World Geomorphological Landscapes

Series Editor

Geomorphology – 'the Science of Scenery' – is a part of Earth Sciences that focuses on the scientific study of landforms, their assemblages, and surface and subsurface processes that moulded them in the past and that change them today. Shapes of landforms and regularities of their spatial distribution, their origin, evolution, and ages are the subject of geomorphology. Geomorphology is also a science of considerable practical importance since many geomorphic processes occur so suddenly and unexpectedly, and with such a force, that they pose significant hazards to human populations. Landforms and landscapes vary enormously across the Earth, from high mountains to endless plains. At a smaller scale, Nature often surprises us creating shapes which look improbable. Many geomorphological landscapes are so immensely beautiful that they received the highest possible recognition – they hold the status of World Heritage properties. Apart from often being immensely scenic, landscapes tell stories which not uncommonly can be traced back in time for millions of years and include unique events. This international book series will be a scientific library of monographs that present and explain physical landscapes across the globe, focusing on both representative and uniquely spectacular examples. Each book contains details on geomorphology of a particular country (i.e. The Geomorphological Landscapes of France, The Geomorphological Landscapes of Italy, The Geomorphological Landscapes of India) or a geographically coherent region. The content is divided into two parts. Part one contains the necessary background about geology and tectonic framework, past and present climate, geographical regions, and long-term geomorphological history. The core of each book is however succinct presentation of key geomorphological localities (landscapes) and it is envisaged that the number of such studies will generally vary from 20 to 30. There is additional scope for discussing issues of geomorphological heritage and suggesting itineraries to visit the most important sites. The series provides a unique reference source not only for geomorphologists, but all Earth scientists, geographers, and conservationists. It complements the existing reference books in geomorphology which focus on specific themes rather than regions or localities and fills a growing gap between poorly accessible regional studies, often in national languages, and papers in international journals which put major emphasis on understanding processes rather than particular landscapes. The World Geomorphological Landscapes series is a peer-reviewed series which contains single and multi-authored books as well as edited volumes.

World Geomorphological Landscapes – now indexed in Scopus® !

Adolfo Quesada-Román
Editor

Landscapes and Landforms of Costa Rica

Springer

Editor
Adolfo Quesada-Román
Facultad de Ciencias Sociales
Escuela de Geografía
Universidad de Costa Rica
Sede Rodrigo Facio
San José, Costa Rica

ISSN 2213-2090 ISSN 2213-2104 (electronic)
World Geomorphological Landscapes
ISBN 978-3-031-64939-4 ISBN 978-3-031-64940-0 (eBook)
https://doi.org/10.1007/978-3-031-64940-0

This Springer imprint is published by the registered company Springer Nature Switzerland AG
The registered company address is: Gewerbestrasse 11, 6330 Cham, Switzerland

Series Editor Preface

Landforms and landscapes vary enormously across the Earth, from high mountains to endless plains. On a smaller scale, nature often surprises us by creating shapes which look improbable. Many physical landscapes are so immensely beautiful that they have received the highest possible recognition—they hold the status of World Heritage properties. Apart from often being immensely scenic, landscapes tell stories which not uncommonly can be traced back in time for tens of million years and include unique events. In addition, many landscapes owe their appearance and harmony not solely to natural forces. For centuries, or even millennia, they have been shaped by humans who modified hillslopes, river courses, and coastlines and erected structures which often blend with the natural landforms to form inseparable entities.

These landscapes are studied by geomorphology—'the Science of Scenery'—a part of Earth Sciences that focuses on landforms, their assemblages, the surface, and subsurface processes that moulded them in the past and that change them today. The shapes of landforms and the regularities of their spatial distribution, origin, evolution, and ages are the subject of research. Geomorphology is also a science of considerable practical importance since many geomorphic processes occur so suddenly and unexpectedly, and with such a force, that they pose significant hazards to human populations and not uncommonly result in considerable damage or even casualties.

To show the importance of geomorphology in understanding the landscape, and to present the beauty and diversity of the geomorphological sceneries across the world, we have launched a new book series *World Geomorphological Landscapes*. It aims to be a scientific library of monographs that present and explain physical landscapes, focusing on both representative and uniquely spectacular examples. Each book will contain details on geomorphology of a particular country or a geographically coherent region. This volume covers a small but nevertheless geomorphologically highly diverse country of Costa Rica. Costa Rican landscapes include volcanic terrains, testifying to ongoing volcanism at a convergent plate boundary, tectonic landforms, high mountains with a unique record of Quaternary glaciation in the tropics, dynamic slopes shaped by landslides, and equally dynamic rivers. Varied coasts complement the general picture. In addition, the landscape of Costa Rica contains many pristine natural areas, but also those with strong anthropic imprints, as demonstrated in a number of chapters in this volume. For some time already Costa Rica is a popular tourist destination, especially for nature-based tourism. The volume convincingly shows that the country would also appeal to geotourists.

The *World Geomorphological Landscapes* series is produced under the scientific patronage of the International Association of Geomorphologists—a society that brings together geomorphologists from all around the world. The IAG was established in 1989 and is an independent scientific association affiliated with the International Geographical Union and the International Union of Geological Sciences. Among its main aims are to promote geomorphology and to foster the dissemination of geomorphological knowledge. I believe that this lavishly illustrated series, which sticks to the scientific rigour, is the most appropriate means to fulfil these aims and to serve the geoscientific community. To this end, my great thanks go to the editor of the volume, Adolfo Quesada-Román, for his initiative and willingness to show the geomorphology of his country to the global geomorphological community. I am grateful to all individual contributors who agreed to participate in this project and delivered interesting stories to read. I am sure the volume will for long remain the prime reference source about geomorphology of Costa Rica.

Wrocław, Poland

Piotr Migoń
Series Editor

Introduction: Landscapes and Landforms of Costa Rica

Welcome to a comprehensive exploration of the diverse and captivating geomorphology that defines Costa Rica. This book, titled *Landscapes and Landforms of Costa Rica*, offers a detailed journey through the country's remarkable terrain, from the towering volcanoes to the intricate karstic formations, providing an in-depth understanding of the processes that have shaped its landforms and landscapes.

The volume is divided into two sections, each contributing unique perspectives to the overarching narrative. The first section, comprising general chapters, lays the foundation by delving into fundamental geological aspects and overarching features of Costa Rican landscapes. The esteemed authors, experts in their respective fields, have meticulously crafted chapters that cover topics ranging from the geotectonic history of Costa Rica to the dynamics of mass movements and the intricate networks of rivers that traverse the country. The interdisciplinary nature of these chapters ensures a holistic grasp of Costa Rica's geological and geomorphological intricacies.

In the second section, the focus shifts to regional studies, offering a more localized examination of specific areas within Costa Rica (Map 1). These chapters delve into the geomorphological nuances of regions, unveiling the distinct features and challenges that characterize each locale. From the upper reaches of the Barva Volcano to the coastal plains of northern Costa Rica, each study unveils the unique interplay of geological processes and human interactions, contributing to the dynamic mosaic of the country's landscapes.

This collaborative effort, published by Springer, brings together a wealth of knowledge from esteemed scholars, researchers, and experts in the field of geoscience in Costa Rica. The culmination of extensive research, fieldwork, and academic collaboration, this book aims to serve as a valuable resource for scholars, students, and enthusiasts interested in the geological and geomorphological wonders of Costa Rica for many years.

As we embark on this journey through the diverse terrains of Costa Rica, we invite readers to delve into the intricate details of its geological history, the dynamic interplay of natural forces, and the profound impact of human activities on this captivating Central American landscape. The chapters within this volume not only provide a scientific understanding of the landforms but also offer a narrative that weaves together the geomorphological past, present, and future of Costa Rica.

Adolfo Quesada-Román

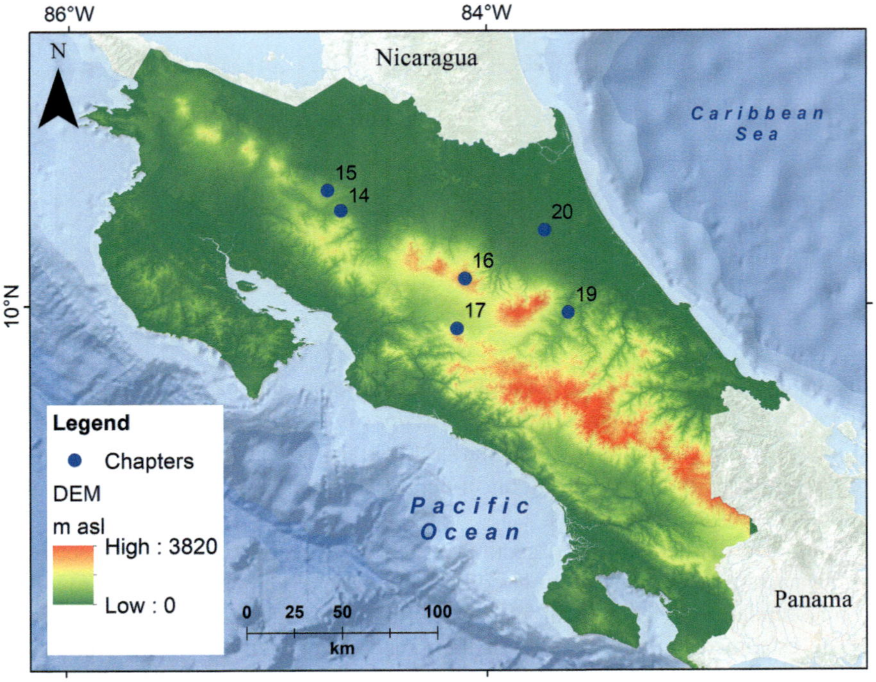

Map 1 Location of the different chapter numbers of particular study areas analysed in this book

Contents

About the Editor

Adolfo Quesada-Román is a geographer with over 15 years of experience, having worked across local and national governments, NGOs, private enterprises, and academic institutions worldwide. His expertise spans disaster risk management, geosciences, geographic information systems (GIS), remote sensing, and geospatial technologies that deliver actionable solutions to environmental, territorial, and natural resource projects. He specializes in geomorphology, applied geography, natural hazard zoning, disaster risk analysis, land-use change, climate change adaptation, and geoheritage conservation. With over 120 indexed publications, he has significantly shaped the fields of Earth Sciences, Geography, and Environmental Sciences, contributing vital knowledge to dynamic, real-world challenges.

Part I

General Chapters

A Synopsis of the Geology, Geotectonic, and Geological History of Costa Rica

1

Guillermo E. Alvarado, Armando Bonilla-Mata, and Guaria Cárdenes-Sandí

Abstract

Costa Rica is a small territory but with a complex geological structure. The continental territory (51,179.92 km^2) is predominantly made up of volcanic rocks, followed by low-grade metamorphic rocks and intrusive rocks. The oldest rocks date back about 200 million years (Ma). The maritime territory (589,683 km^2) is much larger and is made up of volcanic and sedimentary rocks between 30 and 0 Ma old. However, there are minerals formed at other latitudes, transported by ancient rivers and/or by wind as old as 1000 Ma. Rocks older than 75 Ma were formed at other latitudes and below sea level. It is from 75 Ma that the volcanism of an island arc commenced, with multiple stages of formation and destruction, to finally begin to shape the current territory from about 40 Ma, and especially in the last 10 Ma. Costa Rica has volcanic, fluvial, coastal, tectonic, karstic, and mass removal landscapes with many transport and deposition processes, as well as resulting landform diversity. The glacial landscape is more restricted, but of great interest as it is in the tropics. Volcanism and active tectonics, in addition to a wide variability in rainfall (with extremes of up to 11,200 mm/year), and a high contrast of altitudinal relief over short distances and variable roughness, coupled with being surrounded by the Pacific Ocean and the Caribbean Sea, gives it a very particular local climates and a wide variety of landscapes.

G. E. Alvarado (✉)
Centro de Investigaciones Ciencias Geológicas (CICG), Universidad de Costa Rica, San Jose, Costa Rica
e-mail: galvaradoinduni15@gmail.com

A. Bonilla-Mata · G. Cárdenes-Sandí
Escuela Centroamericana de Geología (ECG), Universidad de Costa Rica, San Jose, Costa Rica

Keywords

Geology · Geomorphology · Geotectonic · Morphotectonic · Geological history

1.1 Introduction

Central America, including Costa Rica, has a unique location between the Caribbean Sea and Pacific Ocean (interoceanic) and North and South America subcontinents (intercontinental) (Fig. 1.1). The narrow Panamanian isthmus holds a long and complex geological, biological, and climatic history, which have motivated decades of research. The Isthmus of Panama as natural long-term laboratory encompasses an intricate mosaic of dynamic landscapes shaped by a wide range of processes, such as volcanism, tectonics, fluvial and coastal erosion and sedimentation, hydrothermal, karstic, and glacial processes; coupled with weathering (physical, chemical, and biological transformations) (Fig. 1.2) as a fundamental element of modeling; ending in a physiography characterized by a heterogeneous array of geomorphic and tectonic provinces, each featuring a distinctive landform that documents a unique history of landscape evolution.

The geographic and tectonic position of Costa Rica produces a wide range of dramatic climatic, topographic, and ecological gradients, with seasonal precipitation coupled with variations in slope, wind direction, and orographic precipitation, which result in extraordinarily diversity of microclimates, vegetation cover, landforms, and soil types within a single mountain range (Gómez 1986; Marshall 2007; Alvarado 2021). Due to its proximity to the equator, the country is characterized by not having well-defined seasons, as occurs in other latitudes. The rainfall is very variable, with an annual average of 2900 mm/year and extremes of up to 11,200 mm/year. The temperature varies between − 9 °C and 43 °C, with an annual average ranging between 20.3 °C and 29.4 °C (IMN 2008; Alvarado 2021).

A. Quesada-Román (ed.), *Landscapes and Landforms of Costa Rica*, World Geomorphological Landscapes, https://doi.org/10.1007/978-3-031-64940-0_1

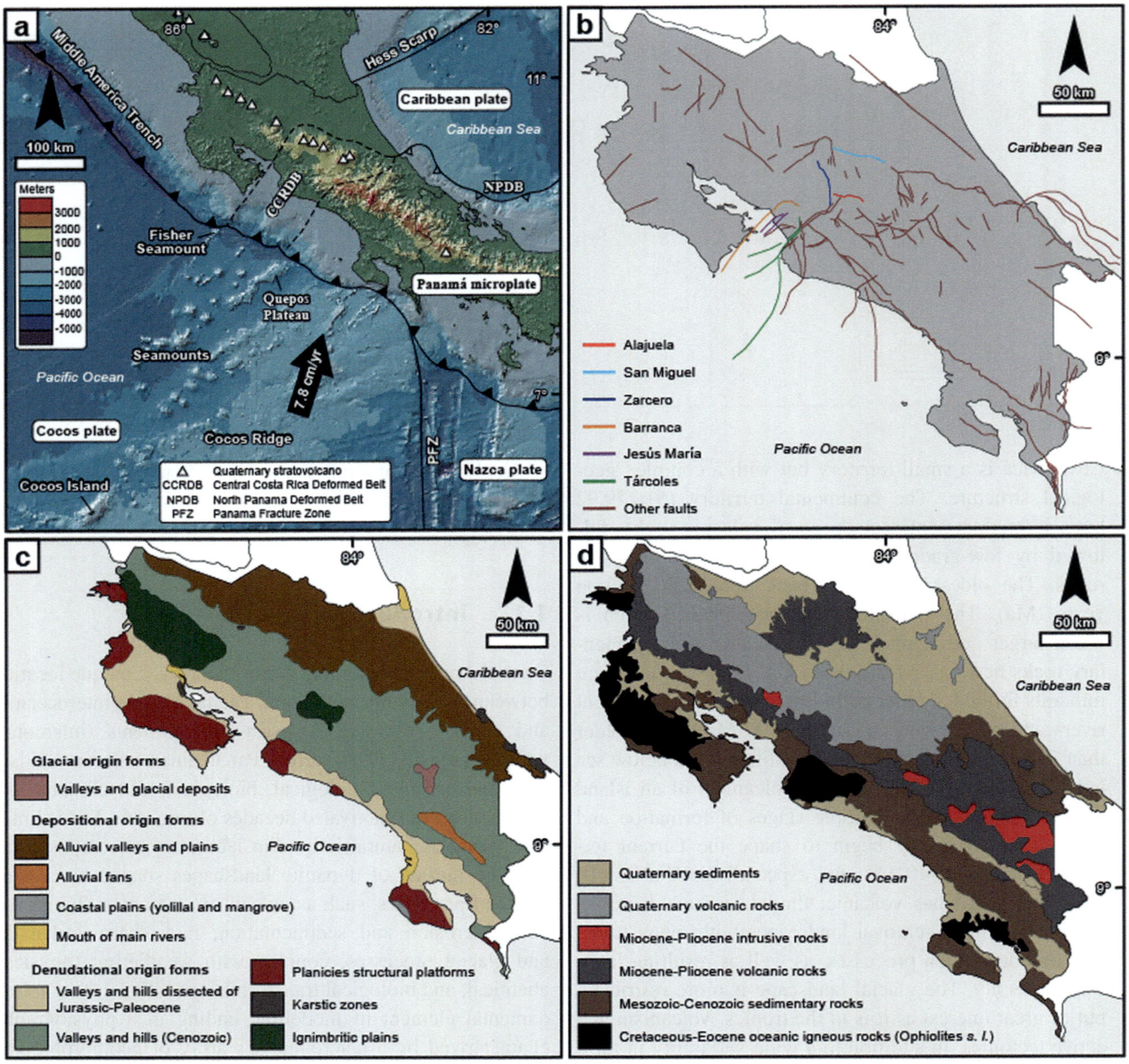

Fig. 1.1 Different simplified types of thematic maps of Costa Rica. **a** Tectonic context, **b** Main active faults, **c** Geomorphology, **d** Geology

The natural and geological history, chiseling the landscapes of Costa Rica, has long interested geographers, biologists, archeologists, civil engineers, and of course, geologists. For example, systematic regional geological investigations in Costa Rica began in the late nineteenth century (Denyer and Alvarado 2000).

The aim of this chapter is to provide a synopsis of the geology of Costa Rica (including the geotectonic and geological history) with particular emphasis on the Quaternary geomorphology (the last 2.58 million years or ~ 2.59 Ma), and focusing on the Holocene (last 11,700 years or 11.7 ka). Those who want to learn more about specific geological topics, lithology, and geochronological periods may consult the original documents and references cited along this chapter (see also Bundschuh and Alvarado 2007).

For the precise age of rocks and landform evolution chronology we used several references (papers, theses, and books), using different techniques (radiometric dating, paleontology, relative ages) (Fig. 1.3).

1.2 Tectonics

The interaction of four tectonic plates (Caribbean, Panama, Cocos, and Nazca) (Fig. 1.1) creates a very complex puzzle responsible for earthquakes and volcanic eruptions.

Fig. 1.2 Different weathering effects. **a** Exfoliation by thermoclastic process (mechanical weathering) producing spheres, **b** Exposed rock blocks (**a** and **b** at Cerro El Indio, Cerro de la Muerte), **c** Cobble-boulder alignment between Cañas and Tilarán, **d** Tafoni, cavities that develop in granular rocks, in this case sandstones, at Papaturros beach, La Cruz, Guanacaste, **e** Spheroidal weathering at Juanilama Creek, Esparza, **f** Pseudobombs at Fila Costeña

The Costa Rican crust has a typical continental signature, being ~ 40 km thick and indicating increase maturity of the arc (Hayes et al. 2013) located at the southern terminus of the Middle America Trench (MAT). The most prominent geotectonic features are: (a) the tectonic trench as a product of the subduction of the Cocos plate under the Caribbean and Panama microplate, (b) subduction of a major topographic seafloor feature called the Cocos Ridge (Cordillera submarina volcánica del Coco), an aseismic submarine volcanic range, extending along much of Southern Costa Rica, where the trench apparently has been plugged by the anomalously buoyant oceanic crust, (c) the Panama Fracture Zone, which defines a triple junction between Cocos, Nazca, and Panama plates offshore southern Costa Rica, (d) the Costa Rican territory, with its well-defined magmatic and sedimentary chains, its older oceanic igneous

Fig. 1.3 Different dating methods applied to establish ages of sediments and landforms in Costa Rica. **a** OSL dating in the Río Corozal fault at Puerto Jiménez, **b** Paleoseismological trench for neotectonic study at Zarcero fault, **c** Morphological recent lava flow at Arenal volcano, a relative dating example, **d** Tephrostratigraphic studies including radiocarbon dating from charcoal and paleosols, useful for the last 45,000 years, **e** Glyptodon fossil from Palmares, useful for relative ages, **f** Precolumbian pottery, useful in Costa Rica for relative dating from the last 4000 years

complexes (also described by some authors as ophiolites), and the back-arc, including a passive plain and non-passive marine margin (Fig. 1.1).

The faulting patterns of Neogene and Quaternary rocks and sediments in Costa Rica may reflect predominantly horizontal shortening perpendicular to the trench, and is consistent with the width, indentation, and subsequent subduction of the submarine buoyant Cocos plate and its Cocos Ridge chain (Montero 1994; Kolarsky et al. 1995), but is also due to the compressive stress generated by the Panama microplate (de Boer et al. 1995), causing an incipient subduction zone of the Caribbean microplate under the Panama microplate (Alvarado 2016; Arroyo and Linkimer 2021; Bourke et al. 2023). The predominant faults show strike-slip movement, followed by thrust (reverse) faults, and, less commonly, normal faults, including composite faults (strike-slip with normal or reverse component, and the other variations) (Figs. 1.4 and 1.5).

1.3 Geology and Geomorphology

Weyl (1971) and Mora (1981a) established the first division of Costa Rica into morphotectonic units, modified by Denyer et al. (2003) (Fig. 1.6), while the most significant

Fig. 1.4 Tectonic structures typical for Costa Rica. **a** Pájaros syncline at Costa de Pájaros, **b** Fault valley in the foothills of the Cordillera de Talamanca, **c** Domino style normal faults with strike-slip component dipoing 50° toward N60°–64° E (rake 70° W), Campos de Oro Arriba, Cordillera de Tilarán, **d** Alkaline sills connecting by dikes through contemporary faults (beside the Reventazón reservoir, right margin), **e** Fracture in the wavecut platform of Montezuma, **f** Small horst near Salitral, Bagaces

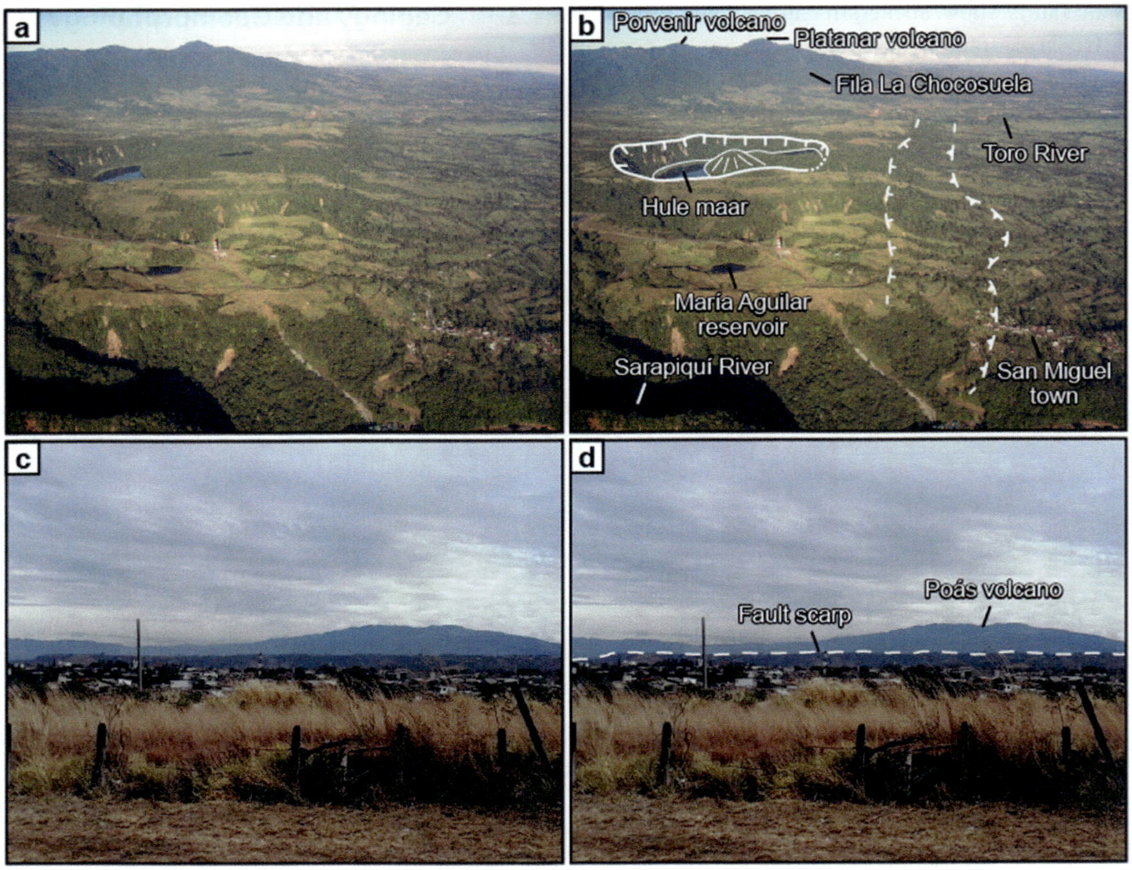

Fig. 1.5 Morphoneotectonic structures related to large fault scarps. **a** San Miguel fault and Hule maar, **b** Geomorphological interpretation of **a**, **c** Alajuela fault scarp, **d** Geomorphological interpretation of **c**

geomorphological studies of Costa Rica were made by Weyl (1980), Madrigal and Rojas (1980), Bergoeing (1978, 1998), Bundschuh and Alvarado (2007), Marshall (2007), and Alvarado (2021). We present here an up-to-date account of the main morphotectonic regions and include several surveys of marine geology. The continental territory of Costa Rica is composed in about 50% of volcanic rocks (subalkaline to alkaline lava flows and pyroclastic rocks), 45% of sedimentary rocks and sediments (sandstone, siltstone, limestone, conglomerates, etc.), 3% of metamorphic rocks (mainly serpentinite and hornfels), and 2% of intrusive rocks (granites to gabbros), but if we include the marine territory, the amount of volcanic and sedimentary sediments increases significantly (Alvarado 2021). The oldest rocks in Costa Rica are of sedimentary origin, with radiolarian fossils of about 200 Ma old, found in the Península de Santa Elena (Tournon and Alvarado 1997).

1.3.1 Submarine Topography from the Abyssal Plain to the Continental Slope in the Marine Pacific Margin

1.3.1.1 The Abyssal Plain and the Middle America Subduction Trench

At the Pacific margin, the Cocos plate is being subducted beneath the Caribbean and Panama plates at a rate of 7.8 cm/year (Protti et al. 2012). On the oceanic plate, three domains can be clearly delineated from seafloor morphology (von Huene et al. 1985, 2000): 1. Northwest of Fisher Seamount, the oceanic crust is 23–22 Ma old and has a relatively smooth surface with normal faulting. It is being subducted opposite the Península de Nicoya, where there is a simple continental slope morphology. 2. In the adjacent area to the southeast, the oceanic crust is approximately 19–13

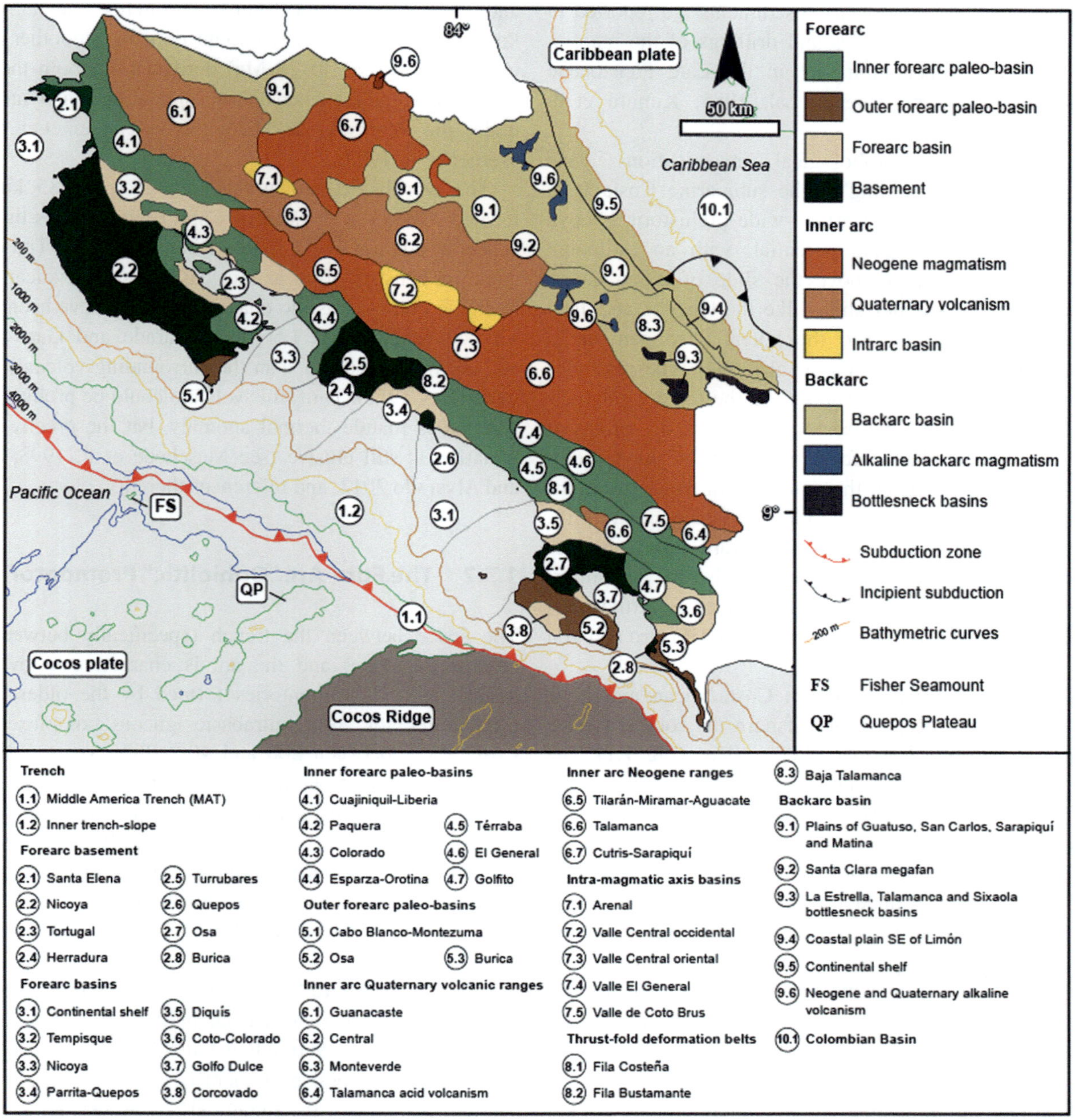

Fig. 1.6 Morphotectonic units of Costa Rica (Modified after Denyer et al. 2003)

Ma old and is covered with seamounts (inactive volcanoes, some still with a crater) that range in size from 1 to 4 km high and 10 to 20 km wide and account for about 40% of the area. In the continental slope, there is more rugged and scars in the slope mark places, where seamounts have subducted. 3. Further to the southeast, the Cocos Ridge uplifts the margin at the Península de Osa (Figs. 1.1 and 1.6).

Seismic reflection data show a 300–600 m thick pelagic sequence of calcareous and siliceous biogenic sediments being inserted into the subduction zone. The sedimentary strata over the Cocos plate have been registered several times by drilling and dredging (McIntosh et al. 1993; Vannucchi and Tobin 2000). The sedimentary sequence can be subdivided into three major units, comprising from top to bottom diatomaceous ooze, followed by silty clay, and then calcareous ooze, of early Middle Miocene (about 16.4 Ma) to Holocene age.

A series of igneous rocks compose the margin wedge, and it is covered by 1–2 km thick slope sediments (Ranero and von Huene 2000). The middle continental slope is characterized by a dominant subsidence from upper bathyal (200–600 m) and even nearshore limestone breccia of Middle Miocene age (16.4 Ma) to its modern abyssal depths of more than 2000 m (Vannucchi et al. 2003). Thick

Pliocene and Pleistocene marine sediments are reported in the seismic reflection images and drillings of the margin, with highest sedimentation rate in the Late Pleistocene (125–164 m/Ma) (von Huene et al. 1985; Kimura et al. 1997).

Comprehensive morphological investigations have detected regions of sliding in the submarine Costa Rica margin, among which the 50 km-wide prehistoric Nicoya slide had a tsunamogenic potential, with an estimated maximum wave height above the slide of 27 m. There are several other smaller slides like the Quepos and Jacó scarps, each of which had the potential to generate a 6–7 m high wave (von Huene et al. 2004). The reason is that the seamounts are transported on the rapidly subducting Cocos plate and have led to a characteristic pattern of upper Caribbean plate deformation. Each seamount creates a broad furrow from the deformation front to its current position, which is clearly indicated by a circular uplift. The uplifted domes are associated seaward with landslides and steep scars. Seamount subduction also has an important role for other geological processes such as earthquakes (Bilek et al. 2003) and subduction erosion (Ranero and von Huene 2000).

The trench along the northern Costa Rican margin is at ~ 4000–5000 m deep, while offshore the southern coast the trench is locally less than 1000 m deep (Fig. 1.1). The buoyant Cocos Ridge and large flanking seamounts in the southern reaches of the Middle America Trench have been proposed as sites of accelerated subduction erosion due to faulting and mass slides (Ranero and von Huene 2000; von Huene et al. 2000). Subduction along the MAT produces much of the seismicity recorded in Costa Rica, including large earthquakes (M_w 7.0–7.7) that have occurred along the Nicoya and Osa segments with an estimated recurrence interval of ~ 10–40 years (Montero 1986), the last ones in 2012 and 1983, respectively.

1.3.1.2 The Cocos Ridge

Located at the Cocos plate, the Cocos Ridge extends from southern Costa Rica to the Galápagos Archipelago and is a 1200 km N45°E long submarine volcanic range, being the longest range (about 815 km in the marine territory) in Costa Rica (Rojas and Alvarado 2012; Alvarado 2021). Adjacent to Costa Rica, it forms a 270–320 km wide, shallow, bathymetric feature that raises 2 to 2.5 km above the ocean floor (Fig. 1.1). The oceanic crust is up to 25 km thick beneath its crest (Walther 2003). The ridge was formed by hotspot volcanism at the Galápagos plume. Werner et al. (1999) determined its age between 13 and 14.5 Ma at the Costa Rican margin.

One of the most astonishing morpho-volcanic features outboard of central Costa Rica are the seamounts (mostly extinct volcanoes), which cover about 40% of the seafloor and range in size from 0.1 km to 2.5 km high and are 10–20 km wide. North of the seamount domain, smoother, older oceanic crust (*ca.* 30–25 Ma) derived from both the East Pacific Rise and the Cocos-Nazca spreading center subducts under the Península de Nicoya (von Huene et al. 1985; Werner et al. 1999).

The worldwide known Isla del Coco (23.85 km^2) is located 496 km southwest of Cabo Blanco and is the only subaerial outcrop of the Cocos Ridge (Fig. 1.1). The actual structure consists of an extinct basaltic shield volcano and trachytic domes (Castillo et al. 1988); they have been dated for between 2.2 and 1.5 Ma (Alvarado and Gans 2012) and are much younger than the surrounding volcanic basement (Hey 1977). Thus, this volcano could be product of an anomalous mantle thermal anomaly, but the precise interpretation is still elusive (see Meschede et al. 1998; Rojas and Alvarado 2012, and references therein).

1.3.2 The Fore-Arc "Ophiolitic" Promontories

The space between the trench (specifically between the continental slope) and the arc is characterized by main peninsulas and promontories formed by the oldest rocks ("ophiolites" or "basic-ultrabasic igneous complexes") of Costa Rica (between 200 and 40 million years old), and relatively younger sedimentary sequences (Figs. 1.1 and 1.7). All these conspicuous geomorphic structures in the Central America land-bridge could be uplift-related features formed by the long-term subduction processes of submarine seamounts and volcanic ranges for the last 40 Ma (Hoernle et al. 2004; Alvarado et al. 2009).

1.3.2.1 Nicoya and Santa Elena Peninsulas

The Península de Santa Elena, built of ultramafic rocks, follows a linear E-W trending fault that bisects the peninsula, forming several prominent bays (Figs. 1.6 and 1.8). The morphology of the rugged shoreline extending southward from the Península de Santa Elena to the northern Península de Nicoya is determined primarily by resistant fault-bounded basement outcrops and steep cliffs formed by the overlying Guanacaste ignimbrites.

Basement rocks exposed at the Península de Nicoya consist of a severely deformed oceanic sequence of Cretaceous basalts, intrusive rocks, and pelagic sediments. Along the margins of the peninsula, a sequence of Upper Cretaceous to Quaternary marine sediments unconformably overlies the Nicoya Complex basement. These sediments include Cretaceous-Paleocene turbidites, Paleocene to Eocene deep- to shallow-water carbonates, Miocene clastic shelves, and a shallowing upward sequence of Pliocene–Pleistocene shelf (sandstones and conglomerates) (Dengo 1962; Jaccard et al. 2001; Denyer et al. 2014) (Figs. 1.1 and 1.6).

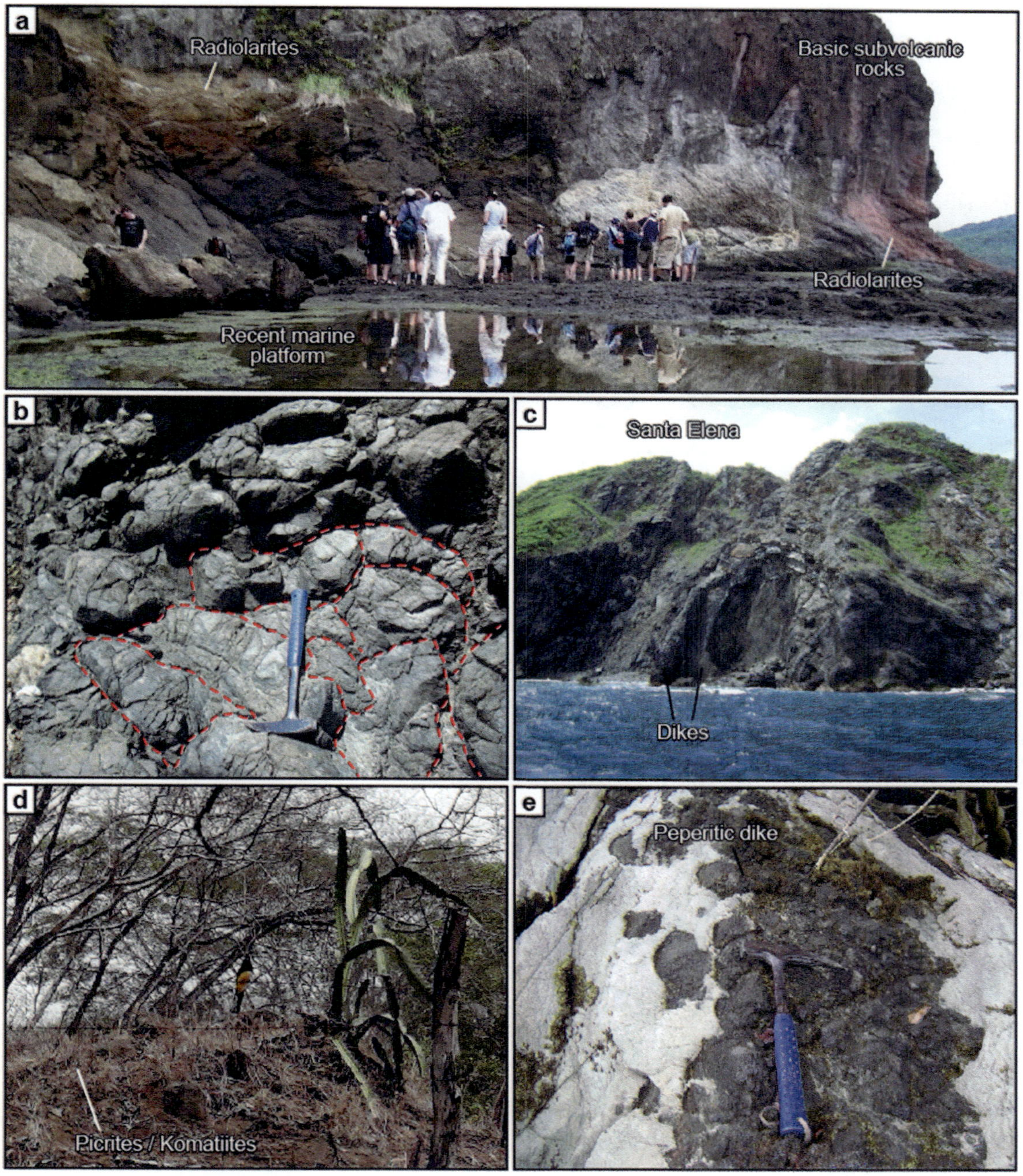

Fig. 1.7 Geology of the forearc ophiolitic promontories of Costa Rica. **a** Basalts and radiolarites in the Nicoya Complex, **b** Pillow basalts (**a** and **b** at Península de Nicoya), **c** Dikes at Península de Santa Elena, **d** Ultramafic lavas at Tortugal, **e** Peperitic dike at Golfito

The Península de Nicoya, located 60 km inboard of the MAT, lies directly above the seismogenic zone. This unique location results in pronounced seismic cycle deformation, which is readily observed along the peninsula's shorelines. Coseismic uplift of > 1 m affected the central Pacific coastline of the Península de Nicoya during the M_w 7.7 subduction earthquakes of 1950 and 2012 (Marshall and Anderson 1995; Protti et al. 2014). Based on the recurrence period of 21.6 ± 11.8 years (Montero 1986) or 40–50 years (Protti et al. 2001), a strong earthquake was predicted to occur since 1991 according to some seismologists, but it finally happened in 2012.

The rugged Pacific coastline of the Península de Nicoya features abundant pocket bays and sandy beaches, bounded by steep, rocky headlands (Arozarena 2024). Uplifted marine terraces and paleo-beach deposits indicate active emergence throughout the late Quaternary (Gardner et al. 2001; Marshall 2007). On the other hand, the peninsula's gulf coast follows a low-relief alluvial plain with extensive mangrove systems (Fig. 1.8). The coastal piedmont

Fig. 1.8 Landforms of Nicoya and Santa Elena peninsulas. **a** Hanging valleys at Península de Santa Elena, **b** Mangrove system at the Península de Nicoya gulf coast, **c** Marine platform and uplift island at Cabuya, **d** Irregular coast with rias at North Pacific coast

along all sides of the Península de Nicoya rises steeply into a mountainous interior highland that reaches over 900 m in elevation. Elevated Holocene shore platforms, beach ridges, and stream terraces record active coastal emergence at the Descartes headland. Valley-fill terraces within coastal embayments extend several kilometers inland, reaching 15–20 m elevation. Streams draining to the Bahía de Salinas are incised into these deposits, exposing uplifted shallow bay to intertidal muds overlain by beach ridge sands and fluvial gravels. Radiocarbon dates obtained for Holocene deposits were used to calculate uplift rates, which show 2.0–3.5 m/ka (Marshall and Vannuchi 2003). The Península de Nicoya coastline has a particular morphology; the western part is an uplift coastline, characterized by co-existence of cliffs and sandy beaches, whereas the eastern part is a subsidence coast, which hosts a well-developed mangrove vegetation (Golfo de Nicoya-Tempisque estuarine system) (Fig. 1.8).

The uplift is recorded by a sequence of Quaternary marine and fluvial terraces at the Península de Nicoya.

High-elevation remnants (400–1000 m a.s.l.) of a Pliocene–Pleistocene marine erosion surface (Cerro Azul surface) are preserved within the mountain block of the peninsula. The lower elevation alluvial terrace (La Mansión surface) occupies interior river valleys at 4–10 m above local base level. Also, two extensive Holocene marine terraces are well developed along 40 km with nearly perpendicular coastlines at the tip of the Península de Nicoya, which extend approximately 1 km inland, they present an impressive feature, covered with up to 2 m of fossiliferous, intertidal sand, and beach rocks (Marshall 1991; Gardner et al. 2001). Thus, several paleo-beach ridges and beachrock strata are visible in the continuous blanket of unconsolidated Holocene beach deposits that covers the wave-cut platform from Cabuya to Montezuma. The upper Cabuya surface is a 1.5 m thick unconsolidated deposit of shell-rich beach sand and gravels that overlies the paleo-platform at elevations between 12.5 and 15.0 m above the highest high tide. This deposit has yielded a calibrated radiocarbon ages of 4190 ± 100 year. B.P. and 4690 ± 220 yr. B.P., respectively (Fig. 1.8). The lower Cabuya surface is the most prominent of these reaches, and it has a height of over 1.5 m continuous for over 500 m. The calibrated radiocarbon ages varied from 2330 ± 70 year. B.P. and 700 ± 90 year. B.P., for elevations of 8.3 and 2.5 m above highest high tide, respectively (Marshall 1991).

A small coastal bluff from Cabuya-Montezuma coast, ranging in height from 0.5 m to over 6 m, is composed of high-energy, conch-rich, pebble-cobble beach deposits, an upward fining sequence of gravels and sand beach, overlain by alluvial gravels, sands, and muds. The estimated age of these deposits is about 650 year. B.P. Similarly, the Isla Cabuya is surrounded at high tide by water, but as the tide falls, it exposes the modern intertidal platform, and the island becomes connected to the mainland and can be reached by foot (Fig. 1.8). Its level surface is covered by a 0.5 m thick unconsolidated beach deposit composed mostly by marine shells and coral fragments of a ^{14}C calibrated age of 490 ± 60 year. B.P. The uplifted wave-cut platform beneath this deposit is about 2 m in elevation above the present highest high tide (Marshall 1991).

This uplift and tilting of Holocene terraces at the southeastern tip of the Península de Nicoya is occurring in response to subduction of seamounts along the projected trend of the Fisher seamount chain, onto the Pacific margin of the Caribbean plate. Uplift rates decrease linearly from a maximum of about 6.0 m/ka near the tip of Cabo Blanco to less 1.0 m/ka along both coastlines (Gardner et al. 2001).

1.3.2.2 The Turrubares Block and Quepos Promontory

The regional basement at Turrubares and Quepos hills is built of basic igneous rocks (basalts, gabbros, volcaniclastic sediments) of Cretaceous to Early Eocene age (*ca.* 65–45 Ma), exposed at the Quepos promontory but mainly in the Herradura block (Hauff et al. 2000; Arias 2003) (Figs. 1.6 and 1.9). The Herradura headland exhibits the highest topographic relief at the Chorotega forearc (> 1700 m). This fault-bounded block exposes Late Cretaceous oceanic basalts, which have been stripped of their sedimentary cover by rapid quaternary uplift and erosion. The differential uplift between the Herradura block and adjacent lower-relief blocks is accommodated by dip-slip along steep, margin-perpendicular faults. Holocene river terraces and wave-cut benches attest to a rapid uplift along the Herradura headland.

1.3.2.3 The Península De Osa and Punta Burica

The Península de Osa is a 62 km-long outer forearc, northwest-trending, high inboard of the subducting aseismic Cocos Ridge (Fig. 1.6). The Península de Osa consists of a narrow coastal piedmont surrounding a northwest-trending mountainous core that narrowly exceeds 700 m elevation. The rocks cropping out along the Península de Osa are a Middle Eocene-Middle Miocene (45–15 Ma) *mélange* dominated by basalts, cherts, and limestones (Fig. 1.10); they formed by the accretion of seamounts. At the end of the Miocene, the subduction of a paleo-Cocos Ridge caused uplift and severe tectonic erosion of the accretionary edifice, allowing for the exhumation of the *mélange*. The *mélange* is overlain by a clastic sequence of Pliocene–Quaternary age, represented by conglomerates, mudstones, and siltstones with fine-grained volcanoclastic turbiditic layers; locally there are megabreccias formed by slumps and calcareous greywacke turbidites (Di Marco et al. 1995; Vannucchi et al. 2006).

Late Pleistocene fossiliferous marine sands unconformably overlie beveled surfaces cut across the competent rocks of the *mélange* basement described above. Presently, exposures of these rocks are found at more than 75 m a.s.l., requiring uplift rates up to 6 mm/yr (Sak et al. 2004).

The Península de Burica juts southward into the Pacific Ocean forming a 25-km-long promontory at the Costa Rica-Panama border (Fig. 1.6). It is an emergent fragment of the Cretaceous-Paleogene oceanic basalt basement, unconformably covered by a Pliocene–Pleistocene sequence of marine sandstone and conglomerate (turbidites) beds (Fig. 1.10). The relationships of sedimentary facies and faunal assemblages indicate that Pliocene subsidence was interrupted by rapid Pleistocene uplift. The Pio-Pleistocene sediments exhibit significant folding and vertical displacement along a prominent north-trending fault valley that bisects the peninsula. Uplifted wave-cut platforms along the Península de Osa coast attest to ongoing deformation (Corrigan et al. 1990) (Fig. 1.10). The elongate N–S form of the Península de Burica is the consequence of the Panama Fracture Zone, also known as a source of large ($M_w \leq 7$) strike-slip earthquakes (Alvarado et al. 2017).

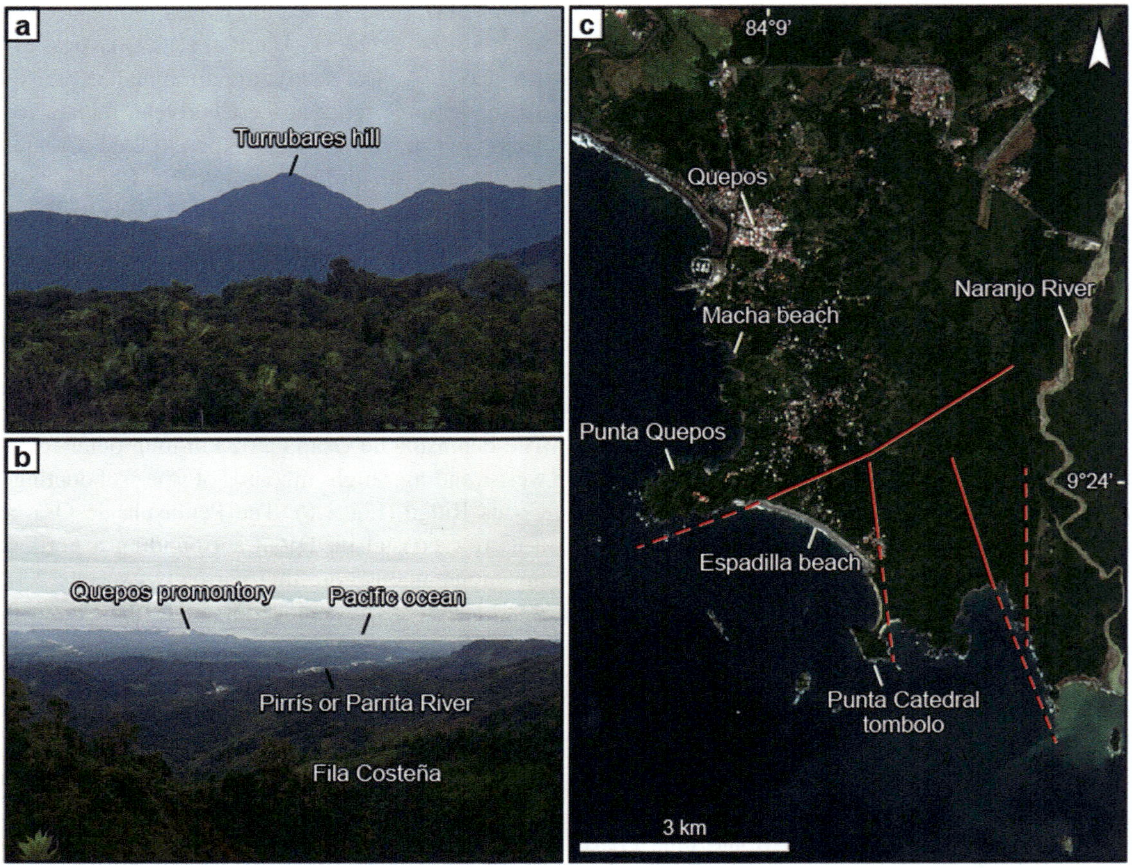

Fig. 1.9 Geomorphological features of Turrubares Block and Quepos Promontory. **a** Turrubares hill, **b** Fila Costeña, Parrita-Quepos in the background, **c** Quepos Promontory (red lines represent local faults)

A relevant feature of Golfo Dulce is its goldfield character, and it has been mined since pre-Columbian times and has produced at least twice as much gold as the entire Tilarán-Aguacate gold province in Northern Costa Rica. The gold occurs in alluvial, colluvial, and beach placers (mainly gravels or conglomerates) of Late Pliocene to Holocene age, overlying the basic igneous complexes of Late Cretaceous to Early Tertiary age (Berrangé 1992). The gold of these placer deposits has their origin in the ancient rocks of Golfito-Burica-Osa and not in the Cordillera de Talamanca as some authors have proposed (Berrangé 1992; Alvarado and Gans 2012).

1.3.3 The Marine Basins Between the Fore-Arc and the Magmatic Arc

1.3.3.1 The Tempisque, Nicoya, and Orotina Basins

At the Península de Nicoya, deep-water sedimentary rocks (calciturbidites and mass flow deposits) of Late Cretaceous age rest unconformably on basalts of the Nicoya Complex (Fig. 1.6). Pelagic limestone clasts with planktonic foraminifera and abundant rudist fragments are frequent in the sedimentary rocks (Denyer et al. 2014; Denyer 2019). The Barra Honda limestone, of Late Paleocene-Lower Eocene age, represents remnants of a formerly continuous carbonate platform from a restricted to open marine environment (Jaccard et al. 2001). Although the Barra Honda limestone exposures on the hill flanks do not exceed 90 m in thickness, caves extend to depths of 200 m, suggesting the development of syn-sedimentary loading and deformation in the central part of the platform, where continuous deposition triggers subsequent weight-subsidence in the center of the hills (Mora 1981b; Denyer et al. 2014).

Offshore of region of the Golfo de Nicoya there has been documented about 3.5 km of Cretaceous-Quaternary sediment extending toward the northeast onshore, limited in the south by a shallow uplift of the basement. The sedimentary sequence typifies a shallowing upward prograding succession of sediments that grades from pelagic sediments to slope and continental deposits. The most recent ones are represented by alluvial fan and estuarine sediments (Barboza et al. 1995).

Fig. 1.10 Geomorphological features of Osa and Burica peninsulas. **a** Slightly inclined sandstones, **b** Fluvial and marine terraces, **c** Uplifted marine terrace (1 ka) (**a, b,** and **c** at Puerto Jiménez), **d** Tilted pillow basalts at Punta Banco

Along 150 km of coastline south of the Península de Nicoya, major trunk rivers draining the inner forearc flow along a system of active, coast-orthogonal faults. These steep faults segment the inner forearc coastline into several fault-bounded blocks with sharply differing Quaternary uplift rates as determined from elevated marine and fluvial terraces. The major rivers follow fault-controlled valleys incised into Neogene-Quaternary nearshore sediments, volcaniclastic debris, and pyroclastic deposits. The Barranca, Jesús María, and Tárcoles faults form the boundaries of the Esparza and Orotina fault blocks (Marshall et al. 2003; Marshall 2007) (Figs. 1.1 and 1.11).

The low-lying Orotina fault block between the Jesús María and Tárcoles rivers is covered by a > 100 m thick Quaternary sequence of lahar-derived debris avalanche deposits, ignimbrites, and their fluvial gravel deposits. During the early Quaternary, a series of eruption-generated debris avalanches, diluted by water inclusion into lahars, descended from the volcanic front into the coastal plain, forming the framework of a 25-km-wide Orotina fan (Alvarado and Gans 2012). Meandering paleo-channels of the Tárcoles River are preserved across the fan surface as inverted topographic ridges of welded tuff overlying river gravels (Marshall et al. 2003).

Up to five late Quaternary alluvial fill terraces (10–260 m elevation) occur along the lower reaches of the fault-controlled Barranca and Tárcoles rivers (Fig. 1.11). The total number of terraces, and the vertical spacing between them, varies along the coast with respect to the magnitude of local tectonic uplift rates. This correlation suggests that terrace generation along this coastline is strongly controlled by the interaction of rock uplift and eustatic sea-level fluctuation (Marshall et al. 2000, 2003).

The most prominent Quaternary sedimentary feature is the Puntarenas sand spit (part of the Golfo de Nicoya-Tempisque estuarine system). This landform appears on maps since the eighteenth century, and it is where the capital city of the province is located. It is 600 m wide, 7 km long, with an average elevation of 3 m a.s.l. Based on historical maps and aerial photographs, the sand bar shows lateral growth, so the origin of the spit could be extrapolated only about 500 years ago, related to sediment transport by marine currents from the Barranca estuarine system (Denyer et al. 2004) (Fig. 1.11).

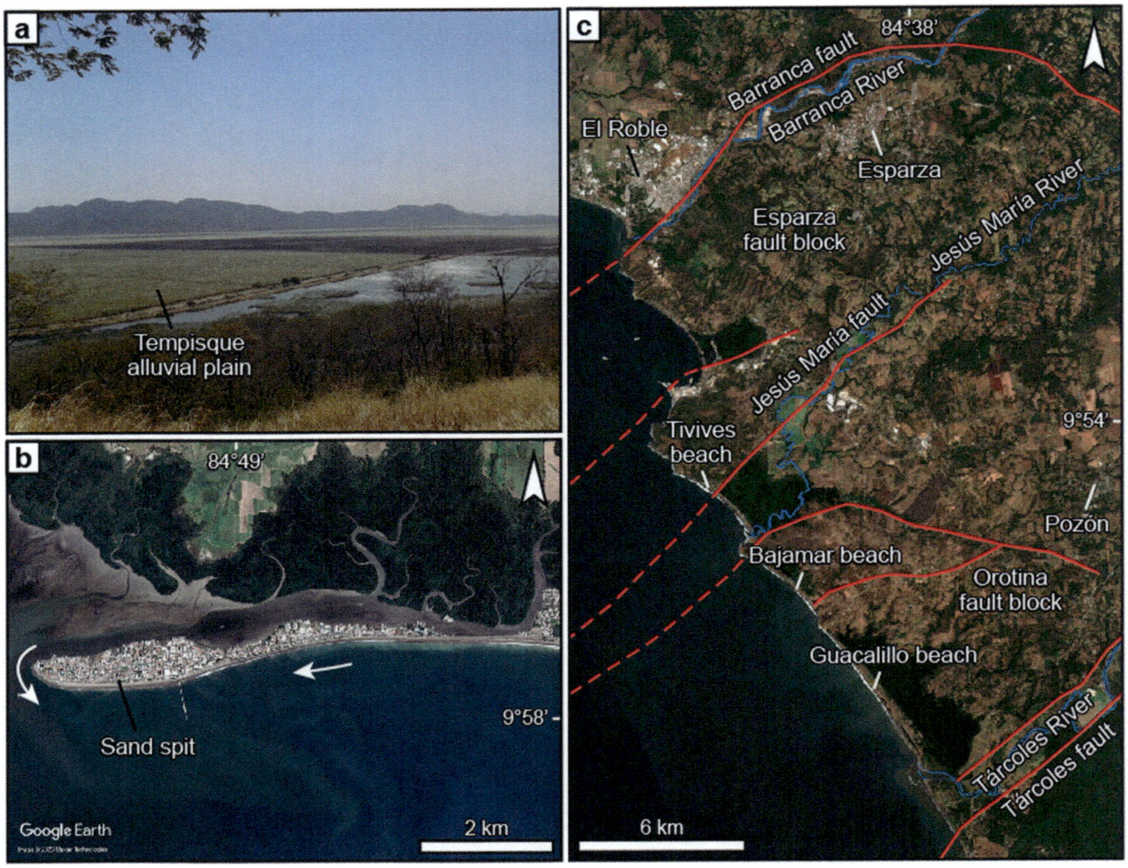

Fig. 1.11 Geomorphological features of Tempisque, Nicoya and Orotina basins. **a** Valle del Tempisque from Taboga farm, **b** Puntarenas sand spit, **c** Esparza and Orotina fault blocks (satellite image of Sentinel-2)

1.3.3.2 The Parrita-Térraba-Golfo Dulce Basins

The Parrita, Térraba, and Golfo Dulce basins in southwestern Costa Rica (Fig. 1.6) include more than 4.5 km of strata ranging in lithology and age from shallow-marine carbonate sedimentation of Middle Eocene, through deep-water Oligocene to Miocene sedimentation (mudstones and sandstones), and shallow-water to subaerial Miocene to Pliocene mudstones and sandstones, followed by Pliocene to Lower Quaternary volcaniclastic rocks and non-marine sands and gravels of Quaternary age (Fig. 1.7). These basins are located offshore and onshore, limited in the south by a shallow uplift of the basement. The sediments represent a shallowing upward prograding succession of sediments that grades from pelagic to slope and continental deposits. The youngest sedimentary rocks correspond with alluvial and deltaic paralic sediments (Barboza et al. 1995).

A wide range of recent sedimentary environments have been recognized along the central Pacific coast of Costa Rica. The continental environment is represented by alluvial deposits such as alluvial plains, active and inactive meandering channels, active braided river channels, colluvial and small lacustrine (lake) deposits. The main rivers, including Tárcoles, Parrita, Savegre and others, are meandering systems with flood plains limited in the north by ranges corresponding with steep dip-slip fault zones. Consequently, the analysis of the morphological and sedimentological changes associated with the local tectonic setting suggests a SE inclination of Parrita area. The alluvial system presents a meandering belt (Parrita River), which migrated SE to the actual alluvial plain. This abnormal situation would have been generated by the SE inclination of the area (Cárdenes 2003) (Fig. 1.12).

The drainage systems in the Parrita coastal plain are deflected around the Quepos highland and exhibit at least four late Quaternary terraces that evidence active uplift (see Marshall 2007). The oldest unit of these terraces is composed of laminated, fine-grained red, and gray clay containing foraminifera and poorly preserved bivalves. The deposit is mottled extensively with purple mottles in a blue-gray matrix, and a minimum thickness of 10 m, and typically forms low flat-topped hills with elevations of 25–35 m. The most recent units are composed of coarse-grained alluvium, highly weathered fluvial deposits, 2.5–30 m thick, distal fine-grained deposits, and buried soils. From regional

Fig. 1.12 Geomorphological features of Parrita and Golfo Dulce basins **a** Parrita alluvial system (satellite image of Sentinel-2), **b** Paleo-beach ridge at Puerto Jiménez, **c** Golfo Dulce basin

correlation and ^{14}C dating, the ages of the terraces are estimated from less than 34,000 years old to about only 400 yr. B.P. Average incision rates (0.5–3.0 m/ka) are estimated from the terrace sequence and apparently reflect regional uplift rates (Drake 1989).

According to our observations, several raised paleo-beach ridges (between 5 and 2) are locally visible in the topography of the continuous blanket of unconsolidated Holocene beach deposits from Savegre to Playa Ballena beaches. One of those was dated using radiocarbon for 5540 ± 70 year. B.P. (Fisher et al. 2004).

Quaternary marine sands, beach ridges, and alluvial gravels along the Península de Osa shorelines record high rates of tectonic uplift (6.5–2.1 m/ka) that decrease along an arc-ward trend from the Peninsula's interior, north-eastward toward the Golfo Dulce (Fig. 1.12). Along the north-eastern coastal piedmont, a sequence of uplifted beach ridges yielded radiocarbon ages ranging from < 1 ka near the modern shoreline to > 30 ka at an elevation of 25 m. Rivers draining the coastal piedmont exhibit two extensive Pleistocene gravel terraces that form a thick alluvial apron across the fault-bounded mountain front. These deposits overlie nearshore marine sediments dated for > 30 ka. Two lower terraces with late Holocene radiocarbon ages occur adjacent to active channels showing a continuous uplift (see Marshall 2007 and references therein).

1.3.4 The Magmatic Front

The magmatic front (volcanic and intrusive) in Central America is not arcuate in plan and hence, not a volcanic arc, but for convenience and tradition, it is classified as one. The magmatic front is the axis of Costa Rica formed by two active volcanic ranges, the Cordillera Central and Guanacaste, and by two extinct ranges, the Tilarán-Aguacate and Talamanca (Fig. 1.6 and 1.13). The latter two ranges do not have enormous stratovolcanoes forms because they are mostly extinct and deeply eroded. There is isolated and local evidence of relatively recent volcanic activity during the Quaternary, as shown as by the colossal Monteverde volcanic plateau (2.2–1.0 Ma), the Perdidos-Chato-Arenal volcanoes (0.1–0 Ma), and the volcanic domes in Talamanca (4–0.1 Ma). For an extended

Fig. 1.13 Geomorphological features of magmatic arc of Costa Rica. **a** Cordillera de Tilarán, Cordillera de Guanacaste in the background, **b** Dike at Cañas, **c** Bella Vista Mine South of the Cordillera de Tilarán at Miramar, **d** View to the N of the Valle Central, Cordillera Central in the background, **e** Jaboncillo hill and "La Bruja" quarry lake at Cerro de la Muerte

description, see Alvarado and Gans (2012) and Alfaro et al. (2018).

The major volcanic centers of Quaternary age are regularly spaced 17–18 km apart, with one exception between Barva and Irazú, which is filled by the Zurquí hills. Arenal volcano is also isolated, being separated by recent volcanic gap from Guanacaste and Cordillera Central by 41 km (Alvarado 2021) (Fig. 1.13).

1.3.4.1 The Cordillera De Guanacaste and the Santa Rosa Ignimbrite Plateau

The Cordillera de Guanacaste occurs in northern Costa Rica (Fig. 1.6). Geologically, it comprises a 90 km-long chain of four major stratovolcanic complexes (Orosí-Cacao, Rincón de la Vieja-Santa María, Miravalles-Zapote, and Tenorio-Montezuma) oriented NW–SE (Fig. 1.13). Rincón de la Vieja is the only volcano active in historical time. The Quaternary volcanoes of Guanacaste, representing basalts to andesites, with rare dacites, grew for about 1 Ma over a regional basaltic to dacitic basement of Late Miocene to Lower Pleistocene age. Two remnant calderas occur along the Cordillera de Guanacaste, the Alcántaro and the Guayabo calderas. An extended fluvio-lacustrine sequence (*ca.* 100 km^2) with a strong volcanic influence (volcanic sandstones and siltstones) and diatomite deposits represents the sediments of an ancient lake that originated within the Alcántaro caldera (Zamora et al. 2004; Alvarado and Gans 2012).

The Santa Rosa plateau consists of an accumulation of pyroclastic flow deposits (ignimbrites) with minor interbedded lava flows and terrigenous and paralic sediments of Lower Pliocene (~ 5 Ma) to Middle Pleistocene age, about 0.6 Ma (Alvarado and Gans 2012) (Fig. 1.1). These units form a broad plateau (2000 km^2) that extends seaward from the base of the volcanic chain, consisting of ignimbrites emitted from pre-neocordillera stratovolcanoes. They form a gently undulating plain that ends in an abrupt 100–150 m high escarpment near the modern Pacific coast. Rivers draining the cordillera have incised deep barrancas and canyons (i.e., Liberia River) into the plateau. Due to tectonic uplift since the Pliocene, several rivers near the Pacific coast drain to the Tempisque River instead flowing directly to the Pacific Ocean (Alvarado and Denyer 2019).

1.3.4.2 The Cordillera De Tilarán and Montes Del Aguacate

The extinct volcanic ranges of Tilarán and Aguacate (100 km long) (Fig. 1.6) consist of densely dissected remnants of stratovolcanoes, andesitic shield volcanoes, and calderas composed of Miocene-Lower Pleistocene basaltic to andesitic lavas, rare dacites, and volcaniclastic rocks (breccias, conglomerates, and tuffs). Hydrothermal alteration (Alvarado and Gans 2012) and deep tropical weathering have destabilized steep slopes of these ranges, resulting in pervasive landsliding (Alvarado et al. 2003).

In the Montes del Aguacate, deeply incised linear canyons have developed along active faults of the Central Costa Rica deformed belt (Fig. 1.1). The Guacimal and Grande de Tárcoles rivers cut deep gorges through the Tilarán and central Aguacate ranges, respectively, connecting rivers of the Valle Central basin with the Pacific coastal plain to the southwest. Along the Tárcoles canyon and many of its tributaries, resistant ignimbrite deposits form level benches and isolated hilltops 50–100 m above the valley floor. Based on late Quaternary isotopic ages for the ignimbrites, the bedrock incision rates range from 0.1 to 0.5 mm/yr (Marshall et al. 2003; Marshall 2007).

An old and large former lake basin ($\geq$ 50 km^2) developed during the Middle Pleistocene in the Palmares and San Ramón area persists today as a depression within the Pliocene volcanic terrain (Alvarado and Gans 2012). Within the basin is a thick sequence (up to 90 m) of lacustrine and fluvial sediments with beds of diatomites, pumiceous conglomerates, tuffaceous siltstones, and sandstones. The lake is drained via the impressive canyon of the Río Grande that flows eastwards. Other diatomite deposits are also known to fill smaller lake basins in Turrúcares and Agua Caliente (Cartago). Small lenticular (5–30 m thick) diatomite and fluvial outcrops are also present along the Santa Rosa and San José rivers, south of Líbano town, and near Peñas Blancas town (Mathers 1989).

Isolated Upper Quaternary volcanic vents located on the northern slope of the Cordillera de Tilarán are Los Perdidos domes (active about 90 ka), and Chato (40–3.4 ka) and Arenal (7–0 ka) volcanoes; Arenal was one of the 16 most active volcanoes in the world between 1968 and 2010 (Alvarado and Gans 2012).

In regions of the old volcanic arc such as the Cordillera de Tilarán, Montes del Aguacate or Crucitas, where different styles and types of hydrothermal alteration (near volcanic focus) occurred in the Neogene, plus an intensive acid leaching of the volcanic rocks, it is possible to generate a multicolor landscape, in some cases with a positive relief due to the existence of a resistant vuggy silica or an iron cap or gossan (Fig. 1.14).

1.3.4.3 The Cordillera Central

The stratovolcanoes complex to andesitic shield volcanoes of the Cordillera Central (Fig. 1.6) forms a 105 km-long volcanic chain including Platanar-Porvenir, Poás, Barva, Irazú, and Turrialba volcanoes. Minor but important volcanic centers include Congo, Cacho Negro, and Hule, along with a dozen parasitic small pyroclastic cones (i.e., Sabana Redonda, Monte de la Cruz, Pasquí, and Armado cones). The Cordillera Central contains the largest volcanoes (elevations ranging from 2000 to 3400 m), in both area and volume, of the entire Central American volcanic front. Their summits exhibit multiple craters and transverse alignments of parasitic cones, and extensive slides occurred on the slopes, generating debris avalanche deposits. Poás, Irazú, and Turrialba have been active in historical times; meanwhile, Barva and Hule were active in the Holocene (Alvarado 2021).

Strong precipitation across the Caribbean side of the range results in greater weathering and erosion, origining fluvial canyons (i.e., La Vieja, Aguas Zarcas, Toro, Sarapiquí, Río Sucio, Toro Amarillo), waterfalls, and more frequent landslides on the humid Caribbean slope (Marshall 2007; Alvarado 2021). Along both flanks of the Cordillera Central prominent fault-propagation-fold scarps along the base of the mountains (Borgia et al. 1990).

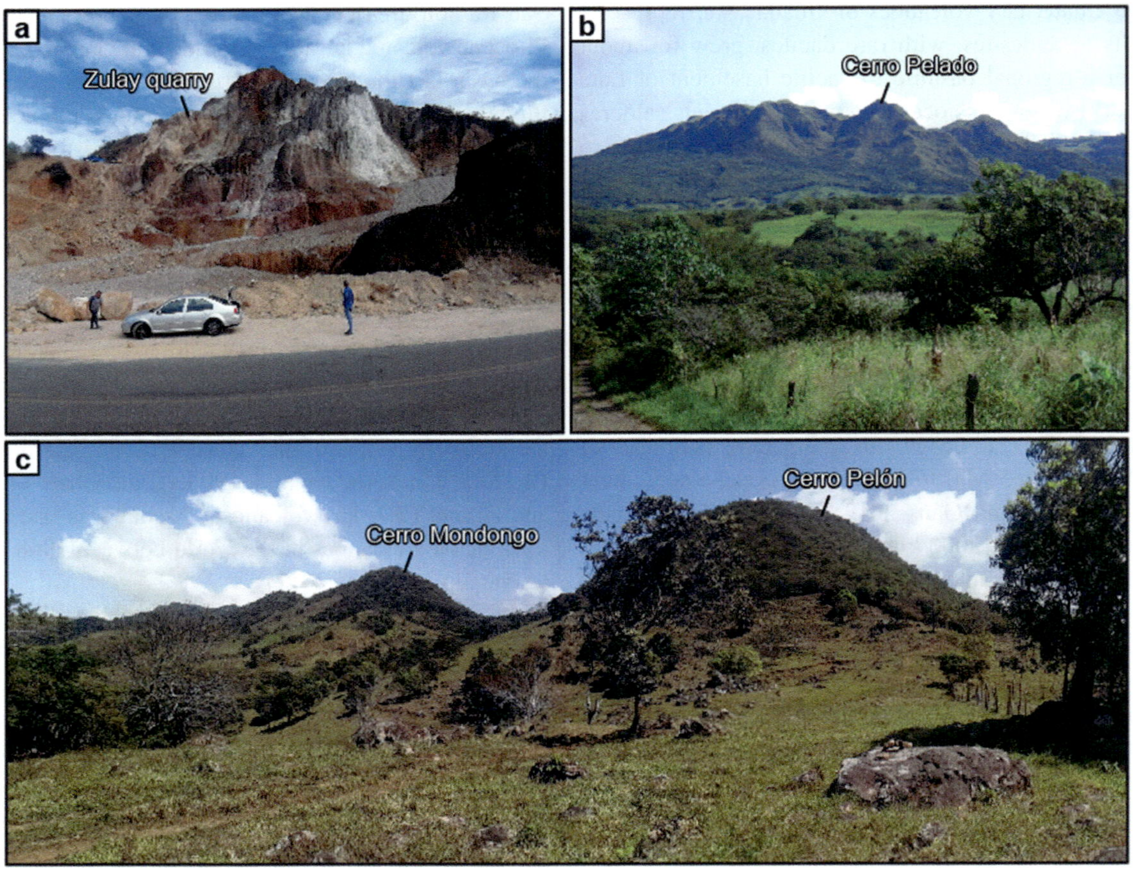

Fig. 1.14 Geomorphological features of Cordillera de Tilarán and Montes del Aguacate. **a** Iron cap (gossan) at Zulay quarry, **b** Intense silicification at Cerro Pelado, Tilarán, **c** Intense silicification at Cerro Pelón-Mondongo, San Mateo

1.3.4.4 The Cordillera De Talamanca

The Cordillera de Talamanca (190 km long) is characterized by elevations over 3500 m and relative heights above 2000 m (Fig. 1.1). It is a conspicuous and unique structure within Central America, as the highest mountain range, and an uplifted inactive segment of the Central American Magmatic Arc (Fig. 1.6). The topography of the Cordillera de Talamanca is asymmetric in cross section. The SW flank is steep, whereas the NE slope is moderately inclined but with profound large scarps. It constitutes the magmatic axis in the southeastern part of Costa Rica, which is composed of intrusive batholiths and stocks of quartz-diorites and monzonites, and subordinate granites and gabbros, which are pervasive intruding sedimentary and volcanic rocks. Sedimentary rocks are predominately volcarenites, breccias, fossiliferous calcarenites, and sandy black shales (Alvarado and Gans 2012; Alfaro et al. 2018).

Most of the intrusive rocks of Talamanca are Miocene in age, between 9 and 11 Ma. There are only a few rocks from the Talamanca intrusion as old as Early Miocene to Late Eocene (19–35 Ma) or as recent as Pliocene (5.3–2.2 Ma) (Alvarado and Gans 2012; Alfaro et al. 2018) (Fig. 1.15). Dacites and andesites produced by the partial melting of hydrated oceanic crust (called adakites) erupted as the latest phase of magmatic activity at the location of previously voluminous calc-alkaline arc magmatism (Abratis and Wörner 2001; Alfaro et al. 2018; Rivera-Flores 2021).

1.3.5 The Intra-Magmatic Axis Basins

1.3.5.1 The Arenal Depression

The Arenal reservoir lake of WNW-ESE trend (28.7 km × 1.1–5 km, 600 m elevation), located between the Arenal active volcano and the eroding remnants of the Monteverde volcanism (2.1–1.0 Ma; Alvarado and Gans 2012), is located within a complex tectonic depression (40 km × 10 km) limited by active strike-slip and normal faults (Camacho 2015) (Figs. 1.6 and 1.16). Several earthquakes of intermediate magnitude ($5 < M_w \leq 6.5$) have affected the region (Montero 1986).

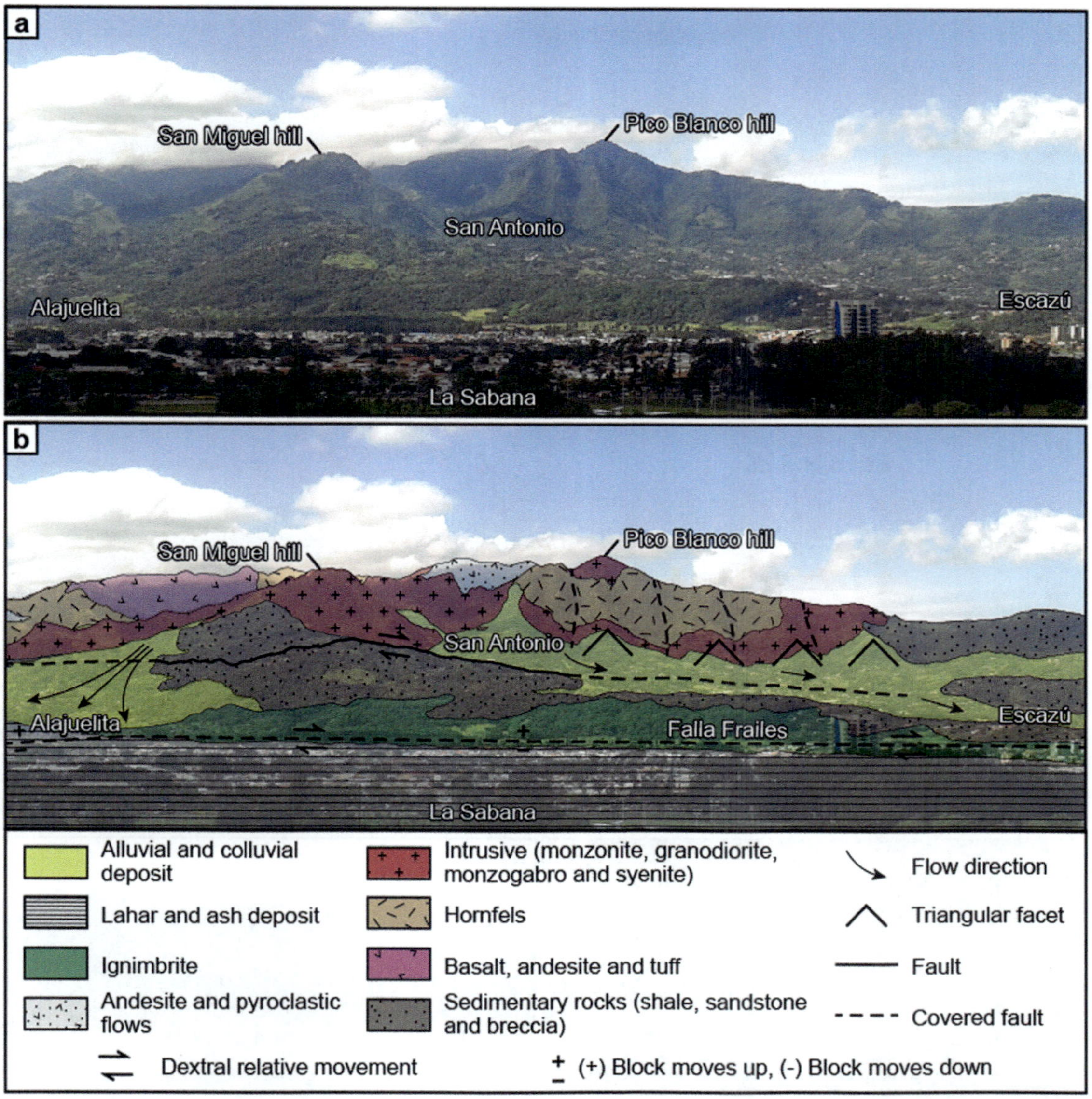

Fig. 1.15 Cerros de Escazú (northwestern foothills of the Cordillera de Talamanca). **a** View from the ICE building, in La Sabana, **b** Geological and geomorphological interpretation of **a** (Modified after Denyer and Arias 1990)

1.3.5.2 The Valle Central

The heartland of Costa Rica is the Valle Central, an elongate E-W trending (*ca.* 70 km × 10 km) basin (600–1500 m elevation) between the active volcanoes of the Cordillera Central and the eroded volcanic remnants of the Aguacate and Talamanca ranges (Figs. 1.6 and 1.16). Thus, the Valle Central is an intermountain valley, filled with a thick accumulation (> 1 km) of volcanic products (andesitic to dacitic lavas, pyroclastic rocks, lahar, and debris avalanche deposits), limited by piedmont faults. The Quaternary migration of the magmatic front shifted volcanism north-eastward to the Caribbean slope and created a new topographic feature—the Valle Central basin (Marshall et al. 2003). The floor of the Valle Central consists of a low-relief upland surface with deeply incised river canyons (i.e., Virilla River) cut into the underlying Quaternary volcanic rocks and less deeply into the Neogene volcanic and sedimentary sequence.

The fault-controlled drainage system of the Valle Central feeds into the Tárcoles gorge, a prominent canyon cut through the eroded highlands of the Aguacate mountains, which provides a link between Valle Central rivers and the Pacific coastal plain below. During the Middle-Late Pleistocene, the Tárcoles River breached the Aguacate drainage divide, leading to progressive capture and re-routing of Valle Central drainage networks toward the Pacific slope (Marshall et al. 2000, 2003; Porras et al. 2012).

Neotectonic and seismic data show that the Central Costa Rica Deformed Belt is a wide diffuse active fault

Fig. 1.16 Intra-arc basins of Costa Rica. **a** View to the W of the Arenal depression seen from the Arenal volcano flank, **b** View to the SW of the Valle Central, Cerros de Escazú in the background, **c** The General and Coto Brus valleys (satellite image of Sentinel-2), **d** NW sector of the Grano de Oro (Moravia de Chirripó) tectonic damming basin at Alta Talamanca

system running through the central part of the country (Montero 2001; Marshall et al. 2000) (Fig. 1.1). Also, paleoseismicity and historical studies indicate the occurrence of intermediate magnitude earthquakes ($M_w \leq 6.5$) and seismic swarms, related mostly to NW dextral strike-slip faults and ENE to NE sinistral strike-slip faults (Denyer et al. 2003; Alvarado et al. 2017).

1.3.5.3 The General and Coto Brus Valleys

The General and Coto Brus valleys are part of an elongate structural basin that stretches for over 90 km along the Pacific slope of the Cordillera de Talamanca and is 10 km wide. This basin (100–1000 m elevation) is separated from the Pacific coastal plain to the southwest by the Fila Costeña (Fig. 1.6), a mountain range (see next topic).

A series of broad alluvial fans, which coalesce along the foot of the Cordillera de Talamanca, forms an extensive piedmont surface on the valley bottom. Tributaries of the General and Coto Brus rivers, which drain the Talamanca highlands, have deeply incised the fan complex, leaving a sequence of terrace remnants along canyon margins. These alluvial surfaces are distinguished from one another based on geomorphic setting, sedimentary texture, and morphologic and chemical characteristics of soils. The oldest geomorphic surfaces coincide with an extensive piedmont upland in the northwestern portion of the Valle de El General. These well-drained upland surfaces exhibit dark red, deeply weathered lateritic oxisols. A series of lower fan surfaces with less-developed soils yields Late Pleistocene to Holocene radiocarbon ages. The youngest alluvial surfaces consist of low-elevation aggradational terraces inset along river canyons and abandoned braided channel bars of the General and Coto Brus rivers (Mora 1979; Kesel 1983).

1.3.6 The Thrust-Fold Deformation Belts

A complex tectonic system of reverse faults (thrust faults) and folds constitutes a deformation range that in the literature is called the thrust-fold deformation belt, a tectonic system present on both sides of the Talamanca, in the Pacific and the Caribbean flanks. In the Pacific flank, it forms part of the forearc system and is called the Fila Costeña and the Fila Bustamante. On the Caribbean side, it is called Baja Talamanca and form part of the back-arc system (Fig. 1.6). Both are grouped here in the thrust-fold deformation belts for its similar tectonic origin and structures.

1.3.6.1 The Fila Costeña and Fila Bustamante Ranges

The Fila Costeña is part of the forearc, a 150-km-long and 15–25-km-wide mountain range with peak elevations of 1100–1400 m that runs NW–SE parallel to the Pacific coast; therefore, its name in Spanish. The Fila Bustamante is morphologically less defined, 64-km-long, and 25–35-km-wide mountain range with peak elevations of 1400–2500 m (Fig. 1.6). Morphometric analysis of mountain front sinuosity and facet development indicates rapid uplift of the Fila Costeña in response to compressional tectonism (Weyl 1971; Wells et al. 1988).

Thrust faults generally bound the exterior and interior of the mountain front and imbricate the Neogene forearc basin sequence of the Térraba basin (Fig. 1.17). Another distinct strike-slip fault (mainly in the eastern part) could be observed running perpendicular to the Fila Costeña. Bedrock has a predominant NW strike and NE dip of 20º-45º (Mora 1979). The sedimentary sequence includes Middle to Late Eocene bioclastic carbonate rocks and a transitional zone to a turbiditic sequence (mudstones and volcaniclastic rocks) of Early to Late Oligocene, indicating a deeper marine environment. The existence of a local Pliocene marine mudstone (Kesel 1983) is one indicator that thrusting within the Térraba basin began after the Pliocene, some time between 5.3 and 1.8 Ma. The sedimentary rocks are intruded by gabbroic and hypabyssal (dolerite, basalt, andesite) dikes and sills of Middle Miocene and Pliocene age (McMillan et al. 2004; Alvarado and Gans 2012). A large lenticular subvolcanic body (15 km long and 700 m maximum thickness) was interpreted as a sill (Sequeira and Chesnel 2022), but rather looks like an elliptical laccolith that laterally changes to a sill.

The coalescence of El General and Coto Brus rivers forms the Grande de Térraba River, which cuts the Fila Costeña, indicating an antecedent river incised during the uplift of the range (Henningsen 1966) (Fig. 1.17). Significant Quaternary deposits, mainly from debris and hypoconcentrated flows, within the Fila Costeña are limited to four discontinuous fluvial terraces up to 100 m above river levels, but their ages are poorly constrained (Bullard 2002).

1.3.6.2 The Baja Talamanca Zone

The Baja Talamanca is a part of the back-arc area (Mora 1981a) (Fig. 1.6). The Panama microplate moves independently of both the Nazca and Caribbean plates, and it is bounded to the north by subduction of the Caribbean plate along a series of fold-and-thrust belts named the Northern Panama Thrust Belt. These faults currently propagate westward toward the coast, generating thrust faulting and earthquakes, including an incipient subduction zone of the Caribbean plate under the Panama microplate below Talamanca (Alvarado 2016; Arroyo and Linkimer 2021) (Figs. 1.1 and 1.6).

A clear and dramatic uplift was observed by the inhabitants during the April 22, 1991 Limón earthquake (M_w 7.7); reaching 4.46 m inland and between 0.5 and 1.85 m in the Caribbean coast (Denyer et al. 2003). Due to the presence of coral material in an ancient, uplifted platform, it was possible to study the Holocene tectonic history. Based on ^{14}C calibrated ages and the current elevation, corrected with eustatic sea-level curves, it was possible to determine an average uplift of about 1.75 m/ka (Denyer 2007).

Fossiliferous and non-fossiliferous sandstones and limestones composed by coral fragments within a sandy matrix represent shallow-marine environments (littoral to sublittoral with patch reefs), in which growth was interrupted by fast sedimentation and/or uplift of alluvial delta fan deposits (Aguilar and Denyer 1994). In the back-arc, the faulting

Fig. 1.17 Geomorphological features of Fila Costeña and Fila Bustamante ranges. **a** Thrust fault at San Vito, **b** Tilted Fila Costeña, Cordillera de Talamanca in the background, **c** Térraba River cutting the Fila Costeña, **d** Tilted alluvial deposits in the Barbilla River, Baja Talamanca

and the thrust and folds system show an NW-trend and near Turrialba a combining dextral strike-slip and reverse faulting (Denyer et al. 2003).

1.3.7 The Back-Arc Region

The back-arc region of Costa Rica is located below the present magmatic front (i.e., Cordillera de Guanacaste, Cordillera Central, and Cordillera de Talamanca) (Fig. 1.6). This region includes the Limón basin, which can be divided into Northern (San Carlos basin) and Southern sub-basins, characterized by different structural settings. The North Limón basin is dominated by extension (normal listric faults) and old compression regime (Porras et al. 2021), a typical of passive continental margin (Parkinson et al. 1998). Instead, the South Limón basin on the Caribbean Sea side has a different structural history defined by the compressional tectonics and shortening due to a heterogeneous pattern of accommodation space and sediment thickness, where several small depocenters have been active since the Middle Miocene and their location changed continuously through time. The hinges were more evident during the Pliocene–Pleistocene, when a succession of piggy-back basins evolved in response of offshore activity of the Limón fold-and-thrust belt. This tectonics causes topographic break at the seafloor that controls the position of recent submarine channels (Brandes et al. 2007). The South Limón basin has a strong compressional regime with several large earthquakes, a topic that is treated in the preceding Baja Talamanca section.

1.3.7.1 The San Carlos-Caño Negro-Tortuguero Plain and Low Hills

The part of the North Limón basin at the Caribbean Sea has a wide and structurally homogeneous depocenter developed in relative tectonic quiescence, particularly active during the Pliocene–Pleistocene times, forming structurally controlled depocenters as a result of pronounced normal faulting (Brandes et al. 2007). Some small earthquakes ($M_w < 5$) are reported by the Parismina nest and along the Hess Scarp (Fernández et al. 1994) (Fig. 1.1).

The regional basement is represented by serpentinized peridotites, Albian siliceous pelagites, and Paleocene to

Middle Eocene turbidites, covered by a thick Miocene to Pliocene volcanic sequence. The recent deposits are composed by Quaternary fine alluvial and palustrine deposits (Gazel et al. 2005; Alvarado et al. 2007), including the Upper Pleistocene-Holocene Santa Clara megafan (Galve et al. 2022) (Fig. 1.18). U–Pb geochronology on detrital zircons from the Neogene sediments of San Carlos basin in Costa Rica yielded the most representative Miocene population, constrained to a maximum depositional age of 6.9 ± 0.2 Ma (Upper Miocene, Rojas-Barrantes et al. 2021), according to previous ages of Middle Miocene to Late Upper Pliocene, mostly Late Miocene (Sen Gupta et al. 1986; Pizarro 1993). The volcanic activity of the evolving island arc is also present since Oligocene to Pleistocene (Alvarado and Gans 2012).

The lowlands encompass an extensive alluvial plain (usually less 100 m in elevation) and a low volcanic relief (less 300 m) that reaches 35–150 km seaward from the base of the Quaternary volcanic range. A series of major rivers traverse the alluvial lowlands, transporting a high sediment load for deposition across broad inland flood plains and a coalescing delta complex at the coast. A sequence of massive alluvial fans has developed along the foot of the volcanic cordillera where the major rivers exit the mountain front. In many cases, modern rivers have incised below extensive upland fan surfaces comprised of thick accumulations of Pleistocene fluvial gravels, capped by well-developed, deep-red, clay-rich soils (Galve et al. 2022). In several locations, the low-relief landscape is interrupted by more prominent hills generated by Lower Miocene to Pliocene volcanism (the Sarapiquí arc, a former remnant arc), and isolated Quaternary basaltic volcanism. The volcanic rocks usually are deeply eroded and were altered into deep-red, clay-rich soils (up to 50 m thick). In some

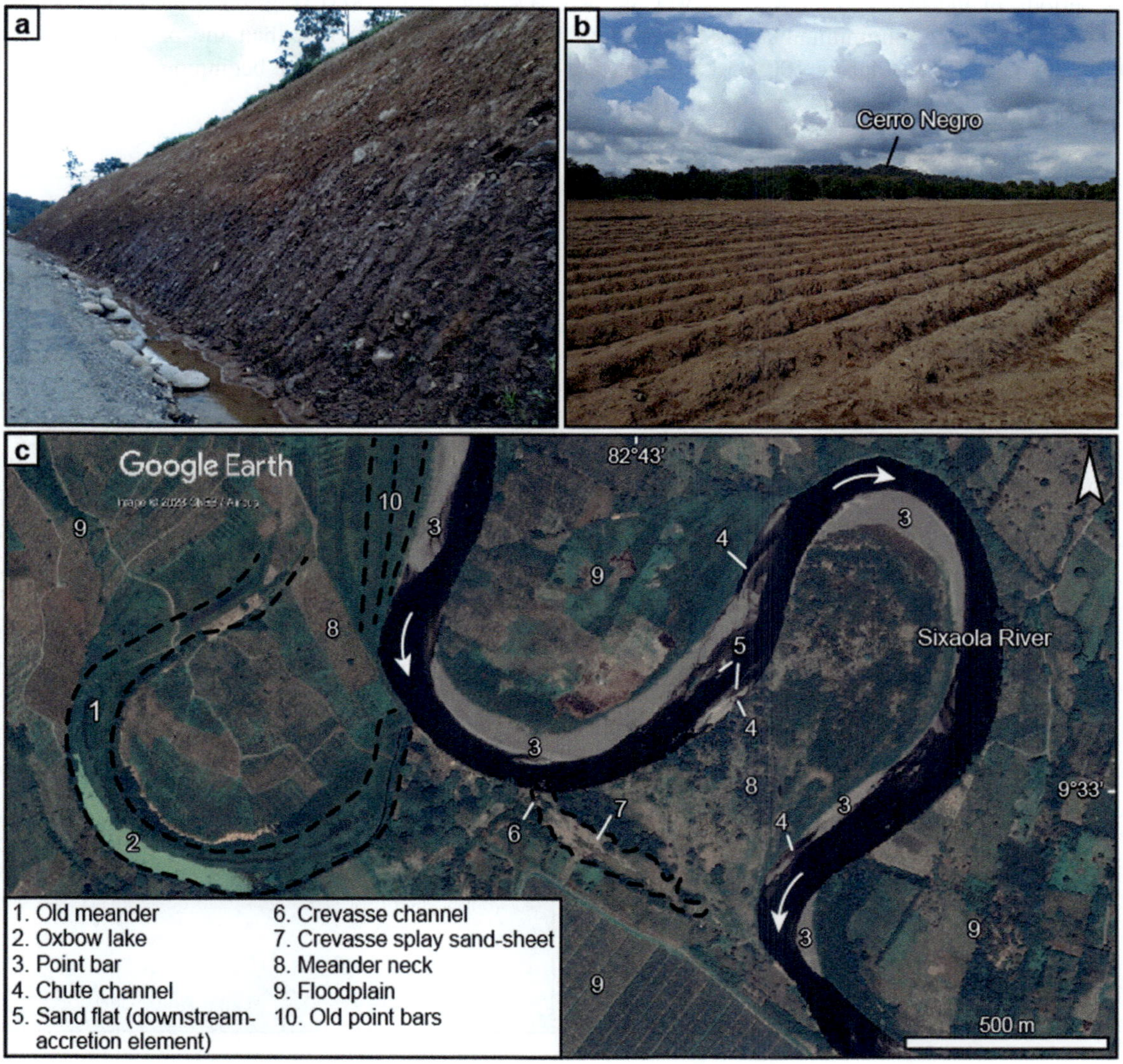

Fig. 1.18 Geomorphological features of Limón basins. **a** Alluvium and lahar on the right bank of the Toro Amarillo River, **b** Santa Clara megafan deposits near Cerro Negro, **c** Features of the meandering fluvial system of the Sixaola River at the Costa Rica-Panama border

geological and geomorphological maps, these features are incorrectly mapped as alluvial terraces, instead as lithological terraces, that means bedrock-controlled benches (Tournon and Alvarado 1997; Gazel et al. 2005).

1.4 Paleoenvironments and Climate Change During the Pleistocene and Holocene

Paleoenvironment and climate remain incompletely understood topics within studies of the geology of Costa Rica, and much of the extant information comprises puzzling fragments and isolated pieces of evidence that often lack good geochronological control (see a summary in Gómez 1986). For a more complete summary of the last 100,000 years for Costa Rica and 36,000 years for the entire Central America region, see Lachniet and Asmerom (2007) and Horn (2007).

Several periods in the last 100 ka were as wet as the Holocene (11–0 ka), with a minimum at *ca.* 60 ka, and several rainfall maxima occurred during 100–65 ka and at *ca.* 37 ka, with a lesser extent during the Holocene (Lachniet and Asmerom 2007).

During Pleistocene glacial epochs, snow accumulated, and glaciers formed on the Cordillera de Talamanca (i.e., Chirripó, Kámuk, and in some instances Cerro de la Muerte) and perhaps on the highest volcanoes (i.e., Irazú and Turrialba) (Fig. 1.19). The equilibrium line altitudes (ELAs) of permanent ice (the Pleistocene snowline) were about 3500 m a.s.l. (Weyl 1971; Orvis and Horn 2000). The lowest reconstructed extent of glacier tongues was at about 3100 m a.s.l. (Orvis and Horn 2000), with a maximum ice thickness of 150–175 m (Lachniet 2007). Pollen in the La Chonta bog indicate that the last glacier interval (*ca.* 50,000–15,600 yr. B.P.) was 7–8 °C cooler than present, and the treeless *paramo* extended down to 2100 m a.s.l. (Islebe and Hooghiemstra 1997; Horn 2007). About 16 ka, after the Last Glacial Maximum, the climate was dry, permitting the formation of sand dunes in the Islas Murciélago (Denyer et al. 2005) (Fig. 1.19), and the ocean level was about 100–120 m less than the present level (i.e., Pinter and Garner 1989). At the beginning of the last deglaciation

Fig. 1.19 Evidence of paleoenvironments in Costa Rica. **a** Small lake in a periglaciar enviroment, Cerro El Indio, Cerro de la Muerte, **b** Flat and ondulated periglaciar morphology at Cerro Las Vueltas, Cerro de la Muerte, **c** Bauxite at Paraíso, Cartago, **d** Fossil sand dune (15.9–17.1 ka) at Islas Murciélago

(15,600–13,000 years), the upper forest limit rose as high as 2700–2800 m a.s.l., indicating a temperature increase of up to 4.6 °C, and precipitation may have increased. The upper forest limit dropped by 300–400 m during the Younger Dryas from 13,100 to 12,300 years, indicating a temperature decline of 2–3 °C. From 12,300 to 11,200 years, the glaciers retreated above 3500 m a.s.l., and subalpine rain forest was gradually replaced by mountain rain forest and the tree line and temperatures rose toward the present-day values (Islebe and Hooghiemstra 1997; Horn 2007).

Also, at least limited savanna is indicated by fossils of grazing mammals, such as horses and mammoths during the Late Pleistocene (Lucas et al. 1997). Unfortunately, the Pleistocene record of mammals in Costa Rica is too diffuse and biased (see above) to allow firm conclusions to be drawn from this rough temporal organization. Clearly, glyptodonts, *Cuvieronius,* and *Equus* were present throughout the Middle-Late Pleistocene, at altitudes at least up to 300, 1200 and 1850 m a.s.l., respectively, and in temporary correspondence with the isolated and rare occurrence of Paleo-Indian (about 13,500–9000 years B.P.) spear points found in Costa Rica.

It is highly probable that the savannas existing in Costa Rica were parkland. Some authors consider, based on the fossil record, that their origin was natural and their existence dates back to the Pleistocene (Kesel 1983; Gómez 1986; Lucas et al. 1997); while other authors suggest that their origin was anthropic (or a combination of both), as it is not clear if pre-Columbian human settlements could have altered the natural ecosystem through deforestation, the use of fire, and agriculture (Artavia and Ávalos 2020).

Analyses of pollen, diatoms, and microscopic and macroscopic charcoal particles in sediment cores from lakes and bogs in the Cordillera de Talamanca suggest wetter conditions during the early Holocene, especially between 7700 and 4800 year B.P., but little change in temperature (Islebe and Hooghiemstra 1997; Horn 2007). A brief early Holocene dry period (8300–8000 years B.P.) is evident in a speleothem record from northern Costa Rica and may be correlated with the high-latitude 8200 year B.P. cold event, but afterward a stable, relatively wetter monsoon climate was established at *ca.* 7600 year B.P. In the late Holocene, a trend toward distinctively drier climates began about 3200 yrs B.P. throughout much of the circum-Caribbean region, including Costa Rica (see Horn 2007).

Extensive aluminum-rich lateritic soils (bauxites) in the Valle de El General, Paraíso of Cartago (Fig. 1.19), and Guápiles (Castillo 1997) should have formed under the following conditions (see Lamey 1966): (a) favorable climatic conditions, as moist and warm (tropical) climate with temperatures exceeding 26 °C most of the time, in which rainfall considerably exceeds evaporation, (b) biological processes can destroy humus faster than the macroflora can produce it, (c) sufficient length of time for pervasive chemical weathering, particular in an alkaline environment, (d) rock located in topographically high and well-drained places, above the level of the permanent water table.

1.5 Brief Summary of the Geological History

The early Caribbean basin began as a narrow seaway between the Pacific and Proto-Caribbean Sea when the last physical connection between North and South America ended about 170 Ma. The next intercontinental connection occurred about 75 Ma ago and lasted for about 6–4 Ma (Bonaparte 1984a, b), when the proto-Antilles were a land-bridge between the Americas in the current position of Central America. Another period of ephemeral land-bridge in the ancestral Antilles also occurred during the Paleocene (Alvarado 1994; Lucas and Alvarado 1994). The emplacement of the oceanic plateaus (the Caribbean Large Igneous Province: CLIP, around 139-69 Ma) between the Americas and the interaction of the Galápagos hotspot track with the Central American volcanic front (between 70 Ma to present) played a fundamental role in the formation of land-bridges (ancestral Antilles) between the Americas during the Campanian, Paleocene, and Pliocene (Central America isthmus) (Hoernle et al. 2002, 2004).

In the Late Cretaceous, an intraoceanic arc formed along the western margin of the Caribbean plate. This arc marked the beginning of the subduction of the Farallon plate in the region. The present territory of Costa Rica emerged above sea-level several times, because of tectonic events, mostly related to subduction. Andesitic calc-alkaline clasts in submarine sediments give evidence that the arc volcanism dates back to the Campanian (Kuijpers 1980). Within the same stage, a regional tectonic uplift, coupled with eustatic sea-level decrease, allowed for the emplacement of a carbonate platform. Similar events occurred during the Late Paleocene-Lower Eocene and Middle-Upper Eocene (Denyer et al. 2000, 2014).

During the Middle Eocene, the areas now occupied by the Tempisque, Nicoya, Parrita, and Térraba basins were part of a major forearc system that spanned along the Pacific margin of Costa Rica. The present structural configuration of the Parrita and Nicoya basins was initiated in the Middle Eocene, related to shear stresses caused by the clockwise rotation of the southern portion of the Costa Rica territory along a main transcurrent fault zone. Such movement displaced the Tempisque and Térraba on the south and north, respectively, developing the half-graben configuration and the system of strike-slip faults of the transtensive basins of Parrita and Nicoya (Barboza et al. 1995).

The northern part of Costa Rica emerged and was subjected to erosion during the Oligocene, whereas in the southern part, sedimentation continued. The modern tectonic configuration began to emerge during the Late Oligocene to Early Miocene (27–25 Ma), when the Farallón plate broke up into the Cocos and Nazca plates separated by the newly formed Galapagos rift zone (Hey 1977; Wortel and Cloetingh 1981). Since the Miocene, shallow-marine deposition was widespread in all Costa Rica.

There are few Oligocene intrusions in Costa Rica; it was not until the Miocene that magmatism began on a wide scale. The Late Oligocene–Miocene (29–8 Ma) was a period of high volcanic activity throughout Costa Rica. In Costa Rica, there has been a 30° counterclockwise (CCW) rotation of the arc from its Middle Miocene position close to the modern volcanic front. This occurred between 15 and 8 Ma and is attributed to deformation in the overriding plate (shortening in the south coeval with extension in the northwest), accompanied by trench retreat in the north (MacMillan et al. 2004; Alvarado and Gans 2012). Alterations in the overriding plate are associated with contractional deformation triggered by plate geodynamics, where changes in the direction of movement of the Cocos plate led to a change of convergence from oblique to orthogonal (Mescua et al. 2017). Subduction of young lithosphere, 25–14 Ma old (von Huene et al. 2000), could have caused migration of volcanism landward in central and north Costa Rica.

Tectonic and magmatic activity during the Middle Miocene to Pliocene in Costa Rica was characterized by uplift, intrusions along the inner arc, extensive volcanism, and folding with reverse and thrust faulting. Since 8 Ma, the arc was parallel to the modern volcanic front but progressively retreated to the northeast in Costa Rica. Adakite restrictive volcanism in Talamanca, represented by small volcanic domes (4.4–0.9 Ma), corresponds in space and time with the subduction of a large scarp associated with a tectonic boundary of southern Panama (MacMillan et al. 2004).

Subduction of the Cocos Ridge produced a shallow ridge indentation (Montero 1994) and affected the southern part of Costa Rica by basement-rooted thrusting, back-arc, and intra-arc compression (Rivier 1985; Kolarsky et al. 1995), volcanic arc extinction (McGeary et al. 1985), and uplift of the coastal belt (Cross and Pilger 1982; Madrigal and Rojas 1980; see also Marshall 2007). The rise of the Cordillera de Talamanca was caused by the subduction of the Cocos Ridge beneath the volcanic front (Rivier 1985; Kolarsky et al. 1995); and the collision of the arc with the South American plate led to the development of Panama microplate (Mann and Burke 1984; de Boer et al. 1995). The age of the Cocos Ridge collision with the Middle America Trench is still, however, a matter of debate. A few authors claim the age of collision of an older Cocos Ridge as early as Middle Miocene, based on geological evidence (Rivier 1985), although the most accepted age at present is around 6.5–5 Ma (see Vannucchi et al. 2003). Where the crustal shortening is larger in the Fila Costeña, the highest mountains of Talamanca (Buena Vista, Chirripó, and Kámuk; 3451, 3821, and 3549 m a.s.l., respectively) are present.

Thus, the Isthmus of Panama began to close 15.2–13 Ma ago (Duque-Caro 1990; Haug and Tiedemann 1998). In Costa Rica, the coast consisted of a shallow-marine embayment during the Neogene. Fossils indicate that the closure of the Isthmus of Panama was almost complete at 3.7–3.6 Ma (Coates et al. 1992). The final closure allowed a land mammal exchange at 2.7 Ma (Marshall 1988), coincident with the glacial-induced sea-level fall during the peak of the Northern Hemisphere ice-sheet growth (Haug and Tiedemann 1998). Most authors (i.e., Coates 1997) assume that the closure took place in the Panama-Costa Rica region ("Panamanian isthmus"), but Gartner et al. (1987) argued that the last interchange of waters between the tropical Atlantic and Pacific was across the Isthmus of Tehuantepec, from the Golfo de Tehuantepec into the Golfo de México. The last significant change in planktonic foraminiferal assemblage occurred at 1.9 Ma and was related to the last closing of the Isthmus of Panama (Keller et al. 1989). The Panama land-bridge permitted the great interchange of terrestrial animals (mastodons, saber-tooth cats, tapirs, ground sloths, armadillos, etc.) and vegetation between North and South America. In several cases, it also worked as a biological filter (or biogeographic frontier) due to differences in topography, vegetation, climate, and geological evolution along the isthmus (see Gómez 1986; Alvarado 1994; Lucas and Alvarado 1994). During the Pliocene, the Panamanian basins were filled with marine and continental clastic sediments (fan and braided delta to alluvial fans).

The beginning of rapid (0.55–0.6 km/Ma) marine subsidence of the Pacific margin of Costa Rica at 6.5–5 Ma is coincident with the arrival of the Cocos Ridge crest at the Costa Rica trench (Vannucchi et al. 2003). The arc volcanism extends the length of Costa Rica coincident with the modern volcanic front (Alvarado and Gans 2012). The Pliocene subduction of the Cocos Ridge about 5.4 Ma led to uplift and increased exhumation of the Cordillera de Talamanca, so that stratovolcanoes are preserved and active to the northwest and east of the Cordillera de Talamanca, whereas in the region of main uplift, they were eroded to deeper crustal levels. This exhumation was characterized by a sudden increase in uplift rate between 5.5 and 3.5 Ma (Gräfe et al. 2002).

The second dramatic subsidence of the Pacific margin occurred very close to the Pliocene–Pleistocene transition (1.8 Ma, Vannucchi et al. 2003), just contemporaneously with the initiation of the Monteverde volcanism (2.2–1.0 Ma). It was followed by the present Middle-Late Quaternary volcanic spine, the Guanacaste and Central ranges, which grew mainly during the last 1 Ma, with major episodes of cone/shield building at 0.98–0.85 Ma (Proto-Cordillera), 0.63–0.40 Ma (Pre-Cordillera), and <0.17 Ma

(Neo-Cordillera) separated by 0.3–0.1 Ma intervals of dormancy (erosion) and/or explosive silicic volcanism (Alvarado and Gans 2012).

In the higher mountains, glaciers and snow caps were present. The initiation of the first glaciations in Costa Rica would have occurred when the Cordillera de Talamanca was uplifted sufficiently high (> 3100 m a.s.l.) and some volcanoes grew above this altitude, but the precise ages of the first glaciations in the present territory of Costa Rica are still unclear, although the last glaciation affected our territory between *ca.* 18 and 11 ka. At the end of the Pleistocene, intensification of continental glaciation resulted in sea-level changes with a stronger amplitude (Haug and Tiedemann 1998). Marine sediments deposited from 11,300 to 9600 years ago contain evidence of an apparent downslope migration of some mountain forest taxa, interpreted to reflect cooler climatic conditions during this period, especially since the deglaciation about 10,000 year BP (Horn 1993). Melting of this ice and snow, together with the volcanic activity in a tropical environment, contributed to the generation of very thick lahars and alluvial deposits.

Finally, as a note of interest, both the oldest and the most recent foreign or exotic volcanic rocks of Costa Rica are found casually in the magmatic arc. One of the oldest rocks still found in Costa Rica corresponds with a very exotic rock, which is the case of sighting and recovering a meteorite in Costa Rica that happened on April 1, 1857, in Heredia. It consists in an H5 Chondrite probably originated in the asteroid 6 Hebe (Soto 2004) or the Aguas Zarcas meteorite, a CM2 Chondrite (Lücke et al. 2019). Another case is the enigmatic precedence of zircons in sediments as old as 1910–265 Ma (Middle Permian to Middle Precambrian). U–Pb geochronology on detrital zircons in sediments of San Carlos basin yielded older ages of Paleozoic (*ca.* 536.5-264.9 Ma) and Proterozoic (*ca.* 1910–552 Ma). Such ages, previously undiscovered in Costa Rica, bring new insights on provenance and indicate multistage zircon recycling from a variety of sources and indicate the influx of recycled sediments from older units (continental and sedimentary rocks), possibly from exhumed Precambrian and Paleozoic continental terranes or blocks from the north of Central America (north of Nicaragua to Guatemala) and probably from southern Mexico (Rojas-Barrantes et al. 2021). In an opposite way, the most recent rocks in Costa Rica are now formed by frequent eruptions (i.e., lava and pyroclastic rocks at Rincón de la Vieja volcano), and also include some travertines in bicarbonate hotsprings, and beachrocks.

Figure 1.20 illustrates the division of geological time starting from the Lower Jurassic period, marking the

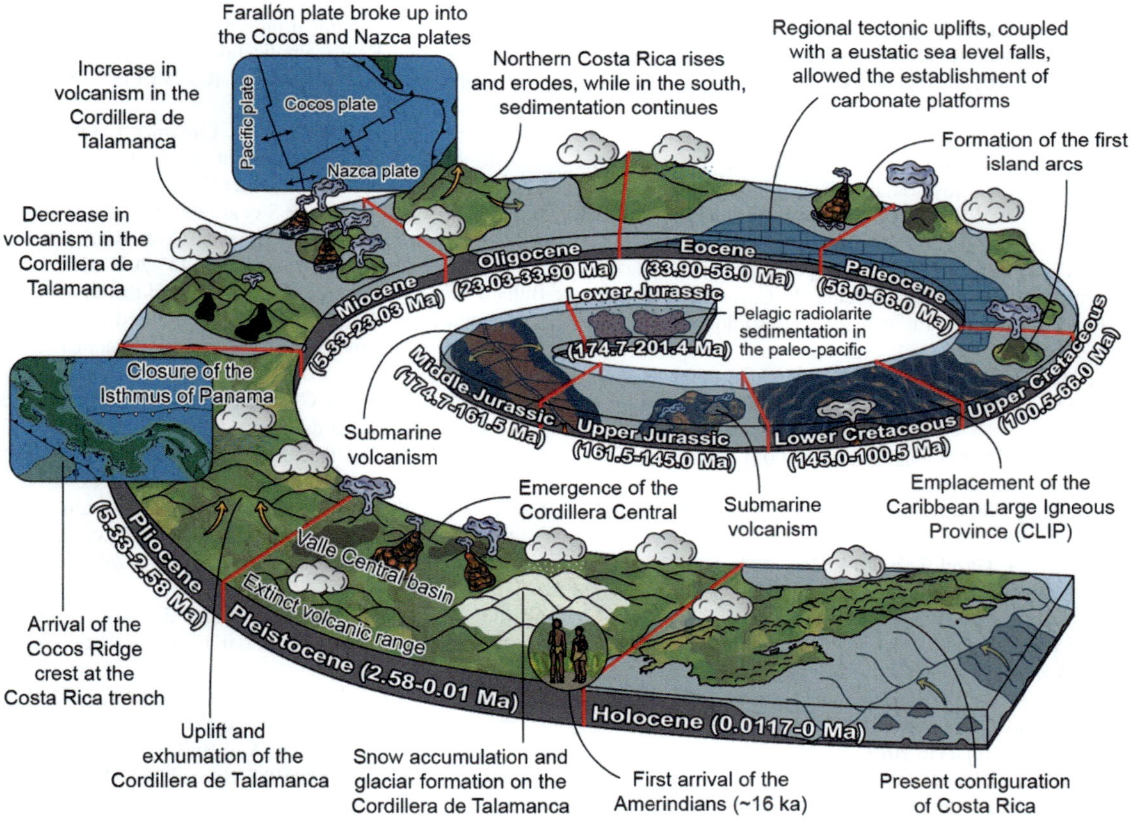

Fig. 1.20 Geological time and some relevant events in the geological history of Costa Rica

commencement of Costa Rica's geological history, up to the present day. It emphasizes some of the most significant events that occurred in each geological epoch.

Acknowledgements The authors thank Piotr Migoń for his valuable comments at the early preliminary draft of this chapter. Our studies have been supported by the Escuela Centroamericana de Geología and Centro de Investigaciones en Ciencias Geológicas at University of Costa Rica, also by the Instituto Costarricense de Electricidad.

References

Abratis M, Wörner G (2001) Ridge collision, slab-window formation, and the flux of Pacific asthenosphere into the Caribbean realm. Geology 29:127–130. https://doi.org/10.1130/0091-7613(2001)029<0127:RCSWFA>2.0.CO;2

Aguilar T, Denyer P (1994) Bioestratigrafía del parche arrecifal de la Quebrada Brazo Seco, Plio-Pleistoceno, Limón, Costa Rica. Rev Geol Am Central 17:55–66. https://doi.org/10.15517/rgac.v0i17.13313

Alfaro A, Denyer P, Alvarado GE, et al (2018) Estratigrafía y petrografía de las rocas ígneas en la cordillera de Talamanca, Costa Rica. Rev Geol Am Central 58:7–36. https://doi.org/10.15517/rgac.v58i0.32669

Alvarado GE (1994) Historia Natural Antigua. Editorial Tecnológica de Costa Rica, Cartago

Alvarado GE (2016) El cinturón deformado del norte de Costa Rica-Colombia: ¿Desde una faja de empuje-plegamiento hasta una zona de subducción incipiente dentro de un bloque o microplaca? In: IASPEI Regional Assembly Latin-American and Caribbean Seismological Commission - LACSC. San José, p 39

Alvarado GE (2021) Costa Rica y sus volcanes. Editorial Universidad de Costa Rica, Editorial Tecnológica de Costa Rica, Editorial Universidad Nacional, San José

Alvarado GE, Denyer P (2019) Estratigrafía volcánica del Neógeno y Cuaternario. In: Denyer P (ed) Perspectiva geológica del noroeste de Costa Rica: historia, evolución y cartografía. Editorial Universidad de Costa Rica, Centro de Investigaciones en Ciencias Geológicas, San José, pp 101–240

Alvarado GE, Gans PB (2012) Síntesis geocronológica del magmatismo, metamorfismo y metalogenia de Costa Rica, América Central. Rev Geol Am Central 46:7–122. https://doi.org/10.15517/rgac.v0i46.1836

Alvarado GE, Mora R, Peraldo G (2003) The June 2000 Arancibia Debris avalanche and block-slide, Costa Rica. Landslide News 14(15):29–32

Alvarado GE, Dengo C, Martens U et al (2007) Stratigraphy and geologic history. In: Bundschuh J, Alvarado GE (eds) Central America: geology, resources and hazards. Taylor & Francis, London, pp 345–394

Alvarado GE, Denyer P, Gazel E (2009) Endeavor research into evolving paradigms around Ophiolites: the case of the oceanic igneous complexes of Costa Rica. Rev Geol Am Central 40:49–73. https://doi.org/10.15517/rgac.v0i40.4186

Alvarado GE, Benito B, Staller A et al (2017) The new Central American seismic hazard zonation: mutual consensus based on up to day seismotectonic framework. Tectonophysics 721:462–476. https://doi.org/10.1016/j.tecto.2017.10.013

Arias O (2003) Redefinición de la Formación Tulín (Maastrichtiano-Eoceno Inferior) del Pacífico Central de Costa Rica. Rev Geol Am Central 28:47–68. https://doi.org/10.15517/rgac.v0i28.7783

Arozarena I (2024) Atlas de playas de Costa Rica con énfasis en la seguridad. Departamento de Física Universidad Nacional, San José, Instituto Costarricense de Turismo

Arroyo IG, Linkimer L (2021) Modelo unidimensional de velocidades sísmicas y características tomográficas tridimensionales del Caribe Sur de Costa Rica One-dimensional seismic velocity model and 3D tomographic features of the Southern Caribbean of Costa Rica. Rev Geol Am Central 65:1–15. https://doi.org/10.15517/rgac.v0i65.46696

Artavia G, Ávalos G (2020) Historia natural del ecosistema de sabana del Valle del Río General, Costa Rica. Rev Geogr161:11–28. https://doi.org/10.35424/regeo.161.2020.861

Barboza G, Barrientos J, Astorga A (1995) Tectonic evolution and sequence stratigraphy of the Central Pacific margin of Costa Rica. Rev Geol Am Central 18:43–63. https://doi.org/10.15517/rgac.v0i18.13523

Bergoeing JP (1978) La fotografía aérea y su aplicación a la geomorfología de Costa Rica. Ministerio de Obras Públicas y Transportes, Instituto Geográfico Nacional, San José

Bergoeing JP (1998) Geomorfología de Costa Rica. Instituto Geográfico Nacional, San José

Berrangé JP (1992) Gold in Costa Rica. Min Mag May 402–407

Bilek S, Schwartz SY, DeShon HR (2003) Control of seafloor roughness on earthquake rupture behavior. Geology 31:455–458. https://doi.org/10.1130/0091-7613(2003)031<0455:COSROE>2.0.CO;2

Bonaparte JF (1984a) Nuevas pruebas de la conexión física entre Sudamérica y Norteamérica en el Cretácico Tardío (Campaniano). In: Actas III Congreso Argentino de Paleontología y Bioestratigrafía. Corrientes, pp 141–149

Bonaparte JP (1984b) El intercambio faunístico de vertebrados continentales entre América del Sur y del Norte a fines del Cretácico. In: Memoria III Congreso Latinoamericano de Paleontología. Oaxtepec, pp 438–450

Borgia A, Burr J, Montero W et al (1990) Fault propagation folds induced by gravitational failure and slumping of the central Costa Rica Volcanic Range: implications for large terrestrial and Martian volcanic edifices. J Geophys Res 95:14357–14382. https://doi.org/10.1029/jb095ib09p14357

Bourke JR, Levin V, Arroyo IG, Linkimer L (2023) Evidence for Caribbean plate subduction in southern Costa Rica. Geology 51:408–412. https://doi.org/10.1130/G50796.1

Brandes C, Astorga A, Back S et al (2007) Fault controls on sediment distribution patterns, Limón basin, Costa Rica. J Pet Geol 30:25–40. https://doi.org/10.1111/j.1747-5457.2007.00025.x

Bullard TF (2002) Geomorphic history and fluvial responses to active tectonics and climate change in an uplifted forearc region, Pacific coast of southern Costa Rica, Central America. Z Geomorphol 129:1–29

Bundschuh J, Alvarado GE (eds) (2007) Central America Geology. Resources and Hazards. Taylor & Francis, London

Camacho JP (2015) Caracterización tectónica y geofísica de la falla Cote, Costa Rica. BSc thesis, Universidad de Costa Rica

Cárdenes G (2003) Evolución de los sistemas sedimentarios costero y aluvial de la región de Parrita, Pacífico Central de Costa Rica. Rev Geol Am Central 28:69–76. https://doi.org/10.15517/rgac.v0i28.7784

Castillo R (1997) Recursos minerales de Costa Rica: génesis, distribución y potencial. Editorial Universidad de Costa Rica, San José

Castillo P, Batiza R, Vanko D et al (1988) Anomalously young volcanoes on old hot-spot traces: I. Geology and petrology of Cocos Island. Geol Soc Am Bull 100:1400–1414. https://doi.org/10.1130/0016-7606(1988)100<1400:AYVOOH>2.3.CO;2

Coates AG (1997) Central America: a natural and cultural history. Yale University Press, London

Coates AG, Jackson JBC, Cronin TM et al (1992) Closure of the Isthmus of Panama: the near-shore marine record of Costa Rica and western Panama. Geol Soc Am Bull 104:814–828. https://doi.org/10.1130/0016-7606(1992)104<0814:COTIOP>2.3.CO;2

Corrigan J, Mann P, Ingle JC (1990) Forearc response to subduction of the Cocos Ridge, Panama-Costa Rica. Geol Soc Am Bull 102:628–652. https://doi.org/10.1130/0016-7606(1990)102<0628:FRTSOT>2.3.CO;2

Cross TA, Pilger RH (1982) Controls of subduction geometry location of magmatic arcs and tectonics of arc and back-arc regions. Geol Soc Am Bull 93:545–562. https://doi.org/10.1130/0016-7606(1982)93<545:COSGLO>2.0.CO;2

de Boer JZ, Drummond MS, Bordelon MJ, et al (1995) Cenozoic magmatic phases of the Costa Rican island arc (Cordillera de Talamanca). In: Mann P (ed) Geologic and Tectonic Development of the Caribbean Plate Boundary in Southern Central America. Geol Soc Am Special Paper 295, pp 35–55. https://doi.org/10.1130/SPE295-p35

Dengo G (1962) Estudio geológico de la región de Guanacaste, Costa Rica. Ministerio de Obras Públicas, Instituto Geográfico de Costa Rica, San José

Denyer P (2007) Tectonically controlled uplifting of the Caribbean coast of Costa Rica. In: Abstracts 20th Colloquium on Latin American Earth Sciences. Kiel, pp 99–100

Denyer P (ed) (2019) Perspectiva geológica del noroeste de Costa Rica: historia, evolución y cartografía. Editorial Universidad de Costa Rica, San José

Denyer P, Alvarado GE (2000) Desarrollo y evolución de la Geología. In: Denyer P, Kussmaul S (eds) Geología de Costa Rica. Editorial Tecnológica de Costa Rica, Cartago, pp 471–492

Denyer P, Arias O (1990) Geología de la Hoja Abra (1:50,000). Escuela Centroamericana de Geología, IGN, San José

Denyer P, Alvarado GE, Aguilar T (2000) Historia geológica. In: Denyer P, Kussmaul S (eds) Geología de Costa Rica. Editorial Tecnológica de Costa Rica, Cartago, pp 155–167

Denyer P, Montero W, Alvarado GE (2003) Atlas tectónico de Costa Rica. Editorial Universidad de Costa Rica, San José

Denyer P, Cárdenes G, Kruse S (2004) Registro histórico y evolución de la barra arenosa de Puntarenas, Golfo de Nicoya, Costa Rica. Rev Geol Am Central 31:45–59. https://doi.org/10.15517/rgac.v0i31.7245

Denyer P, Cortés J, Cárdenes G (2005) Hallazgo de dunas fósiles de final del pleistoceno en las Islas Murciélago, Costa Rica. Rev Geol Am Central 33:29–44. https://doi.org/10.15517/rgac.v0i33.4234

Denyer P, Aguilar T, Montero W (2014) Cartografía geológica de la península de Nicoya, Costa Rica: estratigrafía y tectónica. Editorial Universidad de Costa Rica, San José

Di Marco G, Baumgartner PO, Channell JE (1995) Late Cretaceous-early Tertiary paleomagnetic data and a revised tectonostratigraphic subdivision of Costa Rica and western Panama. In: Mann P (ed) Geologic and Tectonic Development of the Caribbean Plate Boundary in Southern Central America. Geol Soc Am Special Paper 295, pp 1–27. https://doi.org/10.1130/SPE295-p1

Drake PG (1989) Quaternary geology and tectonic geomorphology of the coastal piedmont and range, Puerto Quepos area, Costa Rica. M.Sc. thesis, University of New México

Duque-Caro H (1990) Neogene stratigraphy, paleoceanography and paleobiogeography in northwest South America and the evolution of the Panama seaway. Palaeogeogr Palaeoclimatol Palaeoecol 77:203–234. https://doi.org/10.1016/0031-0182(90)90178-A

Fernández JA, Bottazzi G, Barboza G, Astorga A (1994) Tectónica y estratigrafía de la Cuenca Limón Sur. Rev Geol Am Central, Special Issue Terremoto de Limón 15–28. https://doi.org/10.15517/rgac.v0i0.13391

Fisher DM, Gardner TW, Sak PB et al (2004) Active thrusting in the inner forearc of an erosive convergent margin, Pacific coast, Costa Rica. Tectonics 23:1–13. https://doi.org/10.1029/2002TC001464

Galve JP, Alvarado GE, Pérez-Peña JV et al (2022) River avulsions and megafan development triggered by high recurrent explosive volcanism. Geomorphology 419:108466. https://doi.org/10.1016/j.geomorph.2022.108466

Gardner TW, Marshall J, Merritts D et al (2001) Holocene forearc block rotation in response to seamount subduction, southeastern Península de Nicoya, Costa Rica. Geology 29:151–154. https://doi.org/10.1130/0091-7613(2001)029<0151:HFBRIR>2.0.CO;2

Gartner S, Chow J, Stanton RJ (1987) Late Neogene paleoceanography of the eastern Caribbean, the Gulf of Mexico, and the eastern Equatorial Pacific. Mar Micropaleontol 12:255–304. https://doi.org/10.1016/0377-8398(87)90024-7

Gazel E, Alvarado GE, Obando J, Alfaro A (2005) Geología y evolución magmática del Arco de Sarapiquí, Costa Rica. Rev Geol Am Central 32:13–31. https://doi.org/10.15517/rgac.v0i32.4242

Gómez LD (1986) Vegetación de Costa Rica. Editorial Universidad Estatal a Distancia, San José

Gräfe K, Frisch W, Villa IM, Meschede M (2002) Geodynamic evolution of southern Costa Rica related to low-angle subduction of the Cocos Ridge: constraints from thermochronology. Tectonophysics 348:187–204. https://doi.org/10.1016/S0040-1951(02)00113-0

Hauff F, Hoernle K, van den Bogaard P et al (2000) Age and geochemistry of basaltic complexes in western Costa Rica: contributions to the geotectonic evolution of Central America. Geochem Geophys Geosyst 1:1–41. https://doi.org/10.1029/1999gc000020

Haug GH, Tiedemann R (1998) Effect of the formation of the Isthmus of Panama on Atlantic Ocean thermohaline circulation. Nature 393:673–676. https://doi.org/10.1038/31447

Hayes JL, Holbrook WS, Lizarralde D et al (2013) Crustal structure across the Costa Rican Volcanic Arc. Geochem Geophys Geosyst 14:1087–1103. https://doi.org/10.1002/ggge.20079

Henningsen D (1966) Notes on stratigraphy and paleontology of upper cretaceous and tertiary sediments in Southern Costa Rica. AAPG Bull 50:562–566. https://doi.org/10.1306/5D25B4A9-16C1-11D7-8645000102C1865D

Hey R (1977) Tectonic evolution of the Cocos-Nazca spreading center. Geol Soc Am Bull 88:1404–1420. https://doi.org/10.1130/0016-7606(1977)88<1404:TEOTCS>2.0.CO;2

Hoernle K, van den Bogaard P, Werner R et al (2002) Missing history (16–71 Ma) of the Galápagos hotspot: implications for the tectonic and biological evolution of the Americas. Geology 30:795–798. https://doi.org/10.1130/0091-7613(2002)030<0795:MHMOTG>2.0.CO;2

Hoernle K, Hauff F, van den Bogaard P (2004) 70 m.y. history (139–69 Ma) for the Caribbean large igneous province. Geology 32:697–700. https://doi.org/10.1130/G20574.1

Horn SP (1993) Postglacial vegetation and fire history in the Chirripó Páramo of Costa Rica. Quat Res 40:107–116. https://doi.org/10.1006/qres.1993.1061

Horn SP (2007) Late Quaternary lake and swamp sediments: Recorders of climate and environment. In: Bundschuh J, Alvarado GE (eds) Central America: geology, resources and hazards. Taylor & Francis, London, pp 423–441

IMN (2008) Clima y variabilidad climática en Costa Rica. Instituto Meteorológico Nacional, San José

Islebe GA, Hooghiemstra H (1997) Vegetation and climate history of montane Costa Rica since the last glacial. Quat Sci Rev 16:589–604. https://doi.org/10.1016/S0277-3791(96)00051-0

Jaccard S, Münster M, Baumgartner PO et al (2001) Barra Honda (Upper Paleocene- lower Eocene) and el Viejo (Campanian-Maastrichtian) carbonate platforms in the Tempisque Area

(Guanacaste, Costa Rica). Rev Geol Am Central 24:9–28. https://doi.org/10.15517/rgac.v0i24.8550
Keller G, Zenker CE, Stone SM (1989) Late Neogene history of the Pacific-Caribbean gateway. J South Am Earth Sci 2:73–108. https://doi.org/10.1016/0895-9811(89)90028-X
Kesel RH (1983) Quaternary History of the Río General valley, Costa Rica. Nat Geog Soc Research Reports 15:339–358
Kimura G, Silver EA, Blum P et al (1997) Introduction. In: Proceedings of the Ocean Drilling Program, Initial Reports 170. College Station, Texas, pp 1–17
Kolarsky RA, Mann P, Monechi S (1995) Stratigraphic development of southwestern Panama as determined from integration of marine seismic data and onshore geology. In: Mann P (ed) Geologic and tectonic development of the caribbean plate boundary in Southern Central America. Geol Soc Am Special Paper 295, pp 159–200. https://doi.org/10.1130/SPE295-p159
Kuijpers EP (1980) The geologic history of the Nicoya Ophiolite Complex, Costa Rica, and its geotectonic significance. Tectonophysics 68:233–255. https://doi.org/10.1016/0040-1951(80)90178-X
Lachniet MS (2007) Glacial geology and geomorphology. In: Bundschuh J, Alvarado GE (eds) Central America: Geology, Resources and Hazards. Taylor & Francis, London, pp 171–184
Lachniet MS, Asmerom Y (2007) Central American Rainfall Variations Over the Past 100,000 year From Costa Rican Stalagmites: A Proxy for Cross-Isthmian Water Vapor Transport. Abstracts AGU
Lamey CA (1966) Metallic and industrial mineral deposits. McGraw-Hill Book Company, New York
Lucas SG, Alvarado GE (1994) Role of Central America in land-vertebrate dispersal during the Late Cretaceous and Cenozoic. Profil 7:401–412
Lucas SG, Alvarado GE, Vega E (1997) The Pleistocene Mammals of Costa Rica. J Vertebr Paleontol 17:413–427. https://doi.org/10.1080/02724634.1997.10010985
Lücke OH, Madrigal P, Soto GJ (2019) El condrito carbonáceo Aguas Zarcas: bólido, caída y recuperación de un meteorito extraordinario del tipo CM2. Rev Geol Am Central 61:9–22. https://doi.org/10.15517/rgac.v61i0.40085
MacMillan I, Gans PB, Alvarado GE (2004) Middle Miocene to present plate tectonic history of the southern Central American Volcanic Arc. Tectonophysics 392:325–348. https://doi.org/10.1016/j.tecto.2004.04.014
Madrigal R, Rojas E (1980) Manual descriptivo del Mapa Geomorfológico de Costa Rica (escala 1:20000). SEPSA, Imprenta Nacional, San José
Mann P, Burke K (1984) Neotectonics of the Caribbean. Rev Geophys Space Phys 22:309–362. https://doi.org/10.1029/RG022i004p00309
Marshall JS (1991) Neotectonics of the Nicoya Peninsula, Costa Rica: a look at forearc response to subduction at the Middle America trench. MSc thesis, University of California
Marshall JS (2007) Geomorphology and physiographic provinces. In: Bundschuh J, Alvarado GE (eds) Central America: geology, resources and hazards. Taylor & Francis, London, pp 75–122
Marshall JS, Anderson RS (1995) Quaternary uplift and seismic cycle deformation, Peninsula de Nicoya, Costa Rica. Geol Soc Am Bull 107:463–473. https://doi.org/10.1130/0016-7606(1995)107<0463:QUASCD>2.3.CO;2
Marshall JS, Vannuchi P (2003) Forearc deformation influenced by subducting plate morphology and upper plate discontinuity, Pacific Coast, Costa Rica. Geol Soc Am Abstracts with Programs 35:74
Marshall JS, Fisher DM, Gardner TW (2000) Central Costa Rica deformed belt: kinematics of diffuse faulting across the western Panama block. Tectonics 19:468–492. https://doi.org/10.1029/1999TC001136
Marshall JS, Idleman BD, Gardner TW, Fisher DM (2003) Landscape evolution within a retreating volcanic arc, Costa Rica, Central America. Geology 31:419–422. https://doi.org/10.1130/0091-7613(2003)031<0419:LEWARV>2.0.CO;2
Marshall LG (1988) Land Mammals and the Great American Interchange. Am Sci 76:380–388
Mathers S (1989) Costa Rican diatomite: a review of existing knowledge and future potential. Rev Geol Am Central 10:3–17. https://doi.org/10.15517/rgac.v0i10.12991
McGeary S, Nur A, Ben-Avraham Z (1985) Spatial gaps in arc volcanism: the effect of collision or subduction of oceanic plateaus. Tectonophysics 119:195–221. https://doi.org/10.1016/0040-1951(85)90039-3
McIntosh K, Silver E, Shipley T (1993) Evidence and mechanisms for forearc extension at the accretionary Costa Rica Convergent Margin. Tectonics 12:1380–1392. https://doi.org/10.1029/93TC01792
Meschede M, Barckhausen U, Worm H-U (1998) Extinct spreading on the Cocos Ridge. Terra Nova 10:211–216. https://doi.org/10.1046/j.1365-3121.1998.00195.x
Mescua JF, Porras H, Durán P et al (2017) Middle to Late Miocene Contractional Deformation in Costa Rica Triggered by Plate Geodynamics. Tectonics 36:2936–2949. https://doi.org/10.1002/2017TC004626
Montero W (1986) Periodos de recurrencia y tipos de secuencias sísmicas de los temblores interplaca e intraplaca en la región de Costa Rica. Rev Geol Am Central 5:35–72. https://doi.org/10.15517/rgac.v0i05.12267
Montero W (1994) Neotectonics and related stress distribution in a subduction-collisional zone: Costa Rica. Profil 7:125–141
Montero W (2001) Neotectónica de la región central de Costa Rica: frontera oeste de la microplaca de Panamá. Rev Geol Am Central 24:29–56. https://doi.org/10.15517/rgac.v0i24.8551
Mora S (1979) Estudio Geológico de una parte de la región Sureste del Valle del General, Provincia de Puntarenas, Costa Rica. BSc thesis, Universidad de Costa Rica
Mora S (1981a) Clasificación morfotectónica de Costa Rica. In: Inf Sem Julio-Diciembre. Instituto Geográfico Nacional, pp 35–55
Mora S (1981b) Barra Honda. Editorial Universidad Estatal a Distancia, San José
Orvis KH, Horn SP (2000) Quaternary glaciers and climate on Cerro Chirripó, Costa Rica. Quat Res 54:24–37. https://doi.org/10.1006/qres.2000.2142
Parkinson RW, Cortes J, Denyer P (1998) Passive margin sedimentation on Costa Rica's North Caribbean coastal plain, Río Colorado. Rev Biol Trop 46:221–236. https://doi.org/10.15517/rbt.v46i6
Pinter N, Gardner TW (1989) Construction of a polynomial model of glacio-eustatic fluctuation: estimating paleo-sea levels continuously through time. Geology 17:295–298. https://doi.org/10.1130/0091-7613(1989)017<0295:COAPMO>2.3.CO;2
Pizarro D (1993) Los pozos profundos perforados en Costa Rica: aspectos litológicos y bioestratigráficos. Rev Geol Am Central 15:81–85. https://doi.org/10.15517/rgac.v0i15.13240
Porras H, Cascante M, Granados R, Alvarado GE (2012) Volcanoestratigrafía y tectónica del Valle Central Occidental y las estribaciones de Los Montes del Aguacate a lo largo de la Ruta 27, Costa Rica. Rev Geol Am Central 47:69–93. https://doi.org/10.15517/rgac.v0i47.6489
Porras H, Alvarado GE, Arroyo-Solórzano M et al (2021) La depresión tectónica de Nicaragua en Costa Rica: estructura interna, estilo estructural, edad, extensión y actividad de la cuenca de San Carlos. Rev Geol Am Central 65:1–22. https://doi.org/10.15517/rgac.v0i65.46874
Protti M, Güendel F, Malavassi E (2001) Evaluación del potencial sísmico de la Península de Nicoya. Editorial Universidad Nacional, Heredia

Protti M, González V, Freymueller J, Doelger S (2012) Isla del Coco, on Cocos Plate, converges with San Andres Island, on the Caribbean Plate, at 78mm/year. Rev Biol Trop 60:33–41. https://doi.org/10.15517/rbt.v60i3.28307

Protti M, González V, Newman AV et al (2014) Nicoya earthquake rupture anticipated by geodetic measurement of the locked plate interface. Nat Geosci 7:117–121. https://doi.org/10.1038/ngeo2038

Ranero CR, von Huene R (2000) Subduction erosion along the Middle America convergent margin. Nature 404:748–752. https://doi.org/10.1038/35008046

Rivera-Flores F (2021) Análisis petrográfico y sedimentológico de los materiales de la Formación Suretka, Cuenca Limón Sur, Costa Rica. BSc thesis, Universidad de Costa Rica

Rivier F (1985) Sección geológica del Pacífico al Atlántico a través de Costa Rica. Rev Geol Am Central 2:23–32. https://doi.org/10.15517/rgac.v0i02.10480

Rojas W, Alvarado GE (2012) Marco geológico y tectónico de la Isla del Coco y la región marítima circunvecina, Costa Rica. Rev Biol Trop 60:15–32. https://doi.org/10.15517/rbt.v60i3.28306

Rojas-Barrantes M, Huapaya-Rodriguez S, Solari LA, Žáček V (2021) U-Pb geochronology of detrital zircons from San Carlos Basin, Costa Rica: evidence of Miocene volcanism and implications for the Precambrian and Paleozoic history of the Central American isthmus. J South Am Earth Sci 110:103311. https://doi.org/10.1016/j.jsames.2021.103311

Sak PB, Fisher DM, Gardner TW (2004) Effects of subducting seafloor roughness on upper plate vertical tectonism: Osa Peninsula, Costa Rica. Tectonics 23:1–16. https://doi.org/10.1029/2002TC001474

Sen Gupta BK, Malavassi LR, Malavassi E (1986) Late Miocene shore in northern Costa Rica: benthic foraminiferal record. Geology 14:218–220. https://doi.org/10.1130/0091-7613(1986)14<218:LMSINC>2.0.CO;2

Sequeira C, Chesnel V (2022) Descubrimiento del mayor intrusivo ígneo tabular del sur de Costa Rica (poster). In: 6to. Congreso Geológico UCR. San José

Soto GJ (2004) Meteoritos y meteoros en Costa Rica (verdaderos, posibles y falsos). Rev Geol Am Central 31:7–23. https://doi.org/10.15517/rgac.v0i31.7241

Tournon J, Alvarado GE (1997) Carte géologique du Costa Rica: notice explicative; Mapa geológico de Costa Rica: folleto explicativo, échelle-escala 1 500 000. Editorial Tecnológica de Costa Rica, Cartago

Vannucchi P, Tobin H (2000) Deformation structures and implications for fluid flow at the Costa Rica convergent margin, ODP Sites 1040 and 1043, Leg 170. J Struct Geol 22:1087–1103. https://doi.org/10.1016/S0191-8141(00)00027-4

Vannucchi P, Ranero CR, Galeotti S et al (2003) Fast rates of subduction erosion along the Costa Rica Pacific margin: implications for nonsteady rates of crustal recycling at subduction zones. J Geophys Res 108:1–13. https://doi.org/10.1029/2002jb002207

Vannucchi P, Fisher DM, Bier S, Gardner TW (2006) From seamount accretion to tectonic erosion: formation of Osa Mélange and the effects of Cocos Ridge subduction in southern Costa Rica. Tectonics 25:1–19. https://doi.org/10.1029/2005TC001855

von Huene RE, Azéma J, Coulbourn WT, et al (1985) Initial reports of the deep sea drilling project 67. U.S. Govt. Printing Office, Washington

von Huene R, Ranero CR, Weinrebe W, Hinz K (2000) Quaternary convergent margin tectonics of Costa Rica, segmentation of the Cocos Plate, and Central American volcanism. Tectonics 19:314–334. https://doi.org/10.1029/1999TC001143

von Huene R, Ranero CR, Watts P (2004) Tsunamigenic slope failure along the Middle America Trench in two tectonic settings. Mar Geol 203:303–317. https://doi.org/10.1016/S0025-3227(03)00312-8

Walther CHE (2003) The crustal structure of the Cocos ridge off Costa Rica. J Geophys Res 108:1–21. https://doi.org/10.1029/2001jb000888

Wells SG, Bullard TF, Menges CM et al (1988) Regional variations in tectonic geomorphology along a segmented convergent plate boundary pacific coast of Costa Rica. Geomorphology 1:239–265. https://doi.org/10.1016/0169-555X(88)90016-5

Werner R, Hoernle K, van den Bogaard P et al (1999) Drowned 14-m.y.-old Galápagos archipelago off the coast of Costa Rica: implications for tectonic and evolutionary models. Geology 27:499–502. https://doi.org/10.1130/0091-7613(1999)027<0499:DMYOGP>2.3.CO;2

Weyl R (1971) La clasificación morfotectónica de Costa Rica (América Central). In: Inf Sem Julio-Diciembre. Instituto Geográfico Nacional, pp 107–125

Weyl R (1980) Geology of Central America. Gebrunder Borntraeger, Berlin-Stuttgart

Wortel R, Cloetingh S (1981) On the origin of the Cocos-Nazca spreading center. Geology 9:425–430. https://doi.org/10.1130/0091-7613(1981)9<425:OTOOTC>2.0.CO;2

Zamora N, Méndez J, Barahona M, Sjöbohm L (2004) Volcano—estratigrafía asociada al campo de domos de Cañas Dulces, Guanacaste, Costa Rica. Rev Geol Am Central 30:41–58. https://doi.org/10.15517/rgac.v0i30.7253

Morphotectonic Regions of Costa Rica: A Review and Updated Classification

2

Mario Arroyo-Solórzano and Adolfo Quesada-Román

Abstract

The tectonic and geomorphological dynamics of Costa Rica are primarily driven by the interaction of tectonic plates and erosional and weathering processes that take place in this tropical environment. We focus our study on the history and evolution of the morphotectonic classifications for Costa Rica using geological, morphological, and seismotectonic data. Based on an in-depth review and analysis of previous morphotectonic subdivisions for Costa Rica, a synthesis is proposed in this study. Moreover, we carefully revised the accuracy of boundaries between different morphotectonic units through a detailed evaluation of the adjustment. For this purpose, comparison of geological maps, topography, and in some cases seismic zonation was performed. We first propose a subdivision of major tectonic elements: (1) tectonic plates, (2) the most relevant tectonic-structural elements, and (3) the morphotectonic regions (Arc-Trench sector, Inner-Arc, and Back-Arc). Based on this, we established ten specific morphotectonic units, grouping and encompassing Costa Rica's main reliefs and geomorphological features. These morphotectonic regions and units were formed primarily due to the interaction between the Cocos and Caribbean plates, and the Panama Microplate, in a subduction tectonic regime. Furthermore, is remarkable that the relief and morphotectonic composition of Costa Rica was originated due to historical interaction and evolution of several geotectonic processes, as well as the formation and modification of major tectonic structures or elements generated by the driving force of plate tectonics.

Keywords

Morphotectonics · Tectonic geomorphology · Plate tectonics · Landscape · Geological history · Costa Rica

2.1 Introduction and Background

From a geotectonic perspective, Central America exhibits significant diversity. Specifically, Costa Rica is situated within the Pacific Ring of Fire in a tectonic framework dominated by the subduction of the Cocos Plate beneath the Caribbean Plate (and the Panama microplate) at a rate of 83–89 mm/year (DeMets et al. 2010; Fig. 2.1). The neotectonic environment of Costa Rica is highly complex, with multiple internal forces inducing crustal movements and generating several tectonic structures (Montero and Morales 1990; Montero et al. 1992). Furthermore, the country lies within the Intertropical Convergence Zone (ITCZ), which is characterized by high humidity and temperature, resulting in substantial exposure to these external factors that shape and play a pivotal role in the relief and geomorphology of the country (Veas-Ayala et al. 2023). Based on these factors, it can be affirmed that gross landforms of Costa Rica are the result of the combined tectonic-erosional processes (Mora 1981; Quesada-Román 2021).

Some of the most important tectonic structures include the Panama Fracture Zone (PFZ), the Hess Scarp, as well as the extensive crustal deformation zones proposed as the boundaries between the Caribbean Plate and the Panama microplate: the Northern Panama Deformed Belt (NPDB) and the Central Costa Rica Deformed Belt (CCRDB) (e.g., Adamek et al. 1988; Fan et al. 1993; Marshall et al. 2000; Montero 2001; Fig. 2.1). Additionally, there are the Volcanic Arc Faults (VAF) and the Central American Forearc Block (CAFB), which moves northwestward at a velocity of 11 mm/year (Franco et al. 2012; Montero et al. 2017; Fig. 2.1). Another significant tectonic structure is the Cocos

M. Arroyo-Solórzano (✉)
Helmholtz Centre Potsdam, GFZ German Research Centre for Geosciences and University of Potsdam, Potsdam, Germany
e-mail: marroyo@gfz-potsdam.de

A. Quesada-Román
Escuela de Geografía, Universidad de Costa Rica, San Pedro, San José, Costa Rica
e-mail: adolfo.quesadaroman@ucr.ac.cr

A. Quesada-Román (ed.), *Landscapes and Landforms of Costa Rica*, World Geomorphological Landscapes, https://doi.org/10.1007/978-3-031-64940-0_2

Fig. 2.1 Tectonic framework and morphotectonic regions of Costa Rica (see Table 2.1). The region contained within the dotted lines represent the Central Costa Rica Deformed Belt (CCRDB). The dashed line represents the simplified northeast boundary of the Central American Fore-arc Block along the Volcanic Arc Faults (VAF). CAFB—Central America Fore-arc Block, FS—Fisher Seamount, NPDB—North Panama Deformed Belt, PFZ—Panama Fracture Zone, QP—Quepos Plateau, RSB—Rough-Smooth Boundary

Ridge, a volcanic submarine range that extends 1200 km in length and is 100–350 km wide, which is subducted off the southeastern sector of the Pacific coast of Costa Rica (Alvarado et al. 2017). Table 2.1 summarizes the main tectonic features of Costa Rica described above.

Morphotectonic classifications are part of current trends in geomorphological studies and refer to the grouping of structural-tectonic units of the Earth's crust and their role in the shaping of the present-day relief (Dramis et al. 2011; Velásquez et al. 2015). In their characterization and classification, the aim is to highlight important tectonic structures based on the recognition of geomorphological features and recent tectonic processes (Lin et al. 2013). Numerous morphostructural criteria are considered to establish the hierarchy of the identified units. Some of the main criteria include morphological, morphometric, genetic, geodynamic, and analysis of the tectonic activity (e.g., Golts and Rosenthal 1993; Modenesi-Gauttieri et al. 2002; Fodor et al. 2005; Magaz et al. 2006; Cotilla et al. 2007, Arrowsmith and Zielke 2009; Bathrellos and Antoniou 2009; Ali et al. 2010; Velásquez et al. 2015; Sofia 2020; Arroyo-Solórzano et al. 2021).

Table 2.1 Tectonic setting of Costa Rica (see Fig. 2.1)

General tectonic features	Name
Tectonic plates	Cocos plate Caribbean plate Panama microplate Nazca plate
Main tectonic structures	Mesoamerican trench Panama fracture zone North Panama deformed belt Central Costa Rica deformed belt Central American forearc block Volcanic arc faults Hess scarp Cocos ridge
Morphotectonic regions	Arc-trench sector (Fore-arc) Inner-arc Back-arc

Previous morphotectonic classifications for Costa Rica, refining and detailing possible subdivisions of the country's relief, were based on its geotectonic origin. The very initial approaches stemmed from Dengo (1968) and Weyl (1971) who established the first morphotectonic schemes which were later described from a geomorphic perspective (Bergoeing and Brenes 1977). Subsequently, the first formal classification, aligning with global trends, was proposed by Mora (1981), which was based on terminology provided by Dickinson (1975) using some adaptations. Since then, multiple studies on the internal geodynamics of Costa Rica and research focused on the seismotectonic characterization of the country allowed for further improvement and a better comprehension of the evolution of morphotectonic units and the main geomorphological features in each of them (e.g., Duncan and Hargraves 1984; Morales 1985; Montero and Morales 1990; Montero et al. 1992; Goes et al. 1993; Protti and Schwartz 1994; Di Marco et al. 1995; Kolarsky and Mann 1995; Westbrook et al. 1995; Marshall et al. 2000; Montero 2001; Fodor et al. 2005; Bunduschuh and Alvarado 2007; LaFemina et al. 2009; Kobayashi et al. 2014).

Recent proposals include updated maps that refine and/or modify previously defined structures as the tectonic knowledge of the country improves. Among these are the Tectonic Atlas of Costa Rica (Denyer et al. 2003) and the national geological map (Denyer and Alvarado 2007). Additionally, other studies showed efforts to refine the classification of Mora (1983), providing a comprehensive description of these morphotectonic units and including some marine structures in the descriptions (Alvarado and Cárdenes 2016).

The purpose of this chapter is to focus on, and reinforce, morphotectonic classifications previously developed for Costa Rica, providing a synthesis of the morphotectonic unit classifications to date. It is evident that the main units do not vary significantly from previous proposals, but the map and morphotectonic units presented here incorporate information from detailed digital elevation models and involve combination or revision of previous classifications with regional seismic zoning (Alvarado et al. 2017). The Chap. 1 of this book provides a detailed description of the main structures and morphotectonic units that are discussed here, so the descriptions we present are more concise and briefer, inviting readers to refer to the previously mentioned chapter for more details. A disaggregation and classification of morphotectonic units is presented, which we consider suitable and practical for widespread use, serving as a reference at the national level.

2.2 Summary of the Geotectonic History and Evolution

The geological basement of Costa Rica was formed during the Jurassic-Cretaceous period. Its bedrock originated from intense submarine volcanism along a mid-ocean ridge, where the Farallon Plate was moving northeast, subducting beneath the western edge of both continents (North and South America), and other tectonic blocks were moving eastward from the North American Plate, such as the Chortis Block (Duncan and Hargraves 1984). The crust of the Farallon Plate thickened on its journey as it passed over the Galapagos hotspot, producing the Caribbean Large Igneous Province (CLIP), and the movement of the Farallon Plate displaced rocks originally formed much farther southwest of Costa Rica's current geographic position, including the Santa Elena, Nicoya, Herradura, and Osa peninsulas (Di Marco et al. 1995).

For the Paleocene to early Eocene period, approximately 55 Ma, rock formations indicate the presence of turbid currents on the continental slope, suggesting that Costa Rica was an archipelago at that time. It is during this period that the Costa Rican territory acquired its current position, and the tectonic features began to resemble their present-day configuration (Pindell and Barrett 1990). It is at this point that subduction along the Mesoamerican Trench occurred, defining the existence of the Caribbean Plate in the inter-American tectonic puzzle (Duncan and Hargraves 1984). This led to the appearance of new islands that built the current fore-arc, including the Tempisque depression.

During the early Miocene, approximately 20 Ma, due to subduction processes the Farallon Plate was fragmented, giving rise to the Cocos Plate as an individual tectonic plate, moving northeastward (Di Marco et al. 1995). This event favored the emergence of numerous active volcanoes, whose remnants form the current Talamanca, Guanacaste, Tilarán, and Central Volcanic Ranges, as well as the Central Tectonic Depression and the General Valley. Toward the end of the Miocene (~ 5 Ma), it is inferred that the Cocos Ridge collided with the Mesoamerican Trench, causing significant geotectonic repercussions, such as the uplift of Talamanca (Pindell and Barrett 1990).

In the Pliocene, approximately 3.5 Ma, intense volcanic activity, combined with sedimentary basin infilling and tectonic processes, led to the definitive closure of the Central American Isthmus (Coates 1997). Finally, in the Quaternary period from 1.8 Ma to the present, the current pattern of active volcanoes in Costa Rica began to take shape, particularly those located in the Central and Guanacaste Ranges. During this period, especially in the Last Glacial Maximum, there was also climate cooling resulting in glacier formation and glacial modeling in the highest mountains of Costa Rica (Quesada-Román et al. 2021). Since their melting around 10,000 ka, the country's current relief has been fully configured, including the formation of major plains through exogenous processes (Duncan and Hargraves 1984).

2.3 Previous Morphotectonic Classifications

We present an overview of the main morphotectonic classifications or geodynamic subdivisions upon which the synthesis presented in this chapter is based. In this section, we briefly summarize the morphotectonic subdivision proposed by Mora (1981), the internal geodynamic structures outlined by Montero et al. (1992), and the modifications suggested by Denyer et al. (2003) and Alvarado and Cárdenes (2016). Mora (1981) defined his morphotectonic classification based on four major tectonic units or elements: the Cocos Plate, the Mesoamerican Trench, the Benioff Zone, and the Caribbean Plate. The Cocos Plate and the Mesoamerican Trench are defined in a very general manner as they do not encompass the continental territory of Costa Rica. Regarding the Cocos Plate, Mora mentioned the influence of the Galápagos hotspot on its bathymetric structures such as the Fisher Seamount and the Quepos Plateau, as well as the rough-smooth boundary, plate motion velocities, and the existence of the triple junction with the Nazca and Panama plates through the PFZ (Fig. 2.1). Concerning the trench, he only indicated its extension from the PFZ to the southern coast of Mexico, and the convergence rate between plates.

Regarding the Benioff Zone delimitations, Mora (1981) suggested four segments with distinct characteristics based on their respective dips and depths of seismic activity. Thus, he defined the segments of Guanacaste, Puntarenas, Parrita, and Golfito. This subdivision is not mapped as it refers to deep seismic activity related to the behavior of the subducting slab. However, for further reference, it is nowadays recommended to consult more recent seismic zoning studies that provide more detailed information on these aspects (e.g., Morales 1985; Climent et al. 2008; Benito et al. 2012; Alvarado et al. 2017).

Lastly, Mora (1981) described different morphotectonic structures of the Costa Rican territory by detailing the Caribbean Plate since at that time, the Panama Microplate was not yet recognized as a separate lithospheric block. He subdivided it into three main domains: the Arc-Trench Segment, composed by the outer arc and the intra-trench region; the Inner-Arc, composed by the intra-arc basins and the plutonic-volcanic arc encompassing the country's main mountain ranges; and finally, the Fore-arc Area. Each of these subdivisions encompasses different reliefs and geomorphological features. It is worth noting that this classification served as the basis for subsequent proposals, which are also explained below. Likewise, it is the basis for the synthesis we present here, where the boundaries between each morphotectonic unit have mainly been refined, and in some cases, the respective terminology was updated.

Another significant work is Montero et al. (1992), focusing on the internal geodynamics of Costa Rica. These authors included a highly relevant map that integrated the classifications of Mora (1981) with information on Plio-Pleistocene volcanoes, seismic activity with focal mechanisms, mapped Quaternary faults, as well as gravity anomalies and tectonic stress data. Montero et al. (1992) described and highlighted the following main morphotectonic domains: the Fore-arc, the Fore-arc Basin, the Inner-Arc, the Back-arc Area, and additionally, the Colombia Basin. Furthermore, in their map crustal thickness was added and emphasized where, as expected, an increase is observed from the oceanic area to the Inner-Arc, with thicknesses of up to 40 km in the latter.

Finally, the most recent proposals are the Tectonic Atlas of Costa Rica (Denyer et al., 2003) and the work of Alvarado and Cárdenes (2016). These studies improved and provided detailed descriptions of the morphotectonic map of Costa Rica, adding newly mapped tectonic structures after the first classifications, as well as updating part of their nomenclature. They also included studies of marine geology and structures on the ocean floor of the Cocos Plate. The authors further broke down the morphotectonic regions presented by Weyl (1971) and Mora (1981), assigning different landforms that correspond to the tectonic genesis of each sector. In these new morphotectonic classifications, seven major morphotectonic regions were identified, encompassing different reliefs or geomorphological features of Costa Rica:

- Trench and bathymetric features (mountain ranges and seamounts)
- Fore-arc basement (first uplifted terrains in the country, mainly the peninsulas)

- Fore-arc basins (depositional basins of the first uplifted terrains)
- Magmatic front (main mountain ranges of the country)
- Intra-magmatic axis basins (tectonic depressions within the volcanic arc)
- Thrust-fold deformation belts (recent relevant uplift)
- Back-arc basins (depositional basins and alluvial plains)

2.4 Synthesis of the Morphotectonic Classification for Costa Rica

In this chapter, a map that compiles the main previously described works is presented, seeking more detail and updating the nomenclature of the initial classifications (e.g., Weyl 1971; Mora 1981; Montero et al. 1992), but at the same time providing a synthesis or grouping of morphotectonic domains from more recent morphotectonic classifications (e.g., Denyer et al. 2003; Alvarado and Cárdenes 2016), with the aim of achieving a simplified map for practical use. Furthermore, in delimiting each of the proposed morphotectonic units, we conducted a detailed analysis of relief based on a Digital Elevation Model (DEM) with a 10 m resolution, as well as a review of the overlap and checking the correspondence between geology and previous morphotectonic classifications, based on the geological map of Costa Rica at a scale of 1:400,000 (Denyer and Alvarado 2007). Additionally, the most recent seismic zoning for Central America (Alvarado et al. 2017) was used as an additional and complementary criterion to define the boundaries between units.

Table 2.2 summarizes the grouping, where the Costa Rican territory is divided into three main morphotectonic regions: (1) Arc-Trench sector, (2) Inner-Arc, and (3) Back-Arc (Fig. 2.1). These regions are controlled and defined by tectonic relations and regional stresses resulting from the interaction of main tectonic structures explained in Table 2.1. Based on these main morphotectonic regions, each of them was further disaggregated to establish ten main morphotectonic units, and within each of these units, the main relief systems and geomorphological features that comprise them are presented (Fig. 2.2 and Table 2.2).

2.4.1 Arc-Trench Sector

The transition zone between the Inner-Arc and the subduction zone between the Cocos Plate and the Caribbean Plate/Panama Microplate is included in this section. In addition to the main morphotectonic units within the Costa Rican territory, it also encompasses the Middle American Trench and the domains of seafloor morphology off the Pacific coast of Costa Rica, where notable structures such as the Fisher Seamount, the Quepos Plateau, and the Cocos Ridge stand out. The Cocos Ridge is the longest range in Costa Rica, extending for approximately 815 km in the marine territory (Von Huene et al. 2000) (Fig. 2.1).

2.4.1.1 Fore-Arc (1a)

The Fore-Arc is called the "ophiolitic promontories" zone, where the oldest rocks and geological basement of Costa Rica are found. It mainly consists of basic igneous rocks such as basalts, gabbros, and volcaniclastic sediments of Cretaceous to Early Eocene age (65–45 Ma) (Denyer and Alvarado 2007; Alvarado and Cárdenes 2016). The main geomorphological features of this unit are the peninsulas of

Table 2.2 Grouping and subdivision of morphotectonic units of Costa Rica

Morphotectonic region	Morphotectonic unit	Name
1. Arc-Trench sector	**1a**. Fore-Arc	• Santa Elena, Nicoya, Osa, and Burica eninsulas • Herradura and Quepos promontories • Nicaragua fore-arc
	1b. Intra-Trench tectonic depressions	• Nicoya and Dulce gulfs • Tempisque, Parrita, Diquis and Coto-Colorado valleys
	1c. Recent tectonic uplifts	• Coasts of Liberia, Nicoya, Central Pacific, Dominical, and South Zone
	1d. Thrust fault deformation belt	• Fila Costeña
2. Inner-Arc	**2a.** Plutonic/Volcanic Arc	• Guanacaste and Central Volcanic Mountain ranges • Tilarán, Aguacate, and Talamanca Mountain ranges
	2b. Intra-Arc tectonic depressions	• Central, Orosi, Turrialba, Talamanca and General Coto-Brus valleys, and Arenal Lake
3. Back-Arc	**3a.** Limon South Basin	• De la Estrella valley, Punta Uva, Manzanillo, and Cahuita
	3b. Limon North Basin	• Tortuguero plains, Barra del Colorado, Calero island
	3c. San Carlos Basin	• San Carlos plains
	3d. Nicaragua tectonic depression and back-arc	• Graben and Caribbean of Nicaragua

Fig. 2.2 Morphotectonic units of Costa Rica (see Table 2.2). The tectonic boundaries are the same as Fig. 2.1. AgMR = Aguacate mountain range, AL = Arenal lake, CV = Central valley, CCV = Coto-Colorado valley, CVMR = Central Volcanic Mountain range, DV = Diquis valley, FC = Fila Costeña, GCV = General Coto-Brus valley, GVMR = Guanacaste Volcanic Mountain range, OV = Orosi valley, P = Parrita, Q = Quepos, SE = Santa Elena peninsula, TaV = Talamanca valley, TiMR = Tilarán mountain range, and TV = Turrialba valley

Santa Elena, Nicoya, and Osa, as well as the promontories of Quepos and Herradura (Figs. 2.2 and 2.3a–d).

2.4.1.2 Intra-Trench Tectonic Depressions (1b)

The unit includes tectonic depressions filled with sediments with more than 4.5 km of strata that span various ages. These sediments include shallow-marine carbonate deposits from the Middle Eocene, deep-water sedimentation from the Oligocene to Miocene, and shallow-water to subaerial mudstone and sandstone deposits from the Miocene to Pliocene. They are followed by volcaniclastic rocks of Pliocene to Lower Quaternary age, as well as non-marine sands and gravels of Quaternary age (Denyer and Alvarado 2007; Alvarado and Cárdenes 2016). The major rivers in this area flow through valleys controlled by faults, which have incised into Neogene-Quaternary nearshore sediments, volcaniclastic debris, and pyroclastic deposits blocks (Denyer and Alvarado 2007). The main geomorphological features of this unit are the Nicoya and Dulce gulfs, as well as the Tempisque, Parrita, Diquis (Acuña-Piedra and Quesada-Román 2021), and Coto-Colorado valleys (Figs. 2.2 and 2.3g, h).

Fig. 2.3 Geomorphological features of the Arc-Trench sector (see Fig. 2.2). ***Fore-arc (1a):*** **a** view to the west of Bahia Salinas gulf, Guanacaste; **b** view to the south of Playas del Coco beach, Guanacaste; **c** view to the north of the Papagayo gulf, Guanacaste; **d** view to the south of Manuel Antonio beach, Quepos. ***Thrust fault deformation belt (1d):*** **e** and **f** view to the NE of Fila Costeña range, south of Pérez Zeledón. ***Intra-Trench tectonic depressions (1b):*** **g** view to the SW of the Tempisque valley, Guanacaste; **h** view to the SW of the Dulce gulf, Puntarenas. ***Recent tectonic uplifts (1c):*** **i** view to the south of Dominical beach, Puntarenas; **j** view to the south of Flamingo beach, Guanacaste. **k** view to the south of Guacalillo cliff, Puntarenas; and **l** view to the SW of the Coto-Colorado valley, Puntarenas

2.4.1.3 Recent Tectonic Uplift Areas (1c)

Uplifted marine terraces and paleo-beach deposits indicate active emergence throughout the late Quaternary within or near to the Fore-arc peninsulas (Gardner et al. 2001; Denyer and Alvarado 2007). These uplifts primarily result from significant earthquakes occurring within the seismic cycle (Protti et al. 2014). Additionally, is proposed that the uplift of these terraces is linked to the subduction of seamounts along the projected trend of the Fisher seamount (Gardner et al. 2001). Differential uplift has been identified between the Herradura block and adjacent lower-relief blocks, where dip-slip displacements along steep margin-perpendicular faults accommodate the uplift process. Moreover, ongoing deformation is evidenced by the presence of uplifted

Fig. 2.3 (continued)

wavecut platforms along the coast of the Osa peninsula (Corrigan et al. 1990). The elongated north–south shape of the Burica peninsula is a consequence of the influence of the PFZ on its uplift (Alvarado and Cárdenes 2016). Notable examples include the Dominical beach in Puntarenas, Flamingo beach in Guanacaste, Guacalillo cliff in Puntarenas, and several locations in southern Costa Rica (Figs. 2.2 and 2.3i–l).

2.4.1.4 Thrust Fault Deformation Belt (1d)

In this study, we include this unit as part of the Arc-Trench sector because it is situated between the Inner-Arc and the trench. Other classifications group it as part of the Inner-Arc or as an independent morphotectonic region. The region is characterized by predominantly northwest-southeast-oriented reverse faults, resulting from compressive stresses (Quesada-Román et al. 2023). Morphologically, it is characterized by significant escarpments, notably represented by the Fila Costeña (Figs. 2.2 and 2.3e, f), a mountain range approximately 150 km long and 15–25 km wide, with peak elevations ranging from 1100 to 1400 m. The Fila Costeña runs parallel to the South Pacific coast in a northwest-southeast direction (Morell et al. 2008).

2.4.2 Inner-Arc

We subdivide it in two main units: the Plutonic/Volcanic arc and the Intra-Arc tectonic depressions. The Plutonic/Volcanic arc represents the axis of Costa Rica and consists of two active volcanic ranges: the Cordillera Central and Guanacaste, as well as two extinct ranges, the Tilarán-Aguacate and Talamanca. Moreover, the Intra-Arc tectonic depressions are also present as the flat zones in this region (Fig. 2.2).

2.4.2.1 Plutonic/Volcanic Arc (2a)

This unit comprises the main volcanic and plutonic mountain ranges of the country (Fig. 2.2). Among the volcanic mountain ranges, the Guanacaste Range (90 km long) and the Central Range (105 km long) stand out, composed of complex stratovolcanoes and andesitic shield volcanoes. The volcanoes of Rincón de la Vieja, Arenal, Poás, Irazú, and Turrialba have shown the highest activity in recent decades (Alvarado and Gans 2012). Among plutonic mountain ranges, especially prominent are the Talamanca Range (190 km long, characterized by topographic elevations over 3500 m) and the Tilarán-Aguacate Range (100 km long, consisting of densely dissected remnants of stratovolcanoes). It is suggested that volcanic activity in these ranges ceased in the Pliocene, with the Talamanca Range having the highest altitudes in the country (Alvarado and Gans 2012). Some examples of geomorphic landscapes within this morphotectonic unit include the Central Volcanic Mountain range (Fig. 2.4a), monogenetic cones at Irazú volcano (Fig. 2.4b), the Talamanca Mountain range (Fig. 2.4c), and the Cerro Chirripó hill (Fig. 2.4d).

2.4.2.2 Intra-Arc Tectonic Depressions (2b)

Intermountain and intramountain tectonic depressions (Fig. 2.2) are primarily of tectonic origin, filled with volcanic, fluvial, and hillslope sediments (Mora 1981). Among them, the Arenal Lake stands out (Fig. 2.4h), located between the Arenal volcano and the eroded remnants of the Monteverde volcanoes, and limited by active strike-slip and normal faults. The Central Valley (Fig. 2.4a) is also noteworthy, situated between active volcanoes of the Central Volcanic Mountain range and the eroded volcanic remnants of the Aguacate and Talamanca ranges, and filled with a thick accumulation of volcanic products. Additionally, the General and Coto-Brus valleys (Fig. 2.4g) in southern Costa Rica are notable (Alvarado and Cárdenes, 2016; Camacho et al. 2020). Other minor intra-arc depressions include the Orosi Valley (Fig. 2.4e), Turrialba Valley (Fig. 2.4f), and Talamanca Valley.

2.4.3 Back-Arc

The Back-Arc corresponds to the zone located below the present Inner-Arc, which mainly includes the plains of the northern and Caribbean slopes of the country (Fig. 2.2). This region is built of clastic, marine, and continental sediments, and some alkaline volcanic centers are present. Moreover, active compressional tectonics has been evidenced, mainly in Limón Sur (e.g., Mora 1981; Camacho et al. 2010; Alvarado et al. 2017). This region is composed of the Limón North and Limón South basins, the San Carlos Basin, and the Nicaraguan graben tectonic depression, characterized by different structural settings.

2.4.3.1 Limon South Basin (3a)

The basin is characterized by a strong compressive regime that includes several historical earthquakes (Brandes et al. 2007; Alvarado and Cárdenes 2016; Quesada-Román 2021). Structurally, it belongs to the North Panama Deformed Belt, which is a wide zone of folds and thrust faults on the Caribbean margin of Panama, suggested as a zone of incipient subduction, causing a double subduction in southern Costa Rica (e.g., Camacho et al. 2010; Arroyo & Linkimer 2021; Bourke et al. 2023). The Limón Sur thrust system is characterized by a series of asymmetric anticlines, listric faults, and predominant northwest-striking planes with a detachment level at a depth of 5–7 km in the outer fore-arc (Zamora et al. 2021). The structural boundary between this basin and the Limón North basin is a zone of active seismicity known as Alto Moín (Fig. 2.2), marking the transition between the Central Costa Rica Deformed Belt (Montero 2001) and the NPDB (Protti and Schwartz 1994). The main geomorphological features of this unit are De la Estrella Valley (Fig. 2.5c), Punta Uva, Manzanillo, and Cahuita (Fig. 2.5d).

2.4.3.2 Limon North Basin (3b)

The Limon North Basin presents a relatively straight coastal plain morphology (Fig. 2.2), with beach ridges parallel to the coast (Quesada-Román and Pérez-Briceño 2019). It is a wide and structurally homogeneous depocenter, with normal listric faults that were active during the Quaternary (Brandes et al. 2007; Zamora et al. 2021). This basin is affected by low to moderate magnitude earthquakes, related to the Hess Scarp (Fig. 2.1) and faults in the eastern sector of the Parismina river. Additionally, there is a seismic

Fig. 2.4 Geomorphological landscapes of the Inner-Arc (see Fig. 2.2). ***Plutonic/Volcanic Arc (2a):*** **a** view to the south of the Central Volcanic Mountain range and valley; **b** view to the west of a monogenetic cone, Irazú volcano, Cartago; **c** view to the NE of the Talamanca Mountain range; **d** view to the north of Cerro Chirripó hill, Talamanca. ***Intra-Arc tectonic depressions (2b):*** **e** view to the south of the Orosi valley; **f** view to the west of the Turrialba valley; **g** view to the NE of the General Coto-Brus valley; **h** view to the west of the Arenal volcano and its lake

swarm known as the "Parismina nest," which stands out for being in a low-seismicity zone (Linkimer et al. 2018; Arroyo-Solórzano and Linkimer 2021). The main geomorphological features of this unit are the Tortuguero plains (Fig. 2.5b), Barra del Colorado, and Calero Island.

2.4.3.3 San Carlos Basin (3c)

San Carlos Basin is mostly covered by recent alluvial and volcanic deposits, except for some sites along the San Juan River and in the Venado Fault (Porras et al. 2021). In this basin, tectonic-induced unconformities have been identified within the Neogene succession, associated with different structural styles of extension and compression. The basin has two main morphostructures described by Ballestero et al. (1995) and named by Porras et al. (2021) as the Venado and Machuca structures on the northern and southern margins, respectively, defined by topographic lows filled with volcanic and sedimentary deposits in the middle part of the basin (Porras et al. 2021). A recent study in the San Carlos Basin has revealed Paleozoic ages within detrital zircons, shedding light on a previously unknown aspect of this region. These zircons are believed to have strong connections with northern Central America and are thought to have originated from fragmented formations of the continental crust (Rojas-Barrantes et al. 2021). The main geomorphological features of this unit are the San Carlos plains (Figs. 2.2 and 2.3a).

2.4.3.4 Nicaragua Tectonic Depression and Back-Arc (3d)

It is a complex morphotectonic structure related to an extensional regime, characterized by predominantly normal faults located on the southwest edge of the Nicaragua Lake, with a NW–SE orientation and dipping toward the northeast, forming a graben (Alonso-Henar et al. 2015; Porras et al. 2021). Its tectonic origin is not fully established though, and different models have been proposed to explain the formation of the graben, commonly involving a slab roll-back process and volcanic front migration (Porras et al. 2021). Geomorphologically, it corresponds to lowlands that encompass an extensive alluvial plain, typically at an elevation of less than 100 m, and includes some volcanic relief. Major rivers drain toward the volcanic cordillera, traversing the alluvial lowlands and carrying a high sediment load that is deposited across broad inland floodplains (Alvarado and Cárdenes 2016).

Fig. 2.5 Geomorphological features of the Back-Arc morphotectonic units (see Fig. 2.2). **a** San Carlos plains (2.5a), **b** Tortuguero plains (2.5b), **c** La Estrella River (2.5c), and **d** Cahuita beach (2.5d)

2.4.4 Final Remarks

The main morphotectonic classifications for Costa Rica were revisited, and we present a new map that integrates, synthesizes, and provides an updated representation with sufficient detail and usefulness for its practical and simplified use. We propose a classification that primarily distinguishes between major tectonic elements, including: (1) tectonic plates that interact or have influence over Costa Rica, (2) the largest and most relevant tectonic-structural elements, and (3) the morphotectonic regions (Arc-Trench sector, Inner-Arc, and Back-Arc).

Ten specific morphotectonic units were determined, encompassing Costa Rica's main geomorphological landscapes and features. These units were carefully delineated, and their boundaries defined, involving a review and overlap of previous morphotectonic classifications, analysis of detailed digital elevation models, the geological map of Costa Rica, and even recent seismic zoning. The Arc-Trench sector was subdivided into four units, consisting of the Fore-Arc, Intra-Trench tectonic depressions, recent tectonic uplift zones, and the thrust fault deformation belt. The Inner-Arc has been further divided into two units, namely the Plutonic/Volcanic Arc, which encompasses the country's main mountain ranges, and the Intra-Arc tectonic depressions, which constitute flat intermontane regions and the most densely populated areas of the country. Finally, the Back-Arc was subdivided into the Limon South and North basins, the San Carlos Basin, and the Nicaragua tectonic depression, comprising the country's main plains.

We conclude that relief and morphotectonic composition of Costa Rica owe their origin to the historical interaction and evolution of several geotectonic processes which occurred in the territory, as well as to the formation and modification of some major tectonic structures or elements generated by the driving force of plate tectonics. The morphotectonic regions and units were created primarily due to the interaction between the Cocos and Caribbean plates and the Panama Microplate, and this interaction being the main generator of the tectonic stresses, forces, and the tectonic regime that dominates in the country (subduction). However, through the breakdown of the morphotectonic units into smaller subunits, it can be affirmed that many other structural elements have a significant influence on Costa Rica's relief, such as the North Panama Deformed Belt, the Panama Fracture Zone, the Inner-arc processes and faulting, the Coco Ridge and submarine mounts, the Longitudinal fault, as well as the Nicaragua tectonic depression. On the other hand, structures inferred from seismic or geodetic data, such as Central Costa Rica Deformed Belt and the sliver connecting with the back-arc, do not appear to have a clear influence on the identified morphostructures.

References

Acuña-Piedra JF, Quesada-Román A (2021) Multidecadal biogeomorphic dynamics of a deltaic mangrove forest in Costa Rica. Ocean Coast Manag 211:105770

Adamek S, Frohlich C, Pennington WD (1988) Seismicity of the Caribbean- Nazca boundary: constraints on microplate tectonics of the Panama region. J Geophys Res 93:2053–2075. https://doi.org/10.1029/JB093iB03p02053

Ali SA, Jones B, Buckman S et al (2010) Morpho tectonic units of the Zagros Orogenic Belt, NE Iraq: a modern Morpho tectonic units of the Zagros Orogenic Belt, NE Iraq: a modern analogue for subduction accretion processes analogue for subduction accretion processes Recommended Citation Recommended Citation. Science, Medicine and Health—Papers: part A 16–22

Alonso-Henar J, Schreurs G, Martinez-Díaz JJ et al (2015) Neotectonic development of the El Salvador Fault Zone and implications for deformation in the Central America Volcanic Arc: insights from 4-D analog modeling experiments. Tectonics 34:133–151. https://doi.org/10.1002/2014TC003723

Alvarado GE, Benito B, Staller A et al (2017) The new Central American seismic hazard zonation: mutual consensus based on up to day seismotectonic framework. Tectonophysics 721:462–476. https://doi.org/10.1016/j.tecto.2017.10.013

Alvarado GE, Cárdenes G (2016) Geology, tectonics, and geomorphology of Costa Rica: a natural history approach. In: Kappelle M (ed) Costa Rican Ecosystems, 1st edn. Chicago Scholarship Online, Chicago

Alvarado GE, Gans PB (2012) Síntesis geocronológica del magmatismo, metamorfismo y metalogenia de Costa Rica, América Central. Revista Geológica de América Central 46:7–122

Arrowsmith JR, Zielke O (2009) Tectonic geomorphology of the San Andreas Fault zone from high resolution topography: an example from the Cholame segment. Geomorphology 113:70–81. https://doi.org/10.1016/j.geomorph.2009.01.002

Arroyo IG, Linkimer L (2021) Geometría de la zona sismogénica interplacas en el sureste de Costa Rica a la luz de la secuencia sísmica de Golfito del 2018. Geofísica Internacional 60:51–75

Arroyo-Solórzano M, Linkimer L (2021) Spatial variability of the b-value and seismic potential in Costa Rica. Tectonophysics 814. https://doi.org/10.1016/j.tecto.2021.228951

Arroyo-Solórzano M, Quesada-Román A, Vargas-Bolaños C (2021) Morphotectonic-volcanic units of the northwest sector of Poás volcano, Costa Rica. Investigaciones Geograficas. https://doi.org/10.14350/RIG.60279

Ballestero C LE, Dobrinescu MM, Jager Refmadora Costarricense de Petr GC, et al (1995) An integrated geological and geophysical interpretation of the San Carlos basin, Costa Rica. In: Miller R, Escalante G, Reinermund J, Bergin M (eds) Circum-Pacific council for energy and mineral resources earth science series. Springer, Berlin Heidelberg

Bathrellos GD, Antoniou V (2009) Morphotectonic characteristics of Leas Island during the Quaternary (Ionian Sea, Greece). Geologica Balcanica 1–3:23–33

Benito B, Lindholm C, Camacho E et al (2012) A new evaluation of seismic hazard for the Central America Region. Bull Seismol Soc Am 102:504–523. https://doi.org/10.1785/0120110015

Bergoeing J, Brenes L (1977) Regiones morfográficas de Costa Rica. Informe Semestral, Instituto Geografico Nacional 53–58

Bourke JR, Levin V, Arroyo IG, Linkimer L (2023) Evidence for Caribbean plate subduction in southern Costa Rica. Geology 51:408–412. https://doi.org/10.1130/G50796.1

Brandes C, Astorga A, Back S et al (2007) Deformation style and basin-fill architecture of the offshore Limón back-arc basin

(Costa Rica). Mar Pet Geol 24:277–287. https://doi.org/10.1016/j.marpetgeo.2007.03.002
Bunduschuh J, Alvarado G (2007) Central America: geology, resources, hazards. Taylor & Francis, London
Camacho E, Hutton W, Pacheco JF (2010) A new look at evidence for a wadati-benioff zone and active convergence at the north Panama deformed belt. Bull Seismol Soc Am 100:343–348. https://doi.org/10.1785/0120090204
Camacho ME, Quesada-Román A, Mata R, Alvarado A (2020) Soil-geomorphology relationships of alluvial fans in Costa Rica. Geoderma Reg 21:e00258. https://doi.org/10.1016/j.geodrs.2020.e00258
Climent Á, Rojas W, Alvarado GE, Benito B (2008) Proyecto Resis II Evaluación de la amenaza sísmica en Costa Rica
Coates A (1997) Central America: a natural and cultural history. Yale University Press, New Haven
Corrigan J, Mann P, Ingle JC (1990) Forearc response to subduction of the Cocos Ridge, Panama-Costa Rica. Bull Geol Soc Am 102:628–652. https://doi.org/10.1130/0016-7606(1990)102<0628:FRTSOT>2.3.CO;2
Cotilla MO, Cárdoba D, Calzadilla M (2007) Morphotectonic study of Hispaniola. Geotectonics 41:368–391. https://doi.org/10.1134/S0016852107050032
DeMets C, Gordon RG, Argus DF (2010) Geologically current plate motions. Geophys J Int 181:1–80. https://doi.org/10.1111/j.1365-246X.2009.04491.x
Dengo O (1968) Estructura geológica, historia tectónica y morfología de América Central. ICAITI, Ciudad de Guatemala, Guatemala
Denyer P, Alvarado G (2007) Mapa geológico de Costa Rica [escala 1:400000]. Librería Francesa, San José, Costa Rica
Denyer P, Montero W, Alvarado G (2003) Atlas Tectónico de Costa Rica (2a ed.), 2nd edn. Editorial Universidad de Costa Rica, San José, Costa Rica
Di Marco G, Baumgartner P, Channell J (1995) Late Cretaceous-Early Tertiary paleomagnetic data and a revised tectonostratigraphic subdivision of Costa Rica and western Panama. In: Geologic and tectonic development of the Caribbean plate boundary in southern Central America. Geological Society of America, Boulder, Colorado, USA, pp 1–27
Dickinson W (1975) Plate tectonics and sedimentation. Tectonics and Sedimentation 22:1–27
Dramis F, Guida D, Cestari A (2011) Nature and aims of geomorphological mapping. In: Developments in Earth Surface Processes. Elsevier B.V., pp 39–73
Duncan R, Hargraves R (1984) Late tectonic evolution of the Caribbean region in the mantle reference frame. In: Bonini W, Hargraves R, Shagan R (eds) The CaribbeanSouth American Plate Boundary and Regional Tectonics. Geological Society of America Memoir. Geological Socienty of America, Colorado, USA, pp 81–93
Fan G, Beck SL, Wallace TC (1993) The seismic source parameters of the 1991 Costa Rica aftershock sequence: evidence for a transcurrent plate boundary. J Geophys Res 98. https://doi.org/10.1029/93jb01557
Fodor L, Bada G, Csillag G et al (2005) An outline of neotectonic structures and morphotectonics of the western and central Pannonian Basin. Tectonophysics 410:15–41. https://doi.org/10.1016/j.tecto.2005.06.008
Franco A, Lasserre C, Lyon-Caen H et al (2012) Fault kinematics in northern Central America and coupling along the subduction interface of the Cocos Plate, from GPS data in Chiapas (Mexico), Guatemala and El Salvador. Geophys J Int 189:1223–1236. https://doi.org/10.1111/j.1365-246X.2012.05390.x
Gardner T, Marshall J, Merritts D et al (2001) Holocene forearc block rotation in response to seamount subduction, southeastern Península de Nicoya, Costa Rica. Geology 29:151–154
Goes SDB, Velasco AA, Schwartz SY, Lay T (1993) The April 22, 1991, Valle de la Estrella, Costa Rica (Mw = 7.7) earthquake and its tectonic implications: a broadband seismic study. J Geophys Res 98:8127–8142. https://doi.org/10.1029/93JB00019
Golts S, Rosenthal E (1993) A morphotectonic map of the northern Arava in Israel, derived from isobase lines. Geomorphology 7:305–315
Kobayashi D, Lafemina P, Geirsson H et al (2014) Kinematics of the western Caribbean: Collision of the Cocos Ridge and upper plate deformation. Geochem Geophys Geosyst 15:1671–1683. https://doi.org/10.1002/2014GC005234
Kolarsky R, Mann P (1995) Structure and neotectonics of an oblique subduction margin, southwestern Panama. In: Mann P (ed) Geologic and tectonic development of the Caribbean plate boundary in Southern Central America. Geological Society of America, Colorado, USA, pp 1–349
LaFemina P, Dixon TH, Govers R et al (2009) Fore-arc motion and Cocos Ridge collision in Central America. Geochem Geophys Geosyst 10. https://doi.org/10.1029/2008GC002181
Lin Z, Kaneda H, Mukoyama S et al (2013) Detection of subtle tectonic-geomorphic features in densely forested mountains by very high-resolution airborne LiDAR survey. Geomorphology 182:104–115. https://doi.org/10.1016/j.geomorph.2012.11.001
Linkimer L, Arroyo IG, Alvarado GE et al (2018) The National Seismological Network of Costa Rica (RSN): an overview and recent developments. Seismol Res Lett 89:392–398
Magaz A, Hernández J, Días J, Hernández I (2006) Formación y consolidación de las morfoestructuras septentrionales de la región central del archipiélago cubano y su geodinámica reciente Recibido: 12 de agosto de 2005. Investigaciones Geograficas 7–23
Marshall JS, Fisher DM, Gardner TW (2000) Central Costa Rica deformed belt: Kinematics of diffuse faulting across the western Panama block. Tectonics 19:468–492. https://doi.org/10.1029/1999TC001136
Modenesi-Gauttieri MC, Hiruma ST, Riccomini C (2002) Morphotectonics of a high plateau on the northwestern f lank of the Continental Rift of southeastern Brazil. Geomorphology 257–271
Montero W (2001) Neotectónica de la región central de Costa Rica: frontera oeste de la Microplaca de Panamá. Revista Geológica De América Central 24:29–56
Montero W, Lewis JC, Araya MC (2017) The guanacaste volcanic arc sliver of Northwestern Costa Rica. Sci Rep 7https://doi.org/10.1038/s41598-017-01593-8
Montero W, Morales L (1990) Deformación y esfuerzos neotectónicos en Costa Rica. Revista Geológica de América Central 69–87
Montero W, Paniagua S, Kussmaul S, Rivier F (1992) Geodinámica interna de Costa Rica. Revista Geológica de América Central 1–12. https://doi.org/10.15517/rgac.v0i14.13171
Mora S (1981) Una revisión y actualización de la clasificación morfotectónica de Costa Rica, según la teoría de la tectónica de placas. Informe Julio- Diciembre 1981. Instituto Geografico Nacional 33–55
Morales L (1985) Las zonas sísmicas de Costa Rica y alrededores. Revista Geológica de América Central 69–101
Morell KD, Fisher DM, Gardner TW (2008) Inner forearc response to subduction of the Panama Fracture Zone, southern Central America. Earth Planet Sci Lett 265:82–95. https://doi.org/10.1016/j.epsl.2007.09.039
Pindell J, Barrett S (1990) Geologic evolution of the Caribbean: A plate¬tectonic perspective. In: Dengo G, Case J (eds) The Geology of North America and The Caribbean Region. Geological Society of America, pp 405–432
Porras H, Alvarado GE, Arroyo-Solórzano M et al (2021) La depresión tectónica de Nicaragua en Costa Rica: estructura interna, estilo estructural, edad, extensión y actividad de la cuenca de San

Carlos. Revista Geológica de América Central 65:1–22. https://doi.org/10.15517/rgac.v0i65.46874
Protti M, González V, Newman AV et al (2014) Nicoya earthquake rupture anticipated by geodetic measurement of the locked plate interface. Nat Geosci 7:117–121. https://doi.org/10.1038/ngeo2038
Protti M, Schwartz S (1994) Mechanics of back arc deformation in Costa Rica: Evidence from an aftershock study of the April 22, 1991, Valle de la Estrella, Costa Rica, earthquake (Mw = 7.7). Tectonics 13:1093–1107
Quesada-Román A, Pérez-Briceño PM (2019) Geomorphology of the Caribbean coast of Costa Rica. J Maps 15(2):363–371. https://doi.org/10.1080/17445647.2019.1600592
Quesada-Román A (2021) Review of the geomorphological effects of the 1991 Limón earthquake. Revista Geológica de América Central 65:1–13. https://doi.org/10.15517/rgac.v0i65.46697
Quesada-Román A, Campos N, Granados-Bolaños S (2021) Tropical glacier reconstructions during the Last Glacial Maximum in Costa Rica. Revista Mexicana De Ciencias Geológicas 38:55–64
Quesada-Román A, Umaña-Ortíz J, Zumbado-Solano M, Islam A, Abioui M, Tefogoum GZ, Kariminejad N, Mutaqin BW, Pupim F (2023) Geomorphological regional mapping for environmental planning in developing countries. Environ Dev 48:100935. https://doi.org/10.1016/j.envdev.2023.100935
Rojas-Barrantes M, Huapaya-Rodriguez Parra S, Solari LA, Žáček V (2021) U–Pb geochronology of detrital zircons from San Carlos Basin, Costa Rica: evidence of Miocene volcanism and implications for the Precambrian and Paleozoic history of the Central American isthmus. J South Am Earth Sci 110. https://doi.org/10.1016/j.jsames.2021.103311
Sofia G (2020) Combining geomorphometry, feature extraction techniques and Earth-surface processes research: 1 the way forward 2. Geomorphology 355:107055
Veas-Ayala N, Alfaro-Córdoba M, Quesada-Román A (2023) Costa Rican wetlands vulnerability index. Prog Phys Geogr 47(4):521–540. https://doi.org/10.1177/03091333221134189
Velásquez GDLÁ, Ayala IA, Hernández JR, Garnica RJ (2015) Volcanic-morphotectonic-zoning in the morphostructural analysis of relief: the case of the municipality of Managua, Nicaragua. Investigaciones Geograficas 87:118–140. https://doi.org/10.14350/rig.43549
Von Huene R, Ranero CR, Weinrebe W, Hinz K (2000) Quaternary convergent margin tectonics of Costa Rica, segmentation of the Cocos Plate, and Central American volcanism. Tectonics 19:314–334. https://doi.org/10.1029/1999TC001143
Westbrook G, Hardt N, Heath R (1995) Structure and tectonics of the Panama-Nazca boundary. In: Mann P (ed) Geologic and tectonic development of the Caribbean Plate Boundary in Southern Central America. Geological Society of America, Colorado, USA, pp 1–349
Weyl R (1971) La clasificación morfotectónica de Costa Rica. Informe Semestral Julio-Diciembre, Instituto Geografico Nacional 25–35. San José, Costa Rica
Zamora N, Arroyo-Solórzano M, Porras H, et al (2021) Evaluación del potencial de tsunamis locales con base en análisis sismo-tectónico en el Caribe de Costa Rica. Revista Geológica de América Central. https://doi.org/10.15517/rgac.v0i65.47002

Volcanic Geomorphology of Costa Rica

3

Guillermo E. Alvarado and Yenory Segura

Abstract

Costa Rica has two active volcanic ranges (< 1 Ma) and several volcanically inactive ranges (1.0–29 Ma), where isolated eroded volcanic foci can be recognized. Large features are described and exemplified (i.e., calderas, andesitic shield volcanoes, stratovolcanoes, volcanic domes, and pyroclastic cones), including their constructive basic elements (lava flows, debris fans, volcanic ash fields), among others small characteristic features of the volcanic landscape. Under a tropical climate, the morphology of lava flows with minor surface features (i.e., levees, lava fronts) can well be recognized up to 100 ka, and craters at<350 ka, although the morphologies of cones, domes, calderas, lava platforms, and large lava fields could be preserved for longer (~ 2 Ma). Ignimbrite fans and plateaus could be preserved under dry tropical forest for even longer (< 5 Ma). Volcano-tectonic features are mainly related to calderas and volcanic grabens. Erosional forms and resistant elements (planezes, waterfalls, and volcanic necks) are part of the erosive pattern. The soil development and the denudation rate are more developed on the Caribbean side of the volcanic ranges than on the Pacific side due to differential orogenic rainfall and the presence of areas with hydrothermal alteration. Inversion of relief is particularly well development on ignimbrites. Along the piedmont, it is possible to recognize alluvial, fluvial, and debris avalanche fans and lahars. On the Pacific Ocean, side of Costa Rica several representative marine landforms can recognized as seamounts (30 to 0.6 Ma), the Cocos submarine volcanic range (< 15 Ma), and an isolated subaerial volcanic island (< 2.2 Ma), the Coco Island, which emulates a guyot.

Keywords

Volcanic landforms · Morphochronology · Erosion · Tropical modeling · Geotourism · Geoheritage

3.1 Introduction

Volcanic eruptions and volcanic terrains of Costa Rica attract attention worldwide because of their beautiful and varied landscapes (symmetrical cones, cliffs, waterfalls, craters, and crater lakes) and diverse manifestations of volcanic activity (acidic lakes, hot springs, geysers, fumaroles, eruptions). In addition, several volcanoes and volcanic hills are considered sacred or mystical and/or are national parks. The volcanic rocks represent about 50% of the Costa Rica continental territory (Fig. 3.1), in which two major Quaternary volcanic ranges are present (Cordillera de Guanacaste and Cordillera Central), but there are also volcanic forms in other ranges of Neogene (< 29 Ma) age (Tilarán, Montes del Aguacate and Talamanca ranges, Sarapiquí hills). However, if we include the Coco submarine volcanic range, the presence of volcanic landforms and their features increases significantly in our physiography.

Unfortunately, very little fundamental work about volcanic landforms and surface features (volcanic geomorphology) of Costa Rica has been published in English, and there are also several misinterpretations of the origin of landforms. Hence, this chapter fills, only in part, an evident gap in this respect, and we hope that this chapter will serve as an aid to those engaged in learning or teaching the fundamentals of geology and geography, and the subfield of geomorphology. We also considered tour guides and

G. E. Alvarado (✉)
Centro de Investigación en Ciencias Geológicas, Universidad de Costa Rica, San Jose, Costa Rica
e-mail: galvaradoinduni15@gmail.com

Y. Segura
Independent Geologist, San Jose, Costa Rica

A. Quesada-Román (ed.), *Landscapes and Landforms of Costa Rica*,
World Geomorphological Landscapes, https://doi.org/10.1007/978-3-031-64940-0_3

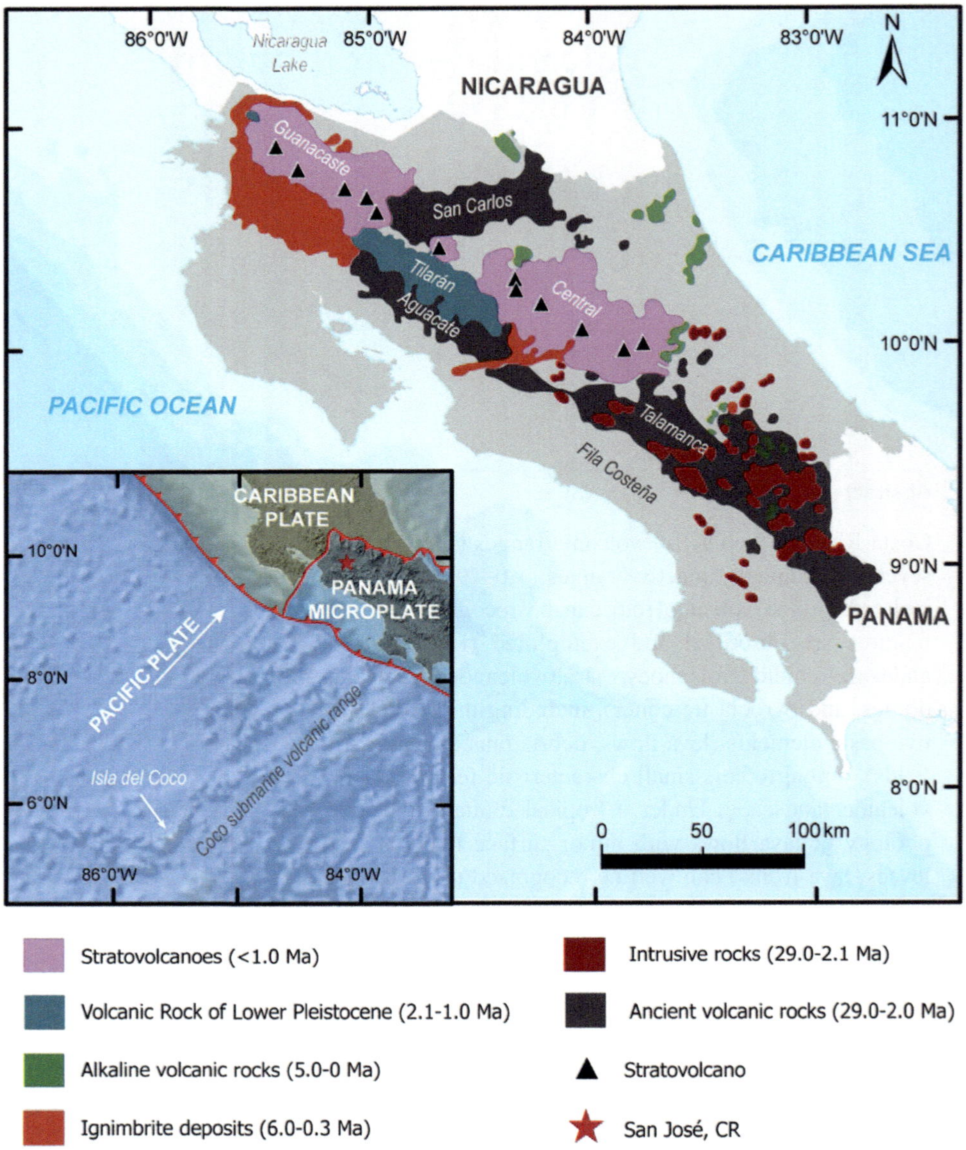

Fig. 3.1 Simplified geological map of the volcanic rocks and simplified tectonic configuration of Costa Rica (based on Denyer and Alvarado 2007)

non-professionals who are interested in volcanic processes as a natural wonder as another audience; therefore, this chapter is designed to reach a wide audience.

Finally, we have added some current topics related to erosion controlled by orogenic rains in tropical environments, morphochronology, volcanic soil development and, at the end, conservation and geotourism in Costa Rica.

For the present chapter, an exhaustive review of the Costa Rican literature was done, including examination of geological and geomorphological maps. Published geologic information was consulted to make first-order evaluations of area, volume, geology, geomorphology, radiometric ages, slopes, and local meteorological conditions.

Problems of nomenclature are often troublesome in volcanology. For terminology, we used the classic textbooks of Green and Short (1971), Macdonald (1972), the review by Alvarado (1990), and the overview paper of Thouret (1999). We have omitted multi-language definitions but have included some commonly used terms derived from languages other than English. The arrangement of the photographs is from large to small features, including variations that characterize the subjects being described. The reader unfamiliar with terms or concepts will be introduced to the topic through a definition.

A range of small- to large-scale features on the major volcanoes were identified using black and white color aerial photographs, in conjunction with topographic maps, satellite images, digital elevation models (DEMs), including LiDAR-derived or drone-derived DEMs. Measurements were taken from aerial photographs, topographic maps, and Google Earth. For some morphometric parameters, we used the topographic maps, Google Earth, and several geological maps. The estimated volume of lava and ejected material in each volcano is based on Alvarado (2021).

Morphometric and morphological characteristics, including the rate of erosional processes based on breaching and rill dissection, are the evidence of different erosional stages. They allowed for a detailed relative temporal classification of volcanic fields. Soil and vegetation cover and their grade of development (thickness, clay content) were also used as relative age indicators, as climatic conditions are quite uniform on both flanks of the volcanic ranges, Pacific, and Caribbean. Results of radiometric dating reported in the literature (^{14}C, K/Ar, ^{40}Ar/^{39}Ar, U/Th, Rb/Sr, U/Pb) were consulted to estimate the age of the volcanic landform (Alvarado and Gans 2012; Alvarado et al. 2020a, b; 2021, 2022, 2023a; Ryan et al. 2022).

3.2 Recent and Old Neogene-Quaternary Volcanic Terrains in Costa Rica

The Quaternary Guanacaste and Central volcanic ranges grow over a Paleogene to Neogene sedimentary and volcanic rocks (Fig. 3.2). Lithology of the volcanic ranges (active and extinct) are mainly interbedded lava flows and pyroclastic rocks of varying composition, mainly from mafic to intermediate (basalts to andesites) composition, with rare acid volcanics center (dacites and rhyolites), with an increasing volume of felsic rocks related to ignimbrite plateaus. Volcanic activity was mainly of Vulcanian and Strombolian type in major volcanoes and cinder cones. Prehistoric Plinian eruptions occurred in almost all the larger volcanoes during the Late Pleistocene and Holocene (except for Irazú volcano), including prehistoric and historic lava flows. Gravity-driven flows can be related either to explosive eruptions (i.e., column collapse, sector failure, and dome or lava flow failure) or to events that occurred during volcanic quiescence (i.e., slope instability, climatic events, and earthquakes). They include generation of pyroclastic density currents (pyroclastic flows and pyroclastic surges), debris avalanches, and volcaniclastic flows (lahars, alluvial deposits), in addition to gravity deposits (landslides, colluvium), which have different dynamics of transportation and emplacement.

The Costa Rican volcanic front results from the subduction of the oceanic Cocos Plate beneath the Caribbean and Panama continental lithospheric plates. The Neogene (Miocene and Pliocene) and Quaternary (Pleistocene and Holocene) volcanic activities are represented by extensive lava and ignimbrite fields and large and small volcanoes that are active, dormant, and extinct. In general terms, our Quaternary volcanic mountain ranges have gone through three major construction periods and are comprised of several constructive phases (one or more edifices in each one). The older, called Protovolcano, is preserved at the base as a portion of the early edifice; the Paleovolcano is usually a deeply eroded edifice that is overlapped by pristine cone edifices termed the Neovolcano, with ranges of intense effusive activity of about 1.61–0.85, 0.74–0.25, and 0.25–0 Ma, respectively. Each constructive phase has alternated with phases of erosion and destruction of the volcanoes associated with the formation of calderas with associated ignimbrites. A period of marked erosion (erosional unconformity) and thick soil formation typically preceded the resumption of volcanic activity. Thus, several earlier volcanoes of possible comparable dimensions were largely obliterated or covered (Alvarado and Gans 2012). The entire volcanic massif is overlain and interbedded with welded and nonwelded ignimbrites. This Quaternary volcanic province, which belongs to the Central America Volcanic Front, is comprised of at least 20 major Quaternary volcanic centers and a few monogenetic edifices and fields. Among these volcanic centers, five volcanoes have experienced eruptive activity since 1700 AD, and at least, five more have demonstrated Holocene eruptive activity. These edifices include the well-known historically active Rincón de la Vieja, Arenal, Poás, Irazú, and Turrialba volcanoes, as well as other lesser-known edifices with Holocene (prehistoric) eruptions such as Miravalles, Chato, Barva, and Hule volcanoes (Fig. 3.2).

There is also a series of volcanic mountain ranges (Cordillera de Tilarán and Montes del Aguacate), low volcanic mountains (Sarapiquí area), and a high-altitude range (Cordillera de Talamanca) composed of a mixture of volcanic, intrusive, and metamorphic rocks (hornfels), where it is possible to recognize the remains of stratovolcanoes, volcanic platforms, volcanic necks, and domes, with different ages and degrees of preservation, and erosion.

3.3 Brief History of Research in Volcano Geomorphology

Volcanic geomorphological curiosities of Costa Rica, particularly the volcanoes, craters, satellite cones, and lava flows, have been visited, drawn, photographed, and described by naturalists since the second half of the nineteenth century (Figs. 3.3 and 3.4). Later, the eruptions of Irazú (1963–1965) and Arenal (1968–2010) attracted the attention of national and foreign geologists, fostering the creation of institutes, schools of geology and geography, and later, seismological and volcanological observatories, increasing the geomorphological descriptions of volcanic landforms; the history being summarized by Alvarado (2021). Healy (1969), Bergoeing (1978), Madrigal and Rojas (1980), and Alvarado (1989, 1990) provided comprehensive presentations of geomorphology of the Costa Rican

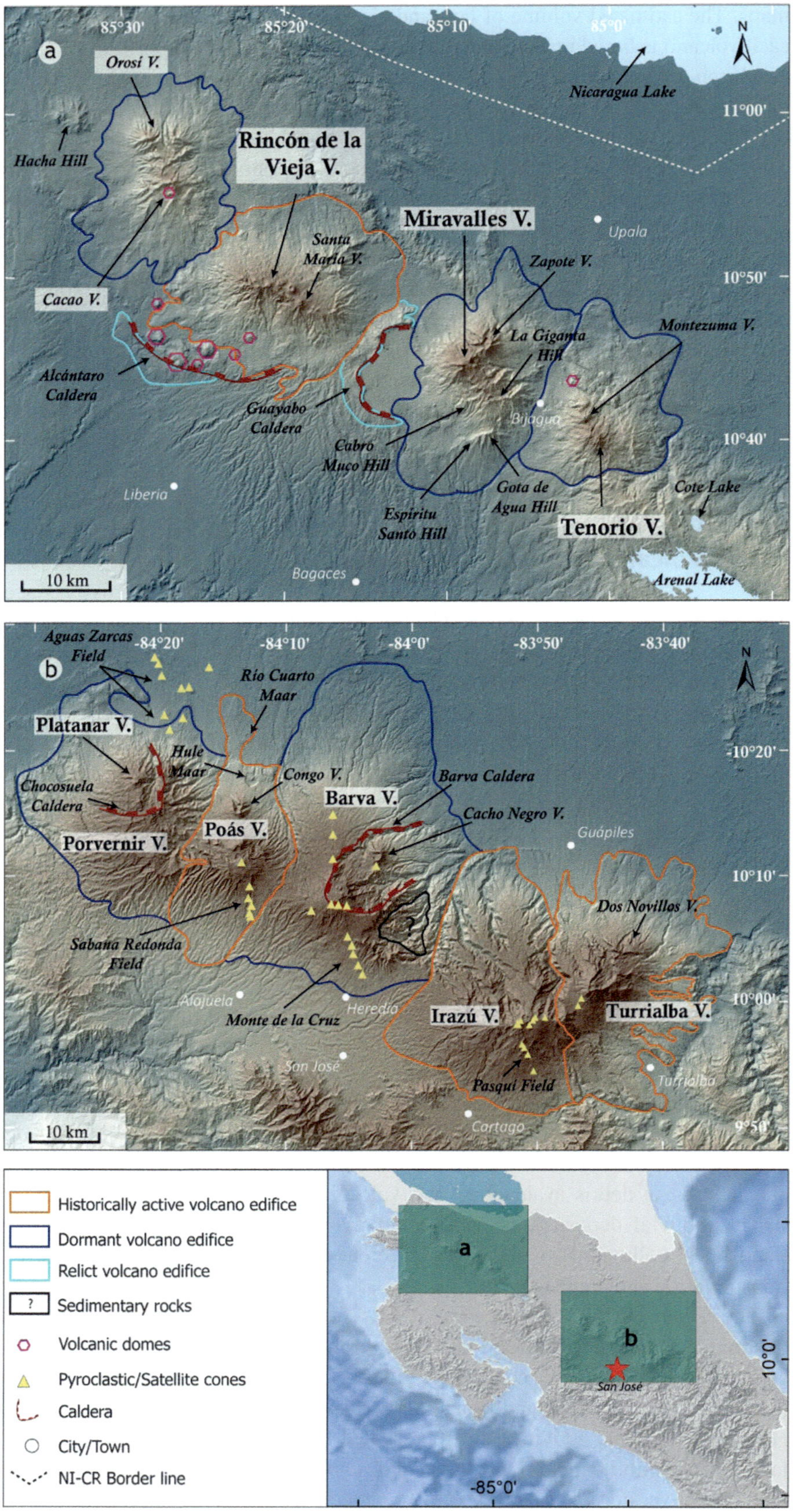

Fig. 3.2 Volcanic edifices of the Cordillera de Guanacaste (**a**) and Cordillera Central (**b**), and their major volcanic structures and approximate limits. A higher grade of erosion (fluvial incision and river density by km^2) reflects the increase in relative age from Neovolcanism, Paleovolcanism, and Protovolcanism in the Quaternary volcanic ranges

Fig. 3.3 First historical drawings of volcanoes and volcanic ranges in Costa Rica. **a** Craters at Irazú volcano in 1858 drawn by Ramón Páez (Meagher 1859a, b, 1860), **b**Turrialba volcano in 1858 (Meagher 1859a, b, 1860), **c** and **d** Volcanic Cordillera Central drawing by Anders S. Oersted made in 1846–1847 and published in 1863 (Oersted 1863), **e** Turrialba in eruption drawn from La Fuente farm at its present location (von Seebach 1892), **f** Turrialba drawn from Irazú volcano and cartoon of the craters (von Seebach 1865)

Fig. 3.4 First historical photographs by naturalists and tourists of Costa Rican volcanoes. **a** Poás crater lake in 1892 (unknown author), **b** Panoramic view of the old volcano morphology at Irazú in 1896 (Vargas 1967), **c** Internal small pit craters at Irazú in 1897 (Tristán 1923), **d** Local tourists visiting the Turrialba summit in 1920 (unknown author), **e** Aerial view of Irazú summit and their craters around 1926 (unknown author). The morphology of the craters at the summit of Irazú changed significantly from volcanic eruptions between 1917 and 1965. **f** First photograph of Arenal volcano by topographers in 1937 (Trejos 1959)

volcanoes. Local studies were carried out in the decade of 1980–1990 for Arenal volcano (Bravo 1982; Borgia et al. 1988) and Tenorio volcano (Bravo et al. 1989), and on old volcanic landforms (Bergoeing 1979; Bergoeing et al. 1979; Alvarado 1989). Since then, several additional studies with different levels of detail have been done at each volcano and volcanic range by several authors, which will be cited in the present chapter. Bergoeing and co-workers also produced several monographs of the volcanic ranges in Costa Rica (Bergoeing 1978, 2007, 2009; Bergoeing et al. 1983, 2010). Martens (2004) paid considerable attention to volcanic pyroclastic cones fields. Elizondo et al. (2019) described historical morphological and structural evolution of volcanic craters. Detailed descriptions of lava fields (morphology, ages, parasitic cones) were made for the volcanoes of Arenal (Borgia et al. 1988; Borgia and Linneman 1990; Soto and Alvarado 2006), Irazú (Alvarado and Vega 2013), Barva (Rojas et al. 2017), Turrialba (Ruiz et al. 2010), Poás (Ruiz et al. 2019), and Miravalles (Castro et al. 2020).

Thus, the period spanning the end of the past century through the present has been largely a period of data gathering. Volcanic morphochronology has advanced in surges in the last decades, particularly with the radiometric techniques of radiocarbon and isotopic dating (Soto and Alvarado 2006; Alvarado and Gans 2012; Alvarado et al. 2020a, b, 2021, 2022; Ryan et al. 2022). The history of landform evolution and a complete compilation of geomorphological references, mainly derived from the examination of spatial relationships between volcanoes and their ages, is summarized in studies focused on Poás (Ruiz et al. 2019; Alvarado et al. 2020b), Turrialba (Ruiz et al. 2010; Alvarado et al. 2020a), Irazú (Alvarado et al. 2021), Rincón de la Vieja (Alvarado et al. 2022), and Arenal volcanoes (Alvarado et al 2023a). Alvarado (2021) developed a summary of the Costa Rican volcanoes and volcanic ranges, including morphometric parameters, which is an important reference on volcanic landforms for the region.

Several interesting and promising recent developments, including the use of LiDAR, radar satellite images, digital maps, and drone photography, have significantly improved quantitative and descriptive interpretation of volcanic landforms (Alvarado et al. 2011; Castro et al. 2020; Granados-Bolaños et al. 2021). In fact, such gathering will continue, now entering the field in a more quantitative and chronological way, and it promises to be increasingly fascinating for the geological and geomorphological history and form interpretation. Such data also provide a base for better hazards studies and territorial planning (Alvarado et al. 2020a, b, 2021, 2022, 2023a; Granados-Bolaños et al. 2021).

3.4 Principal Volcanic Landforms and Surface Features

Volcanic landforms, in contrast to other types of landforms, result from both constructive and destructive forces near simultaneously; thus, volcanic geomorphology needs to be studied regarding their processes of growth and erosion (Thouret 1999).

3.4.1 Basic Volcanic Vents

An eruptive vent could be quasi-circular (crater) or elongated (volcanic fissure) for several hundred meters or kilometers and can be formed by lava effusion and/or accumulation of materials. At one extreme are volcanoes which are products of a single eruption (monogenetic volcano), at the other extreme are long-term volcanoes (thousands to millions of years of activity), which experiences numerous individual eruptions (polygenetic volcano) and construct an edifice (Mercalli 1907; Rittmann 1962; Walker 1965). Orientations of volcanic features (monogenetic cones, small stratovolcanoes, volcanic grabens; see below for descriptions) are usually associated with fissures related to local and regional stress orientations (Nakamura 1977).

A parasitic cone or parasitic crater (adventive cone, lateral cone, secondary cone, satellite cone) is a small crater or cone developed on the flank or foot of a major volcano. A pit crater is a circular volcanic depression with steep walls, and a nested crater is a large crater containing one or more small craters or boccas in its bottom (Mercalli 1097; Rittmann 1962; Green and Short 1971; Francis and Oppenheimer 2004). The form of a volcano is the response to several factors including the eruptive style, magma viscosity, number and location of vents, flank instability, gravitational spreading, tectonics, geography location (under water or ice or subaerially exposed, climate, wind direction), among several other conditions (Lagmay et al. 2000).

In Costa Rica, volcanic craters that contain lakes (crater lakes) are present at several locations throughout the Volcanic Front (Fig. 3.5). A systematic examination of crater lakes in Costa Rica is presented in several papers, including recent examples of eruptions and hydrologic events that involved crater lakes at Rincón de la Vieja and Poás volcanoes (Horn and Haberyan 1993; Kempter and Rowe 2000; Tassi et al. 2009), or cold-water crater lakes at Tenorio, Votos, Barva, Hule and Río Cuarto, and sometimes at Irazú and Turrialba volcanoes. The eruptive events at Rincón de la Vieja and Poás illustrate the role of lake water in influencing eruptive processes and hazardous hydrologic phenomena (lahars).

3.4.2 Polygenetic Volcanoes

3.4.2.1 Stratovolcanoes

Stratovolcanoes are composite stratified cones which consist of both lava flows and beds of pyroclastic material. At the base, the slope generally flattens gradually to near horizontality, principally due to the presence of fluid lava flows, tephra, and fluvial deposits, including lahars (Green and Short 1971; Macdonald 1972). In Costa Rica, typical symmetrical stratovolcanoes (simple cones) to complex stratovolcanoes (volcano complexes) vary in height from 600 to 1700 m, with slopes ranging from about 20° to 34°. Relatively simple cones exist at Arenal, Cacho Negro, and Congo volcanoes (Figs. 3.6, 3.7 and 3.8).

For example, Arenal stratocone, where the last eruption occurred at the beginning of 2010, is now a dormant volcano 1700 m high. The slope of the cone is about 34° near the summit and decreases toward the skirt. The profile reflects incipient erosion (dissected gullies and ravines) and deposition of nearly horizontal lavas and debris fans at the base. Remnants of older lava flow fields are recognized by different surficial morphology, and the grade of forest and soil development (Borgia et al.

Fig. 3.5 Diversity of volcanic craters in Costa Rica. **a** Large crater and internal active crater lake at Poás volcano, **b**Craters and domes at Tenorio volcano, **c** Santa Rosa cone with crater, a parasitic crater of Irazú, **d** Olla parasitic cone with crater at Irazú, **e** Irazú main crater, **f** Barva crater lake

Fig. 3.6 Topographic profile of Arenal and Chato volcanoes. Arenal is a small stratovolcano with a more symmetrical form. Chato has an explosive crater that destroyed its summit and two parasitic foci

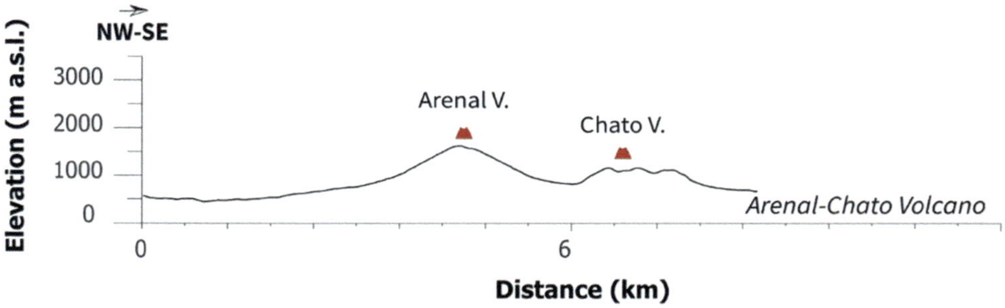

Fig. 3.7 Large volcanoes of Costa Rica. **a** Arenal stratovolcano with a conical shape from the west, **b** twin cone of Arenal volcano from the north, **c** Congo stratovolcano, **d** Cacho Negro stratovolcano and its parasitic cone on the left, **e** complex stratovolcanoes at Miravalles volcano massive, **f** Irazú andesitic shield volcano

1988). The influence of regional tectonic setting on the emplacement, morphology, and potential edifice instabilities at Arenal has been assessed using numerical models (Alvarado et al. 2010).

3.4.2.2 Andesitic Shield Volcanoes

Andesitic shield volcanoes are relatively low, broad volcanoes, many kilometers in circumference, built up by lava flows, with slopes seldom as steep at 15–20° at the summit and 5–10° at the base. The form of the shield is controlled largely by the presence of zones of fissures (rift zones), which have served as the channels for eruption of parasitic cones and lava fields (Green and Short 1971; Macdonald 1972; Alvarado 1990). Most of the volcanoes in the Cordillera Central have a complex morphology and basal breaks in slope, caused by the juxtaposition of the two or three main edifices (Proto, Paleo and Neo) and several smaller flank vents (N–S or NE–SE trend). In Costa Rica, shield volcanoes such as Barva and Irazú vary in height from 2800 to 3200 m, with an angle ranging from about 4° to 27° (Figs. 3.7, and 3.8).

3.4.2.3 Guyot and Seamounts

Costa Rica has thousands of submarine volcanoes (seamounts) in its maritime territorial waters (Fig. 3.9a) from 30 to 0.6 Ma, most of them along the Cocos Ridge (although the local and correct name is Coco), formally known in Costa Rica as the Cordillera Submarina Volcánica del Coco, which is the result of a mantle thermal anomaly that erupted through volcano-tectonic fissures in the oceanic crust. The Cordillera del Coco is a well-defined linear bathymetric ridge (N45°E) that originated from active volcanism of the Galápagos hotspot during the last 15 million Ma. It is the largest geographic feature of Costa Rica, being a volcanic range of 780 km long in its territorial seawaters (Rojas and Alvarado 2012; Alvarado 2021).

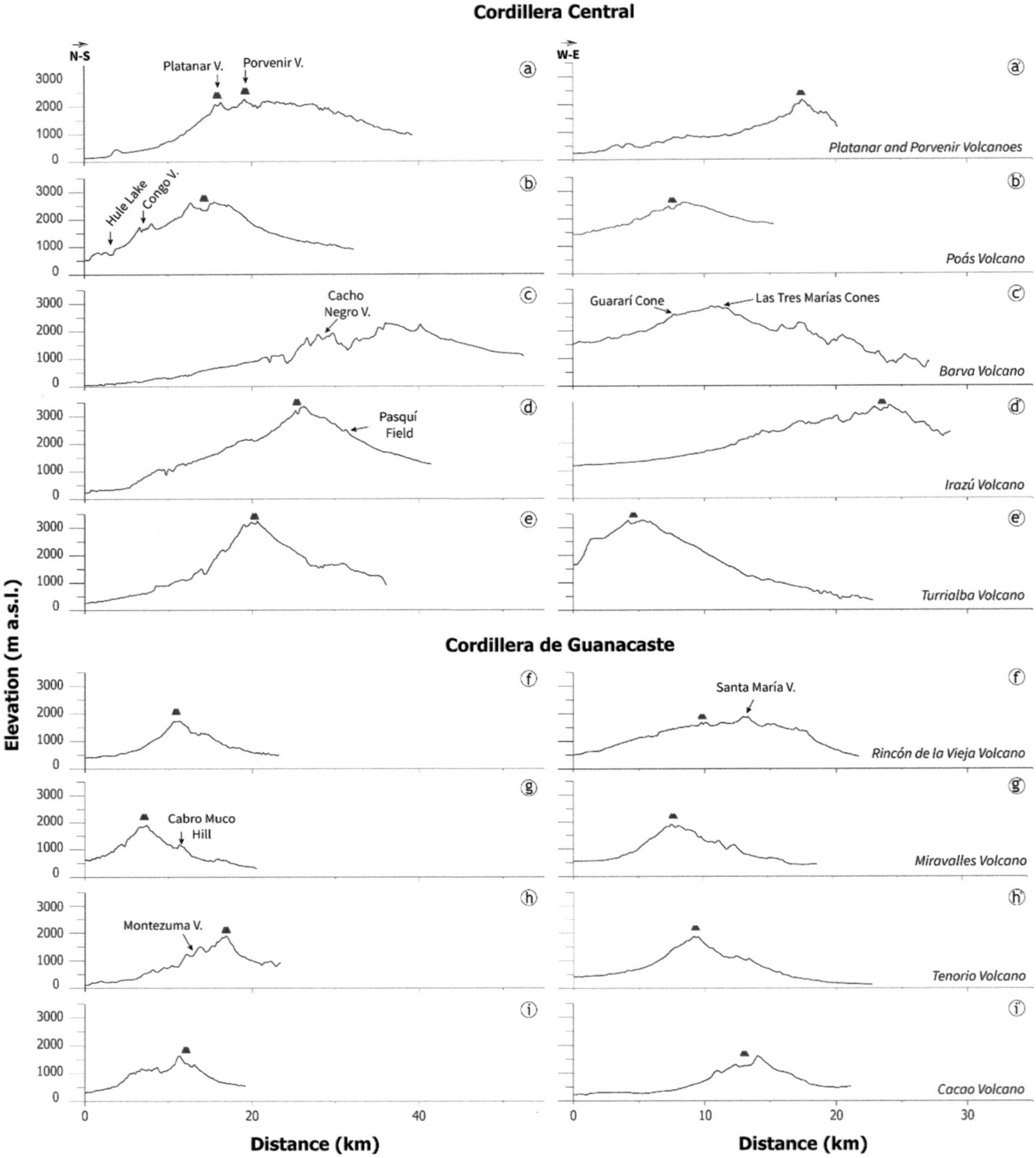

Fig. 3.8 Topographic profiles of Costa Rican volcanoes with parasitic cones. The volcanoes of the Cordillera Central have more asymmetrical profiles and are larger than the volcanoes in Guanacaste, except for Rincón de la Vieja. The rougher topography reflects the occurrence of parasitic cones and old eroded volcanic massifs with a pervasive and highly evolved erosion pattern on the Caribbean side of the volcanic ranges

The Isla del Coco (also known as Cocos Island), in the Eastern Pacific Ocean, is the only true marine island in the country (out of the continental platform; Cortés, 2007). Isla del Coco is part of a submarine shield volcano of complex history, which has erupted several times above sea level during the Lower Pleistocene (2.2–1.5 Ma). It is the subaerial topographic expression of a marine volcano located on the margin of the Cocos Ridge. The island has a rough topography limited by sea cliffs (the highest up to 183 m), only two sand beaches and an emerged area of

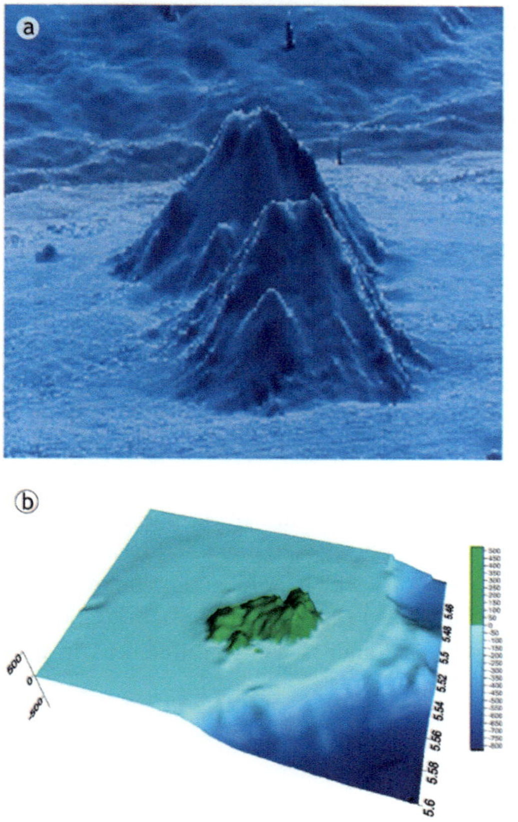

Fig. 3.9 Seamounts and a volcanic island. **a** Twin submarine extinct Miocene volcanoes (seamounts) called Evi volcanoes, located in front of the subduction zone of Costa Rica, **b** Coco island (green color) and submarine flat top of the seamount, resembling a guyot

24 km^2. The highest point of the island is the Yglesias hill (575 m a.s.l.), the remnant of an eroded cone (Rojas and Alvarado 2012). The rocks consist mainly of alkali basaltic lava flows (aa, pahoehoe, blocky lavas) and dikes, minor trachyte lava flows, volcanic domes, and dikes, with subordinate pyroclastic and epivolcanic rocks (Castillo et al. 1988). Colluvial soils and local littoral deposits such as sand and gravel beaches are also present. The island has a juvenile erosive amphitheater with several waterfalls on the sea cliffs that pour directly into the ocean. Their submarine erosive platforms (at the depth of 90–110 m and 183 m) are probably the result of erosion that occurred during last two glacial maxima, in addition to slow subsidence events of the island due to thermal cooling of the volcanic shield and its oceanic crust. These marine platforms give the island a truncated-conical shape (Fig. 9b) that emulates a guyot (Alvarado 2021). The Coco seamount (1900 km^3) is the largest volcano in Central America (Schindlbeck et al. 2016).

3.4.3 Monogenetic Volcanoes

Monogenetic volcanoes are in most cases products of a single eruptive episode. The event may last a few hours or a few years, but the essential point is that once the eruption has ceased, the plumbing connecting the vent to the magmatic source freezes over, so that the volcano never erupts again. Examples are pyroclastic cones, fissure eruptions, domes, and possibly some small shield volcanoes (Mercalli 1907; Rittmann 1962; Walker 1965; Németh 2010). However, a new volcano could be born near the original location several years or thousands of years later, such as the Hule maar (Alvarado et al. 2011), or in very rare examples, the same volcano reactivates, such as the Cerro Negro volcano in Nicaragua (McKnight and Williams 1997). Naturally, the crisp profiles of the cones soften with the age, such that the ratio of height to basal diameter decreases with age (Inbar and Risso 2001).

3.4.3.1 Pyroclastic Cones (Tephra Cones)

Pyroclastic cones consist of a cone-shaped hill composed predominantly of fragmental materials produced by explosive volcanic activity (mainly of Strombolian type, except for a few with phreatomagmatic styles). Slopes of the cones are largely determined by the angle of repose of the ejected material. Fine volcanic ashes form slopes of about 10° near the base of the cone. Coarser fragments build higher slopes at angles of 20–30°. The crater at the top is most typically rather smoothly bowl-shaped or funnel-shaped. The summit crater of the cone may be breached (horseshoe-shaped crater) because a portion of the crater wall has been removed by outflow of lava or explosive activity, predominant wind direction, or prolonged erosion. Asymmetrical cones may result from non-uniform building because of inclined explosive jets that deposit tephra on one preferential side, or because of a strong wind in a persistent direction. Cinder cones are often associated with lava flows (Green and Short 1971; Macdonald 1972). In the case of hydrovolcanism (particularly phreatomagmatic eruptions), the result could be smaller and steeper (20–30°) cones (*tuff cones*), or vents with flat-floored explosion craters with diameters much greater than the heights of surrounding rims (2–10° slopes), usually formed by a single eruption or short-lived eruptions (*maars and tuff rings*). Usually, the maar excavates down the substrate, whereas the tuff ring is built up above it. The crater can be dry or can contain a water-filled lake and is usually surrounded by low tuff rings. The German name maar means "crater lake," and these rarely contain internal cones (Green and Short 1971; Macdonald 1972; Francis and Oppenheimer 2004).

In Costa Rica, pyroclastic cones (Fig. 3.10) formed by Strombolian activity could be developed on large volcanoes in the Cordillera Central or as isolated cones within flat-lying volcanic fields (e.g., Chopo and Chopito cones in Guanacaste). The pyroclastic cones could be composed of scoria ejecta (cinder cones), as most of the pyroclastic cones in Costa Rica (e.g., Tortuguero and Aguas Zarcas field on the Caribbean, Sabana Redonda field at Poás volcano, Monte de la Cruz at Barva volcano, La Laguna and Pasquí-Las Cazuelas field at Irazú), or composed of cannon ball ejecta (e.g., Chopo cone), or even a mixture of different morphology and tephra granulometry (Alvarado and Carr 1993; Gazel and Ruiz 2005; Rojas et al. 2017; Alvarado 2021). The cone can have a symmetrical or asymmetrical form according to the previous slope topography, wind direction or lateral lava flow, or the elongation could result from simultaneous eruptions along a fissure (elongated or linear cinder cone), as the Santa Rosa and Cazuela cones and craters on the southern flank of Irazú (Alvarado and Vega 2013; Alvarado 2021).

Typical maars and tuff rings are the Hule and Río Cuarto lakes. The Hule maar rim is 1.8 km × 2.3 km, and the nearly circular crater at Río Cuarto is ~ 700 m in diameter, with exterior slope angles below 20° (Fig. 3.10). Hule maar has an intra-maar asymmetrical double cinder cone that has been produced by successive eruptions from closely spaced single vents, but also a parasitic small maar crater called Pata de Gallo or Ángeles (Alvarado et al. 2011). There is another unnamed maar, not yet studied, located on the northern flank of Barva volcano and at least two on the northern flank of Tenorio volcano. An asymmetrical tuff cone was formed during the 1963–1965 eruption at Irazú volcano (Alvarado 2021).

Martens (2004) made the most detailed study of the Costa Rican pyroclastic cones. He selected and analyzed morphometric characteristics of 33 pyroclastic cones. The regression obtained for the height and basal diameter ratio (H_{co}/W_{co}) is 0.17, which is in good agreement with data obtained for cones located in other places of the world. The wide dispersion found is due to differences in the factors

Fig. 3.10 Pyroclastic cones. **a** Hule maar caldera and its internal cinder cone on the right, **b** Loma Hermosa cinder cone with the rest of crater (Aguas Zarcas cinder cone field), **c** La Laguna cinder cone at the summit of Irazú, **d** horse shape crater at the Pasquí cinder cone, **e** cinder cone with small 300 ka lava flow on the right (Aguas Zarcas cinder cone field), **f** asymmetrical Chopo pyroclastic cone between Cañas and Tilarán

that control eruptions leading to pyroclastic cones, particularly the ejection velocity and angle and vent diameter. The analysis of the top and basal diameter ratio (W_{cr}/W_{co}) and slope (S) enables the identification of several groups that broadly coincide in terms of their position relative to the volcanic front. The Aguas Zarcas back arc cones possibly have been exposed to erosion for a longer period. There is also a moderate correlation between cone morphology and chemical composition. Cones that contain more silica and less magnesium are higher, and those with more alkalis and less ferrous content tend to be steeper. The geochemical and structural framework of the Aguas Zarcas field and the Tortuguèro cones suggests deep crust faulting reaching the upper mantle.

3.4.3.2 Domes

A volcanic dome is a protrusion of viscous lava forming a steep, dome-shaped mound above the vent, leaving no crater. Deposits may flow for short distances and form tongues of small to moderate dimensions. In ground plan, most domes are more or less circular, or show very short ovals. More rarely, they are elongated as a result of the extrusion of lava through a long fissure vent (Mercalli 1907; Rittman 1962; Green and Short 1971; Macdonald 1972).

A Peleean dome (peléean dome, cumulo dome, tholoid) is a mass of viscous lava protruded from a volcanic vent with a vault-shaped form. In these features, lava scarcely flows but is mostly squeezed up to form the volcanic cupule composed of large angular blocks (central mass of dome) and an apron of angular debris (crumble breccia). The endogenous domes that grow by expansion from within, and exogenous domes which are built by surface effusion, normally from a central summit crater, can be distinguished. A low lava dome (low domes, tortas) is flattish, roughly symmetrical extrusion that has a convex cross-section, instead a coulee is a dome built onto a sloping surface that moved down-slope. A volcanic material which is so viscous or strong as not to deform while emerging from a conduit but is pushed bodily upwards like a piston and emerges with steep sides is an upheaved plug or spine (Fig. 3.11). A plug of near-surface magma, which deforms the surface topography without breaking through, is termed cryptodome (Macdonald, 1972; Blake 1990; Francis and Oppenheimer 2004).

In Costa Rica, there are about 41 lava domes and 12 more structures that could be associated either with domes or bodies with dome-like morphologies (Alvarado 2021). However, all of them are poorly studied and have not been properly classified. Domes may grow in the craters of volcanoes (Poás in 1953, Arenal in 1968, and 2000–2010), near the caldera rim or inside it (Cañas Dulces and Los Perdidos dome fields) or may develop independently of older volcanic edifices (La Cruz and San Miguel domes in

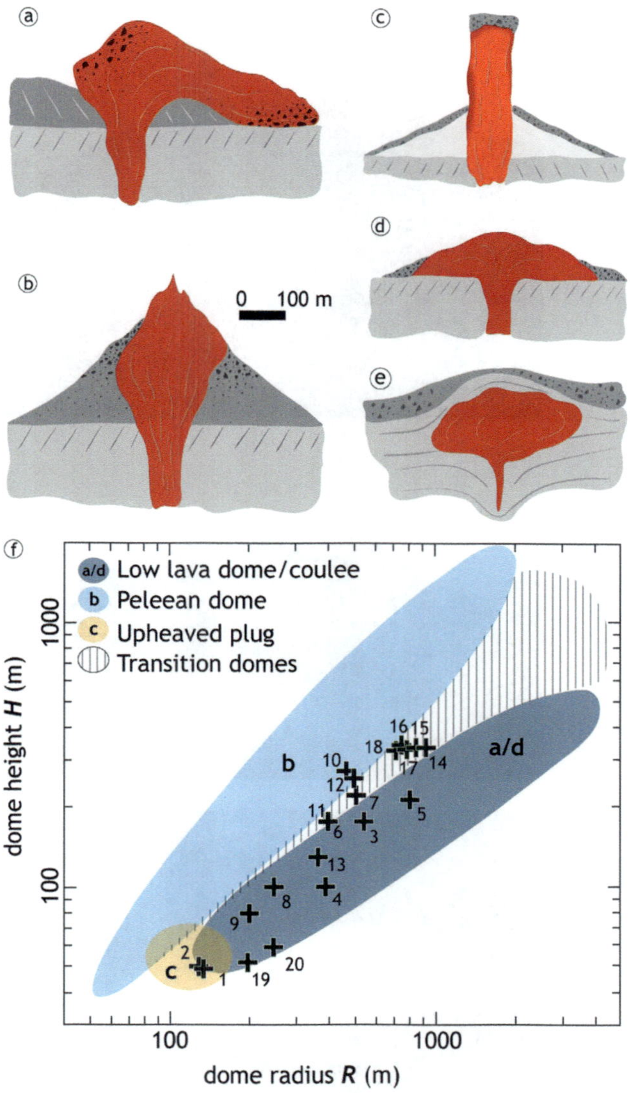

Fig. 3.11 Dome forms (Macdonal 1972; Blake 1990) and classification diagram (modified after Blake 1990). **a** Coulee, **b** Peleean dome, **c** upheaved plug, **d** Low lava dome, and **e** Cryptodome 1) Poás 1953, 2) Espina, 3) Fortuna, 4) Young Góngora, 5) San Miguel, 6) Cacao, 7) Cañas Dulces, 8) Torre, 9) Lori 2, 10) Lori 4, 11) Lori 5, 12) Pelón, 13) Los Perdidos A, 14) San Roque, 15) Old Góngora, 16) La Cruz, 17) Lori 1, 18) Lori 3, 19) Los Perdidos B, 20) San Vicente

Cordillera de Tilarán; Río Lori dome field at Cordillera de Talamanca). Following the morphological classification of Blake (1990), the Costa Rican domes are mostly low lava domes with transition to Peleean domes, though there are some structures classified as coulees domes (Poás) and upheaved plugs (Fig. 3.12). Other old bodies (> 10 Ma) with dome-like morphologies of undetermined origin could correspond to eroded subvolcanic bodies, either acid volcanic necks or true cryptodomes.

In terms of petrochemical composition, the domes consist mainly of rhyolites and dacites (La Cruz and San

Fig. 3.12 Panoramic view of different dome types. Rhyolitic-dacitic Cañas Dulces volcanic dome field: **a** Góngora, **b** San Roque, **c** Fortuna, **d** possible trachyte dome at Isla del Coco, **e** 1953 asymmetrical basaltic andesitic coulee dome at Poás volcano (destroyed during the 2017 eruption), as a natural dam for the crater lake, **f** Dome at the summit of Arenal volcano in 2009

Miguel hills; Cañas Dulces dome field), followed by andesites to basaltic andesites (e.g., Los Perdidos dome field; Tenorio, Arenal and Poás volcanoes). Unusual compositions are also present, like the kersantite lamprophyre at the Bola hill, a dome or volcanic neck at the foot of the Fila Costeña. Although the composition is mainly subalkaline (calcoalkaline), it is possible to find adakitic andesites to dacites in the domes at the top of the Cordillera de Talamanca (Río Lori dome field), or alkaline trachytes on the Cocos Island (Gazel and Ruiz 2005; Rojas and Alvarado 2012; Alvarado and Gans 2012; Alvarado and Denyer 2019).

The age of dome-like structures (perhaps cryptodomes or felsic necks) varies from 22.75–10.9 Ma, such as the Sarapiquí old volcanic arc, or the Cordillera de Tilarán, which varies across the Upper Miocene to Pliocene (7.3–4.40 Ma). During the Late Pliocene to Early Pleistocene (2.80–0.95 Ma), the Río Lori dome field in Talamanca erupted, the only recent volcanism in that range. Regarding neovolcanism distribution, the Los Perdidos domes are Late Pleistocene in age (0.09–0.083 Ma) as are the intra-crater domes at the Cacao and Tenorio volcanoes, although without radiometric age, these are only defined as Late

Quaternary in age. Finally, in historical times, some couleé domes originates such as the 1953 dome on the Poás volcano and the Pelean ones on the Arenal volcano (Alvarado and Gans 2012; Alvarado 2021).

3.4.4 Lava Flows

Lava flows are streams of molten rocks, the morphologies of which are influenced by the composition, crystal content, temperature, cooling story, topography, and gas content (Francis and Oppenheimer 2004). Lava levees are lava gutters or channels produced by overflow of lava onto the flanks of the channel. The formation of such open channels depends on lava temperature, cooling rate, and viscosity; the lava must be neither too viscous nor too fluid and should flow over a moderate gradient (Baker and Harris 1963). A finite quantity of lava that is emplaced during continuous effusion and maintains fluid continuity throughout its lengths is defined as a simple flow or unit flow, while a compound flow or composite flow are usually products of several individual flows that traverse the same path over short time intervals, being overrun by later ones. A lava field describes the tangle of composite flows from a single eruptive event at the same crater. The summation of lava fields of a volcano constitutes the lava armor (Borgia and Linneman 1990; Francis and Oppenheimer 2004).

Lava flows in Costa Rica are mainly of Aa lavas (aa, a'a', or á'ā lava) in which the surface consists of jagged stony clinkers with numerous projections bristling with sharp points and angles, including sharp protrusions and spines (Cervantes and Ángeles lava fields, at Irazú and Barva volcanoes, respectively), or block lavas, which are formed by angular blocks that have sharp edges and even planar surfaces (Gavilán, Río Peje-Pretil, and 1968–2010 lava fields, at Rincón de la Vieja, Miravalles and Arenal volcanoes, respectively). Aa lavas could change to block lavas during the transport. Lava fields commonly form a rough terrain or malpais (Fig. 3.13), although a thin cover of ash makes possible limited agriculture activity or use of these rocks for local construction (Cigolini et al. 1984; Alvarado and Vega 2013; Rojas et al. 2017; Alvarado 2021). Also, lava flows can host lava caves, lava tubes, and collapse lava tubes (see the chapter of Ulloa and Alvarado in this book). Lava flows may cause the obstruction of main drainage lines producing temporary lakes. This was the case at the end of the Pleistocene, when lake formation at Cachí was caused by the Cervantes lava flow, or when the Palma Lake formed at Arenal volcano.

A lava plateau (lava platform) is an extensive lava flow field that builds a plain, including at high altitudes and, consequently, can be named as a plateau. A flat-topped hill capped with a near-horizontal lava flow is called lava mesa (Green and Short 1971; Macdonald 1972). An example of a lava mesa is present near the Sandillal dam (Cañas town), while lava plateaus are part of the Monteverde hills and occur around Tilarán town.

During cooling of thick lava flows, but also within dikes, sills, small- and large-volume pyroclastic density currents and even paleosols, the process can produce cooling structures such as columnar jointed colonnades with a polygonal pattern in cross-section. Other forms include entablatures, or two-tiered columnar jointing, columnar-joint rosettes, spherical joints, platy joints, and slabbies (Fig. 3.14).

3.4.5 Large Volcanic Depressions: Calderas and Volcano-Tectonic Structures

A caldera is a large, bowl-shaped, or flat-floored topographic depression (> 1.5 km in diameter) related to faulting, mostly associated with ignimbrite eruptions and large volumes of lava effusion. Some studies also relate their formation to sector collapse by debris avalanche or even subsequent erosion. Historically, calderas have been classified according to their origin: collapse, explosion, and erosion (Green and Short 1971; Macdonald 1972), but at the present, the terms explosion and erosion calderas are not used anymore. An ignimbrite caldera results from collapse following the eruption of large amount of ash and pumice that mantle the flanks of the volcanic depression and the surrounding region. An avalanche caldera generates an amphitheater-shaped head formed by sector collapse of a volcano producing a volcanic debris avalanche (Leyrit 2000). The term erosional caldera, a topographic depression formed and enlargement by erosion and landslides rather than by direct volcanic processes, is a former and old term, but in some cases can be useful, particularly under tropical conditions. Finally, a volcanic graben (sector graben) is a segment of a cone that may have dropped down between faults forming a graben (Green and Short 1971; Macdonald 1972).

In Costa Rica, well-studied ignimbrite calderas are the Guayabo (Zúñiga et al. 2005; Cigolini et al. 2018) and Alcántaro (Zamora et al. 2004; Molina et al. 2014) calderas in the Cordillera de Guanacaste, whereas minor studied ones include the Chocosuela (Alvarado and Carr 1993; Huapaya et al. 2021) and Barva (Pérez et al. 2006) calderas in the Cordillera Central. Avalanche calderas are present at almost all large volcanoes, but not in our conical stratovolcanoes (Alvarado et al. 2004). Most of the calderas represent different erosional stages. The Hule maar, because it has a diameter of 2 km, could be also considered as an explosion caldera (Alvarado et al. 2011). Volcanic grabens are present at the summits of Tenorio and Turrialba volcanoes (Alvarado 1989; Calvo et al. 2019).

Fig. 3.13 Lava flows. **a** Rural road between levees of a 17 ka Eastern Cervantes lava flow, **b** 1995 blocky lava flow at Arenal volcano; **c** and **d** active front of a blocky lava flow at Arenal volcano in June 1996 and July 1995. The steepness of the front is typical for viscous lavas, **e** aerial view of large 5 ka blocky lava flow at Miravalles volcano and a kipuka ("island of live" surrounded by lava flows), **f** Rough blocky lava morphology at the Caribbean flank of Rincón de la Vieja volcano (*ca.* 5 ka old)

Bergoeing (1998, 2007, 2009), Bergoeing and Brenes (2007), and Bergoeing et al. (1983, 2010, 2019) offered general geomorphological descriptions of the most outstanding features of volcanoes. However, they overestimated the number of caldera structures, many of which have no geological or geomorphological support (see Alvarado 2021).

3.4.6 Primary Morphological Features of Pyroclastic Deposits

Pyroclastic deposits can generate different types of geomorphological landscapes, reflecting the type of eruption, volume, grade of welding, and previous relief.

Fig. 3.14 Cooling structures in lavas. **a** Inclined andesitic columnar joints in andesites at Taboga farm (8 Ma), **b** columnar-joint rosettes near Cabuyal beach (7.8 Ma), **c** old columnar jointed colonnade in basalts at Manuel Antonio (61 Ma), **d** spherical joints within a 0.3 Ma andesitic lava flow at Virilla river, **e** and **f** platy joints and slabbies at 1.17 Ma old Mogote basaltic flow, Guayabo caldera

Ignimbrite platforms are voluminous and very extensive plains containing series of partly to strongly welded ignimbrites (pumice flow deposits, Fig. 3.15). In some cases, they form a broad, flat shield (ignimbrite shield), or even coalescing shields with very low slopes, usually with a caldera at the apex (Macdonald, 1972).

The Arenal pyroclastic density currents (pyroclastic flows and pyroclastic surges) from 1975 to 2010 produced Vulcanian pyroclastic flow fans, where the surface is comprised of numerous overlapping lobate deposits and boulder trains. Some of these were valley-confined basal coarse pyroclastic density currents (blocks and ash flows and bombs and ash flows) corresponding to the deposition of a high-density ground-hugging granular or inertial grain flow, in which the bulk of material is transported during small-volume pyroclastic density currents. But there are also several coarse debris deposits (colluvium and grain flows to hyperconcentrated flow deposits) interbedded with the pyroclastic material (Fig. 3.16).

Impact craters are secondary craters produced by ground impact of volcanic bombs and blocks ejected during an eruption (Fig. 3.17).

3.4.7 Primary Morphological Features of Epivolcaniclastic Deposits

Volcanoes are dynamic structures characterized by continued growth marked with episodes of destruction (caldera collapse) and/or instability that led to failure (e.g. landslides

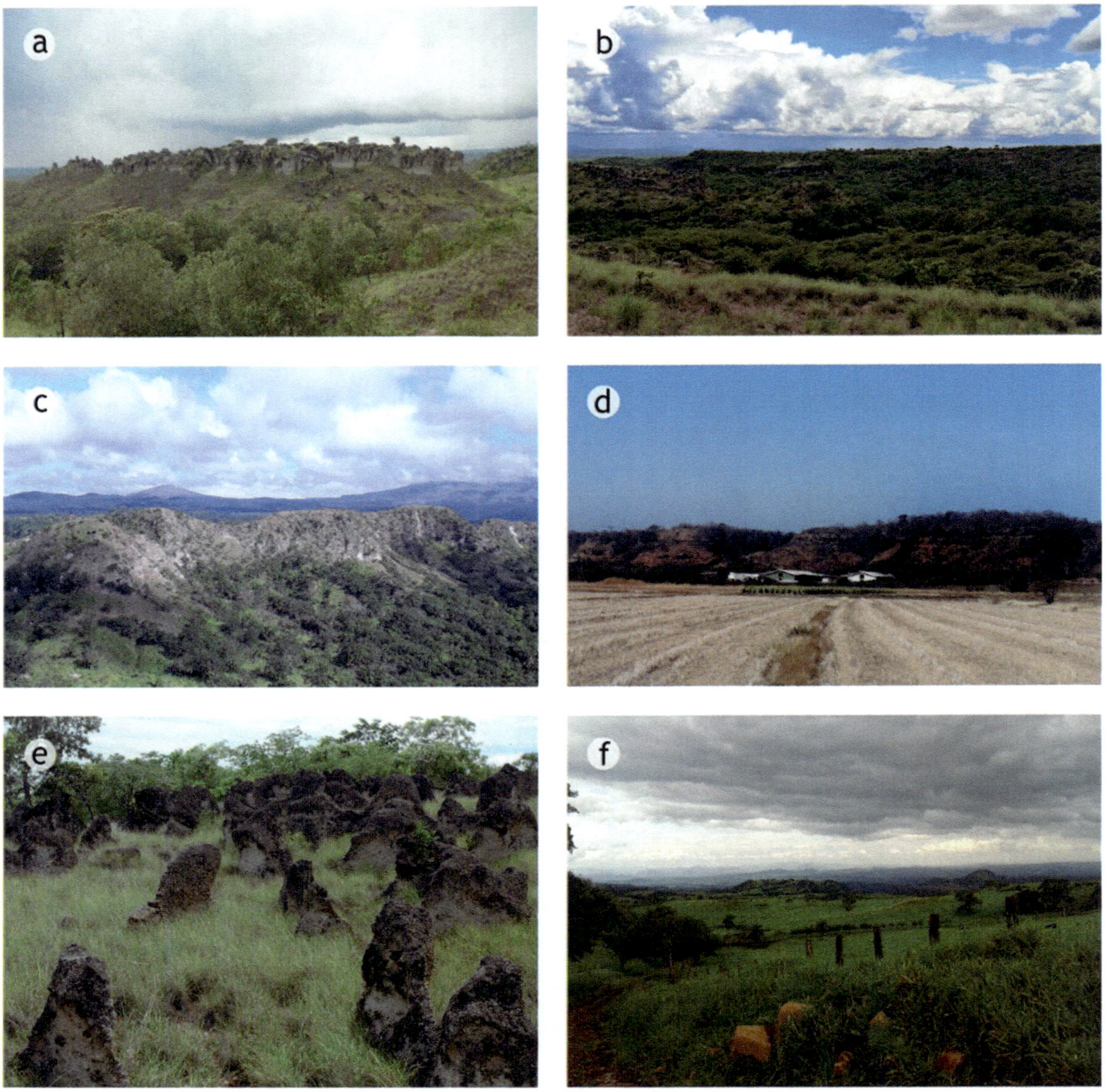

Fig. 3.15 Platforms or volcanic plateaus. **a–c** Ignimbrite platform (volcanic plateau) and lateral cliffs of non-welded pumiceous flows at the southern flank of Miravalles and Rincón de la Vieja volcanos, Guanacaste, **d** plateau underlain by welded ignimbrite (Meseta de Santa Rosa), **e** pinnacles eroded (old fumaroles in the ignimbrites) in non-welded pumice flows (near Santa Fe, Bagaces), and **f** lava plateau near Cañas, from the Meseta de Monteverde

and large avalanches). A striking feature of gravity-driven flows, including lahars, debris avalanches, and pyroclastic flows, are their erosive capability in ravines and canyons, during which they remove all vegetation and topsoil as well as generally smooth and round the topography, producing a wide, denuded corridor. The side walls of ravines and bends in valleys, along which the flows pass, are severely scoured and striated, and bedrock surfaces are conspicuously exposed.

An avalanche caldera is a horseshoe-shaped amphitheater that were formed by major collapse during large flank failures that emplaced a debris avalanche deposits, producing the hummocky topography. Hummocks are small hills with a circular, rectangular, or elliptical basal shape and are a characteristic structure of debris avalanche deposits in volcanic environments (Fig. 3.18). They may be separated by depressions, which can host small lakes. The most extensive gravitacional valleys have fans that are built of volcanic debris avalanche deposits, generally composed of heterogeneous, polymictic breccia, which consists of highly cataclased, metric-size blocks with abundant jigsaw cracks in pebble to clay matrix. Deposits vary from unaltered to highly hydrothermally altered lavas and pyroclastic rocks, yellow-reddish-gray in color. Debris avalanche deposits are present in all the large Costa Rican volcanoes (Alvarado et al. 2004).

Fig. 3.16 Primary landforms produced by Vulcanian pyroclastic eruptions. **a** and **b** 1975 pyroclastic flows at Arenal volcano in the direction of the Tabacón River, **c** pyroclastic fan created during the 1968 Arenal blast eruption, **d** pyroclastic flow at Arenal volcano, an example of Vulcanian eruption

Lahar fans and plains are volcaniclastic deposits with very poor sorting and grain sizes that range from clay to meter-sized boulders, up to 10 m in diameter (Alvarado and Schmincke 1994). Discharge of water from crater lakes that led to the development of lahars has only been documented at Rincón de la Vieja and Poás volcanoes, where acidic water was drained several times during volcanic eruptions via ravines and produced lahars (also hot lahars). Lahars and floods downstream from the volcano have also been recorded in association with earthquakes or phreatomagmatic eruptions (Fig. 3.18).

3.4.8 Hydrothermal and Geochemical Morphologies in Volcanic and Geothermal Environments

In active, dormant, and extinct volcanoes vents of gases and mineral springs (particularly hot or thermal springs) are a common feature called secondary manifestations of volcanism. Around fumaroles, the rocks are attacked by acid gases and condensates. The common products are opal, clay minerals, hydrated iron oxides and sulfates, alunite, and sulfur. Some fumaroles have no connection with deep sources and derive their gases wholly from the lava flow and pyroclastic ash flows on which they occur. In some cases, deposition of secondary minerals has made these features more resistant to erosion and more elevated than the surrounding, less cemented rocks, producing small landforms (Macdonald, 1972). The most relevant geoforms are the following:

- Hydrothermal explosion craters (Fig. 19a). Using detailed digital topography images, it was possible to recognize two craters of this type on the flanks of the Miravalles volcano (Castro et al. 2020).
- Travertine terraces. Calcareous terrace deposits (calcareous tufa) originate by precipitation of calcium carbonate at hot springs. Several examples are known from the Pacific flank of the extinct Cordillera de Tilarán and Montes del Aguacate, as well as from around the Palmera-La Marina geothermal area in San Carlos region, to the north of Platanar volcano (Fig. 3.19b).
- Mud pools and mud volcanos are small pools or pond filled with water and volcanic mud from hot springs and fumaroles, or minor cones, respectively. Numerous

Fig. 3.17 Impact craters. **a** and **b** Volcanic bombs and blocks ejected during the 1968 eruption from the western flank of Arenal volcano produced impact craters, most of which were filled with water (Melson and Sáenz 1968); **c** present morphology of one of the impact crater at Arenal; **d** 2017 impact crater at Poás volcano

examples of mud pools and mud volcanoes occur on the southern flanks of Rincón de la Vieja and Miravalles volcanoes (Fig. 3.19c, d).

- Siliceous sinter cone and travertine cone. Hot springs containing considerable dissolved silica and carbonates may deposit these compounds to form cones built of opal and/or other secondary minerals such as aragonite and calcite. Several examples are on the Pacific flank of the extinct Cordillera de Tilarán and Montes del Aguacate, north Rincón de la Vieja volcano, and around the Palmera-La Marina geothermal area in San Carlos region (Fig. 3.19e).
- Volcanic pavement is a thin shell or armor-plate flat surfaced, developed in volcanic ash with or without fragments of rocks forming a soil surface coating (Alvarado 1990; Jongmans et al. 1996). The shell is composed of fine ash (usually < 3 mm thick), light gray to grayish brown, with primary volcanic crystals and secondary minerals (small spherules of iron hydroxides in amorphous or opal silica cement, gypsum) or even plaster under the action of volcanic acid rain (Jongmans et al. 1996; Alvarado 2021). Examples are present around the craters and flanks of active Costa Rican volcanoes as Rincón de la Vieja, Arenal, Poás, Irazú and Turrialba (Fig. 3.19f).

3.5 Erosion and Preservation of Volcanic Landforms Under Tropical Conditions

On Earth, the preservation of volcanic landforms is largely controlled by the prevailing local climate, including the frequency of heavy rainfall, or the aggressivity of acid conditions (acid rain and fumaroles) that form clays. In the humid environment of Costa Rica, under additional influence of ongoing hydrothermal activity, primary landforms are easily eroded and last for relatively short time compared to analogous features in more benign conditions. The degree of denudation becomes larger for the progressively older volcanoes or volcanic geoforms, and extreme erosion has occurred under wet climates on the Caribbean flank of active or extinct Costa Rican volcanic ranges. The degree of denudation versus age correlation makes it possible to estimate the "erosional" age for different stages of volcanoes, based on morphology, as was described, and grouped in Proto-Paleo and Neo-volcanoes (i.e., Alvarado and Gans 2012).

Fig. 3.18 Landforms produced by volcanic debris avalanche (DA) and lahar deposits: **a** and **b** Conical small blocky hills in DA deposits (a: 9 ka Miravalles DA, b: ~ 1 ka Río Costa Rica DA); **c** 2017 debris flow at Miravalles volcano, related to heavy rainfall associated to Otto hurricane, **d** hyperconcentrated flow deposits in a distal part of debris flow, Miravalles volcano, Otto 2017 hurricane, **e** lahar fan at Taras town (Cartago city) related to 1963–1965 Irazú eruption, **f** hot lahar descending from the Rincón de la Vieja volcano, related to a volcanic eruption

Particularly intense erosion since the Late Pleistocene is responsible for an increase in relative relief, incision of the trunk valleys, headward erosion in the tributaries, and consequently, the origin of picturesque gorges in several volcanoes of the dissected ranges. The tephra fields commonly form a rough terrain (malpais) and show very dense dendritic patterns of rills, gullies, and ravines. Evidence of high denudation rates include dramatic incision on the flanks of the volcanoes in general and particularly in the valleys and canyons that cross the Caribbean side of the Guanacaste, Tilarán, and Central volcanic ranges (i.e., Caño Negro, Peñas Blancas, Aguas Zarcas, Toro, Desagüe, Anonos, Sarapiquí, Sucio, and Toro Amarillo rivers). Galve et al. (2016) measured valley walls up to 1115 m high in the Toro Amarillo River valley (between Irazú and Turrialba volcanoes) and estimated long-term fluvial incision rates of

Fig. 3.19 Landforms related to hydrothermal activity. **a** Hydrothermal explosion crater (Las Hornillas, Miravalles), **b** travertine terraces (La Marina), **c** mud pool, Rincón de la Vieja, **d** Las Hornillas acid hot lake, Rincón de la Vieja, **e** travertine cone, Finca Ande, Cañas, and **f** volcanic pavement, Arenal volcano

c. 5.6 m ka^{-1}. On the Pacific side of the volcanic ranges, there are less developed fluvial canyons such as the Virilla and Guacimal canyons.

Differential erosion of more resistant rock with respect to weaker slope lithologies is very common in volcanic terrains and produces scarps, cliffs, and waterfalls. A triangular wedge of lava on the slopes of a dissected volcano, narrowing upwards to culminate in a broken projection on or near a crater-rim, is a planeze. These occur where the lava flows protect an otherwise non-resistant cone and the planeze are part of the erosion stage of a volcano. Instructive examples of planeze are present on the northern flanks of Irazú (Galve et al. 2022) and Barva volcanoes, but smaller ones exist at other volcanoes such as Arenal.

Denudation of a volcanic cone commonly leaves resistant filling of the former vent standing out as a bold monolithic relief called a volcanic neck (Green and Short 1971). The best examples of volcanic necks are the Pelado hill in the Cordillera de Tilarán and the volcanic old complex of Pelón-Mondongo-Tinajita (Alvarado 2021) (Fig. 3.20).

Another important erosional process is the inversion of relief in a volcanic landscape, which occurs when a

Fig. 3.20 Examples of volcanic necks or eroded Pliocene volcanoes, at present composed by highly hydrothermally altered (silicified) volcanic rocks. **a** Pelado hill near Cañas, **b** Tinajita hill observed from the village of Dulce Nombre, **c** Pelón hill, and **d** Mondongo-Pelón hills, the last three near San Mateo (Montes del Aguacate)

volcanic flow deposit within a valley or depression become more resistant to erosion. As erosion proceeds, the valley floor becomes a ridge or hill bounded by newly formed valleys on each side, excavated in less resistant materials. Thus, relief inversion is a process where a negative landform (e.g., a valley or a lake) becomes a positive one, a common process in volcanic terrains (Pain and Oilier 1995; van Wyk de Vries et al. 2022). In Costa Rica, inversion of relief is less frequently found on lava flows (i.e., Sandillal area, Guanacaste) but is common in lacustrine pumiceous and diatomite lake deposits (Palmares and Turrúcares in the Valle Central; Loma Camastro and Montano in Guanacaste, among others) and ignimbrite plains. In these locations, a protective ignimbrite cap exists and is widespread in several localities in the Santa Rosa plateau (Pacific side of Guanacaste) and along the Río Grande de Tárcoles valley, which contain many examples that should be regarded as a part of ignimbrite landscape evolution (Fig. 3.21).

Based on numerous isotopic studies using radiometric dating over the last decades (Alvarado and Gans 2012), the age of these landforms could be well constrained. For example, the ignimbrite platforms are usually well preserved in the Pacific sector of the Guanacaste range, with ages of up to 5 Ma. On the other hand, in the Caribbean sector, they cannot be easily recognized. Similarly, domes usually maintain their primary shapes with morphochronological ages of up to 2 Ma. The lava flows show their meandering morphology and their different fronts with ages close to 2 Ma (e.g., lava field of the Paleo-Rincón de la Vieja volcano), if they are not covered with later eruptive products. Morphologies of pressure ridges and levees in lava flows can be preserved for approximately 100 ka.

Volcanic craters can preserve their shapes perhaps up to about 350 ka (e.g., Loma Morena cone, part of Aguas Zarcas cinder cone field with an age of 300 ka). Other forms attributed to former craters are usually questionable, since they rather look like landslide-produced amphitheaters (e.g., Mano de Tigre hill in the Cordillera Costeña and Cerro Yglesias on the Cocos Island). Volcanic calderas related to ignimbrites can be better preserved compared to craters, and their shapes can be recognized with ages of up to 2 Ma. Many caldera structures visualized in the Cordillera de Talamanca have a dubious volcanic origin and

Fig. 3.21 Examples of inversion of relief, where the resistant volcanic material deposited in the former valley becomes the high ground forming inverted relief. **a** 5 Ma old andesitic block and ash flow stratified deposits from an old volcanic apron, now exposed as pronounced cliffs of volcanic breccia at the Cruz de Guatuso, **b** former Palmares lake (old lake deposits) and ignimbrite on top form positive relief in the area, **c** Ancient Pliocene fluvial deposit with a 2 Ma old ignimbrite cap at Pelón de la Bajura, Guanacaste, **d** old paleovalley in volcanic breccias filled by Middle Pleistocene ignimbrite near Atenas, **e** 1 Ma old lake fluvial and lake deposit in a high topographic position, with a deep paleosol protected by a 0.6 Ma pumice flows at La Ese (between Bagaces and Guayabo); **f** Puente de Piedra, a natural bridge composed of Middle Pleistocene ignimbrite, near Tacares

rather seem to correspond to large amphitheaters of mega-landslides (debris avalanches), whose sliding planes usually correspond to marine shales that are transformed into clays. One case is the supposed caldera of Barbilla, with an amphitheater visible in the topography. Although there are volcanic rocks in the area with petrographic bimodalism (alkaline and calcalkaline rocks), there is no evidence of ignimbrites that could have come from this alleged "caldera" (Alvarado, 2021), and instead, it corresponds to an eroded anticline (J.P. Galve, verb. comm., 2023).

The geomorphic evolution of fluvial canyons during the last 100 ka has been influenced by volcanic activity, landslides triggered by earthquakes, heavy rainfall including hurricanes, and the erosive mechanism of lahar-related events that have occurred during that time. Frequent strong hydrothermal alteration in those areas produced weak rocks and loose material that favor landslides and erosion (gravitacional slide valleys).

More enigmatic are the forms attributed to domes in the Caribbean sector of Costa Rica from Sarapiquí to the border with Nicaragua. The rocks attributable to these domes have ages between 22.7 and 11 Ma (Alvarado and Gans 2012). Although they have a dome shape and a typical petrography of domes (dacites to rhyolites), it is unknown if they are primary morphologies or exposed cryptodomes by erosion.

3.6 Brief Comments on Volcanic Soil Development

The nature and type of volcanic activity produce and disperse soil particles, which can be classified by grain size and density to create a textural gradient along volcanic slopes. Coarse materials are deposited nearest to the crater, whereas finer grains are deposited further away from the volcano (Alvarado and Mata 2016). The weather and environmental conditions take control of soil-forming processes once the deposition is completed. The United States Department of Agriculture Soil Taxonomy (Soil Survey Staff 1999) describes volcanic soils as andosols, defined by the single property of having pyroclastic parent material, such as volcanic ash, pumice or scoria lapilli. On the other hand, entisols are soils with little or no evidence of the development of pedogenic horizons, Orthents are the most common on recent erosional surfaces (Soil Survey Staff 1999). Costa Rica's high volcanic activity allows for the formation of andisols associated with entisols, for example, on lavas and very recent ash (Alvarado et al. 2014; Alvarado 2022).

In Costa Rica, andosols occupy around 14% of the land area and occur in the NW–SE belt where volcanic activity has influenced landforms and soil development. The area includes the Valle Central, the Cordillera de Guanacaste, and the regions influenced by Barú volcano ashes between Coto Brus and Panama. Some areas are affected by the wind and have fluvio-volcanic deposition such as the Caribbean and Northern zone (Alvarado and Mata 2016), and some areas in the Cordillera de Talamanca where the NE wind spread ashes from the Cordillera Central (Chinchilla et al. 2011).

Orthents are the most common and abundant kind of entisols and are usually found on rocky hillsides with low temperatures and/or particularly erosive precipitation, recent volcanic depositions such as ashes or lava, and the existence of weather-resistant parental material (Alvarado and Mata 2016). Other Orthents are found in low-relief sites of rhyolithic origin in comparatively dry Pacific areas, with less than 1500 mm of precipitation (Alvarado and Mata 2016). The glass-minerals with short halloysitic crystallization range tend to dominate in these kinds of soils. Characteristics are conditioned primarily by climatic variation induced by differences in elevation, and to a lesser extent by mineralogy of the parent material, since most of ash deposits in the country have andesitic composition. This way, the andosols mineral formation sequence is described based on elevation. In the mountain highlands, vitric profiles and humic allophanic soils can be found near the craters. On the slopes, there are allophanic placic soils in the middle, hallositic soils in the middle lower portion, kaolinitic soils in the lower part, and smectitic soils in the distal areas (Alvarado et al. 2014; Alvarado 2022).

3.7 Conclusions and Future Research

Neogene-Quaternary volcanism in Costa Rica is represented by an extensive variability of primary (major and minor) volcanic forms and denudation landforms. Monogenetic cinder cones and domes are infrequent in Costa Rica, opposite to other Quaternary volcanic regions. Cinder cones are controlled by extensional tectonic regimes. Six monogenetic volcanic fields are recognized in Costa Rica: Pasquí-Cazuelas (Irazú volcano), Monte de la Cruz (Barva volcano), Sabana Redonda (Poás volcano), Cañas Dulces (Paleo-Rincón de la Vieja volcano), Aguas Zarcas (at the piedmont of Platanar volcano), and Río Lori.

At least 2 million people are under the shadow of the Costa Rican volcanoes, so, the volcanic geomorphology, including the quantitative reconstruction of recent volcanic activity, gravitational slides and lahars is essential for hazard studies and risk assessment, and for the development of models used in long-term territorial planning.

Intense erosion since the Late Pleistocene is responsible for an increase in relative relief, incision of the trunk valleys, headward erosion in the tributaries, and consequently, the origin of picturesque gorges in several volcanoes and old volcanic terrains of the dissected ranges. Their rejuvenation is ongoing, as attested by impressive waterfalls, some of which are more than 50 m high.

Territorial protection was introduced in 1955, and since then, several national, protected, and reserved parks have increases in number and size. Thus, conservation and tourism are one of the most important goals and economical entries in Costa Rica. National parks ensure proper safeguarding of its natural heritage, including outstanding geomorphology. Some trails also play the role of educational paths, with explicative and interpretative panels and/or printed leaflets available, particularly in the volcanic national parks (see Alvarado et al. 2023b). Fluvial canyons are impressive and are always attractive from a tourist and adventure viewpoint.

The confinement generated by the COVID-19 pandemic though 2020–2021 brought with it the revival of outdoor life and more knowledge of the national parks and surrounding areas. This offers an opportunity to further exploit the beauties in volcanic landscapes in a more sustainable way. For example, the tourist value of Las Cazuelas and Pasquí satellite craters and cones of Irazú volcano, and their beautiful landscapes and viewpoints were not recognized by the local people until 2021 (Alvarado et al. 2023b).

With this study, we aim to draw attention to the importance of volcanic landform features, which form an important attraction, generally producing the highest and most spectacular landscapes. The variability of landscapes offers tourists the opportunity to understand the different processes of geodynamics, and thus being of "geodiversity" and geoheritage interest. This chapter also highlights that volcanic inverted relief is more frequent in ignimbrite environments in Costa Rica, with respect to the widespread existence of inverted lava flows well described around the world (i.e., Pain and Oilier 1995; van Wyk de Vries et al. 2022).

Volcanoes have attracted the attention of human beings for as long as anyone can remember due to their activity, hazards, and processes. But volcanic landforms continue to attract the attention of tourists even if the volcano is dormant, particularly the conical forms and their craters. But why? If it is assumed as true that beauty is related to symmetry and perhaps to the golden number (1.61893…, i.e.,

Fig. 3.22 Golden triangle such as Arenal volcano form (**a**), or snail-like crater similar symmetry as the golden or Fibonacci spiral, such as the Rincón de la Vieja (**b**), could explain their particular appeal to the tourists

Hemenway 2005; Corbalán 2010), then it can be explained that those volcanoes that meet near the golden triangle (internal angles of 36°, 36° and 108°) such as the Arenal volcano (Alvarado 2021), or those craters that due to eruptive and erosive processes emulate a snail with its intrinsic symmetry (golden or Fibonacci spiral), such as the active craters of Rincón de la Vieja, Poás, and Irazú, are more attractive if compared to less symmetrical volcanoes such as complex massifs and shield volcanoes, or the very simple circular craters (Fig. 3.22).

There are still several enigmatic landforms identified on aerial photographs and satellite images that cannot be easily interpreted without detailed fieldwork (geological mapping), such as the semi-depression around Valle Azul between the Platanar and the Arenal volcanoes, or the depression located 12 km to the NE of the Cacao volcano, or those located at Bajo La Rosa, to the WNW of the Irazú volcano.

Finally, it is necessary to undertake more quantitative research on the volcanic subaerial and submarine landforms. These include morphometric measurements of sizes, shapes, and estimate of growth and denudation rates (in the case of subaerial geoforms) based on geochronology, geomorphology, and geological field mapping.

Acknowledgements We thank the Instituto Costaricense de Electricidad (ICE) and Comisión Nacional de Prevención de Riesgos y Atención de Emergencias (CNE) for making available essential data for this research and for the necessary means used in our field surveys in Costa Rica. We thank Adolfo Quesada (UCR) for inviting us to participate in this chapter. Jani Radebaugh and Piotr Migón made very valuable comments on the text and English editing.

Credits of photographs and thanks to: Figs. 5b (Blas Sánchez), 9a (Ernst Flüeh), 9b (Omar G. Lizano), 10c (Yemerith Alpízar), 12d (Jorge Cortés), 12f (Federico Chavarría), 13e (Google Earth), 14d (Laura Castro), 16a and b (William G. Melson, r.i.p.), 16d (Luis Madrigal), 18e (Instituto Geográfico Nacional), 18f (Jean Paul Dinarte).

References

Alvarado A, Mata R (2016) Soils of Costa Rica. In: Kappelle M (ed) Costa Rican ecosystems. University of Chicago Press, Chicago, pp 64–94

Alvarado A (2022) Suelos de Centroamérica. EDUPUC, San José

Alvarado A, Mata R, Chinchilla M (2014) Arcillas identificadas en suelos de Costa Rica a nivel generalizado durante el período 1931–2014: I. historia, metodología de análisis y mineralogía de arcillas en suelos derivados de cenizas volcánicas. Agron Costarric 38:75–106. https://doi.org/10.15517/rac.v38i1.15160

Alvarado GE (1989) Los volcanes de Costa Rica. EUNED, San José

Alvarado GE (1990) Los relieves de las rocas ígneas. ETCR, Cartago

Alvarado GE (2021) Costa Rica y sus volcanes. EUCR, EUNA, ETCR, San José

Alvarado GE, Carr MJ (1993) The Platanar-Aguas Zarcas volcanic centers, Costa Rica: spatial-temporal association of Quaternary calc-alkaline and alkaline volcanism. Bull Volcanol 55:443–453. https://doi.org/10.1007/BF00302004

Alvarado GE, Schmincke H-U (1994) Stratigraphic and sedimentological aspects of the rain triggered lahars of the 1963–1965 Irazú eruption, Costa Rica. Zentralblatt für Geol und Paläontologie, Tl I 1:513–530

Alvarado GE, Gans PB (2012) Síntesis geocronológica del magmatismo, metamorfismo y metalogenia de Costa Rica, América Central. Rev Geológica América Cent 46:7–122. https://doi.org/10.15517/rgac.v0i46.1836

Alvarado GE, Vega AE (2013) La geomorfología de la colada de Cervantes, volcán Irazú (Costa Rica): Descripción de uno de los campos de lava más grandes de América Central. Rev Geológica América Cent 48:99–118. https://doi.org/10.15517/rgac.v0i48.12238

Alvarado GE, Denyer P (2019) Estratigrafía volcánica del Neógeno y Cuaternario. In: Denyer P (ed) Perspectiva geológica del noroeste de Costa Rica: historia, evolución y cartografía. EUCR-CICG, San José, pp 101–238

Alvarado GE, Vega E, Chaves J, Vásquez M (2004) Los grandes deslizamientos (volcánicos y no volcánicos) de tipo Debris Avalanche en Costa Rica. Rev Geológica América Cent 30:83–99. https://doi.org/10.15517/rgac.v0i30.7260

Alvarado GE, Carboni S, Cordero M et al (2010) Stability of the cone and foundation of Arenal volcano, Costa Rica. In: Olalla C, Hernandez LE, Rodriguez-Losada JA et al. (eds) Volcanic Rock Mechanics. Rock Mechanics and Geo-engineering in Volcanic Environments. CRC Press, Taylor & Francis Group, London, pp 135–150

Alvarado GE, Soto GJ, Salani FM et al (2011) The formation and evolution of Hule and Río Cuarto maars, Costa Rica. J Volcanol Geotherm Res 201:342–356

Alvarado GE, Esquivel L, Sánchez BE, Matamoros G (2020a) Actualización del peligro volcánico del Turrialba, Costa Rica. CNE Internal report, San José

Alvarado GE, Esquivel L, Sánchez BE, Alfaro JC (2020b) Actualización del peligro volcánico del Poás, Costa Rica. CNE Internal report, San José

Alvarado GE, Campos-Durán D, Brenes-André J et al (2021) Peligros volcánicos del Irazú, Costa Rica. CNE Internal report

Alvarado GE, Brenes-André J, Venegas L et al (2022) Análisis de peligros volcánicos Rincón de la Vieja, Costa Rica. ICE-CNE Internal report

Alvarado GE, Brenes-André J, Núñez D, et al (2023a) Actualización del análisis del peligro volcánico del Arenal, Costa Rica. ICE Internal report

Alvarado GE, Alpízar Y, Araya R et al (2023b) A new impetus in "geo-literacy" in the national volcanic parks of Costa Rica and surrounding sectors for better risk management. Rev Estud Latinoam sobre Reducción del Riesgo Desastr—REDER 7(2): 6–18. https://doi.org/10.55467/reder.v7I2.121

Baker P, Harris P (1963) Lava Channels on Tristan da Cunha. Geol Mag 100:345–351

Bergoeing JP (1978) La fotografía aérea y su aplicación a la geomorfología de Costa Rica. Ministerio de Obras Públicas y Transportes, Instituto Geográfico Nacional, San José

Bergoeing JP (1979). Geomorfología del sector volcánico Las Nubes-Cabeza de Vaca. Informe semestral IGN (enero-junio), 136–146

Bergoeing JP (1998) Geomorfología de Costa Rica. Instituto Geográfico Nacional, San José

Bergoeing JP (2007) Geomorfología de Costa Rica, 2nd edn. Librería Francesa, San José

Bergoeing JP (2009) Costa Rica, paisajes volcánicos. Ediciones Jadine, San José

Bergoeing JP, Brenes L (2007) Las calderas concéntricas del Volcán Platanar, Costa Rica. Rev Geográfica 141:168–178

Bergoeing JP, Malavassi E, Protti R (1979). Tres posibles edificios volcánicos del sector Cerros del Aguacate. Informe semestral IGN (julio-diciembre), 35–45

Bergoeing JP, Brenes L, Malavassi E (1983) Geomorfología del Pacífico Norte de Costa Rica (Explicación del mapa geomorfológico 1:100 000). MOPT-IGN-UCR, San José

Bergoeing JP, Brenes L, Protti R et al (2010) Atlas geomorfológico del Caribe de Costa Rica. UCR-IGN, San José

Bergoeing JP, Brenes LG, Fernández M (2019) Las calderas de Molejón, Santa Clara y Flores, Costa Rica. Rev Geográfica 153:167–178

Blake S (1990) Viscoplastic Models of Lava Domes. In: Fink J (ed) Lava flows and domes: emplacement mechanism and hazard implications. Springer, Berlin, pp 88–126

Borgia A, Linneman SR (1990) On the Mechanisms of Lava flow emplacement and volcano growth: Arenal, Costa Rica. In: Fink JH (ed) Lava flows and domes: emplacement mechanism and hazard implications. Springer, Berlin, pp 208–243

Borgia A, Poore C, Carr MJ et al (1988) Structural, stratigraphic, and petrologic aspects of the Arenal-Chato volcanic system, Costa Rica: evolution of a young stratovolcanic complex. Bull Volcanol 50:86–105. https://doi.org/10.1007/BF01275171

Bravo J (1982) Geomorfología de la Hoja Fortuna (Escala 1:50 000). Universidad Nacional, Sede Omar Dengo

Bravo J, Barquero J, Fernández E et al (1989) Geomorfología del Complejo Volcánico Tenorio. Bol Vulc 10:11–14

Calvo C, Madrigal K, Merayo F et al (2019) Modelo volcanotectónico del graben cuspidal complejo del Turrialba (Costa Rica) y su relación con los colapsos sectoriales bajo un régimen transpresivo y transtensivo. Rev Geológica América Cent 61:57–77

Castillo P, Batiza R, Vanko D et al (1988) Anomalously young volcanoes on old hot-spot traces: I. Geology and petrology of Cocos Island. Geol Soc Am Bull 100:1400–1414. https://doi.org/10.1130/0016-7606(1988)100<1400:AYVOOH>2.3.CO;2

Castro D, Solís A, Alvarado GE (2020) Evolución geomorfológica del volcán Miravalles, Guanacaste, Costa Rica. Rev En Torno a La Prevención 25:1–12

Chinchilla M, Mata R, Alvarado A (2011) Andisoles, inceptisoles y entisoles de la subcuenca del río Pirrís, región de Los Santos, Talamanca, Costa Rica. Agron Costarric 35:83–107. https://doi.org/10.15517/rac.v35i1.6688

Cigolini C, Borgia A, Casertano L (1984) Intra-crater activity, aa-block lava, viscosity and flow dynamics: Arenal Volcano, Costa Rica. J Volcanol Geotherm Res 20:155–176

Cigolini C, Taticchi T, Alvarado GE et al (2018) Geological, petrological and geochemical framework of Miravalles-Guayabo caldera and related lavas, NW Costa Rica. J Volcanol Geotherm Res 358:207–227. https://doi.org/10.1016/j.jvolgeores.2018.05.013

Corbalán F (2010) La proporción áurea. El lenguaje matemático de la belleza. RBA Coleccionables S.A., España

Cortés J (2007) Coastal morphology and coral reefs. In: Bundschuh J, Alvarado GE (eds) Central America, Geology, Resources and Hazards. CRC Press, London, pp 185-200

Cotton CA (1969) Volcanoes as landscape forms, 2nd edn. Hafner Publishing Company, New York

Denyer P, Alvarado GE (2007) Mapa Geológico de Costa Rica 2007. 1:400,000. San José: Librería Francesa S.A

Elizondo V, Alvarado GE, Soto D (2019) Evolución espacio-temporal de las bocas eruptivas de los volcanes Irazú, Arenal, Turrialba y Poás en tiempo histórico (Costa Rica). Rev Geológica América Cent 61:35–55. https://doi.org/10.15517/rgac.v61i0.40087

Francis P, Oppenheimer C (2004) Volcanoes, 2nd edn. Oxford University Press, New York

Galve JP, Alvarado GE, Pérez-Peña JV et al (2016) Megafan formation driven by explosive volcanism and active tectonic processes in a humid tropical environment. Terra Nov 28:427–433. https://doi.org/10.1111/ter.12236

Galve JP, Alvarado GE, Perez-Pena JV et al (2022) River avulsions and megafan development triggered by high recurrent explosive volcanism. Geomorphology 419:108466

Gazel E, Ruiz P (2005) Los conos piroclásticos de Sabana Redonda: componente magmático enriquecido del Volcán Poás, Costa Rica. Rev Geológica América Cent 33:45–60. https://doi.org/10.15517/rgac.v0i33.4235

Granados-Bolaños S, Alvarado GE (2021) Simulación de lahares en la cuenca alta del río Platanar y sus posibles impactos en Ciudad Quesada (volcán Platanar, Costa Rica. En Torno a La Prevención 26:1–16

Granados-Bolaños S, Quesada-Román A, Alvarado GE (2021) Low-cost UAV applications in dynamic tropical volcanic landforms. J Volcanol Geotherm Res 410:107–143. https://doi.org/10.1016/j.jvolgeores.2020.107143

Green J, Short NM (1971) Volcanic landforms and surface features. Springer, Berlin, Heidelberg

Grosse P, van Wyk de Vries B, Petrinovic IA, et al (2009) Morphometry and evolution of arc volcanoes. Geology 37:651–654. https://doi.org/10.1130/G25734A.1

Healy J (1969) Notes on the volcanoes of the Guanacaste volcanic range, Costa Rica. New Zel Geol Surv

Hemenway P (2005) Divine proportion. Phi in Art, Nature, and Science. Sterling Publishing Co., Inc., New York

Horn S, Haberyan K (1993) Costa Rican Lakes. Natl Geogr Res Explor

Huapaya A, Rodríguez E, Ceciliano A (2021) Mapa Geológico 1:50 000 de la Hoja Quesada 3346-IV. Dirección de Geología y Minas

Inbar M, Risso C (2001) A morphological and morphometric analysis of a high density cinder cone volcanic field—Payun Matru, south-central Andes, Argentina. Zeitschrift für Geomorphol 45:321–343. https://doi.org/10.1127/zfg/45/2001/321

Jongmans AG, Groenesteijn K, Buurman P, Mulder J (1996) Soil surface coatings at costa Rican recently active volcanoes. Soil Sci Soc Am J 60:1871–1880. https://doi.org/10.2136/sssaj1996.03615995006000060036x

Kempter KA, Rowe GL (2000) Leakage of Active Crater lake brine through the north flank at Rincón de la Vieja volcano, northwest Costa Rica, and implications for crater collapse. J Volcanol Geotherm Res 97:143–159. https://doi.org/10.1016/S0377-0273(99)00181-X

Lagmay AMF, van Wyk de Vries B, Kerle N, Pyle DM (2000) Volcano instability induced by strike-slip faulting. Bull Volcanol 62:331–346. https://doi.org/10.1007/s004450000103

Leyrit H (2000) Flank collapse and debris avalanche deposits. In: Leyrit H, Montenat C (eds) Volcaniclastic Rocks, from Magma to Sediments. Gordon and Breach Science Publishers, Amsterdam, pp 111–129

Macdonald GA (1972) Volcanoes. Prentice-Hall, New Jersey

Madrigal R, Rojas E (1980) Manual Descriptivo del Mapa Geomorfológico de Costa Rica (escala 1:200 000), SEPSA. Imprenta Nacional, San José

Martens U (2004) Aspectos morfométricos de los conos de escoria de Costa Rica. Rev Geológica América Cent 30:15–29. https://doi.org/10.15517/rgac.v0i30.7251

McKnight SB, Williams SN (1997) Old cinder cone or young composite volcano? The nature of Cerro Negro Nicaragua. Geology 25:339. https://doi.org/10.1130/0091-7613(1997)025<0339:OCCOYC>2.3.CO;2

Meagher TF (1859a) Holidays in Costa Rica. III.–San José to Cartago. Harper's New Mon. Mag.

Meagher TF (1859b) Holidays in Costa Rica. I.–Puntarenas to San José. Harper's New Mon. Mag.

Meagher TF (1860) Holidays in Costa Rica. II.–San José. Harper's New Mon. Mag.
Melson WG, Sáenz R (1968) The 1968 Eruption of Volcán Arenal, Costa Rica: Preliminary summary of field and laboratory studies. Smithsonian Inst, Center for Short-lived Phenomena
Mercalli G (1907) I Vulcani Attivi Della Terra: Morfologia-Dinamismo-Prodotti-Distribuzione Geografica-Cause. Biblio Bazaar, Milan
Molina F, Martí J, Aguirre G et al (2014) Stratigraphy and structure of the Cañas Dulces caldera (Costa Rica). Geol Soc Am Bull 126:1465–1480. https://doi.org/10.1130/B31012.1
Nakamura K (1977) Volcanoes as possible indicators of tectonic stress orientation—principle and proposal. J Volcanol Geotherm Res 2:1–16. https://doi.org/10.1016/0377-0273(77)90012-9
Németh K (2010) Monogenetic volcanic fields: origin, sedimentary record, and relationship with polygenetic volcanism. In Cañon-Tapia E, Szakács A (eds) What Is a Volcano? Special Paper The Geol Soc Amer 470:43–66. https://doi.org/10.1130/2010.2470(04)
Oersted A. (1863) L'Amerique Centrale. Recherches sur sa flore et sa géographie physique. Resultats d'un voyage dans les etats de Costa Rica et Nicaragua execute pendant les annes 1846–1848. Bianco Luno, Copenhagen
Pain CF, Oilier CD (1995) Inversion of relief - a component of landscape evolution. Geomorphology 12:151–165. https://doi.org/10.1016/0169-555X(94)00084-5
Pérez W, Alvarado GE, Gans PB (2006) The 322 ka Tiribí Tuff: stratigraphy, geochronology and mechanisms of deposition of the largest and most recent ignimbrite in the Valle Central, Costa Rica. Bull Volcanol 69:25–40. https://doi.org/10.1007/s00445-006-0053-x
Rittmann A (1962) Volcanoes and their activity. Interscience Publishers, New York
Rojas V, Barahona D, Alvarado GE (2017) Geomorfología y petrografía de la colada Ángeles y del cono Monte de la Cruz, volcán Barva, Costa Rica. Rev Geológica América Cent 56:17–35. https://doi.org/10.15517/rgac.v0i56.29240
Rojas W, Alvarado GE (2012) Marco geológico y tectónico de la Isla del Coco y la región marítima circunvecina, Costa Rica. Rev Biol Trop 60:15–32. https://doi.org/10.15517/rbt.v60i3.28306
Ruiz P, Mana S, Gazel E et al (2019) Geochemical and geochronological characterization of the Poas Stratovolcano Stratigraphy. In: Tassi F, Vaselli O, Mora Amador R (eds) Poás Volcano: The Pulsing Heart of Central America Volcanic Zone. Springer, Cham, pp 13-43
Ruiz P, Turrin BD, Soto GJ et al (2010) Unveiling Turrialba (Costa Rica) volcano's latest geological evolution through new 40Ar/39Ar, ages. In: American Geophysical Union, Fall Meeting Abstracts
Ryan PC, Alvarado GE, McCanta M et al (2022) The importance of overbank deposits and paleosol analyses for comprehensive volcanic hazard evaluation: the case of Holocene volcanism at Miravalles Volcano, Costa Rica. Nat Hazards 112:413–449. https://doi.org/10.1007/s11069-021-05187-6
Schindlbeck JC, Kutterolf S, Freundt A, Andrews GDM, Wang KL, Volker D, Werner R, Frische M, Hoernle K (2016) Alkalic marine tephra layers at ODP Site 1241-Major explosive eruptions from an oceanic volcano in a pre-shield stage? J Volcanol Geotherm Res 328:96–104
Soil Survey Staff (1999) Soil taxonomy: a basic system of soil classification for making and interpreting soil surveys, 2nd edn. Natural Resources Conservation Service. U.S. Department of Agriculture Handbook 436, Washington
Soto GJ, Alvarado GE (2006) Eruptive history of Arenal Volcano, Costa Rica, 7 ka to present. J Volcanol Geotherm Res 157:254–269. https://doi.org/10.1016/j.jvolgeores.2006.03.041
Tassi F, Vaselli O, Fernandez E et al (2009) Morphological and geochemical features of crater lakes in Costa Rica: an overview. J Limnol 68:193. https://doi.org/10.4081/jlimnol.2009.193
Trejos JF (1959) Geografía de Costa Rica, física, política y económica. Imprenta universal, San José
Tristán J (1923) The activity of volcano Irazú in Costa Rica. Zeitschr Vulkanol 7:93–113
Thouret J-C (1999) Volcanic geomorphology-and overview. Earth Sci Rev 47:95–131
van Wyk de Vries B, Karatson D, Gouard C et al (2022) Inverted volcanic relief: Its importance in illustrating geological change and its geoheritage potential. Int J Geoheritage Park 10:47–83.https://doi.org/10.1016/j.ijgeop.2022.02.002
Vargas J (1967) Belleza y Amenaza del volcán Irazú. Efemérides Costarric 1:1–17
von Seebach K (1892) Über Vulkane Zentralamerikas. Aus den nachgelassenen Aufzeichnungen von Karl von Seebach. Dieterich'sche Verlagsbuchhandlung, Göttingen
von Seebach K (1865) Seebach's Besteigung des Vulkans Turrialba in Costa-Rica. Petermann's Geogr Mitth IX:321–324
Walker G (1965) Some aspects of quaternary volcanism in Iceland. Trans Leicester Lit Philos Soc LIX:25–40
Whitford-Stark JL (1982) Factors influencing the morphology of volcanic landforms: an earth-moon comparison. Earth-Sci Rev 18:109–168. https://doi.org/10.1016/0012-8252(82)90050-2
Zamora N, Méndez J, Barahona M, Sjöbohm L (2004) Volcanoestratigrafía asociada al campo de domos de Cañas Dulces, Guanacaste, Costa Rica. Rev Geológica América Cent 30:41–58. https://doi.org/10.15517/rgac.v0i30.7253
Zúñiga E, Chavarría L, Barrantes M, et al (2005) Geologic model of the Miravalles geothermal field, Costa Rica. In: World geothermal congress 2005. Antalya, Turkey

4 Mass Movements Dynamics and Morphologies in Costa Rica

Adolfo Quesada-Román

Abstract

Landslides are natural processes that commonly affect human activities at every latitude and altitude worldwide. Tropical countries such as Costa Rica deal annually with landslides affecting public and private infrastructure and activities. This chapter provides a brief overview of landslide origin, dynamics, and morphologies applicable to Costa Rica. Coseismic and hydrometeorological dynamics are common landslide triggers, but land use changes are also important. Large landslides are related with volcanic slopes and densely vegetated rural contexts, while smaller but more common mass movements are linked with periurban conditions, especially in densely populated areas with altered land use. On average, during the last five decades every year, 80 landslide events occurred in Costa Rica, causing 5 people died, 9 houses destroyed, and approximately, 1 million US dollars of losses. This chapter suggests more studies dealing with modelling, geomorphological characterization, early warning systems, and continued monitoring of landslides on susceptible sites in Costa Rica.

Keywords

Landslides · Seismic landslides · Rainfall-triggered landslides · Earthquakes · Tropical cyclones

4.1 Introduction

Landslides constitute a common natural hazard worldwide which results in greater socioeconomic impacts in developing and tropical countries than anywhere else. Mountainous areas are characterized by closely coupled fluvial and gravitational processes (Carrión-Mero et al. 2021; Trujillo-Vela et al. 2022). Mass movements starting at higher elevations may impact distant, lower areas directly and through cascading effects, involving rivers (Quesada-Román 2021a), or through the amplification of processes by entrainment of material (Lee et al. 2022). Landslides have been widely related to rainfall intensity-duration relations (Sepúlveda and Petley 2015; Rosi et al. 2021) and to seismic activity (Bommer and Rodríguez 2002).

In Central America, including Costa Rica, this phenomenon is mainly triggered by seismicity and extraordinary rainfall. The Central American region is characterized by moderate to high landslide hazard and risk (Sepúlveda and Petley 2015; Quesada-Román and Campos-Durán 2023). Over 90% of the disasters in Costa Rica are hydrometeorological in nature (DesInventar 2022). About 60% of the national territory is mountainous, where landslides are common (Fig. 4.1). In addition, strong weathering, heavy rains, and seismicity make Costa Rica a susceptible territory for landslides (Quesada-Román 2021b). Furthermore, poor agricultural practices, roads with inadequate slope designs, inefficient water management, and unprotected slope cuts make the country highly vulnerable to landslide events (Vallejos et al. 2012; Quesada-Román 2022a). Most landslides sensu lato in Costa Rica are translational and rotational slides, but debris flows, falls, complex slides, and even rock avalanches have been reported in the past (Vallejos et al. 2012; Quesada-Román and Barrantes 2016a, b; Quesada-Román 2021a).

The subduction of the Cocos Plate under the Caribbean Plate, the occurrence of Panama microplate and volcanism favour tectonic activity on the territory of Costa Rica (Alvarado et al. 2017; Vannucchi and Morgan 2019). This dynamic is responsible for three morphotectonic units: forearc, volcanic front, and backarc (Fig. 4.1; Marshall 2007). The first morphotectonic unit, the forearc, is of Cretaceous-Quaternary age and extends along the Pacific

A. Quesada-Román (✉)
Laboratorio de Geografía Física, Escuela de Geografía, Universidad de Costa Rica, San José, Costa Rica
e-mail: adolfo.quesadaroman@ucr.ac.cr

A. Quesada-Román (ed.), *Landscapes and Landforms of Costa Rica*, World Geomorphological Landscapes, https://doi.org/10.1007/978-3-031-64940-0_4

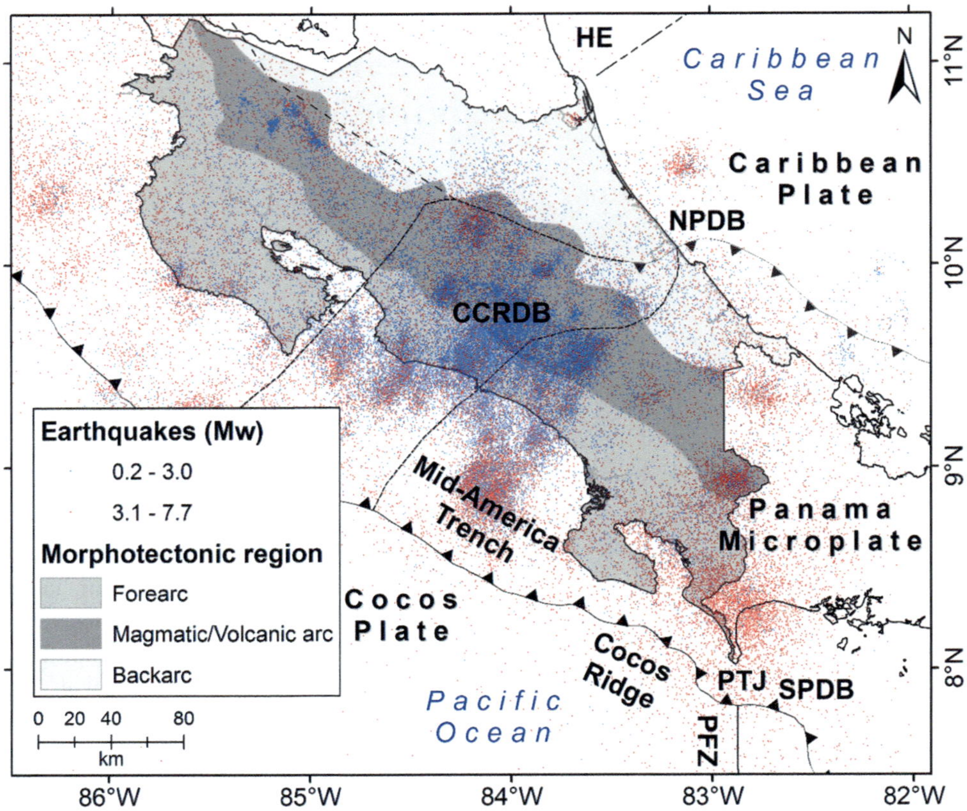

Fig. 4.1 Location, tectonic framework, and morphotectonic units of Costa Rica. CCRDB: Central Costa Rica Deformed Belt; NPDB: North Panama Depression Belt; SPDB: South Panama Depression Belt; PFZ: Panama Fracture Zone; PTJ: Point Triple Junction (for details, see DeMets et al. 2010). Instrumental location of earthquakes from 1975 to 2017 (RSN, 2017)

coast (Denyer and Alvarado 2007). The second morphotectonic unit, the volcanic front, includes the Guanacaste, Tilarán, Aguacate, Central, and Talamanca Mountain ranges composed mainly of volcanic and sedimentary rocks of Paleogene-Quaternary age (Alvarado 2021). Finally, the back arc extends to the Caribbean basin (Quesada-Román and Pérez-Briceño 2019). The climate and precipitation patterns in the back arc are influenced by latitudinal migration of the Intertropical Convergence Zone, ENSO, northeast trade winds, cold fronts, and tropical cyclones (Durán-Quesada et al. 2020). The interaction of trade winds and topography produces dissimilar Pacific and Caribbean climates. The Pacific side presents a bimodal rainfall distribution, while it is difficult to define a dry season for the Caribbean side (Fig. 4.2; Quesada-Román 2022b). This chapter presents a brief overview of landslide origin, dynamics, and morphologies in Costa Rica. Moreover, a bibliometric analysis helped to generate a map indicating the best-studied landslides in Costa Rica.

4.2 Coseismic Landslides

The influence of earthquakes on slope weakening has direct relation with the subsequent rainfall-induced triggering of landslides in mountain landscapes. In fact, the region has a substantial history of earthquake-induced landslides (Bommer and Rodríguez 2002). The seismicity in Costa Rica is particularly abundant in the Pacific coast, especially between the coastline and the Mesoamerican Trench, associated with the interaction between Cocos, Caribbean, and Panama plates (Godínez-Rodríguez et al. 2018). Other major tectonic features include the Panama Fracture Zone (PFZ), the North Panama Deformed Belt (NPDB), and the Central Costa Rica Deformed Belt (CCRDB) (Linkimer et al. 2018). These broad zones of crustal deformation are proposed as the boundaries between the Caribbean plate and the Panama microplate (Marshall et al. 2000). Specific seismic events such as Limón earthquake in 1991 (Hernández et al. 1992; Mora and Mora 1994; Quesada-Román 2016b, Barrantes et al. 2021; Quesada-Román 2021a, b, c; Fig. 4.3a), Cinchona earthquake in 2009 (Alvarado 2010; Barrantes et al. 2011, 2013; Quesada-Román and Barrantes, 2016a, b, 2017; Ruiz et al. 2019; Fig. 4.3b), and Sámara earthquake in 2012 (Peraldo et al. 2014) produced many slope failures (Fig. 4.4), described in numerous publications.

4.3 Landslides Triggered by Rainfall

El Niño-Southern Oscillation temporal and spatial variability heavily influences rainfall oscillations (Méndez et al. 2019). Complex responses (warm or wet) vary in terms of

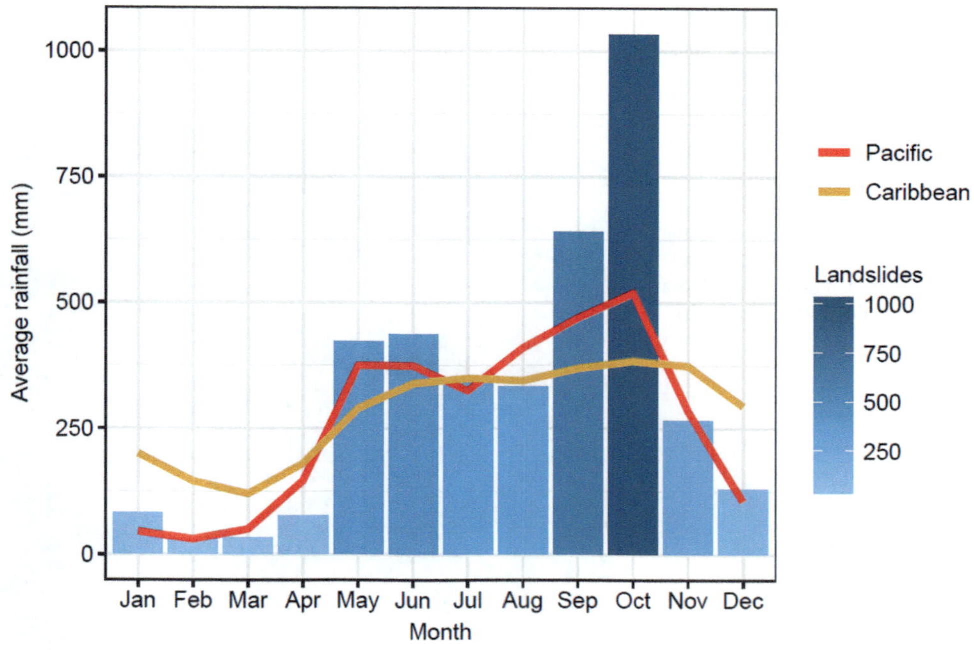

Fig. 4.2 Costa Rican Pacific and Caribbean rainfall trends (based on Méndez et al. 2019) and landslide occurrence by month from 1970 to 2018 (DesInventar 2022)

their signs, magnitudes, duration, and seasonality between water basins draining towards the Pacific and the Caribbean (Waylen and Laporte 1999). These two slopes behave differently and respond in opposite ways to ENSO conditions (Durán-Quesada et al. 2020). On the Pacific side of Costa Rica, El Niño events generally lead to drier conditions whereas La Niña events favour wetter conditions (Quesada-Román et al. 2020). The situation is slightly opposite on the Caribbean side of Costa Rica. Moreover, a statistically significant positive linear trend has been observed in the annual number of intense hurricanes in the Caribbean Sea since the 1970s (Guzman and Jiang 2021). These characteristics facilitate the occurrence of extreme hydrometeorological phenomena related to the ENSO, or more specifically, during the passage of cold fronts and tropical cyclones. Accordingly, the region also has a very high correlation between mean daily rainfall totals and mean daily landslides. Extreme rainfall events trigger landslides in Central America (Campos-Durán and Quesada-Román 2017; Hidalgo 2021), including Costa Rica (Quesada-Román et al. 2021). Therefore, seismic preconditioning and subsequent extraordinary rainfall conditions have occurred several times in Costa Rica, generating landslides (Quesada-Román et al. 2019) (Fig. 4.5).

See Fig. 4.6.

4.4 Landslide Distribution in Space

Disaster database for the period 1970–2018 shows that there have been 3835 events in which 221 people died, 20 were missing, 34 were wounded, 115,498 were affected, 5625 were evacuees, 566 were relocated, 447 houses were destroyed, and 4129 were affected. The reported loss in US$ dollars was 42 million between 1970 and 2018 (DesInventar 2022).

Historical national landslide catalogues for Costa Rica have been done by Mora et al. (1992), Peraldo and Rojas (2000), Arroyo et al. (2011), and Vallejos et al. (2012). The historical occurrence of landslides has been reported along volcanic slopes of the Central and Guanacaste Volcanic Cordilleras (Alvarado et al. 2003, 2004; Siebert et al. 2006). Specific landslide studies in volcanic terrains have been carried out at Poás (Mora and Morales 1992; Morales et al. 1992; Montero et al. 2010; Ruiz et al. 2019; Arroyo-Solórzano et al. 2021; Arroyo-Solórzano et al. 2022), Irazú (Montero and Alvarado 1995; Mora and Pirulli 2017; Fallas et al. 2018), Barva (Salazar et al. 2019; Quesada-Román and Mata-Cambronero, 2020), Platanar (Granados-Bolaños and Alvarado 2021), and Miravalles volcanoes (Quesada-Román et al. 2019; Ryan et al. 2022).

A bibliometric analysis was performed using all the scientific papers and technical reports published in Spanish and English related with landslides in Costa Rica (Fig. 4.7). The results show that the most studied regions in the context of landslides are located mainly surrounding different volcanic regions of the volcanic/plutonic arc, and some exceptions within the Central Costa Rica Deformed Belt (CCRDB) or produced by the Limón 1991 earthquake.

Landslide processes in certain municipalities have received more attention from scientists and decision makers during the last four decades. Some of these areas belong to municipalities characterized by a high hazard, vulnerability, or risk to landslides. For example, Puriscal, a mainly rural

Fig. 4.3 Recent coseismic landslides in Costa Rica. **a** Along the Chirripó River caused by the Limón 1991 earthquake. **b** Around Poás volcano generated by the Cinchona 2009 earthquake. Both examples showed the complexity of resulting morphologies with falls, slides, flows, and even complex movements. Credits: Sergio Mora (**a**), and Guillermo Alvarado (**b**)

municipality in the south of San José province, has been widely studied. A local seismic swarm and an earthquake in 1990, in addition to an altered land caused by intensive agriculture, provoked continuous instability (Mora 1991; Peraldo and Molina 1993; Terlieng et al. 1995; Suárez et al. 2009). A debris flow or lahar such as the one occurred in 1963 in Cartago has been analysed in the past (Mora 1993; Granados-Bolaños et al. 2021). The historical impact of land use change and train line use in Turrialba has been extensively analysed (Mora et al. 1990; Peraldo and Rojas 1998; Segura et al. 2011; Peraldo et al. 2015; Peraldo and Badilla 2015). Other cases have been studied in Pérez Zeledón associated to the impact of the Buenavista earthquake in 1983 as well as the successive tropical cyclones which triggered several landslides (Boschini et al. 1988; Mora and Peraldo 2011; Quesada-Román 2016a; Quesada-Román et al. 2018; Quesada-Román and Zamorano 2018, 2019a, b). During the last decade, special attention has been given to developing studies of landslides which occurred to the south of the GAM municipalities. These

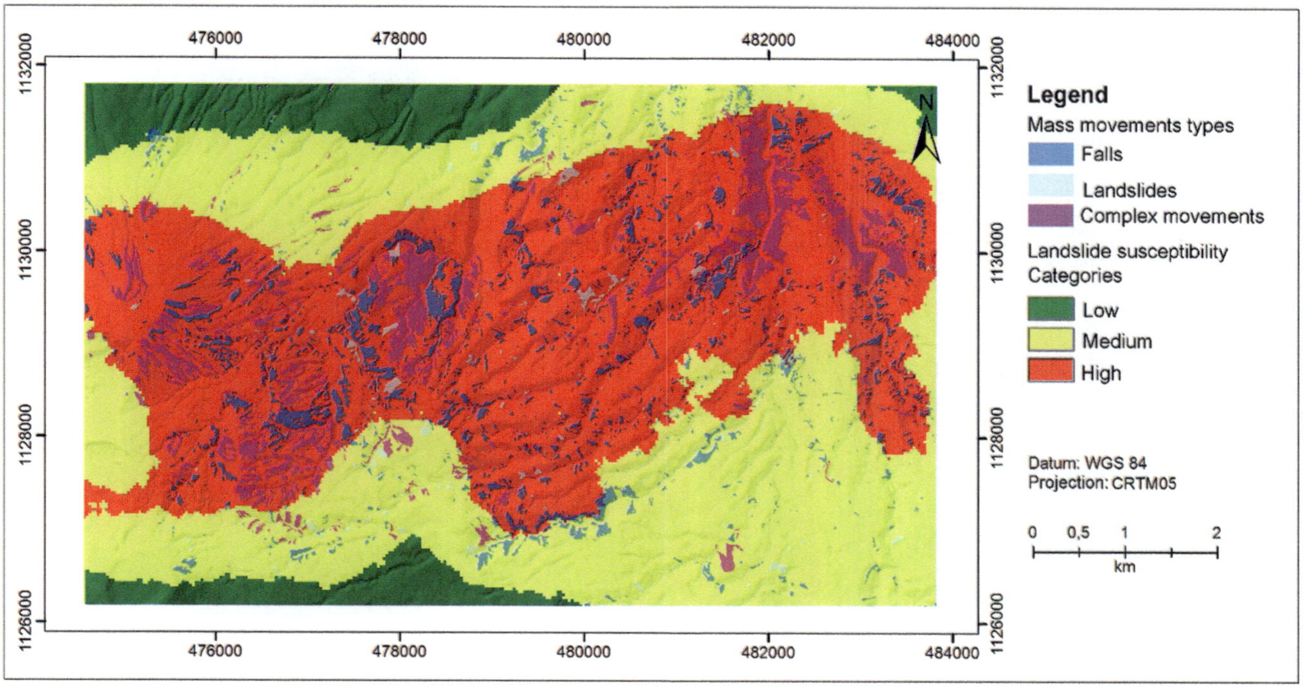

Fig. 4.4 Landslides susceptibility map of the northeastern slope of the Poás volcano by Quesada-Román and Barrantes (2017). The indicated mass movements types are the ones triggered by Cinchona 2009 earthquake

municipalities experience a rapid urban growth over altered volcanic or sedimentary rocks (Arias et al. 2011; Madrigal 2011; Mora and Granados 2014; Quesada-Román 2015; Quesada Román et al. 2020). Other studies have been performed in periurban municipalities of the GAM such as Poás (Barrantes and Quesada-Román 2016a, b), Alajuela, and Moravia-Vázquez de Coronado (Quesada-Román and Feoli-Boraschi 2018; Sequeira-Arguedas 2021).

Addressing the pervasive issue of creep in Costa Rican pastures and its consequential impact on erosion, leading to potential landslides, necessitates a multifaceted approach for sustainable environmental management (Fig. 4.8a). Implementing erosion control strategies, such as the introduction of vegetation cover and the establishment of contour ploughing and terracing, can effectively stabilize the soil within pastures. Concurrently, promoting sustainable livestock management practices, including rotational grazing systems and educational initiatives for responsible grazing, becomes imperative to mitigate overgrazing and soil compaction, both contributors to erosion. To bolster preventative measures, the integration of advanced technologies like satellite imagery, drones, and ground sensors for comprehensive monitoring, coupled with the precise terrain mapping capabilities of Lidar technology, can offer crucial insights into landscape dynamics. Early warning systems based on these technologies can be pivotal for identifying areas prone to creep and erosion, enabling timely evacuation and risk mitigation (Fig. 4.8b). Furthermore, employing modelling and simulation techniques to predict potential landslide scenarios, considering factors like soil composition, land use patterns, and precipitation levels, enhances our ability to develop targeted prevention strategies. In tandem, community engagement and education initiatives empower residents to actively participate in monitoring and adopting sustainable land management practices.

4.5 Conclusions

Costa Rica merges an intense tectonic and hydrometeorological dynamics. Moreover, population is growing without proper land use planning and therefore, increasing their exposure to risk. Bigger landslides of both seismic and hydrometeorological origin are in forested mountain areas affecting small rural towns. However, smaller but more frequent landslides are present in periurban contexts due to the fast urban growth without territorial management. There are some understudied regions such as Fila Brunqueña (southeast cordillera), Cordillera de Talamanca slopes both in the Caribbean and the Pacific sides, the Pacific coastal belt, and Nicoya Peninsula. The study of landslides in Costa Rica needs to expand in terms of modelling, geomorphological description, landslide early warning systems (LAWS),

Fig. 4.5 Rainfall-triggered landslides in the Miravalles volcano slopes during Otto 2016 hurricane. This event showed translational slides in the upper summit of the volcano (**a**), and debris flows in both the Caribbean (Bijagua de Upala) and Pacific foot slopes near Guayabo and La Fortuna de Bagaces (**b**). *Credits* CNE (2017)

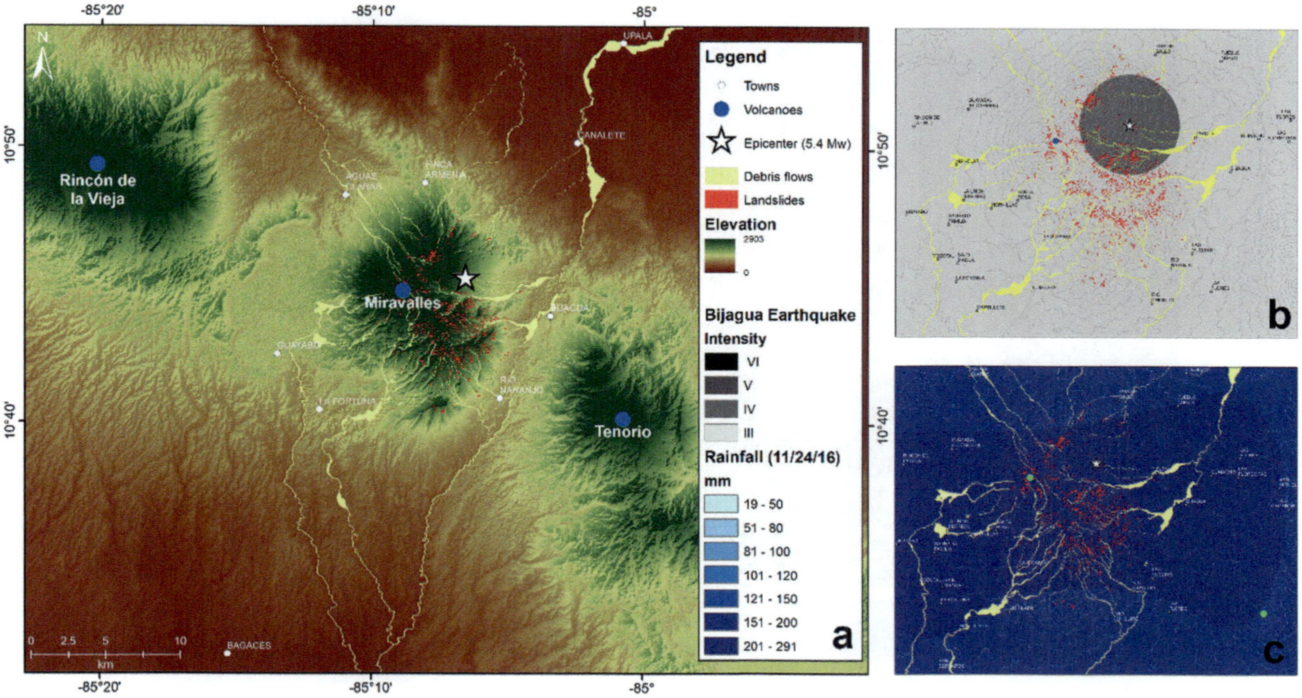

Fig. 4.6 **a** Landslides (red) and debris flows (yellow) triggered by Hurricane Otto around Miravalles volcano. **b** Bijagua earthquake epicentre with its Mercalli Modified Intensity distribution. **c** Rainfall reported around Miravalles volcano during November 24, 2016

Fig. 4.7 Location of the landslides study sites in Costa Rica since mid-1980s

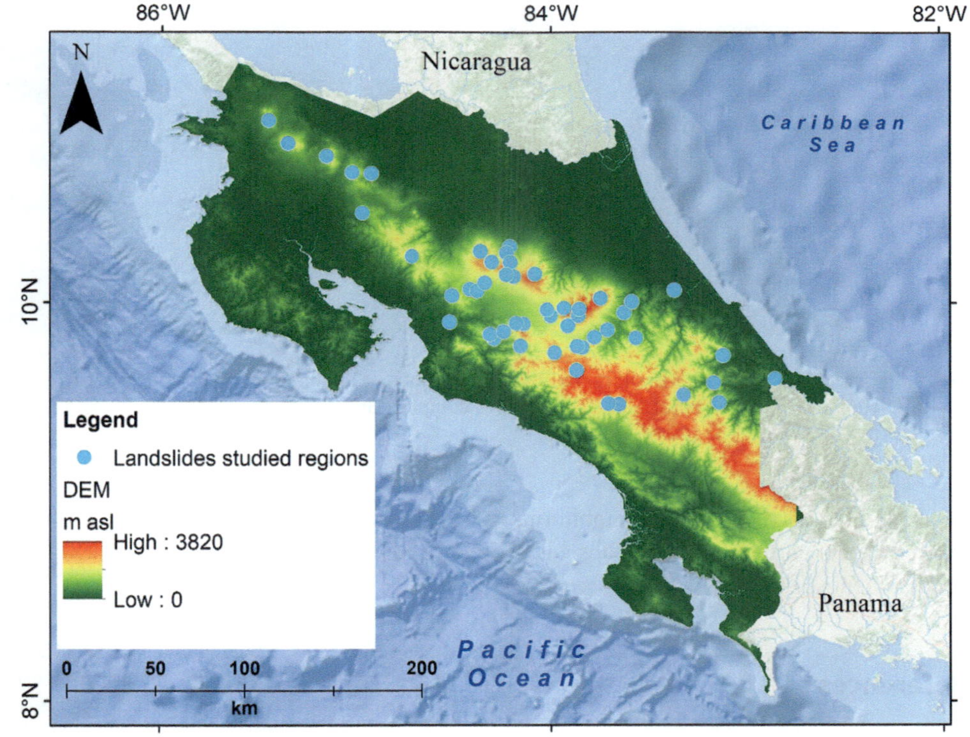

Fig. 4.8 Creep associated with pastures and livestock is another common mass movement type in Costa Rica, this example is from Barú in Pérez Zeledón (**a**). Banderillas (or San Blás) massive complex landslide in the Irazú volcano southern slope near Tierra Blanca, at the bottom Cartago city (**b**). *Credits* The author

and should involve continued monitoring of relationships between local earthquakes and subsequent intense rainfalls at catchment scale.

References

Alvarado GE (2010) Aspectos geohidrológicos y sedimentológicos de los flujos de lodo asociados al terremoto de Cinchona (Mw 6, 2) del 8 de enero del 2009, Costa Rica. Rev Geol De Amér Cent 43:67–96

Alvarado GE, Peraldo G, Mora R (2003) The June 2000 Arancibia Debris Avalanche and block-slide, Costa Rica. Landslide News 14:29–32

Alvarado GE, Vega E, Chaves J, Vásquez M (2004) Los grandes deslizamientos (volcánicos y no volcánicos) de tipo debris avalanche en Costa Rica. Rev Geol de Amér Cent 30:83–99

Alvarado GE, Benito B, Staller A, Climent A, Camacho E, Rojas W, Marroquín G, Molina E, Talavera JE, Martínez-Cuevas S, Lindholm C (2017) The new Central American seismic hazard zonation: mutual consensus based on up to day seismotectonic framework. Tectonophysics 721:462–476

Alvarado GE (2021) Costa Rica y sus volcanes. San José, Costa Rica: EUCR, EUNA, ETCR

Arias M, Fuentes O, Fallas J (2011) Criterios utilizados para la microzonificación de la microcuenca Lajas, Cantón de Escazú, Costa Rica. En Torno a la Prev 7:1–7

Arroyo LN, Salgado D, Moraga CV (2011) Costa Rica: análisis de la incidencia relacionada con inundaciones y deslizamientos por

provincias y cantones, década 1999 al 2009. Rev Geog de Amér Cent (Núm Esp EGAL 2):1–16
Arroyo-Solórzano M, Quesada-Román A, Barrantes-Castillo G (2022) Seismic and geomorphic assessment for coseismic landslides zonation in tropical volcanic contexts. Nat Hazards 114(3):2811–2837. https://doi.org/10.1007/s11069-022-05492-8
Arroyo-Solórzano M, Quesada-Román A, Vargas-Bolaños C (2021) Unidades morfotectónicas-volcánicas del sector noroeste del volcán Poás, Costa Rica. Investigaciones Geográficas e60279. https://doi.org/10.14350/rig.60279
Barrantes G, Barrantes O, Núñez O (2011) Efectividad de la metodología Mora-Vahrson modificada en el caso de los deslizamientos provocados por el terremoto de Cinchona, Costa Rica. Rev Geog de Amér Cent 47:141–162
Barrantes G, Jiménez C, Ocón AJ (2013) Deslizamientos provocados por el terremoto de Cinchona de 2009, Costa Rica. Rev Geog De Amér Cent 51:89–100
Barrantes G, Vahrson WG, Mora S (2021) Cambios geomorfológicos e hidrológicos inducidos por el terremoto (Mw 7, 7) del 22 de abril de 1991 en la provincia de Limón, Costa Rica. Revista Geológica de América Central 65:1–19
Bommer JJ, Rodriguez CE (2002) Earthquake-induced landslides in Central America. Eng Geol 63(3–4):189–220. https://doi.org/10.1016/S0013-7952(01)00081-3
Boschini IM, Alvarado GE, Rojas W (1988) El terremoto de Buenavista de Pérez Zeledón (Julio 3, 1983): Evidencia de una fuente sismogénica intraplaca desconocida en Costa Rica. Rev Geol de Amér Cent 8:111–121
Carrión-Mero P, Montalván-Burbano N, Morante-Carballo F, Quesada-Román A, Apolo-Masache B (2021) Worldwide research trends in landslide science. Int J Environ Res Public Health 18(18):9445
CNE—Comisión Nacional de Prevención de Riesgos y Atención de Emergencias (2017b) Geospatial information of the geomorphic impacts of Hurricane Otto. Unidad de Investigación y Análisis de Riesgo. San José, Costa Rica
DeMets C, Gordon RG, Argus DF (2010) Geologically current plate motions. Geophysical J Int 181(1):1–80
Denyer P, Alvarado GE (2007) Mapa geológico de Costa Rica. Escala 1:400,000. Librería Francesa. San José, Costa Rica
DesInventar (2022) Inventory system of the effects of disasters; Corporación OSSA: Cali, Colombia. Available online: http://desinventar.org. Accessed 22 Oct 2022
Durán-Quesada AM, Sorí R, Ordoñez P, Gimeno L (2020) Climate perspectives in the Intra-Americas Seas. Atmosphere 11(9):959
Fallas M, Prado A, Mora M, Ruiz P, Alfaro E, Soto G (2018) El deslizamiento del 8 de diciembre de 1994 en el volcán Irazú (Costa Rica): aspectos históricos y geomorfología con base en fotografías aéreas históricas y recientes. Rev Geol de Amér Cent 58:55–84
Godínez-Rodríguez K, Arroyo-Solórzano M, Linkimer-Abarca L (2018) Distribución geográfica de los sismos contenidos en el catálogo de la Red Sismológica Nacional de Costa Rica. Rev Geol de Amér Cent 60:161–188
Granados-Bolaños S, Alvarado GE (2021) Simulación de lahares en la cuenca alta del río Platanar y sus posibles impactos en Ciudad Quesada (volcán Platanar, Costa Rica). Revista En Torno a la Prevención 26:1–16
Granados-Bolaños S, Quesada-Román A, Alvarado GE (2021) Low-cost UAV applications in dynamic tropical volcanic landforms. J Volcanol Geoth Res 410:107143
Guzman O, Jiang H (2021) Global increase in tropical cyclone rain rate. Nat Commun 12(1):1–8
Hernández G, Vahrson W, Ruiz A (1992) Deslizamientos producto del terremoto (4–22-91). Mapa publicado. Escuela de Ciencias Geográficas, Universidad Nacional de Costa Rica
Hidalgo HG (2021) Climate variability and change in Central America: What does it mean for water managers? Front Water 2:632739
Lee S, An H, Kim M, Lee G, Shin H (2022) Evaluation of different erosion–entrainment models in debris-flow simulation. Landslides 1–16
Linkimer L, Arroyo IG, Alvarado GE, Arroyo M, Bakkar H (2018) The National seismological network of Costa Rica (RSN): an overview and recent developments. Seism Res Lett 89(2A):392–398
Madrigal J (2011) Análisis general del mega-deslizamiento Ortiga-Potrerillos: Acosta, San José, Costa Rica. En Torno a la Prev 7:8–13
Marshall J, Fisher D, Gardner T (2000) Central Costa Rica deformed belt: kinematics of diffuse faulting across the western Panama block. Tectonics 19:468–492
Marshall J (2007) The geomorphology and physiographic provinces of Central America. In: Bundschuc and Alvarado (eds) Central America: Geology, Resources and hazards. Taylor & Francis, London, p 1436
Méndez M, Calvo-Valverde L, Maathuis B, Alvarado-Gamboa LF (2019) Generation of monthly precipitation climatologies for Costa Rica using irregular rain-gauge observational networks. Water 11(1):70. https://doi.org/10.3390/w11010070
Montero W, Alvarado GE (1995) El terremoto de Patillos del 30 de diciembre de 1952 (Ms= 5, 9) y el contexto neotectónico de la región del volcán Irazú, Costa Rica. Rev Geol de Amér Cent 18:25–42
Montero W, Soto G, Alvarado GE, Rojas W (2010) División del deslizamiento tectónico y transtensión en el macizo del volcán Poás (Costa Rica), basado en estudios neotectónicos y de sismicidad histórica. Rev Geol de Amér Cent 43:13–36
Mora S (1993) Análisis preliminar de la vulnerabilidad económica de la ciudad de Cartago, a causa de los efectos de una avalancha que transite por el río Reventado, Costa Rica. Rev Geol De Amér Cent 15:65–80
Mora R, Granados R (2014) Zonificación de los procesos denudacionales en la micro-cuenca del río Cruz, Escazú, San José, Costa Rica: recomendaciones para mejorar la gestión del territorio. Rev Geol de Amér Cent 50:99–112
Mora R, Morales L (1992) La susceptibilidad a la inestabilidad de laderas en la zona norte de Alajuela y el terremoto de Fraijanes de 1888. Efectos y previsión. Rev Geog de Amér Cent 25–26:311–322
Mora M, Peraldo G (2011) Análisis macrosísmico del terremoto de buena vista de Pérez Zeledón, 3 de julio de 1983 (ms= 6, 1), Costa Rica. Rev Geol de Amér Cent 44:41–70
Mora R, Pirulli M (2017) Deslizamiento Las Torres, volcán Irazú, Costa Rica: definición de su volumen y su dirección preferencial de movimiento. Rev Geol de Amér Cent 56:51–67
Mora R, Cerdas A, Molina F, Vega E (1990) Caracterización geológica del deslizamiento Chiz (Turrialba, provincia de Cartago, Costa Rica): amenaza y prevención. Rev Geol de Amér Cent 11:59–68
Mora S, Alvarado GE, Morales L (1992) Mapa preliminar sintético de amenazas geológicas de Costa Rica. Su importancia en el desarrollo de la infraestructura civil. Rev Geog de Amér Cent 25–26:355–371
Mora S (1991) The Puriscal, Costa Rica landslide and its dynamics. In: Bell D Landslides: Proceedings of VI international symposium on landslides, Christchurch, New Zealand, Balkema, pp 1199–1204
Morales L, Montero W, Madrigal R (1992) El terremoto de Fraijanes de 1888 y el sistema de fallas de Alajuela, implicaciones del peligro sísmico potencial para el Valle Central Occidental. Rev Geog de Amér Cent 25–26:227–240
Peraldo G, Badilla E (2015) El área de deslizamientos de Torito: un proceso histórico de amenaza. En Torno a la Prev 14:7–21

Peraldo G, Molina F (1993) Reconstrucción Histórica del deslizamiento de Santiago de Puriscal. Rev Geol de Amér Cent 16:83–94

Peraldo G, Rojas E (1998) La deslizable historia del ferrocarril al Caribe de Costa Rica. Anua de Est Centroamer 24(1–2):97–128

Peraldo G, Rojas E (2000) Guía para investigadores. Catálogo de deslizamientos históricos para Costa Rica, período 1772–1960. Rev Inf Sem 36:123–171

Peraldo G, Badilla E, Arias O (2014) Efectos morfológicos generados en el cantón de Valverde Vega por el terremoto de Sámara del 5 de septiembre de 2012 (Mw=7,6). En Torno a la Prev 12:13–22

Peraldo G, Badilla E, Soto D, González M, Herra D, Ramos JD, Méndez H, Ceciliano A, Vega A, Van der Laan E (2015) El deslizamiento de Palo Alto, Turrialba, Costa Rica: apuntes para su estudio. En Torno a la Prev 15:7–23

Quesada-Román A (2021a) Landslide risk index map at the municipal scale for Costa Rica. Int J Disaster Risk Reduction 56:102144

Quesada-Román A (2021b) Landslides and floods zonation using geomorphological analyses in a dynamic catchment of Costa Rica. Revista Cartográfica 102:125–138

Quesada-Román A (2021c) Review of the geomorphological effects of the 1991 Limón earthquake. Revista Geológica De América Central 65:1–13

Quesada-Román A (2022a) Disaster risk assessment of informal settlements in the global South. Sustainability 14(16):10261

Quesada-Román A (2022b) Flood risk index development at the municipal level in Costa Rica: a methodological framework. Environ Sci Policy 133:98–106. https://doi.org/10.1016/j.envsci.2022.03.012

Quesada-Román A, Campos-Durán D (2023) Natural disaster risk inequalities in Central America. Papers Appl Geogr 9(1):36–48

Quesada-Román A, Pérez-Briceño PM (2019) Geomorphology of the Caribbean coast of Costa Rica. J of Maps 15(2):363–371. https://doi.org/10.1080/17445647.2019.1600592

Quesada-Román A, Zamorano-Orozco JJ (2019a) Geomorphology of the Upper General River Basin, Costa Rica. J of Maps 15(2):95–101. https://doi.org/10.1080/17445647.2018.1548384

Quesada-Román A, Fallas-López B, Hernández-Espinoza K, Stoffel M, Ballesteros-Cánovas JA (2019) Relationships between earthquakes, hurricanes, and landslides in Costa Rica. Landslides 16(8):1539–1550. https://doi.org/10.1007/s10346-019-01209-4

Quesada-Román A, Ballesteros-Cánovas JA, Guillet S, Madrigal-González J, Stoffel M (2020) Neotropical Hypericum irazuense shrubs reveal recent ENSO variability in Costa Rican páramo. Dendrochronologia 61:125704. https://doi.org/10.1016/j.dendro.2020.125704

Quesada-Román A, Villalobos-Portilla E, Campos-Durán D (2021) Hydrometeorological disasters in urban areas of Costa Rica Central America. Environ Hazards 20(3):264–278. https://doi.org/10.1080/17477891.2020.1791034

Quesada-Román A, Barrantes G (2016a) Procesos de ladera cosísmicos del terremoto de Cinchona (Costa Rica) del 8 de enero de 2009 (Ms= 6,2). Cuad de Geog: Rev Col de Geog 25(1):217–232. https://doi.org/10.15446/rcdg.v25n1.52590

Quesada-Román A, Barrantes G (2016b) Procesos de ladera cosísmicos del terremoto de Cinchona (Costa Rica) del 8 de enero de 2009 (Ms= 6,2). Cuadernos de Geografía: Revista Colombiana de Geografía 25(1):217–232. https://doi.org/10.15446/rcdg.v25n1.52590

Quesada-Román A, Barrantes G (2017) Modelo morfométrico para determinar áreas susceptibles a procesos de ladera. Inv Geog, Bol Inst Geog 94:1–12. https://doi.org/10.14350/rig.57318

Quesada-Román A, Feoli-Boraschi S (2018) Comparación de la metodología Mora-Vahrson y el método morfométrico para determinar áreas susceptibles a deslizamientos en la microcuenca del río Macho, Costa Rica. Rev Geog de Amér Cent 61:17–45. https://doi.org/10.15359/rgac.61-2.1

Quesada-Román A, Mata-Cambronero E (2020) The Geomorphic Landscape of the Barva Volcano, Costa Rica. Phys Geog. https://doi.org/10.1080/02723646.2020.1759762

Quesada-Román A, Zamorano-Orozco JJ (2018) Peligros Geomorfológicos en Costa Rica: Cuenca Alta del Río General. Anuá do Inst de Geoc 41(3):239–251. https://doi.org/10.11137/2018_3_239_251

Quesada-Román A, Zamorano-Orozco JJ (2019b) Zonificación de procesos de ladera e inundaciones a partir de un análisis morfométrico en la cuenca alta del río General, Costa Rica. Inv Geog, Bol Inst Geog 99:1–19. https://doi.org/10.14350/rig.56770

Quesada-Román A, Moncada-López R, Paz-Tenorio JA, Espinoza-Jaime E, Castellón-Meyrat C, Acosta-Galeano N (2018) Las investigaciones sobre movimientos de laderas en Costa Rica, Honduras, México y Nicaragua: enseñanzas desde la academia, las agencias de cooperación y las instituciones públicas. Rev Geog de Amér Cent 60:17–59. https://doi.org/10.15359/rgac.60-1.1

Quesada-Román A (2015) Implicaciones en la gestión del riesgo de desastres y ambiente en el Valle Central en los últimos treinta años (1985–2015). Vigesimoprimero Informe Estado de la Nación en Desarrollo Humano Sostenible. Pavas, Costa Rica: CONARE, Programa Estado de la Nación. 1–39 pp

Quesada-Román A (2016a) Peligros geomorfológicos: inundaciones y procesos de ladera en la cuenca alta del río General (Pérez Zeledón), Costa Rica. Maestría en Geografía con énfasis en Geografía Ambiental, Posgrado en Geografía, Universidad Nacional Autónoma de México. 157 pp. https://doi.org/10.13140/RG.2.1.2731.6080

Quesada-Román A (2016b) Impactos geomorfológicos del Terremoto de Limón (1991; ms=7.5) y consideraciones para la prevención de riesgos asociados en Costa Rica. Rev Geog de Amér Cent 56:93–111. https://doi.org/10.15359/rgac.1-56.4

Quesada-Román A (2017) Los estudios de riesgos naturales y antrópicos a través de cuatro décadas en la Revista Geográfica de América Central (1974 – 2015). Rev Geog de Amér Cent 58:17–43. https://doi.org/10.15359/rgac.58-1.1

Rosi A, Segoni S, Canavesi V, Monni A, Gallucci A, Casagli N (2021) Definition of 3D rainfall thresholds to increase operative landslide early warning system performances. Landslides 18(3):1045–1057

RSN—Red Sismológica Nacional de Costa Rica (2017) The Costa Rica national seismological network catalog during 1975–2017. https://doi.org/10.15517/TC

Ruiz P, Carr MJ, Alvarado GE, Soto GJ, Mana S, Feigenson MD, Sáenz LF (2019) Coseismic landslide susceptibility analysis using LiDAR data PGA attenuation and GIS: the case of Poás volcano, Costa Rica, Central America. In: Poás Volcano. Springer, Cham, pp 79–118

Ryan PC, Alvarado GE, McCanta M, Barca MK, Davis G, de Mendoza LH (2022) The importance of overbank deposits and paleosol analyses for comprehensive volcanic hazard evaluation: the case of Holocene volcanism at Miravalles Volcano Costa Rica. Natural Hazards 112(1):413–449

Salazar M, Alvarado GE, Madrigal J, Monestel Y (2019) Deslizamientos históricos importantes en el volcán Barva (Costa Rica) y su transformación en flujos de escombros volcánicos. Revista Geológica de América Central 61:107–119

Segura G, Badilla E, Obando L (2011) Susceptibilidad al deslizamiento en el corredor Siquirres-Turrialba. Rev Geol de Amér Cent 45:101–121

Sepúlveda SA, Petley DN (2015) Regional trends and controlling factors of fatal landslides in Latin America and the Caribbean. Nat Hazards Earth Syst Sci 15:1821–1833

Sequeira-Arguedas JM (2021) Geología, geomorfología y ocurrencia de deslizamientos en la cuenca alta del Río Virilla, Costa Rica. Revista Geográfica de América Central 66:313–339

Siebert L, Alvarado GE, Vallance JW, de Vries BVW (2006) Large-volume volcanic edifice failures in Central America and associated hazards. In: Volcanic hazards in Central America 412, 1–26

Suárez ÁM, Peraldo G, Badilla E, Obando LG (2009) Zonificación geomorfológica para la evaluación de la susceptibilidad a los deslizamientos en la cuenca del río Viejo, Puriscal, Costa Rica. Rev Geol de Amér Cent 41:55–69

Terlien MT, Van Westen CJ, van Asch TW (1995) Deterministic modelling in GIS-based landslide hazard assessment. In: Geographical information systems in assessing natural hazards. Springer, Dordrecht, pp 57–77

Trujillo-Vela MG, Ramos-Cañón AM, Escobar-Vargas JA, Galindo-Torres SA (2022) An overview of debris-flow mathematical modelling. Earth Sci Rev 232:104135

Vallejos S, Esquivel L, Hidalgo M (2012) Histórico de desastres en Costa Rica (Febrero 1723-Setiembre 2012). In Histórico de desastres en Costa Rica (Febrero 1723-Setiembre 2012). Comisión Nacional de Prevención de Riesgos y Atención de Emergencias (CNE)

Vannucchi P, Morgan JP (2019) Overview of the tectonics and geodynamics of Costa Rica. In: Poás Volcano. Springer, Cham, pp 1–12

Waylen P, Laporte S (1999) Flooding and the El Niño-Southern Oscillation phenomenon along the Pacific coast of Costa Rica. Hydrol Proc 13(16):2623–2638

Fluvial Landscapes of Costa Rica: An Overview of Dynamic Rivers

5

Sebastián Granados-Bolaños, Nicola Surian, Christian Birkel, Guillermo E. Alvarado, Adolfo Quesada-Román, J. P. Galve, and Armando Bonilla-Mata

Abstract

Costa Rica, a small tropical country with an area of 51,179 km^2, has an extensive network of diverse streams controlled by varied environmental and geological conditions. Rugged terrain, active volcanism, and tectonic activity, along with intense precipitation, result in a dynamic drainage system characterized by a wide array of erosive and depositional fluvial landforms. In this chapter, a generalized characterization of Costa Rica's riverscapes is provided, with a focus on channel morphologies and fluvial dynamics within selected reaches across the country. Using a range of remote sensing techniques, including Unmanned Aerial Vehicles (UAVs), hydraulic modeling, and detailed field observations, representative river sections situated within distinct geological contexts are described. The findings illustrate how rivers flow in confined steep volcanic-mountain settings, on alluvial fans, within low-gradient unconfined dynamic reaches, and in urban environments. Moreover, we discuss the intrinsic connection between channel morphology, surface processes, and associated hazards, as most of the disasters occurring in Costa Rica are tied to fluvial environments.

Keywords

Fluvial geomorphology · Channel morphology · River dynamics · Tropics · Costa Rica

5.1 Introduction

Humid tropics are characterized by high rainfall rates that cause frequent and intense hydropeaking events that impact fluvial geomorphology conditions on short timescales (< 1 year) (Chang and Lau 1993). In Costa Rica, an isthmic country, some catchments can receive over 7000 mm of accumulated annual rainfall, and in some cases, the number can exceed over 10,000 mm/yr. Rainfall of this magnitude generates complex rainfall-runoff dynamics and frequently leads to disasters resulting in millions of dollars of losses and the loss of human lives (Sánchez-Murillo et al. 2016).

Costa Rica has approximately 1080 rivers and 2460 streams officially recognized and named by the Geographical Institute of Costa Rica (IGN), summing up to a total length of 138,100 km. In addition, over 6800 km are unnamed streams according to the IGN. Adding a grand total of 144,900 km of superficial drainage in a small country with an area of 51,179 km^2 results in very high drainage densities. Most fluvial environments in Costa Rica are highly dynamic and transport high sediment loads leading to rapid channel changes. In this chapter, we analyze the morphology and dynamics of several river reaches around the country in different environments. The spatial unit of analysis was mainly the river reach scale, which is basically a small (< 5 km length), manageable, and homogeneous river section (Grabowski et al. 2014). Nevertheless, we

S. Granados-Bolaños (✉) · N. Surian
Department of Geosciences, University of Padova, Padua, Italy
e-mail: sebastian.granados@ucr.ac.cr

S. Granados-Bolaños · C. Birkel · A. Quesada-Román
Escuela de Geografía, Universidad de Costa Rica, San José, Costa Rica

G. E. Alvarado
Unidad de Investigación y Análisis del Riesgo, Comisión Nacional de Prevención y Atención de Emergencias (CNE), San José, Costa Rica

J. P. Galve
Departamento de Geodinámica, Universidad de Granada, Granada, España

A. Bonilla-Mata
Escuela Centroamericana de Geología, Universidad de Costa Rica, San José, Costa Rica

A. Quesada-Román (ed.), *Landscapes and Landforms of Costa Rica*,
World Geomorphological Landscapes, https://doi.org/10.1007/978-3-031-64940-0_5

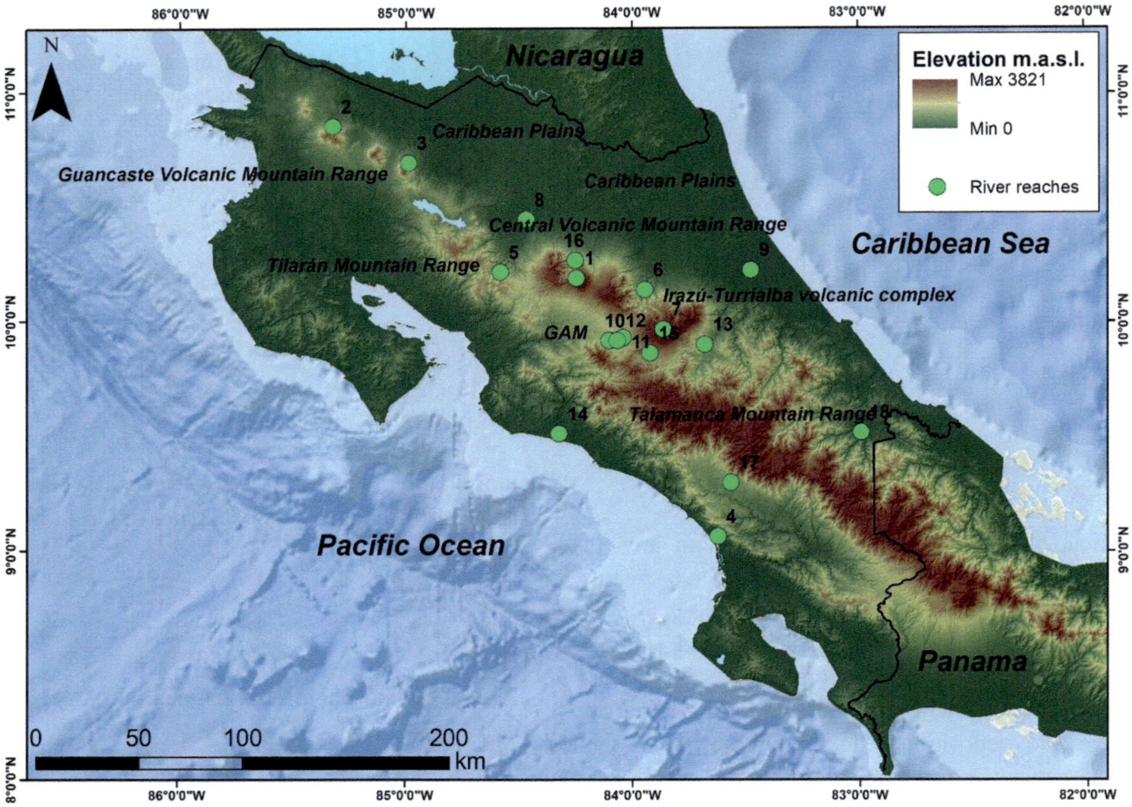

Fig. 5.1 River reaches and important sites mentioned in this chapter: (1) Desagüe River, (2) Zanjonuda River, (3) Celeste River, (4) Pavón River, (5) San Lorencito River, (6) Sucio River, (7) Irazú Volcano Crater, (8) San Carlos River, (9) Reventazón River, (10) María Aguilar River, (11) Negritos River, (12) Ocloro River, (13) Turrialba River, (14) Parrita River, (15) Reventado River, (16) Cinchona-Seco River, (17) Térraba alluvial fan complex, and 18) Talamanca valley alluvial fan complex

also present some larger-scale analysis for selected fluvial systems.

Using a combination of remote sensing techniques, field observations, hydroclimatic monitoring, and bi-dimensional hydraulic modeling, we intend to explain a general context of the fluvial environments of Costa Rica (Fig. 5.1). The use of Unmanned Aerial Vehicles (UAV) was applied in most of the analyzed reaches to give a high-resolution view of landforms and processes (Granados-Bolaños et al. 2021). Overall, our goal is to demonstrate fluvial landform diversity and their highly dynamic processes in Costa Rica.

5.2 The Drainage Network and Catchments of Costa Rica

Costa Rica has two major drainage basins, one draining toward the Caribbean Sea, and the other one to the Pacific Ocean. In the northern part of the country, a series of extrusive volcanic structures (Guanacaste, Tilarán, and Central Volcanic Cordilleras) separates the two major basins; meanwhile in the southern part of the country, the prominent Talamanca Cordillera composed mainly of plutonic and sedimentary formations separates the Pacific from the Caribbean basin.

Numerous differences can be found between both basins regarding the configuration of the drainage network, geology, and fluvial geomorphology of the systems. The Caribbean basin is characterized by younger geological formations; it also concentrates 70% of the Quaternary deposits of the country. In the northern part of the Caribbean basin, headwater catchments are located within volcanic structures, whereas some of the volcanoes show frequent eruptive activity and debris flows (lahars) are common, such as at the Rincón de la Vieja volcano. Meanwhile, in the southern part of the Caribbean basin, rivers form over sedimentary formations composed mainly of Paleocene and Neogene sandstones and limestones from the Talamanca Cordillera. Generally, rivers in the highlands of the Caribbean basin are very confined and typified by high sediment supply, with a pseudo-meandering and step-pool dominant morphology. When streams reach the lower Caribbean plains, river confinement is considerably reduced, and therefore, sinuosity increases, resulting in some cases in meandering and sinuous reaches. Due to the intense and constant rainfall, the Caribbean rivers maintain a relatively constant flow regime all year compared to the more seasonal rivers of the Pacific side. Alluvial fans can be found along the foothills of the volcanoes. For example, a mega-fan can be found on the Caribbean slope with an area of 1440 km^2 and an extension of 40 km from its apex. The constant, intensive, and explosive volcanic activity causes abrupt changes in the river systems over short periods of time (Galve et al. 2016).

The Pacific basin with its older geological formations and less than 30% of the country's Quaternary deposits forces rivers to flow more confined with less sinuosity compared to the Caribbean slope. North Pacific headwater catchments are hosted by volcanic structures, whereas the central and southern Pacific rivers flow over sedimentary formations composed mainly of Oligocene-Pleistocene sandstones and limestones. The largest catchment of Costa Rica is the Tempisque-Bebedero with a combined area of ~ 5000 km^2, located in the northern part of the Pacific side (Fig. 5.1). The Grande de Térraba catchment located in the south is of a slightly smaller size compared to the Tempisque. The Central Pacific basin of Costa Rica is home to most of the population of Costa Rica and only within the Grande de Tárcoles catchment sits over 70% of the Great Metropolitan Area (GAM). Also, an estimated population of over 3 million inhabitants resides within this catchment (Guillen-Montero et al. 2021). Urban rivers that drain the GAM are mainly confined with single-thread morphologies over Quaternary volcanic rocks (Quesada-Román et al. 2021). Unfortunately, this catchment is the most polluted in Costa Rica (Perez-Gómez et al. 2021; Rainwater et al. 2011). On the other hand, in the lowlands of the Grande de Tárcoles catchment river confinement is considerably reduced resulting in a sinuous channel with alternating gravel-sand bars.

5.3 Confined Rivers in Volcanic and Mountain Environments

Over 33% of Costa Rica's territory is composed of Quaternary-Pleistocene volcanic landforms (Fig. 5.2). These dynamic and high sediment source environments play a crucial role in the economy and society of the country. From the energy supply point of view, over 70% of the country's electricity generation comes from hydropower plants located within volcanic formations. Mountain and highland volcanic areas are also the main sources of drinking water for most of the population that live in the central part of Costa Rica (GAM). Also, rivers in volcanic and mountain environments play a key role in the tourist industry of the country. Hundreds of internationally recognized waterfalls can be found within confined volcanic rivers and are visited by thousands of tourists per year (Fig. 5.3). Most of the headwater catchments in volcanic and mountainous areas are protected by strict environmental regulations. Many are defined by the Costa Rican law as National Parks, Biological Reserves, and other protection categories. Therefore, these environments are mostly covered by pristine, dense, and biodiverse forests.

5.3.1 The San Lorencito Headwater Catchment: A Highly Dynamic Stream Under Pristine Forest and on Extinct Volcanic Substrate

The San Lorencito headwater catchment is located in the northern Tilarán Mountain Range. Its 3.2 km^2 area drains to the NE Caribbean and is a tributary to the San Carlos catchment. The catchment forms part of a pristine conservation area known as the Biological Reserve Alberto Manuel Brenes (RBAMB). The landcover is dominated by a dense primary montane rainforest with high humidity conditions throughout the year (> 99%). The catchment has frequent rain events all year with an annual average precipitation of 2830 mm. The geological substrate known as the Monteverde Formation is mainly composed of Pleistocene (1–2.2 Ma) andesitic and basaltic lavas (Alvarado and Gans 2012; Žáček et al. 2012). Pyroclastic flow deposits and breccias can also be found in bedrock channels that drain the hillslopes into the main channel. The elevations of the catchment range from 873 to 1472 m a.s.l. The terrain of the catchment is characterized by steep slopes (> 30%) with frequent landslide and rockfall events during the rainy season (May–December). The San Lorencito River follows a fault line, responsible for its linear and steep channel morphology (~ 22%) flowing from SW to NE.

Despite the relatively small size of the catchment, the San Lorencito River is highly dynamic and experiences frequent and severe channel changes due to extreme rainfall events (Solano-Rivera et al. 2019; Granados-Bolaños et al. 2022). This river has been monitored since 2014 as an experimental natural laboratory for hydrological and geomorphological observations (Birkel et al. 2021). Channel changes have been monitored using high-resolution UAV surveying and hydroclimatic observations to understand the response of the river to rainfall events. The channel of the San Lorencito River reaches mainly shows a step-pool, cascade morphology with few bedrock segments underlain by volcanic breccias. The sediment supply is mainly composed of coarse sediments (boulders, cobbles, and gravel). An extreme rainfall event can change the morphology from cascade to step-pool and vice versa in hours. This phenomenon was observed in a river reach during the rainy season of 2018. The reach has an average slope of 2.8%, a length of 200 m, and an average width of 7.5 m. Figure 5.5 shows

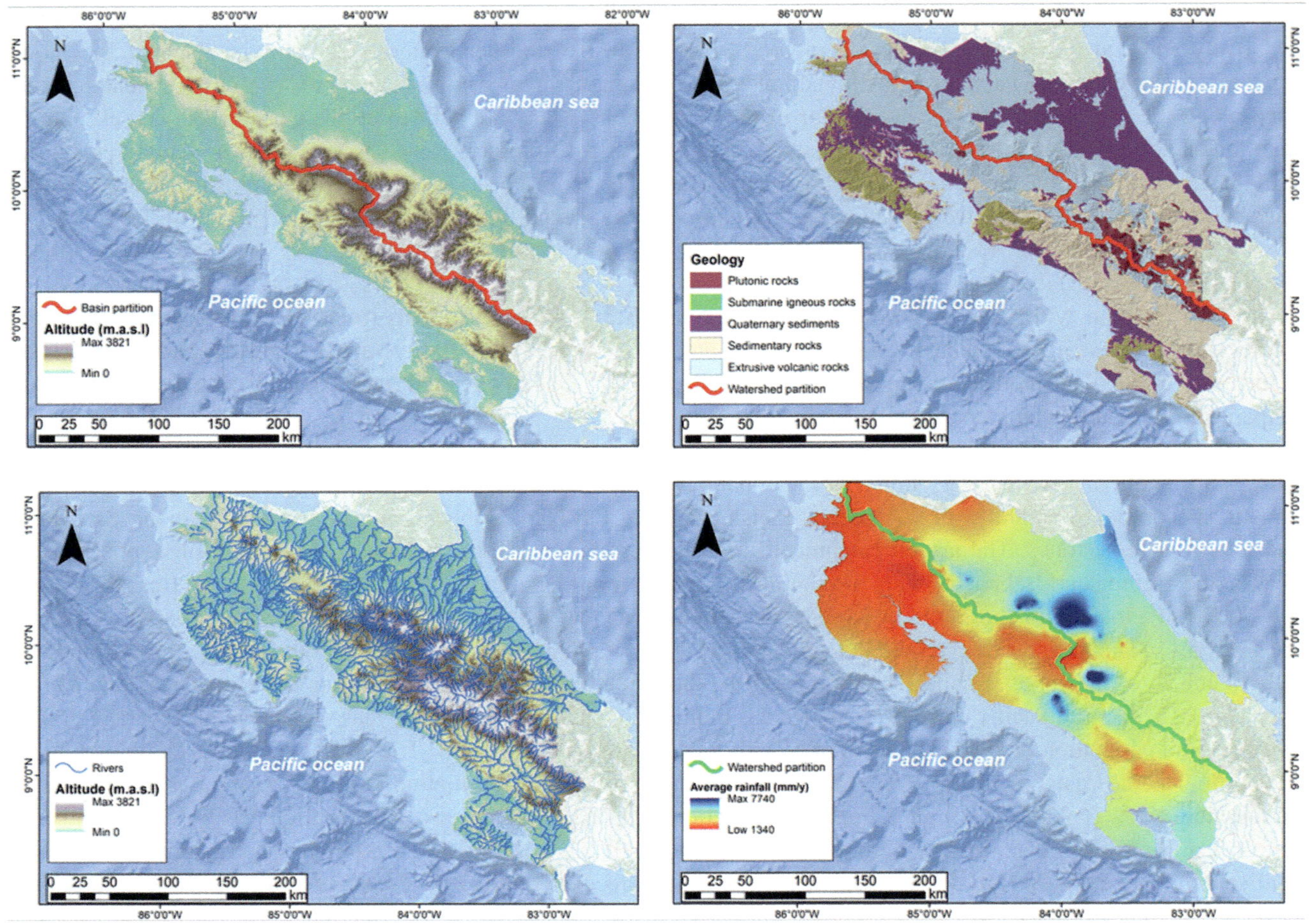

Fig. 5.2 General geological and geographic conditions of Costa Rica. **a** Elevation with the continental divide separating Caribbean and Pacific basins, **b** simplified geology of Costa Rica, **c** Fluvial network of Costa Rica, and **d** average annual precipitation (after Sánchez-Murillo et al. 2016)

the river reach during the rainy season of 2017. The channel was dominated by a step-pool morphology. A cross-section located to the east of the figure shows two dry channels and an active wet channel, as well as a small fluvial island with trees and bushes. The two tributaries shown in Fig. 5.5a, b carry enough sediments to create small alluvial fans at the confluence with the mainstream that frequently causes deflection of the mainstream.

One year later, an accumulated precipitation of 173 mm/13 h caused water levels to rise 2.5 m with an estimated flow of around 300 m^3/s. During this event, multiple landslides and rockfalls occurred incorporating large volumes of debris, vegetation, and soil into the channel. Important morphological changes occurred in the observed reach. Channel width almost doubled its size shifting from 7.5 m to an average of 12 m. Step-pool morphologies almost completely disappeared, and the reach adopted a cascade morphology. Due to the severe bank erosion, riparian vegetation disappeared, resulting in increased channel width. The cross-section located to the east of Fig. 5.5 shows a drastic change in the elevation profile before and after the extreme event. The fluvial island disappeared and was substituted by a large sediment bank surrounded by two channels. After the event, the abandoned channel has become active.

The San Lorencito River is a good example of fluvial geomorphology of the mountain and volcanic areas of Costa Rica. These environments are mostly pristine, protected areas with high rainfall all year. These steep and confined rivers that flow throughout mountain and volcanic environments have very high sediment supply and are extremely dynamic and can have rapid channel changes in hours. Also, there is high connectivity between the hillslopes and the channels. Landslides, rockfalls, and debris flows are frequent and incorporate constant colluvium into the channels and tributaries. The San Lorencito River is no longer affected by volcanic activity because it is located on an extinct volcanic formation, but other streams and rivers are directly affected by volcanic processes, such as the Sucio River, located on the Irazú Volcano.

Fig. 5.3 Waterfalls in volcanic and mountain environments: **a** Desagüe River is located in a restricted area of the Poás National Park, **b** Zanjonuda River in the foothills of the Rincón de la Vieja volcano within the Sensoria private reserve, **c** Celeste River within the Tenorio National Park, and **d** Pavón River is located 15 km southeast from Marino Ballena National Park. Hundreds of accessible waterfalls with legal and safe entries can be found in volcanic and mountain environments of Costa Rica. Waterfalls often play a key role in local economy of rural areas

Fig. 5.4 San Lorencito catchment. **a** The landcover is composed of a primary pristine montane rainforest with minimal human disturbances, **b** steep-pool channel morphology

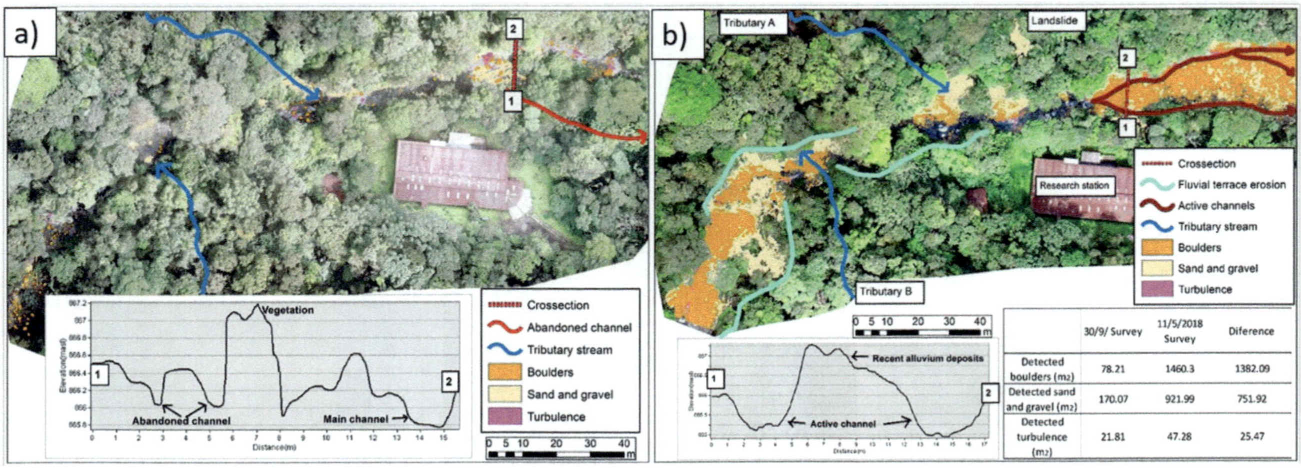

Fig. 5.5 Changes in a reach of the San Lorencito River detected with UAV surveying for **a** October 2017 survey and **b** November 2018 after an extreme rainfall event (173 mm/13 h). The channel morphology changed drastically from step-pool to cascade. Bank erosion caused increased channel width, eliminated vegetation, and a fluvial island located to the east

5.3.2 Confined Rivers of Active Volcanoes: A Braided Reach in the Sucio River

Compared to other rivers in Costa Rica, the Sucio River is particular in terms of its morphology, extreme sediment supply, and dynamics (Fig. 5.6). The word *Sucio* in Spanish means "dirty." The biogeochemical conditions and the extreme amount of dissolved and suspended sediment cause the river's brownish-reddish coloration throughout the year (Arce-Rodríguez et al. 2017).

The Sucio River catchment area is around 191 km^2. The landcover is protected pristine rainforest within the Irazú Volcano National Park, Braulio Carrillo National Park, and the Central Volcanic Cordillera Reserve (Reserva Forestal Cordillera Volcánica Central). The headwater region of the Sucio catchment is located on the northern, unstable flank of the active Irazú volcano. Massive landslides, rockfalls, and debris flows occur frequently in the headwater region of the Sucio River leading to high sediment input into the channel. Figure 5.7 illustrates the dimensions of landslides and unstable hillslopes of the northern flank of Irazú Volcano.

The Sucio River is very steep, with an average slope of 16%. The river descends from a maximum altitude of 2800–400 m a.s.l. in less than 16 km of length. Additionally, the river flows through one of the rainiest places in Costa Rica with annual rainfall that can exceed 7000 mm. Moreover, the Irazú Volcano is considered active. Explosive eruptive activity of the volcano throughout the centuries has caused major morphological changes in the Sucio River bed (Galve et al. 2022). The Sucio River has a particular channel morphology which is not commonly reported in geomorphological literature (Church 2006; Rinaldi et al. 2016; Buffington and Montgomery 2021). Island-braided reaches composed of boulders can be found at the outlet of the catchment located near the 32° National Highway Bridge (Fig. 5.8). The reach we illustrate in this chapter has a length of 5.9 km, an average width of 230 m, and a particularly steep slope of 3.2%.

Figure 5.8a also shows 17 fluvial islands with similar elongated morphology. The largest of the islands measures 2.6 ha and the smallest is 0.09 ha. The average area of the islands is around 1 ha. The largest island is 300 m in length. Meanwhile, the width of the largest island is around 130 m. The vegetation structure and composition of these fluvial islands are different compared to the banks, in terms of its biodiversity and biomass. The height of the vegetation in the largest islands varies from 1 to 14 m. Several trees can be found on the islands which can grow rapidly in this type of environment. A tree count performed with a UAV survey resulted in > 50 units. The average tree height is approximately 6 m.

5.4 Coalescent Alluvial Fan Complexes at the Foothills of the Cordillera of Talamanca

We will now focus on two coalescent alluvial fan complexes. Both are located in the south of Costa Rica, one in the Caribbean basin and the other on the Pacific side. Both are related to the transport of sediments during the deglaciation of the Talamanca Mountain Range before and after the Last Glacial Maximum (LGM) (Quesada-Román et al. 2020a, 2020b, 2022). Due to the predominant mountainous and volcanic environments of the country, alluvial fans are

Fig. 5.6 Oblique aerial image looking upstream of the Sucio River and Hondura River confluence. The Sucio River has an unusual coloration due to its biogeochemical properties and extremely high sediment load. The bridge located on the N° 32 National Highway is one of the few safe places to observe this river under pristine conditions

a common depositional landform. The genesis of these alluvial fans varies according to the environment. Some may be related to explosive volcanism or the occurrence of debris avalanches (Galve et al. 2022), whereas others are related to the deglaciation of the Talamanca Cordillera during the LGM (Bergoeing 2011; Quesada-Román et al. 2021). The vast alluvial deposits constitute a particular environment from geomorphological and geographical point of view since these landscapes play a key role in the human development of the south of the country.

5.4.1 The Talamanca Valley Alluvial Fan Complex and the Indigenous Population

This coalescent alluvial fan complex is formed by the confluence of the Telire, Coen, Lari, and Uren rivers. They come from the highlands of the Talamanca Cordillera, and their headwater catchments include areas over 3000 m a.s.l., meaning that these fluvial systems have experienced glacial and periglacial processes during the last LGM (Quesada-Román and Zamorano-Orozco 2019). The detailed delineation of these four alluvial fans remains unclear due to the lack of geomorphological studies in the area. These alluvial fans are inhabited by the most populous indigenous group of Costa Rica, the Bribris. Over 25% of the Talamanca Bribri indigenous territory is located within these depositional environments, and most of the population is located within these fluvial systems. This 122 km^2 depositional environment is characterized by a complex multi-thread fluvial network that plays a key role in the daily activities of the Bribri population. The rivers are used mainly for transportation, fishing, bathing, and recreational activities (Arias-Hidalgo 2016). The fluvial landscapes are deeply enrooted within the social and cultural identity of the population. Most

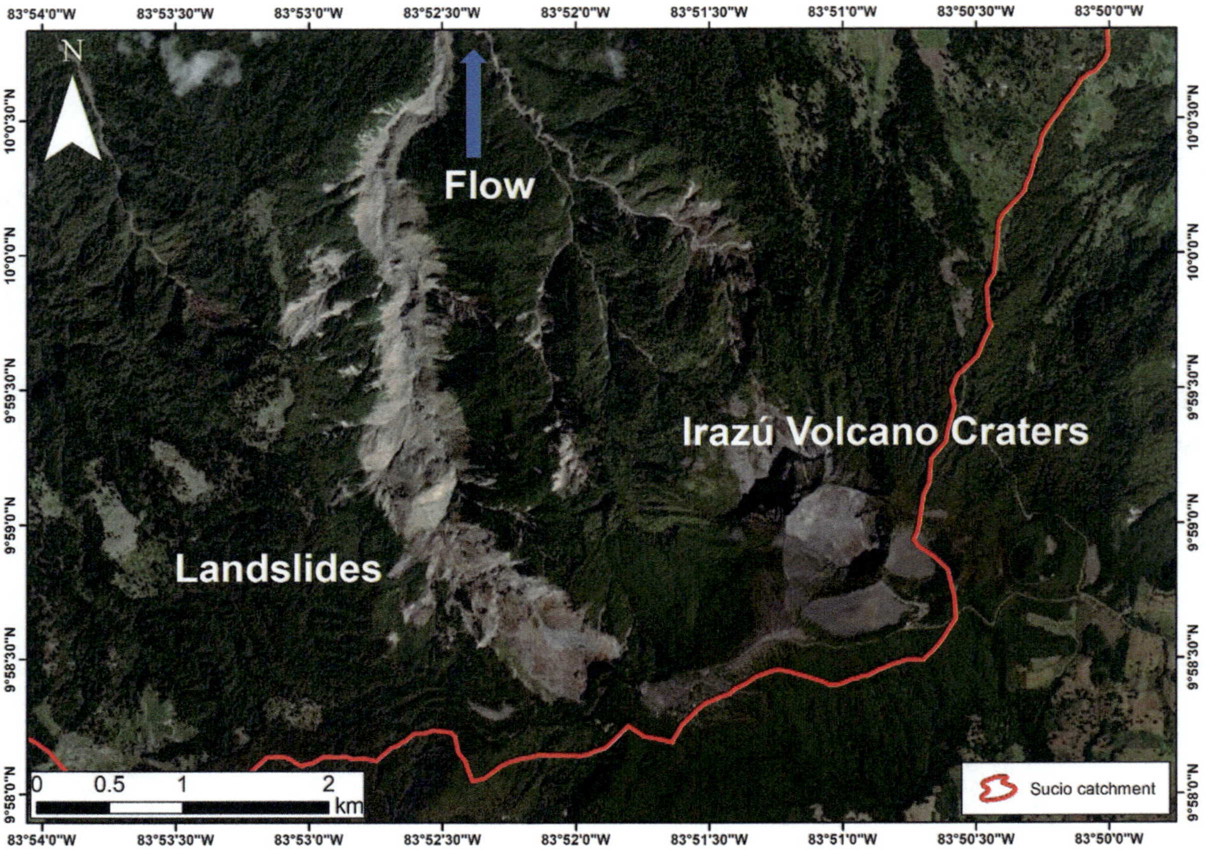

Fig. 5.7 Headwater part of the Sucio River catchment. Landslides and rockfalls deliver material to streams forming colluvial channels. To the east side of the figure, the main craters of Irazú Volcano are located at a maximum altitude of 3432 m a.s.l. These craters are highly dynamic and have experienced severe morphological changes due to explosive eruptions in the last two centuries (Image © CNES/Airbus and © Maxar Technologies, 2023)

people live considerably far away from active channels due to hazard issues since some multi-thread reaches are extremely dynamic and experience frequent lateral channel migrations. Before the rivers arrive at the apex of the alluvial fans, they flow confined within steep step-pool and pseudo-meandering channel morphologies. Once the rivers decrease their slope, a complex river network is developed with over 50 rivers and streams dominated mainly by island-braided gravel-cobble channels. Eventually, all four rivers form the Sixaola River which is one of the southern border rivers that separates Costa Rica from Panama. Due to tectonic uplift processes, the Sixaola River flows confined and incise the Suretka Formation, which is composed of Lower to Middle Pleistocene fluvial-continental conglomerates (Denyer et al. 1994). Once the Sixaola River arrives in the area built of Quaternary alluvial deposits, it flows with a high degree of sinuosity forming meandering channels (Fig. 5.9).

5.4.2 The Térraba Catchment Alluvial Fan Complex, Soils, and Agricultural Land Use

Costa Rica is the largest pineapple exporter in the world (Paniagua-Molina and Solís-Rivera 2020). Most of pineapple production occurs in the northern Caribbean catchments of the country. Nevertheless, the most concentrated pineapple production in the south of the country is located within the Térraba River catchment, which coincides with the most extensive alluvial fan complex of Costa Rica. A series of coalescent alluvial fans follow a NW–SE orientation on the northern side of the Térraba catchment with extensive pineapple plantations. We mapped soils of nine alluvial fans over the total area of 198 km^2 to explore their relationship with agricultural landuse (Quesada-Román and Zamorano-Orozco 2019; Acuña-López et al. 2022). We found that Ultisols are the main soil order with eight types of taxonomic suborders. The most common taxonomic soil

Fig. 5.8 Sucio River. **a** Satellite image of the Boulder island-braided reach. **b** Ground photo of sediment composition. **c** Aerial view of the island-braided reach of Sucio River. An island of 1.7 ha is in the center of the channel. The island vegetation is dominated by 3 or 4 species well adapted to this environment. The islands have elongated morphology and show evidence of intense bank erosion from bankfull events and channel migrations. The surrounding hillslopes that confine the river have an average elevation of 200 m a.s.l., and on the left bank of the river, several fluvial terraces can be found (Image © CNES/Airbus and © Maxar Technologies, 2023)

Fig. 5.9 Talamanca Valley alluvial fan complex. The confluence of the Telire, Coen, Lari, and Uren rivers form a complex multi-thread network where a large amount of the Bribri population of Costa Rica lives. Eventually, these four rivers form the Sixaola River which is the one of the border rivers between Costa Rica and Panama

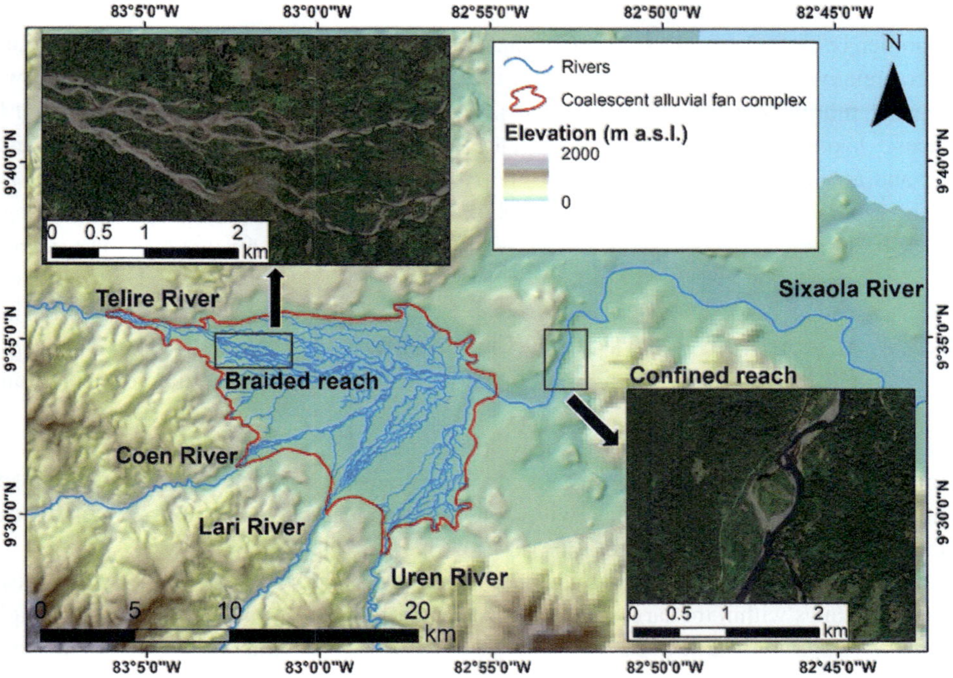

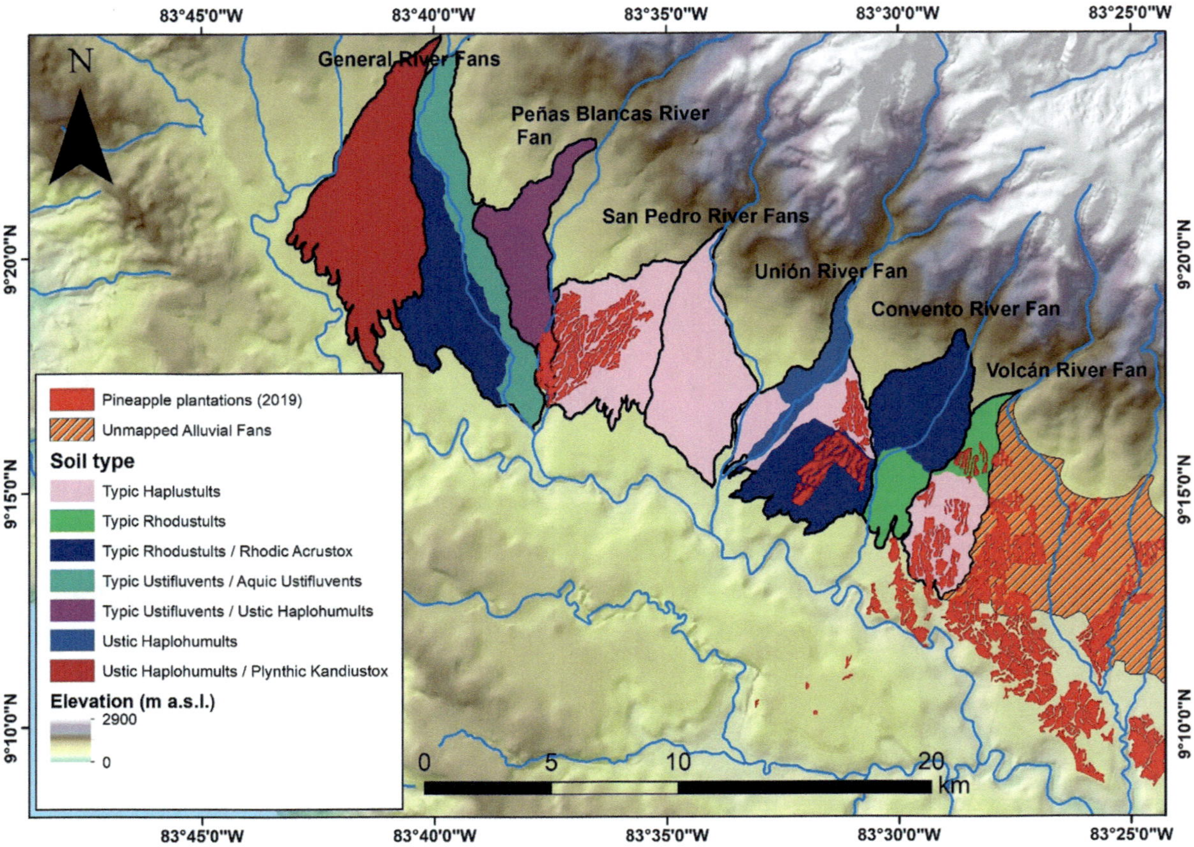

Fig. 5.10 Soils and pineapple plantations of a coalescent alluvial fan complex in the Térraba catchment. These inactive alluvial fans are the major hotspot of pineapple plantations in southern Costa Rica. Intense weathering of these fans has led to the origin of deep well-drained Ultisols presenting optimal conditions for pineapple agriculture

suborders are Typic Haplustults and Typic Rhodustults (Fig. 5.10) (Chinchilla et al. 2011). The first taxonomic soil suborder is located mainly in inactive fans where the river flows relatively confined with a pseudo-meandering steep channel morphology filled with boulders. Meanwhile, the second taxonomic soil suborder is located on intermediate and active alluvial fans where also the river flows with a single-thread pseudo-meandering morphology (Camacho et al. 2020).

5.5 Unconfined Rivers of the Caribbean Plains

Typically, in Costa Rica when rivers leave these steep, confined mountain, or volcanic environments, they transition into single-thread morphologies with sinuous or meandering channels. We now address the case of rivers in the Caribbean plains with unconfined meandering channels and their dynamics within the largest Quaternary deposits of the country.

The Caribbean Plains are located in the NE of the country with the San Juan River as the natural northern border to Nicaragua. The central areas of the plains are heavily used for intense agriculture with pineapple and banana plantations as the predominant land use. To the north, rivers flow toward the Wildlife Reserves Maquenque and Barra del Colorado, and to the east, they flow toward the National Park Tortuguero. Geologically, these plains contain over 70% of Quaternary deposits of Costa Rica. The terrain is characterized by very low relief. Some isolated hills and residual Pliocene and Miocene volcanic reliefs can be found (Alvarado and Gans 2012). Sinuous and meandering channels are the most common in this environment.

5.5.1 Meander Bend Evolution of the San Carlos and Reventazón Rivers

Even though there is less relief in the Caribbean Plains, some rivers are still highly dynamic and can experience important changes in short periods of time. We illustrate the evolution of two meander bends under similar environments and channel morphology. Both reaches flow through dense pineapple fields with scattered secondary forest patches. The average channel width for both reaches is near 75 m. Slope values are also similar, around 0.2%. The sinuosity index (SI) for the Reventazón River is 2.4, while the San Carlos River has a value of SI = 2.7, meaning both rivers are meandering. Both meander bends were analyzed using multitemporal panchromatic satellite imagery to detect changes in the channel and reconstruct neck cutoffs that occurred in the last ten years. In the case of the San Carlos River, the observation period was between 2013 and 2022 and in the Reventazón River between 2014 and 2022.

5.5.2 Point Bar Growth Causes Lateral Channel Migration in the San Carlos River

The San Carlos River catchment has an area of 2632 km^2 and flows into the San Juan River (Costa Rica–Nicaragua border). The catchment drains the Central Volcanic Cordillera at the continental divide at a maximum elevation of 2326 m a.s.l. to the northern Caribbean plains. Mean annual precipitation is around 4000 mm, and spatial variability around 3000–5500 mm is caused by the complex topography (Birkel et al. 2020).

This studied meander bend is located 2 km to the south from Muelle de San Carlos and the river flows through pineapple plantations with scarce forest patches. The reach has evidence of point and scroll bar deposits that record the historic position of the meander along both banks. During the observation period, curvature and lateral channel migration remained relatively stable throughout the reach except for the north point bar. Point bar growth can cause lateral channel migration and increase bank erosion toward the opposite bank (Ikeda, 1989). Figure 5.11 shows how progressive sedimentation at the point bar induced lateral migration that eroded over 80 m of the right bank in less than six years (2013–2016). In the period of 2019–2022, the point bar grew by more than 50 m toward NE, eventually causing rapid channel migration. By 2022, neck cutoff occurred leaving an oxbow lake (Fig. 5.11).

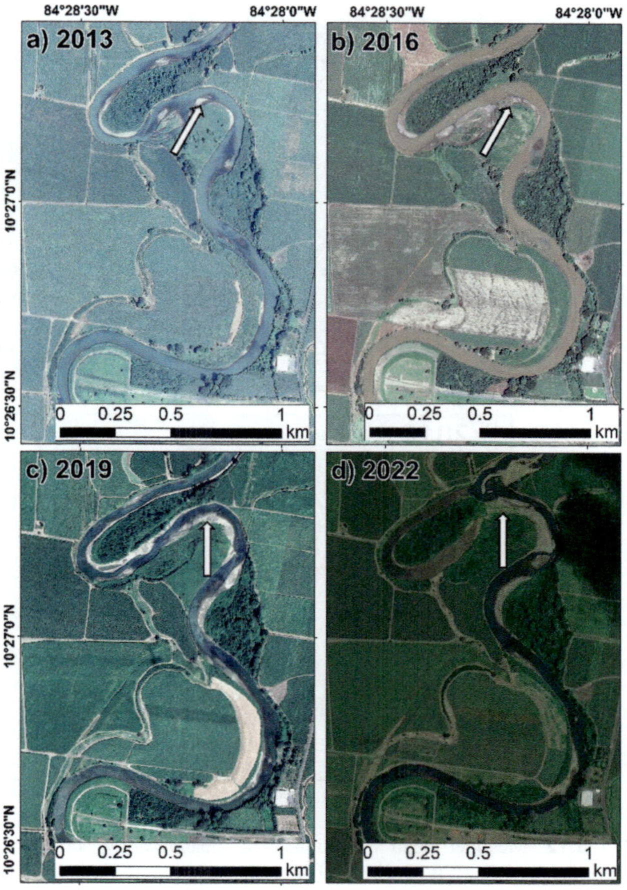

Fig. 5.11 Meander bend evolution in a meandering reach of the San Carlos River: **a** During 2013, the point bar deposit had an average width of 25 m in the inner section of the meander bend, **b** by 2016, the point bar average width increased to 32 m, and the right bank retreated by approximately 52 m, **c** by 2019, image shows an important growth of the inner part of the point bar reaching an average width of 43 m, and **d** by 2022, the neck cutoff occurred leaving and oxbow lake. The point bar increased by 72 m in less than nine years causing a lateral channel migration of 142 m to the north, eroding approximately 2.3 ha. The flow direction is from south to north (Image © CNES/Airbus and © Maxar Technologies, 2023)

5.5.3 Inner Bank Deposition Causes Lateral Channel Migration in the Reventazón River

The Reventazón River catchment with an area of 2819 km^2 drains to the Caribbean Sea and is located in mid-eastern Costa Rica. The headwater region of the catchment drains the Irazú-Turrialba volcanic complex and the Talamanca Cordillera with a maximum altitude of 3463 m.a.s.l. The mid-catchment holds several major hydropower projects operated by the National Electricity Institute of Costa Rica (ICE). This catchment plays a key role in the energy supply of the country. There is a high spatial variability of precipitation rates within the catchment due to topographical irregularities and high relief. Certain areas of the most elevated part of the catchment can reach annual precipitation around 4000–8000 mm, whereas average precipitation in the mid-catchment is around 2000–4000 mm/yr. During 2014–2022 periods, a neck cutoff occurred due to inner deposition in the meander bend that caused lateral migration of the channel. The sediment bar with an area of approximately 0.9 ha caused lateral migration of the

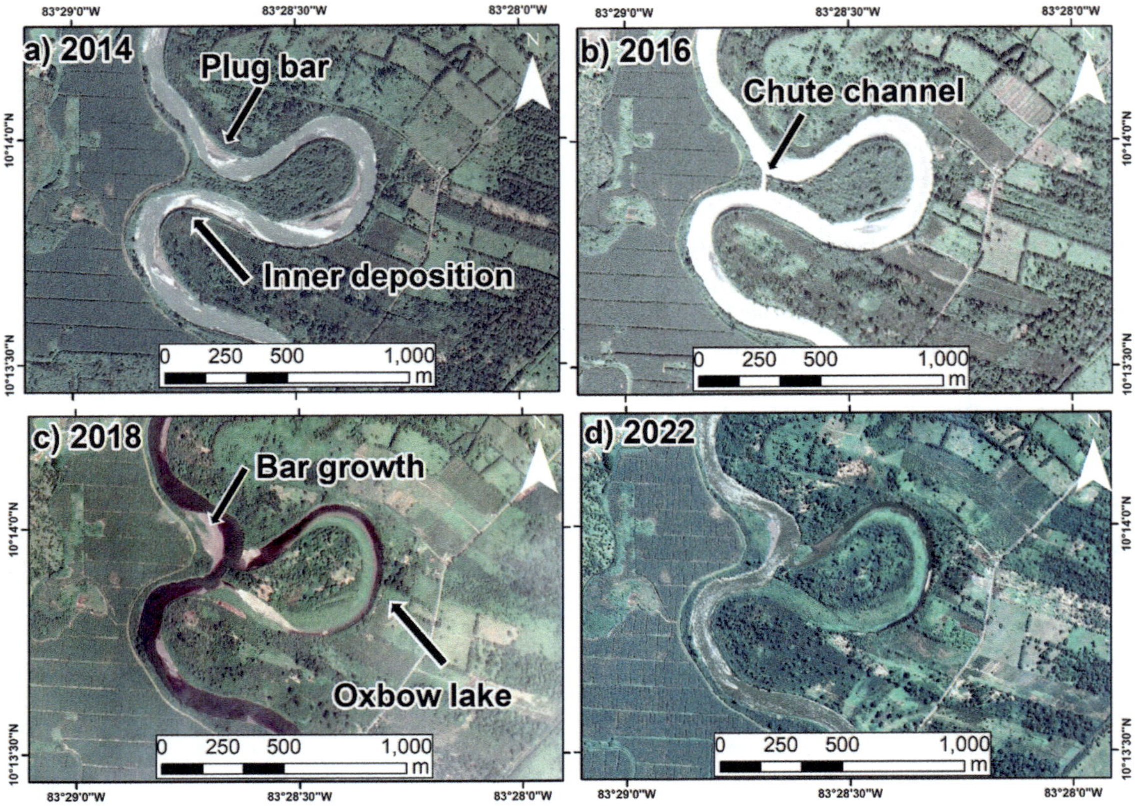

Fig. 5.12 Meander bend evolution in a meandering reach in the Reventazón River: **a** During 2014, two bars led to lateral channel migration, downstream elongation, and headwater incision, **b** by 2016, image shows the river reach under bankfull conditions, a chute channel with a width of 26 m and a length of 64 m was formed, **c** by 2018, the neck cutoff occurred, the oxbow lake enters a transitional lacustrine-terrestrial stage. Also, a large point bar formed, causing lateral migration of the channel toward the NE, and **d** intense vegetation succession takes place within the oxbow lake and bars. A point bar located in the north formed prior to the neck cutoff has caused channel migration of approximately 44 m

channel to the north, intensely eroding the left bank. At the same time, a plug bar with an area of 0.7 ha was deviating the flow southward, toward the cutbank. These two bars caused a combination of two processes: (i) a downstream elongation due to the inner deposition bar and (ii) headwater incision caused by the plug bar. Both lateral migrations eventually lead to a chute channel formation with a width of 26 m and a length of 64 m. During 2016, according to the satellite image, the reach was under bankfull conditions flooding the existing sediment bars (Fig. 5.12). Approximately, one year later the neck cutoff left an oxbow lake. The formation of a point bar of 1.9 ha occurred during the period of 2017–2018, causing lateral migration of the channel to the NE. The channel migrated by approximately 44 m in less than two years (2016–2018). By 2022, the oxbow lake was still in a lacustrine stage (Fig. 5.12).

5.6 Urban Rivers of the Great Metropolitan Area and the Study Case of the Ocloro River: A Stream Flowing Between Volcanic Units

The Great Metropolitan Area (GAM) holds over 50% of Costa Rica's population and is the major economic center of the country. Within the GAM is the capital city of San José, which is the most populous of the country and concentrates the most important political and economic institutions. San José is surrounded by mountains and volcanic structures (Central Volcanic Cordillera and Talamanca Cordillera). The catchment that drains San José city is the Grande de Tárcoles with an area of 2166 km^2. This river is the most polluted in Costa Rica due to high amounts of untreated sewage, industrial waste, agricultural waste, and urban runoff. The GAM rivers flow mainly on Quaternary

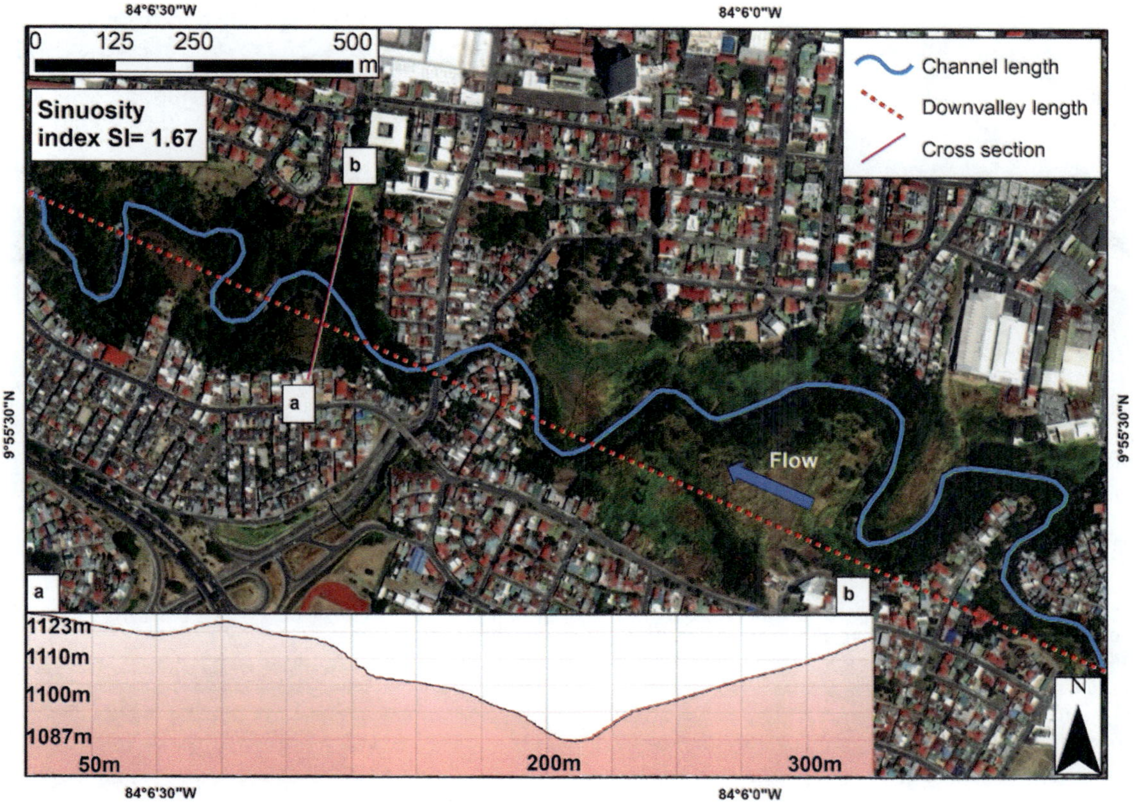

Fig. 5.13 Incised meandering reach of the María Aguilar River with a length of 3.1 km and an average width of 19 m. The river flows in SE-NW direction within the San José Municipality. The lower valley length used to calculate the SI is 1.8 km. Rivers with SI > 1.5 are considered meandering. A cross-section shows the confinement of the channel. The valley depth is approximately 36 m, and the valley width is 350 m. (Image © CNES/Airbus and © Maxar Technologies, 2023)

volcanic substrate composed of lava flows, debris avalanches, pyroclastic flows, and pyroclastic fall deposits (Arias and Denyer 2011). Average annual precipitation in San José averages 2700 mm. Most rivers of San José flow confined through incised deep valleys with single-thread channels varying from straight to meandering (Fig. 5.13). Channel morphology is mainly composed of bedrock, step-pool, plane bed, and pool-riffle reaches. Despite intense anthropic pressure on the river systems of San José, riverine vegetation prevails in many fluvial corridors (Figs. 5.13 and 5.14). These dispersed and fragmentated forest patches provide important ecosystem services to the city, such as regulating air temperature changes, controlling bank erosion, increasing biodiversity, reducing rainfall-runoff rates, and improving the visual scenery of San José.

The Ocloro River flows in the central-east sector of San José city. It has a catchment area of 4.5 km^2 and an average channel slope of 3.8%. Step-pool and plain beds are the most common channel morphologies. The catchment has low vegetation and forest landcover (< 5%) causing high impermeability coefficients and a rapid rainfall-runoff response. The Ocloro River channel has also been artificially modified with culverts, bridges, and bank retaining walls. Floods happen every year in the neighborhood known as Barrio Luján, located between two volcanic geomorphological units that control confinement, slope, and channel morphology of the river. There is an abrupt change of slope and confinement when the river flows out of these volcanic units, which are presumed to be Pleistocene lava flows (Fig. 5.15). Additionally, there are two culverts with an incapable hydraulic capacity to manage the stormflow. Therefore, overbank flow occurs rapidly even in response to moderately intense rainfall.

To reconstruct and illustrate the behavior of this river, flood events were modeled. Within the reach, 2D hydraulic modeling was applied. The hydraulic model was calibrated using observed hydroclimatic data in field (water level sensor and a pluviometer). The high-resolution topography used was generated from UAV flights. The results of the hydraulic modeling validated the hypothesis that several geomorphological units are controlling channel morphology and flow behavior. Also, the hydraulic modeling contemplated geometry of the culverts to understand the impact of anthropic structures in the river. It was concluded that the geomorphological context in which the river flows coupled with poor design and performance of the

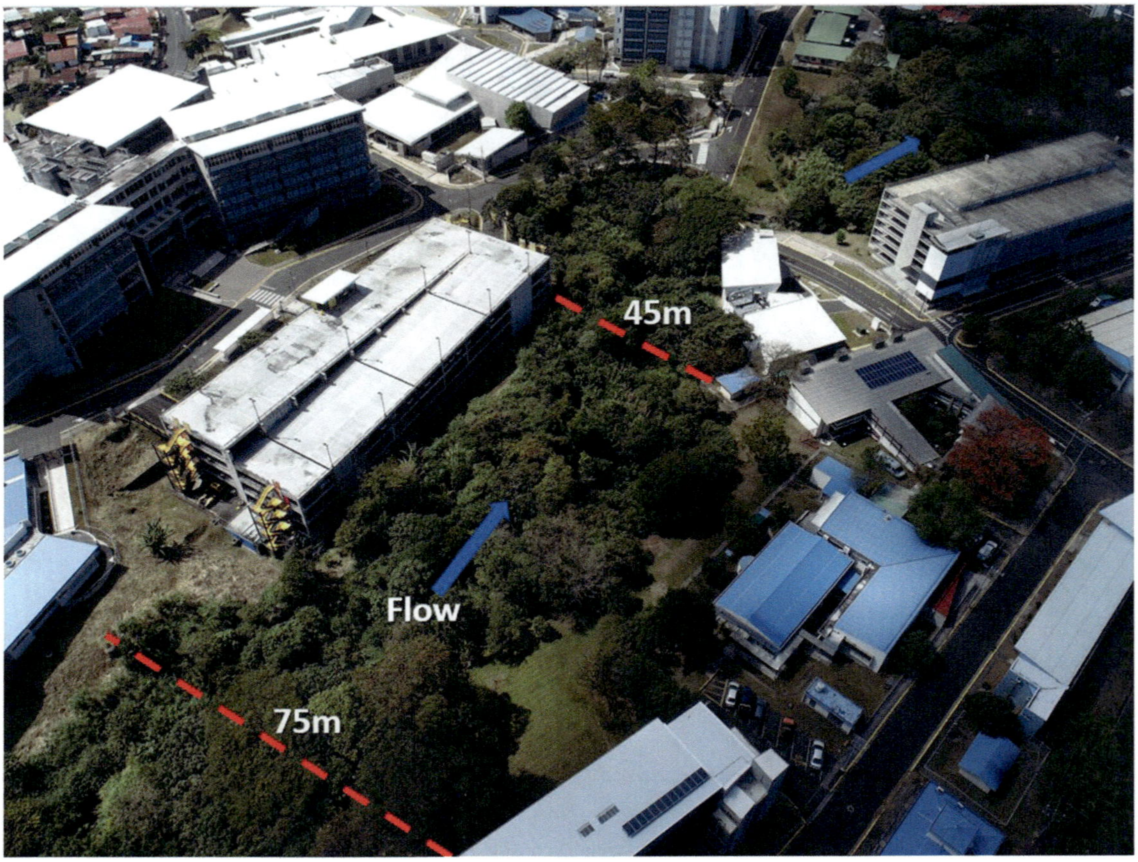

Fig. 5.14 Aerial oblique image of a confined river reach and riparian vegetation of the Negritos River within the University of Costa Rica campus. The reach has a dense secondary forest that follows the channel

hydraulic structures in the river cause frequent floods that affect several residential structures yearly (Fig. 5.16). The Ocloro River is a clear example of a poorly managed river landscape in Costa Rica. Unfortunately, urban rivers of the country are heavily polluted and intervened.

5.7 Fluvial Processes and Hazard in the Tropics

Most disasters in Costa Rica are caused by hydrometeorological events which occur almost every year (Quesada-Román et al. 2021). Floods and landslides cause millions of dollars of losses every year, affecting hundreds of kilometers of transport infrastructure such as roads and bridges. Extreme precipitation events occur frequently in the country affecting both the Caribbean and Pacific basins causing floods, debris flows, and landslides. Recent events such as Hurricane Otto (2016), Hurricane Nate (2017), and Hurricane Eta (2020) caused severe damage to infrastructure and the loss of human lives. For example, Hurricane Nate is considered one of the costliest disasters in Costa Rican history. It caused 14 fatalities and over 560 million dollars in material damage (Vallejos and Bonilla 2017). A large amount of the country's population and public–private infrastructure is exposed to fluvial hazards (Quesada-Román 2022).

We present several examples of fluvial processes and hazards, using the Turrialba and Parrita rivers. Both catchments have high annual rainfall, around 3300 and 2700 mm on average, respectively. Both rivers flow over Quaternary deposits on relatively low-slope areas (< 5%). The Turrialba River presents severe bank erosion and channel widening. Meanwhile, the Parrita River experiences floods frequently and overbank flow affects tens of structures. Also, the Parrita River shows frequent and drastic changes in the channel morphology. On the other hand, we also present two historically important volcanic debris flow events that caused the loss of lives due to complexity of the events.

5.7.1 Severe Bank Erosion Damages Infrastructure in the Turrialba River

The Turrialba catchment is located on the southeastern flank of the Turrialba Volcano, within an altitude range

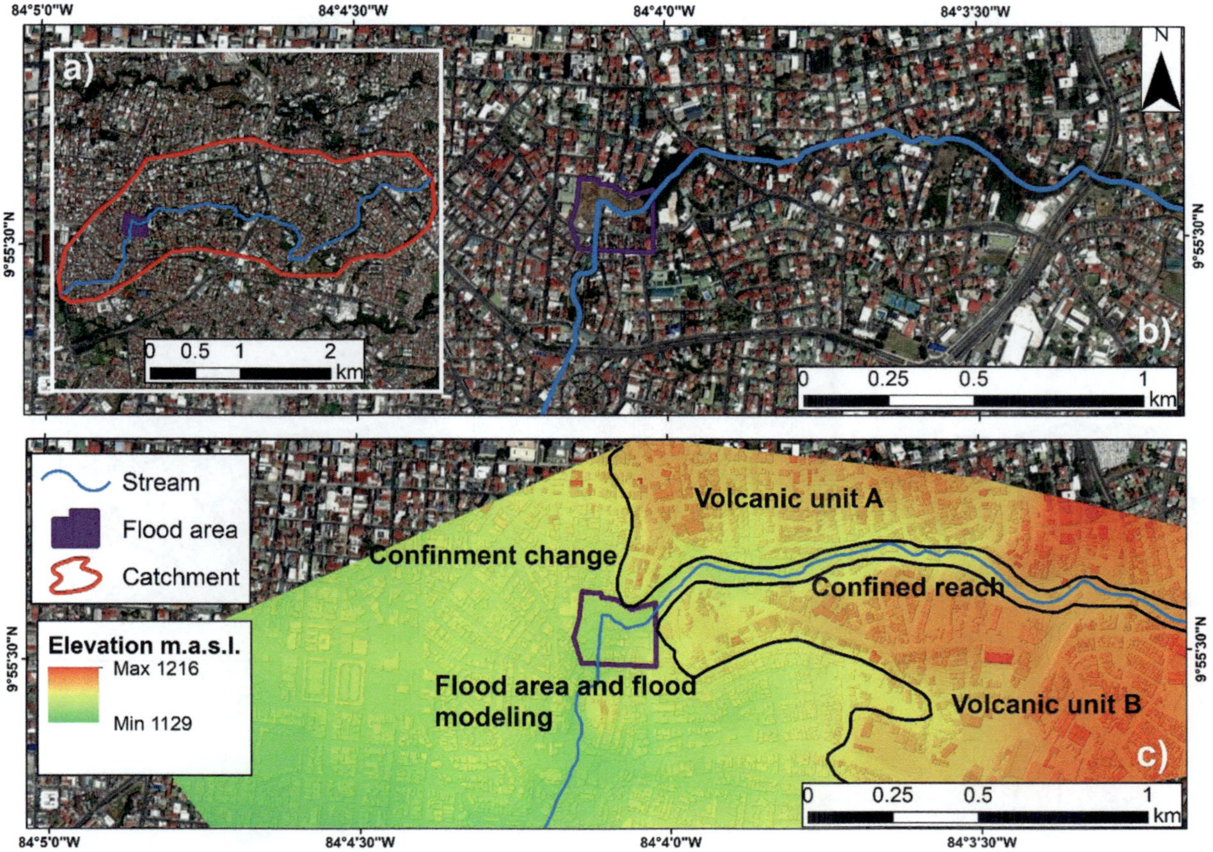

Fig. 5.15 Ocloro River catchment location within the city of San José. **a** Location overview, **b** satellite image of the Ocloro River reach, **c** Ocloro River flows confined between the volcanic units A and B. When the river leaves the confined reach channel morphology and slope change drastically. The purple polygon indicates where floods happen and was reconstructed using 2D hydraulic modeling. (Image © CNES/Airbus and © Maxar Technologies, 2023)

of 2586–1191 m a.s.l. The river flows through the center of Turrialba city with over 26,000 inhabitants. Pseudo-meandering channel morphology dominates the fluvial corridor. The average slope of the Turrialba River is approximately 13%. Nevertheless, the presented river reach has a slope around 2.5%. Available sediment in the channel is composed mainly of coarse materials such as boulders and cobbles. Sediment supply is high within the Turrialba River catchment. One main sediment source is the Coliblanco scarp (avalanche caldera). This caldera rim has steep and unstable hillslopes. Alvarado et al. (2004) suggested that this caldera could have triggered a volcanic debris avalanche that propagated downhill for several kilometers ~ 17 ka ago leaving a massive deposit known as the Angostura Debris Avalanche. The deposit composition is mainly angular lava boulders and subrounded alluvium with matrix of reddish clays and ashes. The Turrialba River incises these deposits and could be a geomorphological explanation regarding the confinement of the channel morphology of the river (Siebert et al. 2006).

In November 2016, an earthquake epicenter was located 4.4 km northwest of Capellades de Alvarado, Cartago. The magnitude was Mw = 5.4 (Campos-Durán and Quintero-Quintero 2020). The earthquake occurred near the Turrialba catchment and caused landslides and rockfalls that led to a debris flow in the Coliblanco Caldera where the main tributaries of the Turrialba River flow. Large amounts of material remain within the channel until today, meaning that there is a high sediment supply available to be transported.

In June 2021, Turrialba city experienced several extreme rainfall events caused by the combined effects of a tropical wave and a low-pressure system. In less than 48 h, over 500 mm of rainfall was registered. This storm caused water levels of the Turrialba River to rise over 3.6 m, severely eroding both banks of the channel and damaging nearby structures (Fig. 5.17). Important amounts of sediments available from the 2016 earthquake were transported from the high catchment downstream causing high sedimentation of the channel in Turrialba city. The event was reconstructed using a combination of techniques. UAV

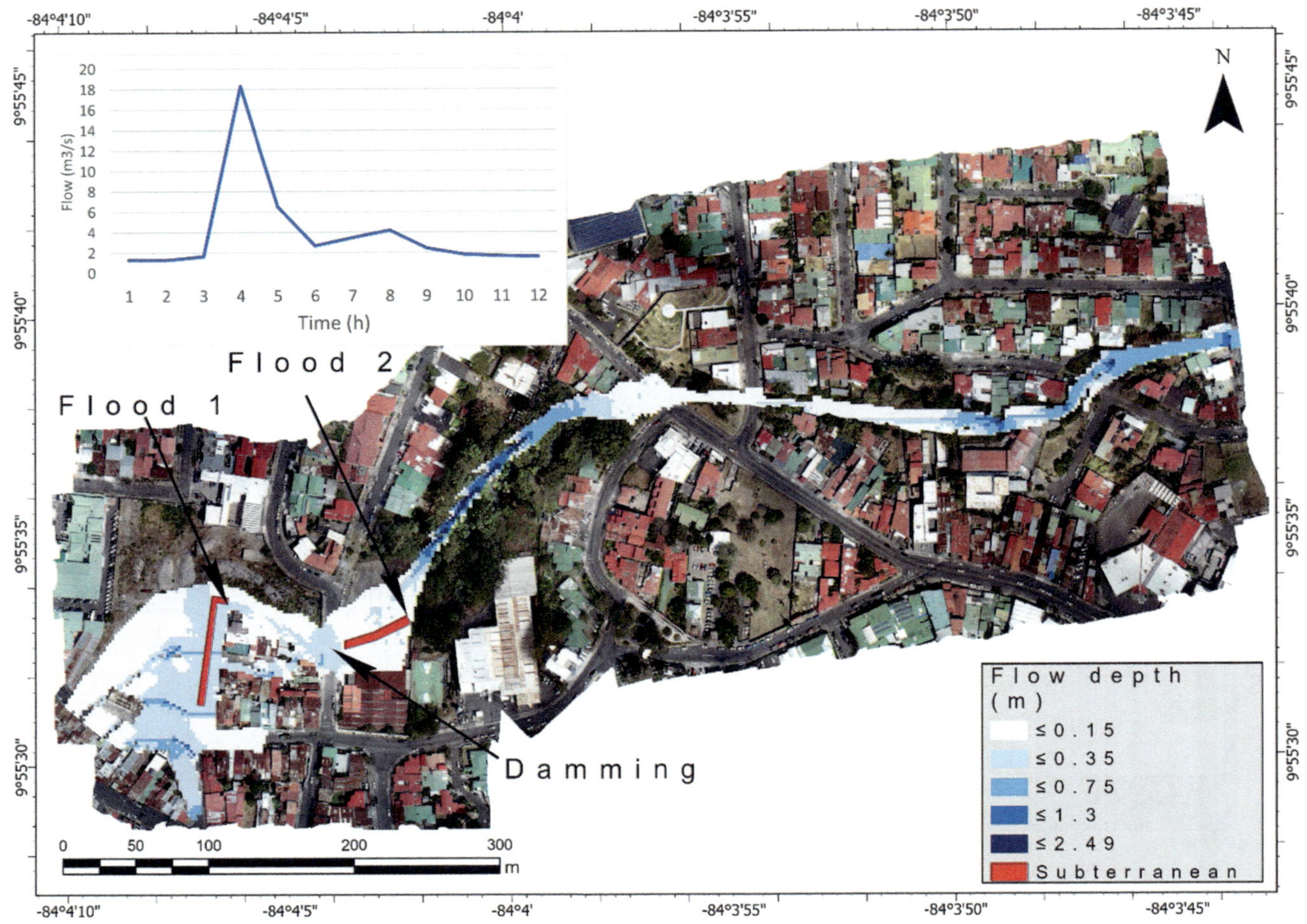

Fig. 5.16 2D hydraulic modeling of a flood in the Ocloro River. The flood simulation time was 12 h with a 15 min temporal resolution, and the maximum stormflow reached 18 m^3/s. The hydrograph was calibrated from observed data in field, specifically a flood that occurred in July 2018. There are two frequently flooded areas. Morphological changes in the channel and poor hydraulic structures cause frequent floods in Barrio Luján. The flood areas reach a maximum flow depth of 0.40 m causing mild structural damage

topographical surveying was performed to obtain high-resolution elevation models for the calibration of a 2D hydraulic model. To generate and calibrate flow hydrographs, flood height marks were collected in the field mainly from tree marks.

A river reach of approximately 1 km was analyzed to reconstruct the flood in terms of flow velocity and depth. Several hydrographs were constructed to calibrate the hydraulic model (Fig. 5.18). Using flood marks on trees, banks, and structures, a spatial database was built to estimate the stormflow. The average flood mark resulted in 3.1 m with a maximum of 3.8 m. The reconstruction of the flood indicated that the maximum flow was around 1000 ± 100 m^3/s.

The event did not cause the Turrialba River to flood the city of Turrialba, since the channel is confined by banks more than 5 m high. Nevertheless, the extreme stormflow caused bank erosion between 10 and 30 m on both banks. The bank retreat caused tens of structures to become inhabitable. Fortunately, there were no human losses. The combination of a high sediment supply, steep slopes, extreme rainfall, and highly erodible banks (debris avalanche deposits) led to a severe increase in channel width.

5.7.2 The Parrita River: A Highly Dynamic River with Frequent Floods

The Parrita River catchment is located in the Central Pacific region of Costa Rica and has an area of 1288 km^2. The geology of the catchment is composed of Cretaceous-Eocene submarine volcanic rocks, Eocene-Miocene sedimentary formations, and Quaternary deposits. The average width of the channel is near 1.1 km. Meanwhile, the average slope is 1%. We will focus on a reach where the river flows unconfined over Quaternary alluvial deposits composed mainly of gravel and cobles (> 50%) (Sandí 2003). There is clear evidence of meandering paleochannels that

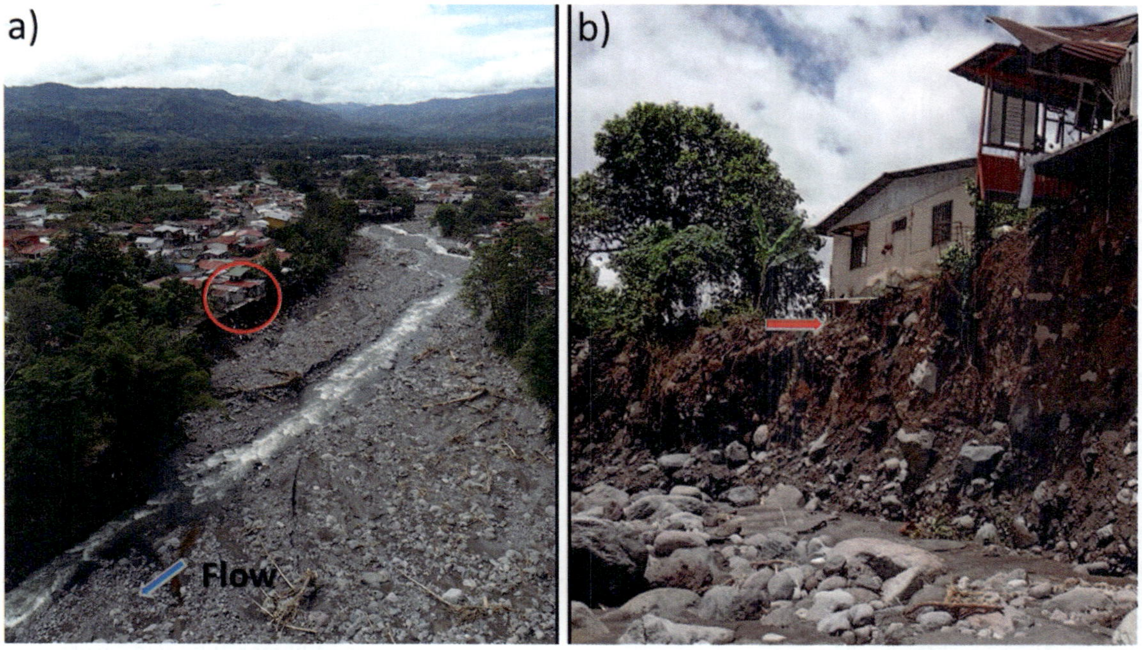

Fig. 5.17 Aerial and ground oblique images of the Turrialba River after the June 2021 rainfall events. **a** Aerial oblique image of the channel, the red circle marks a severely damaged house due to bank erosion, **b** ground photo of the same house

Fig. 5.18 Hydraulic reconstruction of the June 2021 flood event in the Turrialba River. The major flood event caused severe bank erosion causing damage of 28 structures within the reach. The event did not cause overbank flow. The maximum flow depth areas reached 3.8 m

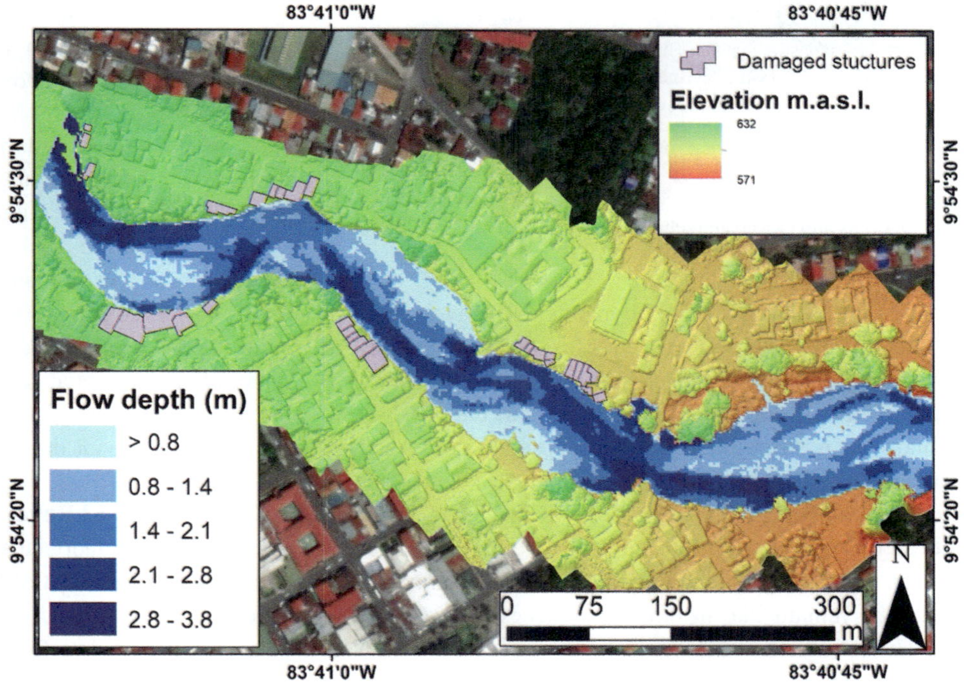

indicate a Quaternary channel migration southeastward, caused by tectonic uplift (Marshall 2007). The Parrita River experiences frequent floods during the rainy season. Flood occurrence is well known and localized by government authorities and inhabitants of Parrita town. Therefore, a series of hydraulic protection structures have been built to protect the town. Nevertheless, these levees are composed mainly of loose and poorly consolidated fluvial material. These are subject to intense erosion during bankfull events. However, they receive constant maintenance by the risk management authorities. Also, due to the levee distribution, they tend to create increased confinement in the river. Precisely in this area, where a bottleneck effect happens, there is frequent overbank flow that leads to floods.

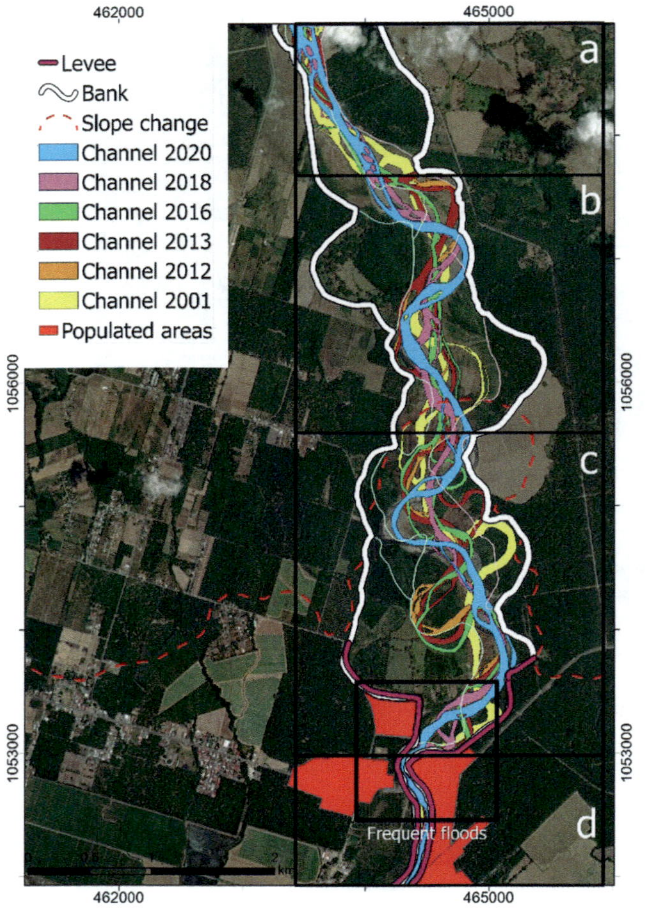

Fig. 5.19 Channel evolution of a Parrita River reach during the period 2001–2020. The most dynamic subreach **c** presents lateral channel migration of over 500 m in less than 8 years. (Image © CNES/Airbus and © Maxar Technologies, 2023)

The Parrita River is highly dynamic and may experience drastic channel morphology changes in short periods of time (< 1 yr). Channel morphology variation was analyzed using panchromatic satellite imagery for the 2001–2020 period. The position of the channel during six observations was mapped in four subreaches to analyze lateral channel migration and other fluvial processes (Fig. 5.19). The first subreach (a) has a relatively stable channel position in comparison with reaches b and c. Despite that the channel width and slope remain like in other subreaches, lateral channel migration, accretion and bank erosion are not so intense. Average lateral channel migration per year is approximately 70 ± 30 m. Subreach b shows considerably more dynamism, with an average lateral channel migration per year around 100 ± 50 m. On the left bank of reach b, bank erosion has been occurring in the last ten years reaching up to 50 m of retreat. Between b and c subreaches, there is a slight slope and elevation change (Fig. 5.19). After slope changes, sinuosity increases drastically and the subreach becomes more dynamic. The channel on some occasions has migrated over 700 m in less than eight years. On some occasions, a neck cutoff may occur in less than one year. Between c and d subreaches, channel confinement is drastically increased due to hydraulic protection levees. Even though the river is highly dynamic in the upper subreaches, bank erosion seems to be artificially controlled by the maintenance of the levees. Fluvial hazard in this reach is related to flooding and overbank flow processes that affect dozens of structures.

5.7.3 The Occurrence of Devastating Lahars in Volcanic Environments

Both Turrialba and Parrita rivers are well known and studied highly dynamic rivers that cause frequent bankfull events. Nevertheless, other rivers that have relatively smaller catchments and less channel morphology variations in short periods (< 10 yrs) can also produce catastrophic events, causing loss of human lives and severe disruption of social and economic actives within their catchment. In this case, we exemplify with two events: (i) the 1963 Reventado River lahar (Fig. 5.19a), and (ii) the 2009 Cinchona earthquake debris flows(Fig. 5.19b, c).

Volcanic debris flows (lahars) can be triggered by multiple mechanisms, such as explosive eruptions, extreme rainfall, earthquakes, landslides, caldera or crater collapses, snow melting, or a combination of multiple processes. Rock fragments carried by lahars make them especially destructive; the abundant liquid allows them to flow over gentle gradients and inundate areas far away from their sources. People in such distal areas commonly neither expect the danger nor anticipate the destructive power of lahars (Thouret et al. 2020). Lahar deposits are common stratigraphic units in the volcanic foothills of Costa Rica (Quesada-Román et al. 2019; Ryan et al. 2022). Here, we briefly explain these historic events for a better understanding of river behavior in Costa Rica.

5.7.3.1 The 1963 Reventado River Lahar

The 1963 Reventado lahar is probably the most studied and notable lahar event of Costa Rica. Located on the Irazú Volcano, the Reventado River has a historical record of lahar events that date from 1724, 1861, 1891, 1928, 1951, and most recently, in 1963. During the 1963–1965 period, the Irazú Volcano entered the strombolian-vulcanian eruptive period that generated over 100 lahar events during the rainy seasons (Alvarado and Schmincke 1994). On the night of December 9, 1963, a 17 h rainfall storm triggered a lahar that severely affected the city of Cartago killing at least 20 persons and damaging over 300 structures located in the proximities of the alluvial fan of the river

Fig. 5.20 Historical photographs of the occurrence of catastrophic lahar events. **a** Aerial photograph of the 1963 lahar event in Cartago city (Instituto Geográfico Nacional de Costa Rica, 1964), **b** severe lateral and vertical erosion in the Seco River trigged by the 2009 Cinchona lahars, and **c** lahar and large wood deposits in the Cinchona River bed after the earthquake, notice massive landslides in the headwater part of the catchment in the background

(Fig. 5.20a). After the event, hydraulic protection structures were constructed with the remaining deposits of lahar, to prevent major damage to the city of Cartago in case of a future event (Ulate and Corrales 1966). Unfortunately, since the construction of these levees 50 years ago, there has been a negligible or complete lack of maintenance of these structures. Furthermore, there has been an increased development of infrastructure in the last decades, accompanied with an absence of urban planning and risk assessment that has caused disordered urbanization within the flood area that the levees were supposed to contain. Construction of informal settlements has also led to a high-density concentration of informal housing on top of the levees, further deteriorating their hydraulic capacity to withstand debris flows (Granados-Bolaños et al. 2021). A high-resolution 2D hydraulic model developed by Granados-Bolaños et al. (2021) estimates that a similar event like the one that occurred in 1963 could affect thousands of persons and hundreds of structures. This steep river has a step-pool channel morphology and has less coarse sediment supply within the channel than most of the rivers analyzed in this chapter (Granados-Bolaños and Surian 2023). Nevertheless, catchment characteristics, such as the geomorphological conditions of the hillslopes, with active massive landslides, intense agricultural landuse, its proximity to the crater areas, and flow direction toward Cartago city makes it one of the most hazardous catchments of the country.

5.7.3.2 The 2009 Cinchona Mw 6.2 Earthquake Triggers Lahars on Poás Volcano

The 2009 Cinchona earthquake (Mw 6.2) with an epicenter located 6.5 km east of the active crater of Poás Volcano triggered several rock falls and landslides (rotational and translational) which in combination with the existing pore water within the landslide mass, aquifer deformation, and mixing of fluvial waters into the moving sediment mass, produced highly a destructive chain of lahars that propagated along several rivers (Sarapiquí, Seco, María Aguilar, Cariblanco, Ángel, La Paz, Mataste, and Tigre) (Alvarado 2010; Quesada-Román and Barrantes-Castillo 2016). The violent event generated extreme water elevations reaching up to 30 m of height, and due to the rapid incorporation of colluvial of material and high-slope rates maximum flow velocities were registered at several hydropower stations with a range of 6–12 km/h. Severe channel morphology change was documented involving both lateral and vertical erosion, also the loss of riparian vegetation that was incorporated as large wood transport (Fig. 5.20b, c).

5.8 Conclusions and Recommendations

Costa Rica has a great and rich diversity of fluvial landscapes with complex dynamics and landforms. Rivers in this tropical part of the world are extremely dynamic due to high-energy input and sediment loads. Depending on the environment and the temporal scale, channel morphology may change drastically in short periods. Streams that flow through volcanoes and mountain catchments are characterized by single thread, steep, and confined channels. These rivers carry high sediment load composed mainly of boulders and coarse particles. Rivers in the Caribbean Plains flow unconfined with sinuous and meandering morphologies. Neck cutoffs and lateral channel migration processes are relatively common in this type of environment. On the other hand, rivers flow incised in urban environments of the GAM, in some streams riparian vegetation is well consolidated, and these dispersed forest patches provide important ecosystem services. The country is exposed to fluvial hazard due to intense rainy seasons or extraordinary events such as tropical cyclones or cold fronts. Population and infrastructure located near highly dynamic rivers have a high probability of being directly or indirectly affected by fluvial processes.

More detailed studies of old (Pliocene and Lower Quaternary) and recent fluvial fans and terraces, including antecedent rivers affected by neotectonics and sea level changes, are necessary. These studies should involve both relative and absolute dating techniques, particularly OSL and 14C dating. Such studies are crucial not only for a better understanding of geological and geomorphological history but also for assessing hazards associated with these processes. The ongoing global warming will alter precipitation patterns and ocean temperatures, which, in turn, will affect the dynamics of fluvial systems in Costa Rica in the near future.

Another crucial aspect for a better understanding of the geomorphological setting of any region is the availability of high-resolution elevation and surface data. Digital terrain models are necessary for analyzing landforms and monitoring-modeling surface processes. In Costa Rica, there are severe limitations in accessing detailed topographical datasets. Therefore, UAV photogrammetry played a significant role in recollecting information for this chapter.

Acknowledgements We would like to thank the multiple collaborators from around the world for this chapter. Acknowledgments to Sol Fernández and the community of the Ocloro River and Fernando Madrigal from the Sensoria Resort. Also, a special thanks to Piotr Migoń for improving the quality of the text. We would also thank the Casa di Risparmio di Padova and the Cariparo Fellowship for the funding of this research.

References

Acuña López S, Ávila Pérez I, Aguilar-Arias H, Hernández Hernández S, Vargas Céspedes A, Vargas Solano Y, Obando-Picado MF, Fallas-Montero E, Jiménez-Rodríguez M, Fernández-Garro J, Calvo-Elizondo Y, Romero-Badilla D, Manrow-Villalobos M, Miller-Granados C (2022) Informe final del paisaje de Cobertura Arbórea presente en un radio de 2 km alrededor de los paisajes productivos del MOCUPP para el año 2019. CENAT-PRIAS https://hdl.handle.net/20.500.12337/8363

Alvarado GE (2010) Aspectos geohidrológicos y sedimentológicos de los flujos de lodo asociados al terremoto de Cinchona (Mw 6,2) del 8 de enero del 2009. Costa Rica. Rev Geol Amér Central 43:67–96

Alvarado GE, Schmincke HU (1994) Stratigraphic and sedimentological aspects of the rain triggered lahars of the 1963–1965 Irazú eruption, Costa Rica. Zbl Geol Palaont Teil 1:513–530

Alvarado GE, Vega E, Chaves J, Vásquez M (2004) Los grandes deslizamientos (volcánicos y no volcánicos) de tipo debris avalanche en Costa Rica. Rev Geol Amér Central (30)

Arce-Rodríguez A, Puente-Sánchez F, Avendaño R, Libby E, Rojas L, Cambronero JC, Pieper DH, Timmis KN, Chavarría M (2017) Pristine but metal-rich Río Sucio (Dirty River) is dominated by Gallionella and other iron-sulfur oxidizing microbes. Extremophiles 21:235–243

Arias O, Denyer P (2011) Estructura geológica de la región comprendida en las hojas topográficas Abra, Caraigres, Candelaria y Río Grande, Costa Rica. Rev Geol Amér Central (12)

Alvarado GE, Gans PB (2012) Síntesis geocronológica del magmatismo, metamorfismo y metalogenia de Costa Rica, América Central. Rev Geol Amér Central 46:7–122

Arias-Hidalgo D (2016) Caracterización de albergues turísticos y percepciones sobre el turismo en el Territorio Indígena Bribri talamanqueño. Trama Rev Cienc Soc Humanidades 5(1):22–32

Bergoeing JP (2011) Los conos de deyección del valle de El General, Costa Rica. Rev Geográfica 21–32

Birkel C, Correa-Barahona A, Martinez-Martinez M, Granados-Bolaños S, Venegas-Cordero N, Gutiérrez-García K, Blanco-Ramírez S, Chavarría-Palma A, Vargas-Arias K, Moore GW, Durán-Quesada AM, Vasquez-Morera J, Soulsby S, Tetzlaff D, Espinoza-Cisneros E, Sánchez-Murillos R (2020) Headwaters drive streamflow and lowland tracer export in a large-scale humid tropical catchment. Hydrol Process 34(18):3824–3841

Birkel C, Correa-Barahona A, Duvert C, Granados-Bolaños S, Chavarría-Palma A, Duran-Quesada AM, Sánchez-Murrillo R, Biester H (2021) End member and Bayesian mixing models consistently indicate near-surface flowpath dominance in a pristine humid tropical rainforest. Hydrol Process 35(4):e14153

Buffington JM, Montgomery DR (2021) Geomorphic classification of rivers: an updated review. Treatise Geomorphol 2:1–47

Camacho ME, Quesada-Román A, Mata R, Alvarado A (2020) Soil-geomorphology relationships of alluvial fans in Costa Rica. Geoderma Reg 21:e00258

Campos-Durán D, Quintero-Quintero R (2020) Intensidades macrosísmicas del sismo de Capellades del 30 de noviembre de 2016 (Mw=5.4) y el contexto sísmico de la región central de Costa Rica. Bol Geol 42(1):57–68

Chang JH, Lau LS (1993) Definition of the humid tropics. Hydrology and water management in the humid tropics, p 71

Chinchilla M, Mata R, Alvarado A (2011) Caracterización y clasificación de algunos Ultisoles de la región de Los Santos, Talamanca. Costa Rica. Agron Costarricense 35(1):59–81

Church M (2002) Geomorphic thresholds in riverine landscapes. Freshw Biol 47(4):541–557

Church M (2006) Bed material transport and the morphology of alluvial river channels. Annu Rev Earth Planet Sci 34:325–354

Denyer P, Arias O, Arias M (1994) Esfuerzos y Paleo-esfuerzos de la cuenca de Limón. Rev Geol Amér Central

Galve JP, Alvarado GE, Pérez-Peña JV, Mora MM, Booth-Rea G, Azanón JM (2016) Megafan formation driven by explosive volcanism and active tectonic processes in a humid tropical environment. Terra Nova 28(6):427–433

Galve JP, Alvarado GE, Pérez-Peña JV, Pérez-Consuegra N, Ruano P, Becerril L, Devoto S, Reyes-Carmona C (2022) River avulsions and megafan development triggered by high recurrent explosive volcanism. Geomorphology 419:108466

Grabowski RC, Surian N, Gurnell AM (2014) Characterizing geomorphological change to support sustainable river restoration and management. Wiley Interdiscip Rev Water 1(5):483–512

Granados-Bolaños S, Surian N (2023) Channel morphology in an active volcanic complex under humid tropical conditions. Copernicus Meetings No. EGU23-3934

Granados-Bolaños S, Quesada-Román A, Alvarado GE (2021) Low-cost UAV applications in dynamic tropical volcanic landforms. Volcanol Geotherm Res 410:107143

Granados-Bolaños S, Surian N, Birkel C, Quesada-Román A, Alvarado GE (2022) UAV applications in humid tropical fluvial geomorphology. Copernicus Meetings No. ICG2022-522

Guillen-Montero D, Núñez-Román OA, Vargas-Bogantes J, Vega-Ramírez LM (2021) Situación de los Sistemas de Información Territorial para la gestión municipal: caso de la GAM, Costa Rica, 2018. Rev Geog Am 66:59–78

Ikeda H (1989) Sedimentary Controls on Channel Migration and Origin of Point Bars in Sand–Bedded Meandering Rivers. In: Ikeda S, Parker G (eds) River Meandering

Marshall JS (2007) The geomorphology and physiographic provinces of Central America. Central Am Geol Resour Hazard 1:75–121

Paniagua-Molina J, Solís-Rivera LR (2020) Effect of "golden pineapple innovation" on Costa Rica's pineapple exports to US Market: an econometric approach. Int J Food Agric Econ (IJFAEC) 8(1128-2020-1528):219–231

Pérez-Gómez G, Alvarado-García V, Rodríguez-Rodríguez JA, Herrera F, Sánchez-Gutiérrez R (2021) Calidad fisicoquímica y microbiológica del agua superficial del río Grande de Tárcoles, Costa Rica: un enfoque ecológico. Cuad Inv UNED 13(1):69–86

Quesada-Román A (2022) Flood risk index development at the municipal level in Costa Rica: a methodological framework. Environ Sci Policy 133:98–106

Quesada-Román A, Barrantes-Castillo G (2016) Procesos de ladera cosísmicos del terremoto de Cinchona (Costa Rica) del 8 de enero de 2009 (Ms=6.2). Cuad Geogr: Rev Colomb Geogr 25(1)

Quesada-Román A, Zamorano-Orozco JJ (2019) Geomorphology of the Upper General River basin, Costa Rica. J Maps 15(2):94–100

Quesada-Román A, Fallas-López B, Hernández-Espinoza K, Stoffel M, Ballesteros-Cánovas JA (2019) Relationships between earthquakes, hurricanes, and landslides in Costa Rica. Landslides 16:1539–1550

Quesada-Román A, Ballesteros-Cánovas JA, Granados-Bolaños S, Birkel C, Stoffel M (2020a) Dendrogeomorphic reconstruction of floods in a dynamic tropical river. Geomorphology 359:107133

Quesada-Román A, Campos N, Alcalá-Reygosa J, Granados-Bolaños S (2020b) Equilibrium-line altitude and temperature reconstructions during the Last Glacial Maximum in Chirripó National Park, Costa Rica. J South Am Earth Sci 100:102576

Quesada-Román A, Campos N, Granados-Bolaños S (2021) Tropical glacier reconstructions during the Last Glacial Maximum in Costa Rica. Rev Mex Cienc Geol 38(1):55–64

Rainwater TR, Millichamp NJ, Barrantes LDB, Barr BR, Bolaños-Montero JR, Platt SG, Abel MT, Cobb GP, Anderson TA (2011) Ocular disease in American crocodiles (*Crocodylus acutus*) in Costa Rica. J Wildl Dis 47(2):415–426

Rinaldi M, Gurnell AM, Del Tánago MG, Bussettini M, Hendriks D (2016) Classification of river morphology and hydrology to support management and restoration. Aquat Sci 78(1):17–33

Ryan PC, Alvarado GE, McCanta M, Barca MK, Davis G, de Mendoza LH (2022) The importance of overbank deposits and paleosol analyses for comprehensive volcanic hazard evaluation: the case of Holocene volcanism at Miravalles Volcano, Costa Rica. Nat Hazards 112(1):413–449

Sánchez-Murillo R, Birkel C, Welsh K, Esquivel-Hernández G, Corrales-Salazar J, Boll J, ... and Araguás-Araguás LJ (2016) Key drivers controlling stable isotope variations in daily precipitation of Costa Rica: Caribbean Sea versus Eastern Pacific Ocean moisture sources. Quat Sci Rev 131:250–261

Sandí GC (2003) Caracterización de los materiales sedimentarios inconsolidados costeros y aluviales actuales en la región de Parrita, Pacífico Central de Costa Rica. Rev Geol Amér Central (29)

Siebert L, Alvarado GE, Vallance JW, van Wyk de Vries B (2006) Large-volume volcanic edifice failures in Central America and associated hazards. In: Volcanic hazards in Central America

Solano-Rivera V, Geris J, Granados-Bolaños S, Brenes-Cambronero L, Artavia-Rodríguez G, Sánchez-Murillo R, Birkel C (2019) Exploring extreme rainfall impacts on flow and turbidity dynamics in a steep, pristine and tropical volcanic catchment. Catena 182:104118

Thouret JC, Antoine S, Magill C, Ollier C (2020) Lahars and debris flows: characteristics and impacts. Earth-Sci Rev 201:103003

Vallejos S, Bonilla WF (2017) Tormenta Tropical Nate y su paso por Costa Rica. Revista en Torno a la Prevención 19:29–37

Žáček V, Vorel T, Kycl P, Huapaya S, Mixa P, Grygar R, Havlíček P, Čech S, Hrazdíra P, Metelka V, Ševčík J, Pécskay Z (2012) Geología y estratigrafía de la Hoja 3646-II Miramar, Costa Rica. Rev Geol Amér Central 47:7–54

Lakes in Costa Rica

6

Sally P. Horn, Gerardo Umaña-Villalobos, and Kurt A. Haberyan

Abstract

Lakes occur throughout Costa Rica, in a variety of geomorphic settings. Nearly every possible mode of lake formation has occurred, involving processes related to volcanism, tectonics, mass movement, fluvial activity, coastal erosion, and glaciation. Humans have also created a diversity of lakes. Based on basal ages on organics in sediment cores, the oldest natural lakes in Costa Rica are the glacial lakes of the Chirripó massif, which date to the late Pleistocene or early Holocene. While these lakes have shown little change over time, lakes elsewhere have changed dramatically due to human and natural disturbances. The great number of lakes in Costa Rica, and their variation in morphometry and limnology, makes them important in the conservation of biodiversity. Lakes are also of economic and tourism importance, and their protection deserves particular attention.

Keywords

Lake formation · Lake morphology · Lake-sediment cores · Limnology

S. P. Horn (✉)
Department of Geography and Sustainability,
University of Tennessee, Knoxville, TN, USA
e-mail: shorn@utk.edu

G. Umaña-Villalobos
Center for Research in Marine Sciences and Limnology,
University of Costa Rica, San José, Costa Rica
e-mail: gerardo.umana@ucr.ac.cr

K. A. Haberyan
Department of Natural Sciences, Northwest Missouri
State University, Maryville, MO, USA
e-mail: khaber@nwmissouri.edu

6.1 Introduction

Costa Rica has hundreds of lakes, distributed from the lowlands of both coasts to the highest peaks. These lakes give character to landscapes, contribute to biodiversity, and provide resources used by people since prehistoric times. Lakes occupy depressions formed by a variety of erosional and depositional processes and can be classified on this basis. For example, volcanic lakes in Costa Rica occupy depressions formed by volcanic explosions or topographic lows within debris or lava flows, fluvial lakes owe their origin to stream dynamics, and glacial lakes resulted from erosion and deposition by glaciers during past periods of cooler climate.

In this chapter, we review the characteristics and geomorphic settings of selected lakes in different regions of Costa Rica, using an adaptation of the regional ecosystem classification in Horn and Haberyan (2016) (Fig. 6.1). The regions are the northern highlands, the southern highlands, the northern Pacific lowlands, the central Pacific lowlands, the southern Pacific lowlands, and the Caribbean lowlands. For each region, we describe major lake-forming processes, the ages of formation of specific lakes where known, and aspects of the limnology of selected lakes, updating and extending earlier reviews by Horn and Haberyan (1993, 2016), Umaña et al. (1999), Haberyan and Horn (1999a), Haberyan et al. (2003), and Umaña (2010). We close this chapter with thoughts on the dynamic nature of Costa Rican lakes in the face of natural and human disturbance, and our hope that these important landscape features can be experienced, valued, and conserved by future generations.

A. Quesada-Román (ed.), *Landscapes and Landforms of Costa Rica*,
World Geomorphological Landscapes, https://doi.org/10.1007/978-3-031-64940-0_6

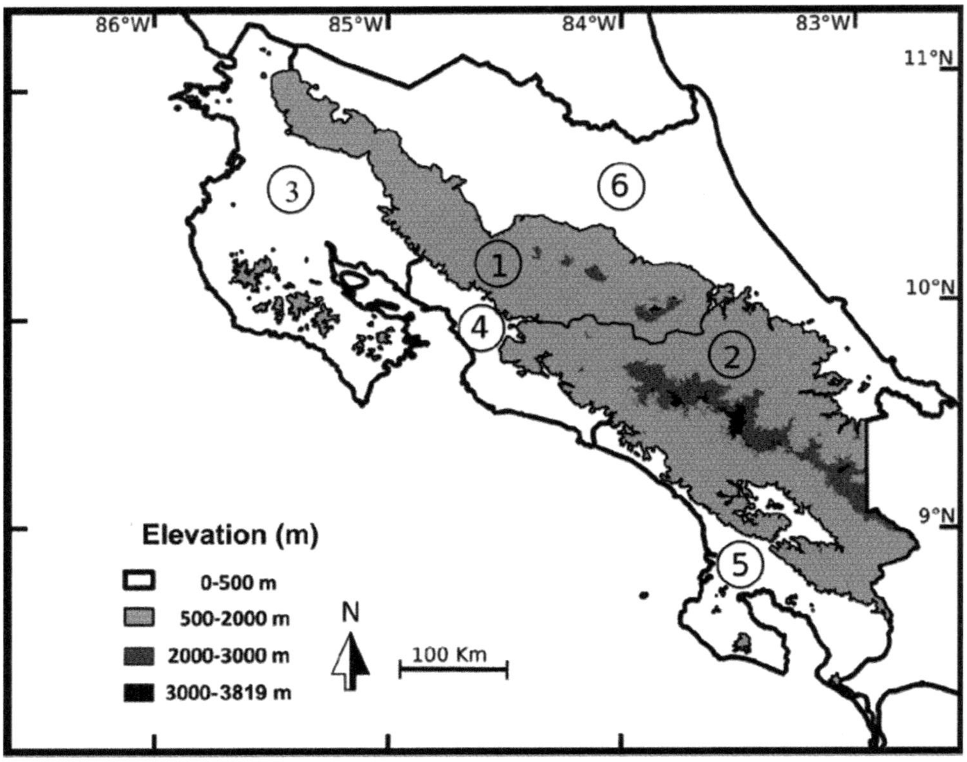

Fig. 6.1 Regions of Costa Rica as used in this chapter. 1: northern highlands, 2: southern highlands, 3: northern Pacific lowlands, 4: central Pacific lowlands, 5: southern Pacific lowlands, 6: Caribbean lowlands

6.2 Lakes of the Northern Highlands: Predominantly of Volcanic Origin

As used here the northern highlands region of Costa Rica include the summits and upper slopes (elevations above 500 m asl) of the volcanic Cordillera de Guanacaste and Cordillera Central, along with the Chato and Arenal volcanoes, the volcanically extinct Cordillera de Tilarán, and the Valle Central. Lakes formed by volcanism are abundant in the northern highlands. Many of the volcanic peaks have inactive craters on their summits or flanks that are occupied by lakes, or lakes dammed by or occupying depressions within lava, lahar, or debris flows. Examples of lakes in craters formed by explosion or collapse include crater lakes on the Rincón de la Vieja and Santa María volcanoes; Cerro Chato Lake, in the crater atop the Chato volcano; Barva and Danta Lakes on Barva volcano; Botos Lake on Poás volcano, and the Hule, Congo, and Bosque Alegre Lakes within a volcanic maar on the northern flank of Poás (Salani and Alvarado 2010; Alvarado et al. 2011) (Fig. 6.2). These crater lakes are exceptionally scenic landforms that are protected within national parks and conservation areas. Some of these crater lakes attract thousands of visitors annually and are well known from a limnological perspective; others are reachable only by hiking off-trail, and limnological data are absent or limited (Baker 1987).

The ages of lakes inside craters can be inferred from radiocarbon dates on organic materials preserved within or underneath lavas and other volcanic deposits associated with crater formation. As the lakes occupy inactive craters, their maximum ages should correspond to the age of the last known volcanic activity. For example, we can infer that Lake Cerro Chato formed around 3500 years ago, based on the estimate by Borgia et al. (1988) that the last activity of Chato volcano occurred about 3500 BP.

Ages of crater lakes and those in other geomorphic contexts can also be determined by dating organic macroremains, such as leaf, wood, or charcoal fragments, at the bases of sediment cores collected from lake basins, if the cores obtained reached the sediments deposited right after lake formation. However, in the case of lakes in volcanic craters, subsequent eruptions from active vents nearby, or the erosion of volcanic deposits on crater walls, can lead to thick tephra layers in sediment profiles of crater lakes that impede efforts to collect cores, and complicate attempts to determine whether dates on organic material at the bases of sediment cores reflect the last eruptive activity from the crater now occupied by the lake. The experience of Horn and students at Lake Barva illustrates this issue. In 1999, they collected the upper 26 cm of organic sediment using a mud-water interface corer with no issue. But after switching to a piston corer to recover deeper material, they were

Fig. 6.2 Crater lakes in the northern highlands of Costa Rica. **a** Barva Lake. **b** Laguna Hule. **c** Google Earth image of Laguna Botos, next to active crater of Poás Volcano (Photos **a** and **b** by G. Umaña and K. Haberyan)

mostly unsuccessful, due to the presence of a thick deposit of lapilli and scoria below the depth penetrated by the mud-water interface corer, that they could not push beyond. The material did not hold well in the core barrel, and in repeated attempts to collect 1-m long sections, they only succeeded in obtaining a short (15 cm) section of this tephra. Leaf fragments picked from the lower 5 cm of the section yielded a radiocarbon date of 170 ± 50 ^{14}C years BP, which calibrates to the period AD 1656–1949 (CALIB software, Stuiver and Reimer 1993; version 8.1). If the tephra in the core is the product of the most recent eruption of the crater occupied by Barva Lake, the date on plant remains indicates that the lake is less than 400 years old. Perhaps, however, the tephra is redeposited, from erosion of material from the sides of the crater, and is older than the age for the leaf fragments. In that case, the date would have no bearing on the age of the lake. While historic accounts mention eruptions of Barva volcano in AD 1770 or 1776, and in AD 1867, Alvarado (2000) suggested that these accounts may be inaccurate. However, he interpreted the less developed vegetation near Barva Lake to indicate that the last eruptive phase of the crater could not be more than a few hundred or thousand years in the past, and he mentioned the presence of fresh-looking lapilli and scoria northeast of the lake that appeared quite recent in age. Titiz and Sanford (2007) dated charcoal fragments below volcanic tephra layers in soil cores collected at 2600 m asl on the Barva summit and obtained results of 170 ± 40 and 140 ± 50 ^{14}C years BP that are in keeping with the date on plant fragments within tephra in the lake-sediment core. These and other lines of evidence suggest a small historic eruption and the possibility that Barva Lake may be a very young lake, but more work is needed to resolve this issue.

Based on radiocarbon dates on sediment cores from Hule Lake and nearby María Aguilar Lake, and by a date obtained by Malavassi et al. (1990) on wood in a pyroclastic deposit, Horn (2001) estimated that Hule Lake formed between ca. 2950 and 2470 cal yr BP. Salani and Alvarado (2010) and Alvarado et al. (2011) used extensive field work and a larger set of radiometric ages along with geoarcheological data to infer that the Hule maar formed about 6200 cal yr BP, since revised by Ruiz et al. (2019) to about 6000 cal yr BP—with both ages indicating that the maar is much older than the lake-sediment cores. These researchers concluded that the pyroclastic deposit dated by Malavassi et al. (1990) was not from Hule but from a later eruption, at about 2800 cal yr BP, of the nearby Pata de Gallo maar; this eruption could also be the source of tephra in María Aguilar Lake and of material in Laguna Hule that prevented the recovery of any older sediments. Studying the mineralogy of tephra in lake-sediment cores could be informative.

María Aguilar Lake (Fig. 6.3), ca. 2 km east of the Hule maar, is one of several lakes in the northern highland region with an origin likely related to flows of lava or volcanic debris, rather than crater formation. It is likely dammed by, or occupies a depression within, a lahar or volcanic debris flow. La Palma and Cedeño Lakes (Fig. 6.3) on the lower north flank of Arenal (at 570 and 610 m asl, respectively) occupy depressions within an area of glowing avalanche deposits and lahars (Alvarado et al. 1988; Zapata and Soto 1990). La Palma Lake is dammed on its western edge by a lava flow dated between AD 1525 and 1650 (Alvarado 2000). Borgia et al. (1990) interpreted the north slope of Lake Cedeño as a remnant of the original steep topography of the ancestral Arenal Valley, before the development of the Arenal and Chato volcanoes. A sediment core 74 cm long recovered from the lake by Horn in 1992 contained highly inorganic sediments that may have originated in part in a recent (post-AD 1968) lahar mapped by Zapata and Soto (1990) as reaching the lake.

Natural lakes formed by non-volcanic processes also occur in the northern highlands, including several lakes we have investigated that were formed by landslides (Haberyan et al. 2003). Arancibia Lake (1260 m a.s.l.) in the Cordillera de Tilarán (Fig. 6.3) has been present since at least ca. 1300 cal yr BP, based on dating of organic macroremains in lake sediments (Horn, unpublished). The lake was not in the path of the major landslide of AD 2000 (Alvarado et al. 2003), but a smaller lake 1 km to the NNW, Lake Arancibita, was buried by that landslide. Gonzalez Lake (710 m elevation on the eastern edge of the Cordillera de Tilarán (Horn and Haberyan 1993) formed in the last 360 years, based on radiocarbon dating of plant fragments in rocks at the base of a lake-sediment core (Umaña et al. 1999). Fraijanes Lake (Umaña-Villalobos 2014) on the south slope of Poás volcano formed from a landslide associated with an earthquake in AD 1888, for which historic documentation exists. Interestingly, the lake formed on the site of a former lake that had been drained to establish a pasture (González Víquez 1910).

The northern highlands are also the site of many hydroelectric reservoirs, including Arenal reservoir in the Cordillera de Tilarán, which is the largest artificial lake in Costa Rica (Umaña 2006). The largest natural lake in Costa Rica, Lago Cote, lies just 3 km north of the reservoir. Though of natural origin, this lake has been modified by humans. Cote Lake originally drained to the Caribbean via an outlet to the Cote River, but in the 1980s, this outlet was dammed and outflow was diverted to the Arenal Reservoir (Umaña and Avilés 2020). The height of the dam was later raised to use the outflow for power generation, resulting in seasonal fluctuations in water level of up to 1.5 m (Umaña and Avilés 2020). The formation of the basin occupied by Lago Cote is uncertain. The roughly circular (sometimes described as heart- or apple-shaped) depression could represent a crater, but the presence of nearby faults indicates that the basin may have been created by tectonic processes, in combination with or independent from volcanism (Alvarado 2000). A sediment core collected from the middle of the lake in 2000 revealed the lake to be at least 4000 years old (Horn et al. in preparation).

Poco Sol Lake, also in the Cordillera de Tilarán, is another highland lake of uncertain origin. It was initially described as occupying an explosion crater, but that was based on an incorrect claim that the lake was over 60 m deep. The actual depth of only 11.5 m and the steep surrounding terrain open the possibility that it was instead formed by a landslide (Horn and Haberyan 2016), though a volcanic origin remains possible (Alvarado 2000). Radiocarbon dates obtained for a short sediment core

Fig. 6.3 Lakes in the northern highlands of Costa Rica. **a** Maria Aguilar Lake, occupying a depression within or created by a lahar or volcanic debris flow on the flank of the Hule crater. **b** Cedeño Lake located within an area of glowing avalanche and lahars from Arenal volcano. **c** Arancibia Lake, formed by a landslide. **d** Gonzalez Lake formed by a landslide. **e** Fraijanes Lake formed by a landslide associated with an historic earthquake. **f** Poco Sol Lake, formed by a landslide or possibly occupying a volcanic crater. (Photos by S. Horn, K. Haberyan, and G. Umaña)

recovered from Poco Sol Lake in 1998 revealed a very high rate of recent sedimentation in the lake. A leaf located 63 cm below the mud-water interface and bulk sediment 97 cm below the mud-water interface both yielded "modern" (post-AD 1950) radiocarbon dates. Calibration using the Calibomb program (Reimer and Reimer 2022) showed the leaf to correspond to either AD 1962 or to the interval AD 1979–1981, and the sediment lower in the core to correspond to either AD 1959 or AD 1986–1989. Assuming an age of ca. 1980 for the leaf and 1959 for the sediment indicates a sedimentation rate of 1.6 cm/year between 1959 and 1980 and 3.5 cm/year since 1980, suggesting the possibility of major erosive events in the watershed or slumping of sediments within the lake.

Basic limnological data are available from all the lakes mentioned above, with the exception of lakes atop the Rincón de la Vieja and Santa Maria volcanoes, and Danta Lake on Barva volcano (see review papers cited in introduction for data and additional references). Cerro Chato, Botos, and Barva lakes have morphologies that are typical of crater lakes: they are nearly circular in outline and relatively deep, and their watersheds are relatively small. Typically, the waters are relatively clear and low in pH and dissolved ions. In these lakes, as well as Hule Lake, the persistence of stratification varies inversely with elevation (Umaña 2010). Hule Lake lies at 740 m asl and is oligomictic, and Cerro Chato Lake (1020 m asl) mixes seasonally, owing in part to the greater depth in these lakes (22 and 18 m, respectively). Botos Lake (2600 m asl) and Barva Lake (2840 m asl) are similar in depth (9 and 7 m, respectively) and mix frequently: Botos is more exposed to wind and is polymictic, while in Barva stratification may be frequent but short-lived (Umaña 2010).

Hule, Congo, and Bosque Alegre lakes are found in a single maar crater, and their morphologies reflect the presence of a secondary volcanic cone which subdivides the crater: Hule Lake and Congo Lake are more semi-circular, and Bosque Alegre Lake is irregular. Of these lakes, only Hule Lake (Fig. 6.2) has been mapped in detail (Gocke 1996). Its steepest slope is adjacent to the crater rim and its bottom is relatively flat (Fig. 6.4). Stratification has been observed in Hule Lake and Bosque Alegre Lake, but Congo Lake has only been sampled once by limnologists and was not stratified at that time (Haberyan and Horn 1999a).

Chemically, the pH of the three Hule lakes is slightly less than neutral (6.4–6.6). The ionic composition of these lakes reflects the rainfall inputs (Tassi et al. 2009). Ionic concentrations are low except for chloride (2.8–9.5 mg L^{-1}), which is probably not of magmatic origin (Haberyan and Horn 1999a). However, it is likely that magma is the source of dissolved carbon dioxide, which is affected by bacteria and archaea (Cabassi et al. 2014) and can accumulate in the deep water of the lake. Hule lake, like Río Cuarto Lake located 6.5 km to the north in the Caribbean lowlands region, does not seem to turn over every year, owing in part to its depth and its sheltered location inside the crater. Between turnovers, its deep water collects carbon dioxide, which has been measured as high as 65 mg L^{-1} (Horn and Haberyan 2016). This carbon dioxide together with hydrogen sulfide (Gocke, 1996) may be released during mixis, which kills fish and produces a rotten egg smell, according to local residents. To date, no human fatalities have been reported, but the volume of degassed carbon dioxide may be sufficient to fill the basin to 2 m (Haberyan et al. 2003) and therefore pose a danger to humans.

Poco Sol Lake is rather deep compared to its area and also likely does not turn over every year; it is strongly stratified (at about 4.5 m) which reduces its biological productivity and thus increases its Secchi disk depth (2.8 m). Hypolimnetic concentrations of CO_2 are elevated, about 31 mg L^{-1} (Haberyan et al. 2003). Other natural lakes in the northern highlands are relatively shallow. Few limnological data are available for Arancibia and Arancibita lakes. Maria Aguilar, Cedeño, and Fraijanes lakes vary in depth (3.1–6.9 m) and likely turn over annually; when present, the thermocline is found between 1 and 2 m. Annual turnover increases their productivity and decreases Secchi disk depth (0.9–2.5 m). La Palma Lake is distinct, being deeper, with a deeper thermocline, and greater values of pH and most of the measured ions (calcium, silica, chloride, sulfur, and total conductivity) (Haberyan et al. 2003).

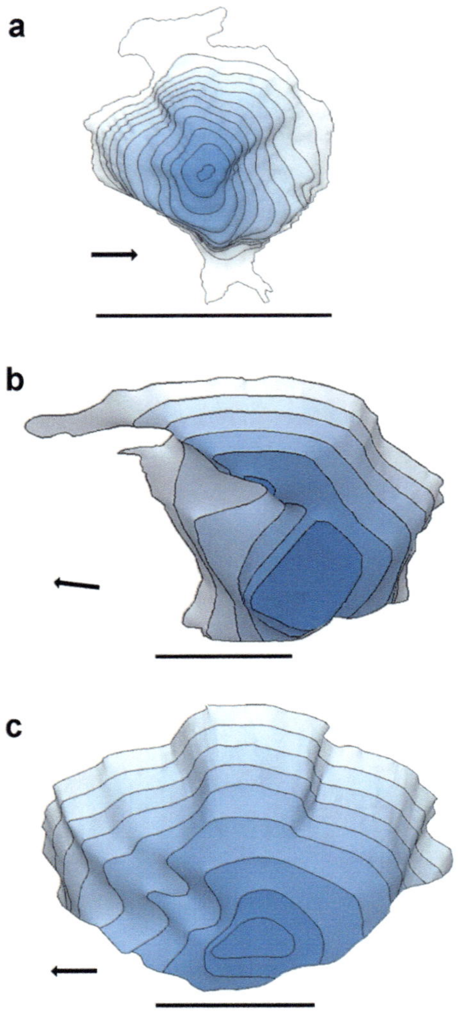

Fig. 6.4 Bathymetry of selected Costa Rican lakes. **a** Chirripó Lake (contour interval 2 m). **b** Hule Lake (contour interval 5 m with additional contour at 22 m). **c** Rio Cuarto Lake (contour interval 10 m with additional contours at 65 and 66 m). Arrows indicate north. Scale bars represent 250 m horizontally across the center of the image. Bathymetry for Chirripó Lake is based on Gocke et al. (1981), for Hule Lake based on Gocke (1996), and for Río Cuarto Lake based on Gocke et al. (1987). These references show similar 3D maps from other perspectives

Long-term observations on lakes in the northern highlands include monitoring over ten years in Lake Cote, plus some repeated sampling in lakes Fraijanes, Barva, Cerro Chato, and Gonzalez. In general, lakes maintain their limnological characteristics through time, which are determined by lake morphometry and local weather conditions (Umaña 2010). In Lake Cote, the data show some hints of warming since the lake was modified for hydropower generation (Umaña 2014).

6.3 Lakes of the Southern Highlands: Glacial Lakes and Some Other Types

The southern highlands of Costa Rica include the upper slopes (elevations above 500 m asl) of the imposing, nonvolcanic Cordillera de Talamanca, composed of a suite of intrusive (mainly granodiorite) and extrusive rocks of Neogene to Quaternary age, and the Fila Costeña, a steep-fronted range positioned subparallel to the southern Pacific coastline that is composed of sedimentary rocks (Bolz and Calvo 2018; Alfaro et al. 2018). Rapid and recent (late Cenozoic) uplift of both mountain ranges is tied to the subduction of the Cocos Plate (Marshall 2007). The Cordillera de Talamanca was high enough during the late Quaternary to have supported small glaciers and ice fields (Orvis and Horn 2000; Lachniet and Seltzer 2002; Cunningham et al. 2019; Li et al. 2019; Potter et al. 2019; Quesada-Román et al. 2019, 2020). Processes of glacial erosion and deposition formed over 30 perennial lakes and ponds on the massif surrounding Cerro Chirripó, whose summit reaches 3819 m asl (Horn et al. 2005). The southern highlands also include several lakes formed by faulting or landslides on the flanks of the range and in the Fila Costeña. Highland lakes formed by human activity include small ponds along the Inter-American highway that may be natural in some cases or formed by road construction.

The glaciated Chirripó massif reaches above treeline and supports páramo vegetation dominated by grasses and shrubs (Kappelle and Horn 2016). Chirripó National Park protects the páramo and surrounding montane forests, accessible only by trail yet still a popular tourist destination (Quesada-Román and Pérez-Umaña 2020). Here, we focus on lakes in two valleys adjacent to Cerro Chirripó, the Morrenas Valley and the Valley of the Lakes, and in the Ditkebi River Valley; other lakes are described in Horn et al. (2005). The Morrenas Valley lies north of the glacial horn that is Cerro Chirripó, includes the headwaters of the Chirripó Atlántico River, and has the most lakes. Orvis and Horn (2000) numbered the lakes to facilitate study (Fig. 6.5). The upper valley has an irregular floor, for two reasons. First, the intrusive igneous bedrock is strongly jointed and has been quarried to different depths by glacial action in the valley, and second, because the bedrock is overlain by an uneven blanket of till and by scattered moraines. As a result, the drainage is also irregular and includes several large lakes primarily created by bedrock quarrying and numerous smaller lakes mainly dammed by till (Horn et al. 2005; Veas et al. 2018). The largest lake in the valley, Morrenas Lake 1 (Fig. 6.6) was described by Hastenrath (1973) as being surrounded by a horseshoe-shaped moraine 20 m high, but bedrock exposed in the till blanket and nearly continuously along the lakeshore indicating strong bedrock control (Horn et al. 2005). This lake is quarried within bedrock, although the water level at the outlet, which flows into Morrenas Lake 2, is controlled by a layer of loose debris. Morrenas Lake 2 is also excavated in bedrock, but at least one moraine crosses a portion of the basin. Morrenas Lakes 3 and 4 (Fig. 6.6) are both dammed on their north sides by a large moraine that overlies a rock step (Horn et al. 2005). Several smaller lakes in the valley, such as Morrenas Lakes 2A and 3A, are dammed by moraines. Morrenas Lakes 3C, 3B, and 3D and some minute ponds not mapped occupy an area of very uneven till that could be ice stagnation topography and could be kettle lakes (Horn et al. 2005).

Radiocarbon ages on lake sediments in cores from Morrenas Lakes 1 and 4 indicate that the lakes formed following deglaciation of the upper valley, after ca. 12,400 cal yr BP but before 9700 cal yr BP (Orvis and Horn 2000). The deepest datable sample from Morrenas Lake 1 consisted of mineral silts with a minor organic component positioned between glacial flour and organic-rich lake sediment, which yielded a radiocarbon date that calibrates to 12,440–11,267 cal yr BP (CALIB software, Stuiver and Reimer 1993; version 8.1). Hastenrath (1973), Orvis and Horn (2000), and Lachniet and Seltzer (2002) interpreted the moraines in the upper valley, near the lakes, and larger moraine down valley to indicate two or more glacial stages, but the results of recent cosmogenic isotope dating (Cunningham et al. 2019; Potter et al. 2019) indicate that even the largest moraines, located several km downvalley from the lakes described, correspond to Marine Isotope Stage 2.

Chirripó Lake, also known as Chirripó Lake 1 and Laguna Grande de Chirripó (Fig. 6.6), is the uppermost of three large lakes in the Valley of the Lakes, the headwaters of the Chirripó Pacífico River immediately to the west of Cerro Chirripó. Chirripó Lake is the largest and deepest lake on the massif; it is a tarn occupying a bedrock basin in the central floor of the upper cirque basin, behind a bedrock step. The upper, southeastern portion of the lake is a shallow bay about 3 m deep, but 80 m from shore the depth increases rapidly, reaching a maximum depth of 22 m

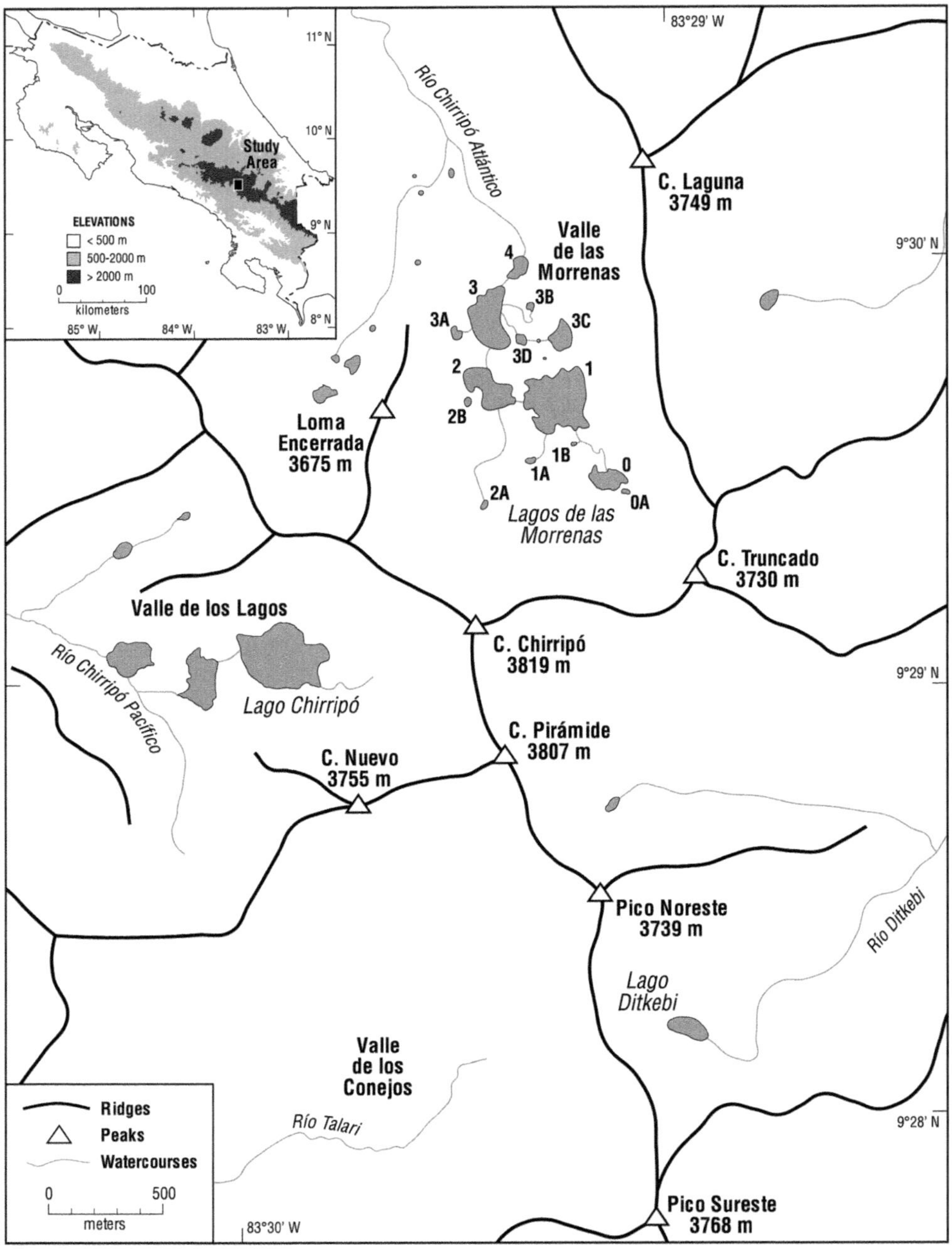

Fig. 6.5 Map showing locations of some of the glacial lakes surrounding Costa Rica's highest peak, Cerro Chirripó, in the Talamanca Mountains. Modified slightly from map in Orvis and Horn (2000)

(Fig. 6.4, Gocke et al. 1981). Chirripó Lakes 2 and 3, downstream of Chirripó Lake, are also both tarns occupying bedrock basins behind glacial steps (Horn et al. 2005).

Ditkebi Lake (Fig. 6.5) is the single lake in the upper Ditkebi River Valley 5 km south of Cerro Chirripó. Quesada-Román et al. (2019) described Ditkebi Lake as a glacial lake dammed by a riegel (a large barrier of resistant rock across a valley), in this case one shaped by glacial abrasion. Ditkebi Lake differs from lakes in the Valle Morrenas and Valle de los Lagos in that it occurs in an area of lavas and pyroclastic deposits, rather than intrusive rocks, based on geological mapping by Wunsch et al. (1999).

The limnology of these glacial lakes reflects their elevation and reduced evaporation, which influences their clarity, temperature, and ionic composition. Only the two largest lakes, Chirripó Lake (7.8 ha, 20.2 m deep) and Morrenas Lake 1 (5.6 ha, 8.3 m deep), have been sampled in any detail. Their alkalinities are low (8–10 mg L^{-1} as $CaCO_3$) but pH is variable (5.4–7.3); although few data are available in different seasons, ions in Chirripó Lake may be 40% more concentrated in the dry season (Jones et al. 1993). These two lakes are not known to stratify, but their surface waters may be slightly warmer during the daytime. In addition, these two lakes have Secchi disk depths of at least 7.5 m (Horn and Haberyan 2016). This clarity led Gocke

Fig. 6.6 Glacial lakes on the Chirripó massif. **a** Panorama of Lakes in the Morrenas Valley, taken from the summit of Cerro Chirripó. **b** Morrenas Lake 4, dammed by the same moraine that forms the north shore of Morrenas Lake 3, which overlies a rock step into which the outlet of Lake 4 has cut. **c** Lago Chirripó, located behind a bedrock step (Photos by B. Watson and G. Umaña)

et al. (1981) to conclude that photosynthesis was likely to be inhibited by the bright conditions above a depth of 5 m. The clarity may allow photosynthesis to occur on the bottom of most of these lakes because they are less than 4 m deep, a depth that is exceeded by only four lakes on the massif (Chirripó Morrenas 1, Morrenas 3, and Ditkebi) (Horn et al. 2005). The bottom of many lakes supports a dense population of *Aulacoseira umanai* (Haberyan and Horn 2020), a diatom that has dominated the Morrenas Lake 1 for the entirety of its record, according to the analysis of a sediment core (Haberyan and Horn 1999b).

The Austrian limnologist Heinz Löffler measured physical properties of Ditkebi Lake in the Ditkebi Valley in August–September 1966. He reported a maximum depth of 7 m, which was less than the Secchi disk depth, and a water temperature of 10.3 °C (Löffler 1972). Like many high-altitude lakes, Ditkebi Lake experiences high precipitation:evaporation ratios that vary with the seasons; some 90% of the inflow occurs during the rainy season (Esquivel-Hernández et al. 2018). Our measurements of water chemistry in March 1998 and March 2001 were similar to values for Morrenas Lake 1, Chirripó Lake, and other lakes in the same valleys, with the exception of Mg^{+2}, which may be related to differing bedrock (Horn et al. 2005). Additional measurements by Wu et al. (2019) in July 2014 found the waters to be extremely dilute (specific conductivity = 0.029 S/cm) and slightly alkaline (pH = 8). They reported a maximum depth of ca. 9 m and a relatively uniform temperature profile with a surface temperature of 10.7 °C that gradually decreased to 10.3 °C at 4 m depth and stayed constant below that. Wu et al. (2019) interpreted this profile to indicate weak stratification, but the definition of stratified requires at least a 1 °C temperature difference in 1 m (Hutchinson 1975), which was not found. The lack of stratification in Ditkebi Lake is consistent with findings for all other glacial lakes on the massif.

Horn measured temperature and oxygen profiles of Morrenas Lakes 1, 2, 3, and 4 in late February 1998 and found them to be highly uniform, as expected, but with water temperatures in these lakes and Chirripó Lake significantly warmer than temperatures measured in Chirripó Lake and Morrenas Lake 1 in previous limnological research by Löffler (1972) in August 1966, Göcke et al. (1981) in March 1979, and Jones et al. (1993) in August 1986 (Horn et al. 2005). We initially interpreted these warm temperatures to be the result of strong El Niño (positive El Nino-Southern Oscillation or ENSO) conditions in 1998 (Horn et al. 1999), but subsequently, we measured high temperatures also under strongly negative ENSO conditions in March 2000 (Horn et al. 2005). In January 2003, under positive ENSO conditions, temperatures in the glacial lakes were lower than we measured in 2000, but still higher than temperatures in 1966, 1979, and 1983. From these observations, and from records of past ENSO events that indicate the earlier, cooler temperatures were measured under neutral to positive ENSO conditions, we concluded that ENSO cycles are unlikely to explain the differences in measurements over the years (Horn and Haberyan 2016). Water temperatures in these clear, high-elevation lakes are influenced by both air temperature and by cloud cover, which blocks insolation during the day and cooling by re-radiation at night. The available temperature measurements from lakes on the massif may indicate changes in air temperatures or cloudiness, perhaps associated with global changes in climate; however, to determine whether lake temperatures are showing evidence of climate change will require continuous monitoring with temperature sensors and data loggers (Horn and Haberyan 2016).

Asunción Pond (3340 m asl) is located at the highest point of the Inter-American highway crossing of the Cordillera de Talamanca, in the Buenavista páramo. The pond is shown on a photograph from 1949, taken during highway construction (Horn 1989), and may owe its origin to road grading at the time. The Tres de Junio Pond, located 15 km NE of Asuncion Pond at 2670 m asl, also lies directly adjacent to the Inter-American Highway and could be artificial, though Standley (1936–1937) described abundant natural ponds and bogs near Tres de Junio Pond prior to highway construction. These areas of swampy terrain, of which only remnants now remain, were described by Weber (1959) as flat, pot-like depressions that held water because of their waterproof, clay bottoms. Both Asunción and Tres de Junio Ponds are cool and shallow, with low pH (< 5.3) and waters low in dissolved ions (Horn and Haberyan 1993; Haberyan et al. 2003). Waters are tannin-stained in Tres de Junio Pond.

Lakes formed by landslides are common at mid-elevations in the southern highlands. Sacabico (1095 m asl) and Dabagri (1040 m asl) lakes on the Caribbean slope of the Cordillera de Talamanca likely formed as a result of landslides, probably recently as indicated by the presence of standing dead trees (Jones et al. 1993). Landslides, possibly associated with faulting, likely explain the origin of Santa Elena Lake (1055 m asl, Fig. 6.7), on the Pacific slope of the Cordillera de Talamanca, and nearby Zoncho Lake (1190 m asl, Fig. 6.7) in the Fila Costeña, but analyses of sediment cores show that these lakes are significantly older than Sacabico and Dabagri lakes. Santa Elena formed about 1800 cal yr BP, and Zoncho Lake formed about 3300 years ago, based on radiocarbon dates on macroremains in lake-sediment cores (Kerr et al. 2020; Taylor et al. 2020). Sacabico and Dabagri lakes are known to stratify and have Secchi disk depths of 1.5 and 2.2 m, respectively, along with elevated nutrient levels (Jones et al. 1993). Santa Elena and Zoncho lakes are not known to stratify and have Secchi disk depths of 1.4 and 1.1 m, respectively (Horn and Haberyan 2016).

6.4 Lakes of the Northern Pacific Lowlands: Lakes Formed by a Debris Avalanche and Fluvial Activity

The northern Pacific lowlands include all areas below 500 m on the Pacific side of the northern Cordilleras, from the Nicaraguan border to Puerto Chomes, Santa Elena Peninsula, and all of the Nicoya Peninsula. Natural lakes in the northern Pacific lowlands are or were concentrated in two areas that constitute "lake districts" (Horne and Goldman 1994), which are areas with multiple lakes with

Fig. 6.7 Lakes formed by landslides in the southern highlands of Costa Rica. **a** Santa Elena Lake. **b** Zoncho Lake, while setting up platform to recover sediment cores (Photos by S. Horn)

the same origin that are valuable for comparative limnological and paleolimnological studies (Horn and Haberyan 2016). One is located on the SW slope of Miravalles volcano, where a debris avalanche about 8300 cal yr BP (Alvarado 2000; Siebert et al. 2006) left a large area of hummocky terrain occupied by more than a dozen lakes and wetlands. The second lake district, now much reduced, is in the lower Tempisque River basin and adjacent Bebedero River basin. In the 1970s, when the aerial photographs were taken that were then used to create the first 1:50,000 scale maps of the region, this area was dotted with lakes formed by fluvial activity. However, most of the lakes shown on these maps have been drained for agriculture. A number of artificial lakes also exist in the northern Pacific lowlands, including farm ponds.

Ages on organic macroremains in lake-sediment cores from the Miravalles lake district are consistent with the age of ca. 8300 cal yr BP reported by Siebert et al. (2006) for the debris avalanche, based on dating of material from geological exposures. The Miravalles lakes are thus among the oldest modern lakes of non-glacial origin in Costa Rica. They occupy depressions between 330 and 570 m asl in the undulating deposit of the debris avalanche, which Siebert et al. (2006) noted to be covered in some areas by pyroclastic-flow deposits, lava flows, and pumice-fall deposits. They range in size from less than one hectare to over 4 ha and are surrounded by pastures and remnant areas of lowland tropical dry forest. San Pablo Lake (Fig. 6.8), the largest of the Miravalles lakes at 4.4 ha, is an elongate, shallow, and turbid lake, with a maximum depth of 2.5 m and a Secchi

Fig. 6.8 Lakes of the Miravalles Lake District, occupying depressions in hummocky terrain formed by a debris avalanche from Miravalles volcano. **a** San Pablo Lake, the largest. **b** Estero Blanco Lake. **c** Martinez Lake, whose sediments contain the earliest botanical evidence of maize agriculture in Costa Rica (Photos by C. Lane and M. Arford)

disk depth of only 0.5 m. The lake is alkaline and has somewhat elevated levels of calcium, sodium, and silica, characteristics shared with other lakes of the district (Horn and Haberyan 2016). Sediment cores from Miravalles lakes contain evidence of changing vegetation and fire incidence related to both climate change and human impacts. Pollen grains of *Zea mays* subsp. *mays*, indicating maize agriculture, are present in late Holocene sediments of six lakes cored by Horn and students. Maize pollen in a core from Laguna Martínez (Fig. 6.8) constitutes the earliest botanical evidence of maize cultivation in Costa Rica, dated between 5590 and 5330 cal yr BP (Arford and Horn 2004).

Agricultural development has greatly modified the seasonal lakes that were originally abundant in the lower basins of the Tempisque and Bebedero rivers. One large lake that has been protected is Mata Redonda Lake, in the Mata Redonda Wildlife Refuge along the lower Tempisque River. Few limnological observations are available. The Taolinga

1:50,000 topographic map produced by the Instituto Geográfico Nacional from aerial photographs taken in 1971 shows an area of open water of about 260 ha, within a large marsh. A report from 1988 (MINAE/SINAC-UICN/ORMA 1988) indicated a maximum depth of 3 m during the wet season and described the site as severely transformed by canalization, agriculture, fire, and contamination.

Sasa et al. (2015) investigated the limnology of seasonal ponds and wetlands in the Lower Tempisque River basin during the period of infilling (June), maximum flooded area (September), and desiccation (January) in 2010–2011. They sampled four locations within Palo Verde National Park (Palo Verde Marsh and Pond, Bocana Pond, Laguna de Nicaragua Marsh) and Tamarindo Pond located on private land outside the park. The sites had high temperatures and transparency of less than 50 cm across all sampling intervals. Median values for pH, conductivity, total dissolved solids, and carbonates were higher during the dry season, while phosphates, nitrites, sulfates, and ammonium were higher during the wet season. The ranges of values reported by Sasa et al. (2015) are consistent with data from Solimar Lake, a permanent artificial lake in the lower basin of the Bebedero River, where Horn and Haberyan (2016) reported a water temperature of 32 °C, low Secchi disk depth (40 cm), and high alkalinity and conductivity.

6.5 Lakes of the Central Pacific Lowlands: Sparse Lakes Formed by Fluvial Activity

The central Pacific lowlands include all areas below 500 m on the Pacific side of the northern and southern highlands, from Puerto Chomes to the mouth of the Naranjo River. Lakes are rare in the central Pacific lowlands. Two oxbow lakes formed by fluvial activity were investigated by Haberyan et al. (2003). Carara Lake (Fig. 6.9) occupies a former channel of the Tárcoles River that was cut off by a change in its course. Limnological data were also collected from the nearby Madre Vieja Lake on the other side of the river. Since sampling in 1997, the Tárcoles River has shifted back into its former channel, destroying the lake. Meanwhile, channel migration downstream may create a new lake (Fig. 6.9). The recent destruction of Madre Vieja Lake and the channel cutoff that may lead to a new lake highlight the dynamic nature of fluvial lakes.

When sampled in 1997, the waters of Madre Vieja Lake and Carara Lake were slightly more concentrated in most ions, compared to other lakes; alkalinities were 202 and 169 mg L^{-1} (as $CaCO_3$), respectively (Horn and Haberyan, 2016). These lakes were turbid (Secchi disk depth less than 0.6 m) with low oxygen levels and elevated carbon dioxide levels, and Carara Lake had elevated sulfur levels (Horn and Haberyan 2016), suggesting extensive decomposition of organic matter, perhaps much of it entering during flood events.

The 1:50,000 topographic maps published by the Instituto Geográfico Nacional show a few additional lakes in the central Pacific region, such as Laguna Ortega near Parrita, and Laguna Sapo near the Jesús María River, 4 km inland of Tivives beach. However, Google Earth imagery suggests that these are no longer open-water lakes, likely because of changes in hydrology due to agricultural development of surrounding areas.

6.6 Lakes of the Southern Pacific Lowlands: Many Landslide Lakes

The southern Pacific lowlands include all areas below 500 m on the Pacific side of the Cordillera de Talamanca from Quepos to the border with Panama. Natural lakes formed by landslides, faulting, and fluvial and coastal processes are present in the southern Pacific lowlands. Vueltas Lake (Fig. 6.10) is a small lake formed by a block slide about 1750 cal yr BP (Horn and Haberyan 2016; Horn unpublished). Los Mangos (Fig. 6.10), Danta (Fig. 6.10), Carse, and Ojo de Agua lakes on the lower slopes of the Fila Costeña south of the General River (Horn et al. 2018) also formed from landslides; basal radiocarbon dates on sediment cores from the first three of these lakes indicate that Los Mangos Lake formed around 4200 cal yr BP (Johanson et al. 2019), while Danta and Carse lakes formed about 800 cal yr BP and 350 cal yr BP, respectively (Johanson et al. 2020). Redonda Lake and Larga Lake (Fig. 6.10) in the town of Buenos Aires in the El General Valley owe their origin to the uplift of the Fila Costeña. This uplift created lakes by reversing the flow of some streams at the distal ends of alluvial fans originating from the Cordillera de Talamanca (Kesel 1983). At least a half dozen lakes existed in the area before the spread of pineapple plantations starting in the 1960s (Artavia and Àvalos 2020), but today only these two may remain, both choked with rooted and floating aquatic plants. Initial radiocarbon dating indicated that they both formed in the late Pleistocene, but may not have accumulated sediments continuously from that time forward (Horn, unpublished). The much larger Sierpe Lake, an open-water area within the extensive Térraba-Sierpe wetland along the southern Pacific coast (Umaña et al. 2015), likely owes its origin to a combination of fluvial and coastal processes (Acuña-Piedra and Quesada-Román 2021).

The landslide lakes have variable morphologies, with depth ranging from shallow to deeper (0.5 m for Los Mangos Lake, up to about 5 m for Danta and Carse lakes),

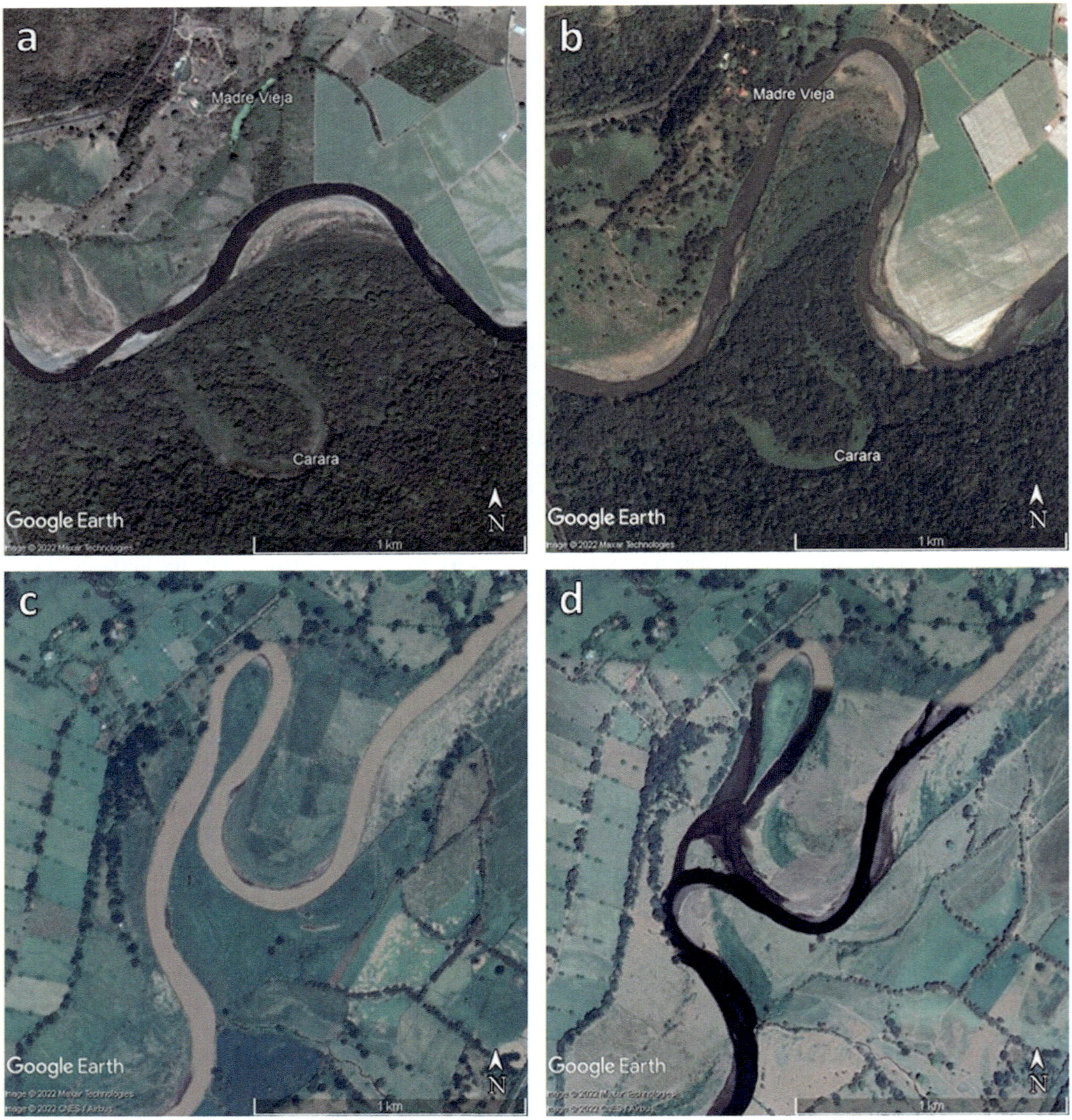

Fig. 6.9 Fluvial lakes of the Central Pacific Lowlands of Costa Rica. **a** Google Earth image from March 2002, showing two lakes occupying former channels of the Tárcoles River, Madre Vieja Lake on the north side of the river, and Carara Lake on the south side. **b** Image from 2014, showing that the river has reoccupied its former channel, destroying Madre Vieja Lake. **c** 2 km downstream, a meander bend narrows from March 2002 to July 2017, when this image was captured. **d** Image from January 2018, following a cutoff that could form a new lake

and these lakes are also quite variable in area (0.2 ha for Ojo de Agua Lake, up to 103 ha for Sierpe Lake). Thermal stratification has been observed in Vueltas and Danta lakes, as well as in Sierpe Lake, despite its relatively shallow depth and extensive surface area. Alkalinities range from 48 to 213 mg L^{-1} (as $CaCO_3$), and pH ranges from circumneutral to somewhat basic (7.1–8.7) (Haberyan et al. 2003; Horn et al. 2018).

Fig. 6.10 Lakes in the southern Pacific lowlands of Costa Rica. **a** Vueltas Lake. **b** Mangos Lake. **c** Danta Lake—all of which are associated with landslides. **d** Larga Lake, in Buenos Aires, which owes its origin to the uplift of the Fila Costeña, which reversed the flow of streams draining alluvial fans from the Talamanca Mountains (Photos by S. Horn)

6.7 Lakes of the Caribbean Lowlands: Of Diverse Origin

The Caribbean lowlands of Costa Rica include all areas below 500 m on the Caribbean side of the cordilleras, from the border with Nicaragua to the border with Panama. The Caribbean lowlands support natural lakes that occupy basins formed by volcanic eruptions, fluvial processes, coastal processes, and landslides. The limnology and sediments of several of these lakes have been investigated.

The deepest natural lake in the country, Río Cuarto Lake (Fig. 6.11), occupies a volcanic explosion crater or maar 6.5 km north of the Hule maar. There are no radiocarbon ages for the formation of the Río Cuarto maar, but Alvarado et al. (2011) estimated an age of 3000–4000 years based on archeological evidence. The low crater walls (< 30 m high) support a narrow ring of remnant tropical wet forest, beyond which the surrounding plain supports cattle pastures and agricultural fields. The lake watershed is only slightly larger than the lake itself, with a few small tributaries arising on the inner crater walls.

Río Cuarto Lake is bowl-shaped, with steep slopes near its shores (Gocke et al. 1987; Fig. 6.4). The great depth of the lake causes its surface waters to be low in nutrients and rather clear (transparency up to 6.1 m); the lake's deepest waters are somewhat elevated in nutrients and are lower in pH (Haberyan et al. 2003). These differences relate to the morphology of the lake basin, which prevents deep mixing every year. Instead, the lake is meromictic, with an upper layer that mixes to a depth of 20 m at least once a year (Umaña and Farah 2018). During these events, transparency drops to < 1 m (Haberyan et al. 2003), and oxygen levels in the upper water column drop because of the upwelling of sulfur-laden water from deeper layers, but mixing does not extend beyond 20 m depth. A three-year program of sampling by Umaña and Farah (2018) showed that the lake has kept its meromictic condition, and that temperatures in the deepest waters have been relatively constant through time. Because of this condition, the lake harbors a dense population of photosynthetic anoxigenic sulfur bacteria (Cabassi et al. 2014) and the chlorophyll maxima is located just below the thermocline at 20–25 m (Umaña and Farah 2018).

Fig. 6.11 Two lakes in the Caribbean Lowlands. **a** Rio Cuarto Lake, which is the deepest natural lake in Costa Rica (66 m). **b** Bonillita Lake, which lies on a terrace of the Reventazón River (Photos by G. Umaña)

Bonilla and Bonillita lakes (Fig. 6.11) are located along the Reventazón River a few km above the upper reaches of the Reventazón reservoir created by the largest hydroelectric dam in Central America, completed in 2016. Bonilla and Bonillita lakes are on fluvial terraces of the Reventazón River and may owe their origin to a combination of landslides, volcanic mudflows (lahars), and stream dynamics. Radiocarbon dates on sediment cores indicate that Bonillita formed about 2700 cal yr BP, but dates on cores from Bonilla are unreliable due to the presence of old carbon in the lake (Northrop and Horn 1996). Nearby lakes include Palo Alto Lake and Azul Lake, near the Guayabo Archeological Site. Palo Alto Lake is located in the upper end of a massive (4 km^2) area of slope instability and mass movement known as the Palo Alto Landslide (Peraldo et al. 2015). The age of the lake is uncertain. Azul Lake formed before 600 cal yr BP, based on an initial test core. Unfortunately, the lake was destroyed during the construction of a small hydroelectric plant (Horn 2006), before a full core could be recovered. Downstream from Bonilla Lake, on the opposite side of the Reventazón River, are the two Lancaster lakes, Lancaster Arriba, and Lancaster Abajo. Dates on sediment cores suggest that both lakes formed in the last 600 years (Umaña et al. 1999; Horn, unpublished), probably in depressions in an irregular landslide body.

The limnology of Bonilla, Bonillita, and Lancaster Abajo lakes is similar in that the lakes are larger (5.0–30.8 ha), rather deep (16.9–27.0 m), and are known to stratify. Alkalinities range from 23 to 97 mg $L-^{1}$ (as $CaCO_3$)

and pH values are circumneutral (6.9–7.2). Lancaster Arriba Lake is shallower (5.6 m). Repeated observations of Bonilla Lake in 1990–1991 (Umaña 1997), 2009–2010 (Umaña 2016), and 2016 (Umaña, unpublished) have shown that the lake mixes irregularly, often showing a thermocline that oscillates between 10 and 15 m deep, and anoxic conditions below 10 m. It accumulates H_2S in its deepest waters and usually shows a peak in chlorophyll at intermediate depths (10–15 m), most probably due to the presence of sulfur bacteria at that depth. The lake can overturn once a year between December and February. In 2009–2010, the whole lake became almost oxygen-deficient during one such event (Umaña 2016) because of the surge of anoxic waters laden with hydrogen sulfide. At the moment, it was completely isothermal, and chlorophyll levels were nearly uniform throughout the water column. The lake stratified back again in the course of a few weeks.

Zent Lake, near the town of the same name, is one of several lakes of fluvial origin on the lower Caribbean slope, with current limnological conditions that appear greatly affected by surrounding banana plantations (Horn and Haberyan 2016). The largest fluvial lakes, on the San Carlos Plain near the border with Nicaragua, were visited only once by one of us (Umaña) in 2004, during the development of a project of the Costa Rican fisheries and aquaculture institution (PREPAC 2005). Those lakes are less than 9 m deep, with low values of Secchi transparency (less than 3 m), and warm, with temperatures above 25 °C, but usually have high oxygen levels even at the bottom. Some sites are known as "lagunas," but they are part of the Tortuguero channel system, like Penitencia, Sierrpe, Caldera, Jalova, and Tortuguero. These water bodies have high conductivities at the surface and slightly saline waters at the bottom, with bottom waters that may be warmer than the surface because of a seepage of salt water from the adjacent Caribbean Sea through the sand bars that separate them from the sea (Quesada-Román and Pérez-Briceño 2019). Most of these lakes are eutrophic, with the exception of En Medio, De Atrás, and La Nueve lakes, in the Barra Colorado region, perhaps because their shape and position protect them from the entrance of water from the main Colorado River.

Gandoca Lake in the Gandoca-Manzanillo Wildlife Refuge near the border with Panama is a sinuous lake that occupies a former channel of the Río Sixaola (Denyer 1998) and is dammed at its seaward end by a sandbar. As in the lakes previously mentioned, this sandbar allows sea water to percolate into the basin while the upper waters of the lake are fresh; the chemocline lies at about 3.5 m. Waters of the epilimnion and hypolimnion are distinct, for example in conductivity (266 vs. 41,100 $\mu S\ cm^{-1}$) and in chloride content (48 vs. 15,400 mg $L-^{1}$) (Haberyan et al. 2003).

6.8 The Dynamic Nature of Costa Rican Lakes and Hopes for the Future

Natural processes and human activity have created a diversity of lakes in Costa Rica, from small ponds created by landslides to enormous reservoirs that provide hydropower, water, recreation, and aquatic habitats that support foodwebs and biodiversity (Veas et al. 2023). Natural and human processes also change and eliminate lakes. Natural change can occur slowly over time, with the natural filling of lakes by material produced within the lake or delivered by runoff or wind. From our work with lake-sediment cores, we know that this can take millennia, even in lake basins with prehistoric and modern agriculture in which we would expect enhanced erosion in watersheds (Horn 2007; Johanson et al. 2019; Kerr et al. 2020; Taylor et al. 2020). Natural change can also occur nearly instantaneously, as in the destruction of Madre Vieja Lake when the Tárcoles River shifted to occupy a former channel, or in the burial of a lake by a landslide, as happened in the case of Arancibita Lake during the major landslide of 2000 (Alvarado et al. 2003; Horn and Haberyan 2016). Human destruction of lakes by draining also tends to occur quickly, to remove water for agricultural or other purposes the landowner has in mind (Veas et al. 2023). Human-induced changes in the flow of nutrients and sediments into lakes and in climate that could greatly alter lakes, or even remove them from the landscape entirely, may already be occurring. Natural climate variability and change have affected Costa Rican lakes in the past, as documented, for example by evidence of lower lake stands associated with past intervals of drier climate (Johanson et al. 2019; Wu et al. 2019). Several lakes and wetlands in Costa Rica have experienced increases in floating aquatic plants such as water hyacinth (*Eichhornea crassipes*) and in rooted macrophytes such as *Typha dominguensis* in recent decades (Horn and Kennedy 2006; Sasa et al. 2015; Horn and Haberyan 2016, Horn, pers. obs.) that may in part reflect changes in nutrient levels, possibly associated with global shifts in the nitrogen cycle or to land uses in watersheds that have increased levels of nitrogen and other nutrients that increase plant growth (Elser et al. 2007). The causes of these increases in floating and rooted macrophytes may vary by site and deserve attention. Increased macrophyte cover can affect lake temperatures, which may also be changing because of global warming. In sporadic measurements in 1998, 2000, and 2003, we found higher temperatures in lakes on Cerro Chirripó than recorded by previous researchers, suggesting possible changes in air temperatures or in daytime or nighttime cloudiness, which also affects water temperatures (Horn and Haberyan 2016). However, securely documenting temperatures in these glacial lakes and others requires systematic, long-term monitoring, which would best be accomplished by deploying arrays of water temperature sensors and data loggers.

Lakes in Costa Rica give character and beauty to landscapes, and create opportunities for reflection and recreation that are enjoyed by Costa Ricans and visitors from around the world. Lakes also provide key ecosystem services and contribute to aquatic and terrestrial biodiversity. Natural and human processes create, modify, and destroy lakes and in so doing change landscapes, ecosystems, ecosystem services, and biodiversity. Many natural lakes in Costa Rica have been drained in the last 50 years, and climate change may be slowly changing other lakes, perhaps in undesirable ways. Our hope for the future is that diverse and beautiful lakes will continue to be available for Costa Ricans and visitors to experience, value, and preserve.

Acknowledgements For their support of our research on lakes and lake histories in Costa Rica, we thank the University of Costa Rica, the University of Tennessee, Northwest Missouri State University, the National Geographic Society, The A.W. Mellon Foundation, the American Association of Geographers, and the U.S. National Science Foundation (grants 9111588, 9512484, 0242286, 0538420, 0825406, 1600376, 1657170, 1660185). We also thank the Costa Rican Ministry of Environment and Energy for permission to sample lakes and local landowners and land managers for access to lakes. Many students and colleagues have worked with us across the years, and we are grateful for their collaboration and friendship.

References

Acuña-Piedra JF, Quesada-Román A (2021) Multidecadal biogeomorphic dynamics of a deltaic mangrove forest in Costa Rica. Ocean Coast Manage 211:105770

Alfaro A, Denyer P, Alvarado GE, Gazel E, Chamorro C (2018) Estratigrafía y petrografía de las rocas ígneas en la Cordillera de Talamanca, Costa Rica. Rev Geol Am Cent 58:7–36

Alvarado GE (2000) Los volcanes de Costa Rica: geología, historia y riqueza natural, 2nd edn. Editorial Universidad Estatal y Distancia, San José, Costa Rica, p 269

Alvarado GE, Matumoto T, Borgia A, Barquero R (1988) Síntesis geovulcanológico del Arenal (Costa Rica): 20 años de continua actividad eruptiva (1968–1988). Boletín Observario Vulcanológico De Arenal 1(1):1–61

Alvarado GE, Mora R, Peraldo G (2003) The June 2000 Arancibia debris avalanche and block slide. Landslide News 14(15):29–32

Alvarado GE, Soto GJ, Sanai FM et al (2011) The formation and evolution of Hule and Río Cuarto maars, Costa Rica. J Volcanol Geoth Res 201:342–356

Arford MR, Horn SP (2004) Pollen evidence of the earliest maize agriculture in Costa Rica. J Latin Am Geogr 3(1):108–115

Artavia G, Àvalos G (2020) Historia natural del ecosistema de sabana del Valle del Río General, Costa Rica. Revista Geográfica 161:11–28

Baker RGE (1987) Notes on the high altitude freshwater lakes of Volcán Rincón de la Vieja National Park, northern Costa Rica. Brenesia 27:47–54

Bolz A, Calvo C (2018) Carbonate platform development in an intraoceanic arc setting: Costa Rica's largest limestone sequence-the fila de cal formation (Middle Eocene to Lower Oligocene). Rev Geol Am Cent 58:85–114

Borgia A, Poore C, Carr MJ, Melson WG, Alvarado GE (1988) Structural, stratigraphic, and peterologic aspects of the Arenal-Chato volcanic system, Costa Rica: evolution of a young stratovolcanic comple6. Bull Volcanol 50:86–105

Borgia A, Burr J, Montero W, Morales LD, Alvarado GE (1990) Fault-propagation folds induced by gravitational failure and slumping of the Costa Rica volcanic range: implications for large terrestrial and Martian volcanic edifices. J Geophys Res 95(B9):14,357–14,382

Cabassi J, Tassi F, Mapelli F, Borin S, Calabrese S, Rouwet D, Chiodini G, Marasco R, Chouaia B, Avino R, Vaselli O, Pecoraino G, Capecchiacci F, Bicocchi G, Caliro S, Ramirez C, Mora-Amador R (2014) Geosphere-biosphere interactions in bio-activity volcanic lakes: evidences from Hule and Rio Cuarto (Costa Rica). PLoS ONE 9:e102456

Cunningham MT, Stark CP, Kaplan MR, Schaefer JM (2019) Glacial limitation of tropical mountain height. Earth Surf Dynam 7:147–169

Denyer P (1998) Historic-prehistoric earthquakes, seismic hazards, and Tertiary and Quaternary geology of the Gandoca-Manzanillo National Wildlife Refuge, Límon, Costa Rica. Rev Biol Trop 46(suppl 6):237–250

Elser JJ, Bracken MES, Cleland EE, Gruner DS, Harpole WS, Hillebrand H, Ngai JT, Seabloom EW, Shurin JB, Smith JE (2007) Global analysis of nitrogen and phosphorous limitation of primary producers in freshwater, marine and terrestrial ecosystems. Ecol Lett 10:1135–1142

Esquivel-Hernández G, Sánchez-Murillo R, Quesada-Román A, Mosquera GM, Birkel C, Boll J (2018) Insight into the stable isotopic composition of glacial lakes in a tropical alpine ecosystem: Chirripó, Costa Rica. Hydrol Process 32:3588–3603

Gocke K (1996) Basic morphometric and limnological properties of Laguna Hule, a caldera lake in Costa Rica. Rev Biol Trop 44(45):537–548

Gocke K, Lehman E, Rojas G, Romero R (1981) Morphometric and basic limnological data of Laguna Grande de Chirripó, Costa Rica. Rev Biol Trop 29:165–174

Gocke K, Bussing W, Cortés J (1987) Morphometric and basic limnological properties of the Laguna de Río Cuarto, Costa Rica. Rev Biol Trop 35:277–285

González Viquez C (1910) Temblores, terremotos, inundaciones y erupciones volcanicas en Costa Rica 1608–1910. Tipografía de Avelino Alsina, p 239

Haberyan KA, Horn SP (1999) Chemical and physical characteristics of seven volcanic lakes in Costa Rica. Brenesia 51:85–95

Haberyan KA, Horn SP (1999) A 10,000 year diatom record from a glacial lake in Costa Rica. Mt Res Dev 19:63–70

Haberyan KA, Horn SP (2020) Discovery of an undescribed species of *Aulacoseira* from highland Costa Rica. Diatom Res 35:289–300

Haberyan KA, Horn SP, Umaña G (2003) Basic limnology of fifty-one lakes in Costa Rica. Rev Biol Trop 51:107–122

Hastenrath S (1973) On the Pleistocene glaciation of the Cordillera de Talamanca, Costa Rica. Zeits Gletscherkunde Glazialgeol 9:105–121

Horn SP (1989) The Inter-American Highway and human disturbance of páramo vegetation in Costa Rica. Yearbook Conf Latin Am Geog 15:13–22

Horn SP (2001) The age of the laguna Hule explosion crater, Costa Rica, and the timing of subsequent eruptions: evidence from lake sediments. Rev Geol Am Cent 24:57–66

Horn SP (2006) Pre-Columbian maize agriculture in Costa Rica: pollen and other evidence from lake and swamp sediments. In: Staller J, Tykot R, Benz B (eds) Histories of maize: multidisciplinary approaches to the prehistory, biogeography, domestication, and evolution of maize. Elsevier, San Diego, pp 367–380

Horn SP (2007) Late Quaternary lake and swamp sediments: recorders of climate and environment. In: Bundschuh J, Alvarado GE (eds) Central America: geology, resources, hazards, vol 1. Taylor & Francis/Balkema. Leiden, Netherlands, pp 427–445
Horn SP, Haberyan KA (1993) Costa Rican lakes: chemical and physical properties. Natl Geogr Res Explor 9:86–103
Horn SP, Kennedy LM (2006) Pollen evidence of the prehistoric presence of Cattail (*Typha*: Typhaceae) in Palo Verde National Park, Costa Rica. Brenesia 66:85–87
Horn SP, Haberyan KA (2016) Lakes of Costa Rica. In: Kappelle M (ed) Costa Rican ecosystems. University of Chicago Press, Chicago, pp 656–682
Horn SP, Orvis KH, Haberyan KA (1999) Investigación limnológica y geomorfológica de lagos glaciares del Parque Nacional Chirripó, Costa Rica. Rev Informe Semestral [Instituto Geográfico Nacional, Costa Rica] 35:95–106
Horn SP, Orvis KH, Haberyan KA (2005) Limnología de las lagunas glaciales en el Páramo del Chirripó, Costa Rica. In Kappelle M, Horn SP (eds) Páramos de Costa Rica. Instituto Nacional de Biodiversidad, Santo Domingo de Heredia, Costa Rica, pp 161–181
Horn SP, Johanson EN, Haberyan KA et al (2018) Initial limnological observations at five small lakes in southern Pacific Costa Rica. UNED Res J 10:119–134
Horne AJ, Goldman CR (1994) Limnology, 2nd edn. McGraw Hill, New York, p 480
Hutchinson GE (1975) A Treatise on Limnology vol 1 part 1. Geography and Physics of Lakes. Wiley, New York, p 540
Johanson EN, Horn SP, Lane CS (2019) Pre-Columbian agriculture, fire, and Spanish contact: a 4200-year record from Laguna Los Mangos, Costa Rica. Holocene 29(11):1743–1757
Johanson EN, Horn SP, Lane CS, Sánchez M, Cecil J (2020) Fire history across the Little Ice Age in southern Pacific Costa Rica. J Paleolimnol 64:1–23
Jones JR, Lohman K, Umaña G (1993) Water chemistry and trophic state of eight lakes in Costa Rica. Verh Internat Verein Limnol 25:899–905
Kappelle M, Horn SP (2016) The páramo grasslands of Costa Rica's highlands. In: Kappelle M (ed) Costa Rican ecosystems. Univ of Chicago, Chicago, pp 492–523
Kerr MT, Horn SP, Lane CS (2020) Stable isotope analysis of vegetation history and land-use change at Laguna Santa Elena in southern Pacific Costa Rica. Veg History Archaeobot 29:477–492
Kesel RH (1983) Quaternary history of Río General Valley, Costa Rica. Natl Geogr Res Repts 15:339–358
Lachniet MS, Seltzer GO (2002) Late Quaternary glaciation of Costa Rica. Bull Geol Soc Am 114(5):547–558
Li Y, Tieche T, Horn SP, Li Y, Chen R, Orvis KH (2019) Mapping glacial landforms on the Chirripó massif, Costa Rica, based on Google Earth, a digital elevation model, and field observations. Rev Geol Am Cent 60:109–121
Löffler H (1972) Contribution to the limnology of high mountain lakes in Central America. Internat Rev Gesamten Hydrobiol 57(3):397–408
Malavassi E, Gill JB, Trimble D (1990) Nuevas dataciones radiométricas del alineamiento Volcánico de Poás (Costa Rica): contribución a la evaluación de peligros volcánicos. 7th Congreso Geológico de América Central, San José, Costa Rica, November 1990. Programa y Resúmenes de Ponencias p 37
Marshall JS (2007) The geomorphology and physiographic provinces of Central America. In: Bundschuh J, Alvarado GE (eds) Central America: geology, resources, hazards, vol 1. Taylor & Francis/Balkema. Leiden, Netherlands, pp 75–122
MINAE/SINAC-UICN/ORMA (1988) Inventario de los Humedales de Costa Rica. Córdoba R, Romero J, Windevoxhel J (eds) MINAE/SINAC/UICN/ORMA. San José, Costa Rica, p 380
Northrop LA, Horn SP (1996) PreColumbian agriculture and forest disturbance in Costa Rica: palaeoecological evidence from two lowland rainforest lakes. Holocene 6:289–299
Orvis KH, Horn SP (2000) Quaternary glaciers and climate on Cerro Chirripó, Costa Rica. Quaternary Res 54:24–37
Peraldo Huertas G, Badilla Coto E, Sota Acosta D, González Artavia M, Herra Herrera D, Ramos García J, Méndez Solórzano H, Ceciliano Chacón A, Vega Arce AE, van der Laan E (2015) El deslizamiento de Palo Alto, Turrialba, Costa Rica: apuntes para su estudio. Rev Entorno Prevención 15:7–23
Potter R, Li Y, Horn SP, Orvis KH (2019) Cosmogenic Cl-36 surface exposure dating of late Quaternary glacial events in the Cordillera de Talamanca, Costa Rica. Quaternary Res 92:216–231
PREPAC (2005) Inventario de cuerpos de agua continentales de Costa Rica con énfasis en la pesca y la acuacultura. Plan Regional de Pesca y Acuicultura Continental–PREPAC. OSPESCA/TAIWAN/OIRSA. INCOPESCA, San José, Costa Rica, p 767
Quesada-Román A, Pérez-Briceño PM (2019) Geomorphology of the Caribbean coast of Costa Rica. J Maps 15:363–371
Quesada-Román A, Pérez-Umaña D (2020) Tropical paleoglacial geoheritage inventory for geotourism management of Chirripó National Park, Costa Rica. Geoheritage 12:12–58
Quesada-Román A, Ballesteros-Cánovas JA, Stoffel M, Zamorano-Orozco JJ (2019) Glacial geomorphology of the Chirripó National Park, Costa Rica. J Maps 15:538–545
Quesada-Román A, Campos N, Alcalá-Reygosa J, Granados-Bolaños S (2020) Equilibrium-line altitude and temperature reconstructions during the last glacial maximum in Chirripó National Park, Costa Rica. J S Am Earth Sci 100:102576
Reimer RW, Reimer PJ (2022) CALIBomb http://calib.org. Accessed 10 Oct 2022
Ruiz P, Mana S, Gazel E, Soto GJ, Carr MJ, Alvarado GE (2019) Geochemical and geochronological characterization of the Poas stratovolcano stratigraphy. In: Tassi F, Vaselli O, Mora Amador RA (eds) Poás Volcano: the pulsing heart of the Central America volcanic zone, Springer Nature, Switzerland, pp 13–43
Salani FM, Alvarado GE (2010) El maar poligenético de Hule (Costa Rica): revisión de su estratigrafía y edades. Rev Geol Am Cent 43:97–118
Sasa M, Armengol X, Bonilla F, Mesquita-Joanes F, Piculo R, Rojo C, Rueda RM, Monrós JS (2015) Seasonal wetlands in the Pacific coast of Costa Rica and Nicaragua: environmental characterisation and conservation state. Limnetica 34(1):1–16
Siebert L, Alvarado GE, Vallance JW, van Wyk de Vries B (2006) Large-volume volcanic edifice failures in Central America and associated hazards. In: Rose WI, Bluth GJS, Carr MJ et al (eds), Volcanic hazards in Central America, Geological Society of America Special Paper 412, pp 1–26
Standley PC (1936–1937) Flora of Costa Rica. Field Museum Nat Hist Bot Ser 18(1–4):1–1571
Stuiver M, Reimer PJ (1993) Extended C-14 database and revised Calib 3.0 C-14 age calibration program. Radiocarbon 35:215–230
Tassi F, Vaselli O, Fernández E, Duarte E, Martinez M, Delgado Huertas A, Bergamaschi F (2009) Morphological and geochemical features of crater lakes in Costa Rica: an overview. J Limnol 68:193–205
Taylor ZP, Lane CS, Horn SP (2020) A 3600-year record of drought in southern Pacific Costa Rica. Quater Res 98:36–48
Titiz B, Sanford RL Jr (2007) Soil charcoal in old-growth rain forests from sea level to the continental divide. Biotropica 39(6):673–682
Umaña G (1997) Basic limnology of Lago Bonilla, a tropical lowland lake. Rev Biol Trop 45:1429–1437
Umaña G (2010) Comparison of basic limnological aspects of some crater lakes in the Cordillera Volcánica Central, Costa Rica. Rev Geol Am Cent 43:137–145

Umaña G (2014) Ten years of limnological monitoring of a modified natural lake in the tropics: Cote Lake, Costa Rica. Rev Biol Trop 62:567–578
Umaña VG (2016) One year weekly dynamics of limnological conditions and phytoplankton in Lake Bonilla, Costa Rica. Rev Biol Trop 64(4):1771–1781
Umaña Villalobos G (2006) Ciclo annual de estratificación y circulación en el Embalse Arenal. Costa Rica. Rev Biol Trop 54(Suppl 1):257–263
Umaña-Villalobos G (2014) Phytoplankton variability in Lake Fraijanes, Costa Rica, in response to local weather variation. Rev Biol Trop 62:483–494
Umaña V. G, Haberyan K, Horn S (1999) Limnology in Costa Rica. In: Wetzel RG, Gopal B (eds) Limnology in developing countries, vol 2, pp 33–62
Umaña Villalobos G, Acuña González J, García-Céspedes J, Agüero Alfaro G, Zúñiga Peña JC, Gómez Ramírez E, Picado Borboza J (2015) Productividad primaria en el Humedal Nacional Térraba-Sierpe, Costa Rica. Rev Biol Trop 63(Suppl 1):9–28
Umaña-Villalobos G, Farah-Pérez A (2018) Revisiting the limnology of Lake Río Cuarto, Costa Rica, thirty-five years later. Rev Biol Trop 66(Suppl 1):S42–S52
Umaña-Villalobos G, Avilés-Vargas L (2020) Plankton variations in Lake Cote, Costa Rica, from 2002 to 2018. Hydrobiologia 847:4177–4190
Veas-Ayala N, Quesada-Román A, Hidalgo HG, Alfaro EJ (2018) Humedales del Parque Nacional Chirripó, Costa Rica: características, relaciones geomorfológicas y escenarios de cambio climático. Rev Biol Trop 66(4):1436–1448
Veas-Ayala N, Alfaro-Córdoba M, Quesada-Román A (2023) Costa Rican wetlands vulnerability inde6. Prog Phys Geog 47:521–540
Weber H (1959) Los páramos de Costa Rica y su concatenación fitogeográfico con los Andes Suramericanos. Instituto Geográfico Nacional, San José, Costa Rica, p 67
Wu J, Porinchu DF, Campbell NL, Mordecai TM, Alden EC (2019) Holocene hydroclimate and environmental change inferred from a high-resolution multi-proxy record from Lago Ditkebi, Chirripó National Park, Costa Rica. Palaeogeogr Palaeocl Palaeoecol 518:172–186
Wunsch O, Calvo G, Willscher B, Seyfried H (1999) Geologie der alpinen Zone des Chirripo-Massives (Cordillera de Talamanca, Costa Rica, Mittelamerika). Profil 16:193–210
Zapata MI, Soto GJ (1990) Lahares antiguos y recientes en el Volcán Arenal, Costa Rica. Boletín Observatorio Vulcanológico De Arenal 3(5):1–28

Main Karstic Landscapes and Caves of Costa Rica

7

Andrés Ulloa, Mariángela Vargas, Alejandro Argüello, Yoselyn Álvarez, Andrea Hidalgo, and Stanimira Deleva

Abstract

Several carbonate rocks present in the geological record of Costa Rica host the karst landscapes of the country. Up to 450 caves have been reported, of which 74% occur in carbonate rocks. This paper presents updated information on the main karstic areas of Costa Rica, with exhaustive information on four karst landscapes located in Tempisque, Zona Sur, Venado of San Carlos, and Limón. Tempisque karstlands host the Barra Honda National Park, the only National Park intended for cave and karst protection in the country. This park exhibits several caves (mostly with vertical passages) and surface karst expressions such as dolines, mogotes, travertine cascades, and karren fields. The Terciopelo cave and the Cuevita are the only ones used for touristic activity. Zona Sur contains the most extensive karstic landscape in the country, with carbonate rock outcrops of about 100 km along Fila Costeña. This area is developed mainly in highly tectonized Eocene limestone. Furthermore, Zona Sur contains the most remarkable dolines in the country and several sinking streams that connect with cave systems. The Corredor River basin includes a cave system of almost 3 km long of surveyed passages. Another important karstic area is Venado of San Carlos, which hosts a karst landscape developed in Miocene carbonate rocks. This region is tectonically affected by Venado Transtensive System, which is a fault system that limits the limestone outcrops, and influences cave patterns. Limestone is overlain by younger volcanic deposits, making superficial karstic expression not very evident. Nevertheless, this area hosts the most extensive cave system known in the country, which consists of three caves that in total have almost 5 km of surveyed passages. Finally, Limón karstlands occur in Pliocene–Pleistocene reef-carbonate rocks in a seismically active area. A LiDAR survey in Limón revealed hundreds of dolines and conical karst as the most remarkable surface karst expressions.

Keywords

Costa Rica · Karst · Carbonate rocks · Karstlands · Limestones · Caves · Karst aquifer · Tempisque · Zona Sur · Venado of San Carlos · Limón

7.1 Introduction

Costa Rica has an interoceanic and intercontinental location in Central America. The country originated as an island arc due to the subduction process of the Cocos Plate under the Caribbean Plate since the Upper Cretaceous. At least, nine episodes of carbonate deposition were documented from the Late Cretaceous to the Holocene, including several geological formations. Ulloa et al. (2011) reviewed the limestones of Costa Rica and related karst areas, with a detailed description of each karst locality. Chesnel and Rodríguez (2021) offered the most recent geological update for carbonate deposits from the Oligocene and Miocene in the Zona Sur of Costa Rica.

A. Ulloa (✉)
Centro de Investigación en Ciencias Geológicas, Universidad de Costa Rica, San José, Costa Rica
e-mail: andres.ulloa@ucr.ac.cr

A. Ulloa · A. Argüello · Y. Álvarez · A. Hidalgo
Escuela Centroamericana de Geología, Universidad de Costa Rica, San José, Costa Rica

M. Vargas
Instituto Clodomiro Picado, Universidad de Costa Rica, Vázquez de Coronado, Costa Rica

S. Deleva
Sede del Sur, Universidad de Costa Rica, Golfito, Costa Rica

National Museum of Natural History-Bulgarian Academy of Sciences, Sofia, Bulgaria

A. Quesada-Román (ed.), *Landscapes and Landforms of Costa Rica*,
World Geomorphological Landscapes, https://doi.org/10.1007/978-3-031-64940-0_7

Carbonated rocks in Costa Rica cover a surface of ~430 km^2, which is less than 1% of its territory (Ulloa et al. 2011). Thus, there are no extensive carbonate platforms as in some neighboring Central American countries such as Guatemala, Nicaragua, Honduras, and Belize, which have karst areas ranging between 5000 and 1,5000 km^2 (Day 2007). Fila Costeña, located in Zona Sur of Costa Rica, contains the most extensive outcrops of limestone in the country, where tilted carbonate rocks crop out for more than 100 km along the range (Mora 1979; Denyer and Alvarado 2007; Chesnel and Rodríguez 2021). Despite its modest size, the Costa Rican karst landscape is represented by various karst forms, including dolines, dry valleys, karst springs, travertine waterfalls, mogotes, and extensive cave systems. The karst lands of Costa Rica are shown in Fig. 7.1 and were delimited based on a regional geological map by Denyer and Alvarado (2007). Mora (1986, 1987) briefly described karstic zones for the national territory according to the limestone age. Ulloa et al. (2011) defined ten karstic zones: (1) Pacífico Norte, (2) Tempisque, (3) Zona Sur, (4) Pacífico Central, (5) Cabo Blanco and Mal País, (6) Turrialba and Pacuare, (7) Talamanca, (8) Venado, (9) Valle Central, and (10) Limón.

Caves are one of the most fascinating karst features, and they serve as "windows through which we can observe and study the subterranean world" (Racovitza 1907). Since prehistory, caves have attracted people's interest as shelters, sacred sites, or artistic expression sites (Goicoechea 2015). However, they constitute a challenge for scientific research because the exploration of caves requires specialized techniques. Most of the caves do not appear on topographic maps, aerial or satellite images, making cave research a priority for a small number of highly specialized researchers (Kambesis 2007). Karstic caves are the most studied and documented karstic elements in Costa Rica. A national database of caves has been carried out by Grupo Espeleológico Anthros (GEA). The information in this database was reviewed, updated, and refined in a new database, for which an exhaustive bibliographic search was carried out. This paper presents a review of the main karstic regions and caves of Costa Rica. In the case of Zona Sur, Tempisque, and Venado of San Carlos, they were selected because they are well-documented based on their karstic phenomena abundance. Limón, on the Caribbean side, has an extensive reef-carbonate platform in which multiple dolines, karst hills, and caves are found. Also, Limón has relevance because it is the most densely populated karst region of Costa Rica.

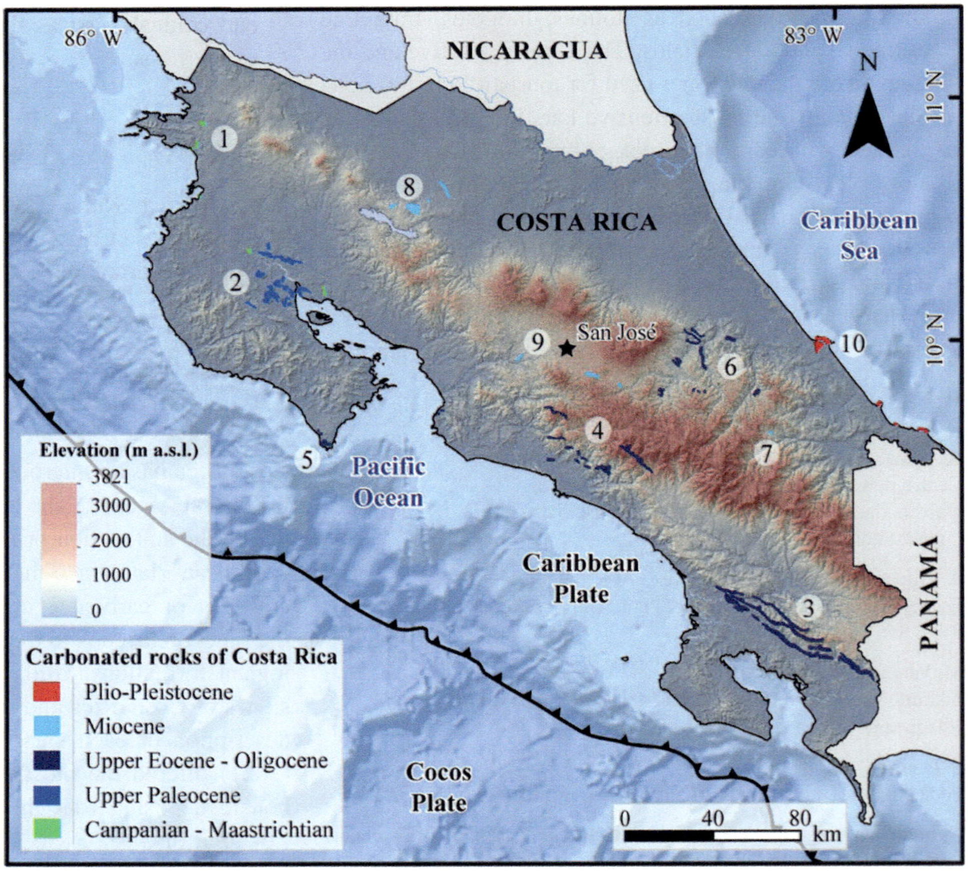

Fig. 7.1 Map of Costa Rica showing a relief model and the main limestone rocks, classified based on the geological age. (1) Pacífico Norte, (2) Tempisque, (3) Zona Sur, (4) Pacífico Central, (5) Cabo Blanco and Mal País, (6) Turrialba and Pacuare, (7) Talamanca, (8) Venado, (9) Valle Central, and (10) Limón. The limestone outcrops are based on the regional geological map from Denyer and Alvarado (2007)

7.2 Brief History of Cave Exploration in Costa Rica

Reports of caves in Costa Rica have appeared since the beginning of the sixteenth century; however, it was in the 1960s when the exploration of Costa Rica's caves began with the creation of Grupo Espeleológico (Goicoechea et al. 2009). One of the first karstic areas to be studied in Costa Rica was the Cerros de Barra Honda, where the discovery of caves and karst (Wells 1974) encouraged the creation of the Barra Honda National Park in 1974. Later, Mora (1981) published a book about the Barra Honda National Park, which focused on geological and karstological aspects. At the beginning of the 1990s, crucial explorations were carried out in the southern zone of the country, as well as in Barra Honda, with the leading promoters of these explorations being international (National Speleological Society, Swiss Society of Speleology, Cave Research Foundation) and national organizations (i.e., Asociación Espeleológica of Costa Rica), who led some of the most multidisciplinary studies regarding caves and karst in Costa Rica (Hempel 1989; Hapka et al. 1992; Peacock and Hempel 1993). In 1995, GEA was created, whose main contribution has been the systematic compilation of speleological information (national and Central American karst registry, using the speleobase software). Since 2007, GEA has been in charge of Espeleo Informe Costa Rica, which includes speleological exploration reports. Nowadays, the University of Costa Rica has multidisciplinary projects focused on geomorphology, speleogenesis, biology, and archeology of the caves.

7.3 Main Karst Landscapes and Caves in Costa Rica

Costa Rica encompasses carbonate rocks deposited since the Late Cretaceous; nevertheless, principal karst localities (e.g., Tempisque, Zona Sur, and Venado of San Carlos) occur in Paleocene–Neogene limestones (Ulloa et al. 2011). Most outcrops of karstic zones in Costa Rica are fault-controlled. Limestone outcrops are generally limited to a few hundred meters, but a few localities have limestone outcrops with a kilometric scale.

The total number of caves registered in Costa Rica is 454, of which ~ 336 correspond to karstic caves. The number of caves increased by ~ 155% since the last published report almost a decade ago (Ulloa et al. 2011), because cave and karst research increased in this period. Based on the speleological plans of 164 caves, there is an estimated of 21.3 km of surveyed underground cave passages in Costa Rica.

One of the most remarkable karst landscapes occurs in the Tempisque karstic zone, where multiple hills, composed mainly of limestone of the Barra Honda Formation, protrude from the alluvial plain of the Tempisque River. Zona Sur is the karstic region with the highest number of reported caves in Costa Rica and the most extensive carbonate outcrops along Fila Costeña. The Corredores river basin includes typical karst dry valleys in tropical forests, with several cave system developed in Fila de Cal Formation. Venado of San Carlos area has discrete carbonate outcrops with an important fault control and hosts the country's longest-known cave system. Karst of Limón has been poorly studied from a karstological point of view, but LiDAR data over the Moín aquifer revealed hundreds of dolines and several karstic hills. Additionally, Limón has relevance because the city is developed on the karstic formation.

7.4 Tempisque Karstlands

This karst zone extends on both sides of the Tempisque River and covers a surface of approximately 80 km^2, being the second largest karst zone in Costa Rica (Ulloa et al. 2011). It extends from Colorado of Abangares to Nicoya. The Barra Honda National Park hosts the best-known karst of this area, and the Tempisque karstlands are developed mainly in the Barra Honda Formation (Upper Paleocene to Lower Eocene) (Denyer et al. 2014).

The limestone rocks of the Barra Honda Formation unconformably overlie the turbidites of the Curú Formation's (Denyer et al. 2014; Denyer 2019). The Barra Honda Formation limestones protrude from the alluvial plain of the Tempisque River, reaching heights up to 481 m a.s.l. above the Tempisque plains, in the Barra Honda hill (Fig. 7.2a). Most of the hills are composed of limestone with evidence of karstification, such as Barra Honda, Corralillo, Quebrada Honda, Taburete, Minisango, Caballito, Rosario, Copal, Carazo, Panamá, Sonzapote, and La Cueva. Morphologically, they represent mogotes (karst hill with steep sides) and conical karst. In addition, there are karren fields, travertine waterfalls, dolines, springs, blind ravines, and caves (Wells 1974; Mora 1978, 1981; Rivier 1983; Hempel 1989; Goicoechea 1989). In the Tempisque karstlands, at least 54 caves have been recognized.

The Palo Verde National Park includes several limestone hills (e.g., Bocana, Piedra Hueca, El Roble, Guayacán, la Cueva del Tigre, Piedra Blanca) made up of carbonate rocks associated with the Barra Honda Formation (Denyer et al. 2013a, 2019). Limestone outcrops show karren fields (Fig. 7.2c), and karren in this area present extremely sharp fluting (Graening 2004). The karst is characterized by conical hills that stands out in relief over alluvial plains, some

Fig. 7.2 Tempisque karst landscape. **a** View of Barra Honda Hill, where it is possible to observe limestone outcrops with karren in the upper slopes. The hill rises above the plain of Tempisque. **b** Travertine cascades near the Barra Honda National Park. **c** Karren in the surrounding of Guayacan hill, in the Palo Verde National Park. **d** Very steep hills considered mogotes in the Palo Verde National Park; on the left is Piedra Hueca hill, and on the right is Bocana Hill

of them presents very steep sides and are classified as mogotes (Fig. 7.2d). Only three caves have been described in this region; the longest cave is Cueva del Tigre, with 180 m of surveyed passages (Graening 2004).

In 1974, the Barra Honda National Park was created by the Costa Rica government (Law N° 5558), in order to protect the karst and promoting tourist visits of the Barra Honda caves. Most of the caves show vertical passages; therefore, many of the caves are unsuitable for tourism because access requires the use of speleological techniques. The park has at least 27 caves, among them two are used for tourism: Terciopelo and La Cuevita caves (Ulloa and

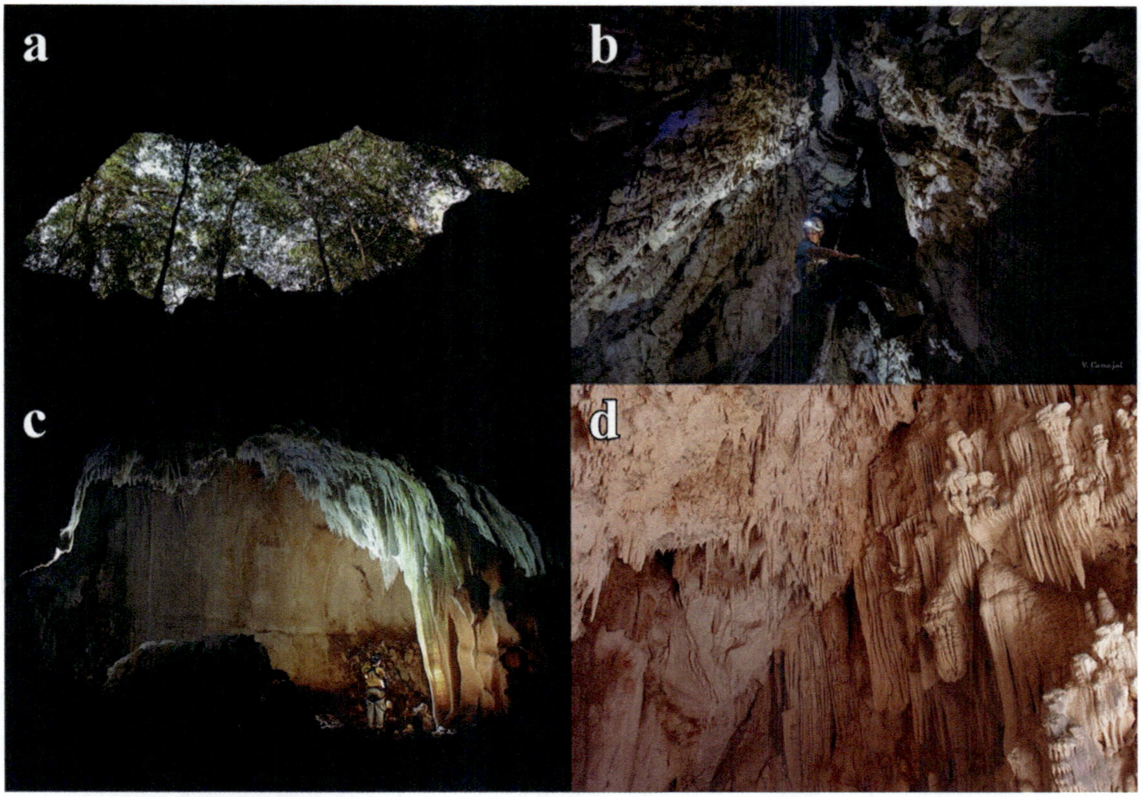

Fig. 7.3 Caves and speleothems in Barra Honda National Park caves. **a** Entrance of Chorotega cave. Photography by Víctor Carvajal. **b** A vertical passage in Chorotega cave. Photography by Víctor Carvajal. **c** Final gallery in La Trampa-Baqueano cave. Photography by Scott Trescott. **d** Speleothems of Terciopelo cave. Photography by Allan Arauz Barrantes

Goicoechea 2013). The caves of the Barra Honda hill present a significant vertical development (Fig. 7.3a–c), with a large number of shafts, most of which are controlled by fractures and faults (Mora 1981). The caves are characteristic of the vadose zone, have vertically dominated passages, and are generally blocked at depth by mud and blocks, so the explorations have not reached water table so far (Hempel 1989). The deepest known cave is Santa Ana-Pozo 110, with 125 m in depth (Hapka et al. 1992). Generally, caves in the area have plenty of speleothems, the most common being stalagmites, stalactites, columns, flowstones, pillars, and calcareous pearls (Mora 1981; Hempel, 1989).

One of the main touristic attractions in the Barra Honda National Park is the Terciopelo cave. This cave is 41 m deep and is home for a wide variety of speleothems (e.g., stalactites, stalagmites, columns) (Fig. 7.3d). The tourist access is enabled through a rigid ladder to facilitate the 18 m descent of the first section of the cave. Secondly, La Cuevita (17.2 m long) is decorated with speleothems and does not require special gear for access. Visiting both caves requires the company of a local guide (Ulloa and Goicoechea 2013).

The geomorphology of Barra Honda National Park is distinctive in that it is an isolated karstic hill with minimal surface drainage. Ephemeral drainages mainly control the hill (Wells 1974; Mora 1978, 1981). Colluvium deposits are widespread in the flanks of Barra Honda hill and composed of varying-sized limestone blocks (Wells 1974; Mora 1981; Denyer et al. 2013a,b).

Wells (1974) described the surface karst as poorly developed "cockpit karst" with iron-red and dark brown soils covering the upper part of the hills, and little to no soil cover in the limestone outcrops. Exposed limestone is characterized by diverse forms of karstification, such as solution ripples, flutes, rinnekarren (rain solution runnels), kluftkarren, schichtfugenkarren, and small shafts (Wells 1974). At least 17 dolines have been identified at the Barra Honda hill, with maximum diameter of up to 291 m and circular to elliptical shapes (Mora 1978, 1981).

Karst springs are the hydrological expression of the karst aquifer. The main ones are located in the surroundings of Cerros de Barra Honda, and some are captured as the water supply for local communities (Mora 1995). Wells (1974) reported karst-spring discharges up to 70 L/s in Barra Honda hill and 130 L/s in Corralillo hill. Between Barra Honda and Corralillo hills, travertine cascades (Fig. 7.2b) are located in the axis of an anticline (Wells 1974). These cascades grew overlying siltstones and mudstone rocks of the Curú Formation. Several small entrances of caves have been reported in this area, but no significant cave development has been found (Hempel 1989).

7.5 Zona Sur Karstlands

Zona Sur is the largest karstic zone in Costa Rica, which extends from the Grande de Térraba River to the border with Panamá (Fig. 7.4), along the Fila Costeña, with a karstic area of approximately 185 km^2 (Ulloa et al. 2011). The extensive middle Eocene–early Oligocene carbonate platforms in southern Costa Rica are related to the Fila de Cal Formation (Malavassi 1961; Mora 1979; Denyer and Alvarado 2007; Chesnel and Rodríguez 2021). Other younger limestones have been reported in the area from the Oligocene (Yuan 1984) and Oligocene to lower Miocene (Chesnel and Rodríguez 2021). Compressive active tectonics (NE-SW) associated with the Cocos Ridge collision affect the sedimentary sequence, with the presence of thrust faults, folds, and some strike-slip faults. As a result of this tectonic regime, thrust faults and folds parallel to the Middle America Trench are common (Denyer et al. 2003; Fisher et al. 2004; Morell et al. 2019). Dextral faults with a normal component, trending N to NNE, are also reported in the area (Arroyo 2001; Chesnel and Rodríguez 2021); these faults are probably younger than thrust faults as inferred from shear and displacement relationships.

Zona Sur is the karstic region with most reported caves in Costa Rica, with a total of 156. Among the main karstic features of the area, the most relevant are dolines, dry valleys, karren, karstic river canyons, springs, travertine waterfalls, and extensive cave systems (Hapka et al. 1992; Peacock and Hempel 1993; Ulloa et al. 2011). Some of the most remarkable areas of Zona Sur karst lands correspond to the Corredores River basin, Carma section, Claro River, and Térraba River.

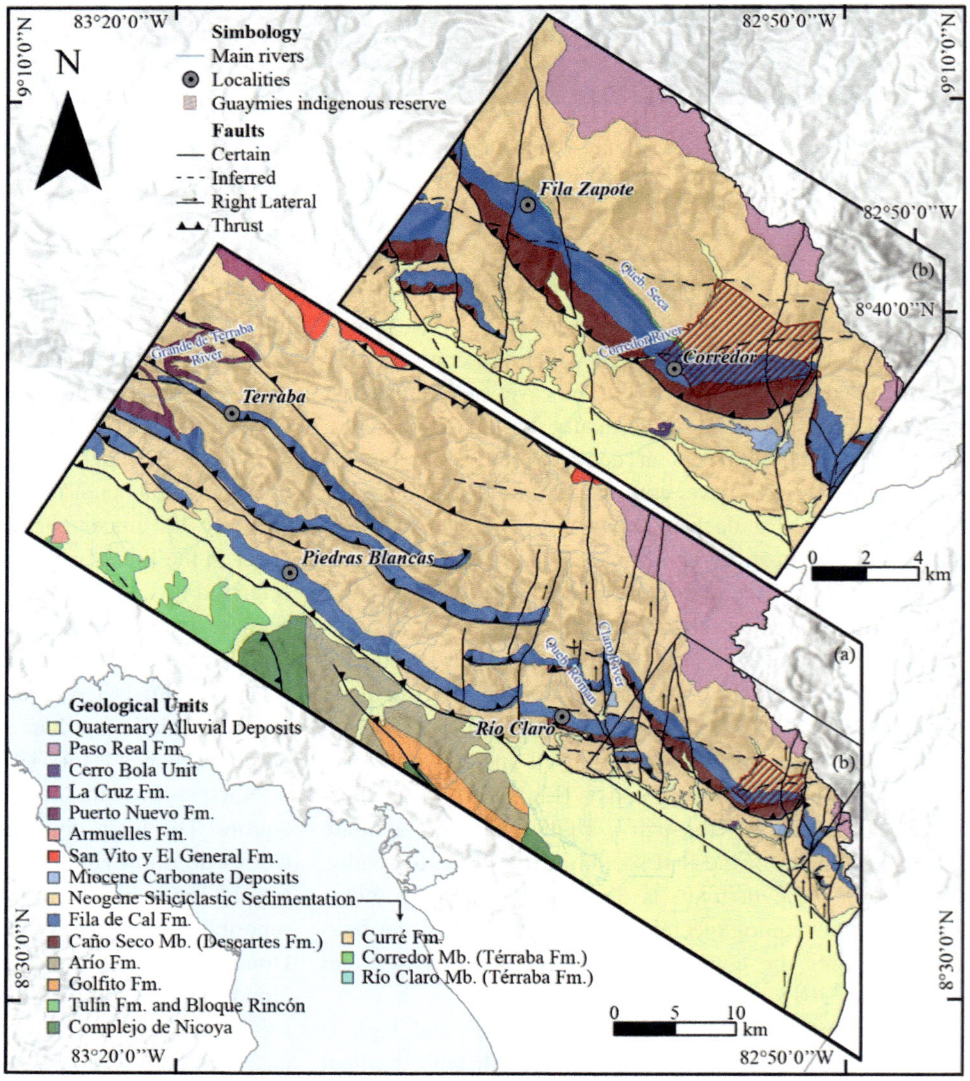

Fig. 7.4 Generalized geological map of Zona Sur. **a** Regional geological map of the Zona Sur, showing the geological units in Fila Costeña, with the main fault traces. Most of the karst is located in Fila de Cal formation Based on the maps by Denyer and Alvarado (2007) and Rodríguez (2020). **b** A detailed geological map of the Corredor River basin showing the geological units and the main faults based on the map by Rodríguez (2020)

7.5.1 Karst in Corredor River Basin

In the surroundings of the Corredor River, near Ciudad Neily, outcrops of a prominent limestone sequence related to Fila de Cal Formation (Rodríguez 2020; Chesnel and Rodríguez 2021), which is highly karstified (Clauser and Hempel 1990; Peacock and Hempel 1993; Ulloa et al. 2011), occur. Peacock and Hempel (1993) reported two main karst systems without surface water flow: (1) La Bruja-Corredores, in the eastern part of Corredor River, and (2) Quebrada Seca, in the western part of Corredor River (Fig. 7.5a). La Bruja-Corredores system is located inside an indigenous reserve of the Guaymies (Ngöbes).

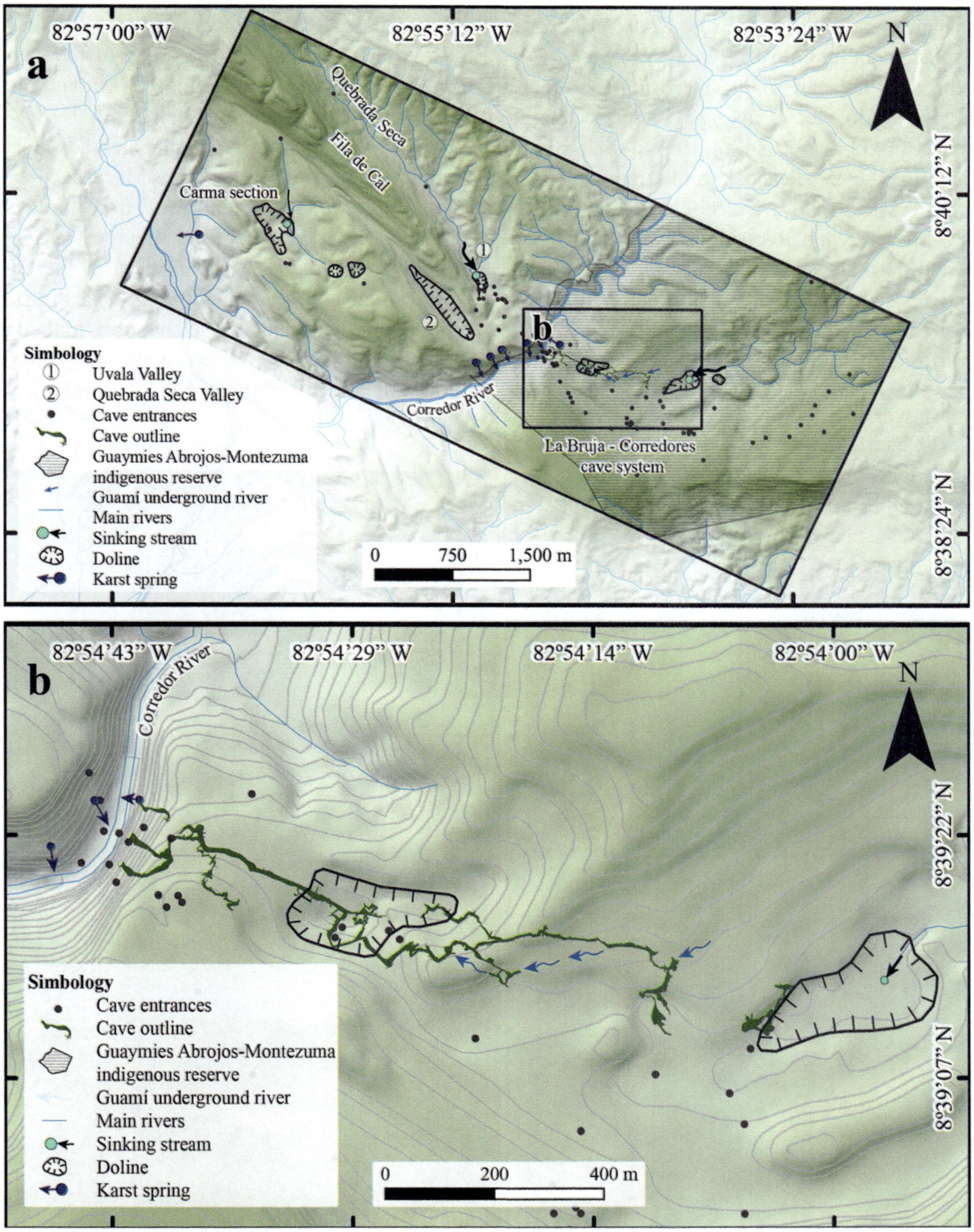

Fig. 7.5 Detailed map of karst features in the Corredor River basin. **a** General map showing Carma, Quebrada Seca, and La Bruja–Corredores cave system with surface water flow paths, sinking streams, caves, and dolines. **b** Detailed map of the La Bruja–Corredores cave system showing the outline of cave passages, with dolines and sinking streams, cave entrances, and karst springs. Main karst elements and cave outlines are based on Peacock and Hempel (1993)

The Corredor River presents a deeply incised karstic canyon (Fig. 7.6b) with a tropical evergreen moist lowland forest in the section where the limestone outcrops occur (Fig. 7.6a). In the NW walls, karst springs related to the Quebrada Seca system are found, while the water coming from the La Bruja–Corredor system comes out through a lower branch of the Corredores cave. Surface geomorphology is characterized by the presence of dolines, karstic hills, canyons, and karst springs. Peacock and Hempel (1993) indicated constant migration of La Bruja Creek up the valley, with a displacement of almost 1.7 km.

Three main caves are part of the La Bruja–Corredores cave system (Fig. 7.5b): La Bruja (221 m), Rectángulo (1411 m), and Corredores (1624 m) for a total of ~3.2 km of mapped passages (Peacock and Hempel 1993). Structural control with a NW trend is evident in the passages of the Corredores and Rectángulo caves. Gilli et al. (1994) suggested normal faulting for NW–SE faults in the Corredores cave based on the shift of the main gallery. In the Corredores cave, there are inactive passages that have an elevation difference of approximately 25 m concerning the regional base level and could be associated with a rapid uplift during the Quaternary that affected the karstic system (Peacock and Hempel 1993; Gilli et al. 1994; Gilli 1996).

Quebrada Seca is located to the north of the Corredor River, where two karstic valleys are found. The first is known as the Chuck's Well Valley or Uvala Valley, and it extends from the surroundings of the Carma cave to the entrance of the Gran Galeria cave. Due to the large number of small dolines in this valley, there is no development of streams and/or superficial rivers (Peacock and Hempel 1993). The second karstic valley corresponds to Quebrada Seca, which is crossed by a creek of the same name that develops superficially for almost 3 km and terminates in a karst sink (Peacock and Hempel 1993).

The karst-sinking stream of Quebrada Seca (Fig. 7.6d) is then found at the base of the MacaMeca cave, which is found approximately 1 km away from the Corredor River karstic canyon (Peacock and Hempel 1993). The underground flow is along a fault line called the Quebrada Seca fault, which has been recognized as a thrust fault and is observed in the Bananal cave and close to the main entrance of Corredores cave. Beyond Quebrada Seca sinkhole, a blind valley with multiple dolines develops within which

Fig. 7.6 Surface karst features of Zona Sur. **a** Fila de Cal foothill, tropical forests cover karst. Photography by Iván Salazar. **b** Karstic canyon in Corredor River, where a cave entrance is located. **c** Doline entrance in the surroundings of Claro River. **d** Quebrada Seca sinking stream in the surroundings of Fila de Cal. Photography by Iván Salazar

Fig. 7.7 Caves in Zona Sur. **a** Entrance of the Gran Galería cave. Photography by Iván Salazar. **b** Siphon in the Carma cave. **c** Wet passage in the Emus cave, near Claro River. Photography by Víctor Carvajal

the Gran Galería (Fig. 7.7a), Bananal, Banano Quemado, Corn Cob, MacaMeca, and other caves are located (Clauser and Hempel 1990).

The Bananal cave is currently one of the longest caves in the country, with 1028.7 m (Peacock and Hempel 1993). It hosts the Lacsa Lake, which is the largest cave-lake known in the country. LACSA lake also connects with the Banano Quemado cave and is related to the underground flow of the Quebrada Seca River (Clauser and Hempel 1990; Peacock and Hempel 1993). The sinking stream of Quebrada Seca has flowed through various caves, from the Gran Galería to its current location (Clauser and Hempel 1990).

7.5.2 Carma Section

This section is located between Fila de Cal and Caño Seco River, to the north from Ciudad Neily. In this area, a ~ 0.5-km diameter doline captures the water and conducts it through the Carma Cave, the main sink in the area. Two surface streams are sinking into the Carma cave, and according to Peacock and Hempel (1993), the water could emerge as springs located ~ 650 m west of the Carma cave entrance. The Carma cave has a surveyed length of 875 m and a depth of 98 m (Peacock and Hempel 1993). This cave reaches water level, and at least two siphons are present in the cave (Fig. 7.7b). Gilli et al. (1994) suggested normal faulting for NW–SE faults (dipping NE) of the Carma cave. Also, they reported fractures affecting the deepest layers of speleothems along a trace of the fault, which could be suggestive of active tectonics in the cave.

7.5.3 Claro River Karst

In the surroundings of Claro River, at least 42 cave entrances have been identified, but the only cave that reaches at least 1 km of surveyed passages is the Emus cave, located in the canyon of the Claro River. The main stream of the Emus cave was dye-traced to a nearby resurgence confirming the hydrological connection (Peacock and Hempel 1993). The Emus cave is the most technically challenging water cave for exploration in the area due to the presence of siphons and waterfalls inside the cave (Fig. 7.7c). This cave presents a characteristic pattern of caves in fractured rocks, developed along thrust faults. The cave includes inactive passages with an elevation difference of approximately 25 m in relation to the regional base level and could be associated with a rapid uplift during the Quaternary that affected the karstic system (Peacock and Hempel 1993; Gilli et al. 1994; Gilli 1996).

Several small dolines (Fig. 7.6c) and small pit-entrance caves a few meters long are found in the surroundings of Claro River, Quebrada Román and Fila Avión, but most of the caves and dolines are smaller than a few tens of meters.

The Quebrada Roman karst area is located in the valley of the Roman River—a tributary of Claro River. The Roman valley area is characterized by karst, covered by agricultural land with patches of broadleaf forest, where cave entrances are of vertical character. Karst landforms are mostly represented by dolines and sinkholes. The karst area was first discovered in January 2016 during an exploratory speleological expedition. Extensive research was performed in March 2016, when 23 caves were discovered and surveyed by an international team of speleologists (Trescott and Deleva 2016). All the caves in this sector have either a completely vertical development or both vertical and horizontal passages. The caves are primarily dry, without known underground streams. The longest cave in the sector is Gran Madre (length = 48 m), which is the only one accessible without ropes. The deepest shaft (35 m) is the Tinamu.

7.5.4 Other Karst Landscapes in Zona Sur

In the margin of the Térraba River, in the Cajón area, a limestone outcrop occurs related to the Fila de Cal Formation. This is a highly tectonic-influenced area, with the presence of thrust faults and chevron folds (Mora 1979). A few caves are recognized in this area, with a total extension of less than 152 m of surveyed passages (CEK 1994; Quesada 2006), presenting some karren and dolines of limited extension.

The Miramar of Abrojo karst is characterized by vertically developed caves. At least 14 caves have been recognized in the area, including the deepest-known Costa Rican cave, i.e., Serpiente Dormida, which reaches a depth of -157 m according to the cave map (Ashbrook 1997). Another karstic area that hosts caves with vertical domain is Fila Zapote. Hapka et al. (1992) described 13 caves in the area and several dolines of metric to decametric size. Most of the caves in Fila Zapote have vertical passages, and the deepest reported cave is Sima Guayabí, with -142 m of depth.

7.6 Venado of San Carlos Karstlands

The Venado of San Carlos karst region is located in San Carlos of Alajuela Province, northern Costa Rica. The local geology is complex due to the variety of lithologies and structural emplacements. The karst landscape has developed in bioclastic limestones (e.g., grainstone, wackestone, and packstone) of the Venado Formation. The Venado Formation dates back from the Upper–Middle Miocene, based on the presence of macrofossils, mainly mollusks, echinoderms (Clypeaster), and ichnofossils (Malavassi and Madrigal 1970); benthic foraminiferal records (Gupta et al. 1986); and shark teeth (e.g., Carcharocles megalodon and Carcharhinus sp) (Argüello et al. 2022). These Miocene geological units are usually covered by Pleistocene volcanic deposits from the Monteverde Formation (e.g., lavas, lahars, tephras) and recent tephra deposits mostly from the Arenal volcano (Argüello 2021; Ulloa et al. 2021). Although this area features a karstic landscape, it is restricted because carbonate rocks are largely covered by volcanic materials. The most common surface karstic landforms are dolines. Ulloa et al. (2021) reported 31 dolines in the Quebrada el Túnel locality, most of which are covered by volcanoclastic deposits (Fig. 7.8a), but also a few collapse dolines have been reported. Other dissolution forms, such as karren (e.g., wandakarren) have been reported, usually exposed in river beds.

The caves in this zone have a predominantly horizontal development, with a pattern controlled by fracture directions, while some passages follow bedding planes. The cave system of Gabinarraca-Menonitas-La Higuera, located in Venado of San Carlos, is the longest in the country, with almost five kilometers of surveyed passages. Three caves are at water level, with some siphons and evidence of significant sedimentation. Argüello (2021) recently described a NW-dextral Venado Transtensive System that limits the limestone outcrops and controls the cave system patterns and most of karstic superficial expressions (Fig. 7.9c). The Gabinarraca cave runs along a NW trend parallel to the fault system, while the passages of Higuera and Menonitas caves present a NE trend (Fig. 7.8b). Hydrological connection by dye tracing (Fig. 7.9f) has been confirmed between the three caves (Ulloa et al. 2021). Each of these three caves is briefly described below.

7.6.1 Gabinarraca Cave

The Gabinarraca Cave, also known as "Cuevas de Venado," is the largest cave in the country, with 2351 m in length (Ulloa et al. 2021) and has been a show-cave since 1976. Guided tours are conducted in the main section of the cave, and visitors receive a speleological experience, visiting the cave using helmets with lights and going through interconnected dry and wet sections along various passages (Ulloa and Goicoechea 2013).

The Gabinarraca cave has five known entrances, the main one corresponding to the water outlet of the local

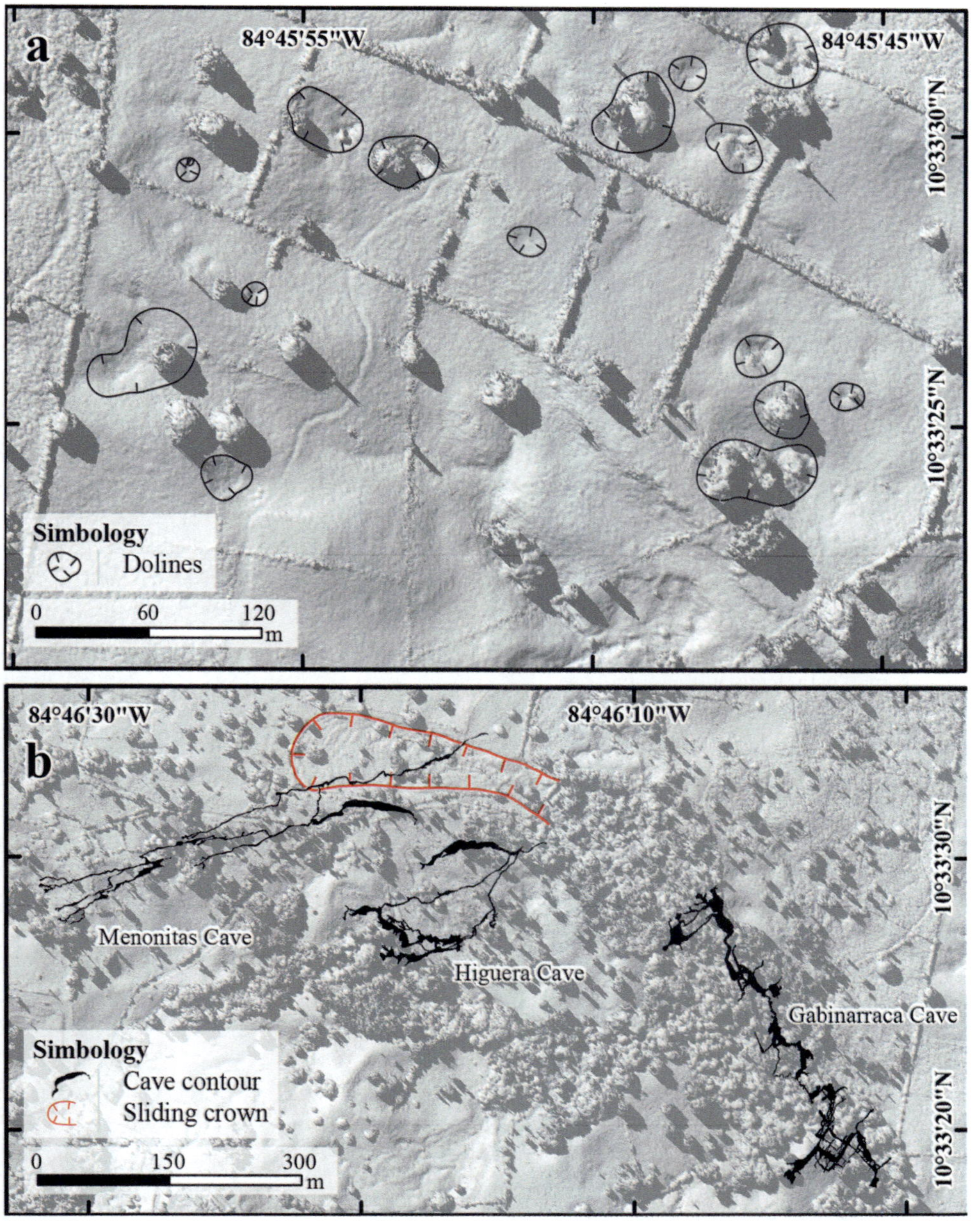

Fig. 7.8 Surface karst manifestation of Venado of San Carlos. **a** Dolines covered by volcanoclastic deposits in a pasture field. **b** Hillshade model of Quebrada el Túnel sector, with the outlines of the caves of Menonitas, Higuera, and Gabinarraca

karstic system (Fig. 7.9b). The cave is structurally controlled, with the main NW-trending galleries following dextral fault traces associated with the Fortuna fault and the Venado Transtensive System. In addition, linear passages related to antithetic fractures are recognized. Therefore, the morphology of cave passages suggests an important structural domain. The sediments in this cave are responsible for paragenetic passages, alluvial notches, and passages filled with sediments. Some cave gallery ceilings have bell holes (Fig. 7.9a) and sinuous paragenetic ceiling channels. The interlocking pattern of many sections of the cave, combined with ceiling channels and paragenesis, suggests that floods and sediment input were involved in the speleogenesis of this cave (Ulloa et al. 2021). Recently, a LiDAR scanning of the main passages has been performed for this cave, which is a useful tool for future detailed analyses such as sedimentation in the cave, speleogenesis, and detailed paragenesis analyses, structural geology, among others.

7.6.2 Menonitas Cave

The Menonitas cave has two known entrances. The stream insurgence is located in Sofocas entrance. The cave has 1620 m of surveyed passages, and the pattern is mainly linear, following fissures and dip direction. This cave is narrow, and most passages converge in a final tributary, a siphon at the end of the surveyed passages. Some sections are characterized by meandric and canyon-type passages (Ulloa et al. 2021). This cave is affected by a neotectonic,

Fig. 7.9 Karst features in Venado of San Carlos. **a** Gallery in the Gabinarraca cave with bell holes in the roof. **b** The main entrance of Gabinarraca cave, which corresponds to the water outlet of the main cave system. **c** Karst dissolution along fractures in the bank of La Muerte River. **d** Menonitas gypsum geodes on the walls. **e** Hola Salvaje gallery of the Higuera Cave. It shows fault traces on the roof and evidence of collapses in the floor. **f** Boca de la Culebra entrance in the Gabinarraca cave, while dye tracing test was performed

showing displacements of passages and speleothems (Argüello 2021).

The cave presents gypsum geodes in the walls (~15 cm diameter; Fig. 7.9d) and secondary gypsum mineralization in alluvial deposits inside the cave. Ca–sulfate minerals with the presence of decimetric-size conducts (like feeders) could be associated with sulfuric acid speleogenesis at some stage of the cave formation. Sediment input is mostly allochthonous and volcanic in composition and is responsible for dissolution niches and paragenesis in some passages (Ulloa et al. 2021).

7.6.3 Higuera Cave

The Higuera cave has 954 m of surveyed passages. This cave has only one known entrance and features underground rivers. The cave pattern is branching, where three tributaries converge into a common subterranean river flow, ending in a siphon with copious amounts of fine sediments (Ulloa et al. 2021). The cave presents fissure passages, where most of the walls are covered by fine sediments (silts and clays). Cave development follows dip direction. This cave includes big chambers (45 m long, 10 m wide), which are structurally controlled with NW preferential orientation and generally (Fig. 7.9e) associated with roof collapses (Ulloa et al. 2021).

7.7 Limón Karstlands

Limón is the most populated city in Costa Rica that is located within a karstic region. Although the presence of caves and karst in the area is known by the locals, there is very little scientific information regarding the caves and karst in the area. The Limón karst region is located near the limit of the Caribbean Plate and the Panamá microplate. This area is tectonically active and one of the biggest earthquakes known in Costa Rica, the Limón Earthquake (Mw 7.7), occurred in this area (Montero et al. 1994). Near the current coastline, there are very young reef-limestones, uplifted after the Limón Earthquake (Fig. 7.10a), that are exposed and subjected to karstic processes. Several limestone outcrops occur in Limón, Moín, Cahuita, Manzanillo, and Brazo Seco, where limestone occupies an approximate area of 20 km^2 (Ulloa et al. 2011). The geology of the area is composed of mixed reef-carbonate and siliciclastic sediments between Moín and Limón. These geological units are part of the Limón Group (lower Pliocene to lower Pleistocene), which is composed of two geological units with limestone reefs; the Quebrada Chocolate Formation (early–late Pliocene), composed by reefal and siliciclastic deposits; and the Moín Formation (late Pliocene–Pleistocene boundary), composed by reefal and fine-grained, mollusk-rich deposits (McNeil et al. 2000).

7.7.1 Karst Geomorphological Features of the Western Limón Zone

In 2018, the Instituto Costarricense de Acueductos y Alcantarillados (AyA) contracted a LiDAR survey for the western sector of Limón (Moín), and an approximate area of 6.3 km^2 was investigated. The aim of this survey was to analyze the vulnerability of a sector of the Moín aquifer. From the analysis of the LiDAR images, the presence of karstic hills (~10 m), similar to cockpits, with a predominant E-W trend, was observed. Also, a total of 438 dolines have been identified in the area by LiDAR survey (Fig. 7.10b). Such dolines represent about 6.5% of the analyzed area, of which 75% have a depth ~5 m, and 24 dolines showed depths over 10 m, with a maximum of 16 m (Stereocarto Centroamérica 2018). The LiDAR of the Moín sector offers a better understanding of karst geomorphology; nevertheless, field characterization is necessary to have a better understanding of the Caribbean karst.

7.7.2 Main Caves of Limón Area

To date, five caves have been located in the northern part of Limón, between Playa Bonita and Playa Piuta (Fig. 7.10c). The El Encanto Cave (also known as Capilla of Portete) is situated next to the road and is probably the most popular cave in the area (Fig. 7.11 a,b). El Encanto cave has two entrances, with 36.1 m of surveyed passages and 512 m^2 of superficial area. Two other caves are located ~300 m westward. The entrances of both caves correspond to water outlets. One cave presents fissure-type passages with an underground river. An estimated length of the explored area of this cave is at least 400 m (Fig. 7.11c), and the cave continues. A proper survey of this cave is needed to elucidate the dimensions of the karst system. Next to this cave, a siphon cave is found, where it is necessary to perform cave diving to explore the underwater passages.

Another cave entrance has been partially explored close to Hotel Oasys del Caribe (Fig. 7.11d). The water level marks in this cave suggest that it is periodically flooded after heavy rains. Preliminary exploration showed that it has at least two sub-horizontal levels, with multiple passages and a length over 300 m. Also, a doline intersects this cave. Trash and probably highly contaminated waters are evident in the cave.

Over the main road to Limón town, one small cave named Al Caribe was found in 2020 during the road-cut construction. This cave is small, ~7 m in length. On the same road, next to the Liceo of Limón, the Gioconda cave was reported with at least 250 m of passages (Goicoechea 1968). Also, small caves (< 10 m) close to the localities of Cahuita, Punta Uva, and Manzanillo have been reported, but accurate descriptions and surveys of these coralline caves are still pending. Seismic effects after the 1991 Limón Earthquake have been disclosed in the caves. Gilli (1996) reported a collapse in the El Encanto cave and suggested reverse direction of water flow and collapse, compared to previous reports from Goicoechea (1968) in La Gioconda cave.

7.7.3 Moín Karstic Aquifer

Moín has a well-developed karstic aquifer with approximately 5 km^2; it is developed in reef-carbonate rocks, where water flows through porous caves in the reef-limestone and

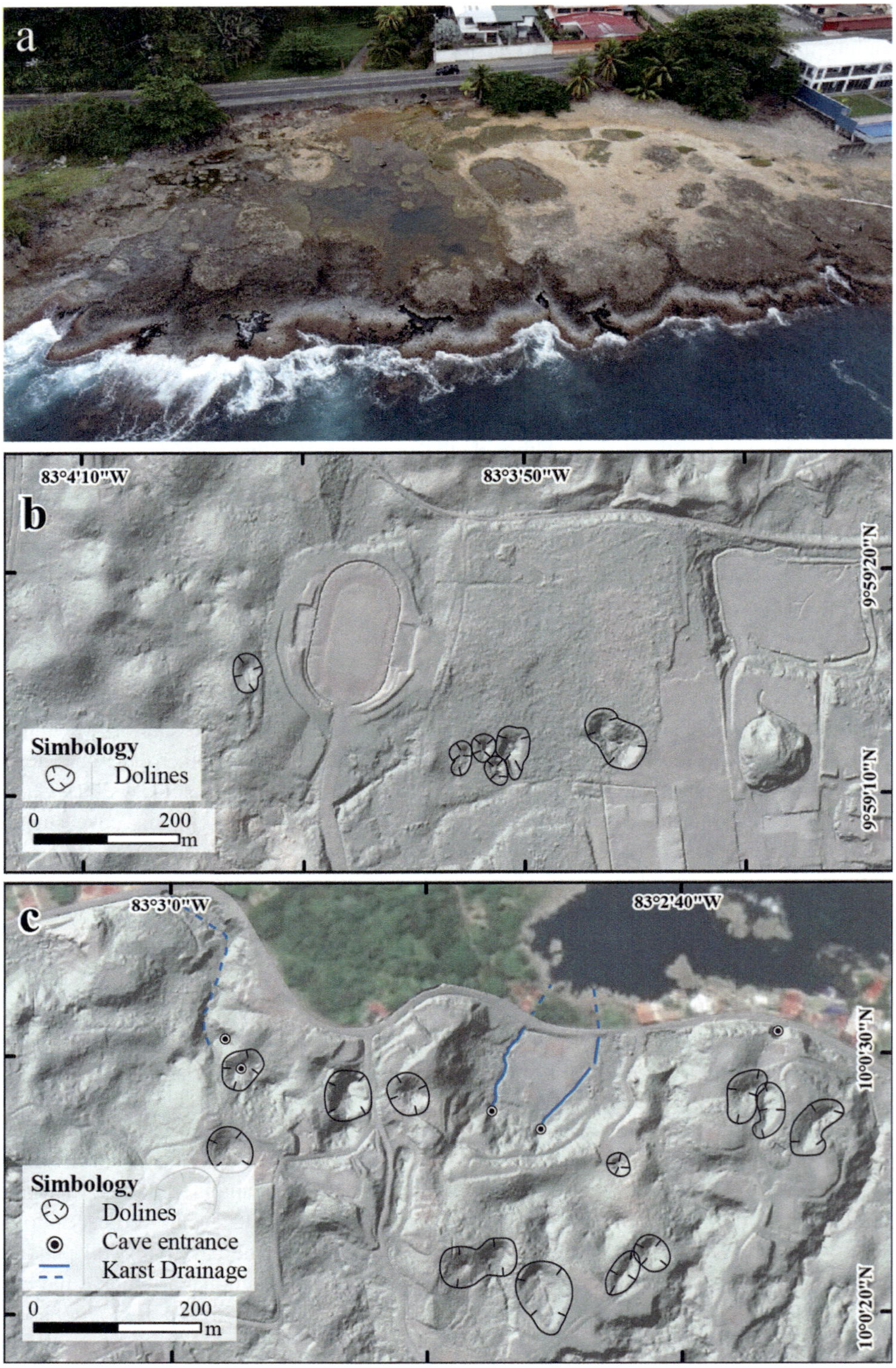

Fig. 7.10 Drone photography and hillshade model from LiDAR topography of the Moín sector, showing various karst features. **a** Uplifted reef platform after Limón earthquake (1991). **b** Dolines and karstic hills near Limón stadium. **c** Dolines and cave entrances of this karst locality between Playa Bonita and Playa Piuta

emerges in Moín karstic springs (ONU 1975; Ramos 2005). Ramos (2005) indicated that karst aquifers in the area present are recharged by infiltration and directly through dolines. Several dolines are reported in the Moín area. Tracing analyses have been performed with fluorescein (ONU 1975) and sodium chloride (Ramos 2005) to analyze the residence time from the sinkholes to the karst springs of Moín. The hydrological results indicated a residence time of less than 40 hours (ONU 1975), which makes this aquifer highly vulnerable to water contamination.

Based on the high vulnerability shown in hydrogeological studies (Ramos 2005), in 2007, the Instituto Costarricense de Acueductos y Alcantarillados (AyA) declared an area of Moín aquifer as protected, with

Fig. 7.11 Main caves of Limón area. **a,b** Cueva el Encanto (or Capilla de Porteté), which is the best-known cave in the area. **c** Fissure passages of a cave located close to Playa Piuta, with an estimated length of at least 400 m. **d** Cave in reef-limestone close to hotel Oasys del Caribe

restrictions to develop industrial activities, use of agrochemicals, fuel storage, cutting of vegetation, and housing estate development. In the Moín sector, there is conflicting land use around the karst. Even though state institutions such as the AyA have issued decrees with policies for the protection of water resources, there exists a precarious situation that poses a significant social and environmental problem. This situation has a detrimental impact on the water quality of the karstic aquifer.

7.8 Future Perspectives

The presence of limestone exposures within tropical climates, abundant water, and active tectonics are processes that promote the development of karst landscapes in several areas of Costa Rica. Numerous karst landforms have been identified such as caves, dolines, sinkholes, karst springs, and karstic hills. In this chapter, we described four key karstlands of Costa Rica; nevertheless, more karst localities require an accurate description and characterization. The most studied karst landforms are caves, where almost 20 km of underground passages have been documented with cave maps. However, only 36% of the documented caves has a cave map or croquis.

Most of the karst features in Costa Rica have limited extent. So, the vast majority of surface karst expressions, such as dolines and blind valleys, are not represented in conventional cartography (e.g., contour levels each 10 m). Therefore, most karstic areas require more detailed cartography or systematic fieldwork to be analyzed. LiDAR technology in karst is a precise tool for the identification of karst features. In Moín, by using LiDAR data, hundreds of sinkholes were recognized in a small study area. The use of this technology in other sectors of the country could lead to a better characterization of the karst.

The scientific documentation of caves and karst in Costa Rica began more than half a century ago. However, this is an emerging line of research that has been poorly covered by the country's academic institutions. Currently, the University of Costa Rica has multiple research projects related to caves and karst. The interdisciplinary study of caves and karst is needed to understand these unique landscapes from the perspective of various sciences such as geology, geomorphology, biology, microbiology, archeology, anthropology, and others.

Acknowledgements We are grateful to Grupo Espeleológico Anthros for sharing essential data about caves in Costa Rica. Carlos Goicoechea Carranza merits posthumous acknowledgment for his kindness in sharing valuable information about caves and karst and his significant contribution to elucidating the history of cave exploration in Costa Rica. UEN Gestión Ambiental of AyA is thanked for bringing LiDAR data of areas with highly hydrogeological vulnerability of Moín aquifer in Limón. We are grateful to Víctor Carvajal, Scott Trescott, Iván Salazar, and Allan Arauz Barrantes for providing photographs used in this paper. Fernando Gázquez is also acknowledged for reviewing the manuscript.

References

Argüello A (2021) Estudio tectónico y kárstico de la región de Venado de San Carlos, Costa Rica. Bachelor´s Thesis, Universidad de Costa Rica, San José, Costa Rica

Argüello A, Laurito CA, Valerio AL, Quesada Q, Ulloa A (2022) Primeros registros fósiles de Condrictios para la Formación Venado, Mioceno Superior, provincia de Alajuela, Costa Rica. Rev Geol Am Central 67:1–9. https://doi.org/10.15517/rgac.v67i0.51593

Arroyo I (2001) Sismicidad y Neotectonica en la region de influencia del proyecto Boruca: hacia una mejor definición sismogenica del Sureste de Costa Rica. Bachelor´s Thesis, Universidad de Costa Rica, San José, Costa Rica

Ashbrook B (1997) Sima Serpiente Dormida (Escala 1:924)

CEK (1994) Caverna La Danta (Escala 1:1000)

Chesnel V, Rodríguez E (2021) Facies analysis of a Bartonian-Aquitanian siliciclastic-carbonate system, Costa Rica. Sediment Geol 417:1–20. https://doi.org/10.1016/j.sedgeo.2021.105884

Clauser L, Hempel JC (1990) 1990 Costa Rica Expedition to Rio Corredores. NSS News 48(10):261–268

Day MJ (2007) Karst landscapes. In: Bundschuh J, Alvarado GE (eds) Central America: geology, resources and hazards. Taylor and Francis, London, UK, pp 155–170

Denyer P (2019) Perspectiva geológica del Noroeste de Costa Rica: historia, evolución y cartografía. Editor Univ Costa Rica, San José, Costa Rica, pp 360

Denyer P, Aguilar T, Montero W (2013a) Mapa geológico de la hoja Talolinga [Escala 1:50 000]. In: Denyer P (ed) Cartografía Geológica de la Península de Nicoya, Costa Rica, Editorial UCR, San José, Costa Rica, pp 216

Denyer P, Aguilar T, Montero W (2013b) Mapa geológico de la hoja Matambú [Escala 1:50 000]. In: Denyer P (ed) Cartografía Geológica de la Península de Nicoya, Costa Rica, Editorial UCR, San José, Costa Rica, pp 216

Denyer P, Aguilar T, Montero W (2014) Cartografía geológica de la Península de Nicoya. Costa Rica. Editorial UCR, San José, Costa Rica, pp 216

Denyer P, Alvarado GE (2007) Mapa Geológico de Costa Rica [Escala 1: 400 000]. Librería Francesa, San José, Costa Rica

Denyer P, Alvarado GE, Montero W, Barrantes G, Chavarría M (2019) Mapa geológico de la hoja Tempisque [Escala 1:50 000]. In: Denyer P (ed) Perspectiva geológica del Noroeste de Costa Rica: historia, evolución y cartografía. Editor Univ Costa Rica, San José, Costa Rica, pp 360

Denyer P, Montero W, Alvarado G (2003) Atlas tectónico de Costa Rica. Editorial de la Universidad de Costa Rica, San José, Costa Rica, pp 64–216

Fisher DM, Gardner TW, Sak PB, Sanchez J, Murphy K, Vannucchi P (2004) Active thrusting in the inner forearc of an erosive convergent margin, Pacific coast, Costa Rica. Tectonics 23:1–13. https://doi.org/10.1029/2002TC001464

Gilli E (1996) Effects des sêismes dans l'endokarst Application aux études de sismicité historique. Rev D'analyse Spat Quant Appliquée 38–39:121–132

Gilli E, Delange P, Larre P (1994) Reconnaissance Sur La Neotectonique Dans Les Karst Du Costa Rica. Centre d'Etude du Karst, Rue Massena, Francia, pp 52

Goicoechea C (1968) Caverna La Gioconda (Escala 1:2525)

Goicoechea C (2015) Proceso Histórico de la Espeleología en Costa Rica. Grupo Espeleológico Anthros, San José, Costa Rica, pp 143

Goicoechea C (1989) El Sistema Kárstico de Barra Honda (y vecindades). NSS, San José, Costa Rica, pp 88

Goicoechea C, Quesada G, Ulloa A (2009) Evolution of speleology in Costa Rica. In: White W (ed) 15th international congress of speleology. international union of speleology, Kerrville, Texas

Graening G (2004) Use of the Costa Rican Cave, Cueva del Tigre, as a classroom for cave ecology and conservation by the organization for tropical studies. NSS News 168–171

Gupta S, Malavassi L, Malavassi E (1986) Late Miocene shore in northern Costa Rica: Benthic foraminiferal record. Geology 14:218–219

Hapka R, Brasey J, Goicoechea C, Jeannin P-Y, Jeannin V (1992) Explorations dans les karsts tropicaux du Costa Rica. Cavernes 1:1–27

Hempel J (1989) Report of the 1982 Expedition to Barra Honda National Park. NSS, Alabama, USA, pp 1–23

Kambesis P (2007) The importance of cave exploration to scientific research. J Cave Karst Stud 69:46–58

Malavassi E (1961) Some Costa Rican larger foraminiferal localities. J Paleontology 35:498–501

Malavassi VE, Madrigal R (1970) Reconocimiento geológico de la Zona Norte de Costa Rica. San José, Costa Rica (Report)

McNeil DF, Coates AG, Budd AF, Borne PF (2000) Integrated paleontologic and paleomagnetic stratigraphy of the upper Neogene deposits around Limon, Costa Rica: a coastal emergence record of the Central American Isthmus. Bull Geol Soc Am 112(7):963–981

Montero W, Pardo M, Ponce L, Rojas W, Fernández M (1994) Evento principal y réplicas importantes del Terremoto de Limón. Rev Geol Am Central esp 993–102. https://doi.org/10.15517/rgac.v0i0.13434

Mora S (1979) Estudio geológico de una parte de la región sureste del Valle del General, provincia de Puntarenas, Costa Rica. Bachelor's Thesis. Universidad de Costa Rica, San José, Costa Rica

Mora S (1986) Consideraciones sobre el karsismo costarricense. San José, Costa Rica

Mora S (1987) Les Regions Karstiques Du Costa Rica Et Leur Conteste Geologique. Karstologia 10:25–32

Mora S (1981) Barra Honda. EUNED, San José, Costa Rica

Mora S (1978) Estudio geológico de los cerros Barra Honda y alrededores, Nicoya, Guanacaste, Costa Rica. San José, Costa Rica

Mora S (1995) Karst localities in Costa Rica and their geologic background. Earth Sci Ser 16:379–386. https://doi.org/10.1007/978-3-642-79476-6_52

Morell KD, Fisher DM, Bangs N (2019) Plio-Quaternary outer forearc deformation and mass balance of the Southern Costa Rica convergent margin. J Geophys Res Solid Earth 124:9795–9815. https://doi.org/10.1029/2019JB017986

ONU (1975) Investigaciones de aguas subterráneas en zonas seleccionadas de Costa Rica (Informe Técnico). New York, USA

Peacock N, Hempel J (1993) Studies in the Rio Corredor basin. NSS Bull J Caves Karst Stud 55:134

Quesada G (2006) Caverna La Danta (Escala 1:667)

Racovitza EG (1907) Essai sur les problèmes biospéologiques. In: Lacaze-Duthiers H (ed) Archives de zoologie expérimentale et générale, Schleicher. Librairie C. Reinwald, Paris, pp 371–488

Ramos V (2005) Estudio hidrogeológico y vulnerabilidad del acuífero Moín, Limón. Master's Thesis. Universidad de Costa Rica, San José, Costa Rica

Rivier F (1983) Síntesis geológica y mapa geológico del área del Bajo Tempisque, Guanacaste, Costa Rica. Inf Semest IGN 1:7–30

Rodríguez EA (2020) Estratigrafía y evolución sedimentológica del sureste de la Fila Costeña, Costa Rica. Bachelor's thesis, Universidad de Costa Rica, San José, Costa Rica

Stereocarto Centroamérica (2018) Informe técnico de procesado de datos. Servicios para el levantamiento, identificación y sistematización de información técnica como un mecanismo de integración de la información bajo el concepto de cuenca, aplicación mediante un estudio fotogramétrico y LiDAR en un sector del acuifero de Moín, Limón. Stereocarto Centroamérica, San José, Costa Rica

Trescott S, Deleva S (2016) Expediciones GEA a Quebrada Román de Río Claro, Enero y Marzo 2016. San José, Costa Rica

Ulloa A, Aguilar T, Goicoechea C, Ramirez R (2011) Descripción, clasificación y aspectos geológicos de las zonas kársticas de Costa Rica. Rev Geol Am Central 45:53–74

Ulloa A, Argüello A, Obando A, Vargas M (2021) Geomorfología y espeleogénesis del sistema kárstico de cuevas de Venado, Costa Rica. Rev Geol Am Central 64:1–31. https://doi.org/10.15517/rgac.v0i64.46619

Ulloa A, Goicoechea C (2013) Geotourism potential of underground sites in Costa Rica. Tourism and Karst Areas 6(1):43–56

Wells S (1974) Report on the Geologic Reconnaisance of the Cerros Barra Honda Karst Costa Rica. San José, Costa Rica

Yuan PB (1984) Stratigraphy, sedimentology, and geologic evolution of eastern Terraba trough Southwestern Costa Rica (tectonics, forearc basin). Doctoral Thesis. Louisiana State University, Baton Rouge, Louisiana Satate, United States

8 Glacial and Periglacial Landscapes in Costa Rica

Adolfo Quesada-Román

Abstract

Tropical regions in the Americas retain evidence of past glacial activity, offering insights into historical climate conditions, particularly during the Last Glacial Maximum (LGM). Focusing on Costa Rica's higher summits, especially the Cordillera de Talamanca, this study identifies distinct features such as glacial cirques, polished bedrock, moraine and till deposits, and glacial-origin lakes. These remnants serve as invaluable indicators of past climatic conditions and contribute to understanding the maximum extent of tropical glaciation during the LGM in Costa Rica and broader tropical America. The Cordillera de Talamanca reveals areas significantly influenced by colder climates during the LGM. This research enhances our ability to reconstruct the interplay between climate and landscape evolution, deepening our understanding of the LGM in Costa Rica. The documented evidence forms a comprehensive narrative of the region's paleoclimatic history, providing a foundation for comparative studies in tropical America. This study advances comprehension of tropical glacial landscapes by revealing nuanced features left by ancient glacial processes. The Cordillera de Talamanca emerges as a key repository of valuable information about past climatic conditions. Navigating through the glacial remnants of Costa Rica's higher summits unravels the mysteries of the LGM and its lasting impact on tropical American landscapes.

Keywords

Last glacial maximum · Central America · Costa rica · Cordillera de talamanca · Paleoglaciers

A. Quesada-Román (✉)
Laboratorio de Geografía Física, Escuela de Geografía, Universidad de Costa Rica, San José, Costa Rica
e-mail: adolfo.quesadaroman@ucr.ac.cr

8.1 Introduction

Past fluctuations of tropical glaciers offer valuable insights into paleoclimate conditions, particularly in regions where alternative evidentiary sources are scarce (Porter 2001; Benn et al. 2005; Mark et al. 2005; Fig. 8.1). This is particularly evident during the Last Glacial Maximum (LGM), when a multitude of glaciers extended beyond 3000 m asl in the tropical regions of America, including Costa Rica, Guatemala, Mexico, and the Andes (Palacios et al. 2020). The high mountains of Central America bear witness to both erosional and depositional glacial activities in the past (Vázquez-Selem and Lachniet 2017; Quesada-Román et al. 2022). Examining glacial paleo-morphologies provides crucial evidence for reconstructing past climate conditions (Quesada-Román 2022).

The historical documentation of glacial activities in the Cordillera de Talamanca dates back to the late nineteenth century when Henri Pittier first described the glacial landscapes of the high peaks (Castillo-Muñoz 2010; Quesada-Román 2016). Over half a century later, Weyl (1955) delved into the glacial landscape of Cerro Chirripó, utilizing aerial photographs and on-site observations. Hastenrath (1973) contributed by mapping numerous glacier cirques and lakes near Chirripó, while other studies utilized large-scale maps based on aerial imagery to describe the glacial morphology of Cerro Chirripó (Bergoeing 1977; Barquero and Ellenberg 1983; Shimizu 1992; Wunsh et al. 1999).

Building upon this foundation, Lachniet and Seltzer (2002) and Lachniet et al. (2005) conducted extensive studies, detailing the expansive LGM glaciers and erosive/depositional morphologies of glacial valleys surrounding Cerro Chirripó. Their mapping efforts covered an area of 35 km^2 around Cerro Chirripó, with the local equilibrium-line altitude (ELA) estimated at 3500 m asl. Additionally, they proposed that the ELA might have descended to 3200 m asl during the Last Glacial Maximum. Recent research by Quesada-Román and Zamorano-Orozco further expounded

A. Quesada-Román (ed.), *Landscapes and Landforms of Costa Rica*, World Geomorphological Landscapes, https://doi.org/10.1007/978-3-031-64940-0_8

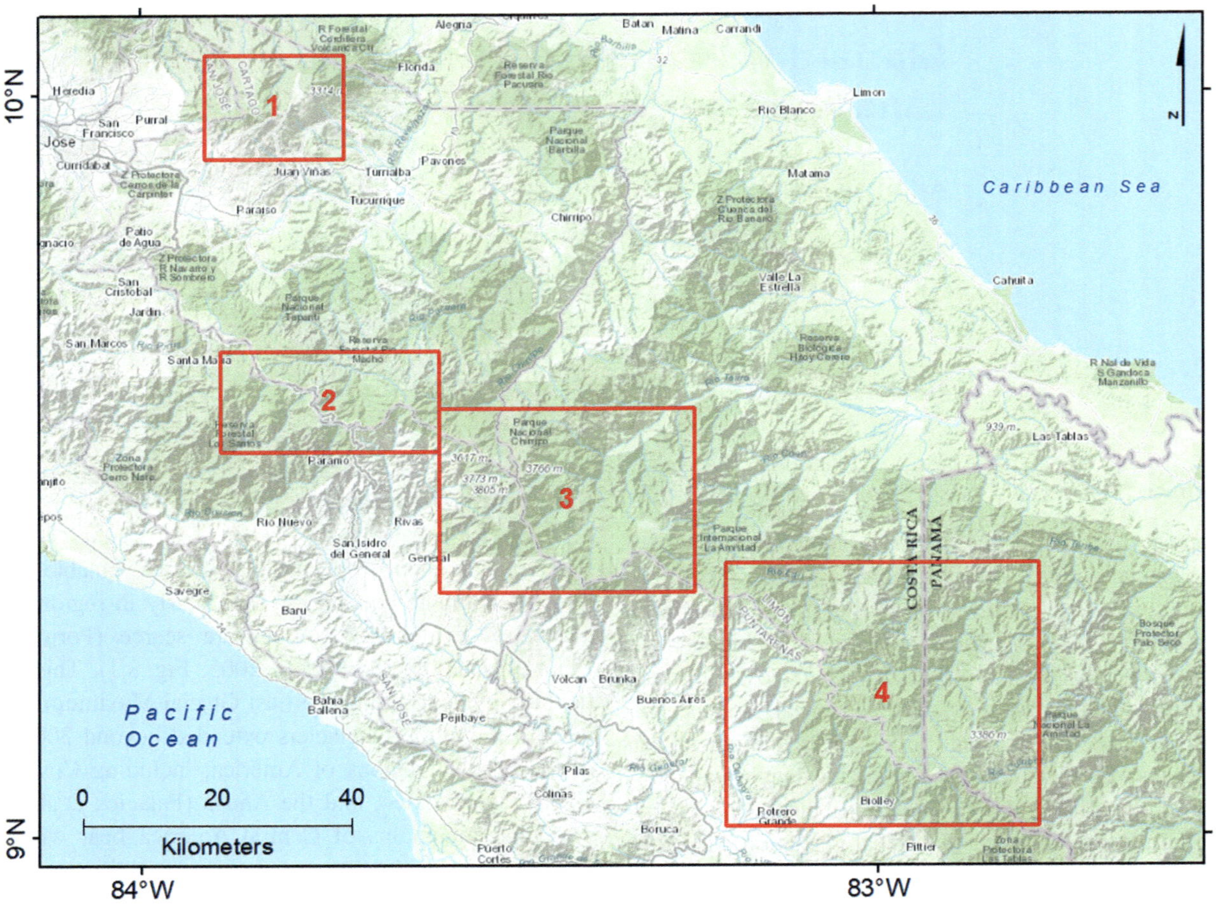

Fig. 8.1 Location of the four main areas with confirmed or probable past glacial activity during the Last Glacial Maximum and/or previous glacial periods in mountains over 3000 m asl of Costa Rica. 1) Irazú and Turrialba volcanoes, 2) From Cerro Vueltas to Cerros Cuericí, 3) From Cerro Chirripó to Cerro Dúrika, and 4) From Cerro Utyum to Cerro Echandi

on the glacial geomorphology of the Upper General River Basin, encompassing Chirripó National Park. Furthermore, Li et al. (2019) meticulously described 22.1 km^2 of surfaces within glacial remnants near Cerro Chirripó. Quesada-Román et al. (2019) expanded the mapping to approximately 90 km^2 at a 1:25,000 scale, encompassing glacial and periglacial landscapes within Chirripó National Park.

In recent decades, beyond the geomorphological characterization of paleoglacial landforms, a significant focus of studies has been directed toward dating these features. Horn (1990) employed charcoal and pollen analyses to pinpoint the end of the last glacial age in the largest lake of Valle de las Morrenas, estimating it to be around ca. ~ 10 ka. Subsequently, Orvis and Horn (2000) determined a deglaciation timeframe ranging from 12.36 to 9.7 ka, concluding at approximately 8580 ± 70 BP. Potter et al. (2019) utilized 36Cl, while Cunningham et al. (2019) employed 10Be cosmogenic nuclides, both converging on the conclusion that the maximum glacial extension in Chirripó National Park aligned with the Last Glacial Maximum (~26.5–19 ka). Consistent with earlier studies, recent research by Quesada-Román et al. (2020a, b) reiterated that the estimated equilibrium-line altitude (ELA) for 31 paleoglaciers during the Last Glacial Maximum stood at 3500 m asl around Cerro Chirripó. Additionally, Quesada-Román et al. (2021) reported a maximum glacier thickness of 178 m and a total volume of 13,863 × 105 m3 for these glaciers during the Last Glacial Maximum.

The Cordillera de Talamanca, characterized by widespread magmatic products, serves as a record of Neogene and Quaternary intrusive and extrusive phases. Generally, three distinct magmatic stages are discernible: 1) Volcanic series predating the Late Miocene, spanning from ~ 17 to 11 Ma; 2) Plutonic events throughout the Middle–Late Miocene, ranging from 12.5 to 7.5 Ma, known as the Talamanca Intrusive Group or Granite-Gabbro Series of Talamanca; and 3) Post-intrusive magmatic pulses occurring from the latest Neogene-Quaternary, dated approximately from ~ 5 to 2 Ma (Alfaro et al. 2018, 2021; Gazel et al. 2019). This chapter provides an in-depth examination of the glacial and periglacial landscapes in the high mountains of Costa Rica, especially those situated above

3000 m asl. These areas have been the focus of previous tropical paleoglacier studies confirming the existence of Last Glacial Maximum extensions (Porter 2001; Benn et al. 2005; Mark et al. 2005; Palacios et al. 2020) and serve as potential sites for further investigation.

8.2 Glacial and Periglacial Regions in Costa Rica

Despite numerous areas exceeding 3000 m asl in Costa Rica, not all exhibit distinct paleoglacial activity imprints from the Last Glacial Maximum (LGM). A prevailing assumption is that regions situated between 3000 and 3500 m asl are more closely associated with paleo-periglacial dynamics, such as frost action and solifluction. In contrast, areas surpassing 3500 m asl are presumed to be more likely to retain evidence of paleoglacial dynamics linked to glacier movement along glacial valleys (Fig. 8.2). However, it is important to note that certain areas may have had glacial imprints, but these could have been effaced by volcanic activity, as observed in the Irazú and Turrialba volcanoes. The Cordillera de Talamanca emerges as the principal regional hub in Costa Rica, encompassing various specific locations exhibiting paleoglacial activity (Quesada-Román et al. 2023). This mountainous region serves as a focal point for understanding the nuanced interplay of paleo-periglacial and paleoglacial dynamics, offering valuable insights into the complex interactions between high-altitude landscapes and past climatic conditions of the country.

8.2.1 Erosional Landforms

The summit areas exceeding 3000 m asl exhibit distinctive features within Central American environments, characterized by the transformative impact of glacial dynamics during the Last Glacial Maximum (Quesada-Román et al. 2019). These high-altitude zones play a pivotal role in shaping the present páramo ecosystem, a critical hydrological and ecological niche of regional Latin American significance (Esquivel-Hernández et al. 2018, 2019, 2022; Veas et al. 2018, 2023; Esquivel-Hernández 2021). The intricate interplay between past glacial processes and the contemporary páramo ecosystem highlights the enduring influence of historical climatic conditions on the current environmental landscape, underscoring the importance of these summit areas in understanding regional ecological dynamics (Quesada-Román et al. 2020a, b).

The erosional paleoglacial landforms in Costa Rica exhibit a variety of features, prominently including arêtes, glacially shaped volcanic slopes, and cirques (Fig. 8.3). The dynamic movement of ice masses has left a lasting imprint on the landscapes of the higher summits in the Cordillera de Talamanca, resulting in the formation of rounded and sharp-crested hills. Two distinct types of volcanic slopes modified by glacial action are evident: (i) subhorizontal surfaces in the lowest parts of glacial valleys and on mountain tops with inclinations less than 15° and (ii) edges of the continental divide with steeper morphologies, characterized by slope angles of up to 36°. The latter morphologies correspond to exposed rocky terrains where glaciers have left behind characteristic scars, p-forms, and striations. A prime example of such a landscape is found at Cerro Chirripó, the highest peak in Costa Rica, soaring to an elevation of 3820 m asl. Various dating procedures applied around Cerro Chirripó, as outlined by Quesada-Román et al. (2021), suggest that equilibrium-line altitudes (ELA) during the Last Glacial Maximum (LGM) were not lower than 3500 m asl. However, the validation of these findings is essential and requires cross-checking at other sites exhibiting probable glacial dynamics, as indicated by geomorphological characteristics like striated rocks, smoothed summits, and the presence of glacial lakes and highland wetlands.

Volcanic slopes shaped by periglacial action constitute extensive surfaces surrounding the glacially shaped hillslopes. According to Orvis and Horn (2000) and Lachniet and Seltzer (2002), temperatures during the LGM in the summits of Talamanca were approximately 9 °C below current temperatures. These slopes underwent various periglacial processes such as weathering, frost action, mass movement, nivation, and frozen ground effects. While periglacial dynamics occurred in several areas during the LGM or previous glaciations, not all have been precisely dated using radiometric methods. These surfaces, albeit rarely found below 3000 m asl, exhibit dominant rocky terrains with sporadic occurrences of whalebacks, striations, p-forms, and scars. Extending over approximately 100 km^2, both glacial and periglacial volcanic slopes are currently covered by páramo vegetation throughout the Cordillera de Talamanca. Over time, paraglacial processes have played a role in the rapid formation of fluvial knickpoints and mass movements in the glacial and periglacial regions during the Holocene.

Cirques, distinctive amphitheater-like morphologies resulting from glacial erosion near mountain peaks, exhibit a clear tectonic control influenced by local faulting with NW and NE orientations. The concave geometry of cirques is a consequence of Last Glacial Maximum (LGM) ice pressure on Miocene plutonic rocks, leading to the abrasion of bedrock. Subsequent frost action and hillslope processes have further shaped these landforms, exemplified in summits like Cerro Urán, Cerro Brunka, Cerro Dúrika, and potentially others within the Cordillera de Talamanca. Cirques along these summits can be categorized into two types: continuous and discontinuous cirques. Continuous cirques, limited in extent and often occupying small areas

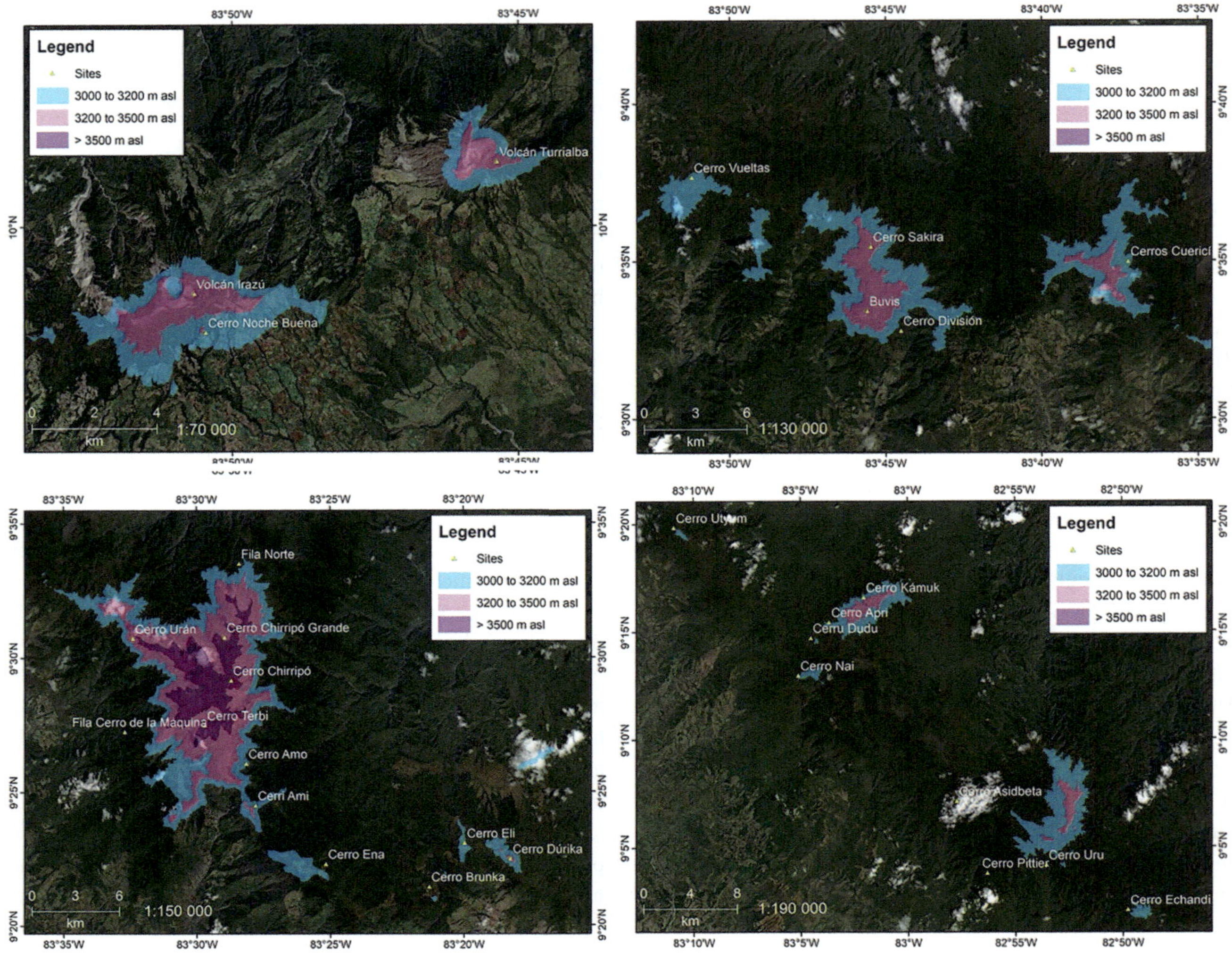

Fig. 8.2 Costa Rican summits over 3000 m asl referenced in Fig. 8.1. The upper left panel shows the Irazú (3432 m) and Turrialba (3329 m) volcanoes. The upper right panel exhibits Cerro Vueltas (Buenavista, 3156 m) (colloquially known as Cerro de la Muerte), Cerro Sakira (3100 m), Buvis (3491 m), and Cerros Cuericí. The lower left panel shows Cerro Chirripó (3820 m asl) and the surrounding higher summits of Costa Rica as well as Cerro Ena, Brunka (3000 m), Eli (3000 m), and Dúrika (3000 m). The lower right panel displays Cerro Utyum (3000 m), Cerro Kámuk (3300 m), and other minor hills as Cerro Pittier (2844 m) and Cerro Echandi (2400 m)

individually, display elongated or arced shapes. Due to their recent age or high altitudinal position, continuous cirques are well-preserved in the landscape, easily recognizable in the field, and persist due to the lack of significant fluvial energy. In contrast, discontinuous cirques, typically taking on a "boomerang-shaped" appearance, exhibit concave slope geometries and are characterized by well-developed fluvial systems. These cirques have undergone substantial post-LGM fluvial erosion, resulting in their discontinuity. The extent of fluvial erosion determines the interruption or erosion of glacial cirques, as fluvial headwaters and tributaries reach basin divides.

Arêtes, narrow and sinuous crests formed when glacial erosion acts on two adjacent cirques, are typically constituted by fresh, exposed bedrock with a convex shape or a slight cover of debris. The presence of arêtes depends on glacial activity and bedrock competence. Sharp (rocky) arêtes are found in areas devoid of vegetation, particularly at the highest elevations of the Cordillera de Talamanca. Examples include the highest summits of Chirripó National Park, Cerro Vueltas, Cerro Asunción, and potentially Cerro Brunka and Cerro Dúrika. Roches moutonnées, characterized by striations, p-forms, scars, and steps, are remnants of glacial abrasion in valley bottoms. These features are particularly evident in the Cordillera de Talamanca, showcasing the lingering impact of past glacial activity. Additionally, riegels, large rock barriers often composed of resistant rock outcrops, are found across valleys, and can create natural dams for lakes. These broad steps, slightly inclined, contribute to the diverse geomorphological tapestry of the region.

Fig. 8.3 Erosional glacial landscapes of the summits of Costa Rica. **a** Cerro Asunción summit (3335 m) heading NE to other Cordillera de Talamanca summits next to the Pan-American Highway nearby Cerro Vueltas (3156 m). **b** Pico Sureste (3760 m) to the left, Cerro Terbi (3760 m), and Cerro Los Crestones (3720 m) to the right in the Chirripó National Park. **c** Cerro Chirripó, the highest summit in Costa Rica with 3820 m. **d** Cerro Dúrika at 3285 m which presumably had paleoglacial influence during the LGM

8.2.2 Depositional Landforms

Morphologies resulting from the accumulation of glacial sediments in the high summits of the Cordillera de Talamanca encompass various landforms, including moraines, glacial valleys transformed by erosion but subsequently sedimented through mass movement processes, predominantly debris falls, till deposits, and glacial lakes. These features, illustrated in Fig. 8.4, are typically located on convex slopes characterized by rocky or smooth morphologies. The interplay of glacial processes and subsequent sedimentation has given rise to a diverse range of depositional landforms in the region. Moraines, prominent features created by the deposition of glacial debris, serve as key indicators of past glacial activity and are found in the high-altitude landscapes of the Cordillera de Talamanca. Glacial valleys, originally formed through erosional processes, have undergone subsequent sedimentation primarily driven by mass movement processes, particularly debris falls. The resulting landscape showcases the intricate interplay between glacial erosion and sedimentation, contributing to the unique geomorphological character of a humid tropical region.

Till deposits, composed of unsorted glacial sediments, further contribute to the high-altitude landscape, reflecting the complex history of glacial dynamics in the Cordillera de Talamanca. Glacial lakes, which initially formed through erosional processes, now act as depositional landforms, adding to the diverse array of features in the higher altitude Costa Rica region. These glacial lakes, situated in convex slopes with rocky or smooth morphologies, are key elements of the landscape and contribute to the overall complexity of the region's geomorphology. The depositional landforms in the Cordillera de Talamanca exhibit a range of characteristics, from ephemeral features that fill only during the rainy season to well-defined formations with considerable widths and lengths. This variability underscores the dynamic nature of the region, shaped by a combination of glacial processes, sedimentation, and mass movement.

Lateral moraines play a significant role in shaping the landscape of Chirripó National Park, particularly on the Caribbean slopes, where lower rainfall totals have resulted in less erosion over time, contributing to the preservation of these moraines (Fig. 8.4). The presence of lateral moraines in the park has been recognized in various studies,

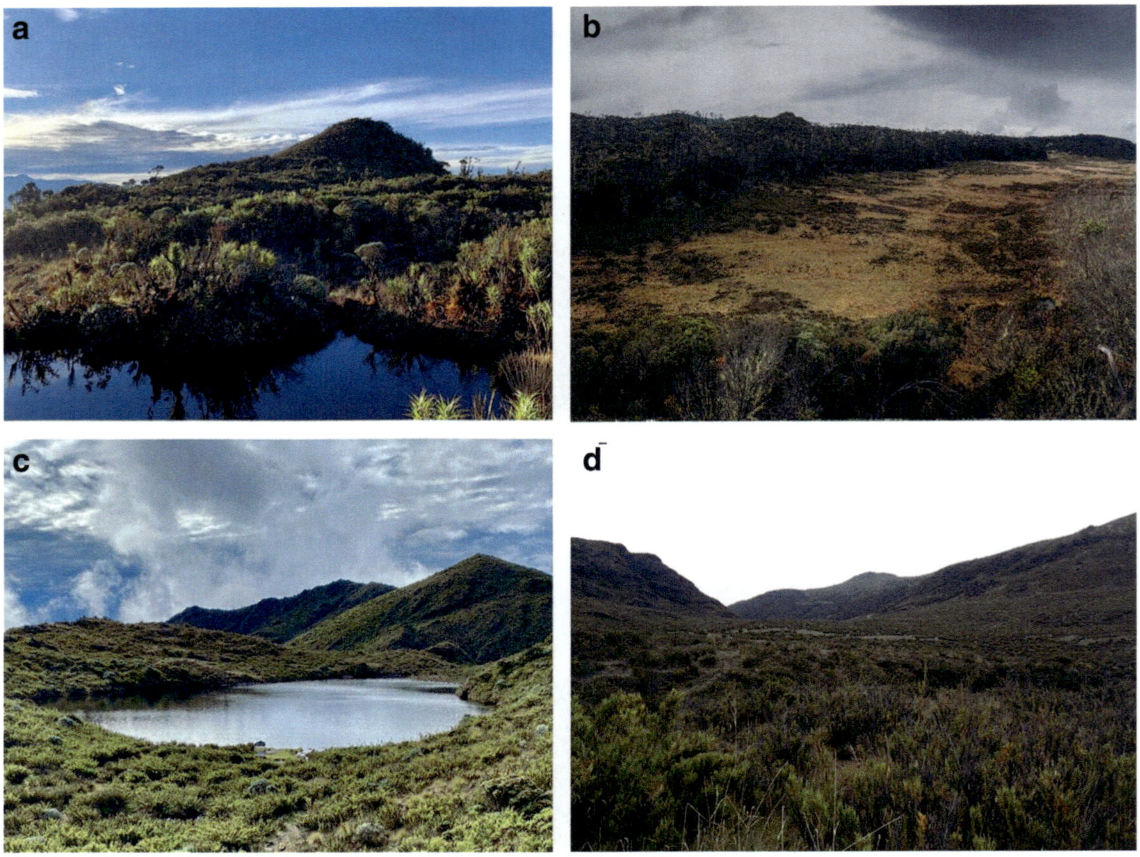

Fig. 8.4 Depositional glacial landforms of Costa Rica. **a** Cerro Utyum (3000 m) and its lacustrine small wetlands. **b** Sabana de Los Leones (~3000 m) in Chirripó National Park. **c** Lago Ditkevi at 3505 m in Chirripó National Park. **d** Subhorizontal till deposits of Valle de los Conejos at approximately 3450 m in Chirripó National Park

including works by Weyl (1955), Hastenrath (1973), Orvis and Horn (2000), Lachniet and Seltzer (2002), and Bergoeing (2017). However, due to the limited resolution of aerial photos examined, often at 1:40,000 scale or less, the details of these moraines have not been fully explored.

Till deposits, common in the region, typically consist of poorly sorted debris lacking stratification, suggesting the alteration of sediment bodies from their original morphology. These materials form mantles in valley bottoms, currently lacking distinct morphology, or are disassembled, as observed in Valle de las Morrenas and Valle de los Conejos in Chirripó National Park. Other regions, such as the summits of Cerro de la Muerte, Brunka, Dúrika, and other subhorizontal areas above 3000 m, may have undergone similar formation processes during the Last Glacial Maximum (LGM).

Glacial cirques in the region support the hypothesis that small bodies of water found in less distinct environments are relict glacial lakes formed from the impoundment of melted glacier waters. These glacial lakes are interpreted as kettle ponds, cirque lakes, paternoster lakes, or ice-marginal lakes, originating from glacial scouring processes that created transversal depressions in the parent material. Despite their erosional origins, the ongoing material deposition resulting from the progressive movement of paleoglaciers, and current fluvial dynamics has transformed these features into accumulation landforms. This dynamic interplay between past glacial processes and current environmental conditions continues to shape the diverse landscape of Chirripó National Park(Fig. 8.5).

8.3 Future Studies

The future exploration of geomorphology in various regions of Costa Rica holds significant potential for advancing our understanding of paleoglacial landscapes. Quesada-Román and Peralta-Reyes (2023) emphasize the need for comprehensive studies, encompassing geomorphological mapping, cosmogenic and non-cosmogenic dating, paleo and current ecological relationships, as well as the promotion of geodiversity and biodiversity. High-resolution methods, incorporating remotely piloted aircraft systems (RPAS) with LiDAR and multispectral cameras, are recommended

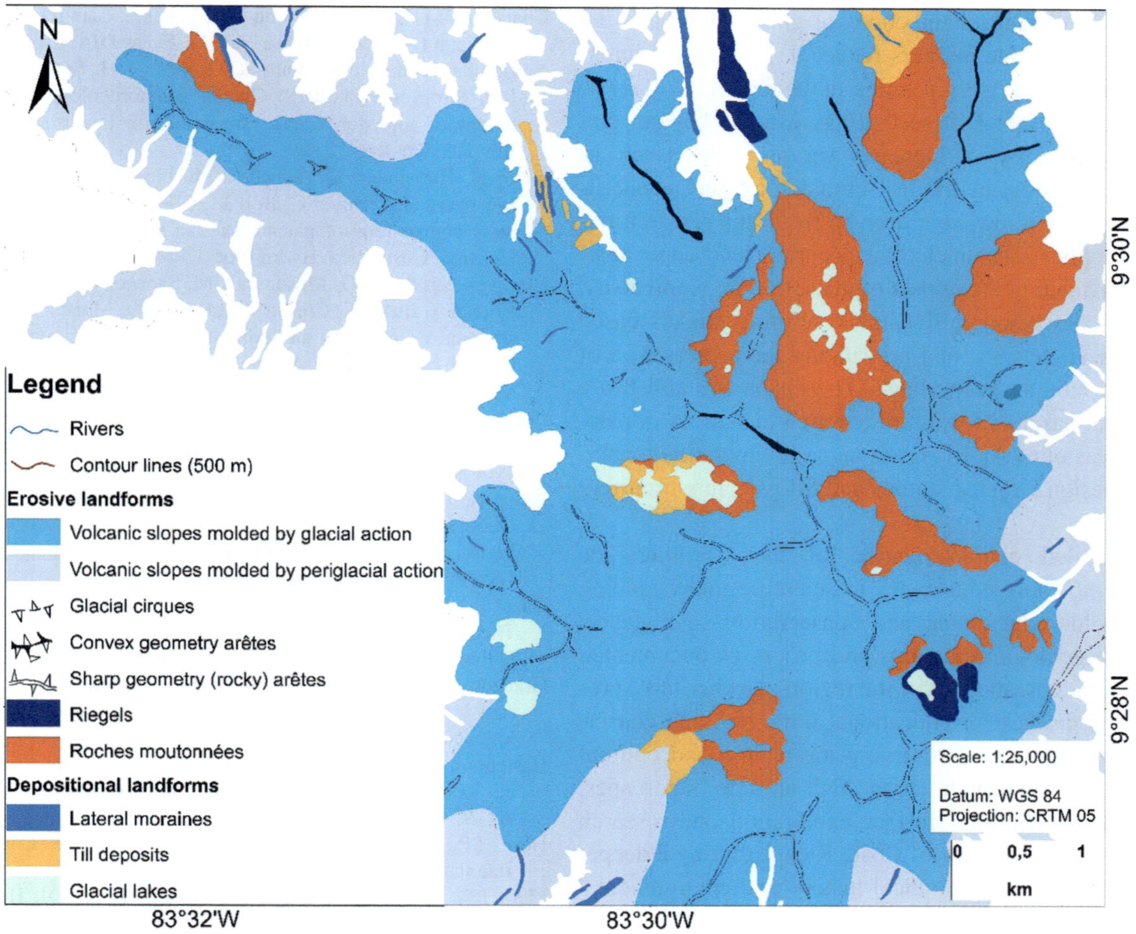

Fig. 8.5 Glacial geomorphological mapping of Chirripó National Park. See many of the explained erosional and depositional landforms which are common in the paleoglacial mountains of Costa Rica

for geomorphological mapping to provide detailed insights into the characteristics of paleoglacial landforms in tropical environments, where weathering and erosion have altered these relicts extensively (Chandler et al. 2018; Granados-Bolaños et al. 2021).

While glacial geoheritage studies are more prevalent in temperate regions, Chirripó National Park in central-south Costa Rica, situated in the Cordillera de Talamanca, stands out as a noteworthy tropical location with significant geodiversity (Quesada-Román et al. 2019). The Strategic Environmental Assessment of UNESCO underscores the importance of La Amistad Biosphere Reserve, to which the park belongs, highlighting the potential for a geopark proposal (UNESCO, 2018). Pérez-Umaña and Quesada-Román (2018) have proposed that Chirripó National Park and La Amistad International Park possess geophysical and cultural elements that warrant consideration for designation as geoparks. Additionally, Quesada-Román and Pérez-Umaña (2020) have conducted an inventory and evaluation of key geosites within Chirripó National Park, paving the way for potential geoconservation and geotourism initiatives. To further expand paleoglacial geoheritage studies, attention can be directed to locations like Cerro Buenavista or Vueltas, Cerro Dúrika, and Cerro Kámuk, where the confirmation of Last Glacial Maximum (LGM) dynamics dating is essential. These efforts contribute not only to scientific understanding but also to the conservation and appreciation of Costa Rica's unique geological heritage.

8.4 Conclusions

The studies conducted since the 1950s represent a transformative era in advancing our scientific understanding of tropical glacial landscapes, particularly in the Cordillera de Talamanca. Technological and methodological advancements in mapping, analytical techniques, and scientific tools have played a pivotal role in unraveling the complexities of the region's glacial and periglacial landforms. This chapter has provided a comprehensive overview of the diverse array of landforms in the Cordillera de Talamanca, ranging from moraines and cirques to aretes. These features not only showcase the dynamic interactions of glacial processes but also serve as valuable markers for reconstructing the intensity and duration of past glaciation in the region.

The significance of these findings extends beyond glacial geomorphology, influencing various fields such as geography, ecology, hydrology, climatology, and geoheritage studies. The knowledge derived from this research has practical applications, including accurate landscape mapping, understanding the impact of glaciation on biodiversity, assessing water resource dynamics, refining models based on past glacial activity, and prioritizing the preservation of the region's unique geoheritage. The chapter establishes a foundational baseline for future research endeavors in paleoglacial areas of Costa Rica, serving as an invitation for scientists to design targeted studies and fill gaps in our current knowledge.

The detailed characterization of the Cordillera de Talamanca not only contributes to scientific understanding but also holds educational and conservation significance. The unique geological features and processes documented in this chapter become part of the region's geoheritage, providing opportunities for educational initiatives and conservation efforts aimed at preserving the natural and cultural heritage associated with these high-altitude landscapes. This chapter encourages further exploration and research in the less-studied regions of Costa Rica, fostering a deeper understanding of the geological history and environmental dynamics in this tropical setting.

Acknowledgements Jason Morales, Germain Esquivel, Danny Gamboa, and Junior Porras for some of the photographs. David Young for his assistance in mapping. Néstor Campos and Soll Kracher for his friendly and very useful revisions.

References

Alfaro A, Denyer P, Alvarado GE, Gazel E, Chamorro C (2018) Estratigrafía y petrografía de las rocas ígneas en la Cordillera de Talamanca, Costa Rica. Rev Geol Amér Cent 58:7–36

Alfaro A, Gazel, E, White, WM., Jicha, B, Rasbury, T (2021) Unravelling the genesis of young continental-arc shoshonites in the Talamanca Cordillera, Costa Rica. Lithos 386

Barquero J, Ellenberg L (1983) Geomorfología del piso alpino del Chirripó en la Cordillera de Talamanca, Costa Rica. Rev Geográ Amér Cent 17–18:293–299

Benn DI, Owen LA, Osmaston HA, Seltzer GO, Porter SC, Mark B (2005) Reconstruction of equilibrium-line altitudes for tropical and sub-tropical glaciers. Quat Int 138–139:8–21

Bergoeing JP (1977) Modelado glaciar en la Cordillera de Talamanca, Costa Rica. Instituto Geográfico Nacional. Informe Semestral 33–44

Castillo Muñoz R (2010) Glaciaciones e Interglaciaciones en Costa Rica: Realidades y enigmas geológicos, San José, Costa Rica: Litografía e Imprenta LIL, S.A. 197

Cunningham M, Stark CP, Kaplan M, Schaefer J (2019) Glacial limitation of tropical mountain height. Earth Surf Dynam 7:147–169

Chandler BM, Lovell H, Boston CM, Lukas S, Barr ID, Benediktsson ÍÖ, Benn DI, Clark CD, Darvill CM, Evans DJA, Ewertowski MW, Loibl D, Margold M, Otto JO, Roberts DH, Stokes CR, Storrar RD, Stroeven AP (2018) Glacial geomorphological mapping: a review of approaches and frameworks for best practice. Eart Sci Rev 185:806–846

Esquivel Hernández G, Sánchez Murillo R, Quesada-Román A, Mosquera GM, Birkel C, Boll J (2018) Insight into the stable isotopic composition of glacial lakes in a tropical alpine ecosystem: Chirripó. Costa Rica. Hydro Proc 32(24):3588–3603

Esquivel Hernández G, Mosquera GM, Sánchez Murillo R, Quesada-Román A, Birkel C, Crespo P, Boll J (2019) Moisture transport and seasonal variations in the stable isotopic composition of rainfall in Central American and Andean Páramo during El Niño conditions (2015–2016). Hydro Proc 33(13):1802–1817

Esquivel Hernández G, Sánchez Murillo R, Vargas Salazar E (2021) Chirripó hydrological research site: advancing stable isotope hydrology in the Central American Páramo. Hydro Proc 35(5)

Esquivel Hernández G, Sánchez Murillo R, Vargas Salazar E, Quesada-Román A (2022) Water stable isotopes reveal the hydrological response of Costa Rican glacial lakes to climate variability. J South Ameri Earth Sci 120:104091

Gazel E, Hayes JL, Ulloa A, Alfaro A, Coleman DS, Carr MJ (2019) The record of the transition from an oceanic arc to a young continent in the Talamanca Cordillera. Geoch Geoph Geosy 20(6):2733–2752

Granados Bolaños S, Quesada Román A, Alvarado GE (2021) Low-cost UAV applications in dynamic tropical volcanic landforms. J Volcanol Geotherm Res 410:107143

Hastenrath S (1973) On the Pleistocene glaciation of the Cordillera de Talamanca. Costa Rica. Zeitschrift Gletscherk Glazialgeol 9(1–2):105–121

Horn SP (1990) Timing of deglaciation in the Cordillera de Talamanca, Costa Rica. Clim. Res. 1:81–83

Mark BG, Harrison SP, Spessa A, New M, Evans DJ, Helmens KF (2005) Tropical snowline changes at the last glacial maximum: a global assessment. Quat Int 138–139:168–201

Lachniet MS, Seltzer GO (2002) Late quaternary glaciation of Costa Rica. Geolo Socie Ameri Bullet 114(5):547–558

Lachniet MS, Seltzer GO, Solís L Kappelle M, Horn SP (2005) Geología, geomorfología y depósitos glaciares en los páramos de Costa Rica. Páramos de Costa Rica. Santo Domingo de Heredia, Costa Rica: Inst ase Biodiver (INBio) 767

Li Y, Tieche T, Horn SP, Li Y, Chen R, Orvis KH (2019) Mapping glacial landforms on the Chirripó massif, Costa Rica, ase don google earth, a digital elevation model, and field observations. Rev Geoló Amér Cent 60:109–121

Orvis KH, Horn SP (2000) Quaternary glaciers and climate on Cerro Chirripó. Costa Rica. Quat Res 54(1):24–37

Palacios D, Stokes CR, Phillips FM, Clague JJ, Alcalá Reygosa J, Andres N, Angel I, Blard PH, Briner JP, Hall BL (2020) The deglaciation of the Americas during the Last Glacial Termination. Earth Sci Rev 203:1031–13

Pérez Umaña D, Quesada-Román A (2018) Una propuesta para la valoración de Geoparques en Costa Rica. Anuár Insti Geocien UFRJ 3:382–394

Porter SC (2001) Snowline depression in the tropics during the last glaciation. Quat Sci Rev 20(10):1067–1091

Potter RS, Li Y, Horn SP, Orvis KH (2019) Cosmogenic Cl-36 surface exposure dating of late Quaternary glacial events in the Cordillera de Talamanca, Costa Rica. Quat Research, pp 1–16

Quesada-Román A (2016) Peligros geomorfológicos: inundaciones y procesos de ladera en la cuenca alta del río General (Pérez Zeledón), Costa Rica. Maestría en Geografía con énfasis en Geografía Ambiental. Posgrado en Geografía. Universidad Nacional Autónoma de México. p 157

Quesada-Román A, Stoffel M, Ballesteros Cánovas JA, Zamorano Orozco JJ (2019) Glacial geomorphology of the Chirripó National Park. Costa Rica J Maps 15(2):538–545

Quesada-Román A, Campos N, Alcalá Reygosa J, Granados Bolaños S (2020a) Equilibrium-line altitude and temperature reconstructions during the Last Glacial Maximum in Chirripó National Park, Costa Rica. J. South Ameri Earth Sci 100:102–576

Quesada-Román A, Ballesteros-Cánovas JA, Guillet S, Madrigal-González J, Stoffel M (2020b) Neotropical hypericum irazuense shrubs reveal recent ENSO variability in Costa Rican páramo. Dendrochronologia 61:125704

Quesada-Román A, Pérez Umaña D (2020) Tropical paleoglacial geoheritage inventory for geotourism management of Chirripó national park. Costa Rica. Geoherit 12(3):1–13

Quesada-Román A, Campos N, Granados Bolaños S (2021) Tropical glacier reconstructions during the Last glacial maximum in Costa Rica. Rev Mex Cienc Geológ 38(1):55–64

Quesada-Román A, Torres Bernhard L, Ruiz Álvarez MA, Rodríguez Maradiaga M, Velázquez Espinoza G, Espinosa Vega C, Toral J, Rodríguez Bolaños H (2022) Geodiversity, Geoconservation, and Geotourism in Central America. Land 11(1):48

Quesada-Román A (2022) Pasado, presente y futuro del clima en el Parque Nacional Chirripó, Costa Rica. UNED Research J. pp 38–71

Quesada-Román A, Peralta-Reyes M (2023) Geomorphological Mapping global trends and applications. Geographies 3(3):610–621

Quesada-Román A, Umaña-Ortíz J, Zumbado-Solano M, Islam A, Abioui M, Tefogoum GZ, Kariminejad N, Mutaqin BW, Pupim F (2023) Geomorphological regional mapping for environmental planning in developing countries. Environ Dev 48:100935

Shimizu C (1992) Glacial landforms around Cerro Chirripó in Cordillera de Talamanka, Costa Rica. J. Geogra [japan] 101:615–621

UNESCO (2018) Evaluación Ambiental Estratégica en sitios Patrimonio Mundial naturales. view/news/strategic_environmental_assessment_in_natural_world_heritage/. Accesed Aug 27 2022

Vázquez-Selem L, Lachniet MS (2017) The deglaciation of the mountains of Mexico and Central America. Cuad Inv Geográfica 43(2):553–570

Veas-Ayala N, Quesada-Román A, Hidalgo H, Alfaro E (2018) Humedales del Parque Nacional Chirripó, Costa Rica: características, relaciones geomorfológicas y escenarios de cambio climático. Rev Biolo Tropic 66(4):1436–1448

Veas-Ayala N, Alfaro-Córdoba M, Quesada-Román A (2023) Costa Rican wetlands vulnerability index. Prog Phys Geogr 47(4):521–540

Weyl R (1955) Contribución a la geología de la Cordillera de Talamanca. Instituto Geográfico Nacional: San José, Costa Rica. p 77

Wunsh O, Calvo G, Willsher B, Seyfried H (1999) Geologie der Alpinen Zone des Chirripó-Massives (Cordillera de Talamanca, Costa Rica, Mittlelamerika. Profil 16:193–210

Coastal and Shallow Marine Geomorphology of Costa Rica

9

Jorge Cortés and Adolfo Quesada-Román

Abstract

Costa Rica has coasts on both the Caribbean Sea and Pacific Ocean side, each with very different geomorphological characteristics. The Caribbean coast is 212 km long. To the north of the port area of Moín-Limón, the coast is rectilinear, with high-energy sandy beaches and coastal lagoons. To the south, fossil reef promontories intersperse with sandy beaches, with only two mangrove forests, numerous seagrass beds, and coral reefs. The Pacific coast is 1254 km long with two enclosed gulfs, Nicoya and Dulce, and open gulfs of Santa Elena, Papagayo, and Coronado. Rocky promontories are made of volcanic or sedimentary rocks, with beaches with many different types of sands in between. Isla del Coco, an oceanic island 500 km offshore in the Pacific, is characterized by basaltic outcrops, many cobble beaches, two sandy beaches, and two bays protected from the predominant southern wave activity.

Keywords

Coral reefs · Mangroves · Beaches · Rocky promontories · Biodiversity

J. Cortés (✉)
Centro de Investigación en Ciencias del Mar y Limnología, Universidad de Costa Rica, San Pedro 11051-2060, San José, Costa Rica
e-mail: jorge.cortes@ucr.ac.cr

Escuela de Biología, Universidad de Costa Rica, San Pedro, San José 11501-2060, Costa Rica

A. Quesada-Román
Laboratorio de Geografía Física, Escuela de Geografía, Universidad de Costa Rica, San Pedro 11051-2060, San José, Costa Rica

9.1 Introduction

Costa Rica, located in the southern end of the Central American Isthmus, has coasts on both the Caribbean Sea and Pacific Ocean sides, each with different geomorphological characteristics (Fig. 9.1). Both the land area and the marine environments of Costa Rica are characterized by a very high biological diversity (Cortés and Wehrtmann 2005; Wehrtmann et al. 2009; Avalos 2019). In this chapter, we describe the main characteristics and geomorphological features of both coastal and shallow marine areas of Costa Rica (Figs. 9.1 and 9.2) and of Isla del Coco, an oceanic island 500 km offshore.

9.2 Caribbean Coast

9.2.1 The Near Shore

The Caribbean coast of Costa Rica is 212 km long and can be divided into two sections. To the north of the port area of Moín-Limón the coast is rectilinear, all the way to the border with Nicaragua. To the South of Limón, toward the border with Panama, there are extensive sandy beaches, separated by river mouths and fossil reef promontories (Parkinson et al. 1998; Cortés 2016a). This coast can be genetically classified into volcanic, fluvial, and coastal landforms.

The volcanic geoforms of the Caribbean coast are comprised in volcanic cones. Cerro Tortuguero, Tortuguerito, and San Francisco are ancient and extinct volcanic edifices of 119, 40, and 25 m high, respectively, composed of pyroclastic deposits and potassium basalts of Quaternary age, less than 1 Ma (Gazel et al. 2011). They exhibit distinct volcanic origin and morphological characteristics, and are being dismantled by fluvial erosion, the construction of artificial channels, and the extraction of pyroclastic materials from the flanks. It should be noted that the Caribbean

A. Quesada-Román (ed.), *Landscapes and Landforms of Costa Rica*,
World Geomorphological Landscapes, https://doi.org/10.1007/978-3-031-64940-0_9

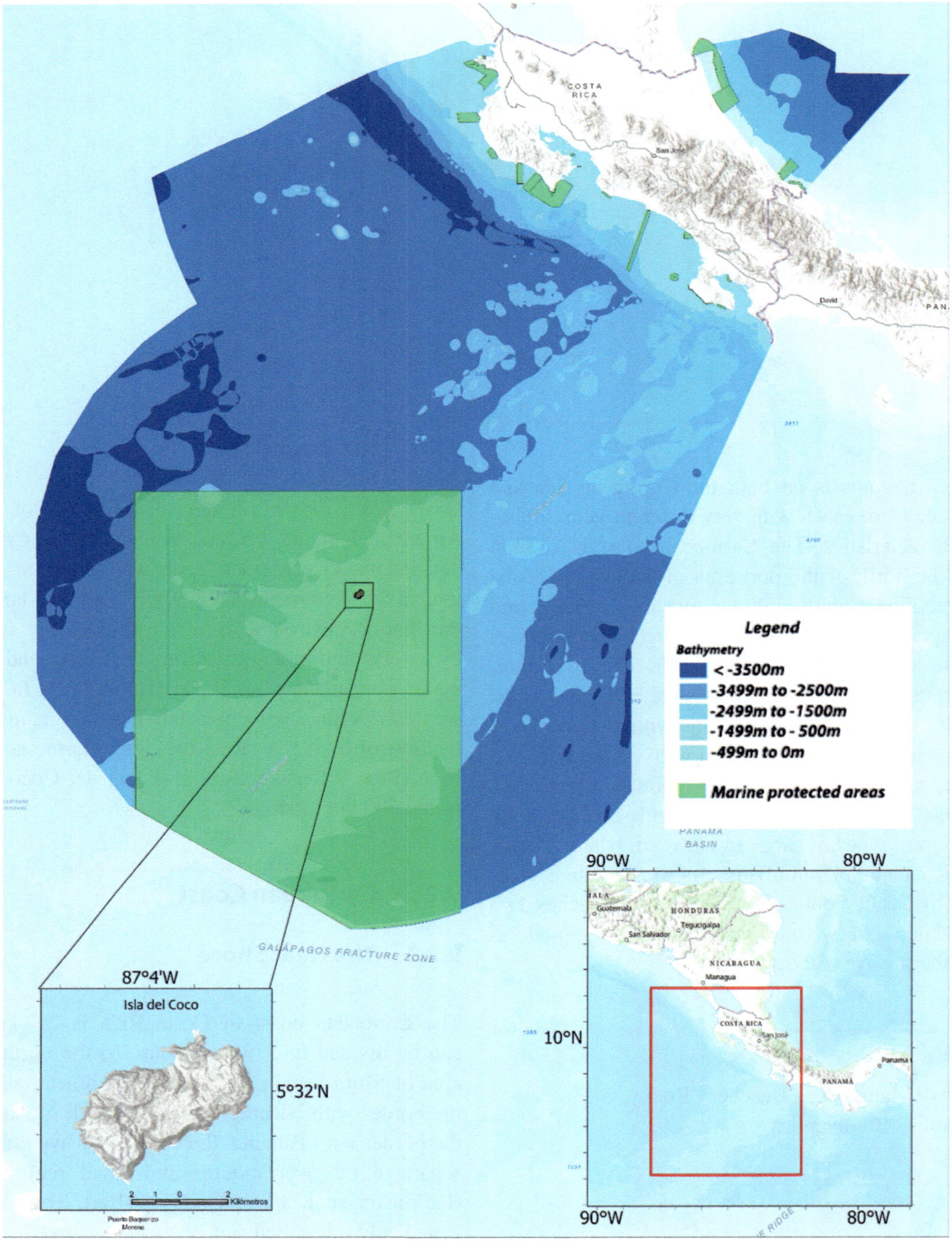

Fig. 9.1 Costa Rica, its Exclusive Economic Zone, marine protected areas, and its location in southern Central America. Isla del Coco geomorphological map

coast includes other examples of this intra-graben volcanism, such as the Lomas del Colorado, the Lomas Azules, the hills of Aguas Zarcas, and the Mercedes Hills (Alvarado 2021).

The fluvial morphologies include rivers, abandoned meanders, alluvial plains, and fluvial islands. Along these lowlands, a complex system of channels marks the transitions of large rivers from the Central Volcanic Cordillera to wide plains (Fig. 9.3). They exhibit extensive systems of natural and artificial channels ending in coastal lagoons (estuaries) in the northern part of the Caribbean (Fig. 9.4). These systems include the Colorado, La Suerte, Parismina, Pacuare, Madre de Dios, and Matina rivers. The Caribbean northern coastal plains were submerged up

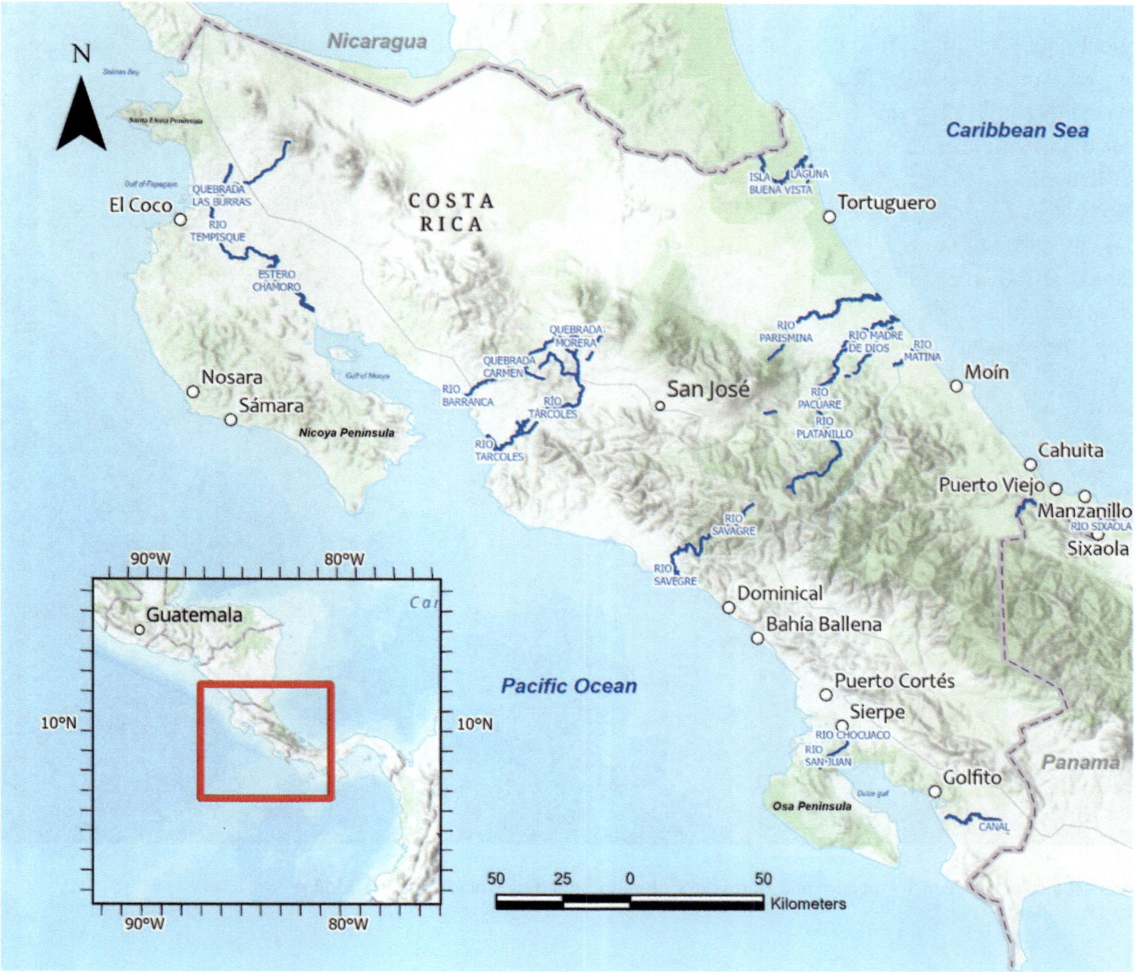

Fig. 9.2 Main coastal geographic landmarks of coastal Costa Rica

to 10 m by the Flandrian transgression until 6 ka and since then emerged due by positive neotectonics (Bergoeing 2017). Abandoned meanders form isolated parts of a river channel in the shape of an arc that formed by cutting the end of a meander and forming a marshy hollow known as an ox-bow lake in floodplains of the Colorado and Parismina rivers (Parkinson et al. 1998). In the northernmost sector of the Caribbean coast of Costa Rica, two segments of fluvial islands of Portillos and Calero are found. In addition, other small fluvial islands are located on the Colorado River and inside the Tortuguero lagoon (Parkinson et al. 1998).

In the Southern Caribbean, the rivers tend to form rapid transitions between the Cordillera de Talamanca, the sedimentary hills, and the narrow alluvial plains (about 2–4 km wide). A last segment of the alluvial plains is in the proximity of the mouth of the Sixaola River, where a marked meandering drainage dominates the landscape.

9.2.2 Coastal and Shallow Marine Landforms

Sedimentary hills are located between the towns of Moín and Sixaola and composed of rocks from the Miocene to Quaternary. Their origin is related to the dynamics of the Central Deformed Belt of Costa Rica and the Depression Belt of Northern Panama, activation of several faults and tectonic uplift (Marshall et al. 2000; Denyer and Alvarado 2007). The sedimentary hills act like a pre-mountain belt between the coastal plains and the Cordillera de Talamanca and are dissected by the rivers of Limoncito, Banano, Vizcaya, Bananito, San Andrés, Estrella, Carbón and Cocles, as well as other minor streams.

The carbonate platforms are segments of marine origin, formed by relatively stable, low inclination Plio-Pleistocene reef limestone rocks that have been gradually raised by slow uplift punctuated by sudden vertical movements due to regional tectonic activity (Cortés 1992) (Fig. 9.5a). The

Fig. 9.3 A coastal river between Tortuguero and Barra del Colorado. Photo by Rocío Córdoba-Muñoz

progressive uplift of old coral reefs due to sudden earthquakes has formed coastal plains adjacent to the coast in the principal cities and towns of Limón, Cahuita, Puerto Viejo and Manzanillo (Cortés et al. 1994a, b; Quesada-Román 2016). Marine abrasion is the prevalent modeling process of these landforms (Fig. 9.5b). The postseismic relaxation favors rapid karstification and could generate problems of subsidence, with the formation of dolines or sinkholes (Quesada-Román 2021).

The northern Caribbean coast is rectilinear with high-energy sandy beaches and coastal lagoons (Fig. 9.6a) (Parkinson et al. 1998; Cortés 2016a). The beaches are important nesting areas, for the Green (*Chelonia mydas*), Hackbill (*Eretmochelys imbricata*) and Leatherback turtles (*Dermochelys coriacea*), and occasionally the Loggerhead turtle (*Caretta caretta*) (Savage 2002; Troëng 2005). Behind the beaches, there is a series of coastal lagoons subject to the ebb and flow of tides and as a result they contain brackish water. These landforms appear as elongated waterbodies of up to 25 km long and 350 m wide, with more than a dozen units along the northern Caribbean region (Fig. 9.6a). Coastal bars originated from the saturation of sediments in the coastal zone due the action of waves and deposition by the coastal currents on surfaces of very shallow slope (Parkinson et al. 1998). The fluctuation of tides, the action of low-energy waves, and the abundant presence of sand contribute to this process. They are closed by ephemeral sand bars and are stable during the dry season. These bars are located between the border with Nicaragua and Moín, reaching a maximum of 82 km long and 1.5 km wide. Beaches can be long (kilometers) and normally narrow (less than 50 m), even with two berms (Fig. 9.6a). Along the channels, there occurs inundated forest (Fig. 9.3) of the palm *Raphia taedigera* and the tree *Pterocarpus officinalis* (Cortés 2016a).

South of Limón, there are extensive sandy beaches, separated by river mouths, and fossil reef promontories (Risk et al. 1980; Cortés 2016a). The beaches are characterized by the presence of white sand, owing to the coastal erosion of coral reefs, such as Cahuita and Manzanillo (Cortés 1992) (Fig. 9.6b); or consist of black sand due to fluvial erosion from the Cordillera de Talamanca (Puerto Viejo and Playa Negra), and the Cordillera Volcánica Central (Parismina, Barra del Colorado and Matina Beach). Some of these beaches have experienced intense erosion during the last years possibly due postseismic relaxation, sea level rise, or human modification (Quesada-Román and Pérez-Briceño 2019).

One of the only two mangrove forests of the Caribbean coast of Costa Rica is located at Moín, and it is highly

Fig. 9.4 Coastal lagoons along the northern Caribbean coast. Photo by Guillermo Alvarado

impacted by the port activities. The other mangrove forest of the Caribbean coast is located in the Gandoca-Manzanillo National Wildlife Reserve (Cortés 1998, 2016a; Coll et al. 2001). Seagrass beds, consisting mainly of turtle seagrass, *Thalassia testudinum,* and manatee seagrass, *Syringodium filiformis*, are present in the lagoons behind the coral reef crest (Fig. 9.7) (Cortés 1998; Paynter et al. 2001; Krupp et al. 2009; Samper-Villarreal et al. 2018). Extant coral reefs are located off the rocky promontories (fossil reefs) all along the coast (Fig. 9.8) (Cortés and Guzmán 1985; Cortés and Jiménez 2003a, b; Cortés 2016a). The coral reefs of the Caribbean coast of Costa Rica have been impacted by sediments originated from the nearby banana plantations (Cortés and Risk 1985).

9.3 Pacific Coast

9.3.1 1The Nearshore

The Pacific coast is 1254 km long and comprises fluvial and coastal morphologies. Lower coastal mountains are present in three Pacific subregions (northern, central, and southern), but they differ considerably in terms of lithology. They vary from Jurassic-Cretaceous intrusive/extrusive igneous rocks domain and to Cretaceous and Holocene sedimentary lithologies (Denyer and Alvarado 2007). These lower mountains near the Pacific coasts are product of interaction between plate subduction processes, lithological control, and differential weathering. Normally, they present a very dense and grouped arrangement of small valleys and clear mountain summits (Bergoeing 2017). Locally they border the sea with the presence of cliffs, more common in the northern section of the Pacific coast and where older igneous rocks crop out. Coastal hills are isolated or grouped promontories with origin similar to the lower coastal mountains, but with a hilly, even hummocky morphology and lower elevation differentiation (Quesada-Román 2018). Normally they constitute the transitional belt between continental mountain regions and flatlands near coasts. Karstic hills are located only in the Nicoya Gulf where elongated elevations built of carbonate sediments stand out in the fluvio-coastal flatlands of the Tempisque River. These units are the result of local faulting and subduction process (Jaccard et al. 2001). The volcanic-coastal altiplanos are situated only in the Papagayo Gulf and are products of ignimbrites deposition from the

Fig. 9.5 **a** Uplifted carbonate platform and coral reef at Punta Piuta after the Limón earthquake of 1991. **b** Fossil reef promontory at Punta Mona. Photos by Jorge Cortés

Fig. 9.6 Diversity of beaches in Costa Rica. **a** High-energy sandy beach at Tortuguero National Park. **b** Low-energy carbonate beach at Cahuita National Park. Photos by Jorge Cortés

Fig. 9.7 Seagrass beds with *Thalassia testudinum*, turtle seagrass, and *Syringodium filiformes*, manatee seagrass, at Cahuita National Park. Photo by Jorge Cortés

Guanacaste Volcanic Cordillera (Marshall 2007; Herrera-Zúñiga and Vargas 2012; Alvarado 2021).

Fluvial landforms include accumulation glacis, fluvial plains, and fluvio-coastal flatlands. Accumulation glacis are gently sloping piedmonts/transitions between the volcanic-coastal altiplanos and the coastline near Papagayo Gulf. They have been molded by physical and chemical weathering and fluvial erosion. Fluvial plains are a widespread type of landscape along the Pacific coast of Costa Rica, reflecting fluvial dynamic of small to large catchments (Quesada-Román et al. 2023). Fluvial plains are located just before the transition to the sea, in estuaries, deltas, coastal plains, lagoons, beaches, or fluvio-coastal flatlands (Fig. 9.9). Normally, the have a low slope of 2–10°. The fluvio-coastal flatlands are intermediate fringes between fluvial and coastal plains with low-angle slopes and a dynamic interplay of fluvial and littoral processes.

Coastal plains are elongated subhorizontal surfaces, tilted toward the ocean. They have been formed during seafloor uplift or sea level decline when abrasive-accumulative shelf surfaces emerged. Cliffs are situated mostly in the northern section of the Pacific coast, especially in the Santa Elena Peninsula and Gulf of Papagayo. Nonetheless, these landforms are typically associated with Cretaceous igneous intrusive rocks, also in the Gulf of Nicoya, Quepos, Dominical, Osa Peninsula, and Dulce Gulf.

Fig. 9.8 Coral reefs diversity. **a** *Acropora palmata* reef in the 1980s, and **b** *Agaricia agaricites* reef in 2012, both at Cahuita National Park. Photos by Jorge Cortés

9.3.2 Coastal and Shallow Marine Landforms

Coastal landforms include deltas, gulfs, estuaries, lagoons, islands, tombolos, arches, abrasion platforms, beaches, and several subtidal ecosystems. Deltas are depositional landforms which owe their morphology primarily to fluvial processes of feeding rivers and to a lesser degree to the action of waves and tides. This condition only happens in the Térraba-Sierpe delta, which results from a complex regional and local tectonic configuration, as well as intense fluvio-coastal dynamics highly dependent of climatic variability (Acuña-Piedra and Quesada-Román 2016). It is one of the best examples of mangrove habitats in Central America (Fig. 9.9). On the Pacific coast, there are two enclosed gulfs, Nicoya and Dulce, and three open gulfs, Santa Elena, Papagayo, and Coronado (Cortés 2016b). Golfo de Nicoya (Fig. 9.10a) is a typical estuary with freshwater entrance in the inner section, becoming deeper and saltier moving out

Fig. 9.9 Fluvio-coastal transitions of Pacific coast of Costa Rica. **a** Coastal plain of the Sierpe and Térraba rivers. **b** Boca Nueva, Térraba-Sierpe National Wetland. Photos by Jorge Cortés

of the gulf (Vargas 2016), while Golfo Dulce (Fig. 9.10b) is a deep (200 m) anoxic basin with coral reefs and mangrove forest along the shallow perimeter (Cortés 1990, 2016b; Hebbeln and Cortés 2001; Quesada-Alpízar and Cortés 2006). These two prominent gulfs are enclosed by the Nicoya and Osa peninsulas. Another important peninsula is in the northern section of the coast, Península de Santa Elena, formed by spectacular outcrops of sedimentary formations and the oldest rocks of the country (Denyer 2019) (Fig. 9.11).

Estuaries are common morphologies all along the Pacific coastline of Costa Rica. They are located in areas under the influence of tides (Lizano 2006) and originate from inundation and expansion of the mouths of large rivers. Some examples are the following river mouths: Coto-Colorado, Esquinas, Savegre, Naranjo, Parrita, Tárcoles, Barranca,

Fig. 9.10 Main gulfs of Costa Rica. **a** Golfo de Nicoya. Photo by Pepe Manzanilla. **b** Golfo Dulce. Photo by Jorge Cortés

the conjunction of Aranjuez (Quesada-Román 2022) and Guacimal, Tempisque, and several small units along Nicoya Peninsula and the northern Pacific coast. Lagoons are natural, shallow water bodies separated from the sea by coastal banks, bars, tombolos, or beaches. Good examples are Pejeperro (Fig. 9.12a), Corcovado Lagoon, Manuel Antonio Lagoon, and the lagoon at Punta Respingue in Santa Elena Peninsula (Fig. 9.12b) (Denyer and Cárdenas 2000).

Impressive examples of tombolos are Punta Respingue in Santa Elena Peninsula, San Juanillo, Punta Catedral, and Puerto Escondido in Manuel Antonio National Park, and the "*whale tail*" in Uvita beach (Fig. 9.13). Arches form within rocky coastal promontories where these are pierced by a tunnel. They are formed due to selective weathering and marine erosion. Some interesting examples are at Playa Ventanas, Marino Ballena National Park (Fig. 9.14a), and Isla San José, Islas Murciélago, Guanacaste Conservation Area (Fig. 9.14b). Abrasion platforms are portions of the coastline leveled by waves and bioerosion of original rocks. Examples are located in Bahía Salinas (Fig. 9.15a) and near other rocky outcrops in the region, between Jacó and Hermosa beaches, near Dominical in the Central Pacific coast (Fig. 9.15b), and in Punta Burica. Beaches are coastal strips, slightly sloping toward the sea, made up of sand, gravel, and pebbles deposited by wave currents. It can be found single or multiple bars, as well as elongated, rocky, and pocket beaches in the Pacific coast of Costa Rica (Cortés 2016b) (Fig. 9.16). The fauna inhabiting some beaches include polychaetes followed by crustaceans and mollusks as the dominant groups (Dexter 1974; Sibaja-Cordero et al. 2019).

Fig. 9.11 Erosive coastal landforms. **a** Volcanic outcrop at the shoreline, Nicoya Complex Formation, Santa Elena Peninsula. **b** Cliffs in sedimentary rocks, Santa Elena Peninsula, Guanacaste Conservation Area. Photos by Jorge Cortés

Rocky promontories of variable size punctuate the Pacific coast and consist of volcanic or sedimentary rocks (Denyer and Cárdenas 2000). Most of rocky formations are being uplifted (Denyer and Cortés 2001) and have an interesting flora and fauna (Sibaja-Cordero and Cortés 2008; Sibaja-Cordero and García-Méndez 2014; Cortés 2016b). On the Pacific coast of Costa Rica, there are approximately 129 islands ranging from 107 m^2 to 43 km^2 (Chira Island). More studies are needed because there are very few publications on their geomorphology, geological origin, biota, or environmental condition (Denyer and Cárdenas 2000; Cortés 2016b). An outstanding feature that has been described are fossil aeolian dunes found on several of the islands of Islas Murciélago in the Guanacaste Conservation Area, Northwest Pacific (Denyer et al. 2005).

Mangrove forests are important coastal ecosystems that modify the coastline by trapping sediments with their roots, fundamental for coastal protection against storms and strong wave action. They are diverse and are important nursery areas for many species, including commercially important species and economically important for tourism. Mangrove forests trap more carbon that any other marine ecosystem and surpass all terrestrial ecosystems (Cannicci et al. 2021; Alongi 2022; Kelty et al. 2022). Mangrove forests are found all along the Pacific coast of Costa Rica, with the largest forest in the Térraba-Sierpe National Wetland (Silva-Benavides 2009; Zamora-Trejos and Cortés 2009; Samper Villarreal et al. 2012; Loría-Naranjo et al. 2014; Cortés 2016b; Acuña-Piedra and Quesada-Román 2021; Veas-Ayala et al. 2023) (Fig. 9.17). Soft sedimentary bottoms range from muddy to sandy substrates. The former are associated with mangrove forest and occur especially in Golfo de Nicoya with a rich fauna, which is an important food source for commercially important species (Vargas 1996, 2016). In muddy to sandy substrate, seagrasses have been found (Samper et al. 2018) (Fig. 9.18a). Sandy substrates are found in many areas along the coast, an example being at Isla del Caño Biological Reserve (Fig. 9.18b).

Subtidal environments in the Pacific include seagrass beds, which are much smaller and ephemeral than their Caribbean counterparts (Samper-Villarreal et al. 2018, 2020) (Fig. 9.18a), coral reefs (Fig. 9.19a, d), and rocky bottoms with abundant algal and animal cover (Suárez et al. 2021) (Fig. 9.19b). Coral reefs of the Eastern Tropical Pacific are small and have few reef-building coral species, but they are important as a refuge for many species (Cortés 2011; Glynn et al. 2017). Another hard substrate subtidal environment is the rhodolith beds, which consist of round, fist-sized calcareous algae (Fig. 9.19c) There are few studies of rhodolith beds, but they have proven to be important for biodiversity and as nursery sites for many species (Solano-Barquero et al. 2022).

9.4 Isla del Coco

Isla del Coco, an oceanic island 500 km offshore on the Pacific Ocean (Fig. 9.1), is characterized by dense forest, basaltic outcrops, many cobble beaches, two sandy beaches, and two bays protected from the predominant southern wave activity (Cortés 2016c) (Fig. 9.20). The Island is the only subaerial exposure of the Coco Submarine Volcanic Range that extends from the Galápagos hotspot to the Osa Peninsula in Southwest Costa Rica (Cortés 2016c; Alvarado 2021). Isla del Coco is part of a submarine shield volcano of complex evolution, which erupted several times during the early Pleistocene (2.2–1.5 Ma). According to Rojas and Alvarado (2012), the island is characterized by alkali

Fig. 9.12 Depositional coastal landforms in Costa Rica. Coastal lagoons: **a** Pejeperro, Osa Península. Photo by Guillermo Alvarado. **b** Respingue, Guanacaste Conservation Area. Photo ACG/Luciano Capelli

basaltic lava flows (aa, pahoehoe, blocky lavas) and dikes, minor trachyte lava flows, volcanic domes, and dikes with subordinate pyroclastic and epivolcanic rocks (Fig. 9.21).

The main coastal features of Isla del Coco are active coastal cliffs, coral reefs, intertidal caves, intertidal and submerged arches, abrasion platforms, surrounding islets, rocky intertidal zones, debris cones, and beaches (Bergoeing 2012; Cortés 2016c) (Fig. 9.20). Rocky intertidal zones are fringes of the coastline leveled by the waves acting upon the original volcanic rocks (Sibaja-Cordero and Cortés 2010) (Fig. 9.22a). Among the main coastal landforms of the island are active cliffs, which are mainly formed by lava flows of even 200 m height (Rojas and Alvarado 2012) (Fig. 9.22b). These morphologies are present all around the island. Active coastal cliffs are alternating with debris cones produced by weathering and mass movements and deposited inside rocky intertidal zones or beaches. Beaches are located mainly in bays with predominant coarse sand, cobbles, and loose rocks (Lizano 2012) (Fig. 9.23). The island is encompassed by intertidal caves

Fig. 9.13 Tombolo at Marino Ballena National Park, Central Pacific coast of Costa Rica. Photo by Rocío Córdoba-Muñoz

and a few arches (Lockwood and Benumof 2000; Cortés 2016c) (Fig. 9.24). Surrounding islets of various sizes are found all around the island.

The shallow subtidal zone consists of sedimentary (Sibaja-Cordero et al. 2012) and rocky bottoms (Sibaja-Cordero et al. 2012), each system with its particular set of organisms (Fig. 9.25a, c). Coral reefs are important shallow water ecosystems and cover extensive areas on the north side of the island, where they are protected from the strong southern swell (Guzmán and Cortés 2007; Alvarado et al. 2020) (Fig. 9.25b). Also in some areas, especially on the north and north-east side, rhodolith beds occur (Fig. 9.25d). Rhodoliths are round calcareous red algae that form extensive beds, associated with a diverse suite of species (Solano-Barquero et al. 2022).

Shallow submerged arches are found around the island (Fig. 9.26), as well as in deep waters, even at depths of 90 m. They are indicators either of a process of subsidence of the island or of a rise in sea level at the end of the last glaciation (Rojas and Alvarado 2012). The limit of the submarine platform around Isla del Coco has variable depth between 110 and 183 m, then it drops-off to several thousands of meters (Lizano 2001, 2012).

9.5 Future Research

The coastal zone is being constantly transformed, either by natural processes, human activity, or a combination of both (Lizano and Salas 2001; Quesada-Román 2021; Graells et al. 2021; Lizano-Araya and Lizano-Rodríguez 2022). These processes are intensifying at an accelerated rate. Tectonics and climate change coupled with sea level rise and changes in storm characteristics are among main drivers of change (Fig. 9.27a). At the same time, alteration of the coastal zone is increasing worldwide, and Costa Rica is no exception (Fig. 9.27b). Improving geomorphological mapping is key to better coastal managing (Quesada-Román and Peralta-Reyes 2023). Knowledge of the present conditions and how they are evolving are necessary to predict future changes.

Marine protected areas (MPAs) are being created as a conservation tool intended to promote healthy and resilient marine ecosystems, protect biodiversity, and provide societal benefits (Grorud-Colvert et al. 2021). MPAs also help in the protection of the coast and the ecosystems present there as well as on offshore. Costa Rica has MPAs on both coasts but more and much larger ones are

Fig. 9.14 Locally controlled and erosive coastal landforms. **a** Tunnel at Punta Ventanas, Marino Ballena National Park. **b** Puerta de la Iglesia (The Church Door) arch at Isla San José, Islas Murciélago, Guanacaste Conservation Area. Photos by Jorge Cortés

needed to ensure real protection of the environment and its organisms (Fig. 9.1). This is important as Ban et al. (2019) has demonstrated that both human well-being and biodiversity conservation can be improved through marine protected areas.

Fig. 9.15 Abrasion platforms: **a** Bahía Salinas, North Pacific coast. **b** Near Dominical, Central Pacific coast. Photos by Jorge Cortés

Fig. 9.16 Common Costa Rican beaches. **a** Isolated sandy beach at Isla del Caño Biological Reserve. **b** Urban beach, Chacarita, Puntarenas. Photos by Jorge Cortés

Fig. 9.17 Mangrove forests of the Pacific. **a** Aerial view of the Térraba-Sierpe National Wetland mangrove forest. Photo by Beatriz Naranjo-Elizondo. **b** Close-up of mangrove forest along the Sierpe River. Photo by Jorge Cortés

Fig. 9.18 **a** Seagrass bed (*Ruppia maritima*) at Iguanita, Beach, Bahía Culebra in the 1990's. **b** Soft sedimentary bottom at Isla del Caño Biological Reserve with Whitetip reef sharks (*Triaenodon obesus*). Photos by Jorge Cortés

Fig. 9.19 **a** Branching coral, *Pocillopora* spp. reef at Isla Leoncillo, Guanacaste Conservation Area. **b** Rocky bottom covered with sea fans, algae of several types and sponges at Isla del Caño Biological Reserve. **c** Rhodolith bed at Isla del Caño Biological Reserve. Photos by Jorge Cortés. Coral reefs of the Pacific. **d** Massive coral, *Pavona clavus* at Tres Hermanas, Marino Ballena National Park. Photos by Jorge Cortés

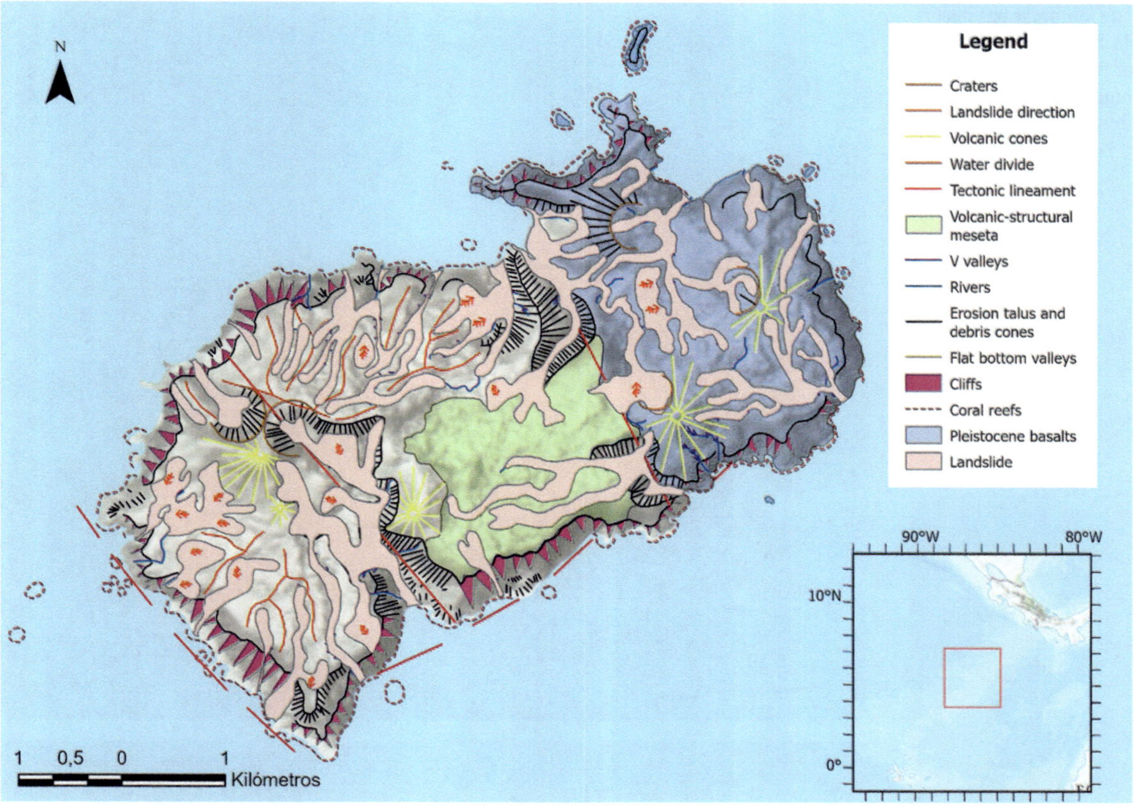

Fig. 9.20 Landforms of Isla del Coco by Bergoeing (2012)

Fig. 9.21 **a** Columnar basalts forming Isla Juan Bautista, southeast of the Island. **b** Massive basalt forming Isla Manuelita. Photos by Jorge Cortés

Fig. 9.22 **a** Rocky intertidal zone at Roca Sucia (Dirty Rock). **b** Basaltic cliffs on the southern side of Isla del Coco. Photos by Jorge Cortés

Fig. 9.23 **a** Beach at Chatham Bay, and **b** detail of rocky section of the beach. Photos by Jorge Cortés

Fig. 9.24 At left hand, one of many intertidal caves and tunnels around the island

Fig. 9.25 **a** Shallow sedimentary bottom at Manuelita Garden. Garden eels (Family Congridae) and corals (*Porites lobata*) in the sandy area and Hammerhead sharks (*Sphyrna lewini*) on the back. **b** Massive coral reef of *Porites lobata*. **c** Shallow rocky bottom with the coral eating sea star, *Acanthaster planci*. **d** Rhodolith bed at Isla Pájara, with coral colonies (*Porites lobata*) and a Whitetip reef shark (*Triaenodon obesus*). Photos by Jorge Cortés

Fig. 9.26 Submerged arch at Roca Sumergida (submerged rock). Photo by Jorge Cortés

Fig. 9.27 **a** Sea level rise is resulting in significant coastal erosion in some areas, as in Cahuita National Park, Caribbean coast. Photo by Jorge Cortés. **b** Transformation of the coastal zone near Playas del Coco, Guanacaste. Photo by Adolfo Quesada-Román

Acknowledgements The logistic and financial support of CIMAR and the Vicerrectoría de Investigación of the University of Costa Rica have been fundamental to study the coastal and marine areas of Costa Rica. We thank David Young who assisted with figures and maps. We greatly appreciate the support received by the Undersea Hunter Group and the Isla del Coco National Park authorities to work there. Many colleagues, students, and assistants have been of great help in the field collecting data and samples, and with enlightening discussions. This chapter is dedicated to the memory of "El Maestro," Ricardo Soto Soto.

References

Acuña FH, Cortés J, Garese A, González-Muñoz R (2020) The sea anemone *Exaiptasia diaphana* (Actiniaria: Aiptasiidae) associated to rhodoliths at Isla del Coco National Park, Costa Rica. Revista De Biología Tropical 68(Supplement 1):S283–S288

Acuña-Piedra JF, Quesada-Román A (2016) Evolución geomorfológica entre 1948 y 2012 del delta Térraba-Sierpe, Costa Rica. Cuaternario y Geomorfología 30(3–4):49–73

Acuña-Piedra JF, Quesada-Román A (2021) Multidecadal biogeomorphic dynamics of a deltaic mangrove forest in Costa Rica. Ocean Coast Manag 211:105770. https://doi.org/10.1016/j.ocecoaman.2021.105770

Alongi DM (2022) Climate change and mangroves. In: Das SC, Pullaiah, Ashton EC (eds) Mangroves: Biodiversity, Livelihoods and Conservation. Springer, Singapore, pp 175–198. https://doi.org/10.1007/978-981-19-0519-3_8

Alvarado JJ, Sánchez-Noguera C, Arias-Godínez CG, Araya T, Fernández-García C, Guzmán AG (2020) Impact of El Niño 2015–2016 on the coral reefs of the Pacific of Costa Rica: the potential role of marine protection. Rev Biol Trop 68(Supplement 1):S271–S282

Alvarado GE (2021) Costa Rica y sus volcanes. EUCR, EUNA, ETCR, San José, Costa Rica

Avalos G (2019) Still searching the rich coast: Biodiversity of Costa Rica, numbers, processes, patterns, and challenges. In: Pullaiah T (ed) Global biodiversity, vol 4: Selected Countries in the Americas and Australia. Apple Academic Press, CRC, Palm Beach, Florida, USA, pp 101–138

Ban NC, Gurney GG, Marshall NA, Whitney CK, Mills M, Gelcich S et al (2019) Well-being outcomes of marine protected areas. Nature Sustainability 2:524–532. https://doi.org/10.1038/s41893-019-0306-2

Bergoeing JP (2012) Geomorfología de la isla del Coco, Costa Rica. Revista Geográfica 151:129–138

Bergoeing JP (2017) Geomorphology and volcanology of Costa Rica. Elsevier, Amsterdam

Cannicci S, Lee SY, Bravo H, Cantera-Kintz JR, Dahdouh-Guebas F, Fratini S et al. (2021) A functional analysis reveals extremely low redundancy in global mangrove invertebrate fauna. In: Proceedings of the national academy of science USA 118:e2016913118. https://doi.org/10.1073/pnas.2016913118

Coll M, Fonseca AC, Cortés J (2001) El manglar y otras asociaciones vegetales de la laguna Gandoca, Limón. Costa Rica. Revista De Biología Tropical 49(Supplement 2):321–329

Cortés J (1990) The coral reefs of Golfo Dulce, Costa Rica: distribution and community structure. Atoll Res Bull 344:1–37

Cortés J (1992) Los arrecifes coralinos del Refugio Nacional de Vida Silvestre Gandoca-Manzanillo, Limón, Costa Rica. Rev Biol Trop 40:325–333

Cortés J (1998) Cahuita and Laguna Gandoca, Costa Rica. In: Kjerfve B (ed) CARICOMP—caribbean coral reef, seagrass and mangrove sites. UNESCO, Paris, pp 107–113

Cortés J (2011) Eastern Tropical Pacific coral reefs. In: Hopley D (ed) The encyclopedia of modern coral reefs: structure, form and process. Springer, Berlin, pp 351–358

Cortés J (2016a) The Caribbean coastal and marine ecosystems. In: Kappelle M (ed) Costa Rican ecosystems. University of Chicago Press, Chicago and London, pp 591–617

Cortés J (2016b) The Pacific coastal and marine ecosystems. In: Kappelle M (ed) Costa Rican ecosystems. University of Chicago Press, Chicago and London, pp 97–138

Cortés J (2016c) Isla del Coco: Coastal and marine ecosystems. In: Kappelle M (ed) Costa Rican Ecosystems. University of Chicago Press, Chicago and London, pp 162–191

Cortés J, Guzmán HM (1985) Arrecifes coralinos de la costa Atlántica de Costa Rica. Brenesia 23:275–292

Cortés J, Jiménez C (2003a) Past, present and future of the coral reefs of the Caribbean coast of Costa Rica. In: Cortés J (ed) Latin American Coral Reefs. Elsevier Science, Amsterdam, pp 223–239

Cortés J, Jiménez C (2003b) Corals and coral reefs of the Pacific of Costa Rica: history, research and status. In: Cortés J (ed) Latin American Coral Reefs. Elsevier Science, Amsterdam, pp 361–385

Cortés J, Risk MJ (1985) A reef under siltation stress: Cahuita, Costa Rica. Bull Mar Sci 36:339–356

Cortés J, Wehrtmann IS (2005) Costa Rica. In: Miloslavich P, Klein E (eds) Caribbean marine biodiversity: the known and the unknown). DEStech Publications, Lancaster, Pennsylvania, pp 169–179

Cortés J, Macintyre IG, Glynn PW (1994a) Holocene growth history of an eastern Pacific fringing reef, Punta Islotes, Costa Rica. Coral Reefs 13:65–73

Cortés J, Soto R, Jiménez C, Astorga A (1992) Earthquake associated mortality of intertidal and coral reef organisms. In: Proceedings of the 7th international coral reef symposium, Guam, vol 1, pp 235–240

Cortés J, Soto R, Jiménez C (1994) Efectos ecológicos del terremoto de Limón. Revista Geológica de América Central, Volumen Especial del Terremoto de Limón, pp 187–192

Denyer P (ed) (2019) Perspectiva geológica del noroeste de Costa Rica: Historia, evolución y cartografía. Editorial Universidad de Costa Rica, San José, Costa Rica

Denyer P, Cortés J (2001) Levantamiento tectónico de la costa pacífica central de Costa Rica. Rev Biol Trop 49(Supplement 2):179–183

Denyer P, Cortés J, Cárdenas G (2005) Hallazgo de dunas fósiles de final del Pleistoceno en las islas Murciélago, Costa Rica. Revista Geológica De América Central 33:29–44

Denyer P, Alvarado GE (2007) Mapa geológico de Costa Rica. Escala 1:400 000. Librería Francesa, San José, Costa Rica

Denyer P, Cárdenas G (2000) Costas marinas. In: Denyer P, Kussmaul S (eds) Geología de Costa Rica. Editorial Tecnológica de Costa Rica, Cartago, Costa Rica, pp 185–218

Dexter DM (1974) Sandy beach fauna of the Pacific and Atlantic coasts of Costa Rica and Colombia. Rev Biol Trop 22:51–66

Gazel E, Hoernle K, Carr MJ, Herzberg C, Saginor I, Van den Bogaard P et al (2011) Plume–subduction interaction in southern Central America: mantle upwelling and slab melting. Lithos 121:117–134

Glynn PW, Alvarado JJ, Banks S, Cortés J, Feingold JS, Jiménez C et al (2017) Eastern Pacific coral reef provinces, coral community structure and composition: an overview. In: Glynn PW, Manzello D, Enochs I (eds) Coral reefs of the Eastern Pacific: persistence and loss in a dynamic environment. Springer Science+Business Media, Dordrecht, pp 107–176

Graells G, Nakamura N, Celis-Diez JL, Lagos NA, Marquet PA, Pliscoff P, Gelcich S (2021) A review on coastal urban ecology: research gaps, challenges, and needs. Front Mar Sci 8:617897. https://doi.org/10.3389/fmars.2021.617897

Grorud-Colvert K, Sullivan-Stack J, Roberts C, Constant V, Horta e Costa B, Pike EP et al. (2021) The MPA guide: a framework to achieve global goals for the ocean. Science 373:eabf0861. https://doi.org/10.1126/science.abf0861

Guzmán HM, Cortés J (2007) Reef recovery 20 years after the 1982–83 El Niño massive mortality. Mar Biol 151:401–411

Hebbeln D, Cortés J (2001) Sedimentation in a tropical fjord: Golfo Dulce, Costa Rica. Geo-Mar Lett 20:142–148

Herrera-Zúñiga P, Vargas AE (2012) Geología de Bahía Culebra, Guanacaste, Costa Rica. Rev Biol Trop 60(Supplement 2):213–223

Jaccard S, Münster M, Baumgartner PO, Baumgartner-Mora C, Denye P (2001) Barra Honda (Upper Paleocene-Lower Eocene) and El Viejo (Campanian-Maastrichtian) carbonate platforms in the Tempisque Area (Guanacaste, Costa Rica). Revista Geológica De América Central 24:9–28

Kelty K, Tomiczek T, Cox DT, Lomonaco P, Mitchell W (2022) Prototype-scale physical model of wave attenuation through a mangrove forest of moderate cross-shore thickness: LiDAR-based characterization and Reynolds Scaling for engineering with Nature. Front Mar Sci 8:780946. https://doi.org/10.3389/fmars.2021.780946

Krupp LS, Cortés J, Wolff M (2009) Growth dynamics and state of the seagrass *Thalassia testudinum* in the Gandoca-Manzanillo National Wildlife Refuge, Caribbean Costa Rica. Rev Biol Trop 57(Supplement 1):187–201

Lizano O (2006) Algunas características de las mareas en la costa Pacífica y Caribe de Centroamérica. Cienc Tecn 24:51–64

Lizano OG, Salas DM (2001) Variaciones geomorfológicas en los últimos 50 años de la Isla de Damas, Quepos, Costa Rica. Revista de Biología Tropical 49(Suplemento 2):171–177

Lizano OG (20019 Batimetría de la plataforma insular alrededor de la Isla del Coco, Costa Rica. Revista de Biología Tropical 49(Supplement 2):163–170

Lizano OG (2012) Rasgos morfológicos alrededor de la Isla del Coco y de sus montes submarinos vecinos, Pacífico de Costa Rica. Revista de Biología Tropical 60(Suplemento 3):43–51

Lizano-Araya MA, Lizano-Rodríguez OG (2022) Creación de escenarios ante el aumento del nivel del mar, para las localidades de Moín y Cahuita, Limón, Costa Rica. Revista Geográfica de América Central 68:103–126. https://doi.org/10.15359/rgac.68-1.4

Lockwood JP, Benumof BT (2000) Investigaciones de las cuevas marinas de la isla del Coco. Informe Puffin Invest. Co., Inc., Datos de campo, p 43

Loría-Naranjo M, Samper-Villarreal J, Cortés J (2014) Structural complexity and species composition of Potrero Grande and Santa Elena Mangrove Forests in Santa Rosa National Park, North Pacific of Costa Rica. Rev Biol Trop 62:33–41. https://doi.org/10.15517/rbt.v62i4.20030

Marshall J (2007) The geomorphology and physiographic provinces of Central America. In: Bundschuh J, Alvarado GE (eds) Central America: geology, resources and hazards. Taylor & Francis, London, pp 75–122

Marshall J, Fisher DM, Gardner TW (2000) Central Costa Rica deformed belt: kinematics of diffuse faulting across the western Panama block. Tectonics 19:468–492

Parkinson RW, Cortés J, Denyer P (1998) Passive margin sedimentation on Costa Rica's north Caribbean coastal plain. Río Colorado. Revista De Biología Tropical 46(Supplement 6):221–236

Paynter CK, Cortés J, Engels M (2001) Biomass, productivity and density of the seagrass *Thalassia testudinum* at three sites in Cahuita National Park. Costa Rica. Revista De Biología Tropical 49(Supplement 2):265–272

Quesada-Alpízar MA, Cortés J (2006) Los ecosistemas marinos del Pacífico sur de Costa Rica: estado del conocimiento y perspectivas de manejo. Rev Biol Trop 54(Supplement 1):101–145

Quesada-Román A (2016) Impactos geomorfológicos del Terremoto de Limón (1991; ms = 7.5) y consideraciones para la prevención de riesgos asociados en Costa Rica. Revista Geográfica De América Central 56:93–111

Quesada-Román A (2018) Cartografía geomorfológica de la cuenca del río Guacimal, Costa Rica. Revista Geográfica Venezolana 59:12–27

Quesada-Román A (2021) Review of the geomorphological effects of the 1991 Limón earthquake. Revista Geológica De América Central 65:1–13

Quesada-Román A (2022) Geomorphology of the Guacimal River catchment, Costa Rica. J Geogr Cartography 5(2):58–67

Quesada-Román A, Peralta-Reyes M (2023) Geomorphological mapping global trends and applications. Geographies 3(3):610–621

Quesada-Román A, Pérez-Briceño PM (2019) Geomorphology of the Caribbean coast of Costa Rica. J Maps 15:363–371. https://doi.org/10.1080/17445647.2019.1600592

Quesada-Román A, Umaña-Ortíz J, Zumbado-Solano M, Islam A, Abioui M, Tefogoum GZ, Kariminejad N, Mutaqin BW, Pupim F (2023) Geomorphological regional mapping for environmental planning in developing countries. Environ Develop 48:100935. https://doi.org/10.1016/j.envdev.2023.100935

Risk MJ, Murillo MM, Cortés J (1980) Observaciones biológicas preliminares sobre el arrecife coralino en el Parque Nacional Cahuita, Costa Rica. Rev Biol Trop 28:361–382

Rojas W, Alvarado GE (2012) Marco geológico y tectónico de la Isla del Coco y la región marítima circunvecina, Costa Rica. Rev Biol Trop 60(Supplement 3): 15–32

Samper-Villarreal J, Cortés J (2020) Seagrass characterization on the southern Pacific coast of Costa Rica: history, vegetation, and environment. Bot Mar 63:429–438. https://doi.org/10.1515/bot-2020-0022

Samper-Villarreal J, Cortés J, Benavides C (2012) Description of the Panamá and Iguanita mangrove stands within Bahía Culebra, north Pacific coast of Costa Rica. Rev Biol Trop 60(Supplement 2):109–120

Samper-Villarreal J, van Tussenbroek BI, Cortés J (2018) Seagrasses of Costa Rica: from the mighty Caribbean to the dynamic meadows of the Eastern Tropical Pacific. Rev Biol Trop 66(Supplement 1):S53–S65

Sánchez-Noguera C, Jiménez C, Cortés J (2018) Desarrollo costero y ambientes marino-costeros en Bahía Culebra, Guanacaste, Costa Rica. Revista De Biología Tropical 66(Supplement 1):S309–S327

Savage JM (2002) The amphibians and reptiles of Costa Rica. University of Chicago Press, Chicago

Sibaja-Cordero JA, Cortés J (2008) Vertical zonation of rocky intertidal organisms in a seasonal upwelling area (Eastern Tropical Pacific). Rev Biol Trop 56(Supplement 4):91–104

Sibaja-Cordero JA, Cortés J (2010) Comparación temporal de la composición y zonación de organismos en el intermareal rocoso del Parque Nacional Isla del Coco, Pacífico de Costa Rica. Rev Biol Trop 58:1387–1403

Sibaja-Cordero JA, García-Méndez K (2014) Variación espacial y temporal de los organismos de un intermareal rocoso, Bahía Panamá, Pacífico Norte. Costa Rica. Revista De Biología Tropical 62(Supplement 4):85–97

Sibaja-Cordero JA, Troncoso JS, Benavides-Varela C, Cortés J (2012) Distribution of shallow water soft and hard bottom seabeds in the Isla del Coco National Park, Pacific Costa Rica. Rev Biol Trop 60(Supplement 3):53–66

Sibaja-Cordero JA, Camacho-García YE, Azofeifa-Solano JC, Alvado-Arranz B (2019) Ecological patterns of macrofauna in sandy beaches of Costa Rica: a Pacific-Caribbean comparison. Estuar Coast Shelf Sci 223:94–104. https://doi.org/10.1016/j.ecss.2019.04.032

Silva-Benavides AM (2009) Mangroves. In: Wehrtmann IS, Cortés J (eds) Marine biodiversity of Costa Rica, Central America. Springer + Business Media, Berlin, pp 123–130

Solano-Barquero A, Sibaja-Cordero JA, Cortés J (2022) Macrofauna associated with a rhodolith bed at an oceanic island in the Eastern Tropical Pacific (Isla del Coco National Park, Costa Rica). Front Mar Sci 9:858416. https://doi.org/10.3389/fmars.2022.858416

Suárez A, Riosmena-Rodríguez R, Cortés J (2021) Checklist of invertebrates and conspicuous fishes in rocky reefs and *Sargassum* beds in the North Pacific of Costa Rica. Revista de Biología Tropical 69(Supplement 2):S180–S193. https://doi.org/10.15517/rbt.v69iS2.48316

Troëng S (2005) Migration of sea turtles from Caribbean Costa Rica: implications for management. Ph.D. dissertation, Lund University. Lund, Sweden

Vargas JA (1996) Ecological dynamics of a tropical intertidal mudflat community. In: Nordstrom KF, Roman CT (eds) Estuarine shores: evolution, environments and human alterations. Wiley, London, U.K., pp 355–371

Vargas JA (2016) The Gulf of Nicoya: Estuarine ecosystem. In: Kappelle M (ed) Costa Rican Ecosystems. University of Chicago Press, Chicago and London, pp 139–161

Veas-Ayala N, Alfaro-Córdoba M, Quesada-Román A (2023) Costa Rican wetlands vulnerability index. Prog Phys Geogr 47(4):521–540

Wehrtmann IS, Cortés J, Echeverría-Sáenz S (2009) Marine biodiversity of Costa Rica: perspectives and conclusions. In: Wehrtmann IS, Cortés J (eds) Marine biodiversity of Costa Rica, Central America. Springer + Business Media, Berlin, pp 521–533

Zamora-Trejos P, Cortés J (2009) Los manglares de Costa Rica: Pacífico norte. Rev Biol Trop 57:473–488

A Costa Rican Biological Puzzle—The Geological, Landscape, and Vicariance History as a Conditioning of the Past and Present Biogeography

10

Ángel Solís, Bert Kohlmann, and Guillermo E. Alvarado

Abstract

The high biological diversity and endemicity for Costa Rica are aspects that have always intrigued naturalists in the past and today. Lamentably, not very much is known about the possible processes that have been active in the past. To understand the highly complex problem of Costa Rica's biodiversity and endemicity and its feasible generation mechanisms, simple proposals or models are established integrating geological and biological lines of evidence that could have favored changes that generated biodiversity and endemicity. This study adduces models integrating geological, paleogeographical, and biological lines of evidence to explain different biodiversity processes operating in Costa Rica. An updated compilation and synthesis of primary information is presented trying to explain the present high levels of organismal biodiversity, distribution, and endemism mechanisms. These possible mechanisms are based on available stratigraphical, radiometric dating, paleontological, paleotectonic and paleogeographical evidence spanning the last 24 Ma to 24 ka. This study amalgamates auspiciously geological and biological processes with very interesting and novel results regarding the coupled evolution and inter-relation between life, geology, and landscape.

Keywords

Biodiversity · Endemicity · Paleoclimate · Paleogeography · Refugia · Southern Central America · Speciation · Tárcoles line

Á. Solís · B. Kohlmann (✉)
BioAlfa Barcoding Project, Santo Domingo de Heredia, Heredia, Costa Rica
e-mail: bkohlmann64@gmail.com

G. E. Alvarado
Centro de Investigaciones, Universidad de Costa Rica, Ciencias Geológicas, San José, Costa Rica

10.1 Introduction

Costa Rica (51,180 km^2) is a country that occupies a bridge position between the Americas and forms a barrier between the Pacific Ocean and the Caribbean Sea, but also, due to its double isthmic and intercontinental conditions, presents a wide variety of biogeographical processes, making it part of the Neotropical region (Savage 2002). Costa Rica is one of the twenty countries in the world with the greatest biodiversity (Groombridge 1992); although it only constitutes 0.03% of the surface of the planet, it is estimated that it possesses about 5% of the world's wealth in this field (about 191,235 described species so far; Avalos 2019). It is considered that only 19% of all species living in Costa Rica have been described (Obando 2002, 2007). The country has more variety of birds than the USA and Canada and has the widest variety of flora and fauna in all Central America (Obando 2002, 2007; Valerio 2006; Avalos 2019). The high biological diversity for such a small area, and particularly its high endemism, are aspects that have always intrigued naturalists in the past and today. Although several probable causes have been proposed, few attempts have been made to explain and understand it (Gómez 1986; Colinvaux 1997; Alvarado and Cárdenes 2016; Kohlmann et al. 2019). This study is the first integrated attempt to try to explain biological concerns that have been in the academy for decades.

Although the pertinent geological and biological information is dispersed, a first approximation can be made using published research that provides useful basic information to attempt a comprehensive study of the paleobiogeography and biodiversity processes in Costa Rica. We begin the study by presenting first the classic geological works that support the evidence of the existence of the first islands in the region and quantify the degree of continental uplift (e.g., uplift of paleobeaches, coral reefs, and raised littoral platforms). This effort is supported by geochronological dating, topographic leveling of the first order, paleontology, paleogeographical, and paleoclimatic aspects, and detailed biological and genetic studies.

A. Quesada-Román (ed.), *Landscapes and Landforms of Costa Rica*,
World Geomorphological Landscapes, https://doi.org/10.1007/978-3-031-64940-0_10

To understand the overly complex issue of Costa Rica's biodiversity and endemicity, we try to establish simple proposals or models integrating geological and biological lines of evidence that could have generated biodiversity. Thus, our work tries to present:

- A synoptic compilation of geological and paleogeographical evidence during the last 24 million (Ma, usage as defined by the International Union of Geological Sciences) years, which could have favored generation of high biodiversity, as well as dispersal and vicariance processes and the creation of iffy endemisms from different areas very close to one another.
- An analysis of the physiography of Costa Rica during the Quaternary (last 2.56 Ma), including processes that must be considered (i.e., volcanism, tectonism) and past climate at the physiographic level (ranges, valleys, and gulfs), and how these mechanisms could have affected the biological distribution and evolution of the different species.
- A reconstruction of Costa Rica's paleogeography during the last glacial period during the last 200,000 years (i.e., ~ 24 ka, close to the last maximum glacial period), including the climate at that time, and how it could have affected the different life zones.

To summarize, this paper intends to study and tries to explain the origin of the horobiota of Costa Rica: "the assemblage of taxa that coexist and diversify in a given area during a period" (Reig, 1981).

10.2 The Costa Rican Scenario

10.2.1 Geography and Geology

The geological history of the present territory of Costa Rica began during the Lower Jurassic as a series of geological events that extended from 200 to 25 Ma, which eventually contributed to the continental constitution of Costa Rica (Fig. 10.1). However, this timeframe remains obscure for the understanding of the recent biogeographic history of the region because these expanses constituted unstable areas that formed isolated islands/archipelagos that emerged due to uplift or volcanic activity during different periods, particularly during the Campanian (ca. 83–71 Ma) and the Middle Eocene (ca. 46 Ma) (Astorga et al. 1995; Hauff et al. 2000). However, the most relevant geological period from the biological point of view is one when the territory slowly began to establish itself or grow to the present geographical subaerial configuration since the Early Oligocene (ca. 30 Ma).

At present, Costa Rica is part of the Central American Isthmus and the westerly exposed area of the Caribbean plate and the microplate of Panama, where the Cocos Plate is subducted under them. In the southern portion, the Panama fracture zone, a transform fault, is the tectonic limit of the Cocos and Nazca plates (Fig. 10.1, inset). The country includes active volcanic mountain ranges of the Cordillera de Guanacaste and the Cordillera Central, and inactive volcanic mountain ranges of Cordillera de Tilarán, Montes del Aguacate, and the Cordillera de Talamanca (Fig. 10.3). The last one comprises mixed igneous-sedimentary-low grade metamorphic rocks and is the highest mountain range in Central America, with elevations up to 3820 m (Figs. 10.1, 10.2 and 10.3). Talamanca has several peaks rising to 3100 and 3820 m.a.s.l. The emplacement of intrusive series at Talamanca (8–12 Ma) constitutes a record of the development of the volcanic arc during the Miocene (Gräfe et al. 2002; Gazel et al. 2009, 2019; Alfaro et al. 2021), although much of the volcanic rocks of these ages have been lost due to erosive processes. Talamanca represents a 180 km gap (Fig. 10.1) in the active andesitic normal arc volcanism between the Barú stratovolcano (Panama) and the Turrialba stratovolcano (Costa Rica). Another important mountain range is the Fila Costeña (Costeña Range), produced by tectonic uplift of sedimentary rocks that, together with Talamanca, enclose the intermontane valley of El General (Alvarado and Cárdenes 2016). Costa Rica has on its Pacific and Caribbean flanks a series of low mountainous areas, alluvial plains, marshy areas, coastal cliffs, mangroves, estuarine deltas, and sandy beach strips. On the eastern part of the Caribbean coast, coral reefs and sandy beaches are present (Cortés 2016a).

At least two-dozen volcanoes were active during the Late Pleistocene (last 126 ka), at least nine volcanoes were active during the Holocene (11 ka), whereas five have been active during historical times since 1700 A.D. to the present, but there are no written records of eruptions before the eighteenth century (Figs. 10.1 and 10.2) (Alvarado and Gans 2012; Alvarado 2021). Multiple active faults occur across the length and breadth of the Costa Rican territory, causing earthquakes of intermediate magnitudes ($5.0 \leq Mw \leq 7.7$) due to the gradual rise and fall of tectonic blocks that have significantly affected the landscape and the drainage network (Denyer et al. 2003a, b).

Lake Arenal has a WNW-ESE trend and has been presently artificially modified and expanded as a hydroelectric reservoir. The lake level today lies at 530 m a.s.l., but initially it used to be at 460 m a.s.l., was 25 km long, and had a maximum width of 4 km. This area lies in a complex tectonic depression (graben) limited by active faults. It is situated between the Arenal volcano (7–0 ka) and the dormant Chato volcano (40–3.5 ka) at one extreme and the eroded

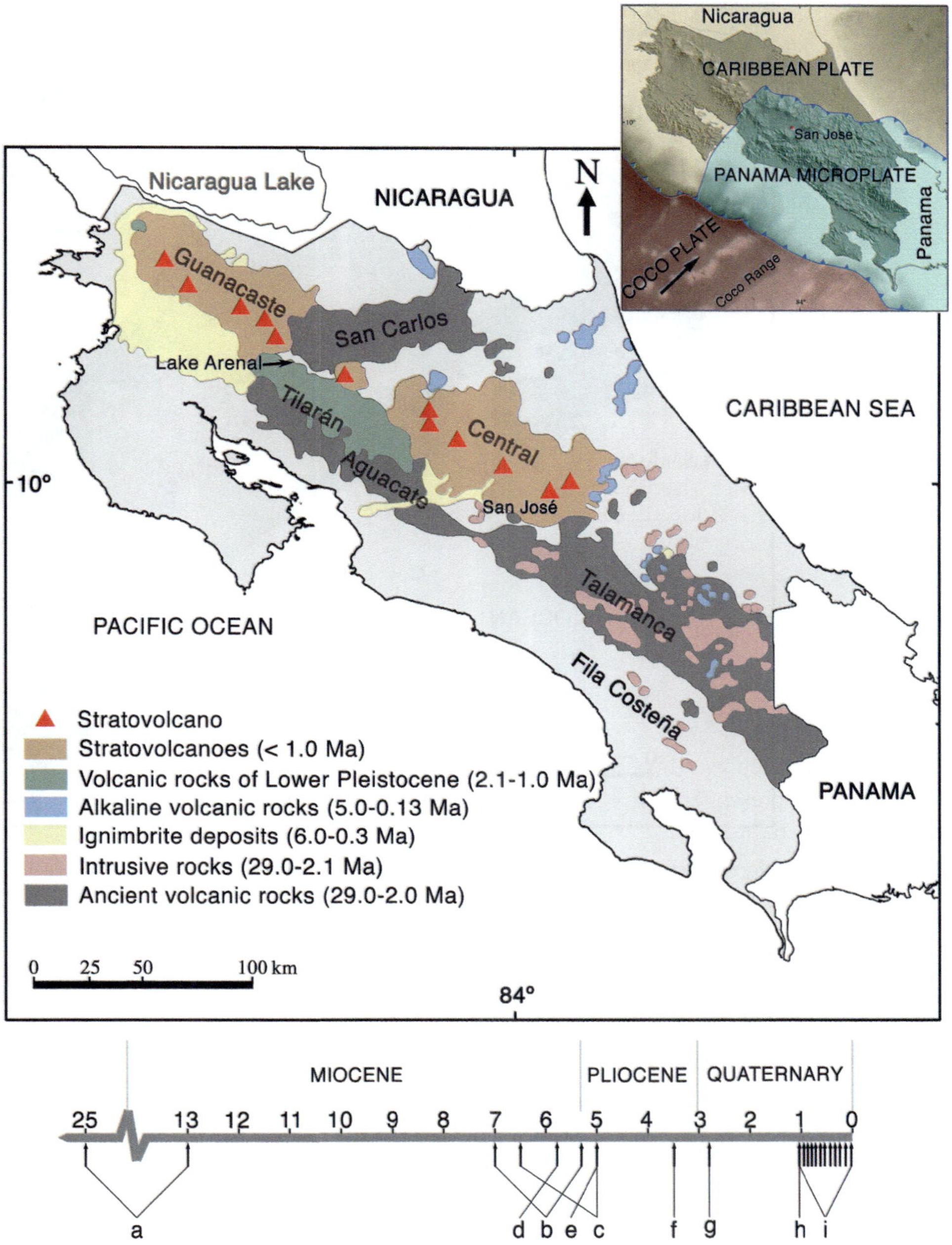

Fig. 10.1 Tectonic configuration of Costa Rica. Simplified geological map of igneous rocks (last 29 Ma), showing the area affected by a pyroclastic flow in Guanacaste (5–0.6 Ma) and Central, Costa Rica (0.8–0.3 Ma) (based on Denyer and Alvarado 2007). Bar below shows the most relevant paleogeographic conditions since about 20 Ma when a bridge started to develop. Chronology of the relevant geological events that determined the paleobiological dynamics of Costa Rica (scale in Ma). **a** Magmatic island arc formation from Nicaragua to Panama and the onset of the development of an archipelago and eventually the Isthmus. **b** Existence of paleovertebrate assemblages composed of fauna derived from North and South America at the southern end of the Fila Costeña. **c** Formation of a peninsula as part of North America is almost complete, but there are still small oceanic connections, one in Nicaragua and the other in Panama. **d** Oceanic connection closure in Nicaragua. **e** Rapid Talamanca uplift began. **f** Volcanism ceased in Talamanca. **g** Closure of the oceanic connection in Panama with the consequent massive exchange of species on the Isthmus. **h** Freshwater lake was connected for the first time to the sea originating the Golfo Dulce. **i** Maximum Ice Age peaks during the last million years that occurred approximately every 100 ka

remnants of the Monteverde volcanic complex at the other (2.1–1.2 Ma; Alvarado and Gans 2012).

10.2.2 Climate

The main factors that determine the climate of Costa Rica are its orography, its latitude, being a land bridge, and its relationship with the latitudinal movement of the Intertropical Convergence Zone (ITCZ) (Waliser and Jiang 2015; Amador et al. 2016a, 2016b; Herrera 2016). Minimum temperature extremes of −8 °C have been recorded above 3000 m altitude, and maximum extremes of 43 °C in the hot, dry lowlands (Fig. 10.4a); average monthly relative humidity varies from 65 to 90%, and rain can occur from 60 to 325 days of the year; the arboreal vegetation limit lies at 3300 m of altitude due to the existence of killing frosts (Herrera 2016).

In Costa Rica, rainfall patterns are controlled by Atlantic air masses from November to March and by the ITCZ from April to October. However, despite its small size, the country presents 95 climate groups, ranging from the sub-moist dry and very warm climate in the lowlands to very wet and cold climate in the mountains (Herrera 2016; Fig. 10.4a). About 51% of the territory corresponds to warm lands below 300 m of elevation; dry periods can range from

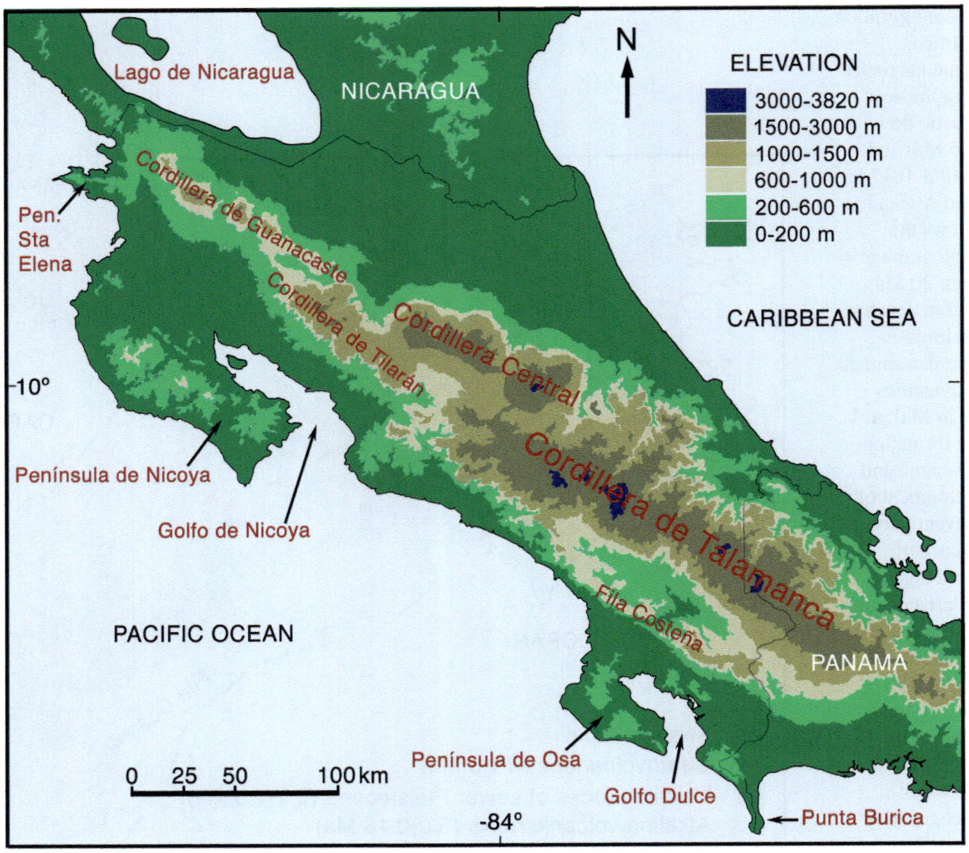

Fig. 10.2 Topography of Costa Rica (modified from Campbell and Lamar 1989)

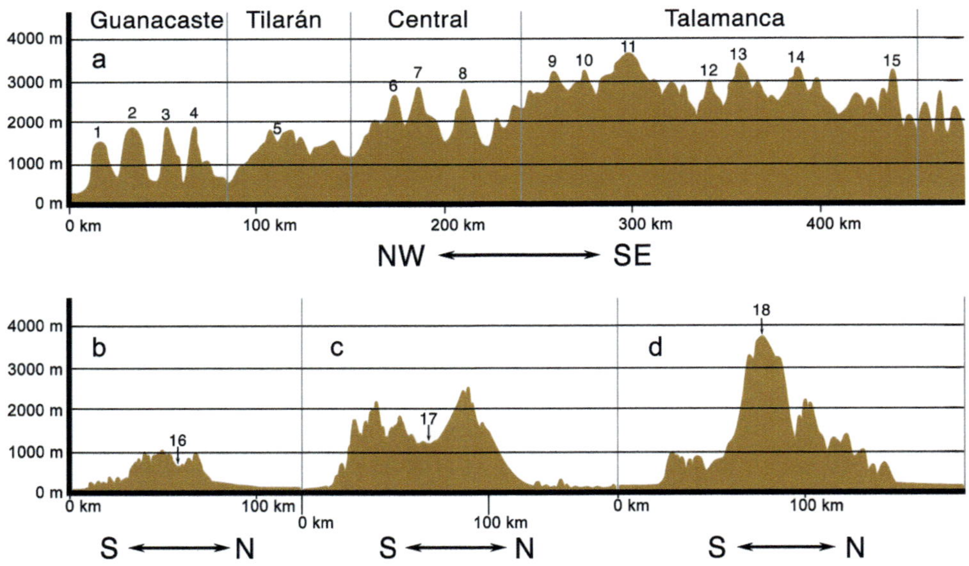

Fig. 10.3 Selected topographic profiles of Costa Rica: **a** Longitudinal mountain and volcano profiles. **b** Profile at the Arenal lake. **c** Profile at the level of the city of San José. **d** Profile at the level of Cerro Chirripó, (1) Orosi-Cacao volcanoes, (2) Rincón de la Vieja volcano, (3) Miravalles volcano; (4) Tenorio volcano, (5) Arenal volcano and Monteverde mountains, (6) Poás volcano, (7) Barva volcano, (8) Irazú-Turrialba volcanoes, (9) Cerro de la Muerte mountain, (10) Cuericí mountain, (11) Chirripó mountain, (12) Durika mountain, (13) Kámuk mountain, (14) Pittier mountain, (15) Cerro Pando mountain, (16) Arenal lake, (17) San José, (18) Chirripó mountain

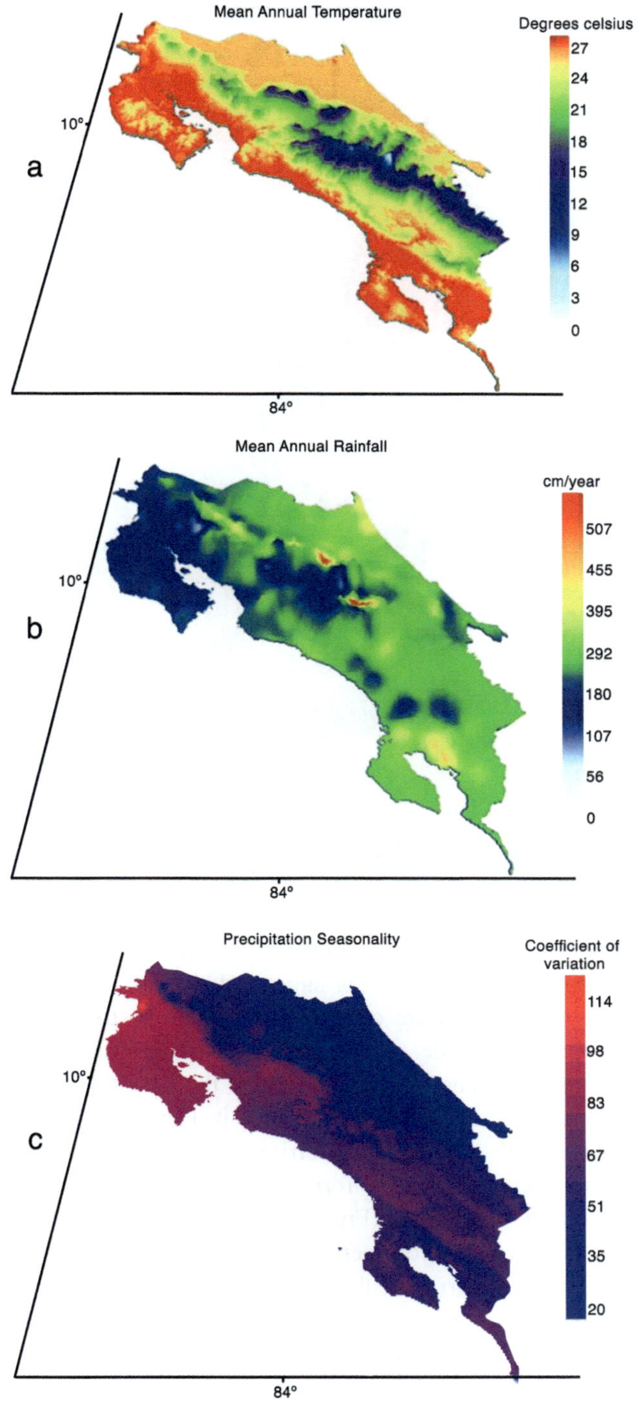

Fig. 10.4 Climate maps of Costa Rica: **a** Temperature. **b** Precipitation. **c** Precipitation seasonality expressed through the coefficient of variation (the standard deviation of the monthly precipitation as a percentage of the annual mean) (modified from Kohlmann et al. 2002)

zero to five months, annual rain can vary from 1300 to 7600 mm, and there is a gradient of 180 to 318 rainy days, monthly thermal variation is less than 4 °C (Fallas, *written comm.*, 2016; Herrera 2016).

Regarding rainfall (Fig. 10.4b), the northern Pacific has a rainfall season extending from May to October with a clear bimodal pattern, where July and August represent midsummer heat (canícula in Spanish). The southern Pacific has a rainy season extending from April to December, with a peak in September (Kohlmann and Wilkinson 2007; Herrera 2016). The Caribbean region resembles this pattern, having a rainy season from May to December, with no pronounced rain peak. Moreover, due to its height and massiveness, Talamanca can generate local rotor winds (Coen 1983) that produce rain patterns able to sustain tropical rain forest in the central and southern Pacific. In contrast, the northern ranges are not capable of this (being minor and less continuous), and an orographic shadow maintains the existence of a dry tropical forest in the northern Pacific (Kohlmann and Wilkinson 2007).

Another important climatic factor is precipitation seasonality, and this can be depicted through the coefficient of variation (Fig. 10.4c). This seasonal variation is influenced primarily by the ITCZ, the Atlantic trade winds, northern cold fronts, hurricanes, and local winds. The Caribbean, with its tropical rain forest, presents an environment with very low seasonality. Seasonality increases in the central-south Pacific's tropical rain forest, becoming marked in the tropical dry forest of the northern Pacific (Fig. 10.4c).

Looking at the topography of Talamanca and Central ranges, one can observe that it is asymmetric in cross-section, whereby the Caribbean flank is longer than the Pacific one, and in Talamanca, the western flank of the range is steeper than the eastern one, although there are large scars and erosive features on the Caribbean side (Fig. 10.3).

10.2.3 Biology

Costa Rican vegetation is incredibly varied (Fig. 10.5). The country has the greatest diversity of plants in the world recording more than 9000 species of flowering plants, about 1640 species of orchids, and around 800 species of ferns. Some of the most extensive vegetation types in Costa Rica are tropical wet forest (10.5% of the total country area), lower montane wet forest (5.9%), and tropical moist forest (5.5%) (Fig. 10.5) (Obando 2002, 2007).

Costa Rica belongs to Middle America, one of the 36 world hotspots, as defined by Mittermeier et al. (2004). What makes this country unique is its species density (number of species per unit of area; Obando 2002; Valerio 2006). Using this last measure, Costa Rica could occupy the first place in global biodiversity (Obando 2002, 2007; Valerio 2006). To give a comparative idea of species density, Costa Rica has 234.8 plant species per 1000 km^2, whereas Colombia, in second place, has only 43.8 plant species per 1000 km^2 (Obando 2007). Costa Rica is considered to have

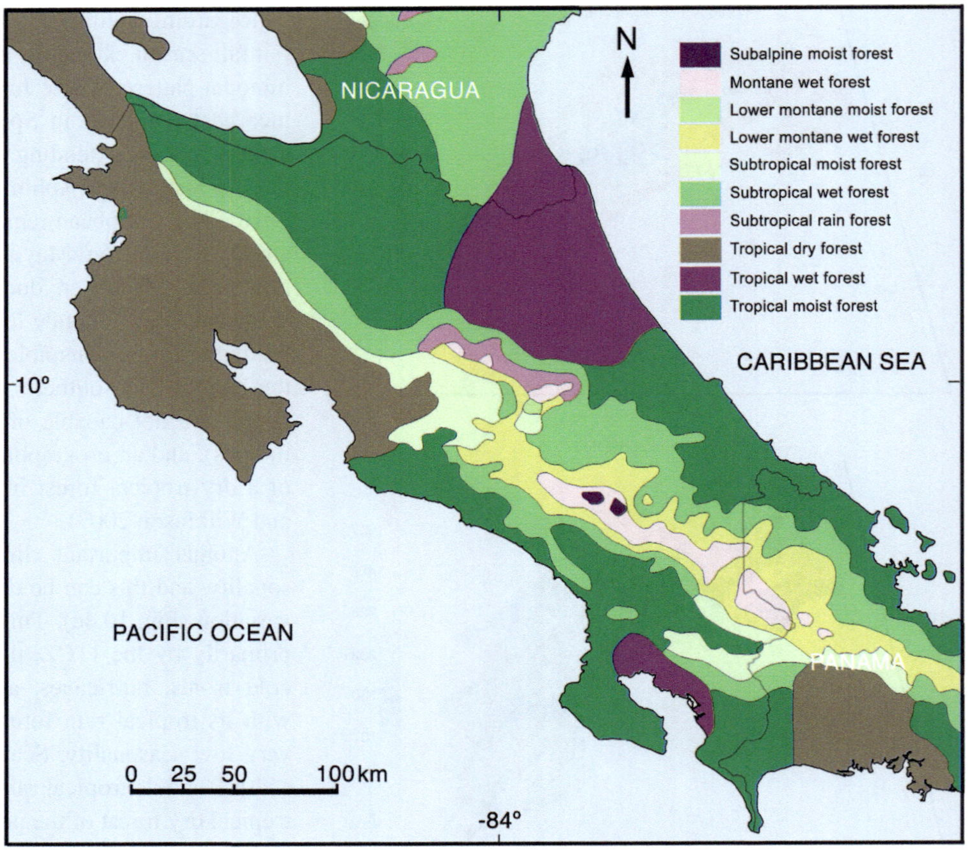

Fig. 10.5 Vegetation zones of Costa Rica (modified from Campbell and Lamar 1989)

a moderate degree of endemism, approximately 1.3%. It is estimated that around 10% of the total plant species are endemics (1102 species), whereas the different vertebrate groups vary from a minimum of 0.7% in birds to 25% for the amphibians (Obando 2007). Using these last two groups, four crucial areas of endemism have been identified for continental Costa Rica: Cordillera Central, Talamanca, the Central Pacific Region, and the Península de Osa; a fifth area has been identified in the Isla del Coco, an oceanic island (isolated from the continental platform) in the Pacific Ocean (Cortés 2016b). From the ecosystem point of view, cloud forests are the most biodiverse and endemic-rich ecosystems (Fig. 10.6) (Obando 2002; Kohlmann 2011).

10.3 The Biological Puzzle

10.3.1 Neogene-Quaternary History of Southern Central America

The composition of the Costa Rican flora and fauna is, to a great degree, the result of the formation and closure of the Central American land bridge. This closure started a gradual interchange of plants and animals from North America via Mesoamerica into South America and vice versa (Stehli and Webb 1985). This biological exchange has given rise to unprecedented biodiversity in Southern Central America.

Geological evidence has suggested that at 25–13 Ma, a magmatic island arc extended from Panama to the present Caribbean side of Nicaragua, called the Talamanca-Sarapiquí volcanic arc in Costa Rica, the Chontales range in Nicaragua, and the Western range in Panama. The configuration of the emerged territory of Costa Rica and Panama began to take a form that gradually approached its current geography (Fig. 10.7; Alvarado et al. 2007). This event can be considered the very early beginning of the Central American Isthmus's closure, even though the definitive closure occurred only around 2.8 Ma (Montes et al. 2015; O'Dea et al. 2016). This arc was almost continuously active, with volcanic activity of variable intensity until the present. Its formation likely began through the origin of underwater volcanoes, which gradually transformed into volcanic islands (< 2 km high), active and inactive, they eventually fused into a volcanic archipelago through tectonics.

Near the actual Panama Canal Zone, the fossil record indicates that at 19 Ma (Early Miocene), terrestrial animals were present, including brachyodont browsing herbivores (oreodonts and rhinoceroses) of open savanna communities (MacFadden et al. 2014). Despite the proximity to

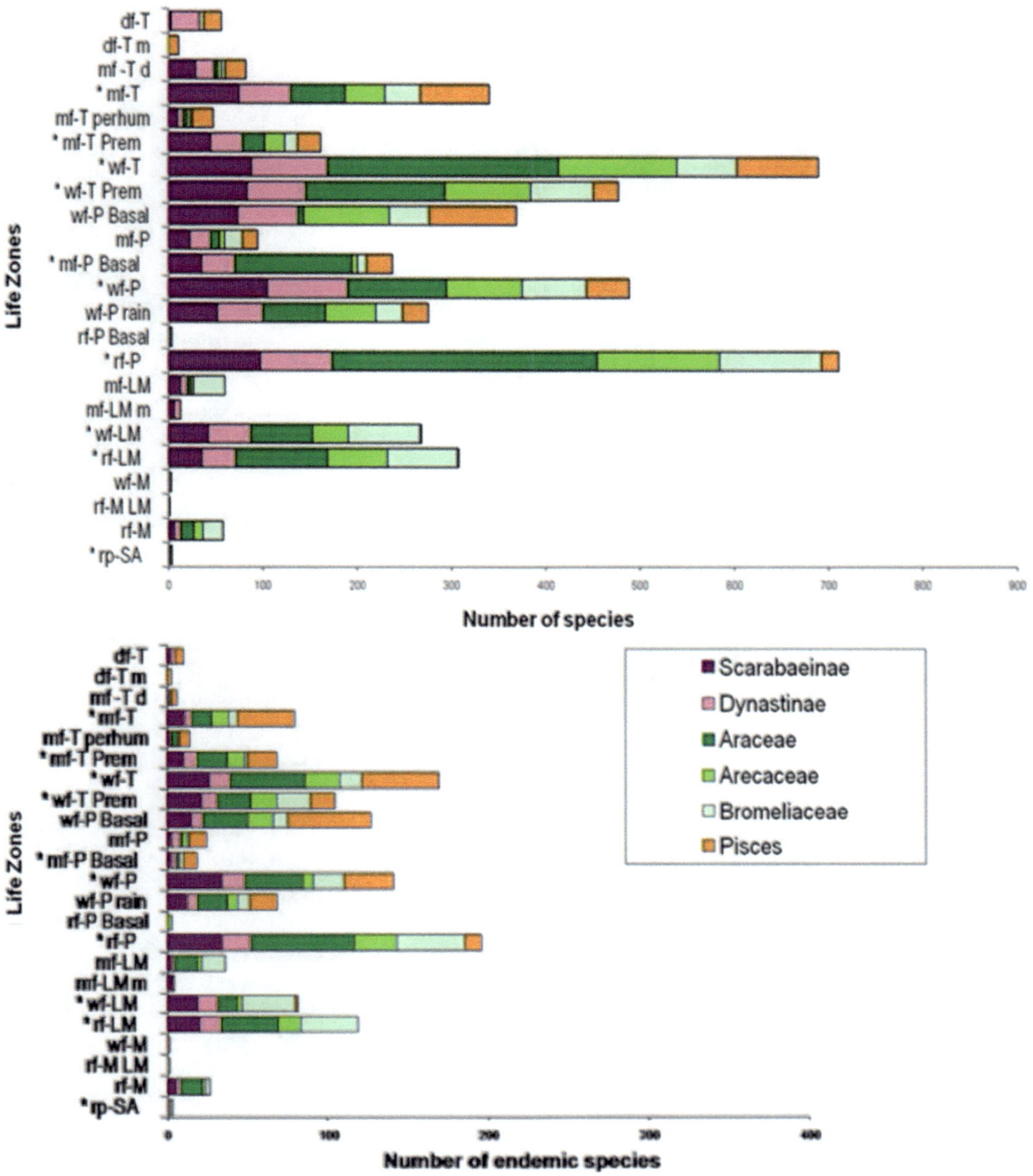

Fig. 10.6 Species number (above) of scarab beetles (Scarabaeinae and Dynastinae), plants (Araceae, Arecaceae, and Bromeliaceae), and fish; the number of endemic species (below) in the different Costa Rican life zones (Kohlmann 2011). Note how the cloud forest (rf-P) is more species-rich and has a higher endemicity than the tropical forest (wf-T). Life zones indicated with a star represent well-collected areas (more than 5 years)

South America, the fauna's taxonomic affinities are North American (MacFadden et al. 2014). Therefore, this portion of land would seem to represent an almost continuous peninsula or an archipelago of nearby islands covering almost all Central America during the entire Miocene (19–5.3 Ma; Kirby and MacFadden 2005; MacFadden et al. 2014; Rincón et al. 2020). Nevertheless, an open, still shallow seaway was likewise present in the Nicaragua oceanic connections to at least the Pliocene (~ 5 Ma), just where are now the Nicaragua (or Cocibolca) and Managua (or Xolotlán) lakes (Funk et al. 2009). Another oceanic connection remained farther south in Panama, called the Atrato Basin (Fig. 10.8) (Barat et al. 2014).

Different chronologies have been proposed for the definitive closure of the Central American Isthmus. Some recent studies suggest that the Isthmus formed about 15 Ma, earlier than the widely recognized age of approximately 2.8 Ma, although other studies are in favor of very recent ages, as recent as 0.65 Ma (see summary in Alvarado et al. 2007; Brikiatis 2023; Montes et al. 2015; O'Dea et al. 2016;), so it is an open and still debatable problem. Baker et al. (2014) and Jaramillo (2018) summarized previous

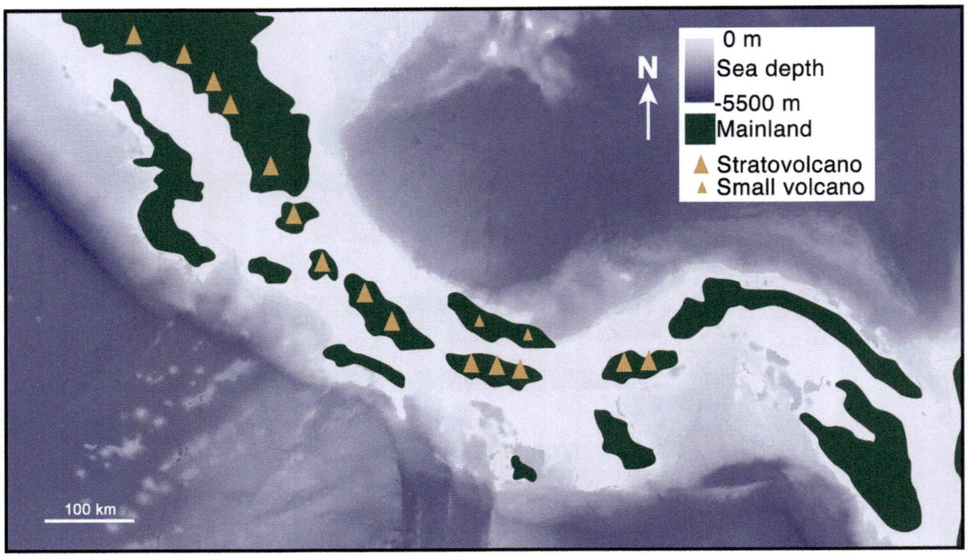

Fig. 10.7 Paleogeographic map of southern Central America during the Middle Miocene (16–11.6 Ma)

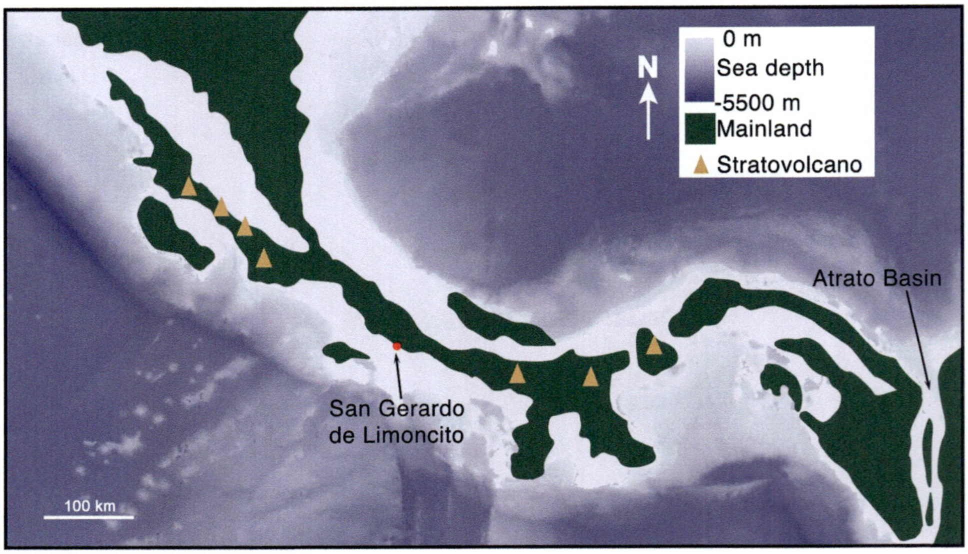

Fig. 10.8 Paleogeographic map of southern Central America during the Early Pliocene (5.3–3.6 Ma)

results and the open questions related to the closure's timing, including how it affected species evolution in this part of the Neotropics and how it has been linked to changing paleoclimate.

During the Late Miocene, volcanism (basaltic to rhyolite) was present since 8 Ma in southern Nicaragua and Costa Rica, which contributed to the amalgamation of volcanic ranges in north and central Costa Rica with Talamanca. This process was followed by the intrusion of intermediate plutons in the paleo-volcanic front; the volcanism in the Talamanca area ceased after 3.5 Ma (Alvarado et al. 2007; Alvarado and Gans 2012). After that, only some isolated small volcanic vents /domes were active until ~ 1 Ma (Alfaro et al. 2021).

Finally, independent lines of evidence that converge upon the formation of the Central American Isthmus sensu stricto around 2.7–2.8 Ma (Marshall 1988; O'Dea et al. 2016) are coincident with the glacial-induced sea-level fall during the peak of the Northern Hemisphere ice-sheet growth (Haug and Tiedemann 1998). The Panamanian land bridge permitted the great interchange of all kinds of terrestrial animals (including mastodons, saber-tooth cats, tapirs, ground sloths, armadillos, etc.) and vegetation between North and South America. In several cases, it also worked as a biological filter due to differences in topography, vegetation, and climate along the Isthmus (Alvarado 1994).

The present biodiversity and ecosystem complexity are derived from this geological history, but various additional climatic and physical factors have played a role such as the mutual interaction of glaciations, differences in seasonal patterns of rainfall, discontinuity of mountain chains (Kappelle 2016; Kohlmann et al. 2019), temperature gradients over short altitudinal ranges (Herrera 2016), and the development of rich mineral soils (Alvarado and Mata 2016).

10.3.2 The Uplift of the Cordillera de Talamanca and Fila Costeña—A Critical Problem

Talamanca was an active volcanic range between 17 and 8.3 Ma, showing a maximum peak in its volcanic activity from 13 to 11 Ma. The Costa Rica volcanic arc in the north, near the Caribbean coast and Nicaraguan border, migrated westward in a 30° counterclockwise rotation from its Middle Miocene (16–11.6 Ma) to the Late Miocene (*ca.* 8 Ma) position, which is the present modern volcanic arc position of the Cordillera de Tilarán and Guanacaste. After that time, the volcanic activity in Talamanca became limited; primarily present as isolated volcanic events between 5 and 0.9 Ma, without building large stratocones. Thus, after 8–6 Ma, Talamanca was a well-defined mountain range that started to be profoundly and rapidly eroded after 3.6 Ma, as determined by the abundance of plutonic clasts in fluvial sediments (produced by the erosion of the plutonic rocks from the root of the mountain range) and by the apatite age measuring methods (Gräfe et al. 2002; Alvarado and Gans 2012; Alfaro et al. 2021).

The impressive rise of Talamanca was caused by the subduction of the Coco Volcanic Submarine Range beneath the volcanic front (Rivier 1985; Kolarsky et al. 1995), and the collision of the arc with the South American Plate led to the development of the Panama microplate (Mann and Burke 1984; De Boer et al. 1995). The age of the Coco Range collision with the Panama Plate at the Middle American Trench is, however, a matter of debate, although the most accepted age at present is around 6.5–6 Ma (Vannucchi et al. 2003; Alvarado et al. 2007; Gazel et al. 2009). These ages are in accordance with isolated indirect evidence based on radiometric and paleontological dating of volcanic and sedimentary rocks, the latter of very shallow marine origin until the early fluvial erosion (molasses) of Talamanca (Table 10.1).

Detailed paleontological studies at the southern end of the Fila Costeña (San Gerardo de Limoncito; Fig. 10.8), near San Vito of Coto Brus, indicate a 5.8 Ma paleovertebrate assemblage ranging from an estuarine environment to a humid tropical forest, already present on the Pacific slope at the base of a modern Talamanca that was forming (Pérez-Crespo et al. 2018; Rincón et al. 2020). The vertebrate fossils are composed by fauna derived from North America, such as mastodons, artiodactyls, peccaries, camels, llamas, and horses (Valerio and Laurito 2008, 2020; Laurito and Valerio 2010, 2016; Pérez-Crespo et al. 2018), and from South America, such as armadillos Pampatheridae, and a Megalonychidae sloth (Laurito and Valerio 2012, 2013). The horse species, the gomphotere, and llama fed on C_3 plants, suggesting they lived in forest clearings not in forest savannas (Pérez-Crespo et al. 2018). Also, in several other small areas of the Fila Costeña there occur shallow marine Lower Pliocene (c. 4 Ma) claystones (Kesel 1983) (Table 10.1), indicating some remnant local marine environments in a coastal range affected by an accelerated uplift rate.

Geomorphologically speaking, at least during Pleistocene and Holocene times (2.6 Ma–to present), the Fila Costeña has been uplifted faster than the Talamanca, and both at a faster speed than during the Pliocene (between 5.3 and 2.6 Ma). This difference in uplift rates also appears to be reflected by different denudation rates in river basins that drain the two ranges (Kesel 1983; Gardner et al. 1987). The estimated average annual denudation rates for the Talamanca and the Fila Costeña at present are about 0.06 and 0.17 mm/yr, respectively (Gardner et al. 1987). However, ten years of continuous global navigation satellite

Table 10.1 Minimum ages of volcanic rocks and sediments that could support the early uplift of Talamanca

Range/evidence	Fila Costeña	Pacific flank of Talamanca	Summit and Caribbean flank of Talamanca (excluding coastal hills)	References
Last marine sediments	~5.8–4 Ma	?	>5.3 Ma	Kesel (1983), Pizarro (1993), Rincón et al. (2020)
Older volcanic rocks without marine sediments on top	4.54–3.59 Ma	≤8.53 Ma	5.78–4.5 Ma	Alfaro et al. (2021), Alvarado and Gans (2012)
Calc-alkaline/alkaline volcanic rocks intruding marine Neogene sediments	6.40–3.51 Ma	?	6.49–5.82 Ma	Alvarado and Gans (2012)
Older molasse (orogenic fluvial sediments)	–	3.87–3.59 Ma	~5.3–3.6 Ma	Alvarado and Gans (2012), Pizarro (1993)

The molasse was dated using paleontology or volcanic rocks at the base systems (GNSS) observations by OVSICORI-UNA (Costa Rican Volcanological and Seismological Observatory), at Cerro de la Muerte, on the summit of the Talamanca Cordillera (measurements between 2011 and 2021, Protti, *written comm.*, 2022), indicates almost no vertical deformation, within the standard errors (-0.2 ± 0.1 mm/y).

10.4 Geological Impacts on Biological Evolutionary Processes

Taken all the geological, biological, and climatological frameworks briefly described before, it is possible to establish a series of evidence lines that could help understand the biological variability of Costa Rica. To test the hypothesis that geology influences geomorphology and climate impacting biological evolution in southern Central America, we review and analyze different animal diversity cases.

10.4.1 First Analyses on Dung-Beetle Biodiversity on the Southern Central American Paleoarchipelago

A fascinating study relating to the older emerged portions of Panama and Costa Rica was done by Howden and Young (1981). These authors considered the following process to have taken place during the Pliocene–Pleistocene (~ 5.2–1.64 Ma), although it could have been older because they did not base their conclusions on any geological facts. Howden and Young (1981) considered these older emerged portions as an island-ship carrying their characteristically evolving species-passengers that initially would come from the surrounding continental masses. They studied the endemicity of dung-beetles (Coleoptera: Scarabaeidae) that occur in Panama (18 species) and in the Panama–Costa Rica border (18 species). They based their study on the island rule, or the island effect, also known as Foster's rule (1964), which is an ecogeographical rule that states that members of a species get smaller or bigger because food resources are limited due to land area constraints or because predation pressure is relaxed due to the absence of some of the predators of the mainland, respectively (Foster 1964). They compared the size of these endemic species against 77 non-endemic dung-beetle species occurring in Panama. Howden and Young (1981) considered that these old areas had had a sufficient length of time to allow a distinctly endemic biota to develop. This isolation or emergence period is not reflected at a generic or tribal level, at least in the Scarabaeinae. However, particularly in the genus *Canthidium* (Fig. 10.9), groups of related species in the Talamanca–Chiriquí area have radiated. Howden and Young (1981) noted that the degree of endemism is higher in small-sized species (under 10 mm) than among genera containing larger species. This size difference may reflect differences in flight behavior and, therefore, dispersal abilities. If the exposed portions of Southern Central America resembled an archipelago or even an isolated island situation, then the endemic elements should mainly reflect this. Matthew's (1966) work on the Scarabaeinae fauna of the West Indies has also shown that a large percentage is composed of small species (under 10 mm), whereas species of moderate to large size (15 mm or more) commonly occur in mainland localities. Darlington (1970) reported the same characteristics for Carabidae (Coleoptera) in the West Indies. If species size does have any relationship with island or habitat isolation, then, if Panama was an archipelago in the recent past, endemic forms should have a smaller average size than the more wide-ranging species. This behavior was shown to be accurate by Howden and Young (1981), at least for *Canthidium* and *Uroxys* in Panama and eastern Costa Rica. Thus, the 36 endemic species of Panama and Panama–eastern Costa Rica showed a statistically different average size of 8.3 mm, while the non-endemic and widespread species showed a value of 10.93 mm.

Following the same line of thought, a similar analysis has been performed in this chapter regarding the two richest dung-beetle genera of Costa Rica, *Canthidium* and *Onthophagus*. The first genus is of South American origin, while the second is of North American (Asian invader) origin. Actually, von Harold (1867) had already indicated that *Canthidium* replaces *Onthophagus* in South America; in other words, it represents its ecological equivalent, thus making this comparison ecologically relevant. The results for the genus *Canthidium* (Fig. 10.9) are statistically significant, where the one-tailed t-test clearly shows that the endemic (Costa Rica-Panama) *Canthidium* species are smaller in size (Fig. 10.9c, d) than the non-endemic species (Fig. 10.9a, b) (Table 10.2). The exact opposite behavior is observed for the genus *Onthophagus* (Table 10.2).

The fact that *Canthidium* presents small endemic species suggests that it invaded and colonized the Costa Rican area at a time when it was still an archipelago during the Miocene. This process of endemicity in *Canthidium* points to a Miocene-Pliocene timing (~23.3–5.2 Ma), older than the Pliocene–Pleistocene process proposed by Howden and Young (1981). However, evidence points to subsequent northern and southern reinvasions by *Canthidium* into Costa Rica. Once Costa Rica had become connected to Nicaragua around 5.3 Ma (Fig. 10.1), species that had island-hopped from south to north and speciated in Nuclear Central America during the Miocene, reinvaded Costa Rica from the north. These are species that Costa Rica shares up to Mexico (Fig. 10.9b). Lastly, there are those species shared with South America which most probably reinvaded Costa Rica recently, spreading northward during the last tropical rain

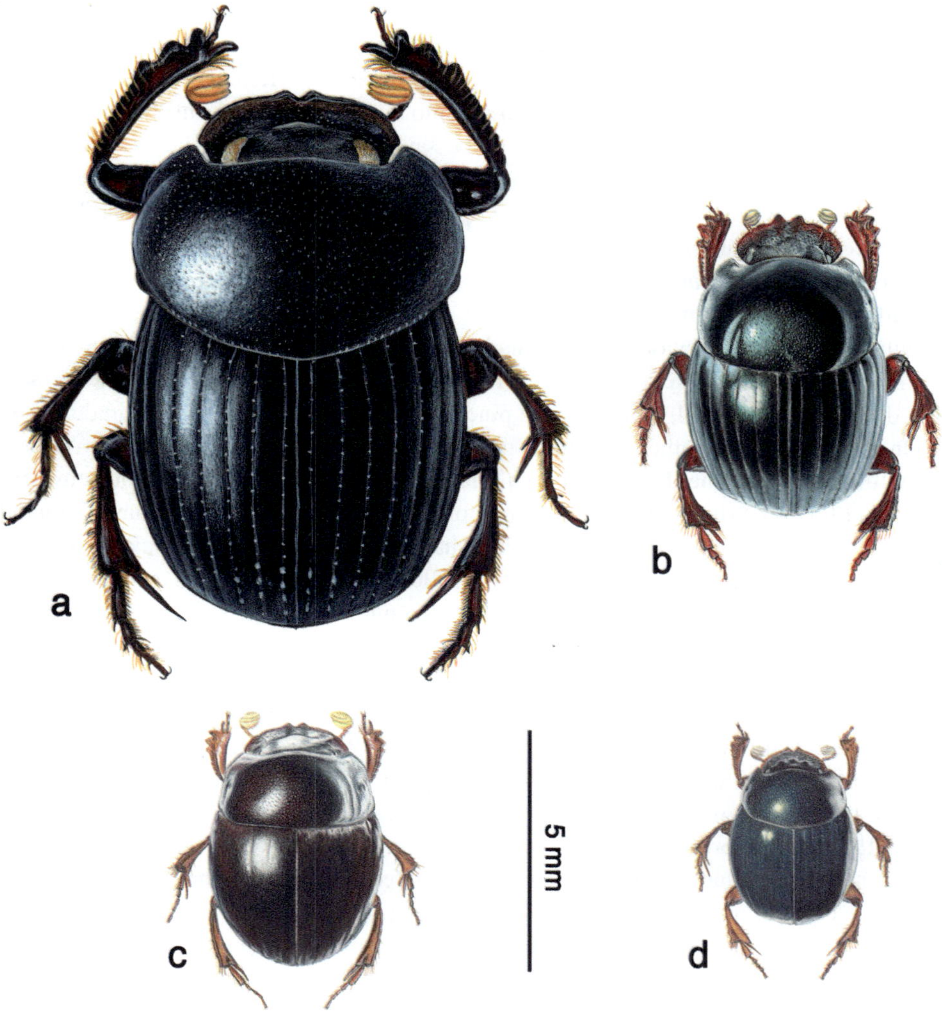

Fig. 10.9 *Canthidium* species inhabiting Costa Rica. **a** *centrale* (shared with North and South America). **b** *laetum* (shared with North America). **c** *planovultum* (endemic, brachypterous). **d** *marielae* (endemic)

Table 10.2 Two independent means one-tailed t-test of the endemic and non-endemic *Canthidium* and *Onthophagus* species of Costa Rica

	Canthidium		Onthophagus	
	Endemic	Non-endemic	Endemic	Non-endemic
Mean	4.34	5.36	7.28	6.24
SD	1.72	2.19	2.37	2.36
S	14	11	24	17
$t_{0.05}$	– 1.8508		– 1.9420	
p	0.0351		0.0279	

Equality of variances (F-test and Levene test) and normality of data (Shapiro–Wilk, Shapiro-Francia, D'Agostino-Pearson, Cramer-von Mises, and Anderson–Darling tests) exist for the present analysis

forest invasion pulse after the Last Ice Age, around 10,000 years ago (Fig. 10.9a) (Toledo 1982; Colinvaux 1997).

Regarding *Onthophagus*, it is known that this phyletic line can follow Foster´s rule. For instance, *Onthophagus massai* living in Sicily (Italy) was considered formerly to be a vicarious species of *O. fracticornis*, a species living in the same country (Baraud 1975). An mtDNA analysis shows it is now considered a small-size individuals population of *O. fracticornis*, following Foster´s rule (Pizzo et al. 2011). Contrariwise, the endemic species of the genus *Onthophagus* in Costa Rica seem to follow the other trend that Foster (1964) predicted, having a greater size than the non-endemic species (Table 10.2). It would therefore seem possible that dung beetles can become smaller or bigger under island environments. Like the case presented for *Canthidium*, *Onthophagus* seems to have elements that have recently invaded Costa Rica. For instance, the mountain fauna (*O. anthracinus*, *O. cyanellus*, and *O. incensus*) is the same as the one distributed along the eastern Mexican mountain ranges. This fact suggests that this mountain fauna is a recent invader, probably mediated by the different glaciation processes during the last 100,000 years.

Barrantes (2009) studied the highland avifauna composition of Costa Rica and Western Panama. This avifauna possesses the highest number of endemic species in Central America. The 37 endemic highland species have affinities to three regions, the Nearctic, Neotropical, and lowland Central America. Of this total, 13 species are of South American origin and 24 of North American origin. Notable is the absence of the rich South American paramo avifauna in the Costa Rican paramos. The same can be said for the highland dung-beetle fauna; no Andean paramo elements are present in the Costa Rican mountains. Like the avifauna case, the Andean dung-beetle fauna could not cross the lowland extensions separating the Andes from Talamanca. Regarding the mountain fauna shared with the Mexican mountains, the three previously mentioned species are not highland inhabitants of Mexican mountains but live in mid-altitude forests, like cloud, pine-oak, and pine forests.

On the same line, another biological process that points in the direction that the Talamanca area was a mountainous island for some time is the phenomenon of brachyptery (Figs. 10.9c and 10.10). Accepted wisdom has proposed that in Scarabaeoidea, the evolution of flightlessness is related to geographically wide areas such as temperate highland forests in the tropics; arid environments, such as deserts or temperate forests at low latitudes in the southern hemisphere; islands; termite nests; and cold regions (Zunino and Halffter 1988; Génier 2000; Scholtz 2000). At present, a very much accepted hypothesis that tries to explain the origin of this phenomenon is by Lawrence et al. (1991), who proposed that wing brachyptery may have a selection value in insects that have adapted to a sedentary, cryptic, or a parasitic way of life, or that live in mountain, island, or high-latitude habitats. The example shown here of the seven brachypterous dung-beetles living in Talamanca (Fig. 10.11) was presented by Kohlmann et al. (2019) as a possible process where brachyptery is correlated with stable mountain environments associated with deeply incised valleys. Kohlmann et al. (2019) also indicated that the degree of dung-beetle brachyptery in Talamanca is the highest known among any mountain ranges in North, Central, or South America. However, several factors that are not necessarily mutually exclusive might be operating at the same time. Living in the Talamanca Island/Range could represent mutually reinforcing processes: a mountain, as well as an island operating process. For example, Frolov and Akhmetova (2020) presented brachypterous orphnines (Coleoptera: Scarabaeidae), where three species of *Aegidium* live as endemics in different islands of the Lesser Antilles, and a fourth one is living at high-altitude on the slopes of the Cotopaxi volcano in the Ecuadorian Andes. We observe here that insularity and mountainous environments can lead to flightlessness in the same genus. On another front, Gillespie and Roderick (2002) indicated a high correlation between endemicity in beetles and brachyptery. All but one (*C. planovultum*, Fig. 10.9c) brachypterous species presented in Fig. 10.11 are endemic to just one mountain range. These authors also mentioned that small ranges characterize species on islands. They also indicate that island remoteness increases the reduction of dispersal abilities among arthropods concerning ancestral colonists. From Fig. 10.11, it is evident that, except for one species, all other brachypterous dung-beetles in Costa Rica have minimal distribution ranges. In sum, the high numbers of brachypterous dung-beetle species, as well as their highly reduced distribution ranges, point to the same conclusion arrived at by Howden and Young (1981) and the previous analysis of *Canthidium* in Costa Rica, that the area of Talamanca might have functioned as an island and a mountain at the same time in the past. A similar mechanism involving high levels of

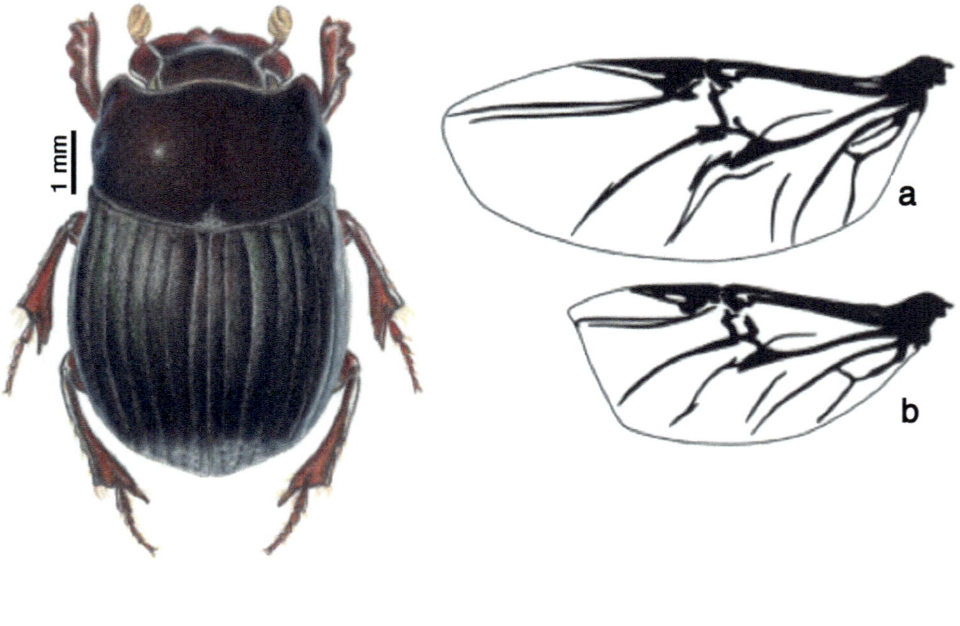

Fig. 10.10 *Ateuchus alutacius*, a brachypterous dung-beetle species (Coleoptera: Scarabaeidae) living in the Fila Costeña, and the only known brachypterous species of this genus. **a** Functional wing of *Ateuchus candezei*. **b** Non-functional wing of *A. alutacius* (taken from Kohlmann and Solís, 2012)

Fig. 10.11 Distribution of the known brachypterous dung-beetle species (Coleoptera: Scarabaeidae) in Costa Rica

endemicity and brachypterism has been reported for the Passalidae (Coleoptera: Scarabaeoidea) by Jiménez-Ferbans et al. (2017) for the Costa Rican mountains, especially the Talamanca.

10.4.2 Formation of Vicariant Species-Pairs

A typical phenomenon observed in Costa Rica is that many closely related species-pairs occur in the Caribbean and the Pacific tropical rain forests (Kohlmann and Wilkinson 2007). This vicariant speciation pattern seems to be mediated by the Talamanca range's gradual emergence, which has divided the original continuous tropical forest into two separate areas: one on the Caribbean slope, and the other one on the Pacific slope (Kohlmann and Wilkinson 2007). This phenomenon has been detected for fishes, amphibians (Fig. 10.12), reptiles (Fig. 10.12), birds, and insects, such as dung beetles (Fig. 10.13; Table 10.3) (Kohlmann and Solís 1997, 2001a, 2001b; Kohlmann and Wilkinson 2007).

10.4.3 Miocene (23–5.3 Ma) Caribbean-Pacific Vicariant Species Formation

Species formation on the Caribbean and Pacific sides has been evident in many cases, but it is still going on in others, even though marked morphological differences are not evident. We can cite the case of two phanaeine species-pairs. *Phanaeus pyrois* Bates has populations with the head and pronotum with a metallic red, green, or blue color on the Caribbean slope, whereas the Pacific populations have a completely black head and pronotum (Fig. 10.13). This situation prompted Arnaud (2002) to establish two different subspecies, *P. pyrois pyrois* on the Caribbean slope and *P. pyrois malyi* Arnaud on the Pacific slope. Subsequently, besides the color difference, differences at the mtDNA level (Solís and Kohlmann 2012) were found, thus prompting both subspecies' elevation to species level by Solís and Kohlmann (2012). However, Edmonds and Zídek (2012) synonymized *Ph. pyrois malyi* with *Ph. malyi*, considering that the color and mtDNA differences were not significant.

Fig. 10.12 **a** *Phyllobates lugubris* from the Caribbean rain forest. **b** *Phyllobates vittatus* from the Pacific rain forest. **c** *Lachesis melanocephala* from the Pacific rain forest. **d** *Lachesis stenophrys* from the Caribbean rain forest. Vicariant amphibian and reptile speciation between the Pacific rain forest and the Caribbean rain forest (Source: Savage 2002)

Kohlmann et al. (2018) re-elevated the subspecies to species level after having found morphological differences between *Ph. pyrois* and *Ph. malyi*. Finally, this taxonomic decision was supported by Gillet and Toussaint (2020), who found in their Phanaeini phylogenetic molecular data analysis that *Ph. malyi* is sister to the *Ph. pyrois-Ph. endymion* species pair. The second case is *Phanaeus beltianus* Bates, from the Caribbean slope, and its recently described sister-species, *Phanaeus changdiazi* Kohlmann and Solís, from the Pacific slope (Fig. 10.13), which seems to be following the same vicariance speciation pattern as the one suggested for the former *Phanaeus* species-pair (Kohlmann and Solís 2001b).

A modified analysis of Costa Rican Phanaeini, based on the work done by Gillett and Toussaint (2020), using a maximum clade credibility chronogram is presented here (Fig. 10.14). This analysis sheds some light on the possible timing when these *Phanaeus* species pairs originated. Using molecular sequences from five gene fragments, they generated a phylogenetic inference coupled to a maximum clade credibility chronogram. Based on this phylogenetic reconstruction, possible timings for speciation events can be proposed. In the case of the *Phanaeus beltianus–Phanaeus changdiazi* sister-species pair, this is 4.5–15 Ma (95% credibility interval), and for the *Phanaeus malyi* + (*Ph. pyrois* + *Ph. endymion*) clade events, this is 7–20.5 Ma and 4.5–15 Ma (95% credibility interval), respectively. Gillet and Toussaint (2020) estimated that the first arrival of *Phanaeus* into Mesoamerica occurred during the Oligocene (34–23 Ma), with diversification taking place during the Miocene (23–5.3 Ma), as exemplified by the species studied above. As can be seen in Fig. 10.7, Talamanca was already an emerged land by the Miocene. If we consider the proposal that Talamanca was already a mountain range by 8–6 Ma, it suggests that this date falls within the 95% speciation-time credibility intervals for the *Phanaeus* species. Thus, Talamanca was likely already acting as a vicariance-producing area since the Miocene for lowland species, separating their continuous distribution into two halves. In the case of the *Ph. malyi* + *(Ph. pyrois* + *Ph. endymion)* clade, the older of the two *Phanaeus* speciation examples presented here, Gillet and Toussaint (2020) considered that this clade [*malyi* + (*pyrois* + *endymion*)] underwent a decrease in body size concerning its sister group (*texensis* + *quadridens*), while the *Ph. beltianus–Ph. changdiazi* clade did not undergo any size change, while other clade members (*amethystinus, amithaon, guatemalensis, lunaris*) living in continental areas increased in size. These results support the reduction of body size or hindrance of size increase, as in the previously studied case of the *Canthidium* evolutionary processes occurring during the Miocene in insular areas of

Fig. 10.13 Example of the proposed formation of vicariant species-pairs between the Caribbean and Pacific rain forests is the uplift of Talamanca and subsequent split into two isolated areas of an initially continuous rain forest. **a** *Phanaeus malyi* from the Pacific. **b** *Phanaeus pyrois* from the Caribbean. The same phenomenon occurs with **c** *Phanaeus beltianus* from the Caribbean, and **d** *Phanaeus changdiazi* from the Pacific

lower Central America. Finally, the divergence time analysis of Gillet and Toussaint (2020) clearly showed that tribal (Phanaeini) composition in Costa Rica is formed of species that branched at different times, ranging from the Oligocene to the Pleistocene (e.g., *Coprophanaeus*, *Sulcophanaeus*).

10.5 Biodiversity Processes

10.5.1 Topographic Control of Winds and the Cordillera de Talamanca

The genesis of sister species could only have occurred if any tropical rainforest were present not only on the Caribbean side but also on the Pacific one. If we compare the present northern Pacific side, we observe that there is currently a tropical dry forest at the base of the Cordillera Central, Tilarán, and Guanacaste (Figs. 10.15 and 10.16). This type of vegetation is primarily the result of the easterly winds. When hitting the Central American mountain chain (Fig. 10.16), these trade winds rise on the Caribbean slope and then sink back to lower altitudes on the Pacific slope of the chain, producing a rain shadow effect. The Pacific coast of the Central American Isthmus, from Mexico to Panama, is warmer and drier than the Caribbean slopes. The rain shadow effect is an almost permanent feature of the Pacific slope of Central America, as can be observed in space photographs (Fig. 10.15), comparatively cloud-free on the Pacific side and with a heavy cloud build-up on the Caribbean side. However, along the Pacific slope of Talamanca, a tropical rainforest exists (Fig. 10.5), not a tropical dry forest.

The macroscale generators of weather conditions and precipitation regimes in Costa Rica are the Intertropical

Table 10.3 Distribution of Caribbean-Pacific species-pairs of different animal groups in Costa Rica (2007)

Dung-Beetles
Canthon aequinoctialis–Canthon sp. nov
Canthon silvaticus–Canthon hartmanni
Deltochilum acropyge–Deltochilum acanthus
Dichotomius annae–Dichotomius amicitiae
Onthophagus cryptodicranius–Onthophagus dicranius
Onthophagus tapirus–Onthophagus sharpi
Phanaeus beltianus–Phanaeus changdiazi
Phanaeus pyrois–Phanaeus malyi
Freshwater Fishes
Astatheros alfari–Astatheros diquis
Bryconamericus scleroparius–Bryconamericus terrabensis
Theraps underwoodi–Theraps sieboldii
Amphibians
Craugastor megacephalus–Craugastor rugosus
Craugastor persimilis–Craugastor stejnegerianus
Dermophis costaricense–Dermophis glandulosus
Hyalinobatrachium talamancae–Hyalinobatrachium vireovittatum
Incilius melanochlorus–Incilius aucoinae
Oedipina gracilis–Oedipina pacificensis
Oophaga pumilio–Oophaga granulifera
Phyllobates lugubris–Phyllobates vittatus
Reptiles
Anolis marsupialis–Anolis humilis
Corallus annulatus–Corallus ruschenbergerii
Dipsas articulatae–Dipsas tenuissima
Lachesis stenophrys–Lachesis melanocephala
Porthidium nasutum–Porthidium porrasi
Sphaerodactylus homolepis–Sphaerodactylus graptolaemus
Sibon annulatis–Sibon dimidiatus
Birds
Amazilia amabilis–Amazilia decora
Cantorchilus nigricapillus–Cantorchilus semibadius
Carpodectes nitidus–Carpodectes antoniae
Manacus candei–Manacus aurantiacus
Pteroglossus torquatus–Pteroglossus frantzii

Caribbean species on the left side, Pacific species on the right side (modified from Kohlmann and Wilkinson 2007)

Convergence Zone (ITCZ), trade winds, and polar depressions (Coen 1983). In the past (Gómez 1986; Herrera 2016), the ITCZ has been invoked as the critical mechanism for explaining greater precipitation values along the southern Pacific in Costa Rica. However, smaller-scale local circulation patterns, called rotor winds, also reveal significant rainfall pattern fluctuations (Kohlmann and Wilkinson 2007).

Along Costa Rica´s Pacific coast, the major exception to the rain shadow precipitation pattern is the Talamanca sector. Coen (1983) discovered a complex wind behavior over the Talamanca, which can be attributed to the significant altitude and massiveness of Talamanca, represented by the existence of air circulations known as rotors. These rotors form most frequently wherever a high, steeped-sided mountain barrier is transverse to regional winds, i.e., easterly winds in this case. The rotors flow in sympathy with the trade winds, producing local diurnal surface winds that blow from the west on both sides of the ranges, i.e., against the easterlies. These trade winds thus have opposite climatic effects on both coasts. The Pacific rotor winds reverse the offshore cold-air flow, bringing air from the sea onshore. This moist air off the ocean is forced to rise over the hills as it blows inland (red arrows, right side of the image, Fig. 10.16), with clouds often developing along the crests of local ridges (curved lines, center right, Fig. 10.16). Inversely, the Caribbean rotor, reinforced by the katabatic cold-air flow off Talamanca (blue arrows, left side of the image, Fig. 10.16), slightly suppresses cloud formation, and therefore rainfall. The rotor winds are vital because they produce rain during the dry season in Talamanca and the Fila Costeña and increase precipitation during the rainy season (Coen 1983). The action of the Pacific rotor explains why the south-central Pacific region of Costa Rica receives higher rainfall than the neighboring Pacific slopes and even more rainfall than the Caribbean (Herrera 2016).

10.5.2 The Biological "Frontier" of the Tárcoles River

An obvious biogeographic frontier is formed by the border of the Pacific dry forest and the Pacific rain forest at the Grande de Tárcoles River (Fig. 10.17), also known as the Tárcoles Line (Kohlmann and Wilkinson 2007). Several groups of organisms (dung beetles, butterflies, amphibians, reptiles, fishes, birds, and vegetation) show a sharp distributional boundary at the Grande de Tárcoles River (Fig. 10.17) (Kohlmann and Wilkinson 2007). Environmental modeling (Fig. 10.18) has shown that regional gradients for climatic variables are coincident with this biogeographic boundary. The clear biogeographical frontier correlates well with the fault system of the Costa Rican-Panama microplate. The origin of the fault system is directly attributable to the effect of the submarine Cocos range. This feature has raised land surface altitudes far above the average values of mountain ranges in the southern Central American region, in the form of the Talamanca Cordillera and the Fila Costeña. Topographic barriers created by the uplift of this range have generated dependent climatic zones (see above). Rotor wind circulations associated with Talamanca enhance precipitation only in the Talamanca sector, on the central and south Pacific facing slopes of Costa Rica. The climatic model (Fig. 10.18) shows an apparent increase in precipitation and a decrease in temperature in the border area, with associated trends of decreased solar radiation and evapotranspiration and an increase in soil moisture. This distinct climatic change might be acting as a present-day climatic/ecological filter, as was suggested by Kohlmann and Wilkinson (2007). It is

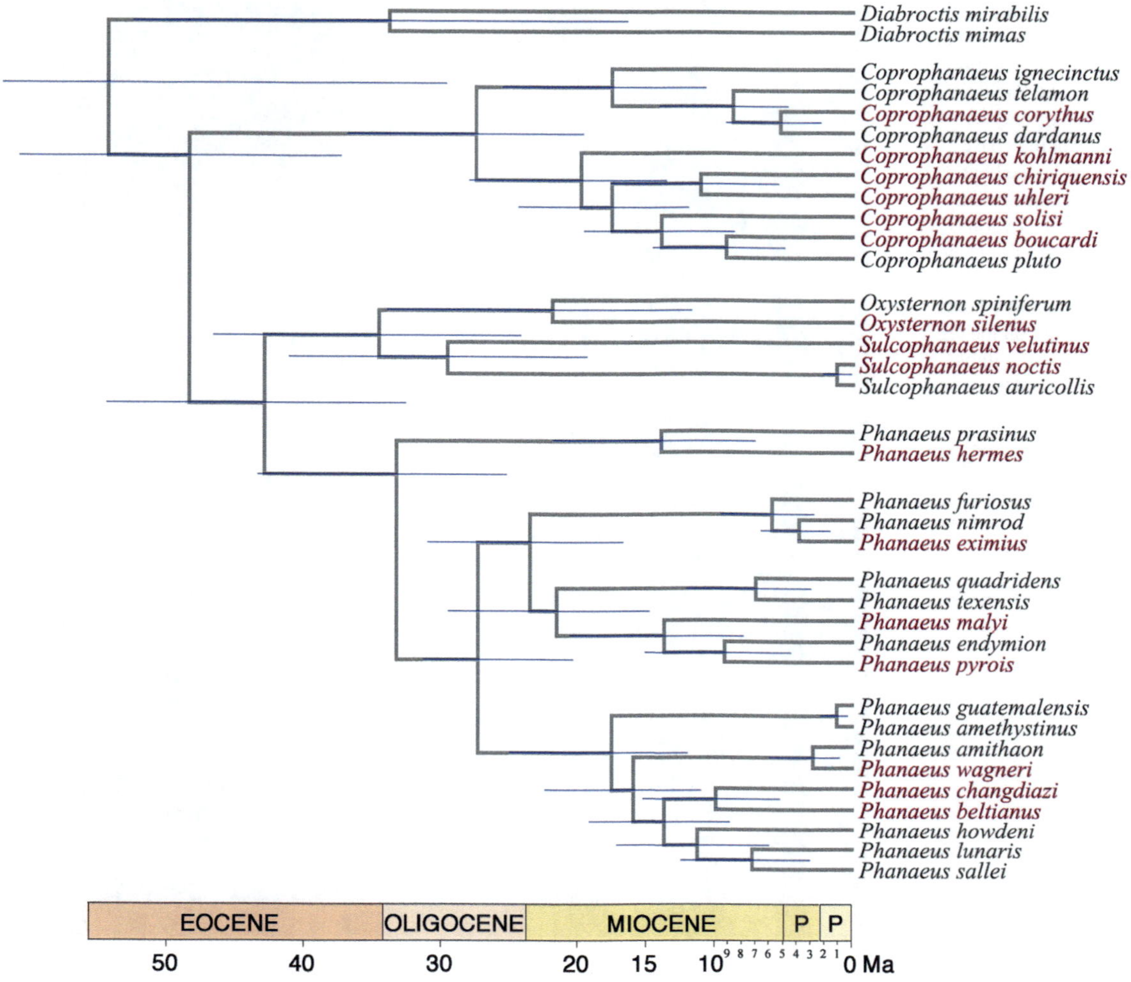

Fig. 10.14 Divergence times of Phanaeini scarab beetles from Costa Rica (Costa Rican species in red). Maximum clade credibility chronogram derived from the best BEAST analysis as selected based on comparison of marginal likelihood estimates (modified from Gillett and Toussaint 2020). The blue lines represent a 95% credibility interval

interesting to note that there is a mirror image of this barrier in the southern distributional limit of the Pacific rain forest (Fig. 10.5), where Talamanca subsides in height (Fig. 10.2), and an orographic shadow maintains a tropical dry forest in the north Pacific of Panama.

Another adjuvant mechanism that might have been operating in unison with this climatic filter is a volcanic one (Fig. 10.1, ignimbrite deposits). Geochronological studies (Alvarado and Gans 2012; Schindlbeck et al. 2016) and geological studies (Denyer et al. 2003a; Pérez et al. 2006; Thiele and Vargas 2017) have shown that the Central Valley has been repeatedly affected by significant volcanic events. All these events (pyroclastic flows, volcanic debris avalanches, and lahar deposits) are associated with the predecessors (proto- and paleovolcanoes) of the currently active and dormant volcanoes, mainly Poás, Barva, and Irazú (Fig. 10.19).

Pyroclastic density currents (pyroclastic flows, Fig. 10.20) have speeds of about 500 km/h, temperatures between 300 and 800 °C, and thicknesses ranging between 5 m and more than 80 m. Its deposits could have covered in the past large areas between 250 and 820 km^2, or more. These deposits are usually accompanied by ash clouds (dilute pyroclastic density currents or pyroclastic surges) that leave thinner deposits (decimeters to centimeters) at elevated temperatures (100–400 °C). However, they usually do not leave deposits since these are rapidly eroded. These pyroclastic flows do not only destroy the areas where we find their sedimentary signatures today, flowing downslope like a river. The accompanying deadly ash clouds can rise more than 500–1000 m above the ground level across which they flow (i.e., ancestral Tárcoles River). The flows are carried by the winds toward the adjacent hills, affecting everything around. Therefore, the area of impact is more extensive, forests disappear, and erosion becomes intense. These events occur in a few days, devastate, seal, and sterilize the ground through which they run. They cause the pyrolysis of forests, drastic changes in river systems,

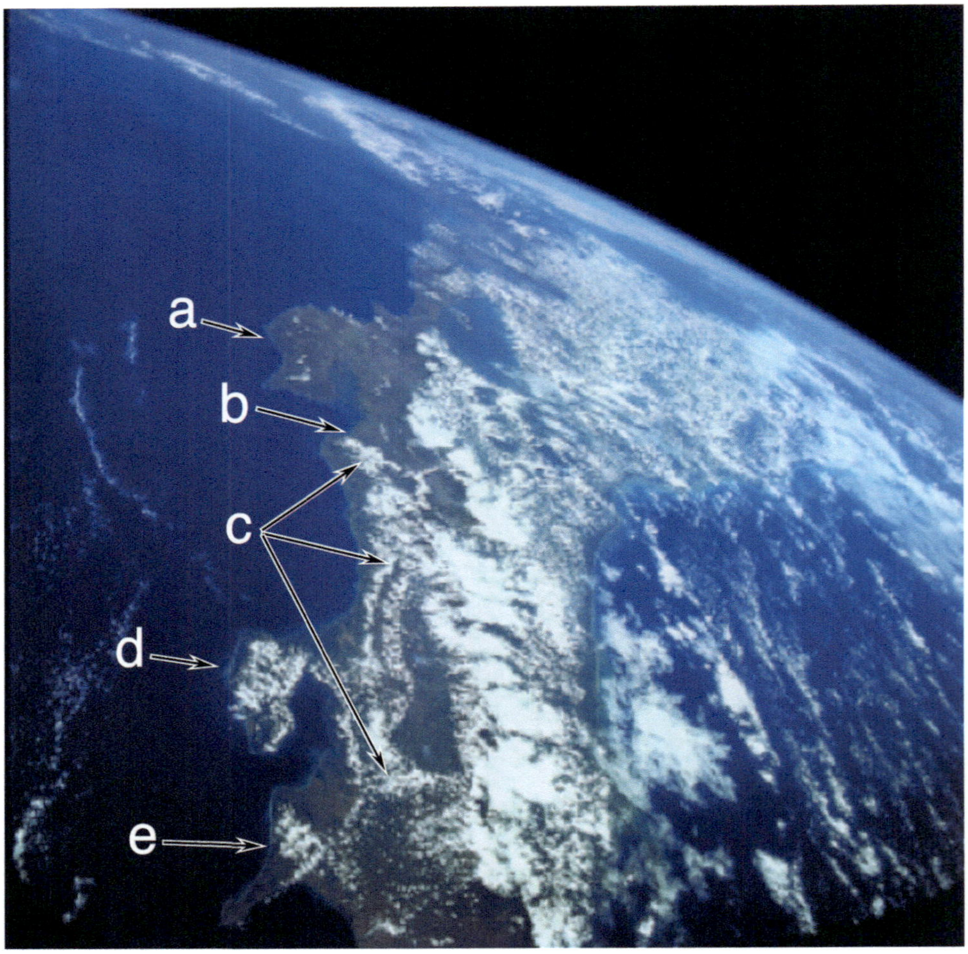

Fig. 10.15 View looking northwest as seen from the Space Shuttle. Typical cloud conditions over Costa Rica. The Pacific coast. **a** Peninsula of Nicoya. **b** Note how the Pacific coast cloud cover ends precisely at the Grande de Tárcoles basin. **c** Pacific coast cloud cover has developed only in the Talamanca sector due to rotor wind activity. **d** Peninsula of Osa, and **e** Punta Burica are prominent on the left side, the first (Nicoya) without cloud and the second (Osa) well-covered with cloud. The easterlies form a cloud cover over the different ranges as they ascend (taken from Kohlmann and Wilkinson 2007)

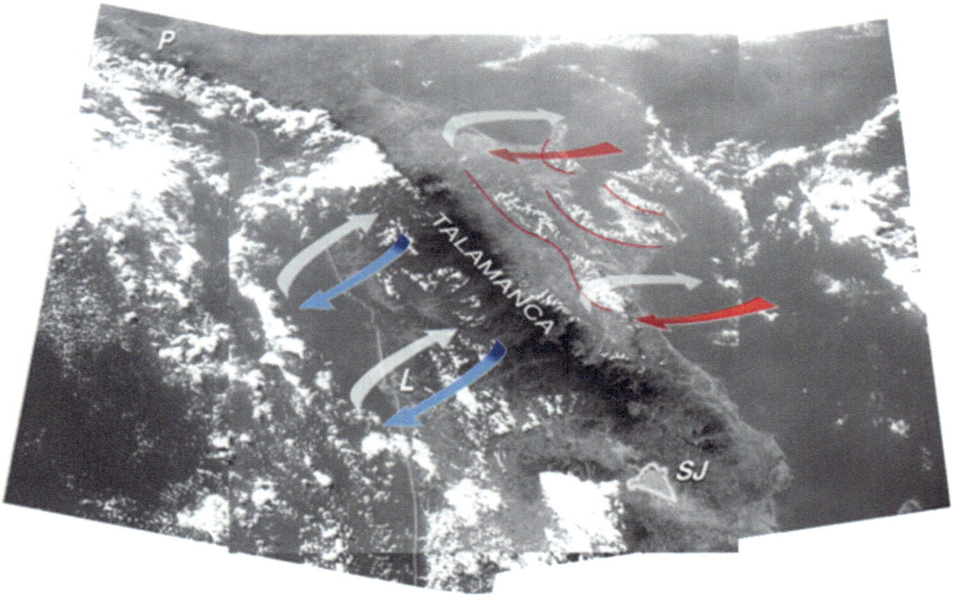

Fig. 10.16 Mosaic view, looking south, as seen in Space Shuttle images. The mechanism of *rotor* winds (blue-gray-red arrows) as secondary circulations within a regional circulation. Rotors in Costa Rica result in moist, onshore winds on the Pacific coast in the Fila Costeña–Talamanca sector (red arrow) and offshore winds in the Caribbean coast (blue arrow), a circulation assisted by the drainage of cold mountain air from high altitudes in Talamanca (taken from Kohlmann and Wilson 2007). Clouds formed by the Pacific onshore rotor, producing clouds and higher rainfall along ridge crests (thin red lines) in the Fila Costeña–Península de Osa zone. L, Limón city; SJ, San José city; P, Panama (modified from Kohlmann and Wilkinson 2007)

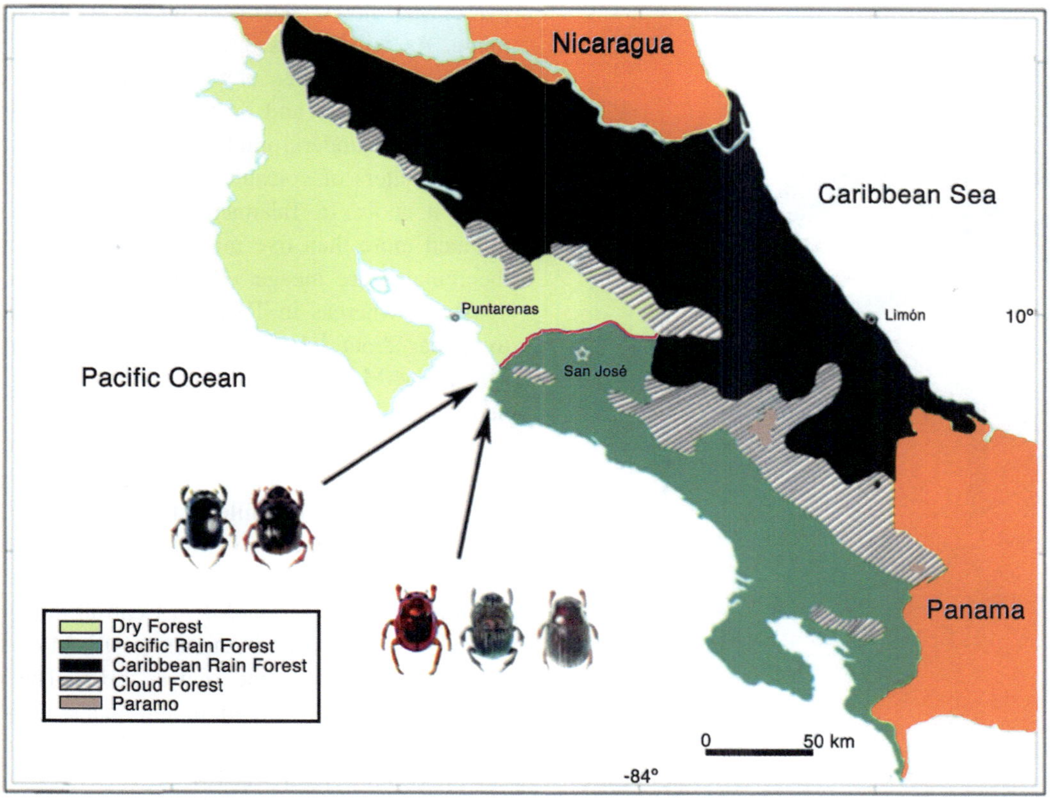

Fig. 10.17 There is a very abrupt biogeographical frontier running along the line of the Grande de Tárcoles River (red line between yellow and green areas), where vegetation types and species distributions abut along a very narrow (2–5 km) transition zone. The arrows indicate the meeting ground along the Grande de Tárcoles River valley with some examples of typical dry forest dung-beetle representatives (pair): *Dichotomius yucatanus* (Bates) (left) and *D. annae* Kohlmann and Solís (right); and typical Pacific rain forest dung-beetles (triplet): *Canthon humboldti* Solís and Kohlmann (left), *Phanaeus changdiazi* Kohlmann and Solís (center), and *Pedaridium bradyporum* (Boucomont) (right) (taken from Kohlmann and Wilkinson 2007)

change the local climate (drop in temperature and radiation), and several decades to centuries are required for the soil to develop again.

Additionally, from time to time, volcanoes present instability on their slopes and partially collapse, developing mega-landslides (volcanic debris avalanches) that also cover several dozen to more than a hundred square kilometers down the slope. These mega-landslides destroy everything in their path, leaving deposits of up to almost 180 m in thickness, but usually between 5 and 50 m (Alvarado et al. 2004; Alvarado and Gans 2012; Thiele and Vargas 2017). After the collapse, a series of mudflows (lahars) occur downstream.

Thus, if one observes the map of volcanic events (Fig. 10.19) and the line marked in blue, one can identify that in the Orotina Plateau there is a fill of volcanic deposits associated with such very destructive events, involving pyroclastic flows or ignimbrites (including co-ignimbrite ash-cloud surges), volcanic debris avalanches, and lahars, starting 2 million years ago and forming intermittently (1.91, 1.6–1.5, 1.1–0.9, 0.6, 0.4–0.3 Ma, Alvarado and Gans 2012; Schindlbeck et al. 2016). These volcanic events were channeled several times from the Central Valley along the Grande de Tárcoles Canyon, which has served as a significant drain or river drainage of the Central Valley and spread out in the form of a fan to the current Orotina Plateau. Everything on its path was destroyed (Figs. 10.19 and 10.20). If we accept that these events could have caused the local disappearance of certain species on active volcanoes, but not in Talamanca or Tilarán, the mechanism could have functioned perfectly as an intermittent filter for insect (and other species) dispersion between 1.91 and 0.3 Ma. After the last event 0.3 Ma ago, ensues a volcanic lull, through which organisms could again freely advance until they reached the Tárcoles River today. This intermittent volcanic filter should have slowed down or delayed the dispersion of living organisms on both sides of the Tárcoles River.

10.5.3 Biodiversity and Endemicity

The uplift of Talamanca generated many habitats that did not exist before, such as a glacier landscape, alpine

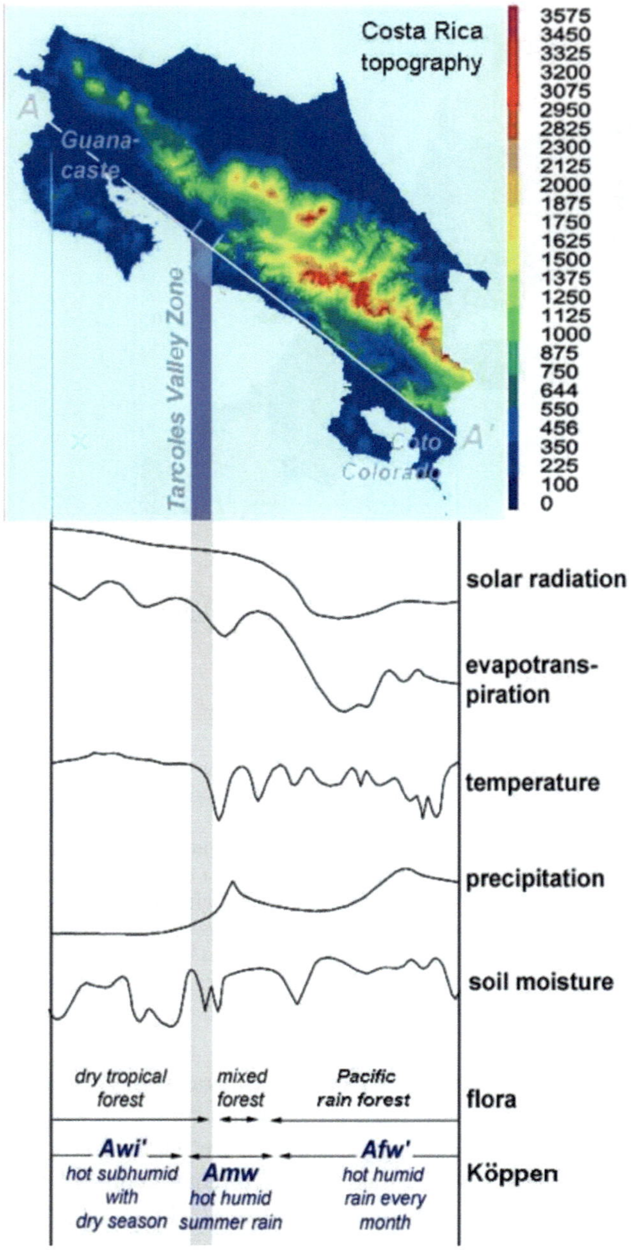

Fig. 10.18 Environmental trends along the Pacific coast of Costa Rica from the Guanacaste (A) to the Coto Colorado (A') area. Some of the significant environmental changes in the coastal transect A–A' are centered at the Tárcoles Valley (Kohlmann and Wilkinson 2007)

meadows (*páramos*), cloud forests, and montane forests. Its slopes and very rough relief were also a speciation engine due to many microenvironments and microclimates derived from the topographic complexity of the range and other neighboring ranges, such as the Fila Costeña and Fila Matama. Besides, the southern part of Costa Rica was, in general, lacking significant effusive eruptions since at least the last 3.5 Ma.

The explosive and devastating volcanism, affecting tens of square kilometers due to eruptions with associated pyroclastic flows, has been different in each mountain range. As will be discussed later, pyroclastic flows (also known as ignimbrites or pyroclastic density currents) are highly destructive and do not leave life where they flow. Not even life underground is saved due to their coverage of several meters of volcanic material, which turns into rock when it cools. In Talamanca, this type of volcanism terminated more than five million years ago, and in general, the remains are meager. In the Fila Costeña, there is no evidence, whereas in Tilarán, explosive volcanism ceased to appear about 1.1 million years ago, and since the last 2 Ma in the Montes del Aguacate, and the Arenal tectonic depression.

On the other hand, it has remained persistent in the Cordillera de Guanacaste and Central for several million years. In the Cordillera de Guanacaste, the evidence is available for 8 Ma ago eruptions, but they became more frequent between 5 and 0.6 Ma, with multiple events every million years. In the Cordillera Central, the evidence dates to at least 0.8 Ma and extends to 0.3 Ma (Alvarado and Gans 2012; Alvarado and Denyer 2019). The Valle Central is of a more recent formation, and its age goes back to less than 1 Ma (Alvarado et al. 2007; Alvarado and Gans 2012; Alvarado and Cárdenes 2016).

Concerning mountains, Rahbek et al. (2019) indicated that they contribute disproportionately to terrestrial biodiversity, especially in the tropics, calling this phenomenon "Humboldt's Enigma", the disproportional mountain contribution to Earth's biodiversity. Mountain regions represent about 25% of all land area, containing more than 85% of the world's species of amphibians, birds, and mammals. Rahbek et al. (2019) considered that the complex climatic characteristics of the rugged mountain terrain likely play a key role in generating and maintaining biological diversity.

This situation is apparent in the case of Costa Rican dung beetles when one compares the lower altitudes (0–1000 m) comprising 40,314 km^2 or 79% of the continental territory of the country, *versus* the middle altitudes (> 1000–2000 m) comprising 7951 km^2 or 15.6% of the continental territory, and the highlands (> 2000 m) encompassing 2773 km^2 or 5.4% of the continental territory (Table 10.4; Kohlmann et al. 2007). Although lower altitudes present a higher number of dung-beetle species (164) than the middle ones (117) or the highland ones (30), once one considers the area, the results present a hump-back behavior; the lower altitudes have a mean number of species per 100 km^2 of 0.76; whereas the middle altitudes present a value of 3.07, and the highlands show a value of 2.45. Interestingly, the dry Pacific slope, having the least surface at a middle altitude (1495 km^2), shows a maximum value of 4.94, indicating that this area is a veritable biodiversity powerhouse. This area defies conventional island biogeography rules because it is neither an old area (Talamanca is

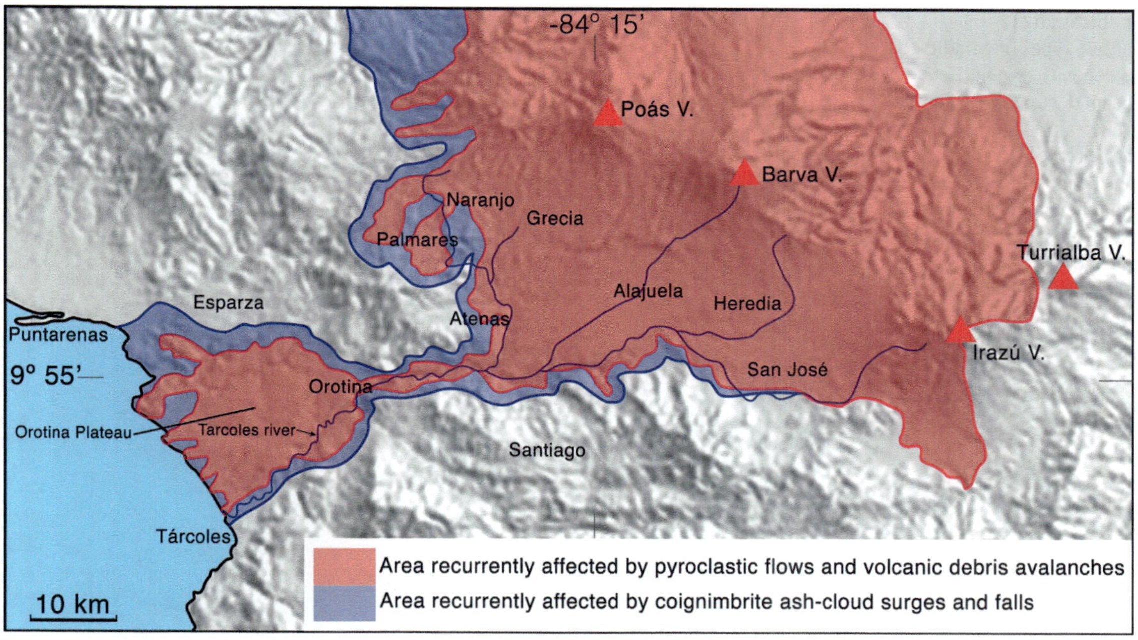

Fig. 10.19 Pyroclastic flows (ignimbrites) and volcanic debris avalanche deposits, including probable co-ignimbrite ash-cloud surges dispersion along the Tárcoles River and fanning out onto the Orotina Plateau

older) nor a very extensive one (the wet Pacific slope has a greater surface (2451 km^2); finally, it is also drier than the Caribbean slope (Table 10.4).

A similar situation occurs with the mean number of endemic species per 100 km^2 (Table 10.4). Lowlands present a value of 0.13 per 100 km^2, whereas middle altitudes have a value of 0.81, and highlands show one of 0.52. Again, the dry Pacific slope shows a maximum value of 1.47 for middle altitudes. There is a hump-shaped behavior for both mean values: the number of species and the number of endemics per 100 km^2. This situation suggests a gradual vertical niche pattern of the different dung-beetle species, where high altitudes start to become a barrier. It is interesting to note also that the Caribbean slope presents the lowest values on both counts, suggesting a less intense biodiversity/speciation activity than the Pacific side, probably a reflection of a recent (10 ka) plant and animal invasion by South American rain forest elements (Colinvaux 1997; Toledo 1982).

If we concentrate on the number of endemic dung-beetle species, it is evident that the Costa Rican mountain ranges share many endemics with the Panamanian mountain ranges. Talamanca shares many endemics with other Costa Rican ranges, and finally, Talamanca, and to a lesser degree Guanacaste, they are essential areas of endemic speciation (Table 10.5). Table 10.6 indicates a high level of regional endemism (34%) for three beetle groups, Cerambycidae, Scarabaeoidea, and Tenebrionidae. Some of these endemic species represent mountain species, but others are endemic lowland species. For example, in the Scarabaeinae subfamily, there are 183 species in Costa Rica, 83 of them or 45% are endemic species (Costa Rica, Nicaragua-Costa Rica, Costa Rica-Panama, Nicaragua-Costa Rica-Panama); of these endemic species, 43 or 52% are from the mountains or mountain slopes, and 40 species or 48% are found in the lowlands. Those 40 species possibly had an origin in which the mountains did not intervene, but they probably originated from isolation factors or from speciation in the archipelago that existed before the union of the isthmus.

10.5.4 The Physiography of Costa Rica During the Last Glacial Maxima

Our planet has had many glacial events in the past with some periodicity (Waliser and Jiang 2015). In Costa Rica, snow caps and glaciers occurred in the highlands rising above 3100 m a.s.l. However, evidence of each glacial may be erased by the subsequent glaciation and hence, studies about the glacial periods have focused on the last glacial maximum (LGM) (Lachniet 2007). Cunningham et al. (2019) and Potter et al. (2019) confirmed with cosmogenic nuclides (36Cl and 10B) that the maximum glacial extension age in the Chirripó National Park corresponded with the LGM (~26.5–19 ka). The Cerro Chirripó (Chirripó Hill) (3820 m.a.s.l.) and its neighboring peaks were covered by glaciers and snow in an approximate area of 35 km^2, while Cerro de la Muerte (3475 m.a.s.l., 2 km^2), the Cuerící (3394 and 3345 m.a.s.l., 2 km^2) and the Kámuk (3549 m.a.s.l., 2 km^2) were most likely covered with snow or even ice, but

Fig. 10.20 Illustrative examples of two types of Middle Pleistocene deposits associated with significant destructive volcanic events, along the paleo-valley of the Tárcoles River (photographs taken on Road 27). **a** Volcanic megabreccia, the product of a mega-slide (volcanic debris avalanche), is several tens of meters thick. **b** Deposits of various pyroclastic flow events, which descended at high temperature and speed from the ancient volcanoes of the Central range

there is no clear demonstration of the existence of mobile glaciers in the strict sense, with the lower snow limits reaching 3200–3260 m a.s.l. (Hastenrath 1973; Lachniet and Seltzer 2002). Clear evidence of glacial cirques occurs above 3500 m.a.s.l. (Quesada-Román et al. 2020), while glaciers descended to 3100 m.a.s.l. at the most (Lachniet 2007).

A retrospective projection was made for the first time to show how the geographic map of current Costa Rica would have looked like during the last glacial maximum at about 24 ka ago (Fig. 10.21). Thus, the most accepted lowering of the sea level for these periods was taken. Following the literature, a compilation of the additional uplift and subsidence velocities of tectonic blocks was regarded (Table 10.7). Thus, the retreat or decline in the sea level was added or subtracted from the rise of the continental block, using current topographic and ocean bottom maps. In the end, a mosaic or jigsaw map was built using ground

Table 10.4 Number of species and endemics of Scarabaeinae by altitude level in continental Costa Rica (modified from Kohlmann et al. 2007)

Slope	Altitude level meters	Area in km^2	% Total area	Species number	Mean species/100 km^2	Endemic species number	Mean Endemic species/100 km^2
Caribbean	0–1000	18,513	36.3	111	0.6	26	0.14
	1000–2000	4005	7.8	67	1.67	14	0.35
	>2000	1740	3.4	6	0.34	1	0.06
	Subtotal Caribbean	24,258	47.5	131	0.54	34	0.14
Wet Pacific	0–1000	9767	19.1	85	0.87	14	0.14
	1000–2000	2451	4.3	64	2.61	15	0.61
	>2000	930	1.8	11	1.18	5	0.53
	Subtotal Wet Pacific	13,145	25.8	121	0.92	26	0.2
Dry Pacific	0–1000	12,034	23.6	97	0.81	16	0.13
	1000–2000	1495	2.9	74	4.94	22	1.47
	>2000	103	0.2	6	5.82	1	0.97
	Subtotal Dry Pacific	13,632	26.7	112	0.82	36	0.26
	Total	51,038	100	177	0.35	73	0.14

Table 10.5 Number of endemic dung-beetle species living in the different Costa Rican ranges that are shared and not shared with Panama, and the number of endemic dung-beetle species living in the different Costa Rican ranges that are unique to that system

Group	Mountain system				
	Guanacaste	Tilarán	Central	Talamanca	Fila Costeña
Shared	25	26	27	41	11
Not shared	18	16	16	29	4
Unique	5	1	1	14	1

Table 10.6 Three well-known Coleoptera insect groups from Costa Rica (Tenebrionidae, Cerambycidae, and Scarabaeoidea) represent 2529 known species. Of these, 852 species (34%) are only found in Costa Rica, with some of them also present in Nicaragua and Panama

Group	Species	Endemics	%
Cerambycidae	1090	351	32.2
Scarabaeoidea	928	414	44.6
Tenebrionidae	511	87	17.0
Total	2529	852	33.7

and underwater contour lines to model the coastline and the height of the main hills in the respective time slices (Fig. 10.21a).

During the last local glacial maximum (LLGM), the sea level dropped around 120 m (Spratt and Lisiecki 2016). Because recent ages attribute the existence of maximum glaciers in the Cerro Chirripó to 23–25 ka, in correspondence to the LLGM (Potter et al. 2019), a time of ~ 24 ka will be fixed in the present work to model the sea-level descent (Fig. 10.21a) (Rabineau et al. 2006), although most recent investigations place it similarly around 120 m (Spratt and Lisiecki 2016). During both descents, a large part of the marine platform, now submerged, was exposed.

Potter et al. (2019) estimated the age of the LLGM at 25–23 ka for the Cerro Chirripó with a reconstructed ELA depression of 1317–1536 m and an associated cooling of ~ 7–9 °C (Table 10.7, Fig. 10.21a). Palynological studies have indicated that the treeless paramo extended down to 2100 m altitude during the last glacier interval, whereas it is distributed from 3300 to 3819 m at present (Islebe and Hooghiemstra 1997; Horn 2007). At the end of the last deglaciation (15.6–13 ka), the upper forest limit rose to 2700–2800 m (3100–3300 m is the present-day forest limit of subalpine tropical rain/elfin cloud forest), indicating a temperature increase of up to 4.6 °C (Islebe and Hooghiemstra 1997; Horn 2007). Subsequently, the upper forest limit dropped 300–400 m from 13.1 to 11.2 ka, indicating a temperature decline of 2–3 °C (Horn 2007). From 12.3 to 11.2 ka, the glaciers retreated above 3500 m, and mountain rainforests gradually replaced the subalpine tropical rain forest, and temperatures rose toward present-day values (Islebe and Hooghiemstra 1997; Horn 2007).

However, neotectonic studies have found evidence of continental uplift of Costa Rica during the last thousands of years using topographic leveling measures of the first order

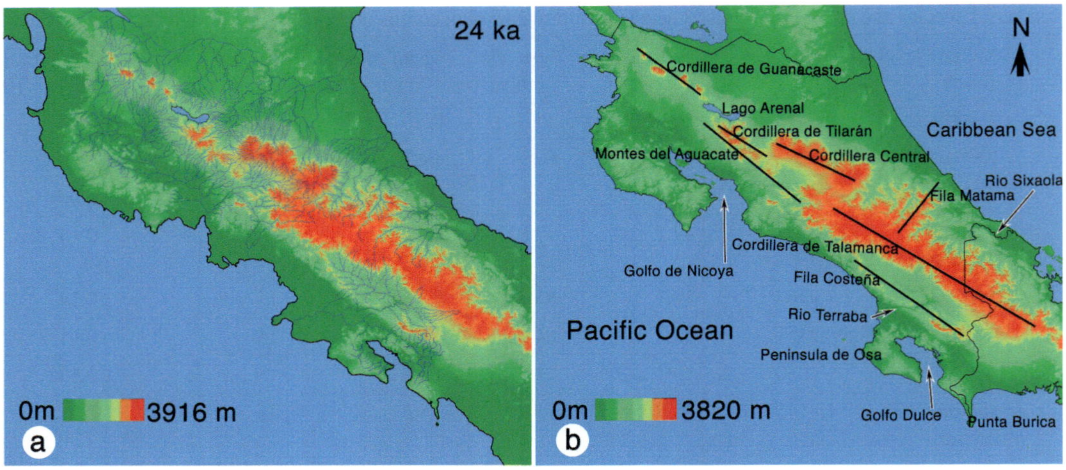

Fig. 10.21 **a** Paleogeographic situation of Costa Rica at 24 ka, and **b** at present. Solid black lines indicate the different orographic systems. The Pacific coast shows more significant changes than the Caribbean one and the terrestrial area was more extensive. This figure is based on a sea-level descent of 120 m according to the differential tectonic uplift (in blocks), using the information presented in Table 10.7

and GPS focused on marine terraces, paleobeaches, raised erosion platforms, and recent paleontological research (Table 10.7).

The most significant uplift occurred in the Península de Osa, in a sector that connects directly with the Pacific open sea, and at Punta Burica, both related to the subduction of the Coco oceanic submarine volcanic range (Sak et al. 2009; Morell et al. 2011; Gardner et al. 2013), followed by the Cabo Blanco (Blanco Cape) (Gardner et al. 2001). Starting from both sectors, the uplift rates decrease along with a generalized tilt (Marshall and Anderson 1995; Sak et al. 2004, 2009). In other Pacific and northern Caribbean coast sectors, tectonic uplift was significantly lower, or there is even subsidence (see Table 10.7). However, strong seismic uplift has also been recorded in the last thousands of years (Denyer 1998; Denyer et al. 2003a, b); the last and most dramatic event corresponding to the extreme case of an uplift was caused by the 1991 Limón earthquake (April 22, 1991, Ma 7.6), whose landmark on the coast was an uplift of 1.85 m, while inland 4.46 cm was recorded (Denyer et al. 2003a, b).

A rough contour of the national territory at 24 ka ago (Fig. 10.21a), considering the descent of the sea level, as well as tectonic uplift, subsidence, or relatively stable areas, suggests that the size of the territory of what is now called Costa Rica would have increased in area by 11,961 km^2, at 24 ka ago, which represents an increase of the territory concerning the present one of 51,180 km^2, of 23%. This difference is caused by differential tectonic uplift (see Table 10.7). Significant changes happened on the Pacific coast, where the Golfo de Nicoya (Nicoya Gulf) disappeared and changed to tropical lowlands (Fig. 10.21a). The Térraba River formed a significantly larger delta and estuarine environment, extending to lowland areas. The extended tropical lowlands should have affected climatic conditions (temperature, precipitation, and wind currents) of the Fila Costeña and Talamanca.

The embayment of Golfo Dulce (Dulce Gulf) (ca. 201 m depth) at present has a water circulation pattern of a typical anoxic gulf with flux stratification within the water column (Hebbeln and Cortés 2001). At 24 ka, the gulf was a narrower and shallower gulf than at present (Fig. 10.21), but although it was not wholly closed or dry, water circulation was more restricted because of a topographic shallow marine barrier near the entrance of the gulf.

On the Caribbean coast, changes were minor in both periods. Terry (1941) had already noticed submarine canyons in the Panamanian marine platform, including the mouth of the Sixaola River. These deep canyons were the product of submarine erosion by the sediment transport from the main rivers and not because of a rise in sea level or tectonic subsidence.

The tectonic depression of Lake Nicaragua (45 m maximum depth, 13 m medium depth, and 37.7 m a.s.l.) is a hydrologically open lake with continuous fluvial drainage into the Caribbean Sea since the Holocene. However, at 24 ka, the San Juan River had to change its equilibrium profile due to the extreme lowering of the sea level. Thus, Lake Nicaragua probably and temporarily remained as an isolated lake without a hydraulic connection to the Caribbean Sea or as an area occupied by shallow swampy lakes and marshes. At least during the Late Pleistocene, it was connected to the smaller Lake Managua (20 m maximum depth, 9.5 m medium depth, and 39 m a.s.l.) in the north, which should also have been a smaller lake or had almost dried up. Most of the flow of the San Juan River did not originate at that time from Lake Nicaragua but was fed instead by a river that came down from the volcanic ranges in Costa Rica.

Table 10.7 Sectors where the rate of tectonic uplift and subsidence have been determined on both coasts of Costa Rica

Place	Uplift rate	Comment	References
(mm/year)			
Tamarindo	≤ 1.0 (~1.6)	Rough estimation	Gardner et al. (1992)
Punta Guiones	0.5–1.6		Sak et al. (2009)
Montezuma-Cabuya	1.7–4.5		Marshall and Anderson (1995) Bee (1999)
Cabo Blanco	3.3–4.0		
	4.1–7		Bee (1999)
	6		Gardner et al. (2001)
Golfo de Nicoya	− 0.5		Marshall and Anderson (1995)
Central mountains (Cordillera Central and Valle Central)	1–2		Miyamura (1975)
Caldera	1.0–2.0		Fisher et al. (1998); Sak et al. (2009)
	2.5		Miyamura (1975)
Herradura	0.4–2.1		Denyer and Cortés (2001)
Esterillos Este	0.1–4.4 (2.1)		Sak et al. (2009)
Quepos	0.5–3.0 (1.6)		Gardner et al. (1992)
Playa Ballena	0.7–1.3		Fisher et al. (2004); Sak et al. (2009)
Península de Osa	1.7–8.5 (4.0)	Faulted blocks (~ 5 km) with differential movements produce very variable estimations of uplift and subsidence. In addition, tilting toward the NE	Gardner et al. (2013)
	1.5–6.5 (NW Osa)		Sak et al. (2004, 2009)
	1.5–8.8 (NE Osa)		
	2.1–6.5		Gardner et al. (1992)
	0.2–0.94		Corrigan et al. (1990)
	4.2–5.6 (S Puerto Jiménez)		Gardner et al. (1987)
	1.7 (10 km W Puerto Jiménez)		
Punta Burica	2.1–19.3 (8.3)	N-S faulted blocks produce very variable estimations of uplift	Gardner et al. (1987)
			Morell et al. (2011)
	0.7–8.0 (2.7)		Gardner et al. (1992)
	1.2 (4.7)		Corrigan et al. (1990)
Limón Centro	0.05		McNeill et al. (2000)
Limón Sur	1.8		Denyer (1998)

10.6 Consequences of Paleogeographical Changes

10.6.1 Speciation at Golfo Dulce

Golfo Dulce has been transforming itself from a smaller to a bigger gulf, to cover 320 km^2. It is a tectonic depression considered to have fjord-like features (Hebbeln and Cortés 2001; Kohlmann et al. 2002) and is also an anoxic basin on the Pacific coast of the Americas, approximately 50 km long and 10–15 km wide. Several rivers discharge freshwater into its basin. Bathymetrically it is characterized by a 215 m-deep inner basin, which is held by an effective 60 m sill, and a shallow valley around 80 m deep (Fig. 10.22), extending southward to the mouth of the gulf (Bessesen and Galbreath 2017). This topography prevents free exchange between the deeper waters of the inner basin and adjacent coastal water masses (Svendsen et al. 2006).

Golfo Dulce is the only area where a particular sea-snake subspecies lives, *Hydrophis platurus xanthos* (Fig. 10.22). This subspecies differs from the nominotypical one, *H. p. platurus,* because it is smaller in size, has a nearly uniform yellow coloration that contrasts with the black and yellow stripes and tail spots or bands typical of the species, and because it has a different hunting behavior, nocturnal ambush *versus* diurnal hunting. Bathymetrics appear to restrict genetic flow between this allopatric population and conspecifics in the broader Eastern Pacific (Bessesen and Galbreath 2017).

Bessesen and Galbreath (2017) considered that the insulated bathymetrics and restricted water currents within the Golfo Dulce interrupted transit and reproductive flow with the wider oceanic population in the past. It is known that subspecific or even specific differentiation can occur rapidly under circumstances of intense selection and/or genetic drift. It is therefore likely that this population inside Golfo Dulce has experienced genetic isolation sufficiently long so that it could well-represent an intrinsically genetically isolated taxon of recent origin.

This proposal begs the question of a possible mechanism/time for this speciation process. We assume that the Punta Burica terrain began to shift eastward between one and two million years ago along a fault line separating it from the Peninsula de Osa, forming a freshwater lake first connected to the sea around one million years ago (Kohlmann et al. 2002). This lake connected and disconnected later intermittently with the ocean during different glacial periods, as sea levels rose and dropped, creating a small and recurrent isolated habitat ideal for speciation processes.

Paleogeographic reconstructions of 24 ka (Fig. 10.21) show that the area of the Golfo Dulce was much smaller than today, and thus genetic isolation and drift could have played

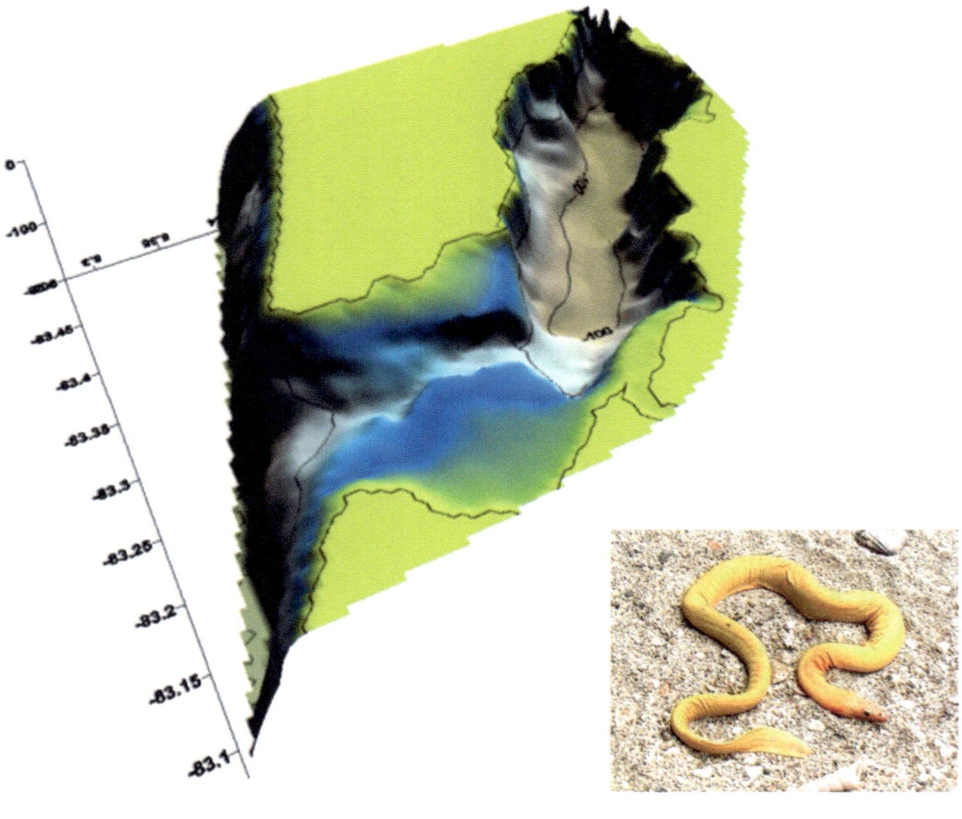

Fig. 10.22 3D bathymetrics of the Golfo Dulce (Source: Morales-Ramírez et al. 2015). It is evident that the gulf mouth is shallow and constricted and is deepest at its innermost end (yellow = land surface; blue = shallow gulf mouth entrance; light brown = deepest floor of the gulf). Inset: *Hydrophis platurus xanthos*, a new subspecies of sea-snake (Elapidae: Hydrophiinae) from Golfo Dulce (Source: Alejandro Solórzano/La Nación)

an essential role in the differentiation of the sea-snake population. On the other hand, the mouth of the gulf also had a smaller breadth, suggesting that the sill could have acted more efficiently in preventing the free exchange of water masses between the open sea and the inner gulf basin. Therefore, it is conceivable that the sea-snake subspecies living in Golfo Dulce originated because of these paleogeographic processes.

10.6.2 Discontinuous Distributions

A phenomenon observed repeatedly is the discontinuous distribution of many tropical rain forest species from the Central Pacific coast to the Península de Nicoya. We present here some examples from dung-beetle species in Fig. 10.23. These species live in deep valleys or ravines of the Península de Nicoya, where the microclimate is suitable for maintaining tropical rain forest conditions.

Note the discontinuous distribution of these species today (Fig. 10.23). However, suppose the paleogeographic reconstruction presented in this study is considered (Fig. 10.21a). In that case, it becomes evident that this modern broken distribution was a continuous one during the last glacial events and that the concomitant sea-level increase interrupted this distribution continuity, isolating many species in the Península de Nicoya.

Another example of a discontinuous distribution pattern is the one at the northern and southern borders of the Pacific rain forest (Fig. 10.5). The climatic barrier present at the Tárcoles River (Figs. 10.17 and 10.18) repeats itself as a mirror image in the southern area in Panama. Here, the rotor effect disappears, and the dry tropical forest reappears. At least three dung-beetle species (*Agamopus lampros, Deltochilum lobipes, Dichotomius yucatanus*) (Fig. 10.17) present in the dry tropical forest of the Guanacaste region suddenly reappear in the dry tropical forest of Panama (Fig. 10.2). These species probably extended their range during the last glacial periods when dry open vegetation existed during the Pleistocene (Colinvaux 1997), or they could represent older distributions when the Cordillera de Talamanca had not yet attained its contemporary height, no rotor winds were generated, and an orographic shadow maintained a continuous dry tropical forest corridor. The dry tropical forest is considered an old type of ecosystem that originated in Mexico's northwest area during the Miocene (Axelrod 1975, 1979), it can be traced back at least as far as 20 million years before present and is seemingly associated with the Tertiary North American flora (Pérez-García et al. 2012), expanding into Central America around 2.5 Ma (Graham and Dilcher 1995; Becerra 2005).

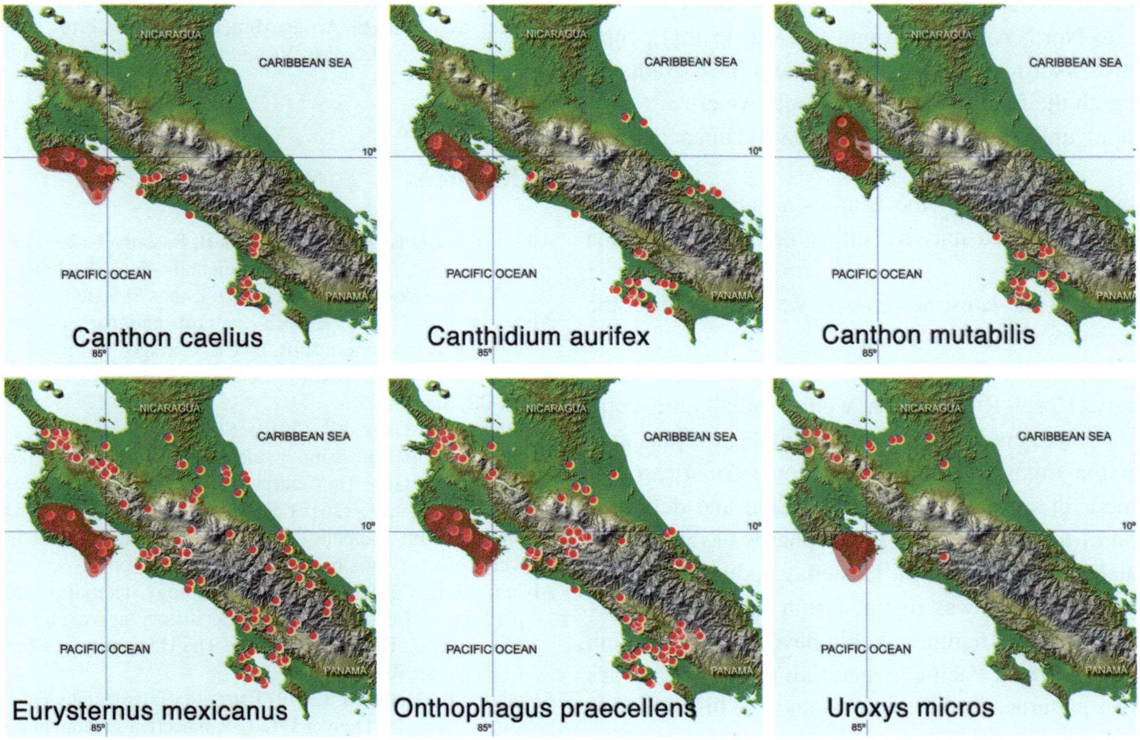

Fig. 10.23 Distribution of six different dung-beetle species (Coleoptera: Scarabaeidae) in Costa Rica (Solís and Kohlmann, 2012). Observe the discontinuous population distribution on the Península de Nicoya (areas highlighted in red)

10.7 Conclusions

Faunal exchange in different grades of intensity between the Caribbean Sea and the Pacific Ocean, and between North America and South America since 15 Ma to the present, represents one of the most critical events in the diversification and extinction of late Cenozoic and modern species in North America, South America, and the Caribbean region (Baker et al. 2014; Jaramillo, 2018). Whether the Isthmus of Panama was continuous and closed at 13–10 Ma ago, or if it was a peninsula of North America with several shallow and often dry seaways (*i.e.*, present Lake Nicaragua, Panama Canal, Atrato Basin), it is still a matter for open and strenuous discussions. The previous period of a biogeographical archipelago is of utmost importance since the species that managed to colonize these islands spreading from both continents could have been the base of the evolution of endemic species in these islands, and eventually, when the isthmian connection got established, formed part of the biodiversity of the Costa Rica of today.

Although there is still a heated debate about the date of the closure of the isthmus and the uplift of Talamanca, from our biogeographical point of view, the following sequence of events significantly influenced the establishment of terrestrial species in Costa Rica:

- The islands of volcanic origin existed since > 13 Ma, and they were slowly coalescing into a big island about 6 Ma ago, and subsequently, as a single peninsular mass connected to North America around 5.3 Ma. In this peninsular area, North American fauna would have gradually mixed with the fauna generated in South America.
- The rapid uplift of the Cordillera de Talamanca about 5 Ma.
- The final connection of the isthmus around 2.8 Ma ago in Panama led to a massive influx of South American fauna.
- The arrival of the latest newcomers was mediated by the last glaciation around 23–25 ka.

In summary, Costa Rican fauna would initially be composed of species generated *in situ* (archipelagic), plus species from the mixture of the latter North American and South American species, during the gradual and definitive connection of the isthmus. The initial phases of rapid uplift of the Talamanca and the Fila Costeña produced a physiographical region that was distinct from the tropical lowlands, and the climate regimes should have changed on both flanks, Caribbean and Pacific, generating profound changes in the forest patterns, fluvial systems, and wildlife distribution. Starting at about 1000 m of elevation, one could suppose Caribbean populations losing contact with Pacific populations, thus generating speciation by vicariance. On the other hand, with the elevation of the Cordillera de Guanacaste, Tilarán, and Central generated by volcanism and that of Talamanca and the Fila Costeña due to tectonic upheavals, climatic isolation was produced in the higher parts generating endemic species. How many millions or thousands of years would be necessary for each speciation event to occur?

Then came the times of glaciations that possibly reordered the fauna of the lowlands by increasing the area due to the descent of the sea, as well as by climatic changes in those parts that led to the existence of vast savannas, allowing for the entry of fauna typical of this type of environment and that today still coexists with us in open areas generated by human beings. However, glaciations also reorganized the fauna of the highlands, allowing for interchange between local mountain peaks and between mountain ranges, including interchanges with the higher parts of Nicaragua.

In conclusion, the exuberant geology, among other processes, is one of the main drivers of the incredible biodiversity Costa Rica has, as exemplified by the discovery of new species every year, for instance, ten new orchid species have been described every year during the last 20 years (Rodríguez 2022). The present study agrees with Croizat's metaphor (1958, 1964): "Earth and life evolve together"; there is, without doubt, a dynamic interaction between biology and geology.

Acknowledgements The authors thank sincerely for revising the text to Alfonsina Arriaga-Jiménez, Jorge Cortés, Piotr Migoń, Matthias Rös, and Lorenzo Vázquez-Selem who helped improve this text.

References

Alfaro A, Gazel E, White WM, Jicha B, Rasbury T (2021) Unravelling the genesis of young continental—arc shoshonites in the Talamanca Cordillera, Costa Rica. Lithos 386–387

Alvarado GE (1994) Historia natural antigua: los intercambios biológicos interamericanos. ETCR, Cartago

Alvarado GE (2021) Costa Rica y sus volcanes. EUCR, EUNA, ETCR, San José

Alvarado GE, Gans PB (2012) Síntesis geocronológica del magmatismo, metamorfismo y metalogenia de Costa Rica. América Central Rev Geol Am Central 46:7–122

Alvarado A, Mata R (2016) Soils of Costa Rica: an agroecological approach. In: Kappelle M (ed) Costa Rican ecosystems, 1st edn. The University of Chicago Press, Chicago, pp 64–93

Alvarado GE, Cárdenes G (2016) Geology, tectonics, and geomorphology of Costa Rica: a natural history approach. In: Kappelle M (ed) Costa Rican Ecosystems. The University of Chicago Press, Chicago, pp 30–63

Alvarado GE, Denyer P (2019) Estratigrafía volcánica del Neógeno y Cuaternario. In: Denyer P (ed) Perspectiva geológica del noroeste de Costa Rica: historia, evolución y cartografía. EUCR-CICG, San José, pp 101–238

Alvarado GE, Vega E, Cháves J, Vázquez M (2004) Los grandes deslizamientos (volcánicos y no volcánicos) de tipo debris avalanche en Costa Rica. Rev Geol Am Central 30:83–99

Alvarado GE, Dengo C, Martens U, Agjuilar BJ, T, Bonis SB, (2007) Stratigraphy and geologic history. In: Bundschuh J, Alvarado GE (eds) Central America: geology, resources and hazards. Taylor and Francis, London, pp 345–394

Amador JA, Rivera ER, Durán-Quesada AM, Mora G, Sáenz F, Calderón B, Mora N (2016a) The easternmost tropical Pacific. Part I: a climate review. Rev Biol Trop 64:S1–S22

Amador JA, Rivera ER, Durán-Quesada AM, Mora G, Sáenz F, Calderón B, Mora N (2016b) The easternmost tropical Pacific. Part II: seasonal and intraseasonal modes of atmospheric variability. Rev Biol Trop 64:S23-57

Arnaud P (2002) Phanaeini. Les Coléoptères du Monde/The Beetles of the World, 28. Hillside Books, Canterbury

Astorga A, Fernández JA, Barboza G, Campos L, Obando J, Aguilar A, Obando L (1995) Sedimentary basins of Costa Rica: Late Mesozoic-Cenozoic evolution and hydrocarbon potential. In: Miller RL, Escalante R, Renemund JA et al (eds) Energy and Mineral Potential of the Central American-Caribbean Regions. Springer, Berlin, pp 85–94

Avalos G (2019) Still searching the rich coast: biodiversity of Costa Rica, numbers, processes, patterns, and challenges. In: Pullaiah T (ed) Global biodiversity, selected countries in the Americas and Australia, vol 4. Apple Academic Press, Boca Ratón, pp 101–138

Axelrod DI (1975) Evolution and biogeography of Madrean-Tethyan sclerophyll vegetation. Evolution 62:280–334

Axelrod DI (1979) Age and origin of Sonoran Desert vegetation. Occ Papers Calif Acad Sci 132:1–74

Baker PA, Fritz SC, Dick CW (2014) The emerging field of geogenomics: constraining geological problems with genetic data. Earth-Science Rev 135:38–47

Barat F, de Lépinay BM, Soisson M, Müller C, Baumgartner P, Baumgartner-Mora C (2014) Transition from the Farallon plate subduction to the collision between South and Central America: geological evolution of the *Panama isthmus*. Tectonophysics 622:145–167

Baraud J (1975) *Onthophagus massai*, nouvelle espèce paléarctique (Coleoptera : Scarabaeoidea). Bulletin De La Société Linnéenne De Lyon 44:292–295

Barrantes G (2009) The role of historical and local factors in determining species composition of the highland avifauna of Costa Rica and Western Panamá. Rev Biol Trop 57:333–349

Becerra JX (2005) Timing the origin and expansion of the Mexican dry tropical forest. PNAS 102:10919–10923

Bee B (1999) Rapid Quaternary uplift of marine terraces: Cabo Blanco to Montezuma area, Peninsula de Nicoya, Costa Rica. Geol Soc Amer Bull 89:168–171

Bessesen BL, Galbreath GJ (2017) A new subspecies of sea snake, *Hydrophis platurus xanthos*, from Golfo Dulce, Costa Rica. Zookeys 686:109–123

Brikiatis L (2023) Evidence for a recent (0.65 Ma) formation of the Isthmus of Panama. Mari Scie Res Ocean 6:25–68

Campbell JA, Lamar WW (1989) The venomous reptiles of Latin America. Cornell University Press, Ithaca

Coates AG, Jackson JBC, Collins LS (1992) Closure of the Isthmus of Panama: the near-shore marine record of Costa Rica and western Panama. Geol Soc Amer Bull 104:814–828

Coen E (1983) Climate. In: Janzen D (ed) Costa Rican natural history. The University of Chicago Press Chicago, pp 35–46

Colinvaux P (1997) The history of forests on the Isthmus from the ice age to the present. In: Coates AG (ed) Central America. Yale University Press, New Haven, A natural and cultural history, pp 123–136

Corrigan J, Mann P, Ingle JC (1990) Forearc response to subduction of the Cocos Ridge, Panama-Costa Rica. Geol Soc Amer Bull 102:628–652

Cortés J (2016a) The Pacific Coastal and Marine Ecosystems. In: Kappelle M (ed) Costa Rican ecosystems. The University of Chicago Press, Chicago, pp 97–138

Cortés J (2016b) Isla del Coco: coastal and marine ecosystems. In: Kappelle M (ed) Costa Rican ecosystems. The University of Chicago Press, Chicago, pp 162–191

Croizat L (1958) Panbiogeography. Volumes 1, 2a, 2b. Published by the author, Caracas

Croizat L (1964) Space, time, form: the biological synthesis. Published by the Author, Caracas

Cunningham M, Stark CP, Kaplan M, Schaefer J (2019) Glacial limitation of tropical mountain height. Earth Surf Dyn 7:147–169

Darlington Jr PJ (1970) Carabidae on tropical islands, especially the West Indies. Biotropica 2:7–15

De Boer JZ, Drummond MS, Bordelon M, Defant MJ, Bellon H, Maury R (1995) Cenozoic magmatic phases of the Costa Rica island arc (Cordillera de Talamanca). In: Mann P (ed) Geologic and tectonic development of the Caribbean Plate Boundary in Southern Central America. Geol Soc Am 295:35–55

Denyer P (1998) Historic-prehistoric earthquakes, seismic hazards, and Tertiary and Quaternary geology of the Gandoca-Manzanillo Natural Wildlife Refuge, Limón, Costa Rica. Rev Biol Trop 46. Supl 6:237–250

Denyer P, Alvarado GE (2007) Mapa Geológico de Costa Rica (1:400,000). Librería Francesa San José

Denyer P, Cortés J (2001) Levantamiento tectónico de la costa del Pacífico central de Costa Rica. Rev Biol Trop 49(Supl 2):179–183

Denyer P, Aguilar T, Alvarado GE (2003a) Geología y Estratigrafía de la Hoja Barranca, Costa Rica. Rev Geol Am Central 29:105–125

Denyer P, Montero W, Alvarado GE (2003b) Atlas Tectónico de Costa Rica. UCR, San José

Edmonds WD, Zídek J (2012) Taxonomy of *Phanaeus* revisited: Revised keys to and comments on species of the New World dung beetle genus *Phanaeus* MacLeay, 1819 (Coleoptera: Scarabaeidae: Scarabaeinae: Phanaeini). Insecta Mundi 0274:1–108

Fisher DM, Gardner TW, Marshall JS, Sak P, Protti M (1998) Effect of subducting seafloor roughness on fore-arc kinematics, Pacific coast, Costa Rica. Geology 26:467–470

Fisher DM, Gardner TW, Sak P, Sánchez JD, Murphy K, Vanucchi P (2004) Active thrusting in the inner forearc of an erosive convergent margin, Pacific coast, Costa Rica. Tectonics 23, TC2007

Foster JB (1964) The evolution of mammals on islands. Nature 202:234–235

Frolov AV, Ahmetova LA (2020) Description of a new brachypterous scarab species, *Orphnus brevialatus* (Coleoptera: Scarabaeidae: Orphninae) from East Africa, with notes on flightlessness in the orphnines. Zootaxa 4750:425–431

Funk J, Mann P, McIntosh K, Stephens J (2009) Cenozoic tectonics of the Nicaraguan depression, Nicaragua, and Median Trough, El Salvador, based on seismic-reflection profiling and remote-sensing data. GSA Bull 121:1491–1521

Gardner TW, Back W, Bullard TF (1987) Central America and the Caribbean. GSA Centennial, Sp. 2:343–402

Gardner TW, Verdonck D, Pinter NM, Slingerland R, Furlong K, Bullard TF, Wells S (1992) Quaternary uplift astride the aseismic Cocos Ridge, Pacific Coast, Costa Rica. Geol Soc Am Bull 104:219–232

Gardner TW, Marshall JS, Merritts DJ, Bee B, Burgette R, Burton E, Cooke J, Kehrwald N, Protti M, Fisher D, Sak P (2001) Holocene forearc block rotation in response to seamount subduction, southeastern Peninsula de Nicoya, Costa Rica. Geology 29:151–154

Gardner TW, Fisher DM, Morell KD, Cupper ML (2013) Upper-plate deformation in response to flat-slab subduction inboard of the aseismic Cocos Ridge, Osa Peninsula, Costa Rica. Lithosphere 5:247–264

Gazel E, Carr MJ, Hoernle K, Feigenson MD, Szymanski D, Hauff F, van den Bogaard P (2009) Galapagos-OIB signature in southern Central America: mantle refertilization by arc-hot spot interaction. Geochem Geophys Geosyst 10:1–32

Génier F (2000) *Dichotomius comarapensis* sp. nov., une nouvelle espèce bolivienne de scarabée brachyptère (Coleoptera: Scarabaeidae, Scarabaeinae). Fabreries 25:25–31

Gillespie RG, Roderick GK (2002) Arthropods on Islands: colonization, speciation, and conservation. Ann Rev Entomol 47:595–632

Gillet CPDT, Toussaint EFA (2020) Macroevolution and shifts in the breeding biology of the New World scarab beetle tribe Phanaeini (Coleoptera: Scarabaeidae: Scarabaeinae). Biol J Linn Soc XX:1–22

Gómez LD (1986) La Vegetación de Costa Rica. EUNED San José

Gräfe K, Frisch W, Villa I, Meschede M (2002) Geodynamic evolution of southern Costa Rica related to low angle subduction of the Cocos Ridge: Constraints from thermochronology. Tectonophysics 348:187–204

Graham A, Dilcher D (1995) The Cenozoic record of tropical dry forest in northern Latin America and the southern United States. In: Bullock SH, Mooney HA, Medina E (eds) Seasonally dry tropical forests. Cambridge University Press, Cambridge, pp 124–145

Groombridge B (1992) Global biodiversity: status of the earth's living resources. World Conservation Monitoring Centre (WCMC). Chapman and Hall, London

Hastenrath S (1973) On the Pleistocene glaciation of the Cordillera de Talamanca, Costa Rica. Z Gletscherk Glazialgeol 9:1051–1121

Hauff F, Hoernle K, van den Bogaard P, Alvarado G, Garbe-Schönberg D (2000) Age and geochemistry of basaltic complexes in Western Costa Rica: contributions to the geotectonic evolution of Central America. Geochem Geophys Geosyst 1. https://doi.org/10.1029/1999GC000020

Haug GH, Tiedemann R (1998) Effect of the formation of the Isthmus of Panama on Atlantic Ocean thermohaline circulation. Nature 393:673–676

Hebbeln D, Cortés J (2001) Sedimentation in a tropical fjord: Golfo Dulce, Costa Rica. Geo-Mar Lett 20:142–148

Herrera W (2016) Climate of Costa Rica. In: Kappelle M (ed) Costa Rican ecosystems. The University of Chicago Press, Chicago, pp 19–29

Horn SP (2007) Late Quaternary lake and swamp sediments: recorders of climate and environment. In: Bundschuh J, Alvarado GE (eds) Central America. Geology, Resources and Hazards. Taylor and Francis, London, pp 423–441

Howden HF, Young OP (1981) *Panamanian Scarabaeidae: Taxonomy, distribution and habits (Coleoptera, Scarabaeidae).* Contr Amer Ent Inst 18

Islebe GA, Hooghiemstra H (1997) Vegetation and climate history of montane Costa Rica since the last glacial. Quatern Sci Rev 16:589–604

Jaramillo C (2018) Evolution of the Isthmus of Panama: Biological, Paleoceanographic and Paleoclimatological Implications. In: Hoorn C, Perrigo A, Antonelli A (eds) Mountains, Climate and Biodiversity. John Wiley and Sons Ltd, Hoboken, p 32–38

Jiménez-Ferbans L, Reyes-Castillo P, Schuster JC (2017) The passalid beetles (Coleoptera: Passalidae) from Costa Rica, with the description of two new species of *Passalus*. Rev Mex Biodiv 88:608–616

Kappelle M (2016) Costa Rica's ecosystems: setting the stage. In: Kappelle M (ed) Costa Rican ecosystem. The University of Chicago Press, Chicago, pp 3–16

Kesel RH (1983) Quaternary history of the Río General Valley, Costa Rica. Nat Geo Soc Res Rep 15:33–58

Kirby MX, MacFadden BJ (2005) Was southern Central America an archipelago or a peninsula in the middle Miocene? A test using land-mammal body size. Paleogeogr Paleoclimatol Paleoecol 22:193–203

Kohlmann B (2011) Biodiversity conservation in Costa Rica—an animal and plant biodiversity atlas. In: Pavlinov O (ed) Research in biodiversity—models and applications. InTech Publishers Rijeka, pp 203–222

Kohlmann B, Solís A (1997) El género *Dichotomius* (Coleoptera: Scarabaeidae) en Costa Rica. G Ital Ent 8:343–382

Kohlmann B, Wilkinson MJ (2007) The Tárcoles Line: biogeographic effects of the Talamanca Range in lower Central America. G Ital Ent 12:1–30

Kohlmann B, Solís A (2001a) El género *Onthophagus* (Coleoptera: Scarabaeidae) en Costa Rica. G Ital Ent 9:159–261

Kohlmann B, Solís A (2001b) Description of a new *Phanaeus* from Costa Rica. Besoiro 6:9–11

Kohlmann B, Solís A (2012) New species and revalidations of scarab beetles (Coleoptera: Scarabaeidae: Geotrupinae and Scarabaeinae) from Costa Rica and Panama. Zootaxa 3193:28–52

Kohlmann B, Arriaga-Jiménez A, Rös M (2018) Dung beetle vicariant speciation in the mountains of Oaxaca, Mexico, with a description of a new species of *Phanaeus*. Zookeys 743:67–93

Kohlmann B, Solís A, Alvarado G (2019) Description of *Onthophagus humboldti* and *Uroxys bonplandi*, two new scarab beetles (Coleoptera, Scarabaeidae, Scarabaeinae) from Costa Rica, with notes on tropical mountain brachyptery and endemicity. Zookeys 881:23–51

Kohlmann B, Wilkinson J, Lulla K (2002) Costa Rica from Space/Costa Rica desde el Espacio. Neotrópica Foundation Publishers, San José

Kohlmann B, Solís A, Elle O (2007) Biodiversity, conservation, and hotspot atlas of Costa Rica: a dung beetle perspective (Coleoptera: Scarabaeidae: Scarabaeinae). Zootaxa 1457:1–34

Kolarsky RA, Mann P, Montero W (1995) Island arc response to shallow subduction of the Cocos ridge, Costa Rica, In: Mann P (ed) Geologic and Tectonic Development of the Caribbean Plate Boundary in Southern Central Americ. Geol Soc Am Spec Pap 295:235–262

Lachniet MS, Seltzer GO (2002) Late quaternary glaciation of Costa Rica. Geol Soc Am Bull 114:547–558

Lachniet MS (2007) Glacial geology and geomorphology. In: Bundschuh J, Alvarado GE (eds) Central America: geology, resources, and hazards. Taylor and Francis, London, pp 171–184

Laurito C, Valerio AL (2010) Los caballos fósiles de la formación Curré, cantón de Coto Brus, Costa Rica. Museo Nacional de Costa Rica, San José

Laurito C, Valerio AL (2012) Primer registro fósil de *Pliometanastes* sp. (Mammalia, Xenarthra, Megalonychidae) para el Mioceno superior de Costa Rica, América Central. Una nueva pista para la comprensión del Pre-Gabi. Rev Geol Am Central 47:95–108

Laurito CA, Valerio AL (2013) *Scirrotherium antelucanus*, una nueva especie de Pampatheriidae (Mammalia, Xenarthra, Cingulata) del Mioceno Superior de Costa Rica, América Central. Rev Geol Am Central 49:45–62

Laurito CA, Valerio AL (2016) Camellos laminos del Mioceno tardío (Henfiliano temprano) de la formación Curré, San Gerardo de Limoncito, Cantón de Coto Brus, Provincia de Puntarenas, Costa Rica. Rev Geol Am Central 54:7–55

Lawrence JF, Nielsen ES, Mackerras IM (1991) Skeletal anatomy and a key to orders. In: The Insects of Australia, 2nd edn. CSIRO Division of Entomology, Canberra, pp 3–32

MacFadden BJ, Bloch JI, Evans H, Foster DA, Morgan GS, Rincon A, Wood AR (2014) Temporal calibration and biochronology of the Centenario Fauna, Early Miocene of Panama. J Geol 122:113–135
Mann P, Burke K (1984) Neotectonics of the Caribbean. Rev Geophys Space Phys 22:309–362
Marshall LG (1988) Land mammals and the great interchange. Sci Am 76:380–388
Marshall JS, Anderson RS (1995) Quaternary uplift and seismic cycle deformation, Península de Nicoya, Costa Rica. Geol Soc Amer Bull 107:463–473
Matthews EG (1966) A taxonomic and zoogeographic survey of the Scarabaeinae of the Antilles (Coleoptera: Scarabaeidae). Mem Amer Ent Soc 21
McNeill DF, Coates AG, Budd AF, Berne PF (2000) Integrated paleontologic and paleomagnetic stratigraphy of the upper Neogene deposits around Limon, Costa Rica: a coastal emergence record of the Central American Isthmus. Geol SocAmer Bull 112:963–981
Mittermeier RA, Robles Gil P, Hoffman M, Pilgrim J, Brooks T, Goettsch Mittermeier C, Lamoreux J, da Fonseca GAB (2004) Hotspots Revisited. CEMEX, Mexico-City
Miyamura S (1975) Recent crustal movements in Costa Rica disclosed by relevelling surveys. In: Pavoni N, Green R (eds) Recent crustal movements. Tectonophysics, vol 29, pp 191–198
Montes C, Cardona A, Jaramillo C, Pardo A, Silva JC, Valencia V, Pérez-Angel LC, Rodríguez-Parra LA, Ramírez V, Niño H (2015) Middle Miocene closure of the Central American Seaway. Science 348:226–228
Morales-Ramírez A, Acuña-González J, Lizano O, Alfaro E, Gómez E (2015) Rasgos oceanográficos en el Golfo Dulce, Pacífico de Costa Rica: una revisión para la toma de decisiones en conservación marina. Rev Biol Trop 63:131–160
Morell KD, Fisher DM, Gardner TW, La Femina P, Davidson D, Teletzke A (2011) Quaternary outer fore-arc deformation and uplift inboard of the Panama Triple Junction, Burica Peninsula. J Geophysical Res 116:B05402
Obando V (2002) Biodiversidad de Costa Rica. Estado del Conocimiento y Gestión. Instituto Nacional de Biodiversidad, Heredia
Obando V (2007) Biodiversidad de Costa Rica en Cifras. Editorial INBio, Heredia
O'Dea A, Lessios HA, Coates AG et al (2016) Formation of the Isthmus of Panama. Sci Adv 2:e1600883
Pérez W, Alvarado GE, Gans PB (2006) The 322 ka Tiribí Tuff: stratigraphy, geochronology, and mechanisms of deposition of the largest and most recent ignimbrite in the Valle Central, Costa Rica. Bull Volcanol 69:25–40
Pérez-Crespo VA, Laurito CA, Arroyo-Cabrales J, Valerio AL, Morales Puente P, Cienfuegos-Alvarado E, Otero FJ (2018) Feeding habits and habitat of herbivorous mammals from the Early-Late Hemphillian (Miocene) of Costa Rica. Acta Paleontologica Polonica 63:645–652
Pérez-García EA, Meave JA, Cevallos-Ferriz SRS (2012) Flora and vegetation of the seasonally dry tropics in Mexico: origin and biogeographical implications. Acta Bot Mex 100:149–193
Pizarro D (1993) Los pozos profundos en Costa Rica. Aspectos litológicos y bioestratigráficos. Rev Geol Am Central 13:81–85
Pizzo A, Mazzone F, Rolando A, Palestrini C (2011) Combination of geometric and genetic approaches applied to a debated taxonomical issue: the status of *Onthophagus massai* (Coleoptera, Scarabaeidae) as an endemic species vicarious to *Onthophagus fracticornis* in Sicily. Zoology 114:199–122
Potter R, Li Y, Horn SP, Orvis KH (2019) Cosmogenic Cl-36 surface exposure dating of late Quaternary glacial events in the Cordillera de Talamanca. Costa Rica Quat Res 92:1–16
Quesada-Román A, Campos N, Alcalá-Reygosa J, Granados Bolaños S (2020) Equilibrium-line altitude and temperature reconstructions during the Last Glacial Maximum in Chirripó National Park, Costa Rica. J South Amer Earth Sci 100:102576
Rabineau M, Berné S, Olivet J-L, Aslanian D, Guillocheau F, Joseph P (2006) Paleo sea levels reconsidered from direct observation of paleoshoreline position during Glacial Maxima (for the last 500,000 yr). Earth Planetary Sci Lett 252:119–137
Rahbek C, Borregaard MK, Colwell RK, Dalsgaard B, Holt BG, Morueta-Holme N, Nogues-Btavo D, Whittaker RJ, Fjeldså J (2019) Humboldt's enigma: What causes global patterns of mountain biodiversity? Science 365:1108–1113
Reig O (1981) Teoría del origen y desarrollo de la fauna de vertebrados tetrápodos de América del Sur. Ameghiniana 2:131–140
Rincón A, Valerio AL, Laurito C (2020) First fossil record of a Megatheriidae-Megatherrinae in the Early Hemiphillian (Late Miocene) from San Gerardo de Limoncito, Curré Formation, Costa Rica. Rev Geol Am Central 62:1–24
Rivier F (1985) Sección geológica del Pacífico al Atlántico a través de Costa Rica. Rev Geol Am Central 2:23–31
Rodríguez I (2022) Ciencia otorga 'cédula tica' a miles de especies cada año. La Nación (Sección El País). Domingo 21 de agosto de 2022:4–5
Sak PB, Fisher DM, Gardner TW (2004) Effects of subducting seafloor roughness on upper plate vertical tectonism: Osa Peninsula, Costa Rica. Tectonics 23:TC1017
Sak PB, Fisher DM, Gardner TW, Marshall JS, LaFemina PC (2009) Rough crust subduction, forearc kinematics, and Quaternary uplift rates. Costa Rican Segment Middle Am Trench Geol Soc Am Bull 121:992–1012
Savage JM (2002) The Amphibians and Reptiles of Costa Rica. The University of Chicago Press, Chicago
Schindlbeck JC, Kutterolf S, Freundt A, Alvarado GE, Wang K-L, Straub SM, Hemming SR, Frische M, Woodhead JD (2016) Late Cenozoic tephrastratigraphy offshore the southern Central American Volcanic Arc. 1. Tephra ages and provenance. Geochem Geophys Geosyst 17:4585–4604
Scholtz CH (2000) Evolution of flightlessness in Scarabaeoidea (Insecta, Coleoptera). D Ent Zeit 47:1–28
Solís A, Kohlmann B (2012) A checklist and distribution atlas of the Scarabaeinae (Coleoptera: Scarabaeiadae) of Costa Rica. Zootaxa 348:21–32
Spratt RM, Lisiecki LE (2016) A late Pleistocene sea level stack. Clim past 12:1079–1092
Stehli FG, Webb SD (eds) (1985) The Great American Biotic Interchange. Plenum Press, New York
Svendsen HR, Rosland R, Myking S, Vargas JA (2006) A physical-oceanographic study of Golfo Dulce, Costa Rica. Rev Biol Trop 54:147–170
Terry RA (1941) Notes on Submarine Valleys off the Panamanian Coast. Geo Rev 31:377–384
Thiele K, Vargas AE (2017) Actualización volcanoestratigráfica del piso del Valle Central, de Costa Rica, específicamente la denominada Formación Colima y sus miembros. (B. Sc. dissertation) Escuela Centroamericana de Geología, San José
Toledo VM (1982) Pleistocene vegetation changes in tropical Mexico. In: University C (ed) Prance GT. Press, New York, pp 93–111
Valerio CE (2006) Costa Rica. Ambiente y Sociedad. Instituto Nacional de Biodiversidad, Heredia
Valerio AL, Laurito CA (2008) Dental remains of immature individuals of *Rhynchotherium blicki* (Frick 1933) of San Gerardo de Limoncito, Coto Brus, Costa Rica. Rev Geol Am Central 39:87–91

Valerio AL, Laurito CA (2020) Primer registro de pecaríes fósiles (Artiodactyla, Tayassuidae) para el Mioceno Superior (Hh3: Hemphilliano tardío) de Costa Rica, América Central. Rev Geol Am Central 62:25–47
Vannucchi P, Ranero CR, Galeotti S, Straub SM, Scholl DW, McDougall-Ried K (2003) Fast rates of subduction erosion along the Costa Rica Pacific margin: Implications for nonsteady rates of crustal recycling at subduction zones. J Geophysical Res 108:25–11
von Harold E (1867) Zur Kenntnis der Gattung *Canthidium* und ihrer nächsten Verwandten. Coleopterologische Hefte 1:1–61
Waliser DE, Jiang X (2015) Tropical meteorology and climate/inter-tropical convergence zone. In: Pyle J, Zhang F (eds) Encyclopedia of atmospheric sciences. Academic Press, Cambridge, pp 121–131

Agricultural Landscapes of Costa Rica

11

Manuel E. Camacho-Umaña, Adolfo Quesada-Román, Mario Villatoro-Sánchez, Bryan Alemán-Montes, Rafael Mata, Carlos Henríquez-Henríquez, Jesús Céspedes-Rivera, Manuel Céspedes-Rivera, and Alfredo Alvarado

Abstract

Agricultural activities and introduction of new crops result in significant changes in society, culture, and most evident in short-term, the transformation of landscapes. For instance, the introduction of coffee (*Coffea arabica* L.) in Costa Rica and its further extension into the Central Valley during the 1800s led to substantial changes in socio-economic, geophysical, and infrastructural aspects of Costa Rican society. More recently, banana (*Musa* spp.), pineapple (*Ananas comosus* (L.) Merr.), sugarcane (*Saccharum officinarum* L.), rice (*Oryza sativa* L.), and oil palm (*Elaeis guineensis* Jacq.) plantations were established in specific regions of Costa Rica, where geomorphic features and agroecological conditions are considered suitable for their extensive production and economic revenue, affecting the evolution of demography and landscapes. The chapter summarizes essential information about landforms, soils, and climatic conditions in areas that have extensive agriculture as their main economic activity. This study highlights the regions and soils dedicated to cash crops and agricultural activities of Costa Rica. For instance, coffee, pineapple, and sugarcane production develop within highly weathered soils (Ultisols and Oxisols). Banana and oil palm production are mainly located in the coastal plains, where soil genesis is mostly associated with alluvial processes (Entisols and Inceptisols). Finally, horticultural crops like potatoes (*Solanum tuberosum* L.) and onions (*Allium cepa* L.) were mainly identified within the volcanic piedmonts, where volcanic ash soils (Andisols) are dominant.

Keywords

Costa Rica · Agriculture · Soils · Landforms · Tropical crops

11.1 Introduction

At around 2000 BC, the Pre-Columbian inhabitants of Costa Rica transitioned from hunting and gathering (10,000–2000 B.C.) to early agriculture (2000–300 B.C.), with "slash and burn" as the main system for cultivating tubbers, roots, and seeds (Corrales 2000). This transition transformed these early societies as they began to group into tribes and establish settlements scattered throughout the country, eventually leading to higher levels of organization and territorial expansion (Corrales 2000).

After the arrival of the Spaniards on the Caribbean coast of Costa Rica in 1502 A.D., several expeditions were conducted throughout the country, introducing cattle, swain, and basic grains such as barley (*Hordeum vulgare* L.), wheat (*Triticum aestivum* L.), and oats (*Avena sativa* L.), as well as bananas (*Musa paradisiaca*) and plantains (Sáenz 1960). Some of these expeditions led to the establishment of villages, cities, and regions where certain crops were predominantly developed such as corn (*Zea mays* L.).

Furthermore, the landscapes changed drastically, especially in the sector *Valle Central* with the introduction of coffee (*Coffea arabica* L.) around the year 1800 in Costa Rica. Coffee cultivation flourished since 1808 when the contemporaneous Costa Rican governor, Tomás de Acosta, imported seeds for its distribution and established coffee plantations as a source of job and economic activity, further

M. E. Camacho-Umaña (✉) · M. Villatoro-Sánchez · B. Alemán-Montes · R. Mata · C. Henríquez-Henríquez · J. Céspedes-Rivera · M. Céspedes-Rivera · A. Alvarado
Centro de Investigaciones Agronómicas, Universidad de Costa Rica, San Pedro, Costa Rica
e-mail: MANUEL.CAMACHOUMANA@ucr.ac.cr

A. Quesada-Román
Laboratorio de Geografía Física, Escuela de Geografía, Universidad de Costa Rica, San Pedro, Costa Rica

A. Quesada-Román (ed.), *Landscapes and Landforms of Costa Rica*,
World Geomorphological Landscapes, https://doi.org/10.1007/978-3-031-64940-0_11

supported by other leaders (Gutiérrez 1994). This last facilitated Costa Rica's coffee export to Panama in 1820, resulting from increased coffee cultivation and high production, leading to further exports to Europe and the United States of America.

Following this expansion of coffee cultivation, the need for a port on the Costa Rican Caribbean coast with a direct connection to the Central Valley to transport coffee and facilitate shipping to Europe led Minor C. Keith (a newcomer in railroad business), to move to Costa Rica in 1871 and further lead the construction of the "Atlantic Railroad," connecting the Reventazón valley with San José (Jones and Morrison 1952). Minor C. Keith was responsible for the introduction of some banana (*Musa* spp) varieties, starting this business in some tributary areas within central "Huetar Atlántica" Region in 1872 (Jones and Morrison 1952). After a slow start and some improvements in the cropping system (plant breeding, nutrient management, pest control, and fruit transportation systems), the banana crop expanded across the Caribbean lowlands of Costa Rica, covering a substantial portion of the Limón province (Soto-Ballestero 1994). This activity has been growing up to date, transforming the socio-economic conditions and, most evidently, the landscape, while promoting Costa Rica internationally due to the quality and the productivity of this crop.

Currently, pineapple (*Ananas comosus* (L.) Merrill) has become the main exported crop of Costa Rica. Pineapple plantations were first recorded in the 1930s when the Costa Rican Banana Company (Compañía Bananera de Costa Rica) introduced cultivars that were highly adapted to the conditions, displacing native cultivars (Castro 1994). After 1983, the company Pineapple Development Corporation of Costa Rica (PINDECO) intensified activities in the southern region and increased exporting fresh pineapple to the USA and Europe, leading to an exponential growth of the cultivated area with pineapple. For instance, in 1984, the total cultivated area with pineapple in Costa Rica was around 2,500 ha, whereas the reported area in 2020 was 44,000 ha (SEPSA 2023). This significant increase in cultivated land has caused dramatic changes in the landscapes of different regions of Costa Rica, transforming not only the socio-economic patterns and population dynamics within the corresponding communities, but also the biophysical environment.

The dynamics of landscapes is highly influenced by inner earth forces (e.g., tectonics and volcanism), other environmental factors (e.g., rainfall, winds, and temperature), and the living organisms developing within them. As mentioned before, agriculture is a powerful changing force acting on landforms, modifying their features and processes in a relatively short period. Therefore, assessing this activity and its diverse variants within the landscapes becomes fundamental to understand and address their evolution and potential fate.

The present chapter addresses the landscapes of Costa Rica where agricultural activities represent a considerable land extension of diverse cash crops under a diversity of climatic conditions and landforms. Socio-economic features, main soil orders, and productivity associated with these crops will also be described. Our findings will represent fundamental knowledge that could be transferred to other tropical landscapes with similar features and climate, where the same crops are currently expanding or have potential to do so.

11.2 Landscapes with Extensive Agricultural Crops in Costa Rica

The territory of Costa Rica can be subdivided into six socio-economic regions (hereafter planning regions): (1) Chorotega, (2) Pacífico Central, (3) Brunca, (4) Huetar Norte, (5) Huetar Caribe, and (6) Central (Fig. 11.1). Within each region, specific cash crops are extensively cultivated due to suitable climatic conditions, soils, and landforms, but some other crops are also observed in several regions. We will describe in detail the main agricultural activities within each of these six regions.

Additionally, and aiming to emphasize the role of the landscape within the current agricultural systems and commodity crops developing in each planning region, we are including the most recent soil orders map of Costa Rica at 1:200,000 scale (Fig. 11.2) as well as their distribution within the regions (Fig. 11.3).

11.2.1 Chorotega Region in the Northwestern Costa Rica

The Chorotega Region extends to cover most of Península de Nicoya (Fig. 11.1), including the municipalities of La Cruz, Liberia, Cañas, Carrillo, Santa Cruz, Hojancha, Bagaces, Abangares, Tilarán, and Nadayure. This region presents contrasting landforms, including a volcanic range with four main volcanic structures outcropping from the southeast to northwest, along with an adjacent volcanic plateau (piedmont sector). Additionally, there is an extensive lowland dissected by several rivers (Bergoeing 2017), which are included into three main watersheds: (1) Ríos de la Península de Nicoya, (2) Río Tempisque, and (3) Río Bebedero (IMN 2011). Lithology is diverse, comprising both sedimentary and igneous rocks distributed throughout these landscapes. Basalts, basalt-agglomerates, gabbros, dolerites, cherts, diorites, and siliceous limestones are prevalent, covering most of the Peninsula (Denyer et al. 2014).

Climatic conditions within the Chorotega Region are predominantly described as rainy with monsoon influence along

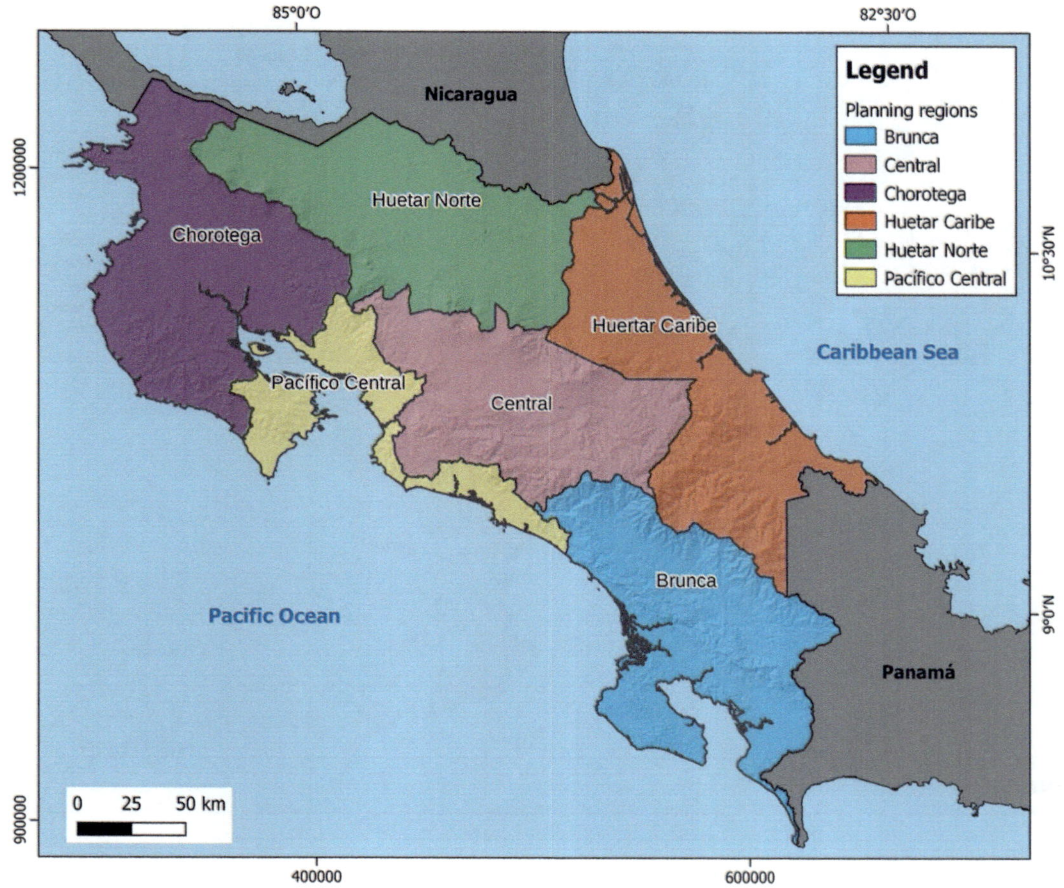

Fig. 11.1 Territory of Costa Rica and its six planning regions

the coastal areas and dry within the inner Peninsula. The average total annual precipitation values are 2321 ± 181 mm and a drought period lasts for 245 ± 18 days. Air temperature average values of 26 ± 2 °C (Solano and Villalobos 2001). Nevertheless, the driest part of the region reports an annual precipitation of only 1800 mm and just 99 days with rainfall. These conditions limit the growth and development of most plants and vegetation, promoting the growth of adapted plants.

However, due to the variability in landforms and geologic materials, most of the vegetation covering this region can be classified according Holdridge's life zone system (Holdridge 1967) as Premontane moist forest basal transition (mf-P Basal), followed by Tropical moist forest (mf-T), Tropical dry forest (df-T), Tropical moist forest dry transition (mf-T d), and Premontane Tropical wet forest (wf-T P). In less extensive areas covering the volcanoes and hills associated, Premontane rain forest (rf-P) and Tropical wet forest Premontane belt transition (wf-T P) occur.

This diversity of landforms and geologic materials explains the main soil orders (Soil Survey Staff 1999) observed in this region, including Entisols, Alfisols, Andisols, Vertisols, Ultisols, Mollisols and Inceptisols, including their associations with other soil orders (Mata et al. 2022). Entisols represent approximately 25% of the total area in the Chorotega Region (Fig. 11.3), followed by Alfisols (19%) and Andisols (14%).

Within these soil orders, various forestry and agricultural activities take place, whereas their respective extents vary due to physical (landforms and climate) and human conditions (socio-economic and cultural aspects). For example, an important activity commonly found in Alfisols of the Chorotega Region is forestry involving quality tropical hardwood such as teak (*Tectona grandis* L.f.), which has been extensively planted in the piedmont and the hills of Guanacaste province due to its high productivity and economic value (Fernández-Moya et al. 2014). Another example of agricultural activities comes from the highlands of Hojancha (southwestern part of the region), where these lands are normally cultivated with coffee (*C. arabica* L.) and orange trees (*Citrus sinensis* L.).

Soils within these landforms are classified as Alfisols, characterized by reddish color, with fine-textured subsoil horizons, and high base status deep soils (Buol et al. 2011). As result of these agriculture and forestry activities, some of land-modifying processes observed like cattle-creeping

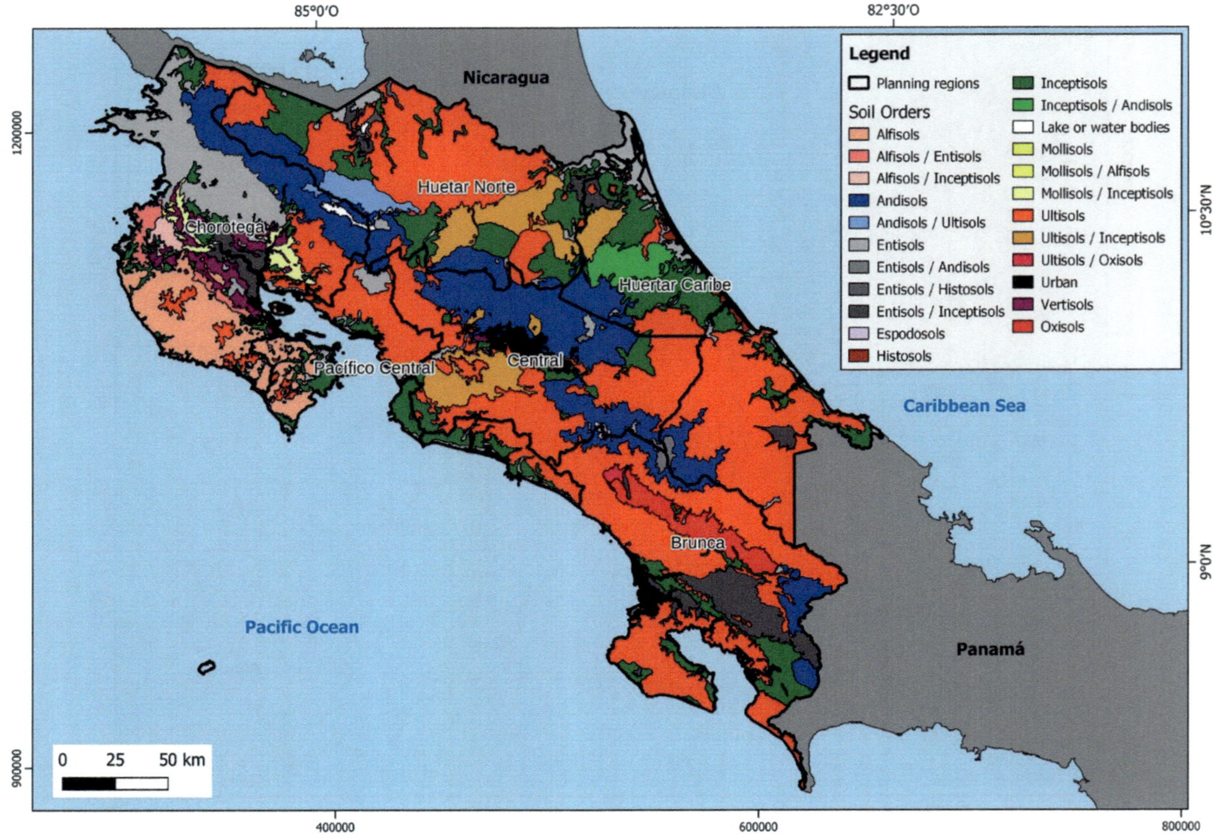

Fig. 11.2 Distribution of the primary soil orders within the six planning regions of Costa Rica. Adapted from Mata et al. (2022)

channels, or erosion processes and landslides, have modeled the landscapes (Fig. 11.4).

These soils are commonly found in regions with ancient geologic materials and landforms supplemented with sufficient nutrients within the parent materials (Buol et al. 2011), a characteristic that aligns with the basaltic materials present in this region (Denyer and Gazel 2009; Denyer et al. 2014). Coffee and orange typically cover a substantial portion of the areas within these hilly landscapes (Fig. 11.4). However, they are being surpassed by other land uses such as pastures, citrus plantations, and secondary forest (Fig. 11.4a–c). In fact, coffee plantations have experienced a decline in coverage, decreasing from 948 ha in 2012 to 791 ha in 2018 (ICAFE 2019). This trend is in favor of orange cultivation systems, despite some efforts to keep the coffee afloat as an economic and social-cultural activity within this sector (Fig. 11.5). Nonetheless, the variability of weather has made more challenging to maintain the land cover with this crop, allowing for other activities such as cattle production using drought-resistant grass species like *Brachiaria brizantha* (Fig. 11.4d).

Within the sector "La Bajura" (a term in Costa Rican Spanish used to describe a lowland), the common soil orders are Vertisols and Mollisols. Generally, Vertisols soils are dark and clayey with shrinking-swelling properties due to drying–wetting cycles (Buol et al. 2011), typically found in least elevated areas. On the other hand, Mollisols soils are deep and dark, friable, with relatively fertile surface horizons and excellent physical–chemical properties (Buol et al. 2011). They are commonly found in association with Alfisols and Inceptisols soils (Mata et al. 2022).

Despite their divergent pedogenesis and contrasting soil physical properties, both Vertisols and Mollisols are considered as fertile soils, with desirable chemical properties such as high base saturation and low acidity, making them suitable for certain agricultural activities. In fact, Vertisols and Mollisols soils assessed in the Chorotega Region are typically used for cultivating cash crops belonging to the *Poaceae* family (Fig. 11.6), including sugarcane (*Saccharum officinarum* L.) and rice (*Oryza sativa* L.).

The cultivation of both sugarcane and rice experienced an increase during the mid-nineteenth century, primarily propelled by new technologies in soil tillage, plant breeding, and government policies that incentivized farmers and landowners (Murillo 1994; Marchena-Sanabria 2015). These two crops play a pivotal role in the socio-economic activities of this region, with nearly 35,000 ha dedicated

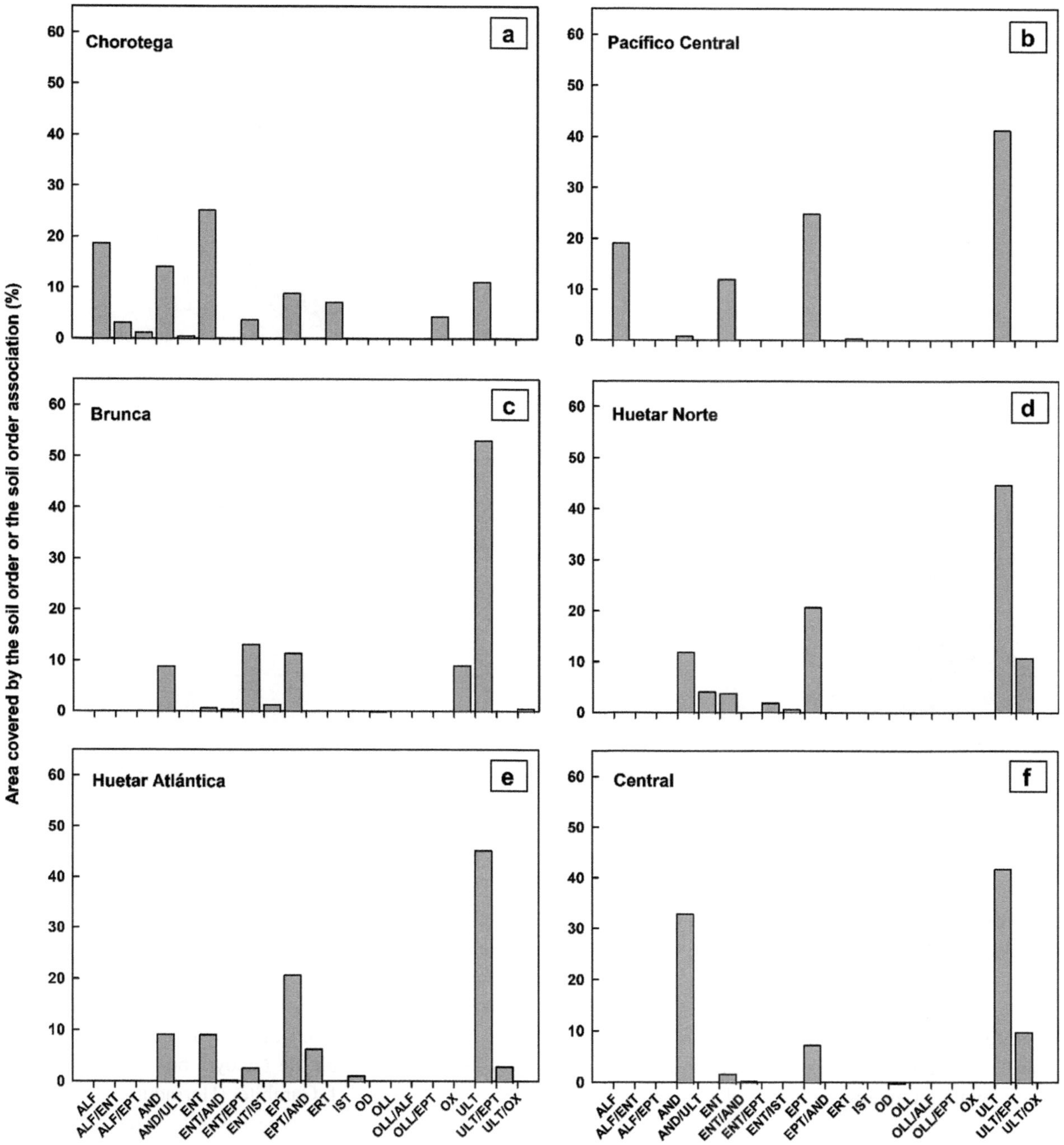

Fig. 11.3 Land coverage of the soil orders or soil associations within the six planning regions of Costa Rica: **a** Chorotega, **b** Pacífico Central, **c** Brunca, **d** Huetar Norte, **e** Huetar Caribe, and **f** Central. Adapted from data available in Mata et al. (2022). *ALF* Alfisols, *AND* Andisols, *ENT* Entisols, *EPT* Inceptisols, *ERT* Vertisols, *IST* Histosols, *OD* Spodosols, *OLL* Mollisols, *OX* Oxisols, *ULT* Ultisols (Soil Survey Staff 1999)

to sugarcane cultivation (Angulo et al. 2020), and an additional 18,206 ha allocated for rice cultivation (SEPSA 2023).

Rice fields and sugarcane plantations dominate most of the "La Bajura" area, competing with cattle farms (pastures), water bodies, and an emerging trend in real estate project development (Fig. 11.7). Moreover, in 2021, the Chorotega Region harvested 96,841 t of rice grains, constituting 63% of national total rice production (SEPSA 2023). Additionally, the sugarcane yield reported for nearby farms during the 2019–2020 season was 2,355,412 t, representing 58% of the total production in Costa Rica (Angulo et al. 2020). Most of these aforementioned landforms have been modified by continuous cycles of agriculture, creating ground creep, raised beds, levees, and dikes due to the tillage and land preparing (Fig. 11.6).

Fig. 11.4 Agricultural landscapes and crops within the hilly sector of the Chorotega Region: **a** Typical landscape with pastures, crops, and forest, **b** citrus (*Citrus* spp.), **c** coffee (*Coffea arabica* L) under shadow, and **d** pastures (*Brachiaria brizantha* (A.Rich))

11.2.2 Pacífico Central Region in the Central Western Coast of Costa Rica

The Pacífico Central Region extends from the southern side of Península de Nicoya and continues along the western central Pacific coast of Costa Rica. It includes piedmont areas with mountains and low elevations (<1500 m asl), comprising the municipalities of Puntarenas, Esparza, Montes de Oro, Quepos, Parrita, Garabito, San Mateo, and Orotina (Fig. 11.1). The geomorphology in this region presents contrasting landforms, including hills within two piedmont sections corresponding to divergent mountain ranges (Tilarán and Talamanca) that stretch in NW–SE direction, and coastal plains (lowlands) dissected by several rivers (Bergoeing 2017), included within the following watersheds: (1) Ríos de la Península de Nicoya (southern sector), (2) Río Abangares, (3) Río Barranca, (4) Río Jesús María, (5) Grande de Tárcoles, (6) Río Tusubres, (7) Río Parrita, (8) Río Damas, (9) Río Naranjo, (10) Río Savegre, and (11) Río Barú (IMN 2011).

Geologic materials observed within the region are diverse, consisting of rocks and soil parent materials of both sedimentary and igneous origins and with different ages (Bergoeing 2017). For instance, the Nicoya Complex (the main geologic unit aforementioned in Sect. 2.1) covers most of Península de Nicoya, but it can also be found in the highlands of Garabito (Herradura Terrain), associated with other pillow-type basaltic materials (Denyer and Gazel 2009), dated from the Maastrichtian to Lower Eocene (Arias 2003). In the northwestern sector (Barranca and Cerros del Aguacate), there mainly occur volcanic rocks from the Miocene-Pleistocene associated with sedimentary materials (alluvial terraces, alluvial deposits) from the Quaternary (Denyer et al. 2003). Additionally, coastline areas are constantly affected by intense erosion–deposition processes caused by both fluctuating tides, waves, or inland flows (Cárdenes and Obando 2005).

Climatic conditions within the Pacífico Central Region are generally described as rainy with monsoon influence, with total annual precipitation values of 3578 ± 237.8 mm

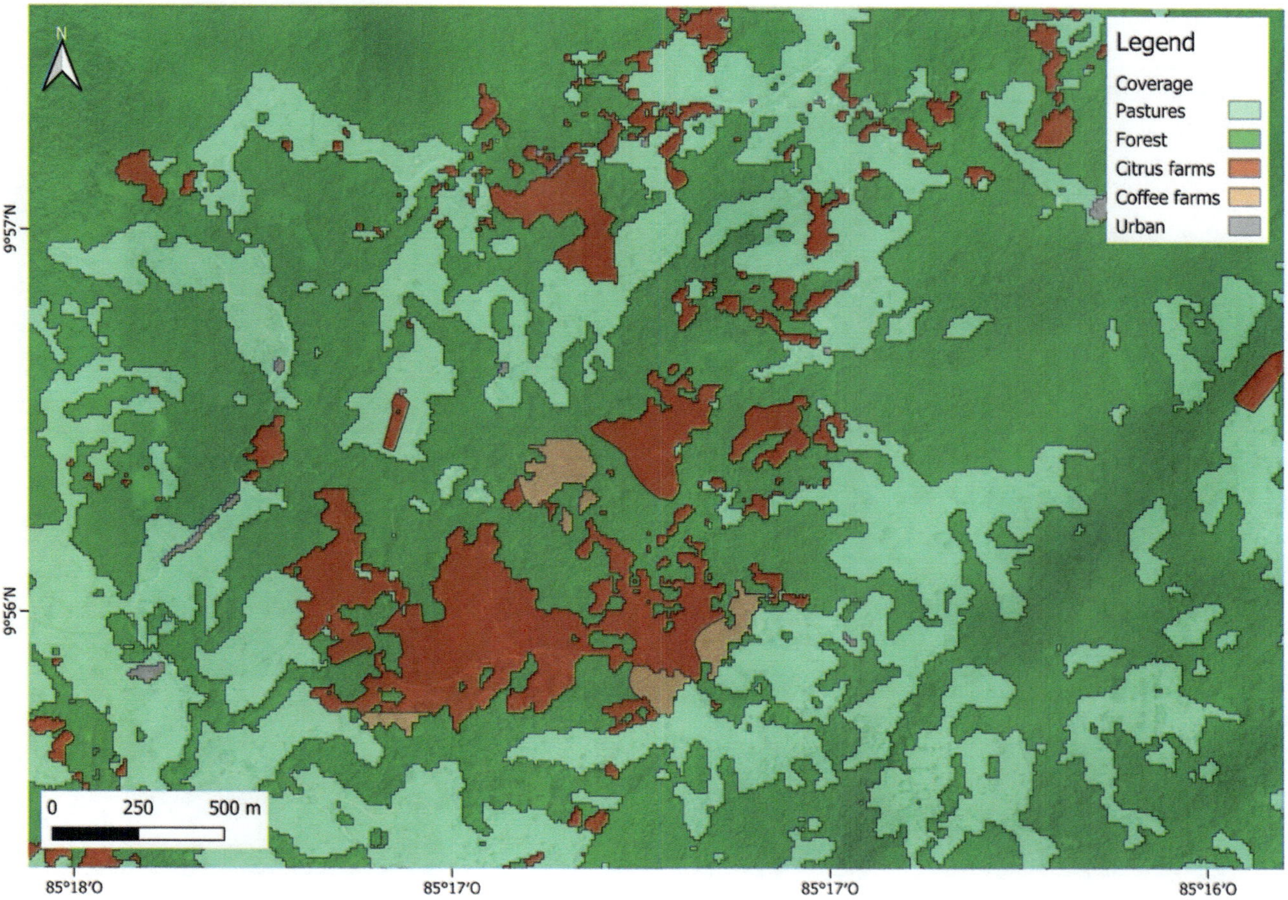

Fig. 11.5 Land coverage of a typical landscape within the hilly sector of the Chorotega Region. Land cover map generated using machine learning technics and supervised classification from PlanetScope Images (Planet Lab 2022)

and a dry period lasting 150 ± 7 days, with average air temperature values of 25 ± 2 °C (Solano and Villalobos 2001). These conditions facilitate the growth and development of vegetation classified according Holdridge's life zone system (Holdridge 1967) as Tropical moist forest (mf-T), Premontane wet forest basal transition (wf-P Basal), Tropical wet forest (wf-T), Tropical moist forest perhumid transition (mf-T perhum), Tropical moist forest dry transition (mf-T d), and Premontane wet forest (wf-T P).

The main soil orders mapped in this region (Mata et al. 2022), from the most to the least abundant, are Ultisols, Inceptisols, Alfisols, and lastly Entisols, including their associations with other soil orders. Ultisols represent around 42% of the total area of the Pacífico Central Region (Fig. 11.3), followed by Inceptisols (25%) and Alfisols (19%). Ultisols and Inceptisols dominate the Central Pacific region of Costa Rica, with Ultisols covering mostly the piedmont region, while Inceptisols extend along the coastal plains (Fig. 11.2). In this regard, Ultisols are described as clayey-reddish highly weathered soils, with low content of bases and desirable soil physical properties (West et al. 1997). These soils are commonly found in the highlands, hills, and piedmont areas, where landform stability prevents further soil-forming processes that would transform them into other soil orders (Buol et al. 2011). On the other hand, Inceptisols are described as weakly profile-developed soils, with some subsoil development and no diagnostic properties established (Skorupa et al. 2017). These last soils are typically found developing from materials accumulated through fluvial processes, including deposition within estuarine or delta rivers (Buol et al. 2011). Under proper soil management, both soil orders are considered suitable for agriculture development.

The Pacífico Central Region has a diversity of agricultural activities that have significantly influenced not only the idiosyncrasy of its habitants and the economy but also transformed the landscape and land cover. For instance, oil palm plantations were established in 1943 in Parrita municipality and extended to neighboring counties and planning regions of Costa Rica, becoming an intensive industrial activity aimed at meeting international demand for vegetable oil (Vargas et al. 2020). Currently, this crop covers a total of 75,700 ha of cultivated areas within the Costa Rican territory (SEPSA 2023).

Fig. 11.6 Agricultural landscapes and primary crops within the lowland sector of the Chorotega Region: **a** Typical landscape with *Poaceae* crops, **b** sugarcane (*Saccharum officinarum* L.), **c** rice (*Oryza sativa* L.), **d** extensive sugarcane farms and pastures (*B. brizantha*)

In the Pacífico Central Region, 10,129.71 ha and 6121.10 ha have been planted in Quepos and Parrita municipalities, respectively, while a total of 227.18 ha are reported for Puntarenas and Garabito counties (MOCUPP 2021a). These oil palm plantations have become an integral part of the landscape in this planning region (Fig. 11.8), sharing soils, and landforms with other activities like pastures and cattle farms and the fruit crop papaya (*Carica papaya* L.) (Fig. 11.8). Another significant economic activity in the Pacífico Central Region is rice production. For instance, during the 2020–2021 season, a total of 5730 ha was cultivated with rice (CONARROZ 2021), resulting in a yield of 22,067 t of grain (SEPSA 2023).

These values were mainly taken from Parrita municipality, where 2746 ha were planted with rice, yielding a total of 11,030 t. This activity has been in decline over time and competing with emerging crops like papaya and banana (Fig. 11.9), which has significant implications for landscape transformation, management practices, and subsequent physical and social-economic consequences into the region (Fig. 11.9).

In this context, the cropping system selected will need specific practices that might be at odds with another system. For instance, within the same landform or landscape (the alluvial lowlands of Parrita), it is possible to find banana, oil palm, papaya, and rice fields (Fig. 11.10a) coexisting over the same soil order Inceptisols (Fig. 11.10b). However, the requirement for levees to retain ponded water (for rice cropping system) or drainages to remove the excess water and lower the water table (for banana, oil palm, or papaya cropping systems) will impact land water dynamics and other related processes such as erosion or sediment and solute transport, which in turn affects the shores and aquifers (Fig. 11.10c, d). These erosion processes have been enhanced by agricultural practices like plowing and leveling to create rice beds or drainages for oil palm production, modifying this landscape and its corresponding landforms (Fig. 11.10).

Hence, this diversity in cropping systems and the corresponding management has directly influenced not only the transformation and evolution of the landscape but also the socio-economic dynamics within the region. For instance, Fig. 11.11 illustrates the distribution of different soil uses

Fig. 11.7 Land coverage of a typical landscape within the lowland sector of the Chorotega Region. Land cover map generated using machine learning technics and supervised classification from PlanetScope Images (Planet Lab 2022)

within the same landscape in the lowlands of Parrita. It clearly highlights the dominant presence of oil palm plantations, followed by rice farms and mangrove forests. These land coverages are situated within an alluvial landscape that is actively shaped by the meanders and discharge of the Parrita River.

11.2.3 Brunca Region in the Southeastern Side of Costa Rica

This region extends from the southern side of Península de Burica (Burica Peninsula) and continues along the northwestern Pacific coast of Costa Rica. It encompasses diverse topography, including piedmont areas with ancient alluvial fans and the highest mountains of Costa Rica (> 3500 m asl), covering the municipalities of Pérez Zeledón, Buenos Aires, Osa, Golfito, Coto Brus, and Corredores (Fig. 11.1).

The geomorphology in this region presents a range of landforms, including Quaternary alluvial fans formed from sediments deposited after glacial periods correlated with the period MIS 6 (200,000–130,000 years ago) and the period MIS 1 (110,000–12,000 years ago) and forming those landforms (Quesada-Román and Zamorano-Orozco, 2019). These alluvial fans extend along the entire Pacific piedmont of Talamanca Mountain Range (Bergoeing 2011), covering a substantial area within the region. The Talamanca Mountain Range rises from southeast to northwest, housing the Costa Rica's highest peaks such as Cerro Chirripó (3819 m asl), Cerro de los Ventisqueros (3812 m asl), Cerro Pirámide (3807 m asl) and Cerro de las Lagunas (3761 m asl) (Bergoeing 2017).

These areas over 3000 m asl were molded by glacial and periglacial processes during the Last Glacial Maximum producing erosional and depositional glacial landforms such as aretes, glacial valleys, glacial lakes, and moraines (Quesada-Román et al. 2019, 2020, 2021). In addition to these diverse landforms, the Pacific Coast within the Brunca Region features the "Fila Brunqueña Costera," a mountain range composed of sedimentary materials from the Miocene, outcropping parallel to the shoreline (Bergoeing 2017). Two primary rivers, Río General and Río Coto Brus, dissect the lowlands of this region, eventually merging to form the Costa Rica's largest river and watershed, the Río

Fig. 11.8 Agricultural landscapes and crops along the Pacífico Central Region: **a** Typical landscape with oil palm (*Elaeis guineensis* Jacq.), **b** pastures (probably *B. brizantha*) and oil palm, **c, d** papaya (*Carica papaya* L.) cultivated within sedimentary terrains

Grande de Térraba. Four watersheds are included in this planning region: (1) Ríos de la Península de Osa, (2) Río Esquinas, (3) Río Coto Colorado, and (4) Río Corredor (IMN 2011).

Geologic materials observed in this extensive region of Costa Rica are remarkably diverse and likely the most intricate found in the country. For instance, the southern area encompasses the Península de Osa and Golfo Dulce (Dulce Gulf), which developed over the Rincón Block and the Golfito Terrane (formerly classified as Nicoya Complex). These two Miocene oceanic complexes are composed of basalts, dolerites, breccias, and interbedded pelagic sediments (Denyer and Gazel 2009). Adjacent to these ancient materials, sedimentary rocks (shales, sandstones, limestones, and conglomerates) emerged during the Miocene due to orogenetic processes (Bergoeing et al. 2010). Moving northwestward, Quaternary alluvial fans deposited along the piedmont region exhibit varying geologic ages and are primarily composed of conglomerates and igneous rock blocks immersed in a red-clayish matrix, highly weathered (Alvarado et al. 2009; Obando & Kussmaul 2009; Bergoeing 2011).

Finally, within the northern side of the Brunca Region, intrusive materials (classified as Comagmatic group Talamanca) predominantly consist of gabbro, monzonites, diorites, and granites. These intrusions cut through lava and other sedimentary strata, dating back to the Late Miocene (12–7.8 Myr.) (van Uffelen 1993; Alvarado et al. 2009; Alfaro et al. 2018). Climatic conditions within the Brunca Region are highly contrasting, described as rainy with monsoon influence, excessively rainy (with a reduction in rainfall from February to April), and characterized by low temperature and a very short drought period (Solano and Villalobos 2001). In general, the total annual precipitation reported for this region is 3836 ± 321 mm during a rainy period of 194 ± 9 days, with average air temperatures of 25 ± 1 °C (Solano and Villalobos 2001). However, the subregions of Valle del Coto Colorado and Península de Osa have reported total annual precipitation values exceeding 4200 mm, with over 190 days with rainfall (> 1 mm day$-^{1}$).

Vegetation within the Brunca Region can classified according Holdridge's life zone system (Holdridge 1967) as follows: Tropical wet forest (wf-T), Premontane wet forest

Fig. 11.9 Crop diversity within the lowlands of Pacífico Central Region: **a** Bananas (*Musa* AAA) and papaya, **b** banana farms with drainage, **c** rice fields, and **d** extensive crops such as rice and bananas within the same landform

basal belt transition (wf-P Basal), Premontane wet forest (wf-P), Premontane rain forest (rf-P), Lower montane rain forest (rf-LM), and Tropical moist forest (mf-T). The primary soil orders described and mapped in this region (Mata et al. 2022), from the most to the least abundant, are Ultisols, Inceptisols, Oxisols, Andisols, and Entisols, including some associations among them. Specifically for this landscape, Ultisols cover approximately 53% of the total area of the Brunca Region (Fig. 11.3c), followed by Inceptisols (12%) and Oxisols (10%). Despite Ultisols being reported as dominant in this region (Fig. 11.2), a revision of soil taxonomic orders assigned to specific geomorphic units within the piedmont (early and intermediate alluvial fans) has been proposed after performing a thorough determination of sand and clay mineralogy (Camacho et al. 2020).

This revision might lead to an increased coverage of Oxisols within this region, especially in landforms described as Quaternary alluvial fans (Camacho et al. 2020; 2021). Oxisols are characterized as highly weathered soils, with low-activity clays and desirable physical properties (Buol and Eswaran 2000). These soils are commonly found in stable landforms in the Tropics, where prolonged pedogenetic processes and limited weatherable minerals prevent further soil evolution. Despite their inherent low natural fertility, extensive crops like soybean, wheat, corn, and coffee can thrive with technified management practices (liming and fertilization) implemented by farmers and landowners (Buol et al. 2011). In fact, the Brunca Region's piedmont hosts extensive agricultural activities such as sugarcane and pineapple cultivation (Fig. 11.12). These activities not only serve as primary socio-economic drivers but also play a crucial role in shaping population dynamics, despite a significant number of people who emigrate due to limited economic opportunities and unemployment (Gatica-López 2017).

Historically, sugarcane cultivation has been a vital socio-economic activity within the region (Figs. 11.12 and 11.13), transitioning from low-input systems to extensive and technified cultivation that transformed the landscape and local society (Durán 2006). In this context, Miranda (1985), while studying the soil use change in the General Viejo community over six years, reported a substantial increase in sugarcane areal from 722 ha in 1973 to 1083 ha in 1979.

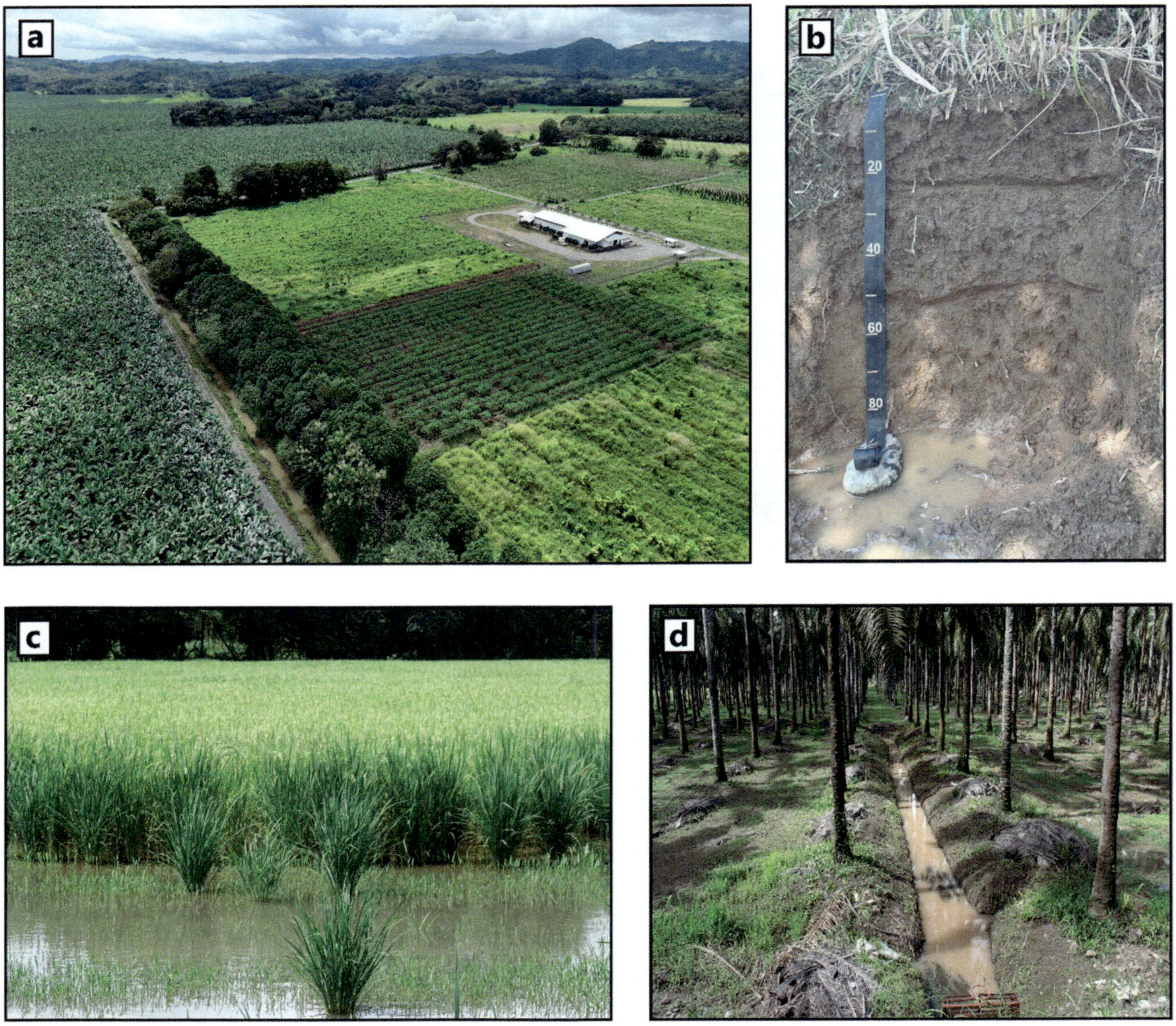

Fig. 11.10 Effect of cropping system and management practices on the agricultural landscape: **a** Distribution of banana, papaya, and rice within the same landform, **b** typical soil profile within this landscape in Parrita (*Aquic Haplustepts*), **c** waterlogged rice fields, and **d** drainage design for oil palm production

The coverage of this crop has continued to expand, reaching 4205 ha in 2019 for the entire region. Within this, 2910 ha is in Pérez Zeledón municipality, and 1295 ha in Buenos Aires municipality, contributing to a total yield of 295,167 t coming from 2312 producers, representing 43% of Costa Rica's sugarcane producers (Barrantes and Chaves 2020). Coffee production in the Brunca Region is primarily concentrated in Pérez Zeledón and Coto Brus municipalities (Fig. 11.13). Historically, coffee cultivation began in Pérez Zeledón when immigrant trailblazers from counties of Dota, Tarrazú, and Desamparados (Central Valley) arrived in the mid-nineteenth century, seeking improved socio-economic conditions (Miranda 1985). Similarly, in the 1940s pioneer families from the Central Valley moved to what is now Coto Brus municipality, establishing coffee farms due to optimal altitude and volcanic ash soils (Quesada-Román 2013).

Over time, coffee cultivation in both counties has expanded, with 8093 ha identified in Coto Brus municipality in 2012, increasing to 8727 ha in 2018. In Pérez Zeledón municipality, the coffee-planted area reached 13,116 ha in 2012 and expanded to 13,315 ha in 2018 (ICAFE 2019). In addition, a *fanega* represents 400 L of ripe coffee cherries, and the minimum volume to obtain 46 kg of coffee beans is considered as the official unit in Costa Rica. These extensions translated to total production of 159,767 *fanega* in Coto Brus and 212,759 *fanegas* in Pérez Zeledón in 2018 (ICAFE 2022). Interestingly, coffee cultivated areas in the

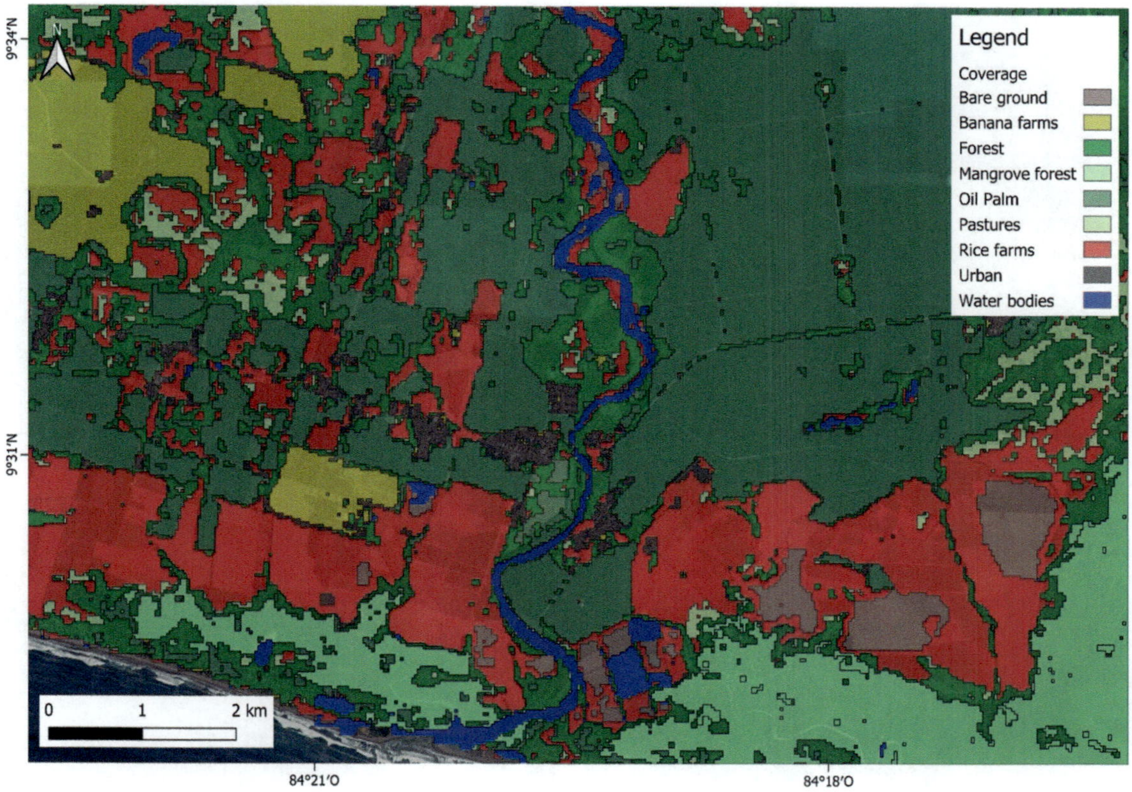

Fig. 11.11 Land coverage and soil use within the Pacífico Central Region. Land cover map generated using machine learning technics and supervised classification from PlanetScope Images (Planet Lab 2022)

last decades have grown in altitudes over even 1800 m asl and have been established on slopes above 40° (Quesada-Román et al. 2022). More recently, ICAFE (2022) reported a total production of 152,656 *fanegas* for Coto Brus municipality and 217,849 *fanegas* for Pérez Zeledón municipality. Coffee harvesting in the region is typically done by hand, requiring a significant labor force, often supplemented by immigrants from Panama (Quesada-Román and Díaz-Bolaños 2019).

Starting in 1983, pineapple cultivation surged in the Brunca Region to meet global demand for the tropical fruit (Castro 1994). This expansion of pineapple farming extended throughout Costa Rica, increasing from 1000 ha in 1984 to 37,660 ha in 2014 (Morales-Abarca 2018). By 2018, Buenos Aires municipality cultivated about 8700 ha with pineapple (Figs. 11.13 and 11.14), yielding a total 630,000 t of fruit. Meanwhile, Pérez Zeledón municipality contributed 2000 ha to produce 153,000 t (INFOAGRO 2023). However, pineapple planting areas decreased significantly in 2019, with only 6871 ha in Buenos Aires and 1200 ha in Pérez Zeledón municipalities (MOCUPP 2021b). Despite these fluctuations, pineapple has become a significant contributor to Costa Rica's economy, representing over 1 billion USD in fresh-fruit exports in 2021. The Brunca Region contributed 8800 ha to this industry (21% of the total pineapple-planted area in Costa Rica), supporting around 6160 direct employees (CANAPEP 2023).

These three crops—sugarcane, coffee, and pineapple—dominate the landscape of Pérez Zeledón and its surroundings (Fig. 11.15), particularly in the piedmont sector characterized by Quaternary alluvial fans. Pineapple is the most common element in the landscape, followed by sugarcane and coffee. As mentioned earlier, highly weathered soils (Ultisols and Oxisols) are commonly associated with these crops, subject to various soil management practices to maximize the yields, including liming, subsoiling, and organic addition (Molina 1998; Bertsch 1998; Camacho et al. 2015).

In the southeastern sector of the Brunca Region, abrupt variations in agricultural and socio-economic activities have resulted in landscape changes. For instance, in 1920, the current Osa and Golfito municipalities were areas where smallholders cultivated bananas, rice, and other grains (Royo 2004). Throughout its history, this region has been closely linked to extensive banana farming through "Compañía Bananera de Costa Rica," a subsidiary of the United Fruit Company, (UFC). Operations ceased in 1985 due to factors such as reduced farm productivity, market crises, and high operational costs that disadvantaged the region against other Central American areas

Fig. 11.12 Primary crops along the Talamanca piedmont sector within the Brunca Region: **a**, **b** Typical sugarcane fields, **c** pineapple (*Ananas comosus* (L.) Merr.), **d** sugarcane and pineapple fields within Quaternary alluvial fans

(Abarca-Jiménez 2015). Following this shutdown, banana farms gradually transitioned to oil palm under the United Fruit Company´s management, and subsequently, smallholders embraced oil palm production, forming associations and cooperative corporations (Fig. 11.16). However, these entities faced competition for land, resources, and operational power until the consolidation of major oil palm enterprises like Agricultural Services and Development (ASD) or “Palma Tica” (formerly UFC) established dominance within oil palm business (Clare 2012).

This activity continues to grow and expand the agricultural frontier in the Southern Brunca Region (Salas-González 2020). For instance, MOCUPP (2021a) reported that 48,160.77 ha were dedicated to oil palm in 2019, representing 5.25% of the Brunca Region’s total area. This area accounts for 68% of the total national oil palm cultivation area (MOCUPP 2021a). For instance, Corredores municipality possesses the largest oil palm-planted area, with 24,521.72 ha, followed by Golfito municipality with 12,480.2 ha, and Osa municipality with 10,969.97 ha.

Rice, another historically significant crop (Fig. 11.17), is also prominently cultivated within the Southern Brunca Region (Royo 2004). However, rice cultivation areas have fluctuated over the past decade and notably decreased over the last four years (Fig. 11.18).

For instance, the rice cultivation area was 6707 ha in the 2017–2018 growing season, dropping to 5650 ha in the 2020–2021 season (SEPSA 2023). This decline in cultivation areas led to reduced production, from 28,202 t of rice in 2017–2018 to 21,009 t of rice in 2020–2021. Similar trends of declining rice production and yields were observed in Costa Rica from 2003 to 2014, attributed to factors such as variation in climatic conditions, land conversion to more profitable activities like cattle production or oil palm (Sánchez and Vega 2018).

Land reconversion to oil palm (Fig. 11.19) was particularly notable in Corredores municipality, the primary political-administrative unit for oil palm cultivation. Here, rice cultivation area and total yield decreased by 16% and 32%, respectively, relative to 2019–2020 season (SEPSA 2023).

Fig. 11.13 Crop diversity along the Talamanca piedmont sector within the Brunca Region: **a** Intercropping system with sugarcane and coffee blocks, **b** typical pineapple fields (*Photo courtesy of S. Marín*), **c**, **d** extensive sugarcane cultivation within typical highly weathered soils

The predominant soils in this area are the Inceptisols, which are suitable for either or both crops under proper management practices (Fig. 11.19).

11.2.4 Huetar Norte Region in the Northern Side of Costa Rica

The Huetar Norte Region extends to cover the northern side of Costa Rica, bordering the San Juan River (the international boundary with the neighboring country of Nicaragua) and the Central Region of Costa Rica to its southern side. Its eastern sector includes the Guanacaste Volcanic Range, which encompasses five volcanoes (1659–2028 m asl), hills, and adventitious cones within the piedmont sector, as well as lowlands and plains that cover most of the entire planning region (Fig. 11.1).

The Huetar Norte Region includes the municipalities and districts of San Carlos, Upala, Los Chiles, Guatuso, Sarapiquí (Heredia), Peñas Blancas (San Ramón), Río Cuarto, Sarapiquí (Alajuela). This region presents contrasting landforms, including Quaternary volcanic cones and craters in the highlands, large alluvial fans formed from sediments deposited during the Pleistocene as part of the piedmont sector, extensive floodplains located in the northern part and crossed by several rivers (including Río Zapote, Río Frío, Río Pocosol, Río Peñas Blancas, and Río San Carlos), and lacustrine estuaries in the lowest areas (Bergoeing and Protti 2006; IMN 2011).

Floodplains belong to the Northern Plains of the Nicaraguan Graben which formed due to continental uplift and sediment deposition within this part of the tectonic depression of Nicaragua during the Quaternary, where sediments mainly originated from the Guanacaste and Central volcanic ranges (Bergoeing 2017).

On the other hand, geologic materials commonly found in this vast region are quite diverse. For instance, within the eastern sector of this region, diverse volcanic rocks are present: (1) dacitic porphyritic mounts, (2) quarzitic blocks, (3) blocks and ashes fluxes with pyroclastic-felsic materials, and (4) vitric welded-tuffs (Vargas and Alfaro 1992; Gazel et al. 2006). Much of these volcanic materials are dated to the Miocene. Porras et al. (2021) studied the San Carlos watershed, which is a significant area within the Huetar

Fig. 11.14 Landscapes and landforms of the Talamanca piedmont sector within the Brunca Region: **a** Talamanca highlands in the background and pineapple fields within the piedmont (*Photo courtesy of S. Marín*), **b** typical soil profile within these old alluvial fans in Pérez Zeledón (*Anionic Acrustox*), **c** sugarcane and pineapple fields within Quaternary alluvial fans, and **d** typical view of sugarcane fields

Norte Region, and described the general stratigraphy, which includes: (1) peridotites and pelagic rocks as the basement (Cretaceous, Albian-Cenomanian), (2) intercalated sandstones, lutites, and limonites (Paleocene), (3) tholeiitic basalts and dacitic material, and rhyolites (Oligocene–Miocene), (4) sandstones and limestones with volcanic materials (Miocene-Pliocene), (5) alkaline volcanic materials (Pleistocene), and (6) andesites and basaltic rocks covered by fluvial and alluvial deposits (Pleistocene-Holocene).

Climatic conditions within the Huetar Norte Region are described as highly rainy, with a decrease in rainfall during February, March, and October, and temperature values not dropping below 22 °C (Solano and Villalobos 2001). In general, the total annual precipitation reported for this region is 3264 ± 204 mm during a rainy period of 206 ± 8 days, with average air temperature values of 24.6 ± 1 °C (Solano and Villalobos 2001). These climatic conditions account for the vegetation within the Huetar Norte Region, which can be classified according Holdridge's life zone system (Holdridge 1967) as Tropical wet forest (wf-T), Premontane wet forest basal transition (wf-P Basal), Tropical moist forest (mf-T), Tropical wet forest premontane transition (wf-T P), Premontane rain forest (rf-P), and Tropical moist forest perhumid transition (mf-T perhum).

The main soil orders described and mapped in this region (Mata et al. 2022), from the most to the least abundant, are Ultisols, Inceptisols, Andisols, and lastly Entisols,

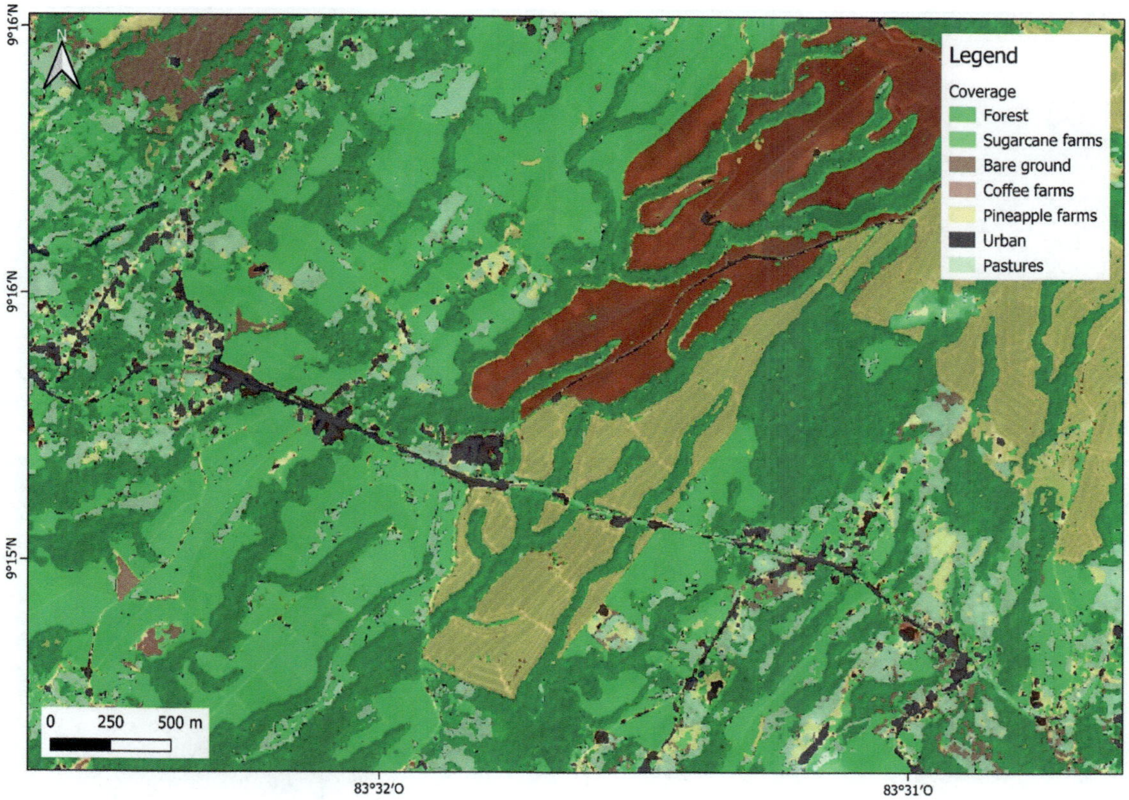

Fig. 11.15 Land coverage and crops observed within the Brunca Region. Land cover map generated using machine learning technics and supervised classification from PlanetScope Images (Planet Lab 2022)

including some associations among them. Specifically, Ultisols represent around 45% of the total area of the Huetar Norte Region (Fig. 11.3), followed by Inceptisols (21%) and Andisols (12%) (Fig. 11.2). Similar to what was reported for the Brunca Region, some Ultisols (dominant within the Huetar Norte) could be reclassified as Oxisols after determining sand and clay mineralogy accurately (Camacho et al. 2020). Indeed, these Oxisols have been reported in Crucitas de Cutris and San Rafael, both within San Carlos municipality (Camacho et al. 2021; NCSS 2023). Oxisols and Ultisols are highly weathered soils that normally have fertility limitations (low base status, low contents of phosphorus, high acidity) but they possess desirable physical properties like high drainage capacity, soil structure, and high porosity, which under proper management make them suitable for agriculture (West et al. 1997; Buol and Eswaran 2000).

Among the agricultural crops that are adapted to these weathered soils, pineapple, orange, sugarcane, and cassava (*Manihot esculenta* Crantz) stand out (Fig. 11.20). In addition to these crops, red and black beans (*Phaseolus vulgaris* L.), or other activities like grasslands and tree species like *Vochysia guatemalensis* Donn. Smith and *Gmelina arborea* Roxb (Moya 2004; Camacho et al. 2016) have been traditionally cultivated in this region, not only on Ultisols and Oxisols, but also Inceptisols. Black and red beans have been cultivated for decades in this sector of Costa Rica. In this regard, Faure and Samper (2004) described the colonization of northern Costa Rica during nineteenth century by trailblazing families and small producers, who initiated small farms with subsistence crops like beans, rice, corn, and roots vegetables. Despite the considerable decrease in planting areas for rice and beans due neoliberal policies established during the 1980s, they still occupy a significant area (Fig. 11.21).

Indeed, according to SEPSA (2023), the areas cultivated with beans reported for 2022–2023 corresponded to 4979 ha, with a total production of 4815 t within the Huetar Norte Region. These values are taken from the municipalities or districts of Upala, Guatuso, Los Chiles, Pocosol, and Santa Rosa, and constitute 33% of the total national production. Similar numbers have been reported for rice-crop areas, with 3666 ha and a total production of 14,520 t, accounting for 9.5% of total production in Costa Rica. Despite these last values of planted land being considerable, they were nearly three times higher during 1999–2000 (Faure and Samper 2004). This decrease is due to the conversion of land to more profitable crops like pineapple or citrus (Fig. 11.22), whose acreage has increased dramatically since the 1970s as industrialized activities (Arias et al.

Fig. 11.16 Extensive oil palm fields observed in the southeastern sector of the Brunca Region: **a–c** Aerial pictures taken in Guaycará, Puntarenas, and **d** close-up of a plantation

2018; Obando 2020). For instance, in 2021, SEPSA (2023) reported that 9,626 ha was planted with oranges within the Huetar Norte Region (Fig. 11.23), with a total production of 144,390 t. These values are distributed within three municipalities: Los Chiles, Upala, and San Carlos.

In addition, pineapple has become the most planted crop within this region and represented 68% of total pineapple-covered land in Costa Rica in 2018 (MOCUPP 2021b), with a total of 44,138.52 ha. These extension values are mostly distributed within six counties: San Carlos (18,509.51 ha), Los Chiles (8742.74 ha), Sarapiquí (6199.70 ha), Río Cuarto (5665.27 ha), Upala (3383.16 ha), and Guatuso (1510.91 ha). Pineapple production as an industrial activity requires a considerable number of workers involved in the agronomic management, fruit harvest, post-harvest practices, and packing, which provide a crucial source of income for families living in these areas. Some communities, however, are striving to find alternative job sources to avoid dependence on this industry (Obando 2020).

11.2.5 Huetar Caribe Region in the Caribbean Coast of Costa Rica

Moving to the Caribbean coast of Costa Rica, the Huetar Caribe Region covers the entire eastern coast, bordering the San Juan River delta on the northern side and the Sixaola River in the southeastern sector (the international boundary with the neighboring country of Panama). The eastern sector of the Huetar Caribe Region borders the Caribbean Sea, while the western sector borders with the Talamanca and the Central Volcanic Range. In addition, the eastern sector encompasses the piedmont region of Irazú and Turrialba volcanoes (1200–2200 m asl), the eastern piedmont of Talamanca hills (600–1400 m asl), and lowlands and alluvial plains that cover most of the whole planning region (Fig. 11.1).

The Huetar Caribe Region includes the counties of Limón, Pococí, Siquirres, Talamanca, Matina, and Guácimo. This region presents two main sectors: (1) coastal plain mostly

Fig. 11.17 Extensive rice fields observed in the southeastern sector of the Brunca Region: **a**, **b** Aerial pictures taken in Guaycará, Puntarenas, **c**, **d** aboveground biomass and panicles almost ready for harvest

composed of lowlands, and (2) piedmont with elevations and isolated hills, including volcanic cones of Quaternary ages (Bergoeing 2017; Quesada-Román and Pérez-Briceño 2019). Both sectors are dissected by several rivers, including Río Sixaola, Río La Estrella, Río Banano, Río Bananito, Río Moín, Río Matína, Río Madre de Dios, Río Pacuare, Río Reventazón, and Río Tortuguero (IMN 2011).

Geologic materials supporting these landforms are quite diverse, including volcanic and sedimentary materials from the Miocene to the Quaternary. For instance, within the northern sector of this region, volcanic materials crop out such as those building Miocene volcanic mounts (Lomas del Cerro Cocorí and Lomas de Sierpe). Moreover, alkaline basalts occur within dissected volcanic cones from the early Pleistocene (Cerro Coronel and Cerro Tortuguero). Other formations include sediments and materials deposited after the Holocene transgression and volcanic-sedimentary materials deposited as alluvial fans and lahars (Battistini and Bergoeing 1984; Bergoeing 2017).

Climatic conditions within the Huetar Caribe Region are mainly described as highly rainy (especially in the mountains) with no clear dry period throughout the year (Solano and Villalobos 2001). In general, the total average annual precipitation reported for the whole region is 3525 ± 494 mm with an average rainy period of 206 ± 14 days, along with average air temperature values of 23.0 ± 1 °C (Solano and Villalobos 2001). Similar to other planning regions, these climatic conditions can explain the vegetation within this region, which can be classified according Holdridge's life zone system (Holdridge 1967) as Tropical wet forest (wf-T), Premontane wet forest basal transition (wf-P Basal), Lower montane rain forest (rf-LM), Premontane rain forest (rf-P), Tropical wet forest premontane transition (wf-T P), and Tropical moist forest (mf-T).

The soils in this region (from most to the least abundant) are Ultisols, Inceptisols, Entisols, and Andisols, including some associations among them and other soil orders (Mata et al. 2022). Specifically, Ultisols represent around 45% of the total area of the Huetar Norte Region, followed by Inceptisols (21%) and Andisols (10%) (Fig. 11.3). Ultisols are mostly located in the southeastern sector (Talamanca Piedmont) and within the northern sector (hills and volcanic cones from the Quaternary) (Fig. 11.2). For instance, Ultisols are commonly found in Cerro Tortuguero, Lomas

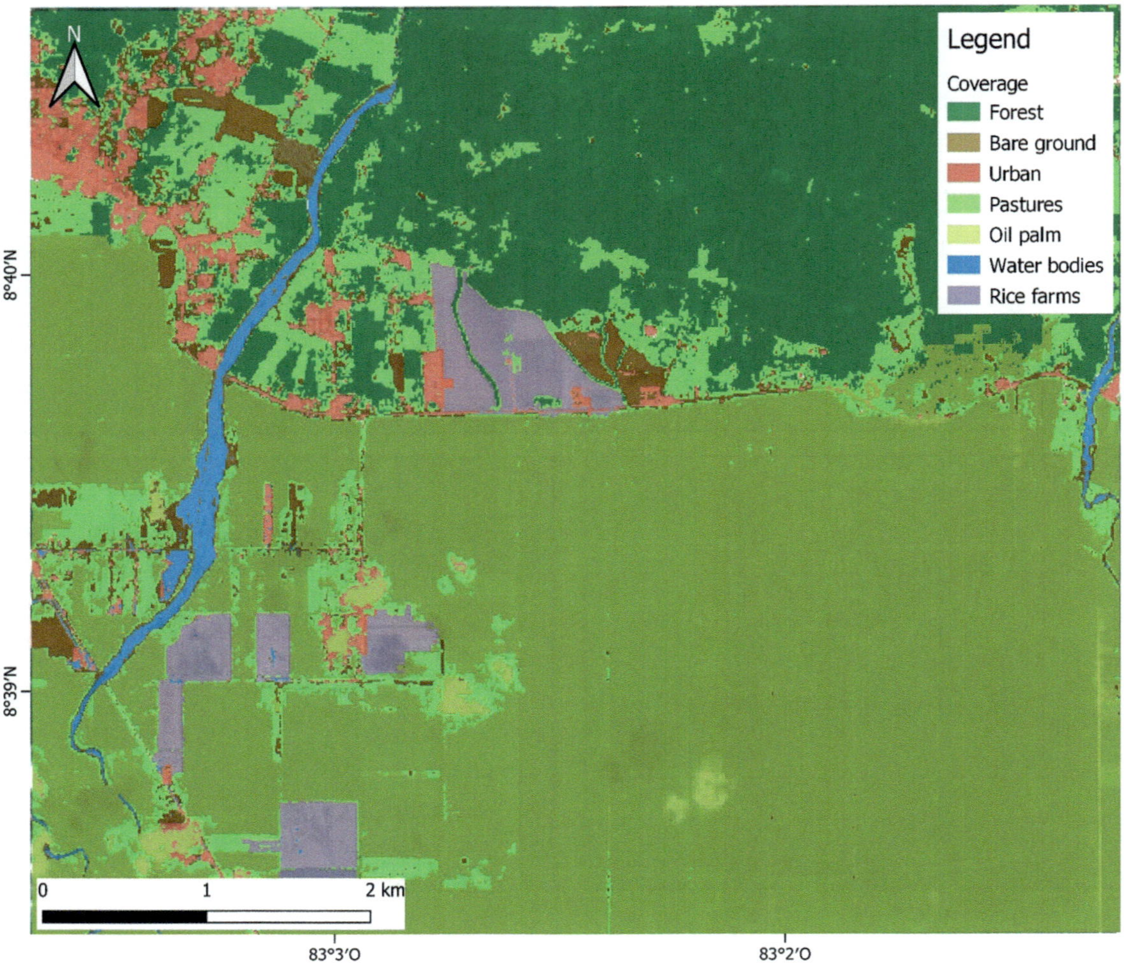

Fig. 11.18 Land coverage and soil use in the southeastern sector of the Brunca Region. Land cover map generated using machine learning technics and supervised classification from Sentinel-2 Images (ESA 2023)

de Sierpe, and Cerro Coronel, with soils in the latter being reclassified as Oxisols (Camacho et al. 2021). Inceptisols and Entisols are more abundant within the northern sector, commonly found within the areas of fluvial deposition as described by IMN (2011). Finally, volcanic ash-derived soils are mostly located within the northern side of Irazú and Turrialba volcanoes and adjacent piedmont region. As well, in some sectors, Andisols developed in association with Inceptisols (Mata et al. 2022).

In this regard, Arias et al. (2010) studied the genesis and mineralogy of soils planted with bananas within the eleven watersheds of the Caribbean side of Costa Rica. They mainly identified soil orders of Inceptisols, Andisols, and their associations, which were separated into two main zones according with their clay minerals and soil genetic pathways: (1) northeastern sector with volcanic ash-derived and medium fertility soils, and (2) southeastern sector with limestone-derived and fertile alluvial soils, physically separated by the rivers Vueltas-Silencio-Parismina. Due to this geomorphologic and pedologic diversity (and subsequent difference in fertility), several cash crops have been developing within the Huetar Caribe, with the most abundant being banana and pineapple extending across the region, as well as papaya and palm-heart (*Bactris gasipaes* Kunth) (Morales-Abarca 2018; Mora-Urpí 2002; Arias et al. 2010; Fallas-Corrales and van der Zee 2020).

Historically, the banana crop has been the main socioeconomic activity within the Huetar Caribe Region. In this regard, Minor C. Keith moved to Costa Rica in 1871 to join his brother's railroad construction business, aiming to connect the Costa Rican Central Valley with Puerto Limón (Limón Port). This project was as ambitious as challenging, where after 42 km of railroad already built, around 4000 lives (including Minor's three older brothers) were taken due to accidents or diseases (Jones and Morrison 1952).

Then, Mr. Keith took charge of this ambitious project, under several troublesome factors like tough climatic conditions and funding cuts that delayed and interrupted construction works several times. During these setbacks, the businessman found a new activity to keep getting profit

Fig. 11.19 Land reconversion in the southeastern sector of the Brunca Region: **a** Aerial picture of rice fields surrounded by oil palm in Guaycará, Puntarenas, **b** typical soil profile within an rice-palm agricultural landscape (*Andic Humudepts*), **c** rice field with plants near to be harvested, and **d** oil palm plants with oil seeds ready (plant is 3–4 years old)

within the agriculture, introducing banana rhizomes (variety "Gros Michel") from the region of Aspiwall, Panamá (currently named as Colón) (Jones and Morrison 1952). Initially, the banana plantations were set up in 1972 within the two valleys: (1) "Valle de Matina" and (2) "Valle de Zent all within the current Matina municipality, but this business further extended to other sectors of the Huetar Caribe Region (Montero and Viales 2014). The banana business and planted area grew dramatically until 1941 when yields started to decrease due soil exhaustion and the presence of two new plant diseases in the region: Panamá disease (*Fusarium cubense*) and Sigatoka (*Cercospora musae*) (Jones and Morrison 1952). Despite these last challenges, bananas have become the second most important cash crop in Costa Rica and currently appear among the first three most exported goods of Costa Rica (Fig. 11.24). During 2021, fresh banana export represented 1.03 billion USD (6.60% of total exported goods) and increased to $1.39 billion USD (7.78% of total exported goods) in 2022 (INEC 2022b, 2023).

According to SEPSA (2023), the areas planted with banana during 2022 corresponded to 41,059 ha and total production within the Huetar Caribe Region was 2,228,499 t. These values are taken from the municipalities Matina, Pococí, Siquirres, Sarapiquí, Limón, Guácimo, and Talamanca and represented 98% of the total national banana production. On the other hand, and despite the criticism about the crop's impact on soils, water pollution,

Fig. 11.20 Crop diversity within the landscapes of Huetar Norte Region: **a** Sugarcane and orange (*Citrus sinensis* L.) fields separated by the road, **b** typical orange field almost ready to harvest, **c** divergent cropping systems within the same landform such as cassava (*Manihot esculenta* Crantz) and beans (*Phaseolus vulgaris* L.), and **d** cassava plants

and sustainability, pineapple production, and land coverage grew dramatically between 2000 and 2017 (Blanco-Obando 2020) and currently represents the second most planted crop within this region (Fig. 11.24). In this regard, the Huetar Caribe Region represents 19% of total pineapple-covered land assessed in Costa Rica during 2018, with a total of 12,303.97 ha (MOCUPP 2021b). Pineapple plantations are mostly distributed within three main counties: Siquirres (2675.12 ha), Guácimo (3006.65 ha), and Pococí (6680.38 ha) (Fig. 11.25). Although the agriculture-covered land is dominated by banana and pineapple in this planning region (Fig. 11.25), there are other extensive crops with economic relevance like papaya and palm-heart (Fig. 11.26). For instance, the palm-heart which originally was extracted from the forests as a wild plant has been planted in monocultures within the Caribbean part of Costa Rica since 1978 (Mora-Urpí 2002).

In subsequent years, there was a huge increment in heart-palm production. In 2002, around 12,000 ha were planted with this crop within the Huetar Caribe Region, becoming the second more extensive crop in this planning region (Ares et al. 2002). However, the planted area started to decrease due to the increment in demand of land for other uses or more profitable crops such as pineapple and banana (Figs. 11.24 and 11.25). In 2014, a total of 4618.0 ha was planted with heart-palm, where the counties Sarapiquí, Pococí, and Siquirres accounted for 2138.3, 974.4, and 492.9 ha, respectively (INEC 2015). Currently, only 3606.3 ha have been planted with heart-palm in the whole territory of Costa Rica (INEC 2022a).

Another important crop within the Huetar Caribe Region is papaya. In 2014, a total of 1393 ha was planted with this permanent crop in Costa Rica, where 515 ha corresponded to the counties of Pococí and Guácimo (INEC 2015). Most of papaya plantations have been established by smallholders and low-input farmers (Fig. 11.27), who struggle with plant diseases and poor nutrient management or fertilizers (Fallas et al. 2020). This could reinforce the increment trend of pineapple-covered land in detriment of other more reliable crops. Figure 11.27 shows the distribution of the area covered with papaya and heart-palm within the western sector of the region and highlights the small coverage

Fig. 11.21 Extensive black and red bean fields in Huetar Norte Region: **a**, **b** Aerial pictures taken, respectively, in Upala and San Antonio de Caño Negro, Alajuela; **c**, **d** aboveground biomass and phenological stages of beans

of papaya while comparing with other land uses such as banana, pastures, or heat-palm. All these activities have been modifying the landscapes within the Huetar Caribe Region along the time. For instance, drainage systems and raised beds designed for banana fields have impacted the surface water and sediment movement.

11.2.6 Región Central in the Central Sector of Costa Rica

The central sector of Costa Rica is known as Región Central (Central Region) and encompasses the central valley along with its surrounding mountains and volcanoes. This region is bordered by the Talamanca Mountain Range and its foothills in the southwestern to southeastern sector, and to the north by the Central Volcanic Range. Additionally, the northwestern sector borders the Tilarán Mountain Range. Notable features include the inner piedmont region of Poás, Barva, Irazú, and Turrialba volcanoes (with elevations ranging from 2500 to 1200 m asl), and the northern piedmont of Talamanca hills (with elevations ranging from 1800 to 900 m asl) that encircle this planning region (Fig. 11.1).

The Región Central comprises over 40 municipalities, with Turrialba covering the eastern part. The northern border is formed by the municipalities of Heredia, Alajuela, Poás, Sarchí, and Zarcero. The western border is defined by the municipalities of San Ramón, Atenas, and Turrubares, and the southern sectors are encompassed by Puriscal, Acosta, Aserrí, Tarrazú, and Dota. This region showcases two primary landform types: (1) the central tectonic depression that includes the cities of San José, Heredia, and Alajuela, also known as "Meseta Central" (Central Plateau), and (2) piedmont with foothills and isolated hills, encompassing the four volcanic cones mentioned above (Bergoeing 2017). Furthermore, several watersheds cover this region, including those of Río Grande de Tárcoles, Río Parrita, Río Naranjo, Río Savegre, Río Chirripó, Río Sarapiquí, Río Moín, and Río Matina (IMN 2011).

Geologic materials found within Región Central are both diverse and complex, consisting of igneous (plutonic or volcanic) and sedimentary materials from the Paleocene-Pliocene and Quaternary periods. For instance, Denyer and

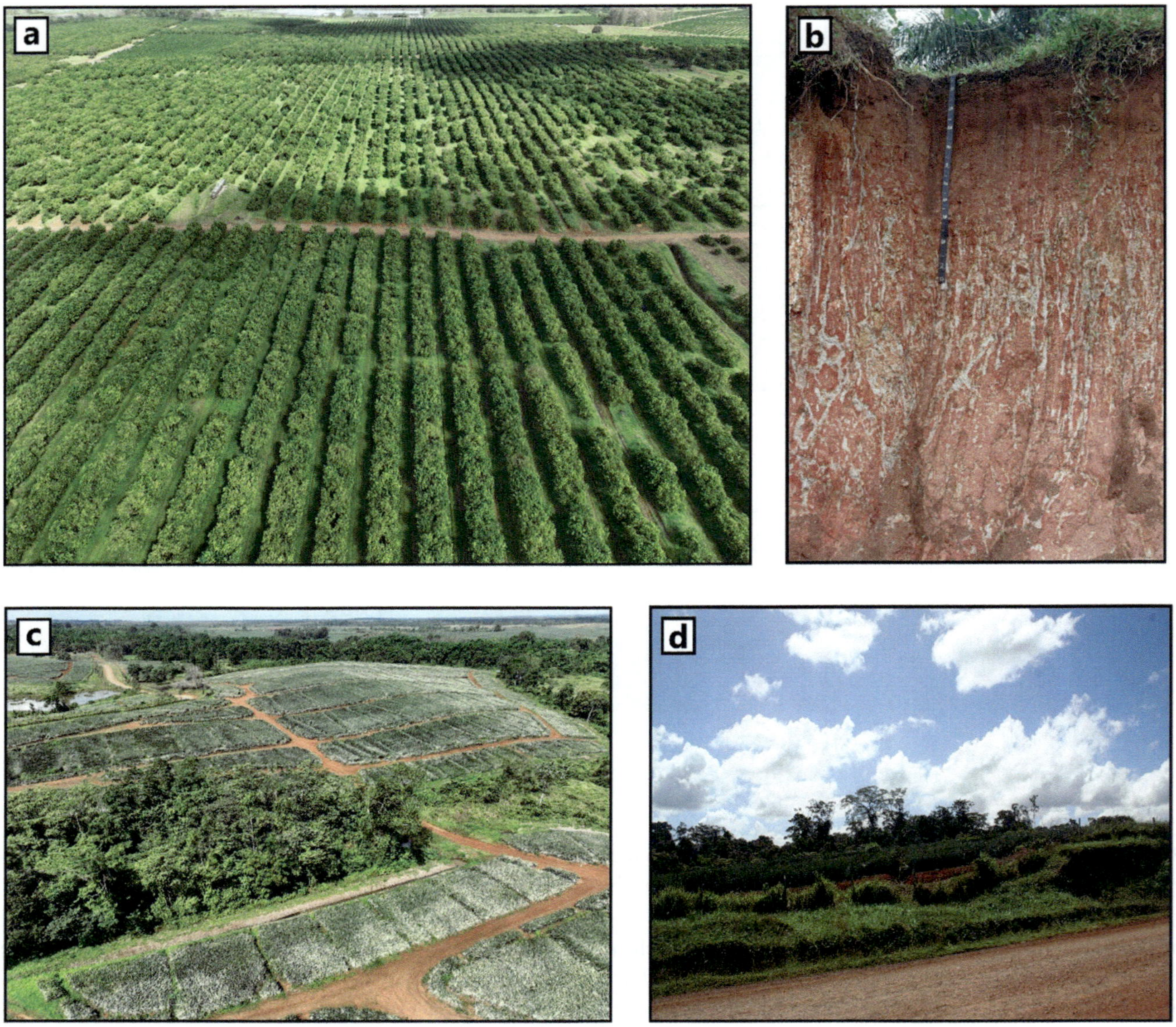

Fig. 11.22 Land reconversion and crops in Huetar Norte Region: **a** Aerial picture of extensive orange fields near of Finca Palmira, Alajuela, **b** highly weathered soil profile (*Plinthic Kandidults*) typically present in this planning region, **c** aerial picture of extensive pineapples fields near of El Concho de Cutris, Alajuela, and **d** pineapple plants at the same location

Arias (1991) examined the stratigraphy of the southeastern sector of the Región Central and described the basement as an ophiolitic oceanic complex (Complejo de Nicoya) overlain by Paleocene-Pliocene materials. These materials range from limestones (Parritilla Formation) and breccias (Caraigres Formation) to volcanoclastic rocks (Pacacua Formation), also including sedimentary materials like mudstones and sandstones associated with limestone (San Miguel and Peña Negra Formations). Further volcanic events, combined with deformation during the Miocene-Pliocene, contributed to additional geological materials. Later on, deposition of ashes and mud flow materials completed the filling of the central depression (Denyer and Arias 1991). Additionally, the volcanoes located within the Volcanic Central Range have significantly impacted the landscape due to their historical activity, producing volcanic ash, lava flows, extrusive rocks, and lahars (Casertano et al. 1983; Alvarado and Schmincke 2013; González et al. 2015).

Weather conditions across the Región Central vary among its sectors. The northern and northeastern sectors, especially in the mountains, are characterized as highly rainy throughout the year, while the south and southwestern sectors are drier, experiencing water deficits for at least five months (Solano and Villalobos 2001). In general, the average annual precipitation for the entire region is 2262 ± 280 mm, with a mean rainy period lasting 134.6 ± 6 days. The average air temperature is 19.0 ± 2 °C (Solano and Villalobos 2001). These climatic conditions

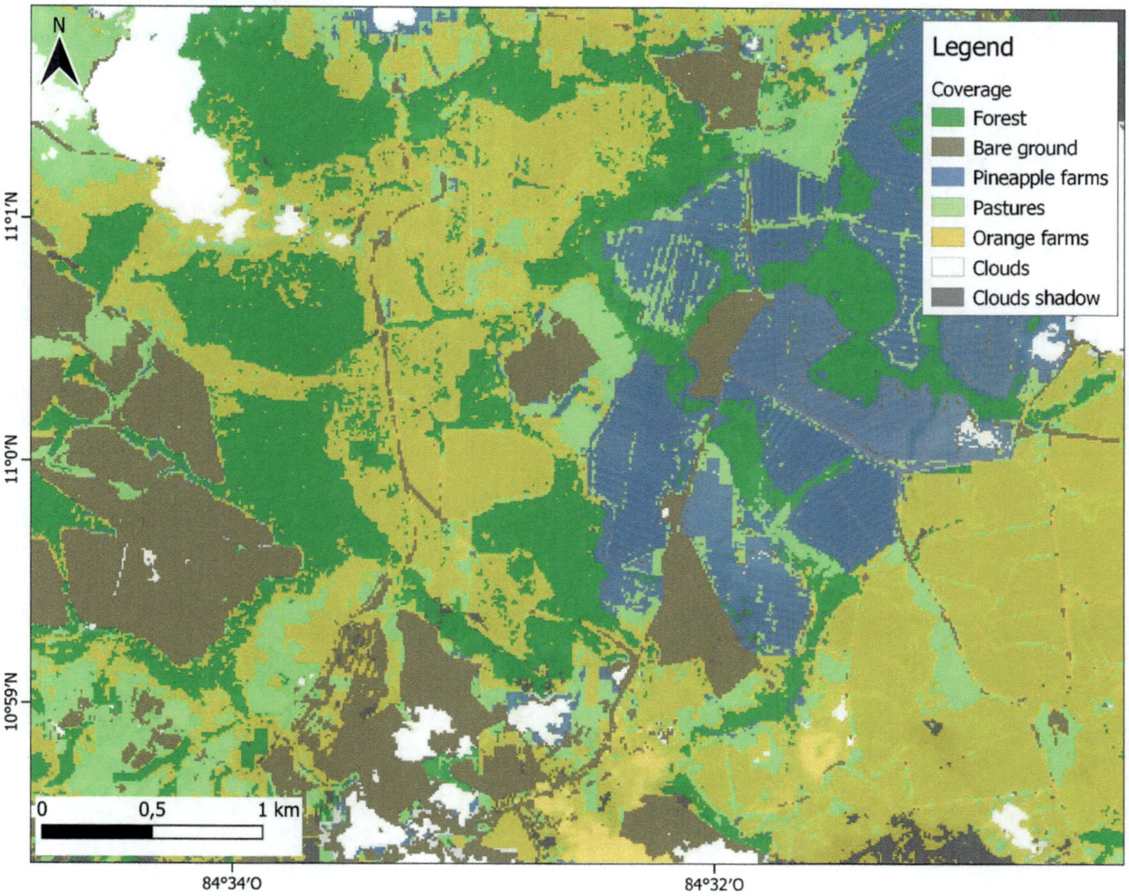

Fig. 11.23 Land coverage and soil use in the southeastern sector of the Huetar Norte Region. Land cover map generated using machine learning technics and supervised classification from Sentinel-2 Images (ESA 2023)

influence vegetation cover, which can be classified according to Holdridge's life zone system (Holdridge 1967) into Tropical wet forest (wf-T), Premontane wet forest (mf-P), Lower montane wet forest (wf-LM), Lower montane rain forest (rf-LM), Premontane rain forest (rf-P), and Tropical wet forest premontane transition (wf-T P).

Soils found in this region (from most to the least abundant), are Ultisols, Andisols, and Inceptisols, including some associations among them and other soil orders (Mata et al. 2022). Ultisols represent approximately 42% of the total area of the Región Central (Fig. 11.3), followed by Andisols (33%) and Inceptisols (8%) (Fig. 11.2). Ultisols are commonly found in the southeastern sector, in the hills and piedmont areas, while Andisols is prevalent around the volcanic piedmonts and foothill slopes.

Despite the pedological divergence and contrasting morphological, chemical, and physical properties of these soils (Buol et al. 2011), they have been found suitable for the development of agriculture with annual and perennial crops, adapted to local physiographic and climatic conditions. For instance, horticultural crops such as potatoes (*Solanum tuberosum* L.), onions (*Allium cepa* L.), carrots (*Daucus carota* Hoffm subsp. sativus), and crucifers (*Brassica* spp.) are primarily cultivated in the highlands and piedmont regions of volcanoes (mainly Irazú and Turrialba).

Extensive coffee plantations began in the "Central Plateau" of Costa Rica and expanded into other sectors within the Talamanca piedmont. Historically, coffee was a driving force for economic growth and social development within Región Central (Molina-Jiménez 1993; Gutiérrez 1994). The introduction of coffee to Costa Rica was first recorded in 1808 during the Tomás de Acosta's government, who imported *Typica* coffee seeds with the aim of establishing a profitable economic activity (Gutiérrez 1994). The expansion of coffee plantations encompassed the entire "Central Plateau" between 1830 and 1840. Subsequently, coffee spread to the western sector (Alajuela and San Ramón) in 1850–1860 and finally to the eastern sector (Reventazón Valley and Turrialba) in 1890 (Montero 2014). Immigrations to unexplored areas and land clearing facilitated the expansion of coffee plantations to southern sectors of this planning region such as "Los Santos" or "Valle del General" (Miranda 1985). Currently, coffee plantations play a pivotal role in several sectors of the Región Central's

Fig. 11.24 Extensive crops observed in the northwestern sector of Huetar Caribe Region: **a** Aerial pictures of extensive banana fields taken in Los Angeles de Cariari, Pococí; **b** banana plantation; **c** pineapple plantation, **d** aerial picture of extensive banana and pineapple fields established within the same landscape

landscape (Fig. 11.28). In 2018, coffee plantations covered an area of 27,953 ha in sector "Los Santos" (Fig. 11.29), 21,992 ha in sector "Valle Occidente" (Western Central Plateau), and 13,321 ha in "Valle Central" (Central Plateau) according to ICAFE (2019). These three sectors reported yields of 732,608, 345,549, and 208,975 *fanegas* (a unit formerly explained), respectively, in 2018 (ICAFE 2022).

Furthermore, the northern sector of Región Central primarily features horticultural crops like potatoes and onions (Fig. 11.30), as well as crucifers, which provide a significant income and employment opportunities. Historically, indigenous varieties of potato were cultivated by the native people in the Cartago highlands during the colonial period. Later, in 1950, onions were introduced, diversifying regional production systems (Ramírez et al. 2008). In 2022, the Irazú and Turrialba piedmonts featured a total of 2155 ha planted with potatoes, predominantly distributed across the municipalities of Turrialba, Oreamuno, Alvarado, and Cartago (Fig. 11.31), with a reported yield of 58,693 t. In the same municipalities, onion-covered land encompassed a total area of 1015 ha (Fig. 11.31 and Fig. 11.32), which obtained a total production of 34,141 t (SEPSA 2023).

The same crops were observed in the northern sector of Región Central, notably in Zarcero municipality. Here, potato- and onion-covered areas amounted to 665 and 106 ha, respectively, with a production of 16,188 and 3592 t, respectively (SEPSA 2023). This consistency in crops across the volcanic piedmont region may arise from similarities in soils and climatic conditions (Solano and Villalobos 2001; Mata et al. 2022). Another significant crop within Región Central is sugarcane, predominantly planted in two main sectors of this planning region: (1) the northwestern Central Valley, covering municipalities such as San Ramón, Naranjo, Grecia, and Palmares, and (2) the eastern Central Valley, encompassing the municipalities of Jiménez and Turrialba (Fig. 11.33). Historically, the northwestern sector witnessed increased development in the nineteenth century due to immigration from the provinces of Alajuela, Heredia, and San José. This influx of people aimed to

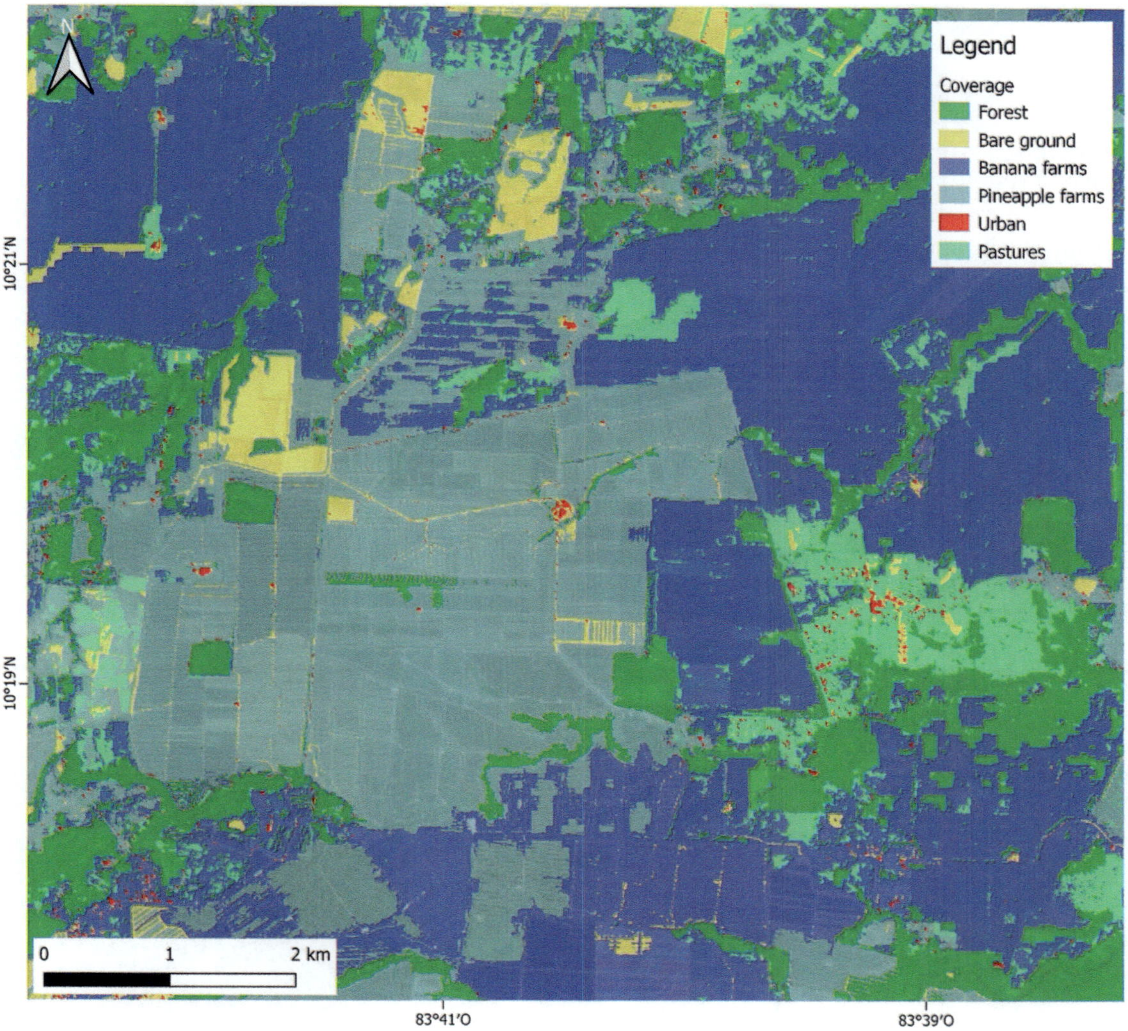

Fig. 11.25 Land coverage and soil use in the northwestern sector of Huetar Caribe Region. Land cover map generated using machine learning technics and supervised classification from PlanetScope Images (Planet Lab 2022)

expand agricultural endeavors, establish settlements, and cultivate these landscapes with the crops of coffee and sugarcane (Samper-Kutschbach 1986; Montero 2014).

Conversely, Turrialba and Jiménez municipalities have been traditionally known as "Campiña azucarera" (sugarcane countryside), with sugarcane production intensifying after 1890 when the railroad connected San José to Limón Port (Calderón and Chaves 2020). Currently, sugarcane occupies 4721 ha (Fig. 11.33), which represents a total yield a total of 225,438 t of fresh cane (Calderón and Chaves 2020). Within this same sector (municipalities of Jiménez and Paraíso), a non-traditional crop called "chayote" (*Sechium edule* (Jacq.) Sw.) is cultivated (Fig. 11.33). In 1999, chayote covered around 700 ha in Paraiso municipality (Saenz and Ruben 2004), which decreased to 407 ha in the same municipality by 2019 (INFOAGRO 2023).

11.3 Conclusions

Costa Rica exhibits diversity not only in its geologic, geomorphic, and climatic features, but also in agricultural crops and associated human activities that represent tradition and heritage within the regions assessed. For instance, coffee and sugarcane were staple crops in Región Central, while pineapple and oil palm dominated in different sectors of the Brunca Region, and banana farms are common along the Huetar Caribe.

Historically, the agricultural frontier underwent significant transformations due to substantial immigration from Región Central to colonize lands within the other five planning regions. These settlers initially planted subsistence crops (e.g., rice, beans, and tubers) and were involved in

Fig. 11.26 Other extensive crops observed in the western sector of Huetar Caribe Region: **a** Papaya (*Carica papaya* L.); **b** palm-heart (*Bactris gasipaes* Kunth); **c** aerial picture of extensive papaya; **d** aerial picture of heart-palm plantations. *Photos A and C courtesy of. R. Fallas and B. Alemán*

cattle rearing. The landscapes and their agricultural systems have continuously evolved, influenced by changes in political, economic, and social factors over the last century such as land reconversion, immigration, government administration shifting, settlement, and environmental policies.

These activities have not only shaped the physical environment but also the social and economic dynamics throughout the country. They have contributed to forging distinct identities and idiosyncrasies within each planning region of Costa Rica.

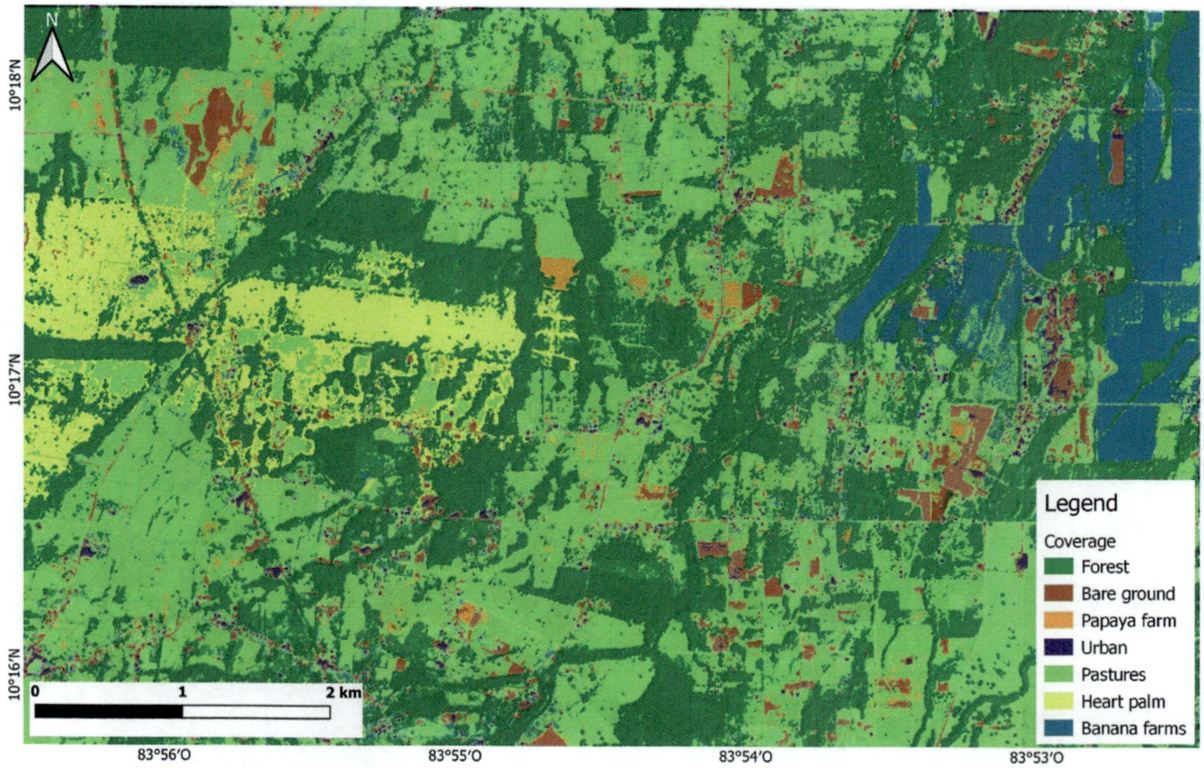

Fig. 11.27 Land coverage and soil use in the western sector of Huetar Caribe Region. Land cover map generated using machine learning technics and supervised classification from PlanetScope Images (Planet Lab 2022)

Fig. 11.28 Coffee (*Coffea arabica* L.) as extensive crop within the southern sector of Región Central: **a** Aerial pictures of extensive coffee fields covering the mountains and sloped areas of Los Santos, **b** just-bloomed coffee plant, observed in La Legua de Aserrí, San José; **c** aerial picture of coffee plantations taken in León Cortés municipality, San José, **d** harvested coffee cherries and official measuring machine (known as "Guillotina"). *Photos B and D courtesy of. M.E. Camacho and A.M. Peavey, respectively*

Fig. 11.29 Coffee (*Coffea arabica* L.) landscape and its distribution within the hilly landforms of "Los Santos" (southern Región Central): **a**, **b** Aerial pictures of coffee fields taken in Aserrí municipality, San José, **c**, **d** aerial pictures of coffee fields taken in León Cortés municipality, San José

Fig. 11.30 Horticultural extensive crops in the northern sector of Región Central: **a** Aerial picture of potato (*Solanum tuberosum* L.) and onion (*Allium cepa* L.) fields in Cartago, **b** typical soil profile (*Humic Haplustands*), **c**, **d** potato and onion plants at the same location. *Photo B courtesy of. M. Sandí*

Fig. 11.31 Highlands of Cartago (Northern Regiõn Central) and their agricultural landscapes: **a**, **b** Sloped landforms cultivated with onion, **c**, **d** slopes cultivated with potato (two contrasting crop stages)

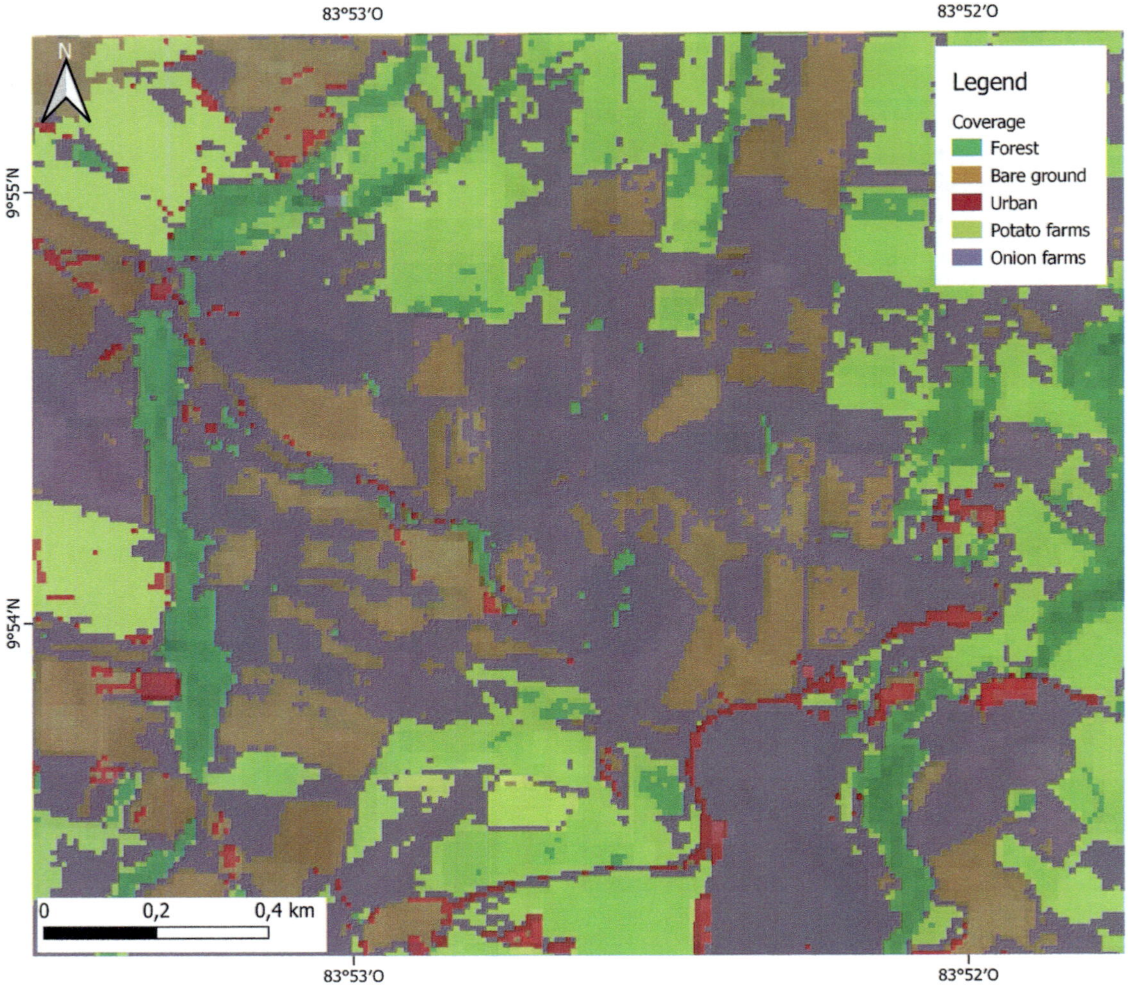

Fig. 11.32 Land coverage and soil use in the northern sector of the Región Central. Land cover map generated using machine learning technics and supervised classification from Sentinel-2 Images (ESA, 2023)

Fig. 11.33 Other extensive crops in the northeastern sector of Región Central: **a** Aerial picture of sugarcane fields (*Saccharum officinarum* L.) in Santiago de Paraíso, Cartago, **b** typical soil profile developed from a lava flow (*Typic Udorthens*), **c** aerial picture of fields cultivated with chayote (*Sechium edule* (Jacq.) Sw.) in Santiago de Paraíso, Cartago, and **d** sugarcane plant observed within this same location

References

Abarca-Jiménez G (2015) Contexto histórico del cese del enclave bananero en la zona sur de Costa Rica (1972–1985). Revista Universidad En Diálogo 5(2):187–205

Alfaro A, Denyer P, Alvarado GE, Gazel E, Chamorro C (2018) Estratigrafía y petrografía de las rocas ígneas en la Cordillera de Talamanca, Costa Rica. Revista Geológica De América Central 58:7–36

Alvarado GE, Schmincke HU (2013) The 1723 AD violent strombolian and phreatomagmatic eruption at Irazú Volcano, Costa Rica. Revista Geológica De América Central 48:41–61

Alvarado GE, Barquero R, Taylor W, Mora M, Peraldo G, Salazar G, Aguilar T (2009) Geología de la Hoja San Isidro, Costa Rica. Revista Geológica De América Central 40:111–122

Angulo A, Rodríguez M, Chaves MA (2020) Guía técnica cultivo caña de azúcar: Guanacaste. Liga Agrícola Industrial de la Caña de Azúcar (LAICA)-Departamento de Investigación y Extensión de la Caña de Azúcar (DIECA). Technical document. 78 p. Available online in: https://servicios.laica.co.cr/laica-cv-biblioteca/index.php/Library/download/jieyzwRDmVvUeWJGLRfYLzXbibjfLZNW

Ares A, Boniche J, Quesada JP, Yost R, Molina E, Smyth TJ (2002) Estimación de biomasa por métodos alométricos, nutrimentos y carbono en plantaciones de palmito en Costa Rica. Agronomía Costarricense 26(2):19–30

Arias O (2003) Redefinición de la formación Tulín (Maastrichtiano-Eoceno inferior) del Pacífico Central de Costa Rica. Revista Geológica De América Central 28:47–68

Arias F, Mata R, Alvarado A, Serrano E, Laguna J (2010) Caracterización química y clasificación taxonómica de algunos suelos cultivados con banano en las llanuras aluviales del Caribe de Costa Rica. Agronomía Costarricense 34(2):177–195

Arias LD, Retana JM, Torres D, Peña L (2018) Los nuevos frutos de oro. Aparato productivo de piña y naranja en la Zona Norte, Costa Rica (1974–2015). Revista Rupturas 8(1):93–121

Barrantes JC, Chaves MA (2020) Guía técnica cultivo caña de azúcar: Región Sur. Liga Agrícola Industrial de la Caña de Azúcar (LAICA)—Departamento de Investigación y Extensión de la Caña de Azúcar (DIECA). Technical document. 73p. Available online in: https://servicios.laica.co.cr/laica-cv-biblioteca/index.php/Library/download/ENAErFDGjylcsJjFYfkSeJlDQdcDiGbC

Battistini R, Bergoeing JP (1984) Geomorfología de la costa Caribe de Costa Rica. Revista Geográfica 99:167–188

Bergoeing JP (2011) Los conos de deyección del valle de El General, Costa Rica. Revista Geográfica 150:21–32

Bergoeing JP (2017) Geomorphology and volcanology of Costa Rica, 1st edn. Elsevier, Amsterdam, The Netherlands, p 280

Bergoeing JP, Brenes LG, Fernández Arce M, Ureña FM (2010) Geomorfología de la cordillera Costeña y de los abanicos aluviales en el piedemonte meridional de la cordillera de Talamanca. Revista Geográfica 148:165–179

Bertsch F (1998) La fertilidad de los suelos y su manejo. ACCS, San José, Costa Rica, 157p

Bergoeing JP, Protti R. (2006). Geomorfología Paleo-Lacustre del Sur del Lago de Nicaragua. Revista Geográfica 139: 27-38

Blanco-Obando EE (2020) Cultivo de piña y conflictos socio-ambientales en la región Atlántico/Caribe, Costa Rica, 1990–2017. Athenea Digital. Revista de pensamiento e investigación social 20(3):e2421

Buol SW, Eswaran H (2000) Oxisols. Adv Agron 68:151–195. https://doi.org/10.1016/S0065-2113(08)60845-7

Buol SW, Southard RJ, Graham RC, McDaniel PA (2011) Soil genesis and classification, 6th edn. Wiley, Chichester, UK, p 543

Calderón G, Chaves MA (2020) Guía técnica cultivo caña de azúcar: Turrialba. Liga Agrícola Industrial de la Caña de Azúcar (LAICA)—Departamento de Investigación y Extensión de la Caña de Azúcar (DIECA). Technical document, 95p. Available online in: https://servicios.laica.co.cr/laica-cv-biblioteca/index.php/Library/download/LgtmpPwEZTIDQoalVkesZIxqTYKHjwlI

Camacho ME, Mata R, Forsythe W (2015) Labranza mecanizada de pasturas mediante tres implementos en un Ultisol y sus implicaciones físicas e hidropedológicas. Agronomía Costarricense 39:101–115

Camacho ME, Alvarado A, Fernández-Moya J (2016) Vochysia guatemalensis Donn. Smith, an alternative species for reforestation on acid tropical soils. New for 47:497–512

Camacho ME, Quesada-Román A, Mata R, Alvarado A (2020) Soil-geomorphology relationships of alluvial fans in Costa Rica. Geoderma Reg 21:e00258

Camacho ME, Mata R, Barrantes-Víquez M, Alvarado A (2021) Morphology and characteristics of eight Oxisols in contrasting landscapes of Costa Rica. CATENA 197:104992

Cámara Nacional de Productores y Exportadores de Piña (CANAPEP) (2023) Statistics. Available online from: https://canapep.com/estadisticas/

Cárdenes G, Obando LG (2005) Índice de erosión-sedimentación costera (IE-SC): una aplicación en la costa del Pacífico Central de Costa Rica. Revista Geológica De América Central 32:33–43

Castro Z (1994) Cultivo de la piña. In: Cortés G (ed) Atlas agropecuario de Costa Rica. EUNED, San José Costa Rica, pp 193–284

Clare P (2012) Poder y medio ambiente. La palma aceitera en el Pacífico costarricense, 1950–2007. Historia Agraria: Revista De Agricultura e Historia Rural 57:135–166

Corrales F (2000) Más de diez mil años de historia precolombina. In: Botey AM (ed) Costa Rica, estado, economía, sociedad y cultura: desde las sociedades autóctonas hasta 1914, 1st edn. Editorial de la Universidad de Costa Rica, San José Costa Rica, pp 25–66

Corporación Arrocera Nacional (CONARROZ) (2021). Informe Anual Estadístico 2021-2022. PDF document consulted on February 10th Available online in: https://conarroz.com/userfile/file/Informeanual estad%C3%ADstico21-22final.pdf

Denyer P, Arias O (1991) Estratigrafía de la Región Central de Costa Rica. Revista Geológica De América Central 12:1–59

Denyer P, Aguilar T, Alvarado GE (2003) Geología y estratigrafía de la hoja Barranca, Costa Rica. Revista Geológica De América Central 29:105–125

Denyer P, Aguilar T, Montero W (2014) Cartografía geológica de la península de Nicoya. Editorial de la Universidad de Costa Rica San José, Costa Rica

Denyer P, Gazel E (2009) The Costa Rican Jurassic to Miocene oceanic complexes: origin, tectonics and relations. J S Am Earth Sci 28(4):429–442

Durán ND (2006) Evolución de la producción azucarera en el distrito de El General, Pérez Zeledón, entre la segunda mitad del siglo XIX y la década de 1970. Revista De Historia 53–54:63–98

ESA (2023) Harmonized Sentinel-2 MSI: MultiSpectral Instrument, Level-2ª. Contains modified Copernicus Sentinel data (2022). Copernicus Global Land Cover Map, 2015

Fallas-Corrales RA, van der Zee SEATM (2020) Diagnosis and management of nutrient constrains in papaya. In: Srivastava AK, Hu C (eds) Fruit crops: diagnosis and management of nutrient constraints. Elsevier, Amsterdam, The Netherlands, pp 607–628

Faure G, Samper M (2004) Veinte años de apertura económica: el porvenir comprometido de la agricultura familiar en el norte de Costa Rica. Anu Estud Centroam 30(1):7–26

Fernández-Moya J, Alvarado A, Miguel-Ayanz S, Marchamalo-Sacristán M (2014) Forest nutrition and fertilization in teak

(Tectona grandis Lf) plantations in Central America. NZ J Forest Sci 44(1):1–8

Gatica-López GA (2017) Los costos de la migración desde Costa Rica. Revista Rupturas 7(2):33–72

Gazel E, Alvarado GE, Obando J, Alfaro A (2006) Geología y evolución magmática del arco de Sarapiquí, Costa Rica. Revista Geológica de América Central 32:13–31

González G, Mora-Amador R, Ramírez C, Rouwet D, Alpizar Y, Picado C, Mora R (2015) Actividad histórica y análisis de la amenaza del volcán Turrialba, Costa Rica. Revista Geológica De América Central 52:129–149

Gutiérrez G (1994) Caficultura costarricense. In: Cortés G. (Ed) Atlas agropecuario de Costa Rica. EUNED, San José Costa Rica, pp 275–286

Holdridge RL (1967) Life zone ecology. Tropical Science Center, San José, Costa Rica, 206p

Instituto del Café de Costa Rica (ICAFE) (2019) Actualización Área Cafetalera 2017–2018. Gerencia Técnica, CICAFE. Heredia, Costa Rica. 62p. Available online in: https://www.icafe.cr/wp-content/uploads/cicafe/documentos/Actualizacion-Area-Cafetalera-2017-2018.pdf

Instituto del Café de Costa Rica (ICAFE) (2022) Informe sobre la actividad cafetalera de Costa Rica. Heredia, Costa Rica, 68p. Available online in: https://www.icafe.cr/wp-content/uploads/informacion_mercado/informes_actividad/actual/Informe%20Actividad%20Cafetalera.pdf

Instituto Meteorológico Nacional (IMN) (2011) Estudio de Cuencas Hidrográficas de Costa Rica. Análisis biofísico, climático y socioeconómico, 724p. Available online in: http://cglobal.imn.ac.cr/documentos/publicaciones/EstudioCuencas/EstudioCuencasHidrograficasCR.pdf

Instituto Nacional de Estadísticas y Censos (INEC) (2015) VI Censo Nacional Agropecuario 2014. San José, INEC, 110p. PDF document consulted on February 10th. Available online in: https://admin.inec.cr/sites/default/files/media/reagropeccenagro2014-002_1_2.pdf

Instituto Nacional de Estadística y Censos (INEC) (2022a) Encuesta Nacional Agropecuaria 2021. Resultados generales de la actividad agrícola y forestall, 75p. PDF document consulted on February 10th. Available online in: https://admin.inec.cr/sites/default/files/2022-09/reagropecENAAGR%C3%8DCOLA2021-01.pdf

Instituto Nacional de Estadística y Censos. (INEC) (2022b) Resumen de resultados. Datos definitivos Comercio Exterior, annual 2021. PDF document consulted on February 10th. Available online in: https://admin.inec.cr/sites/default/files/2022-11/ComercioExterior_ResumenResultadosDefinitivos_2021.pdf.pdf

Instituto Nacional de Estadística y Censos (INEC) (2023) Resumen de resultados. Datos Preliminares Comercio Exterior, Anual 2022. PDF document consulted on February 10th. Available online in: https://admin.inec.cr/sites/default/files/2023-03/coCOMEX2022-preliminar.pdf

Jones CF, Morrison PC (1952) Evolution of the banana industry of Costa Rica. Econ Geogr 28(1):1–19

Marchena-Sanabria J (2015) El nacimiento de las corporaciones azucareras en Guanacaste, 1890–1970. Diálogos Revista Electrónica De Historia 16(2):83–119

Mata R, Rosales A, Sandoval D, Vindas E, Alemán B (2022) Mapa de Órdenes de Suelos de Costa Rica, 2022. Esc. 1:200.000. Universidad de Costa Rica, Centro de Investigaciones Agronómicas. Costa Rica, San José

Miranda M (1985) Cambio en el uso del suelo en General Viejo de Pérez Zeledón. Revista Geográfica De América Central 2(17–18):99–121

Molina E (1998) Encalado para la corrección de la acidez del suelo. ACCS. San José, Costa Rica, 45p

Molina-Jiménez I (1993) Los pequeños y medianos caficultores, la historia y la nación. Costa Rica (1890–1950). Caravelle (1988–), pp 61–73

Mora-Urpí J (2002) Presente y futuro del palmito en Costa Rica. Agronomía Costarricense 26(2):95–100

Monitoreo de Cambio de Uso en Paisajes Productivos (MOCUPP) (2021a) Informe: Monitoreo del estado de la palma aceitera en Costa Rica para el año 2019. San José, Costa Rica. PDF document consulted on February 10th, 2023. Available in: https://mocupp.org/informes/

Monitoreo de Cambio de Uso en Paisajes Productivos (MOCUPP) (2021b) Informe: Monitoreo del estado de la piña en Costa Rica para el año 2019. San José, Costa Rica. PDF document consulted on February 10th, 2023. Available in: https://mocupp.org/informes/

Montero A (2014) Una aproximación a los cambios en el paisaje en el Valle Central de Costa Rica (1820–1900). Historia Ambiental Latinoamericana y Caribeña (HALAC) Revista de la Solcha 3(2):276–309

Montero A, Viales R (2014) "Agriculturización" y cambios en el paisaje. El banano en el Atlántico/Caribe de Costa Rica (1870–1930). Historia Ambiental Latinoamericana y Caribeña (HALAC) Revista de la Solcha 3(2):310–338

Morales-Abarca LF (2018) Producción y rendimiento del cultivo de la piña (*Ananas comosus*) en Costa Rica, periodo 1984–2014. E-Agronegocios 4(2):1–15

Murillo JI (1994) El cultivo del arroz en Costa Rica. In: Cortés G (ed) Atlas agropecuario de Costa Rica. EUNED, San José Costa Rica, pp 87–108

Moya R (2004) Wood of *Gmelina arborea* in Costa Rica. New for 28:299–307

NCSS (National Cooperative Soil Survey) (2023) National cooperative soil survey characterization database. Consulted on February 10th, 2023. Available in http://ncsslabdatamart.sc.egov.usda.gov/

Obando A (2020) Acciones y omisiones del estado costarricense en la Expansión piñera: el caso de la zona norte-norte de Costa Rica Anuario Centro de Investigación y Estudios Políticos 11:22–55. https://doi.org/10.15517/aciep.v0i11.42226

Obando LG, Kussmaul S (2009) Geología de la hoja Buenos Aires, Costa Rica. Revista Geológica De América Central 41:123–137

Planet Labs (2022) Planet explore. Available online: https://www.plantet.com/expl. Accessed on 10 Aug 2023

Porras H, Alvarado GE, Arroyo-Solórzano M, Durán P, Echandi E (2021) La depresión tectónica de Nicaragua en Costa Rica: estructura interna, estilo estructural, edad, extensión y actividad de la cuenca de San Carlos. Revista Geológica De América Central 65:347–369

Quesada-Román A, Díaz-Bolaños R (2019) Impactos ambientales de la colonización agrícola en Coto Brus, Costa Rica (1940–2018). Revista Geográfica De América Central 63:171–203

Quesada-Román A (2013) Condición de uso de la tierra del distrito San Vito, Coto Brus, Puntarenas. Reflexiones 92(1):47–64

Quesada-Román A, Zamorano-Orozco JJ (2019) Geomorphology of the Upper General River Basin, Costa Rica. J Maps 15(2):95–101

Quesada-Román A, Pérez-Briceño PM (2019) Geomorphology of the Caribbean coast of Costa Rica. J Maps 15(2):363–371

Quesada-Román A, Stoffel M, Ballesteros-Cánovas JA, Zamorano-Orozco JJ (2019) Glacial geomorphology of the Chirripó National Park, Costa Rica. J Maps 15(2):538–545

Quesada-Román A, Campos N, Alcalá-Reygosa J, Granados-Bolaños S (2020) Equilibrium-line altitude and temperature reconstructions during the Last Glacial Maximum in Chirripó National Park, Costa Rica. J S Am Earth Sci 100:102576

Quesada-Román A, Campos N, Granados-Bolaños S (2021) Tropical glacier reconstructions during the Last Glacial Maximum in Costa Rica. Revista Mexicana De Ciencias Geológicas 38(1):55–64

Quesada-Román A, Quirós-Arias L, Zamora-Pereira JC (2022) Interactions between Geomorphology and Production Chain of High-Quality Coffee in Costa Rica. Sustainability 14(9):5265

Ramírez L, McHugh A, Alvarado A (2008) Evolución histórica y caracterización socioeconómica de la cuenca media del río Reventado, Cartago, Costa Rica. Agronomía Costarricense 32(2):53–72

Royo A (2004) La Ocupación del Pacífico Sur Costarricense por parte de la Compañía Bananera (1938–1984). Revista electrónica de historia, Diálogos vol 4, no 2

Sáenz A (1960) Reseña Histórica Agropecuaria y Experimental Agrícola de Costa Rica. Revista De La Universidad De Costa Rica 21:5–58

Saenz F, Ruben R (2004) Export contracts for non-traditional products: Chayote from Costa Rica. J Chain Netw Sci 4(2):139–150

Salas-González DM (2020) Cambios en la superficie sembrada de Palma aceitera en el cantón de Osa, Puntarenas. Período 2014–2018. Revista Geográfica De América Central 65:93–120

Samper-Kutschbach M (1986) Uso de la tierra y unidades productivas al finalizar el siglo XIX: Noroeste del Valle Central, Costa Rica. Revista De Historia 14:133–177

Sánchez Y, Vega MF (2018) Situación del mercado del arroz en Costa Rica: una mirada a la realidad. Revista ABRA 38(56):1–22

Secretaría Ejecutiva de Planificación Sectorial Agropecuaria (SEPSA) (2023) Boletín Estadístico Agropecuario 33. Serie Cronológica 2019–2022. Electronic document consulted on February 10th, 2023. Available in: http://www.sepsa.go.cr/

Sistema de Información del Sector Agropecuario Costarricense (INFOAGRO) (2023) Reporte de Área y Producción. Estadísticas agropecuarias. Electronic document consulted on February 10th, 2023. Available in: http://www.infoagro.go.cr/

Skorupa AL, Silva SH, Poggere GC, Tassinari D, Pinto LC, Nilton CURI (2017) Similar soils but different soil-forming factors: converging evolution of Inceptisols in Brazil. Pedosphere 27(4):747–757

Solano J, Villalobos R. (2001). Aspectos fisiográficos aplicados a un bosquejo de regionalización geográfico climático de Costa Rica. Tópicos meteorológicos y Oceanográficos 8(1): 26-39

Soil Survey Staff (1999) Soil taxonomy: a basic system of soil classification for making and interpreting soil surveys, 2nd edn. Natural Resources Conservation Service. U.S. Department of Agriculture Handbook 436

Soto-Ballestero M (1994) El banano. In: Cortés G (ed) Atlas agropecuario de Costa Rica. EUNED, San José Costa Rica, pp 149–160

van Uffelen JG (1993) A geological, geomorphological and soil transect study of the Chirripó massif and adjacent areas, Cordillera de Talamanca. Tesis de Maestría, Wageningen University, Wageningen, Costa Rica, p 72

Vargas F, Alfaro A (1992) Presencia de serpentinitas, basaltos alcalinos y rocas volcánicas ácidas en la zona norte- Atlántica de Costa Rica. Revista Geológica de América Central 14:105–107

Vargas C, Vargas Y, Acuña JF, Miller C (2020) Monitoreo anual del paisaje productivo de palma aceitera para Costa Rica al año 2018 (línea base). Ambientico 276:49–54

Vargas Y, Vargas C, Miller C (2021) Monitoreo del estado de la palma aceitera en Costa Rica para el año 2019. Monitoreo de Cambio de Uso en Paisajes Productivos (MOCUPP). Informe técnico. San José, Costa Rica, 88p. Available at https://mocupp.org/cultivo-de-palma-aceitera/

West LT, Beinroth FH, Sumner ME, Kang BT (1997) Ultisols: characteristics and impacts on society. Adv Agron 63:179–236

12 Anthropogenic Geomorphology of Costa Rica

Guillermo E. Alvarado and Adolfo Quesada-Román

Abstract

Anthropogenic geomorphology examines the impact of human activities on the physical landscape. This chapter focuses on the ways in which different human actions have shaped the terrain in the country, creating landforms and modifying the pace of external geodynamic process since Pre-Columbian time ($\geq$ 4.5 ka ago). Developing countries, including Costa Rica, face the challenge of balancing economic growth and sustainable development while addressing the impacts of anthropogenic geomorphology. Human activities such as urbanization, deforestation, and resource extraction, road building, and large-scale agriculture and tourism can cause alterations to the natural landscape and lead to soil erosion, changes in hydrologic regimes, and loss of biodiversity. Costa Rica has implemented policies to protect its natural resources and promote sustainable development, such as reforestation and low-impact tourism activities. The principal landscapes generated by anthropogenic geomorphology in Costa Rica could be classified in three classes determined by their geomorphic expression in (i) excavated, (ii) leveled, and (iii) artificial landforms. Thus, anthropogenic geomorphology is a critical issue for developing countries like Costa Rica, and effective management strategies are necessary to balance economic growth and environmental protection.

Keywords

Pre-Columbian altered landscape · Anthropogenic geomorphology · Urban geomorphology · Developing countries · Environmental issues

12.1 Introduction

Anthropogenic geomorphology could be defined as the study of the role of humans in creating landforms (human-made landforms) and modifying the operation of geomorphological processes such as weathering, erosion, transport, and deposition. In addition, it predicts the consequences of disruptions to natural equilibria and proposes ways to prevent harmful impacts (Sherlock 1922; Szabó 2010; Goudie 2010).

In Costa Rica and globally, human activities are equally significant as other geomorphic factors in shaping the landscape. This is due to the rapid population growth and increased demand for land, natural resources, and energy, leading to widespread modifications of the environment (Hall 1984; Santillana 2008; Vargas 2014). Geologists, geomorphologists, geographers, civil engineers, biologists, and archeologists have recognized anthropogenic alterations of landforms in Costa Rica and deep-time human environmental impacts, which have occurred in Pre-Columbian time, particularly after the last glaciation (Cooke 2005; Piperno 2011), but extensively since the second half of the twentieth century (Flores 2001; Vargas 2014).

Literature review of the anthropogenic geomorphology of Costa Rica can enhance the knowledge of what happened between the Pre-Columbian, through colonial, to the present eras. The subject of this chapter is therefore, for the first time in the Costa Rican geomorphic literature, to provide description of a large and ever-widening range of surface artificial landforms, extremely diverse in origin and in purpose, created by the operation of human society.

G. E. Alvarado
Centro de Investigaciones Ciencias Geológicas, Universidad de Costa Rica, San José, Costa Rica

A. Quesada-Román (✉)
Escuela de Geografía, Universidad de Costa Rica, San José, Costa Rica
e-mail: adolfo.quesadaroman@ucr.ac.cr

A. Quesada-Román (ed.), *Landscapes and Landforms of Costa Rica*, World Geomorphological Landscapes, https://doi.org/10.1007/978-3-031-64940-0_12

12.2 Brief History in Costa Rica

Slight and early anthropic landform modifications in the contemporary Costa Rican territory could have started as the first human arrive during the Paleoindian Period (~15 ka b.P.) and/or early Holocene (11.7–8.2 ka b.P.), but there is no direct evidence in Costa Rica, even though food production might had taken place propitiated by selective clearing and plant selection since the early part of the Holocene in tropical environments (Cooke, 2005; Piperno 2011). Studies on microfossils, such as pollen, phytoliths, and starch grains, indicate adaptative processes in neotropical areas of the Americas driven by food production and plant domestication along the Archaic Period (10–4 ka b.P.). A wide variety of cultigens made its way to the southern part of Central America in that period, including corn and manioc. Based on documented findings in Panamanian territory between 8.6 and 5.7 ka b.P. (Piperno 2011), the presence of cultigen can be expected for Costa Rica. In fact, early maize agriculture in Costa Rica came back at 5.5 ka b.P. (Arford and Horn, 2004).

It is likely that raise in number, size, and time of permanence in occupation sites happened during the Archaic Period. However, available data for Costa Rica have archeologist to suggest more sedentary lifestyle starting at about 4.5–4.0 ka b.P. (Hoopes 1994; Hurtado de Mendoza 2004), and intentional forest fires at least since 3 ka b.P. (Clement and Horn 2001). Those data show a tangible increase in the quantity of sites and cultural remains in the time interval 300 b.P.–300 A.D., as a reflection of population growth. Pre-Columbian villages with stone constructions, marked walking trails, causeways, and hydrologic works have been documented for later periods, from 1500 to 800 years in the past (Snarskis 1981; Sheets and McKee 1994; Hurtado de Mendoza 2004; Fernández and Alvarado 2006; Peytrequín and Aguilar 2007; Benfer 2012; Corrales 2016; Vázquez and Rosenswig 2017; Solís et al. 2019; Núñez 2020; Hurtado de Mendoza and Alvarado 2021; Salgado et al. 2021; Vázquez et al. 2022).

Few Pre-Columbian interregional routes of human movement, definable by marks of pedestrian use or constructed features, have been identified in Costa Rica (Vázquez et al. 2003, 2021). Long indigenous routes interconnecting territories count with one report only, which was deduced by ethnohistorical correlation and in large part articulated to waterways (Cavallini 2011; Salgado et al. 2016). It has been hypothesized that indigenous roads were taken advantage of in Colonial times, intervened for circulation of carts and afterwords for transportation with motor vehicles (Vázquez 2014). Pathways leading to modern-day indigenous communities, particularly in regions of Talamanca (Fig. 12.1), could be related to long-lasting communication roads, used in Pre-Columbian times and with linkage to archeological sites but, to this regard, only very preliminary information is known (Vázquez, personal communication 2023).

The earliest formalized communication so far identified in Costa Rica belongs to societies on the Arenal-Tilarán area (Fig. 12.1), in the form of narrow pathways used in the bracket A.D. 500–1300 (McKee et al. 1994; Sheets and Sever 2007). Radiocarbon dating of the trails along that extended archeological trajectory was agency by stratified tephra from the Arenal volcano intercalated with paleosols. Pathways became eroded when people walked single line, in processional mode passing through villages, cemeteries, and water springs. This prehistorical behavior attests for the resilience of human communities to vulcanism, reoccupying its territory and having handy a mobilization strategy as a response to natural disasters (Sheets 2012). Investigators of the Arenal-Tilarán's archeology have proposed interactions and exchanges through the network of trails, not only within an intraregional sphere but further away with the northeastern Caribbean lowlands, the northwest territory of Guanacaste-Nicoya, and the central highlands of Costa Rica (Sheets and Sever 2007; Benfer 2012). The erosion by repetitive use resulted in paths entrenched 2 m or deeper. The cultural standard developed over centuries was that the preferred way of entering a special place was by an entrenched path. Thus, people approaching a special place would have a restricted view of their surroundings. Upon entering a site, the view would dramatically open. Through cognitive experiences like that, people created and perpetuated social memory across their landscapes after generations of processional use (Sheets 2008, 2011).

Also, beach dikes on the coast of Guanacaste (Culebra Bay) and Quepos made by Pre-Columbian inhabitants with local stones indicate that they functioned mainly as traps forming ponds for fish, crustaceans, turtles, and mollusks (Figs. 12.1 and 12.2d). In the case of Culebra Bay, the erection of these traps dates between 800 and 1350 AD (Vázquez et al. 2019).

As indicated by archeological survey, differentiated rank societies never developed in the Arenal-Tilarán area (Sheets and Sever 2007). Also, few architectural examples attributable to chiefdoms are recorded for Guanacaste-Nicoya (Creamer and Hass 1985; Lange 1992; Vázquez 2023). By contrast, a series of chiefdoms did develop in other territories of the current Costa Rica, interpreted upon remains of architectural villages and differentiation of salient features. These indicators can be traced back to about A.D. 600 (Vázquez et al. 2003; Corrales 2016; Solís et al. 2019; Salgado et al. 2021). Demonstrative infrastructure of residential centers has been identified in diverse geographic setting, among them: the slope of a volcano, foothills, alluvial

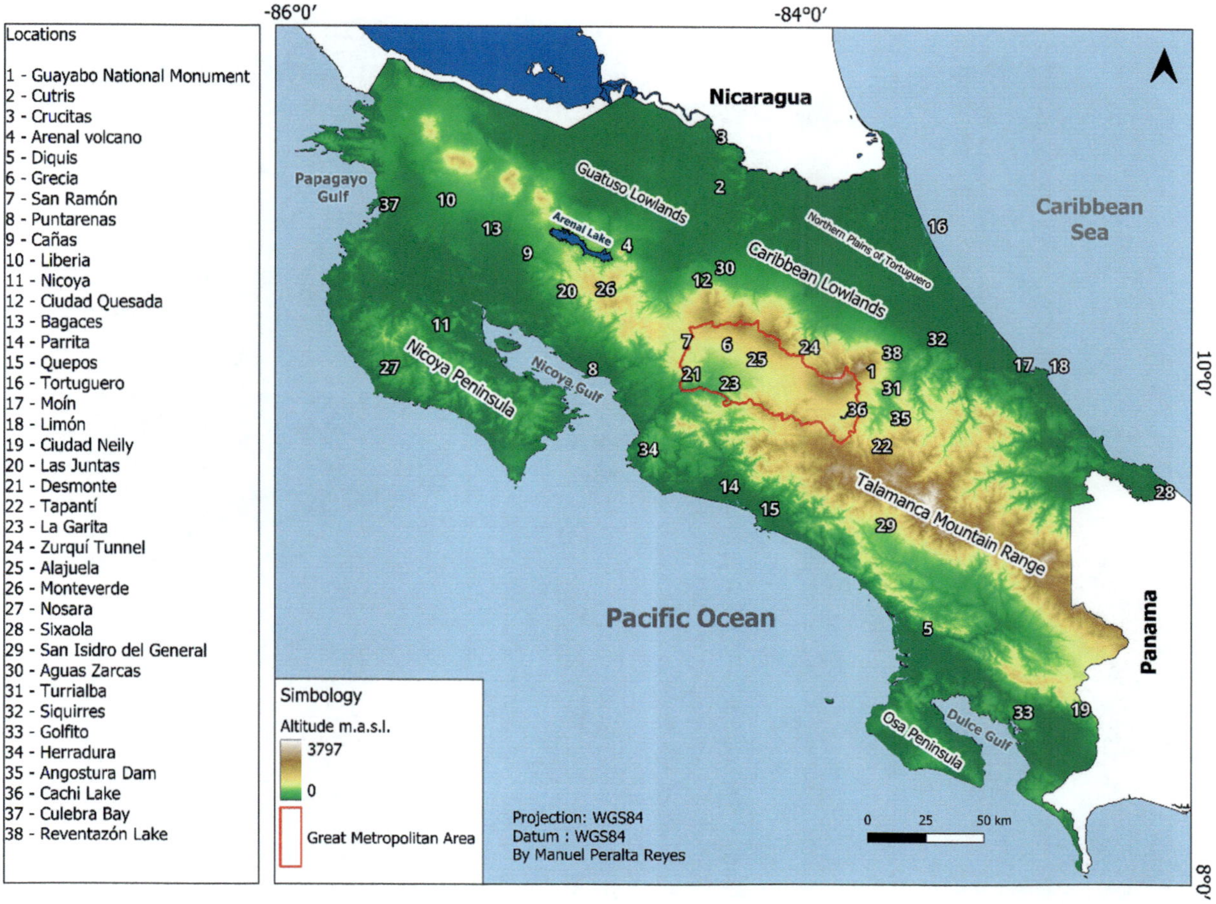

Fig. 12.1 Main locations of Costa Rica studied in this chapter

plains, intermontane valleys, and mountain ranges. The basal architectural structures of such centers were built mostly out of cobble rocks with earthen fills (Hurtado de Mendoza 2004; Hurtado de Mendoza and Alvarado 2021).

It is important to point out that, starting at around 500–300 B.C., considerable energy expenditure was dedicated by Pre-Columbian peoples of Costa Rica to funerary architecture. Archeology has shown that since those centuries, house construction demanded use of copious quantities of forest materials, together with clever skills and even the employment of cobble rocks. In addition, building of graves entailed elaborated standards in, for example, shaft-and-chamber tombs and corridors for aligning interments. Time progressed with the advent of carefully layout stone covers, which incorporated volcanic and sedimentary slabs, as well as rock mounts or tumuli. At the northwestern territory, in the Cañas-Liberia and other region, dozens of huge concentrations of rocks for protecting burial have been recorded by archeological research (Guerrero and Solís 1997; Fernández and Alvarado, 2006).

In Costa Rica, Pre-Columbian villages with monumental architecture and attributes of ceremonial centers are known. Architectural cores in these sites are from small settlement to villages, mostly from 1 to 10 ha. Some sites that come to the point are distributed in the Central Caribbean Watershed, among them: Guayabo (Aguilar 1972; Fonseca 1996; Hurtado de Mendoza 2004; Alarcón 2019), Las Mercedes (Vázquez and Chapdelaine 2008; Vázquez and Rosenswig 2017), Anita Grande (Vázquez et al. 2022), and Nuevo Corinto (Salgado et al. 2021). As an example, the Guayabo archeological site, designated National Monument in1973, shows foundations of dwellings in conjunction with spaces for social interaction, as well as the arrival, and assembly of visitors (Figs. 12.1 and 12.2). These characteristics suggest ceremonial activities linked to the residence at the village of a regional sociopolitical leader (Fonseca 1996; Hurtado de Mendoza 2004). A built road interconnects additional architectural sites, with Guayabo as a hob within a regional system of salient settlements (Acuña 1987; Vázquez et al. 2021). The cusp of construction works is dated at AD 900–1200 upon radiocarbon determinations obtained in the main site (Alarcón 2019). Guayabo includes an aqueduct, a channelized creek, water collectors, and two causeways, among other features that in sum required tons of cobble rocks. It holds the designation of Civil Engineering World Heritage, awarded in 2009 by

Fig. 12.2 Guayabo National Monument. **a** Aerial view of the Turrialba volcano in the top of the image, **b** ceremonial monuments, **c** stony Pre-Columbian road (almost 7 m wide) and extending for at least 7 km, **d** fish trap in Papagayo, Guanacaste

the American Society of Civil Engineers (ASCE), a most prestigious institution (Fig. 12.2).

Another case to mention is the Cutris site, a monumental site hub of the largest network of Pre-Columbian villages known for the northeastern territory of Costa Rica (Fig. 12.1). Cutris is composed of over 70 stone architectural structures, most of them interpreted as elevated dwelling foundations. Four sunken roads constructed with side embankments from the artificial earth movements, some of them huge, radiate out of Cutris toward satellite sites at 6–9 km away. The occupational history of Cutris, from 1500 B.C. to A.D. 900, is similar in extended trajectory when compared to other monumental sites of the country. However, Cutris gave ceramic clues of a rise in high infrastructural proportion at A.D. 600–900 years, earlier to the analogical development in remarkable villages as Guayabo and Las Mercedes of the Central Caribbean (Vázquez et al. 2003). A similar panorama is interpreted for the southern territory of Costa Rica, where the early monumental complexity of Cutris together with its road infrastructure does not have documented correlates. That territory has rendered the stone spheres sculpted out of cobble rocks, usually taken from the riverbeds (Corrales 2016).

Also, beach dikes made by cobbles are consistent with pond traps for catching fish, crustaceans, turtles, and mollusks. These costal features have been documented in

Guanacaste, at Bahía Culebra, and the Central Pacific seashore in Quepos (Figs. 12.1 and 12.2d). The dikes were made by Pre-Columbian inhabitants with rocks readily available scoured from costal cliffs. They were placed in backwaters influenced by tidal flow with low energy of the surf. In the case of Culebra Bay, the erection of raps is dated to between 800 and 1350 A.D., in view of large seafood meadows absent in previous archeological phases of that region (Vázquez et al. 2019).

During the Colonial Period, the Spanish colonizers used these same patterns of movement for their own transportation and communication (Hurtado de Mendoza 2004; Benfer 2012; Salgado et al. 2016). As the Costa Rican population rose after the Conquest in the sixteenth century, but particularly after the Independence process (1821–1823), new lands and resources were exploited. In addition, new technologies were gradually adopted, which increased the impact of humans. At that time, the first landforms were produced by direct anthropogenic actions, associated with extensive farming (cultivation, grazing, and horticulture), human-made wildfire clearing, quarrying of volcanic and sedimentary rocks, building of several new roads, and gold mining. Moreover, in the twentieth and twenty-first centuries, hydrological interference (e.g., the building of dams, reservoirs, and channels), garbage deposition mountains, touristic complexes, including recreation lakes and gulf camps, and open mines were developed. On the marine littoral, for example, the construction of breakwaters, protection walls, and dikes were common (Hall 1984; Santillana 2008; Vargas 2014).

These landform modifications have produced indirect anthropogenic impact involving acceleration (or deceleration) of natural processes of erosion and sediment transport. One of the main environmental changes are land cover/use modification (e.g., cutting, bulldozing, burning, and grazing) that have accelerated rates of soil erosion in some regions and sedimentation in others. Moreover, other related processes include landslides triggered by human actions, the change of hydraulic regimes, and sediment concentration in rivers. These can affect the rate of erosion and sedimentation in the fluvial systems which, in turn, changes the dynamics of sand beaches and coral reefs.

12.3 Landforms and Processes

The effects of human activity on geomorphic processes, particularly those related to denudation/sedimentation, have been investigated at a global scale covering the past few centuries (Goudie 2020). There is evidence from different parts of the world showing that certain geomorphic processes are experiencing acceleration, especially since the mid-twentieth century (Goudie 2022). This suggests that a global geomorphic change is taking place, largely caused by anthropogenic landscape changes (Tarolli et al. 2019). Sedimentation has increased considerably in most regions and in all kinds of sedimentation environments (Moragoda and Cohen 2020). Although the link between denudation and sedimentation is not direct and unequivocal, it is safe to assume that if sedimentation rates increase in different regions during a given period, denudation must have increased too, even though their magnitudes could be different (Owens 2020). This augmentation, particularly marked from the second half of the last century onwards, appears to be determined mainly by land surface changes, in conjunction with climate change. Cendrero et al. (2022) suggested that the changes are clear because:

(a) There is evidence at a global scale of a growing response of geomorphic systems to socioeconomic drivers, such as Gross Domestic Product density, which is a good indicator of the human potential to cause such impacts.
(b) Land-use/cover changes enhance effects of climate change on global denudation/sedimentation and landslide/flood frequency and appear to be a stronger controlling factor.
(c) There is a global geomorphic change which is especially evident since the "great geomorphic acceleration" that began in the middle of the twentieth century and constitutes one of the characteristics of the proposed and informal most recent geologic time period, the Anthropocene.

The principal landscapes generated by anthropogenic geomorphology in Costa Rica could be classified, following Szabó (2010) and Castro et al. (2021), in three classes determined by their geomorphic expression: (i) excavated, (ii) leveled, and (iii) artificial landforms.

12.3.1 Excavation Landforms

Direct human-driven denudation (through activities involving excavation, transport, and accumulation of geologic materials) has increased worldwide by a factor of 30 between 1950 and 2015, representing a tenfold increase of per capita effect. Human-induced denudation, indirectly caused by alterations of the land surface, is currently at least ten times greater than denudation resulting from natural processes alone. Mass movement have also shown a clear intensification representing an important contribution to denudation, sediment generation, and landscape evolution (Cendrero et al. 2022). The frequency of hazardous events and disasters caused by intense rainstorm events has increased significantly in Costa Rica since the mid-twentieth century, often triggering more and frequent slope movements.

12.3.1.1 Quarries and Open-Pit Mines

The Bellavista gold mine has a quarry, mining lakes, and flumes. Most recently, there has been chaotic and illegal open-mining development in Crucitas (north of Costa Rica), which encourages manual extraction that is scattered and irregular (Ching 2014). The intensity of this mining produces several small holes, tunnels, and triggers deforestation, whose dimensions are relatively small, but with a strong environmental impact. The most evident landscape modifications are quarries of volcanic rocks (mainly lava and pyroclastic deposits) that are modified through excavations, blasting, and piling of waste material (Fig. 12.3). The result is a mosaic of excavated land represented by artificial cliffs, small lakes, terraces, and pits of sterile rocks, which increase the roughness of the terrain. There are also hundreds of quarries and open-pit mines in Costa Rica with legal authorization (Fig. 12.4).

12.3.1.2 Road Cuts

Anthropogenic slopes, such as cuttings and embankments for transportation and irrigation systems, can greatly impact regional slope–area relationships. These slopes often have a rectilinear shape and limited gradient range, particularly for embankments (26–32 °) (Tarolli et al. 2013; Brown et al. 2017). The first rural roads connecting villages and towns in the Valle Central were established in the sixteenth century. However, acceleration of international trade between 1844 and 1846 led to the construction of new roads for ox carts between Cartago, Valle Central, and the Puntarenas harbor. Additionally, secondary trails were established in coffee regions like Grecia and San Ramón. This road expansion continued through the end of the nineteenth century and early twentieth century, reaching places like Cañas, Liberia, Nicoya, and Quesada. The Interamerican road was built after World War II (Hall 1984; Flores 2001). Most roads in the country are concentrated in the Greater Metropolitan Area (GAM), but there are several critical highways that connect other regions, such as the Interamerican Highway stretching from Nicaragua to Panama, Route 32 connecting to the Caribbean coast, and Route 27 linking to the Central and South Pacific. Their building involved excavation of deep trenches into bedrock (Fig. 12.5).

Road cuts in Costa Rica are characterized by local but frequent slides (rock falls, rocks slides, and landslides) because of the recent volcanic and sedimentary geologic history, which also involved hydrothermal alteration weakening the rock, and the occurrence of thick soils which become unstable under heavy tropical conditions in the rainy season (May–December). The roads located on unstable slopes and in areas with high rainfall annual rates are commonly affected by mass movement processes (Fig. 12.6), as are old roads which are used by four-wheel drive vehicles, mules, ATVs, and motorcycles. There are also particular tracks used for four-wheel drive vehicles competitions, which also present a problem. Costa Rica is the number two in roads density in Central America (Fig. 12.7). The country is very well connected although it is divided by a mountain system that cross Costa Rica from NW to SE and hilly landscapes are more common than flatlands.

12.3.1.3 Artificial Channels

In the Caribbean lowlands near the homonymous coastline, artificial channels were constructed between the mouth of the Parismina River and Moín beach to connect the estuaries of five rivers: Pacuare, Madre de Dios, Matina, Vueltas, and Moín. These channels span a distance of 82 km, with a width of 1.5 km, and reach a length of 30 km from Tortuguero to Moín (Quesada-Román and Pérez-Briceño 2019). In the Northern Pacific region, near Cañas and Bagaces municipalities, there are about 255 km of artificial channels supplied by waters of Lake Arenal (Fig. 12.8a). In the Central Pacific, near the municipalities of Parrita

Fig. 12.3 Quarries and open-pit mines. **a** Miramar gold mine and **b** Cerro Chopo quarry in Cañas

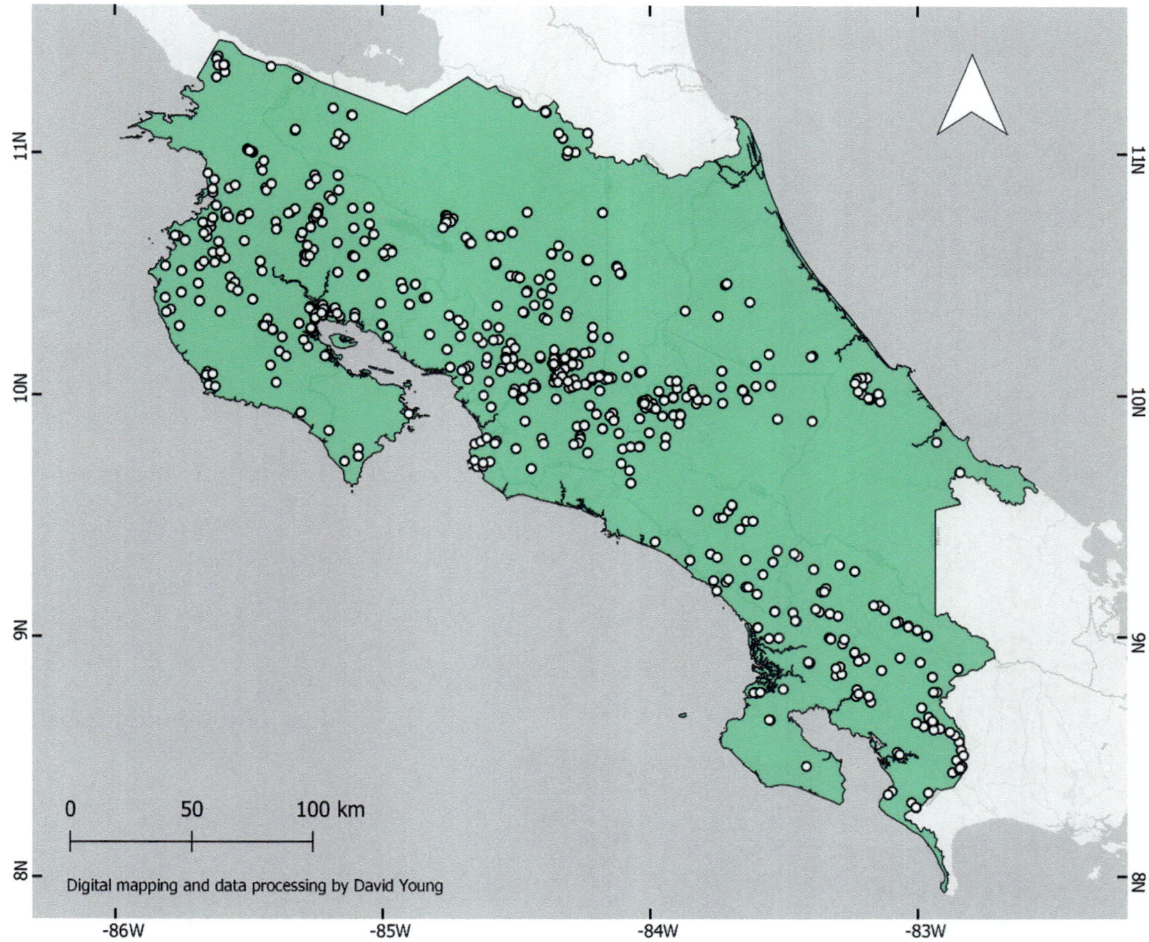

Fig. 12.4 Quarries and open-pit mines with legal authorization in Costa Rica. *Source* Dirección General de Geología y Minas. It is observed that in the Talamanca Mountain Range, Osa Peninsula, and the Northern Plains and Tortuguero, there are very few quarries and open-pit mines

and Quepos (Fig. 12.8b), the development of banana fields during the 1940s and following decades created extensive areas of lineal riverscapes (Royo 2004). In the municipality of Corredores in the Southern Pacific region, artificial channels have been built to modify the 380 km^2 of fluvial plains for the purpose of creating banana fields for the United Fruit Company (Zúñiga-Arias 2014). These modified riverscapes today are used mostly for rice, cattle, and oil palm croplands (Quesada-Román et al. 2023).

12.3.1.4 Recreational Excavated Lakes

Costa Rica is renowned for its stunning natural landscapes, which include many recreational small lakes that are perfect for various outdoor activities (Haberyan et al. 2003). These lakes boast magnificent views of their surroundings, making them ideal for picnics, fishing, kayaking, and swimming (Marcouiller and Prey 2005). Urban areas, like Parque de la Paz and La Sabana, also boast several such lakes. Recreational lakes offer numerous benefits to both local communities and visitors. These include water supply, economic development, environmental protection, outdoor recreation, scenic beauty, and flood control (Horn and Haberyan 2016). They provide vital resources for local communities, including irrigation, supply of drinking water, and hydroelectric power.

12.3.1.5 Treatment Ponds

Treatment ponds in Costa Rica are artificial water bodies used for wastewater treatment. The geomorphology of these ponds is largely influenced by the type of wastewater treatment process used, water chemistry, and the rate of sediment deposition (Vietz et al. 2016). For example, anaerobic ponds are shallower and tend to have a more complex bottom profile than aerobic ponds, which are deeper and have a more uniform bottom (Alfaro-Chinchilla et al. 2019). The environmental implications of treatment ponds can be both positive and negative. On the positive side, these ponds can improve water quality and help to prevent pollution of nearby rivers and lakes, thereby preserving the health of aquatic ecosystems. Additionally, they can serve as a source of irrigation water for agriculture, reducing the need for pumping water from rivers and other sources.

Fig. 12.5 Road cut in Route 27, km 47. Intense hydrothermal alteration and terracing to stabilize the slope are observed

However, treatment ponds can also have negative environmental impacts. For example, they can emit pollutants, such as nitrogen and phosphorus, which can contribute to eutrophication and the growth of toxic algae blooms (Gołdyn et al. 2015). They can also release pathogens and heavy metals, posing risks to human health and the environment. Furthermore, accumulation of sediment in these ponds can reduce the volume of water available for treatment, leading to the need for regular maintenance and cleaning. Therefore, it is crucial to implement effective management strategies for treatment ponds to minimize their negative impacts and maximize their positive impacts (Shilton 2006). This can include regular monitoring of water quality, proper maintenance and cleaning, and the use of appropriate treatment processes to minimize the release of pollutants. Additionally, the use of constructed wetlands and other sustainable wastewater treatment technologies can help to minimize environmental impacts of treatment ponds (Varma et al. 2022).

12.3.1.6 Tunnels, Galleries, and Ventilation Shafts

In Costa Rica, underground human-made structures such as mines, street and train tunnels, and ventilation shafts (Fig. 12.9) are in a way equivalents of caves and cave passages, subterranean rivers, lava tubes, sinkholes, vertical shafts, and volcanic pipes in natural geomorphology. Extensive mining activities in Costa Rica started about 1821 with the exploitation of gold and silver, but initial landscape modification was negligible, including only tunnels and subterranean galleries. However, this activity favored the settlement of sparsely populated towns and villages such as Desmonte, Las Juntas de Abangares, and Líbano. Las Juntas de Abangares grew through mining over time (1890-present), giving this town a particular cultural and geoheritage value (García 1984; Bundschuh et al. 2007; Castillo 2009).

The born economy generated first by gold mining and later by coffee production promoted the construction of railroads, including several train tunnels to the Pacific

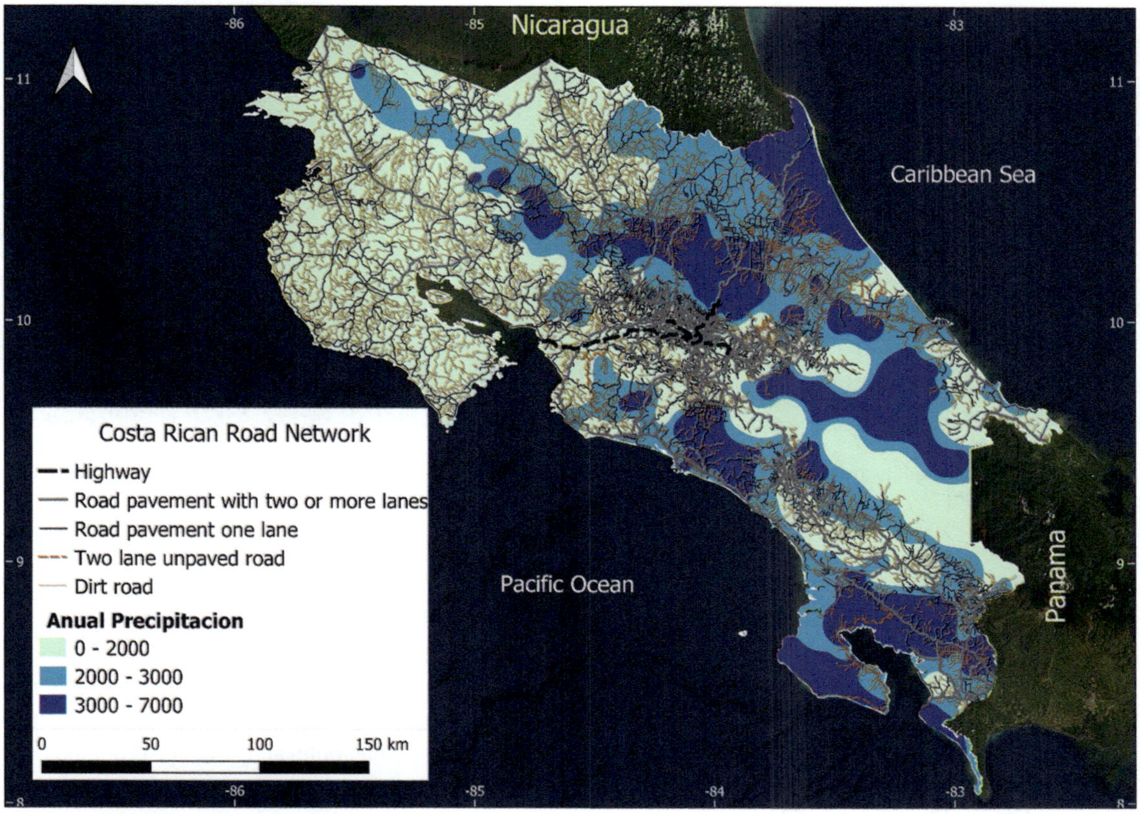

Fig. 12.6 Road network of Costa Rica clearly shows the high density of roads, particularly in the GAM (Greater Metropolitan Area). At the same time, national parks and reserves have very low density, even zero roads, such as Braulio Carrillo, La Amistad National Park (including Chirripó, Quetzales, and Tapantí), Corcovado, and Tortuguero

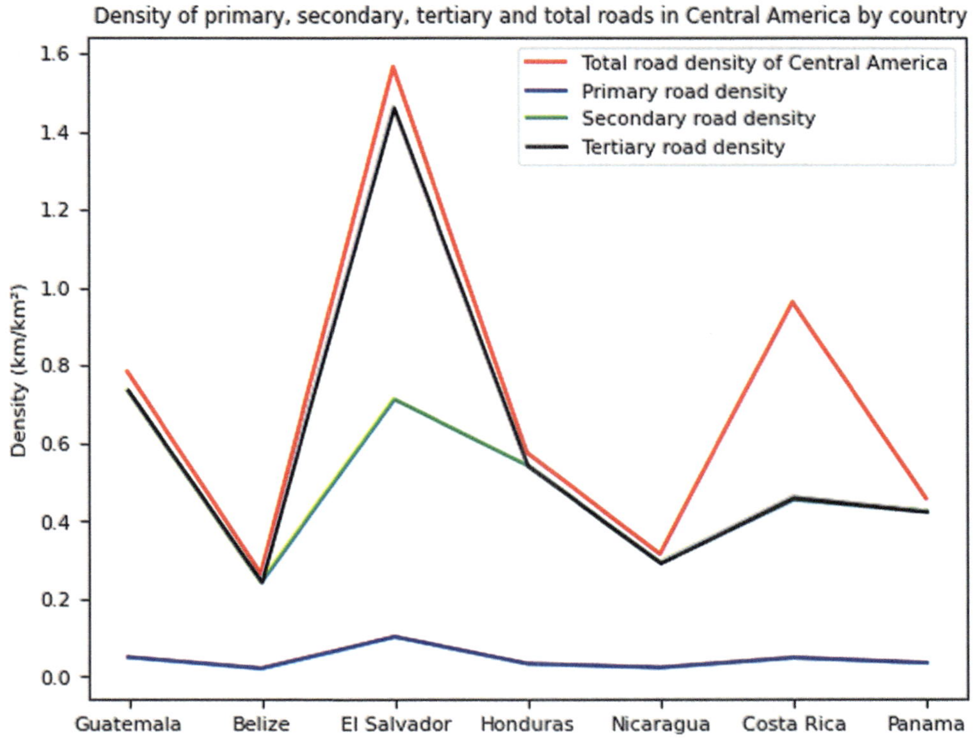

Fig. 12.7 Road density (primary, secondary, tertiary, and total roads) in Central America based on Open Street Maps. Despite being the fifth largest country in Central America, Costa Rica has the second highest density of roads in the region

Fig. 12.8 Artificial channels. **a** Cañas irrigation channels, **b** Parrita oil palm fields and their artificial channels to irrigate these extensive croplands (Credit: Drone pilot 506)

Ocean (1897–1910) such as the Cambalache and Carballo tunnels, and to the Caribbean Sea (1870–1890), with several tunnels near Peralta. Another less known old small train tunnel, now used by cars, is located in Chase (national route 801), at the Panama boundary, and was constructed in 1914 (45 m long, 4.25 m wide, and 4.45 m high) by Chiriquí Land Company. The very known Zurquí street tunnel (562 m long, 12 m wide, and 10 m high) was inaugurated in 1984.

More impressive and extensive was the construction of tunnels, galleries, and vertical shafts for hydroelectrical projects by the Costa Rican Electrical Institute (ICE) since 1950. According to the summary compiled by David Núñez, in total 117 km of underground human-made structures was constructed, including 26 vertical shafts (1078 m total length), 44 galleries (13,436 m total length), and 68 tunnels (102,821 m total length); the oldest tunnel is La Garita, and the longest one is Tapantí (14.9 km long).

12.3.2 Terrain Leveling

12.3.2.1 Airfields

The previous international airport was La Sabana in San José, with international operation between 1940 and 1955, and for local airplanes still during the first years of the decade of 1960. Now, Costa Rica is well connected through air transportation (Fig. 12.10), with its four main international airports: (a) the Juan Santamaría (also known for the pilots as Coco airport) in Alajuela, in operation since 1958 and with a landing track of 3.1 km long, (b) the Daniel Oduber in Liberia, in operation since 1975 with a landing track of 2.7 km long, (c) the Tobías Bolaños in San José, in operation since 1975 with a landing track of 1.6 km, and (d) Limón, an international airport but with irregular operation since end of the decade of 1960 past century and with a landing track of 1.7 km These airports serve as major hubs for both international and domestic flights.

In addition to these main airports, there are many airfields located in the Caribbean coast, particularly near the extensive pineapple and banana croplands. These airfields provide crucial access to the agricultural sector and allow for efficient crop dusting of extensive croplands. Furthermore, there are numerous airfields located near the Pacific coast, typically associated with large private ranches, crop dusting, and tourist beach destinations. These airfields provide convenient access to these properties and destinations, allowing owners and visitors to arrive and depart quickly and easily. The presence of these airfields is also a key factor in the development of tourism industry, attracting more visitors and helping to drive local economic growth (Fig. 12.11). Geomorphologically, associated landforms modifications are localized as roads, landing tracks, and other structures associated to the airport size and dynamic (Pijet-Migoń and Migoń 2018).

12.3.2.2 Urban Areas

Urban development is also associated with significant anthropogenic landform modifications (Thornbush 2015). Urban areas of Costa Rica are characterized by their unique geomorphology and environmental issues that stem from the region's complex topography and climate (Quesada-Román et al. 2021a). Costa Rica is in a tropical zone with a diverse landscape that includes mountainous areas, volcanic ranges, and coastal plains. The urban areas of the country, including the capital city of San José, are situated in valleys and basins, surrounded by rugged terrain and high mountain ranges. This topography makes it difficult to manage the flow of water and to prevent runoff and flooding in the cities (Quesada-Román 2022). Geomorphology of the region

Fig. 12.9 Tunnels. **a** Gold tunnel at Tres Hermanos Mine in 1975, Juntas de Abangares; **b** the former Costa Rican president, Cleto González (third from right to left), possibly at the Cambalache tunnel along the Pacific railroad, around 1909; **c** Zurquí pilot tunnel in 1980 cutting volcanic breccias and two volcanic dykes (~5 m thick); **d** coal mine at Tablazo (Aserrí), 1922. **e** Railroad tunnel at Chase (Talamanca, beside Sixaola River, Costa Rica, and Panama border), built by the United Fruit Company in 1914, now used by cars; **f** Reventazón hydroelectric dam tunnel made by ICE, 2012

also affects local climate, leading to high levels of humidity and temperature, which can exacerbate environmental issues such as air and water pollution. Additionally, the densely populated urban areas of Costa Rica are surrounded by areas of high conservation value.

Modern transformation of deforested land and urban spatial development in Costa Rica started in the sixteenth century, but increased significantly after 1821 (Hall 1984; Vargas 2014). At least 7165 km^2 (14% of the country) is now urbanized, in most cases leading to widespread

Fig. 12.10 Historical photographs of main airports of Costa Rica. **a** Predominance of vegetation around the old Sabana airport. **b** Construction of water infrastructure of the Juan Santamaría airport in Alajuela. **c** La Sabana airport years after expanding its size. **d** La Sabana airport with surrounding vegetation at the old airport. **e** Juan Santamaría airport starting operations in 1958. **f** Construction of ICE building (about 1971) nearby La Sabana airport in San José. Credit: Costa Rican Aeronautic Museum

removal of natural vegetation, creation of entirely artificial surfaces, and heavily modified drainage systems (Fig. 12.12; Gobierno de Costa Rica 2018). Impermeability of urban constructions (houses, edifices, and pavement roads) increases the volume of water runoff and discharge to the urban rivers. This in turn increases garbage transportation to the reservoirs and littoral beaches. These processes affect marine littoral hydrodynamics and sediment transport, resulting in coastal engineering and management interventions (e.g., groins, navigation structures, storm surge barriers, and dredging).

12.3.2.3 Croplands

Since the period of independence in 1821–1823, croplands in Costa Rica have undergone significant evolution and change. During the early years of independence, agriculture was the primary economic activity in the country, and much of the land was devoted to crops such as coffee, sugarcane, pastures, and tobacco (León 1948). Strong transformation of rural and urban areas promoted deforestation and agricultural development, first around the Central Valley and later in other areas (Hall 1984; Vargas 2014). The literature demonstrates that agricultural land use typically accelerates erosion tenfold to 100-fold and that physical erosion in old belts is not as rapid compared to young mountain belts having equivalent relief. Possibly, the most vulnerable lithologies have been removed by earlier erosion (Stallard 1995). This could be the case of older terrains along the Pacific coast of Costa Rica, with rocks spanning the Jurassic to the Eocene interval, compared to the Neogene–Quaternary magmatic front in the central column of the country.

In the twentieth century, the country underwent a period of modernization and industrialization, which led to a shift away from traditional agricultural practices and an increase in urbanization (Astorga et al. 2005). This resulted in the reduction of cropland, as forests and wetlands were cleared to make way for urban development and industrial activities (Fig. 12.13; Rosero-Bixby and Palloni 1998). However, in recent decades, there has been a resurgence of interest in sustainable agriculture and crop production in Costa Rica. The country has become a leader in eco-friendly and sustainable agriculture, with many farmers embracing organic farming methods and promoting agroforestry, which integrates trees and crops in a sustainable system (Nanni et al. 2019). Additionally, there has been an increase in the production of specialty crops such as organic coffee, cocoa, and fruits, which have become important export products (Ferreira et al. 2017).

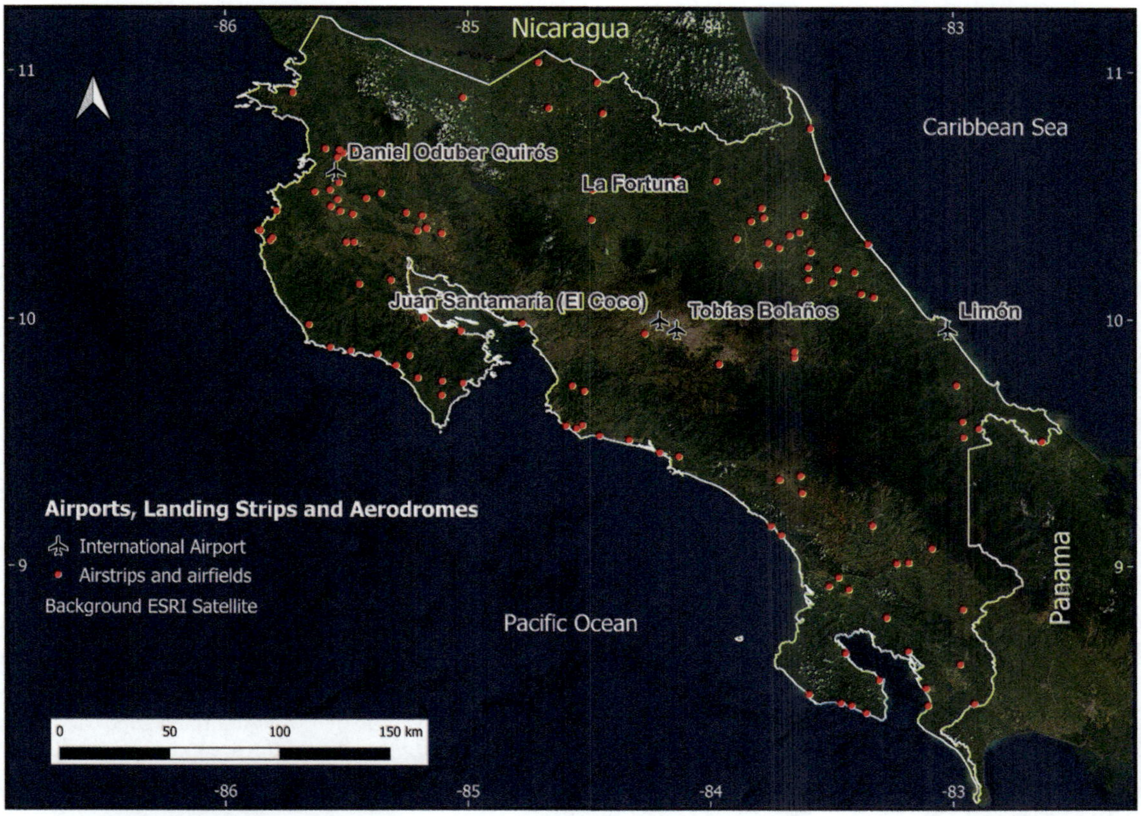

Fig. 12.11 Airports, landing strips, and aerodromes of Costa Rica

Overall, the evolution of croplands in Costa Rica has been marked by periods of expansion and contraction, as well as a shift toward more sustainable and eco-friendly agricultural practices (Fig. 12.14). Today, agriculture continues to play an important role in the national economy, providing livelihoods for many rural communities and supporting the development of sustainable tourism and conservation initiatives.

12.3.3 Artificial Landforms

12.3.3.1 Lake Reservoirs

Human activity can influence fluvial processes indirectly through land-use changes that modify flow discharge and sediment loads and river system connectivity. These approaches show that human impacts may either increase or decrease fluvial activity. River catchment developments (e.g., reservoir construction) may reduce coastal sediment supply (Syvitski et al. 2005). There are thousands of lakes in Costa Rica, with a total area of lacustrine wetlands of 166.3 km^2, and 113.3 km^2 of them are artificial (68% of the total; Veas-Ayala et al. 2022). The latter area includes at least 13 lake reservoirs of different sizes and depths (~5% of the territory). A summary of some of the principal lake reservoirs created for hydroelectrical energy is supplied by Segura (1998).

One of the most famous recreational lakes in the country is Lake Arenal, surrounded by verdant forests and crystal clear waters ideal for fishing and boating. Visitors can rent kayaks or take guided tours to admire its scenic beauty. In addition to being great for outdoor activities, Costa Rica's recreational lakes are also a heaven for nature lovers and adventure seekers (Quesada-Román and Pérez-Umaña 2020). Lake Arenal, for example, is situated in the northern highlands of Costa Rica, in the provinces of Guanacaste and Alajuela, near to the Arenal-Tilarán Conservation Area and is close to the Arenal volcano and Monteverde cloud forest (Fig. 12.15a, b). The lake is approximately 30 km long and almost 5 km at its widest point, making it the largest lake in Costa Rica at 87.8 km^2. Its depth varies between 30 and 60 m depending on the season and location (Castillo-Barahona 2015). Another important hydroelectric project was Reventazón with a lake reservoir of 8 km long and 7 km^2. Reventazón is the largest hydropower plant in Central America since 2017, and this project has been classified as an example of international good practice in hydropower sustainability, according to the International

Fig. 12.12 Greater Metropolitan Area (GAM in Spanish) where roughly 75% of the population of Costa Rica live in only 14% of the territory (Photograph credits: Ramón Masís-Campos). In the background is the Barva volcano

Hydropower Association. ICE (Costa Rican Electricity Institute), a government entity who oversees all the larger hydropower plants, has implemented watershed management programs to improve water quality, increase forest area, and control erosion (Fig. 12.15c). Through reforestation and Payments for Ecosystem Services (PES), the forest has increased significantly, which means less sediment volume lost than expected from the basins and more energy generation at the different plants.

These lakes also attract tourists, boosting local economic growth and creating job opportunities in the tourism industry. They help prevent soil erosion, reduce the impact of droughts, and provide habitats for wildlife. Moreover, recreational lakes offer a wide range of water-based activities and promote a healthy and active lifestyle, while also adding to the scenic beauty of the surrounding areas and providing opportunities for hiking, picnicking, and bird watching. Additionally, they can be used to store excess water during heavy rainfall, reducing the risk of flooding in nearby communities.

A geographic investigation indicates that dam impact in the tropics may be very different from that documented in temperate environments. Consequently, theories developed for temperate areas regarding expected dam impacts may not apply to tropical regions. This has important implications for hydrology, geomorphology, and ecology (Laurencio 2005). Therefore, the experience obtained at ICE in those fields is very important for other tropical and developing countries. In general, there is a clear link between urban development, denudation and sediment, and garbage transportation in Costa Rican rivers. Thus, if sedimentation rates increase in different regions, their magnitudes could affect the reservoirs and the beaches in different ways.

Fig. 12.13 Different croplands of Costa Rica. **a** banana fields in Parrita, **b** a mix of rice and oil palm surrounding Térraba-Sierpe wetland, **c** sugar cane plantation in Cañas, and **d** pineapple plantations in Buenos Aires (Credits to Manuel Camacho, Néstor Veas, and Daniel Quesada)

12.3.3.2 Garbage Disposal Hills

Three major garbage disposal hills are located in San José. The first one, operating in the early 1960 and perhaps before, was located in Pavas and is now a sport center of the Municipalidad de San José. The second one was Río Azul (La Unión, Cartago) that operated between 13 August 1973 and July 2007 (Fig. 12.16; Mora and Mora 2003). The area of the Pavas anthropogenic hill is now difficult to estimate but at least more than one hectare, and the Río Azul is considerably larger, 40 ha. Nowadays, the operating garbage dump is located in La Carpio, eastwards San José. There are other minor garbage disposal hills in the GAM or main cities of the country. In the Torres River catchment in San José, Quesada-Román et al. (2021b) reported that active quarries, old quarries, escarpments, and terraces have become clandestine garbage deposition sites for nearby informal settlements and municipalities that in the past deposited their solid waste in these areas.

12.3.3.3 Earth and Rockfill Dams

Sangregado dam, part of the Arenal-Corobicí-Miguel Dengo hydropower plants, is an earth core rockfill dam with a height of 66 m, crest length of 1012 m, crest width of 8 m, base width of 553 m, and rockfill volume of 4.66×10^6 m^3. Reventazón has a concrete face rockfill dam with a fill volume of 9×10^6 m^3 (loose materials), 130 m high and 535 m base width. Sandillal (part of Miguel Dengo hydropower plant) is a core rockfill dam with 470 m long. Other earth dams are smaller and mostly related to gold mines or local constructions.

12.3.3.4 Levees

Fluvial protection measures are implemented to prevent damage from floods along rivers, including the construction of levees. Levees are embankments built along the sides of rivers to contain water and prevent it from overflowing onto nearby land (Gupta and Ahmad 1999). These

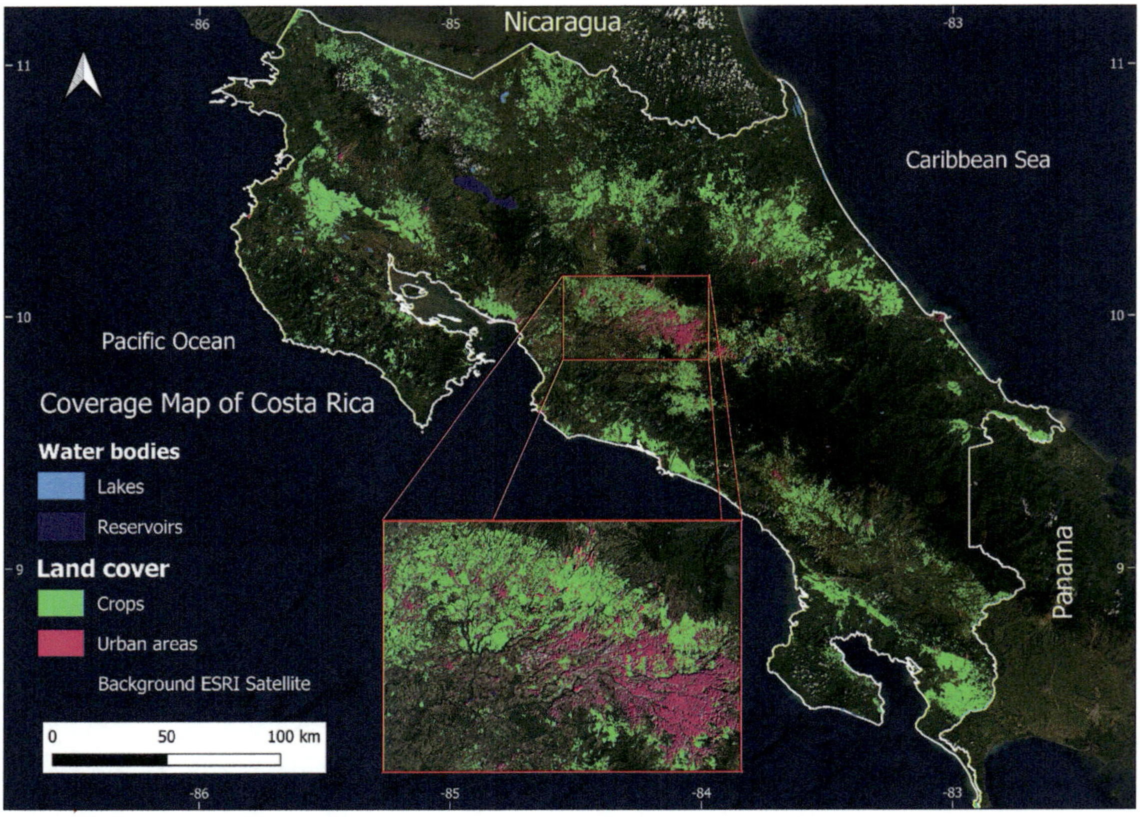

Fig. 12.14 Water bodies and land cover of Costa Rica. The box indicates the GAM

structures are designed to withstand high water levels and protect nearby communities, infrastructure, and crops from damage (Hudson 2017). Several examples exist to protect towns as Parrita, Sixaola, Nosara, Ciudad Neily, San Isidro del General, Turrialba, Siquirres, Aguas Zarcas, and several others (Fig. 12.1). In addition to levees, other measures such as the creation of retention basins, river channelization, and the improvement of drainage systems are used in Costa Rica to manage the flow of water and reduce the risk of flooding. These measures help to protect against the negative impacts of floods and promote sustainable development in the country. However, it is important to note that fluvial protection measures should be integrated into a comprehensive approach to river management that considers the needs of the environment, the local community, and other stakeholders (Bark et al. 2021).

12.3.3.5 Coastal Management

Coastal management to address sea level rise and coastal erosion is an important issue in Costa Rica, where many communities and infrastructure are located near the coast. Sea level rise, which is a result of climate change, can lead to the loss of coastal land and erosion of beaches and cliffs, while also increasing the risk of flooding and other coastal hazards.

To address these challenges, the national and local governments, the private sector, and NGOs have implemented several coastal management measures, including the construction of seawalls and groins, and the creation of artificial reefs (Barrantes-Castillo et al. 2020; Arias-Godínez et al. 2021). These measures help to protect coastal communities and infrastructure, while also promoting sustainable use of coastal resources. In addition, the government is working with local communities, NGOs, and other stakeholders to implement integrated coastal management (ICM) programs that address increasingly complex coastal issues (Calleja-Apéstegui and López-Arias 2022). These programs often involve the development of land-use planning policies and the implementation of ecosystem-based approaches to coastal management, such as the restoration of natural habitats and creation of green infrastructure (Yaney-Keller et al. 2019).

It is important to note that while coastal management measures can help to reduce the risk of coastal hazards, they should be integrated into a comprehensive approach to coastal management that considers the needs of the environment, local communities, and other stakeholders (Acuña-Piedra and Quesada-Román 2021). This will help to ensure that coastal management practices are sustainable and effective in addressing challenges posed by sea level rise and coastal erosion.

Fig. 12.15 **a** Artificial lakes. **b** Lake Arenal reservoir at the bottom of the Arenal volcano. **c** Reventazón reservoir in 2016

Marinas in Costa Rica, such as Marina Papagayo, Marina Los Sueños in Herradura, Marina Quepos, and Marina Golfito, have had a significant impact on the country's economy and tourism industry (Fig. 12.17). However, it is important to carefully consider and mitigate any negative environmental impacts (e.g., damaging coral reefs, disturbing wildlife habitats, and water pollution) that may result from their development and operation (Radulovic 2022).

12.4 Conclusions and Future Research

At least since Holocene, there is indirect, direct, and extensively evidence available of early Pre-Columbian anthropic modification of landscape and vegetation in Costa Rica, particularly at 4.5 ka associated with the introduction of sedentarism and agriculture and its extensive development about 3 ka b.P. This includes building of the first villages, monuments, roads, stone fishing traps, and trails. Later, particularly in the nineteenth and early twentieth centuries, mining of volcanic rocks contributed to local landscape modification, among several other small and large infrastructure constructions.

Following the work of Castro et al. (2021), in the case of Las Juntas de Abangares, landscapes that were altered through gold mining over time have a cultural and heritage value and, therefore, compose the mining-related geoheritage. Major modification of the landscape has occurred since the second half of the twentieth century and the first decades of the twenty-first century, particularly in relation to the development of agriculture and pastures, quarries, building of dams, reservoirs, new roads, and urban expansion.

Multidisciplinary research approach, involving geologists, civil engineers, archeologists, geomorphologists, and geographers, is fundamental to evaluate the extent and long-term consequences of anthropogenic alterations at the landscape scale. Human-made landforms will continue growing in future, and these actions are inducing cascading

Fig. 12.16 Example of artificial accumulation landform is the Río Azul garbage disposal hills, now under WPP enterprise control (Credit: Ramón Masís-Campos)

Fig. 12.17 Yacht marinas and associated engineering coastline modifications. **a** Papagayo, **b** Quepos

environmental changes. These results should also serve to promote the implementation of socioeconomic tasks, including a better territorial planning, environmental protection, and nature conservation.

LiDAR technology and similar applications have become a fundamental tool to aid in mapping anthropogenic structures, including Pre-Columbian trails and other cultural interventions to create a three-dimensional geospatial database to document the true extent of humans as geomorphic agents in Costa Rica. The next goal is quantification of the degree of anthropogenic surface transformation within the concept of hemeromorphy (Rózsa et al. 2020). This methodology and its interpretation make the comparison of landscape units possible according to their anthropogeomorphological transformation, independently of the intensity and quality of their geomorphological processes. One of the best ways to evaluate the effect of human activities is through studies of paired watersheds. Such studies simultaneously address effects of human activities as well as the importance of denudation in smaller basis as described by Milliman and Syvitski (1992).

Future challenges in anthropogenic geomorphology in developing tropical countries like Costa Rica are numerous. One of the most evident trends is deforestation and conversion of forests to agricultural land, urban areas, and other uses. This can lead to soil erosion, changes in local water balances, and impacts on biodiversity. These countries are often more vulnerable to the impacts of rising temperatures, changes in precipitation patterns, and sea level rise.

Developing tropical countries are also more vulnerable to disasters such as hurricanes, earthquakes, and volcanic eruptions, which can have significant impacts on geomorphological processes and the built environment. An increasing demand for infrastructure in the Global South can lead to changes in the natural landscape, including the construction of roads, bridges, and other structures, which can have significant impacts on local geomorphology and hydrology. Undeveloped countries usually lack the resources to properly manage and mitigate the impacts of human activities on the Earth's surface (Campos et al. 2022; Quesada-Román and Peralta-Reyes 2023). This includes a lack of funds, trained personnel, and technology to monitor and manage changes in the landscape. In contrast to the anthropic disturbances that have shaped Costa Rica's landforms and landscapes since prehistoric times, contemporary alterations exhibit a commendable shift due to cultural and educational level of citizenship, and large extensions of natural continental and marine protected areas, which is characterized by a profound respect for nature and meticulous environmental planning, mitigating the impact of strong transformations.

Acknowledgements The authors thank Hugo Rodríguez, Manuel Peralta, Tatiana Ramírez, and David Young for the photographs search, spatial analyses, and mapping; Drone pilot 506, Manuel Camacho, Néstor Veas, Daniel Quesada, and Ramón Masís-Campos for their photographs; Russell Lee Losco, Laura Calderón, Silvia Salgado, Ricardo Vázquez, and David Núñez for their useful friendly review.

References

Acuña A (1987) Relación entre asentamientos precolombinos al norte de Guayabo de Turrialba en la Fase La Cabaña. Revista De Ciencias Sociales 35:43–52

Acuña-Piedra JF, Quesada-Román A (2021) Multidecadal biogeomorphic dynamics of a deltaic mangrove forest in Costa Rica. Ocean Coast Manag 211:105770

Aguilar CH (1972) Guayabo de Turrialba arqueología de un sitio indígena prehispánico. Editorial Costa Rica, San José

Alarcón G (2019) Desarrollo de la jerarquización social precolombina en Guayabo de Turrialba, vertiente del Caribe Central de Costa Rica. Doctoral dissertation, Programa de Maestría y Doctorado en Estudios Mesoamericanos, Universidad Nacional Autónoma de México, México City

Alfaro-Chinchilla MC, Villalobos-Sandí N, Pérez-Salazar R, Scholz C (2019) Sistemas de lagunaje como tecnología alternativa para el tratamiento de aguas residuales: vigencia de esta técnica en Costa Rica. I Congreso Internacional de Ciencas Exactas y Naturales, Universidad Nacional de Costa Rica. Heredia, Costa Rica

Arford MR, Horn SP (2004) Pollen evidence of the earliest maize agriculture in Costa Rica. J Lat Am Geogr 3(11):108–115

Arias-Godínez G, Jiménez C, Gamboa C, Cortés J, Espinoza M, Beita-Jiménez A, Alvarado JJ (2021) The effect of coral reef degradation on the trophic structure of reef fishes from Bahía Culebra, North Pacific coast of Costa Rica. J Coast Conserv 25(1):8

Astorga P, Bergés AR, FitzGerald V (2005) The standard of living in Latin America during the twentieth century 1. Econ Hist Rev 58(4):765–796

Bark RH, Martin-Ortega J, Waylen KA (2021) Stakeholders' views on natural flood management: implications for the nature-based solutions paradigm shift? Environ Sci Policy 115:91–98

Barrantes-Castillo G, Arozarena-Llopis I, Sandoval-Murillo LF, Valverde-Calderón JF (2020) Critical beaches due to coastal erosion in the Caribbean south of Costa Rica during the period 2005–2016. Revista Geográfica De América Central 64:75–102

Benfer AK (2012) Interregional landscapes of movement and the La Unión Archaeological District of Northeastern Costa Rica. MA thesis in Anthropology, Univ Kansas

Brown AG, Tooth S, Bullard JE, Thomas DSG, Chiverrell RC, Plater AJ, Murton J, Thomdycraft VR, Tarolli P, Rose J, Wainwright J, Downs P, Aalto R (2017) The geomorphology of the Anthropocene: emergence, status and implications. Earth Surf Proc Land 42:71–90

Bundschuh J, Birkle P, Finch R, Day M, Romero J, Paniagua S, Alvarado GE, Bhattacharya P, Tippman K, Chaves D (2007) Geology-related tourism for sustainable development. Central America Geol Resour Hazards 2:1014–1098

Calleja-Apéstegui F, López-Arias F (2022) Inventario de humadales para protección costera (IHPC): una herramienta para la gestión costera en Costa Rica. Ingeniería 32(2):32–51

Campos N, Quesada-Román A, Granados-Bolaños S (2022) Mapping mountain landforms and its dynamics: study cases in tropical environments. Appl Sci 12(21):10843

Castillo A (2009) La Guerra del oro. Tierra y minería en Abangares 1890–1930. EUCR, San José

Castillo-Barahona R (2015) Canal del Sur-Cañas, Guanacaste. Avance en la construcción del canal de riego. Boletín Estructuras 1:1–3

Castro PTA, Nascimento ST, de Paula SF (2021) Classification of geo-mining heritage based on anthropogenic geomorphology. J Geol Surv Brazil 4(1):13–20

Cavallini C (2011) El camino precolombino del Sitio Arqueológico Alto del Cardal C-304 AC. Cuadernos De Antropología 21:1–8

Cendrero A, Remondo J, Beylich A, Cienciala P, Forte L, Golosov V, Gusarov A, Kijowska-Strugala M, Laute K, Li D, Navas A, Sodati M, Vergari F, Zwoliriski Z, Dixon J, Knight J, Nadal-Romero E, Placzkowska E (2022) Denudation and geomorphic change in the Anthropocene; a global overview. Earth Sci Rev 233:104186

Ching EW (2014) Costa Rica: the first Latin American country free of open-pit gold mining. In: The Earth charter, ecological integrity and social movements, pp 256–269, Routledge

Clement RM, Horn SP (2001) Pre-Columbian land-use history in Costa Rica: a 3000-year record of forest-clearance, agriculture and fires from Laguna Zoncho. Holocene 11:419–426

Cooke R (2005) Prehistory of native Americans on the Central American land bridge: colonization, dispersal, and divergence. J Archaeol Res 13(2). https://doi.org/10.1007/S10804-005-2486-4

Corrales F (2016) La Gran Chiriquí: una historia cada vez más profunda. Canto Rodado 11–2016:27–58

Creamer W, Haas J (1985) Tribe versus chiefdom in lower Central America. Am Antiq 50(4):738–754

Fernández P, Alvarado GE (2006) Artesanos y piedras. Herramientas y escultura precolombina en Costa Rica. Craftsmen & stones. Pre-Columbian stone tools and sculpture in Costa Rica. Fundación Museos del Banco Central

Ferreira GFC, Fuentes PAG, Ferreira JPC (2017) The successes and shortcoming of Costa Rica exports diversification policies. Background paper to the UNCTAD-FAO commodities and development report

Flores E (2001) Geografía de Costa Rica. EUNED, San José

Fonseca O (1996) Historia antigua de Costa Rica: surgimiento y caracterización de la primera civilización costarricense. Editorial de la Universidad de Costa Rica, San José

García G (1984) Las minas de Abangares. Historia de una doble explotación. EUCR, San José

Gobierno de Costa Rica (2018) Política Nacional de Desarrollo Urbano 2018-203. San José, Costa Rica, pp 28

Gołdyn B, Chudzińska M, Barałkiewicz D, Celewicz-Gołdyn S (2015) Heavy metal contents in the sediments of astatic ponds: influence of geomorphology, hydroperiod, water chemistry and vegetation. Ecotoxicol Environ Saf 118:103–111

Goudie A (2020) The human impact in geomorphology–50 years of change. Geomorphology 366:106601

Goudie A (2022) The impacts of humans on geomorphology. Geol Soc London Mem 58(1):121–134

Goudie A (2010) Foreword. In: József Szabó J, Lóránt Dávid L, Lóczy D (eds.) Anthropogenic geomorphology. A guide to man-made landforms, pp v–vi

Guerrero, JV, Solís F (1997) Los pueblos antiguos de la zona Cañas-Liberia del año 300 al 1500 después de Cristo. Museo Nacional de Costa Rica San José

Gupta A, Ahmad R (1999) Geomorphology and the urban tropics: building an interface between research and usage. Geomorphology 31(1–4):133–149

Haberyan KA, Horn SP, Umaña G (2003) Basic limnology of fifty-one lakes in Costa Rica. Rev Biol Trop 51(1):107–122

Hall C (1984) Costa Rica: una interpretación geográfica con perspectiva histórica. EUCR

Hoopes JW (1994) The Tronadora Complex: Early Formative ceramics in northwestern Costa Rica. Lat Am Antiq 5(1):3–30

Horn SP, Haberyan KA (2016) Lakes of Costa Rica. In: Kappelle M (ed) Costa Rican ecosystems, pp 656–682

Hudson PF (2017) Fluvial depositional processes and landforms. Int Encycl Geogr People Earth Environ Technol 1–9

Hurtado de Mendoza L, Alvarado GE (2021) Notas sobre los factores naturales y culturales en el desarrollo sociopolítico prehispánico en el extremo noroeste de Guanacaste, Costa Rica. Revista de Biología Tropical 69 (Suppl. 2): S15–S28. https://doi.org/10.15517/rbtv69iS2.48739

Hurtado de Mendoza L (2004) Guayabo: historia Antigua de Turrialba. Litografía e Imprenta Lil S. A., Tibás, Costa Rica

Lange FW (1992) The search for elite personages and site hierarchies in Greater Nicoya. In: Lange FW (ed) Wealth and hierarchy in the intermediate area. Washington DC, Dumbarton Oaks, pp 109–139

Laurencio LR (2005) The distribution of dams in Costa Rica and their hydrologic impacts. M.Sc. thesis, Graduate Studies of Texas A&M University

León J (1948) Land utilization in Costa Rica. Geogr Rev 38(3):444–456

Marcouiller DW, Prey J (2005) The tourism supply linkage: recreational sites and their related natural amenities. J Reg Anal Policy 35(1):23–32

McKee BR, Sever TL, Sheets PD (1994) Prehistoric footpaths in Costa Rica: remote sensing and field verification. In: Sheets PD, McKee BR (eds) Archaeology, volcanism, and remote sensing in the Arenal region, Costa Rica. University of Texas, pp 142–157

Milliman JD, Syvitski JPM (1992) Geomorphic/tectonic control of sediment discharge to the ocean: the importance of small mountainous rivers. J Geol 100:525–544

Mora R, Mora R (2003) Reseña histórica del Relleno Sanitario de Río Azul y consideraciones sobre los metales pesados tratados en él y los presentes en nuestros hogares. Rev Reflexiones 82(2):47–58

Moragoda N, Cohen S (2020) Climate-induced trends in global riverine water discharge and suspended sediment dynamics in the 21st century. Glob Planet Change 191:103199

Nanni AS, Sloan S, Aide TM, Graesser J, Edwards D, Grau HR (2019) The neotropical reforestation hotspots: a biophysical and socioeconomic typology of contemporary forest expansion. Glob Environ Change 54:148–159

Núñez Y (2020) Intercambio interregional y configuración arquitectónica en Lomas Entierros, Pacífico Central de Costa Rica. Vínculos 40(1–2):35–66

Owens PN (2020) Soil erosion and sediment dynamics in the Anthropocene: a review of human impacts during a period of rapid global environmental change. J Soils Sediments 20:4115–4143

Peytrequín J, Aguilar M (2007) Los indicadores arqueológicos de un modo de vida cacical en el sitio Agua Caliente. Vínculos 30(1–2):57–81

Pijet-Migoń E, Migoń P (2018) Landform change due to airport building. In: Urban geomorphology. Elsevier, Amsterdam, pp 101–111

Piperno DR (2011) The origin of plant cultivation and domestication in the New World tropics: patterns, process, and new developments. Curr Anthropol 52(S4):S453–S470

Quesada-Román A (2022) Disaster risk assessment of informal settlements in the Global South. Sustainability 14(16):10261

Quesada-Román A, Peralta-Reyes M (2023) Geomorphological mapping global trends and applications. Geographies 3(3):610–621

Quesada-Román A, Pérez-Briceño PM (2019) Geomorphology of the Caribbean coast of Costa Rica. J Maps 15(2):363–371

Quesada-Román A, Pérez-Umaña D (2020) State of the art of geodiversity, geoconservation, and geotourism in Costa Rica. Geosciences 10(6):211

Quesada-Román A, Villalobos-Portilla E, Campos-Durán D (2021a) Hydrometeorological disasters in urban areas of Costa Rica. Central Am Environ Hazards 20(3):264–278

Quesada-Román A, Castro-Chacón JP, Feoli-Boraschi S (2021b) Geomorphology, land use, and environmental impacts in a densely populated urban catchment of Costa Rica. J S Am Earth Sci 112:103560

Quesada-Román A, Umaña-Ortíz J, Zumbado-Solano M, Islam A, Abioui M, Tefogoum GZ, Kariminejad N, Mutaqin BW, Pupim F (2023) Geomorphological regional mapping for environmental planning in developing countries. Environ Dev 48:100935
Radulovic A (2022) Smart technology applied in the management of yachting marinas. Int J Marit Eng 164(A1):81–90
Rosero-Bixby L, Palloni A (1998) Population and deforestation in Costa Rica. Popul Environ 149–185
Royo A (2004) La ocupación del pacífico sur costarricense por parte de la compañía bananera (1938–1984). Diálogos, Revista Electrónica De Historia 4(2):1–29
Rozsa P, Incze J, Balogh S, Novák TJ (2020) A novel approach to quantifying the degree of anthropogenic surface transformation—the concept of hemeromorphy. Erdkunde 74(1):45–57
Salgado S, Ibarra E, Mesén R (2016) De tierras altas a tierras bajas: propuesta y exploración de una ruta prehispánica y colonial en el Caribe de Costa Rica. Vínculos 37(1–2):133–161
Salgado S, Hoopes JW, Aguilar M, Fernández P (2021) Nuevo Corinto: a chiefly village in Northeastern Costa Rica. J Caribb Archaeol 21:2–32
Segura B (1998) Presas: guía práctica de diseño. Instituto Costarricense de Electricidad. Departamento de Investigación y Desarrollo. ETCR, Cartago, Costa Rica
Santillana (2008) La Enciclopedia del Estudiante. Geografía de Costa Rica (Tomo 20). Santillana S.A., Peru
Sheets PD (2008) Memoria social perdurable a pesar de desastres volcánicos en el área de Arenal. Vínculos 31(1–2):1–26
Sheets PD (2011) Pilgrimages and persistent social memory in spite of volcanic disasters in the Arenal area, Costa Rica. Anc Mesoam 22:435–435
Sheets PD (2012) Responses to explosive volcanic eruptions by small to complex societies in ancient Mexico and Central America. In: Sheets PD, Cooper J (eds) Surviving sudden environmental change: answers from archaeology. University of Colorado Press, Boulder, pp 43–63
Sheets PD, McKee BR (1994) Archaeology, volcanism, and remote sensing in the Arenal region, Costa Rica. The University of Texas Press, Austin, Texas
Sheets PD, Sever TL (2007) Creating and perpetuating social memory across the ancient Costa Rican landscape. Remote Sens Archaeol 161–184
Sherlock RL (1922) Man as a geological agent: an account of his actions on inanimate nature. HF and G Witherby, London
Shilton A (2006) Pond treatment technology, vol 5. IWA Publishing
Snarskis MJ (1981) The archaeology of Costa Rica. In: Benson E (ed) Between continents/between seas: precolumbian art of Costa Rica. Harry N. Abrams, New York, pp 15–84
Solís F, Herrera A, Guerrero JV (2019) La Fábrica un sitio temprano con arquitectura en el noroeste del Valle Central, Costa Rica. San José, Museo Nacional de Costa Rica
Stallard RF (1995) Tectonic, environmental, and human aspects of weathering and erosion: a global review using a steady-state perspective. Annu Rev Earth Planet Sci 23:11–39
Syvitski JP, Vörösmarty CJ, Kettner AJ, Green P (2005) Impact of humans on the flux of terrestrial sediment to the global coastal ocean. Science 308(5720):376–380
Szabó J (2010). Anthropogenic geomorphology: subject and system. In: Szabó J, Dávid L, Lóczy D (eds) Anthropogenic geomorphology. A guide to man-made landforms, pp 3–10
Tarolli P, Calligaro S, Cazorzi F, Dalla Fontana G (2013) Recognition of surface flow processes influenced by roads and trails in mountain areas using high-resolution topography. Eur J Remote Sens 46:176–197
Tarolli P, Cao W, Sofia G, Evans D, Ellis EC (2019) From features to fingerprints: a general diagnostic framework for anthropogenic geomorphology. Prog Phys Geogr 43(1):95–128
Thornbush M (2015) Geography, urban geomorphology and sustainability. Area 47:350–353
Vargas G (2014) Geografía de Costa Rica. EUNED, San José
Varma VG, Jha S, Raju LHK, Kishore RL, Ranjith V (2022) A review on decentralized wastewater treatment systems in India. Chemosphere 134462
Vázquez R (2023) Arquitectura demostrativa y asamblearia en el contexto paisajístico del sitio arqueológico Guastomatal, Mansión de Nicoya, Costa Rica. Cuadernos De Antropología 33(2):1–21
Vázquez R, Chapdelaine C (2008) Arquitectura, Caminos Empedrados y Cronología del Sector Principal del Sitio Las Mercedes-1, Caribe Central de Costa Rica. Vínculos 31:27–77
Vázquez R, Rosenswig RM (2017) El sitio arqueológico Las Mercedes: surgimiento de un importante centro sociopolítico en Línea Vieja, vertiente Caribe Central de Costa Rica. Canto Rodado 11–2016:101–133
Vázquez R, Guerrero JV, Sánchez JC (2003) Cutris: Descripción, cronología y afiliación de un centro arquitectónico con caminos monumentales en la Llanura de San Carlos, Costa Rica. Vínculos 28(1–2):149–174
Vázquez R, Solís F, Herrera A, Hardy ET (2019) Trampas prehispánicas para pesca litoral de la Bahía Culebra, Noroeste de Costa Rica. Vínculos 39(1–2):1–26
Vázquez R, Sánchez JC, Massey H (2021) El gran camino de Guayabo en Turrialba, Costa Rica: amplitud exploratoria, asentamientos relacionados, esfera de interacción y análisis comparativo. Vínculos 41(1–2):97–132
Vázquez R, Rosenswig RM, Brenes J, Alfaro I, Ramírez M (2022) Desarrollo derivativo en la arquitectura cacical de Línea Vieja: dos complejos vecinos del sitio arqueológico Anita Grande, Caribe Central de Costa Rica. Cuadernos De Antropología 32(2):1–44
Vázquez R (2014). Turrialba, una encrucijada: arqueología regional y rutas antiguas de comunicación en un valle del trópico húmedo de Costa Rica (11.000 a.C.–1600 d.C.). Doctoral dissertation, University at Albany, State University of New York
Veas-Ayala N, Alfaro-Córdoba M, Quesada-Román A (2022) Costa Rican wetlands vulnerability index. Prog Phys Geogr 47(4):521–540
Vietz GJ, Walsh CJ, Fletcher TD (2016) Urban hydrogeomorphology and the urban stream syndrome: treating the symptoms and causes of geomorphic change. Prog Phys Geogr 40(3):480–492
Yaney-Keller A, Santidrián Tomillo P, Marshall JM, Paladino FV (2019) Using unmanned aerial systems (UAS) to assay mangrove estuaries on the Pacific coast of Costa Rica. PLoS ONE 14(6):e0217310
Zúñiga-Arias AY (2014) Ríos, cultivos e inundaciones: historia reciente del cantón de Corredores, Costa Rica, 1950–2010. Ambiente y Desarrollo 18(34):77–92

13 Geoheritage, Geoconservation, and Geotourism Promotion in Costa Rica

Dennis Pérez-Umaña and Adolfo Quesada-Román

Abstract

Tourism became one of the most important sectors of the economy in Costa Rica since the end of the 1980s. It is from the massive arrival of international tourists attracted by the ecotourism offer based on a robust national system of conservation areas as well as from a rapid adaptation of both public and private sectors that investments in projects to improve the conditions of access, dissemination, and diversification of tourist attractions throughout the country have been increasing steadily. This work shows the main geoheritage and geodiversity characteristics of Costa Rica and focus on an evaluation of the 28 national parks (NP) of the total protected areas of the National System of Conservation Areas (SINAC) for a landscape and aesthetic, cultural, didactic, scientific, and tourism evaluation as potential geoparks. The evaluation determined that 13 of the NPs obtained a high rating, 11 were catalogued as medium, whilst four obtained a low rating. This initiative is an original and novel contribution for Costa Rica since it determines a baseline in the study of geosites and geoparks in the country. In addition, it shows the weaknesses and strengths of each national park in its objective to attract more visitors in the future.

Keywords

Geosites · Geoheritage · Geoconservation · Geoparks · National system of conservation areas · National parks

D. Pérez-Umaña
Comisión Nacional de Prevención de Riesgos y Atención de Emergencias, Frente a Aeropuerto Internacional Tobías Bolaños, San José, Costa Rica

A. Quesada-Román (✉)
Laboratorio de Geografía Física, Escuela de Geografía, Universidad de Costa Rica, 2060 San José, Costa Rica
e-mail: adolfo.quesadaroman@ucr.ac.cr

13.1 Introduction

Costa Rica attracts more than 2 million tourists per year, who bring an investment of about 2.1 billion dollars to the country (Sada 2015; Hunt and Harbor 2019). Since the colonial period, ecosystems and geoforms present in the country had been used for intensive and extensive agricultural purposes. In the 1960s, Costa Rica's environmental policy changed to reverse this phenomenon and subject large areas of land of water, ecosystem, and landscape importance under conservation (Zeledón 1999; Quesada-Román and Pérez-Umaña 2020a; Acuña-Piedra and Quesada-Román 2021).

The country has a National System of Conservation Areas (SINAC), which divides the country into conservation areas and administers protected areas according to their management categories, which include forest reserves, protected zones, national parks, biological reserves, wildlife refuges, wetlands, natural monuments, marine reserves, and marine management areas. Every existing regulation in the country refers to conservation focused on biotic elements, but there is a gap in protection of certain abiotic resources, such as geological and geomorphological elements; in fact, the situation is similar to other Latin American countries (Palacio 2013; Rosado-González et al. 2020; Quesada Román et al. 2022). For this reason, projects have emerged to identify abiotic elements that should be protected such as the Global Indicative List of Geological Sites (GILGES) and IUGS Geological Heritage Sites, which were promoted by the International Union of Geological Science (IUGS) and the International Association for the Conservation of Geological Heritage (ProGEO), as well as the Geomorphosites Working Group within the International Association of Geomorphologists (Palacio 2013; Herrera-Franco et al. 2021).

A geosite is defined as a landform, geological formation, particular soil, or group of fossils that has an importance for understanding the Earth's history, which receives

A. Quesada-Román (ed.), *Landscapes and Landforms of Costa Rica*, World Geomorphological Landscapes, https://doi.org/10.1007/978-3-031-64940-0_13

additional significance from the population according to their perception and exploitation, these added values being aesthetic, cultural, historical, and socio-economic (Panizza 2001; Reynard & Panizza 2005; Zouros 2007; Ilies and Josan 2008; Mucivuna et al. 2019). Geosites, due to their attributes, can become important tourist attractions as they integrate several elements of human and natural history. Through the elaboration of geotourism maps showing local geomorphology, as well as respective geosites and facilities (such as trails and access), an adequate development of local geotourism is facilitated for the benefit of nearby communities (Reynard 2008; Santos et al. 2019).

Several geosite assessment methodologies have been developed in different countries (Feuillet and Sourp 2010; Reynard and Brilha 2018). These different geosite assessment methodologies allow recognising tangible values that are of varying interests, as well as values that give the geosite a higher significance, both natural and social (González et al. 2014). These methodologies even allow comparing different geosites with similar conditions, such as the study by Pérez-Umaña et al. (2018) who made a comparison of geosites in the volcanoes Poás in Costa Rica, Paricutín in Mexico, and Teide Pico Viejo in the Canary Islands, confirming that the same methodology can be implemented to compare geosites in volcanic areas with similar conditions in the world and provide an input to propose geoparks.

Geoparks are well-defined territories with a certain number of sites of geological interest (including geomorphological; Quesada-Román and Peralta-Reyes 2023) of variable size, which are complemented by ecological, archaeological, historical, cultural, and economic interests, meriting protection that highlights their importance, so that to be named a geopark it must contain geosites (Voth 2008; Medina 2012; De Jesús 2014; Brilha 2016; Gordon 2019).

In 2004, the Global Geoparks Network was founded by UNESCO following the example as the European Geoparks Network, and later the Asian Geoparks Network was created in 2007 (Palacio 2013), both of which form part of the Global Geoparks Network. Currently, the largest number of Geoparks are in Europe and Asia, whilst in Africa, there are only 2 UNESCO Global Geoparks. In the Americas, there are 17 UNESCO Global Geoparks: five in Canada and five in Brazil, two in Mexico, and one each in Uruguay, Ecuador, Peru, Chile, and Nicaragua (Global Geoparks Network 2017).

Latin America has great potential for studies of geosites and for the creation of geoparks. However, the theme is still incipient despite the great contributions that geosite assessments could give to already known tourist sites, as well as to sites that could become more valuable based on their natural and/or cultural characteristics that are still unknown to most of the population. This allows scientific information and cultural values to give more relevance to the geosites of a given area (Pérez-Umaña and Quesada 2018; Rosado-González et al. 2020).

This work first shows the general geoheritage and geodiversity aspects of Costa Rica. Secondly, it is more focused on the protected areas of the National System of Conservation Areas (SINAC), because they have an organised and consolidated administrative structure, management plans, as well as the infrastructure to cater for growing tourist groups, attracted by their scenic, cultural, didactic, scientific, and touristic attractions. There are no people living inside the natural protected areas, but there are numerous villages in the surrounding areas whose local economy depends on the national parks (Pérez-Umaña and Quesada-Román 2018a, b). Some national parks have marine and insular areas, which were not contemplated because the valuation carried out considers those cultural values related to elements in continental areas. The exception is Cocos Island National Park, which is located 535 km from Cabo Blanco on the Nicoya Peninsula (Zúñiga 2008), because it has unique geological, cultural, historical, geological, and ecological characteristics of tourist interest (Díaz-Bolaños et al. 2012).

The procedure to identify potential geoparks using national parks in Costa Rica has been based on the geosite evaluation methodology of Santos et al. (2020), which takes into consideration a series of elements that add value to the geological/geomorphological richness. The evaluation was carried out taking into consideration the significance and impact of the national parks on Costa Rican society, taking as a reference cultural attributes, visitation data, and research carried out in them, obtaining this information through the bibliographic review.

Subsequently, the values were summed up to determine the potential of the territories to be proposed as geoparks according to their cultural and tourism values and the geodiversity they possess. The results of this summation are shown by means of cartography, which serve as input in this proposal. The Natural Jenks statistical method was used in ArcMap 10.3 to group the sum values into three ranges (high, medium, and low). The SINAC Conservation Areas were used in the discussion, as these are the administrative boundaries of the national parks. The conservation areas are denominated: Central (ACC), Central Pacific (ACOPAC), Arenal-Huetar Norte (ACA-HN), Arenal-Tempisque (ACA-T), Guanacaste (ACG), Tempisque (ACT), La Amistad Pacífico (ACLAP), La Amistad Caribe (ACLAC), Tortuguero (ACTo), Osa (ACOSA), and Coco Marine (ACMC).

The results of evaluation of potential geoparks in Costa Rica were validated taking as reference access, visitation, attractions, cultural manifestations, sense of national, regional, or local belonging. It was also analysed and discussed why certain national parks have received higher scores than the others from a landscape and/or aesthetic, cultural, didactic, scientific, and touristic point of view. In Costa Rica, tourism represents an important part of Gross Domestic Product and is responsible for approximately 4–5% of its annual economic growth (Quesada-Román et al. 2022). Therefore, the development of initiatives and proposals that promote greater knowledge about tourist attractions of the national parks and its dissemination based on their natural and cultural values is a necessity for the country. The objectives of this chapter are to (i) depict the geodiversity of the country in a general perspective, and (ii) identify the national parks of Costa Rica that have the capacity to be also named UNESCO Global Geoparks given their natural and social characteristics, and based on the geosites that can be used as tourist attractions.

13.2 Geoheritage and Geodiversity of Costa Rica

Costa Rica exhibits a diverse range of geomorphic environments, including fluvial, coastal, and volcanic landscapes, with additional contributions from glacial and karstic formations (see Fig. 13.1). The country is geographically divided into two basins: the Pacific and the Caribbean, separated by a continuous chain of ranges with an NW–SE orientation, influencing precipitation patterns in each basin. Generally, the Caribbean region experiences annual means of up to 3000 mm, whilst the Pacific region receives between 1500 and 3000 mm.

Active Quaternary volcanism is observed in Guanacaste, the Central volcanic ranges, and the Arenal tectono-volcanic complex. The Talamanca Range exhibits Miocene–Quaternary volcanism, preserving volcanic forms eroded by fluvial and glacial dynamics since the Last Glacial Maximum. Non-active volcanic ranges like Tilarán and Aguacate exist, alongside active volcanic ranges with dynamic volcanoes such as Rincón de la Vieja, Arenal, Poás, Irazú, and Turrialba (refer to Fig. 13.3). Some massifs, despite their inactivity, retain original morphologies amid high erosion and weathering rates. Various volcanic landforms, concealed by dense forests, include waterfalls often found in lava flows or faults, showcasing volcanic and fluvial interactions. Examples of such features are abundant in both basins at the foothills of volcanic ranges. In both basins, fluvial and volcanic mountain landscapes experience high erosion and weathering rates, primarily due to rainfall and seismicity, resulting in frequent mass movement activities such as landslides, rockfalls, debris flows, and lahars (Campos et al. 2022). These processes contribute sediment to the fluvial system, impacting both natural landscapes and human activities.

The Northern Caribbean region consists mainly of Quaternary deposits eroded from the Guanacaste and Central volcanic ranges. This region exhibits abrupt changes with knickpoints, forming impressive waterfalls, wide eroded alluvial fans, extensive alluvial plains, river–estuarine channels in Tortuguero and Barra del Colorado, coastal lagoons, coastal bars, dynamic long beaches, and

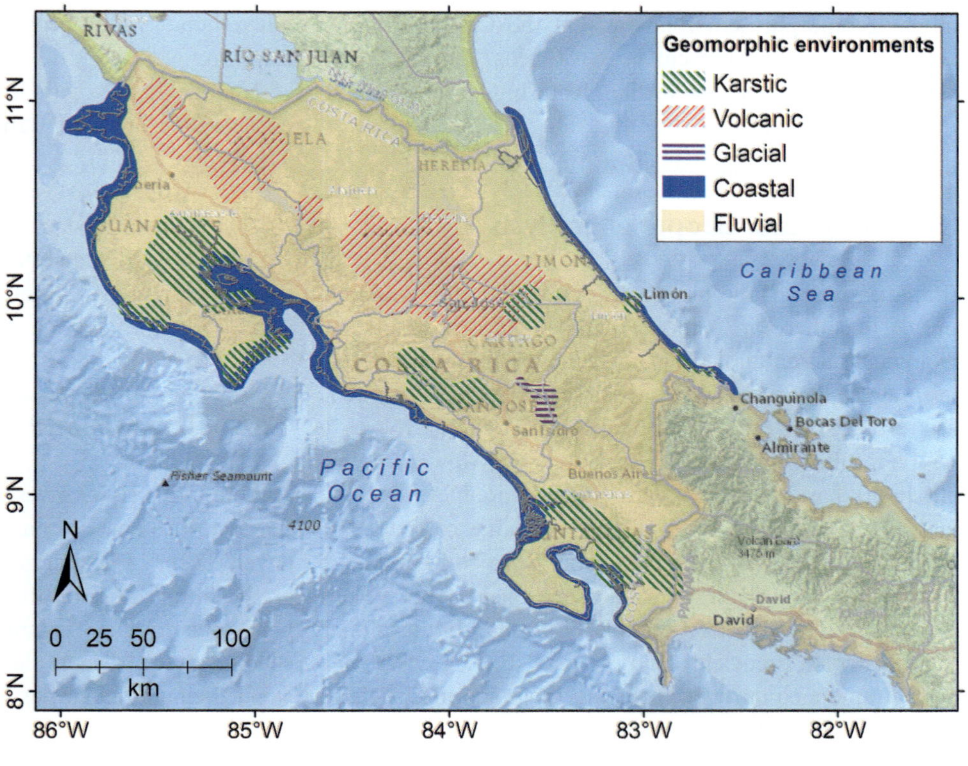

Fig. 13.1 Main geomorphic environments depicting the general geodiversity of Costa Rica. Modified from Quesada-Román and Pérez-Umaña (2020a, b)

islands. Conversely, the Southern Caribbean region, adjacent to Talamanca Range foothills, features a narrow coastline transition dominated by hills, coral reefs, islands, beaches, and carbonate platforms.

The Pacific basin's intricate topography, characterised by lithological diversity and rapid base level changes from active and relict volcanic ranges, fosters the formation of massive alluvial fans, dynamic valleys, or tectonic depressions (e.g. General Valley, Central Volcanic Depression, and Turrialba), and various levels of alluvial terraces leading to the alluvial plains (i.e. Tempisque, Tárcoles, Parrita, and Térraba) with meanders extending to the coast. The Pacific coastal landforms are more diverse than their Caribbean counterparts, featuring gulfs, bays, coves, sea stacks, arches, tombolos, beaches, deltas covered by mangroves, and numerous islands. The country also has a diverse composition of wetlands divided into palustrine, lacustrine, and estuarine (Veas-Ayala et al. 2023).

During the Last Glacial Maximum, periglacial and glacial processes shaped the highest mountains of Costa Rica (above 3000 m) in the Talamanca Range, leaving relict erosional glacial landforms. These include volcanic slopes modified by glacial and periglacial action, glacial cirques, arêtes, riegels, roches moutonnées, as well as depositional landforms such as lateral moraines, till deposits, and glacial lakes. Although these formations are now relict, they are presently covered by páramo vegetation, harbouring several endemic fauna and flora species. Additionally, karst landforms are found along Fila Brunqueña, the Gulf of Nicoya, and certain parts of Nicoya beaches on the Pacific side, where mogotes, caverns, and sinkholes define this unique geomorphic environment. On the Caribbean side, karst landscapes are present in San Carlos, Turrialba, and the Southern Caribbean coastline, featuring carbonate platforms, islands, and caves.

13.3 Potential Geoparks Analysis

The assessment of the proposal developed in this research reveals that there are national parks that have significant geological and cultural values, whereas some obtained values indicate that there are areas that can be better exploited. The results of the evaluation are shown in Table 13.1.

The results were classified in three ranges (low, medium, and high) based on the sum of their variables to compare all the national parks. In the low value category are located those whose final assessments were between 0.310 and 0.545, these being Guanacaste, Diriá, Palo Verde, Tortuguero, Carara, La Cangreja, Barbilla, Internacional La Amistad, Corcovado, and Piedras Blancas National Parks. In the medium value category those scoring between 0.546 and 0.762 were included, corresponding to Santa Rosa, Miravalles Jorge Manuel Dengo, Tenorio Volcano, Marino Las Baulas, San Lucas Island, Juan Castro Blanco, Braulio Carrillo, Manuel Antonio, Tapantí-Macizo de la Muerte, and Los Quetzales National Parks. National parks with scores between 0.763 and 0.973 were grouped in the high value category, including Rincon de la Vieja, Arenal Volcano, Barra Honda, Poás Volcano, Irazú Volcano, Turrialba Volcano, Cahuita, Marino Ballena, Chirripó, and Cocos Island National Parks (Fig. 13.2).

The Conservation Area with the highest number of national parks belonging to the high value category is the Central Conservation Area (ACC) where three national parks (Poás Volcano, Irazú Volcano, and Turrialba Volcano) obtained high scores. The Conservation Areas Guanacaste (ACG), Tempisque (ACT), Arenal-Huetar Norte (ACA-HN), Osa (ACOSA), La Amistad Pacífico (ACLAP), La Amistad Caribe (ACLAC), and Coco Marine (ACMC) have only one national park scoring high values each one. Other Conservation Areas do not have national parks in this category.

The Central Conservation Area (ACC) is the area with the highest number of national parks with a medium rating, with a total of three protected areas, namely Braulio Carrillo, Tapantí-Macizo de la Muerte, and Los Quetzales National Parks. Arenal-Tempisque (ACA-T) and Central Pacific (ACOPAC) Conservation Areas also stand out with two national parks of medium rating. The rest of the conservation areas have at least one national park with average score, except for the Osa (ACOSA) and La Amistad Pacífico (ACLAP), La Amistad Caribe (ACLAC), Tortuguero (ACTo), and Coco Marine conservation areas, where none of their protected areas have been evaluated as medium.

National parks with low ratings are distributed in the conservation areas of Central (ACC), Guanacaste (ACG), Tempisque (ACT), Arenal-Tempisque (ACA-T), Central Pacific (ACOPAC), Osa (ACOSA), La Amistad Pacífico (ACLAP), La Amistad Caribe (ACLAC), and Tortuguero (ACTo). Arenal-Huetar Norte (ACA-HN) and Coco Marine conservation areas do not have protected areas with a low score.

13.4 Tourism and Geosites in National Parks

Although this study gives important evidence of the potential that different national parks have in Costa Rica for the development of geoparks through the dissemination of scientific information on the geosites that each national park hosts due to their natural and cultural characteristics, a specific study of each national park is necessary to understand their particularities in even greater detail. However, although not all national parks have high value, they have a

Table 13.1 Assessment of the proposal in Costa Rica's national parks

National parks	Scientific value	Ecological assessment	Cultural assessment	Aesthetic assessment	Average
Chirripó	0.893	1	1	1	0.973
Poás Volcano	1	0.75	1	1	0.938
Irazú Volcano	1	0.75	1	1	0.938
Cocos Island	0.857	1	1	0.750	0.902
Marino Ballena	0.750	1	1	0.833	0.896
Arenal Volcano	0.893	0.75	1	0.917	0.890
Turrialba Volcano	0.821	0.75	1	0.917	0.872
Rincón de la Vieja	0.821	0.75	1	0.917	0.872
Cahuita	0.893	1	0.75	0.667	0.827
Barra Honda	0.857	1	0.5	0.917	0.818
Miravalles Jorge Manuel Dengo	0.714	0.75	0.75	0.833	0.762
Juan Castro Blanco	0.536	1	0.75	0.667	0.738
Tenorio Volcano	0.607	0.75	0.75	0.833	0.735
Manuel Antonio	0.893	0.5	0.75	0.667	0.702
Braulio Carrillo	0.607	1	0.5	0.667	0.693
Santa Rosa	0.893	0.25	1	0.500	0.661
Los Quetzales	0.500	0.75	0.75	0.583	0.646
Tapantí-Macizo de la Muerte	0.571	0.25	0.75	0.833	0.601
Marino Las Baulas	0.679	0.75	0.25	0.667	0.586
San Lucas Island	0.643	0.25	0.75	0.667	0.577
Corcovado	0.679	0.5	0.25	0.750	0.545
Tortuguero	0.607	0.75	0.25	0.500	0.527
Guanacaste	0.500	0.5	0.5	0.583	0.521
Palo Verde	0.750	0.25	0.25	0.667	0.479
Internacional La Amistad	0.536	0.25	0.25	0.667	0.479
Carara	0.429	0.25	0.5	0.583	0.440
Piedras Blancas	0.393	0.25	0.25	0.583	0.369
Barbilla	0.417	0.25	0.25	0.500	0.354
Diriá	0.321	0.25	0.25	0.500	0.330
La Cangreja	0.321	0.25	0.25	0.417	0.310

high annual visitation due to natural conditions, scenery, or ease of access that make them attractive to local and foreign tourists (Table 13.2).

Table 13.2 shows national parks with the highest annual visitation in Costa Rica, highlighting that one includes a volcano and two are coastal national parks. This high visitation is due to its short distance from San José and a cultural root with them (Alvarado et al. 2023; Mora et al. 2024). It should be noted that three of the nine most visited NPs obtained a high rating, whilst one obtained a medium rating (Manuel Antonio NP). Their high number of visitors indicates that there is good access to these national parks, which is used by the local people to have businesses to earn income from the sale of food and souvenirs. However, during the period between 2012 and 2016, Poás Volcano National Park was the second most visited national park in Costa Rica, but in 2017, it was closed due to eruptive activation which put tourists at risk. This situation was beneficial for other national parks that attracted more tourists during 2017 (Pérez-Umaña et al. 2019). It was until 2019 that Poás Volcano National Park was positioned as the fourth most visited national park in the country with 147,230 visitors and 85,114 visitors in 2020 (SINAC 2021).

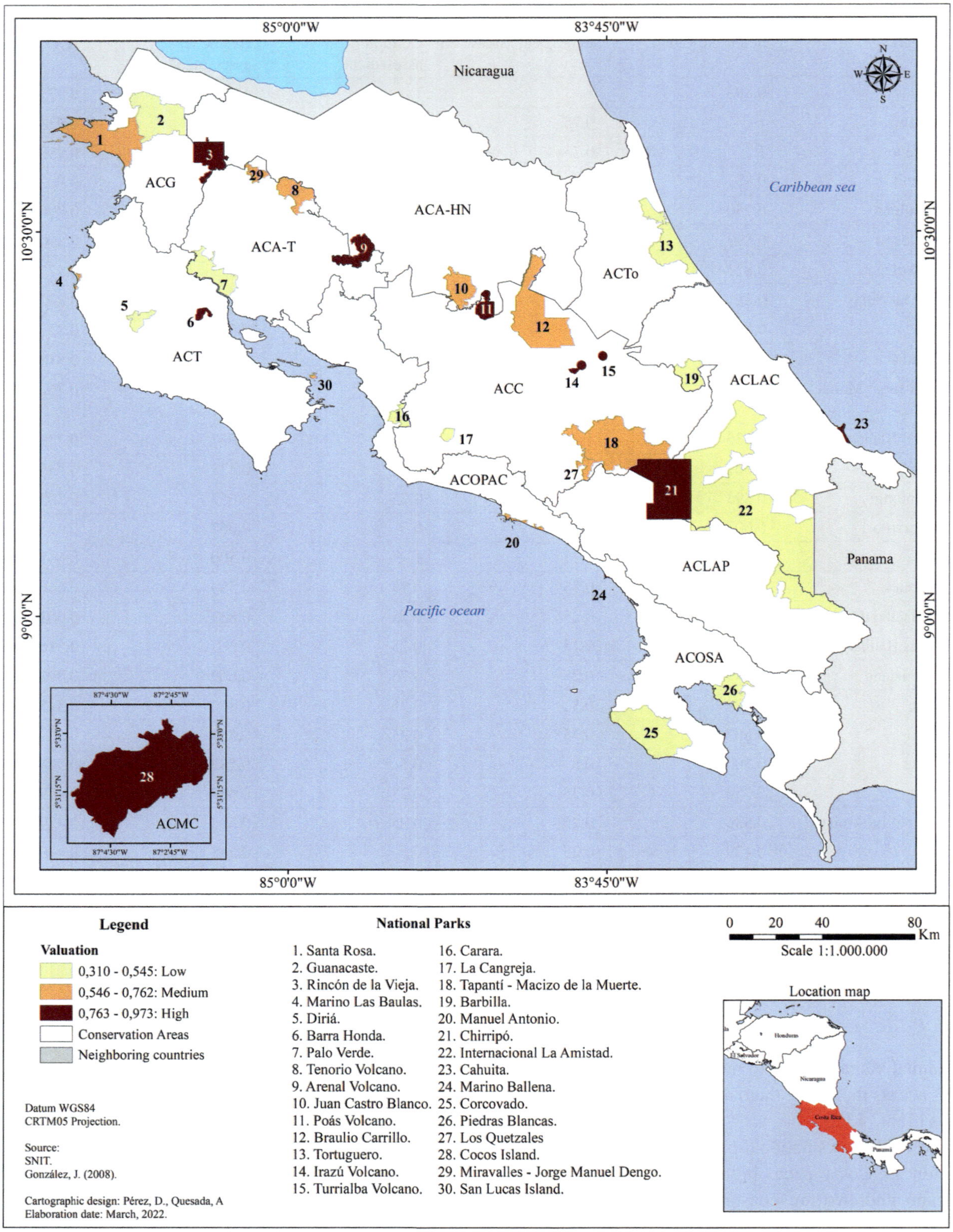

Fig. 13.2 Results of the assessment of national parks proposed as geoparks in Costa Rica

Table 13.2 Most visited national parks in Costa Rica, 2017–2020

National park	Visitors 2017	Visitors 2018	Visitors 2019	Visitors 2020
Manuel Antonio	475,052	524,835	513,050	224,762
Irazú Volcano	343,254	422,947	352,034	188,786
Marino Ballena	167,054	165,875	171,562	127,106

In 2020, there was a significant reduction in visits to national park by almost 50% as a result of restrictions imposed by the Costa Rica government to prevent the spread of COVID-19 and to protect the health of visitors to protected areas (Mora 2020). Some national parks continued to receive visitors, but their numbers were considerably reduced as shown in Table 13.2.

Most of the studied protected areas have at least one road entrance, such as Poás, Irazú, and Barva volcanoes being the only volcanoes in Central America to have a road that reaches their craters (Quesada-Román and Mata-Cambronero 2021; Pérez-Umaña 2023). An important aspect is that not all these protected areas have a paved road, but they do have land access to enter. There are national parks that are only accessible by sea, such as Cocos Island and San Lucas Island National Parks. In addition, some national parks have access by sea or air such as Tortuguero and Corcovado National Parks, but this does not reduce the interest of tourists, as well as to walk their trails and get to know the national protected area. Since a geopark is composed of geosites in its interior that serve as tourist attractions, it is necessary to list recognised sites of natural and/or cultural interest that serve as input for this proposal. These geosites should be sites that tourists can access and are representative of Costa Rica's national parks (Table 13.3).

13.5 Cultural Values in National Parks

Geoparks may have ancient cultural objects or cultural values within their territories, which contribute to the preservation of intangible values, such as cultural heritage (Porshnov et al. 2019). In other words, a geopark contains not only geological and geomorphological heritage, but also cultural heritage. These cultural manifestations are related to everything that people relate to these territories, as for example Pérez-Umaña (2017, 2023) describes that the Poás volcano has been inspiration for many artists as well as it is the place that is inspiration for many tales and legends. Another example is provided by Pérez-Umaña et al (2020) who state that the Irazú volcano had a place where indigenous rituals were performed, believing that it was the place where their gods lived. Vargas (2012) pointed out that the SINAC has functions such as managing, promoting, facilitating, and participating in the development of sustainable tourism in protected areas based on responsible administration, planning, and management practices. These functions are the basis for the formulation of this proposal, as the existence of infrastructure facilitates the establishment of geoparks through the NP and subsequently the other protected areas of the country.

The geosites are binding elements in this proposal, as they allow to establish which attractions are available for the geoparks. In the case of Costa Rica, due to its tectonic, geological, geomorphological, climatic, and biological dynamics, there are diverse geomorphic environments ranging from volcanic, fluvial, coastal, to even glacial of the past, which have been modified by human action over the last thousands of years.

13.6 Examples in National Parks

Amongst sites of volcanic origin are several active and inactive volcanoes, one of its most striking geosites being the Main Crater in the Poás Volcano National Park (Fig. 13.3a). It is a spectacular feature and probably one of the largest craters in the world (Salguero 2003), originated by a strong Plinian eruption 3500 years ago. It has been an important inspiration for various artistic and literary works (Pérez-Umaña, 2017, 2023). Other volcanic geosites of interest are located in Irazú Volcano (Fig. 13.3b), with a summit at 3432 m asl where the craters of Diego de la Haya, Playa Hermosa, and Principal Crater are located. Also, from the summit, one can see the Pacific and Caribbean coasts and the Turrialba Volcano, which is a volcano that borders the Irazú volcano (Pérez-Umaña et al. 2020; Alvarado 2021).

In the Guanacaste Volcanic Range in northern Costa Rica, Orosí Volcano (Fig. 13.3c) is located within Guanacaste National Park, surrounded by tropical dry forest and with a fascinating eruptive history (Vargas-Sanabria and Quesada-Román 2018; Quesada-Román and Vargas-Sanabria 2022). Arenal Volcano (Fig. 13.3d) is one of the most striking volcanoes due to its conical shape, with an age of 7000 years, being one of the youngest volcanoes in Costa Rica (Alvarado 2021).

Costa Rica has a small segment of glacial geomorphology from the Last Glacial Maximum in the Talamanca Range (Orvis and Horn 2000; Quesada-Román et al. 2019). It is in

Table 13.3 Geosites of interest in Costa Rica's national parks

National park	Geosites
Poás Volcano	Main Crater, Botos Cone, Von Frantzius Cone, Congo Volcano
Irazú Volcano	Craters, Cerro Cabeza de Vaca, Active landslide, Pyroclastic Cone
Chirripó	Crestones, Cerro Chirripó, Las Morrenas, Cerro Ventisqueros
Santa Rosa	Peña Bruja, Naranjo Beach
Cahuita	Punta Cahuita, Vargas Port
Cocos Island	Cocos Island, Cerro Iglesias, Wafer Bay, Punta Presidio
Braulio Carrillo	Barva Volcano, Cacho Negro Volcano, Cerro Zurquí, Sucio River
Turrialba Volcano	Craters, Boquetes
Arenal Volcano	Arenal Volcano, Cerro Chato, Lava Flows, Arenal Lake Viewpoint
Manuel Antonio	Punta Catedral, Manuel Antonio Beach, Espadilla Sur Beach, Punta Serrucho
Tortuguero	Cerro Tortuguero, Tortuguero Canals, Barra de Tortuguero, Cerro Tortuguerito
Rincón de la Vieja	Rincón de la Vieja Crater, mud volcanoes, Santa María Crater, Las Pailas
Corcovado	Corcovado Lagoon
Marino Ballena	Punta Uvita, Uvita Beach, Piñuela Beach, Ballena Beach
Marino Las Baulas	Grande Beach, Ventanas Beach, Langosta Beach
Internacional La Amistad	Cerro Kamuk, Cerro Pittier, Sabanas Esperanza, Silencio Valley
Tapantí-Macizo de la Muerte	Cerro de la Muerte, Cerro Asunción, Cerro Jaboncillo, Tapanti Waterfall
Tenorio Volcano	Rio Celeste Waterfall, Los Teñideros, Montezuma Volcano, Cerro La Carmela
Palo Verde	Palo Verde Lagoon, Cerro Catalina, Cerro Guayacán
Los Quetzales	Source of the Savegre River
Barra Honda	Caves, Mirador Nacaome, Cerro Barra Honda
Carara	Meandric Lagoon, Entierro Hills
Diriá	Brasil Waterfall, River sources
Guanacaste	Orosí Volcano, Cacao Volcano, Pedregal Site, Orosilito Volcano
Juan Castro Blanco	Platanar Volcano, Volcán Viejo, Río Segundo Caldera
Barbilla	River sources
Piedras Blancas	River sources
La Cangreja	Cerro La Cangreja
San Lucas Island	Cocos Beach
Miravalles Jorge Manuel Dengo	Miravalles Volcano

this area where the highest mountain in Costa Rica is located, Cerro Chirripó (Fig. 13.4a), with an altitude of 3820 m asl, located within the Chirripó National Park. Also, within this national park are located Los Crestones (Fig. 13.4b), intrusive rocks that were exposed by glacial erosion at an altitude of 3720 m asl. The national park hosts impressive paleoglacial landscapes such as valleys, moraines, lakes, and glacial cirques (Fig. 13.4c; Quesada-Román and Pérez-Umaña 2020b). Several glacial cirques have developed within the summits that nowadays indicate impressive paleostage dynamics during the Last Glacial Maximum, 20,000 years ago (Quesada-Román et al. 2020c, 2021).

Coastal areas are amongst the most popular tourist attractions in Costa Rica. An example is Manuel Antonio National Park, located in the Central Pacific. Punta Catedral, a tombolo bordered by sandy beaches and forest (Fig. 13.5a), is one of the main attractions of this park, whereas Punta Serrucho (Fig. 13.5b), a rocky formation of large cliffs resulting from constant marine abrasion, is another one.

In the southern Caribbean of Costa Rica, the Cahuita National Park is located, with Punta Cahuita (Fig. 13.5c). This place is one of the few coral reefs in the Costa Rican Caribbean that provide sediments for the formation of white sand beaches (Herrera 1992; Zúñiga 2008; Quesada-Román and Pérez-Briceño 2019). Also, in the Marino Ballena National Park in the Central Pacific of Costa Rica is located Punta Uvita (Fig. 13.5d), an interesting geosite, which is a sandy tombolo that connects an islet with the beach (Zúñiga

Fig. 13.3 Volcanic geosites in Costa Rica's national parks. **a** Main Crater, Poás Volcano National Park *Source* (Poás Volcano National Park Facebook). **b** Playa Hermosa Crater and Sapper Hill, Irazú Volcano National Park. **c** Orosí Volcano, Guanacaste National Park. **d** Arenal Volcano, Arenal Volcano National Park

2008). From this site, wide landscapes with a low swell touching the beaches of the tombolo as well as the nearby mountains of the Fila Costeña can be observed.

The geosites of some national parks in Costa Rica strengthen the proposal to establish geoparks in the country. Some characteristics such as scientific, cultural, historical, economic, and scenic beauty are determining factors in this proposal. Visitors to the national parks in Costa Rica can enjoy a variety of sites and features, from exuberant landscapes such as on top of the Turrialba and Irazú volcanoes, with interesting geological and geomorphological characteristics.

Since their origins, the national parks proposed in this study have had a cross-cutting approach to risk management. In 1955, with the approval of the Organic Law of the Costa Rican Tourism Institute N°1917, the first national parks in Costa Rica were created in the volcanic massifs of the main mountain ranges (Zeledón 1999). This becomes one of the first risk management actions in the country because it delimited the areas with risk potential and put a stop to expansion of an agricultural frontier and the development of infrastructure into areas where people would put their lives at risk (Alvarado et al. 2021).

Although Costa Rica has worked on the protection of natural resources through the protection of biodiversity, which is why there are so many national parks in the country (Pérez-Umaña 2020), it is within the objectives for the creation of a protected area that the justification needs to involve a risk management approach. The same is true for environmental education; the entire environmental approach is visualised transversally so that citizens know that the existence of a national park is because it poses a threat to the population, as in the case of volcanoes, and that it is a site that should be taken care of from an environmental point of view. In other words, the creation of a protected area in Costa Rica provides knowledge of how to take care of the environment and stay away from dangers, working with resilience and adaptation to natural threats and providing full environmental education to society.

Another aspect to consider is the impact that national parks have on the local economy. Pérez-Umaña (2019)

Fig. 13.4 Geosites of glacial origin in Chirripó National Park. **a** Cerro Chirripó. **b** Los Crestones. **c** Glacial landscape seen from Cerro Chirripó (*Source* courtesy of Melissa Agüero Rojas)

mentioned that the existence of a figure that protects the local geoheritage would be an additional benefit for neighbouring communities who would see an increase in their income from tourists visiting these places. Currently, as demonstrated by Pérez-Umaña (2021), who reported that through social networks information about site values and access to sites of interest is disseminated and nearby populations have economic income by offering to the public sales of home-made food from the region or other natural products to visitors.

Fig. 13.5 Geosites in coastal national parks. **a** Punta Catedral, Manuel Antonio National Park. **b** Punta Serrucho, Manuel Antonio National Park. **c** Punta Cahuita, Cahuita National Park (*Source* Pucci et al. 2014). **d** Punta Uvita, Ballena Marine National Park

13.7 Conclusions

Costa Rica has gained global recognition for its environmental initiatives developed over the past three decades. These efforts have led to the creation of numerous protected areas, establishing the country as a premier ecotourism destination worldwide. The unique geodiversity of Costa Rica, shaped by the interplay of tectonics, volcanism, tropical climate, and diverse vegetation, encompasses a variety of geological environments, including volcanic, fluvial, coastal, karstic, and glacial landscapes.

In this proposal, 30 national parks of Costa Rica were evaluated and analysed because they are protected areas with a significant and ideal administrative and territorial organisation to be proposed as geoparks. In addition, several geosites of interest were identified inside national parks that strengthen the proposal. They would become important tourist attractions of these potential geoparks. The evaluation found that 10 of Costa Rica's national parks were rated high, 10 were rated medium, whilst 10 were rated low. The evaluation shows that there are national parks that can improve to become geoparks such as better promotion of national parks as a tourist attraction or increasing accessibility, which would allow for greater number of visits by tourists.

Scientific and didactic evaluation work together; therefore, the more scientific research is carried out in these national parks and the more these results are disseminated, the better the general knowledge about the protected areas will be. As far as landscape appreciation is concerned, this can be increased if new viewpoints open to the public that tourists can use as new attractions in national parks, obviously hand in hand with management plans and sustainable development (Quesada-Román et al. 2023).

In the national parks which scored high, efforts should be made to ensure that they maintain the rating obtained.

Local, regional, and national population's attachment to these national parks can be improved, expanded, and new sites with wide landscapes that attract tourists can be found. People have a sense of identification with some national parks, but this must be improved so that the national population identifies with the national parks. Similarly, the tourist, scientific, and didactic structure should be improved to increase the number of visitors and to ensure that the results of studies are widely disseminated and known.

This initiative is a proposal in Costa Rica, a country that depends on the tourism sector, analysing landscape, cultural, didactic, scientific, and tourism variables to strengthen an existing tourism product. The aim is to improve the tourism offer in such a way that the visitor will not only have a place for recreation but will also be able to receive knowledge given the added social and cultural values. The proposal of geoparks accompanied by some geosites enriches Costa Rica in terms of tourism and economy, because by promoting these geoparks, new geosites can be sought to diversify the tourism offer that should be divulged (Quesada-Valverde and Quesada-Román 2023).

References

Acuña-Piedra J, Quesada-Román A (2021) Multidecadal biogeomorphic dynamics of a deltaic mangrove forest in Costa Rica. Ocean Coast Manag 211:105770

Alvarado G (2011) Los volcanes de Costa Rica: geología, historia, riqueza natural y su gente. Editorial UNED, San José, Costa Rica

Alvarado G (2021) Costa Rica y sus volcanes. EUCR, EUNA, ETCR, San José, Costa Rica

Alvarado G, Esquivel L, Sánchez B (2021) Mejoras en el paradigma del manejo de la gestión del riesgo en los parques nacionales volcánicos de Costa Rica, entre el año 2000 y el 2020. Revista Geológica De América Central 64:1–15

Alvarado GE, Alpízar Y, Araya R, Esquivel L, Sánchez B (2023) Un nuevo impulso a la "geoalfabetización" en los parques nacionales volcánicos de Costa Rica y sectores aledaños para una mejor gestión del riesgo. Revista Estudios Latinoamericanos Sobre Reducción Del Riesgo De Desastres REDER 7(2):6–18

Brilha J (2016) Inventory and quantitative assessment of geosites and geodiversity sites: a Review. Geoheritage 8:119–134

Campos N, Quesada-Román A, Granados-Bolaños S (2022) Mapping mountain landforms and its dynamics: study cases in tropical environments. Appl Sci 12(21):10843

De Jesús J (2014) Geomorfositios en el volcán Paricutín, Michoacán, México. Disertación de Licenciatura en Geografía, UNAM

Díaz R, Alfaro E, Ugalde A (2012) Diccionario histórico-toponímico de la Isla del Coco, Costa Rica. Rev Biol Trop 60:1–14

Feuillet T, Sourp E (2010) Geomorphological heritage of the Pyrenees National Park (France): assessment, clustering, and promotion of geomorphosites. Geoheritage 3:151–162

Global Geoparks Network (2017) Geoparks members list. https://www.unesco.org/en/iggp/geoparks?hub=67817. Accessed 10 Diciembre 2023

González M, GonzálezM SE (2014) Lugares de interés geomorfológico, geopatrimonio y gestión de espacios naturales protegidos: el Parque natural de Valderejo (Álava, España). Revista De Geografía Norte Grande 59:45–64

Gordon J (2019) Geoconservation principles and protected area management. Int J Geoheritage Parks 7:199–210

Herrera W (1992) Mapa Guía de la Naturaleza Costa Rica. Editorial Incafo, Santo Domingo de Heredia, Costa Rica

Herrera-Franco G, Montalván-Burbano N, Carrión-Mero P, Jaya-Montalvo M, Gurumendi-Noriega M (2021) Worldwide research on geoparks through bibliometric analysis. Sustainability 13:1175

Hunt A, Harbor L (2019) Pro-environmental tourism: lessons from adventure, wellness and eco-tourism (AWE) in Costa Rica. J Outdoor Recreat Tour 28:100202

Ilies D, Josan N (2008) Some theoretical aspects regarding the genesis of geosites. GeoJournal Tourism Geosites 1:7–12

Medina W (2012) Propuesta metodológica para el Inventario Geológico en Argentina. Universidad de Minho, Braga, Portugal, Disertación de Maestría en Patrimonio Geológico y Geoconservación

Mora A (2020) MINAE y SINAC cerrarán parques nacionales para prevenir propagación de COVID-19. https://delfino.cr/2020/03/minae-y-sinac-cerraran-parques-nacionales-para-prevenir-propagacion-de-covid-19. Accessed 1 April 2022

Mora, M.M., Peraldo Huertas, G. & Cerdas Prado, K. (2024). Percepciones en torno al volcanismo en Costa Rica (siglos XIX al XXI): De lo místico a lo profano, entre la curiosidad y el morbo. Revista de Estudios Latinoamericanos sobre Reducción del Riesgo de Desastres REDER 8(1):91–114. https://doi.org/10.55467/reder.v8i1.141

Mucivuna V, Reynard E, Garcia M (2019) Geomorphosites assessment methods: comparative analysis and typology. Geoheritage 11:1799–1815

Orvis K, Horn S (2000) Quaternary Glaciers and Climate on Cerro Chirripó, Costa Rica. Quatern Res 54:24–37

Palacio J (2013) Geositios, geomorfositios y geoparques: importancia, situación actual y perspectivas en México. Investigaciones Geográficas, Boletín Del Instituto De Geografía UNAM 82:24–37

Panizza M (2001) Geomorphosites: concepts, methods and examples of geomorphological survey. Chin Sci Bull 46:3–6

Pérez-Umaña D (2017) Evaluación del potencial turístico de Geomorfositios en el Parque Nacional Volcán Poás. Disertación de Licenciatura en Ciencias Geográficas con énfasis en Ordenamiento del Territorio, Escuela de Ciencias Geográficas, Universidad Nacional de Costa Rica

Pérez-Umaña D, Quesada-Román A (2018) Una Propuesta para la Valoración de Geoparques en Costa Rica. Anuário do Instituto de Geociências 41(3):382–394. https://doi.org/10.11137/2018_3_382_394

Pérez-Umaña D (2019) Geodiversidad y geopatrimonio en las áreas silvestres de Costa Rica. Ambientico 271:59–65

Pérez-Umaña D (2020) Un lugar para establecer una ruta geoturística: Los Quemaderos del volcán Poás. https://delfino.cr/2020/09/un-lugar-para-establecer-una-ruta-geoturistica-los-quemaderos-del-volcan-poas

Pérez-Umaña D (2021) Las redes sociales, una herramienta para la promoción del geoturismo en Costa Rica. https://delfino.cr/2021/04/las-redes-sociales-una-herramienta-para-la-promocion-del-geoturismo-en-costa-rica

Pérez-Umaña D (2023) Inventory and assessment of geoheritage in Poás Volcano National Park, Costa Rica. Geoconservation Res 6(2): 1–13. https://doi.org/10.57647/j.gcr.2023.0602.21

Pérez-Umaña D, Quesada-Román A (2018b) Metodología para la valoración y evaluación de geomorfositios en Costa Rica. Revista Geográfica De América Central 60:117–135

Pérez-Umaña D, Quesada-Román A, De Jesús J, Zamorano J, Dóniz J, Becerra R (2018) Comparative analysis of geomorphosites in volcanoes of Costa Rica, Mexico, and Spain. Geoheritage. https://doi.org/10.1007/s12371-018-0313-0

Pérez-Umaña D, Quesada-Román A, Tefogoum G (2020) Geomorphological heritage inventory of Irazú volcano, Costa Rica. Int J Geoheritage Parks 8:31–47
Porshnov D, Burlakovs J, Kriipsalu M et al (2019) Geoparks in cultural and landscape preservation context. Rural Environ Eng 1:154–159. https://doi.org/10.22616/rrd.25.2019.023
Pucci S, Pucci G, Gamboa, J (2014) Costa Rica Aérea: Retratos de un país inédito. Editorial IINDEF, San José, Costa Rica
Quesada-Román A, Ballesteros-Cánovas J, Stoffel M, Zamorano-Orozco J (2019) Glacial geomorphology of the Chirripó National Park, Costa Rica. J Maps 15:538–545
Quesada-Román A, Pérez-Briceño P (2019) Geomorphology of the Caribbean coast of Costa Rica. J Maps 15:363–371
Quesada-Román A, Pérez-Umaña D (2020a) State of the art of geodiversity, geoconservation, and geotourism in Costa Rica. Geosciences 10:211
Quesada-Román A, Pérez-Umaña D (2020b) Tropical paleoglacial geoheritage inventory for geotourism management of Chirripó National Park, Costa Rica. Geoheritage 12:1–13
Quesada-Román A, Campos N, Alcalá-Reygosa J, Granados-Bolaños S (2020) Equilibrium-line altitude and temperature reconstructions during the last glacial maximum in Chirripó National Park, Costa Rica. J S Am Earth Sci 100:102576
Quesada-Román A, Mata-Cambronero E (2021) The geomorphic landscape of the Barva volcano, Costa Rica. Phys Geogr 42(3):265–282
Quesada-Román A, Vargas-Sanabria D (2022) A geomorphometric model to determine topographic parameters controlling wildfires occurrence in tropical dry forests. J Arid Environ 198:104674
Quesada-Román A, Peralta-Reyes M (2023) Geomorphological mapping global trends and applications. Geographies 3(3):610–621
Quesada-Román A, Umaña-Ortíz J, Zumbado-Solano M, Islam A, Abioui M, Tefogoum GZ, Kariminejad N, Mutaqin BW, Pupim F (2023) Geomorphological regional mapping for environmental planning in developing countries. Environ Dev 48:100935. https://doi.org/10.1016/j.envdev.2023.100935
Quesada-Valverde ME, Quesada-Román A (2023) Worldwide trends in methods and resources promoting geoconservation, geotourism, and geoheritage. Geosciences 13(2):39
Reynard E (2008) Scientific research and tourist promotion of geomorphological heritage. Geogr Fis Din Quat 31:225–230
Reynard E, Panizza, M. (2005) Geomorphosites: definition, assessment and mapping. An introduction. Géomorphol Relief Process Environ 3:177–180
Reynard E, Brilha J (2018) Geoheritage: assessment, protection, and management. Elsevier
Rosado-González E, Sá A, Palacio-Prieto J (2020) UNESCO global geoparks in Latin America and the Caribbean, and their contribution to agenda 2030 sustainable development goals. Geoheritage 12:1–15
Ross Y (2014) Costa Rica Parques Nacionales Fronteras Naturales. Producciones del Río Nevado, San José, Costa Rica
Sada M (2015) The curious case of Costa Rica can an outlier sustain its success? Harvard Int Rev Summer 2015:11–12
Salguero M (2003) Volcanes de Costa Rica. Editorial Costa Rica, San José, Costa Rica
Santos D, Reynard E, Mansur K, Seoane J (2019) The specificities of geomorphosites and their influence on assessment procedures: a methodological comparison. Geoheritage 11:2045–2064
Santos D, Mansur K, Seoane J, Mucivuna V, Reynard E (2020) Methodological proposal for the inventory and assessment of geomorphosites: an integrated approach focused on territorial management and geoconservation. Environ Manage. https://doi.org/10.1007/s00267-020-01324-2
SINAC (Sistema Nacional de Áreas de Conservación) (2018) SINAC en números: Informe Anual Estadístico 2012. San José, Costa Rica
SINAC (Sistema Nacional de Áreas de Conservación) (2019) SINAC en números: Informe Anual Estadístico 2013. San José, Costa Rica
SINAC (Sistema Nacional de Áreas de Conservación) (2021) SINAC en números: Informe Anual Estadístico 2014. San José, Costa Rica
Vargas G (2012) Geografía de Costa Rica. Editorial UNED, San José, Costa Rica
Vargas-Sanabria D, Quesada-Román A (2018) Influencia geomorfológica en la vulnerabilidad a incendios forestales en el Área de Conservación Guanacaste, Costa Rica. Revista De Ciencias Ambientales 52:1–15
Veas-Ayala N, Alfaro-Córdoba M, Quesada-Román A (2023) Costa Rican wetlands vulnerability index. Prog Phys Geogr 47(4):521–540
Voth A (2008) Los Geoparques y el Geoturismo: Nuevos conceptos de valorización de recursos patrimoniales y desarrollo regional. In: XI Coloquio Ibérico de Geografía. Departamento de Geografía de Universidad de Alcalá, España
Zeledón R (1999) Código ambiental. Editorial Porvenir, San José, Costa Rica
Zouros N (2007) Geomorphosite assessment and management in protected areas of Greece Case study of the Lesbos island—coastal geomorphosites. Geographica Helvetica 62:169–180
Zúñiga R (2008) Guía de Parques Nacionales de Costa Rica. Editorial Jomagar, Madrid, España

Part II

Regional Studies

14 Biogeomorphological Evolution of Arenal Stratovolcano: A Field Laboratory to Investigate Relationships Between Volcanic Activity, Erosion, and Vegetation Succession

Diego Martínez-Atienza and Guillermo E. Alvarado

Abstract

Arenal is a small Holocene twin stratovolcano (~33 km^2, ~9 km^3, 1684 m a.s.l., ≥ 7 ka), almost symmetrical but with some slope irregularities because it is composed of overlapping lava fields (lava armour) in the central area, with a medium to distal zone comprised of tephra, and hyperconcentrated flow deposits on its flanks (tephra and epi-volcaniclastic apron). We present an up-to-day summary of the prehistorical and historical processes, based on morphometric, morphogenesis, and morphochronological information, the last one obtained by radiocarbon dating and historical records. The eruptive foci during the last 3 ka reflect local structural controls that produced different lava fields, with different grades of vegetation and soil development. The difference in age between the zones of the lava fields has an important role in the successional stage of the vegetation community and, therefore, on the erosion rate. Lower the average erosion rate and smaller range of variation over the years are recorded on older flows. The historical lava fields (1968–2010) have shown that the average erosion rate on the complete lava field has decreased from 50.77 t ha^{-1} y^{-1} to 40.99 t ha^{-1} y^{-1} to 40.99 t ha^{-1} y^{-1} to 40.99 t ha^{-1} y^{-1} in 2010–2020. The results confirm that plant recolonization plays an important role in erosion control at Arenal volcano. At present, the areas of greatest vulnerability to erosion processes are in the gullies and rockfall channels in the upper lava field but they will stabilize in the near future.

Keywords

Volcanic landforms · Soil loss · Vegetation colonization · RUSLE · Erosion rate · NDVI

14.1 Introduction

The Arenal volcano has become famous among volcanologists because of its sudden awakening after a long period without previous historical eruptive record, and, after that, it experienced a long eruptive period, from 1968 to 2010. Moreover, it is known for hot springs at the northern flank volcano and for its tourist facilities. Therefore, the La Fortuna town and the Parque Nacional Volcán Arenal (created in 1991) are among the most visited natural areas of Costa Rica that receive around 1.3 million visitors per year (Fig. 14.1).

At Arenal, tropical climate allows for the existence of very humid tropical forests typical for the region. However, persistent volcanic activity was unfavourable for forest development and exerted considerable stress. These types of events occur with a certain frequency over time, which results in physical, chemical, and ecological changes that occur after the event called ecological disturbances (White 1979). Volcanic activity generates changes in biotic and abiotic components of the natural system, and the landscape is transformed during and after a volcanic eruption. The effects of these changes will be related to the nature of the event itself. In some cases, the landscape will experience abrupt changes in altitude and slope, among other variables. At the same time, the edaphic units may be covered by volcanic products (tephras, lava flows, and lahars), which in turn will influence chemical and structural composition of the soil, as well as being modern unaltered materials. Therefore, syn-volcanic and post-volcanic events affect the biotic component, and more specifically, the vegetation community is drastically affected. These changes

D. Martínez-Atienza (✉)
Departamento de Ciencias Ambientales, Universidad de Castilla-La Mancha, Toledo, España
e-mail: diego.martinezatienza@hotmail.com

G. E. Alvarado
Escuela de Química, Universidad Nacional, Heredia, Costa Rica

A. Quesada-Román (ed.), *Landscapes and Landforms of Costa Rica*, World Geomorphological Landscapes, https://doi.org/10.1007/978-3-031-64940-0_14

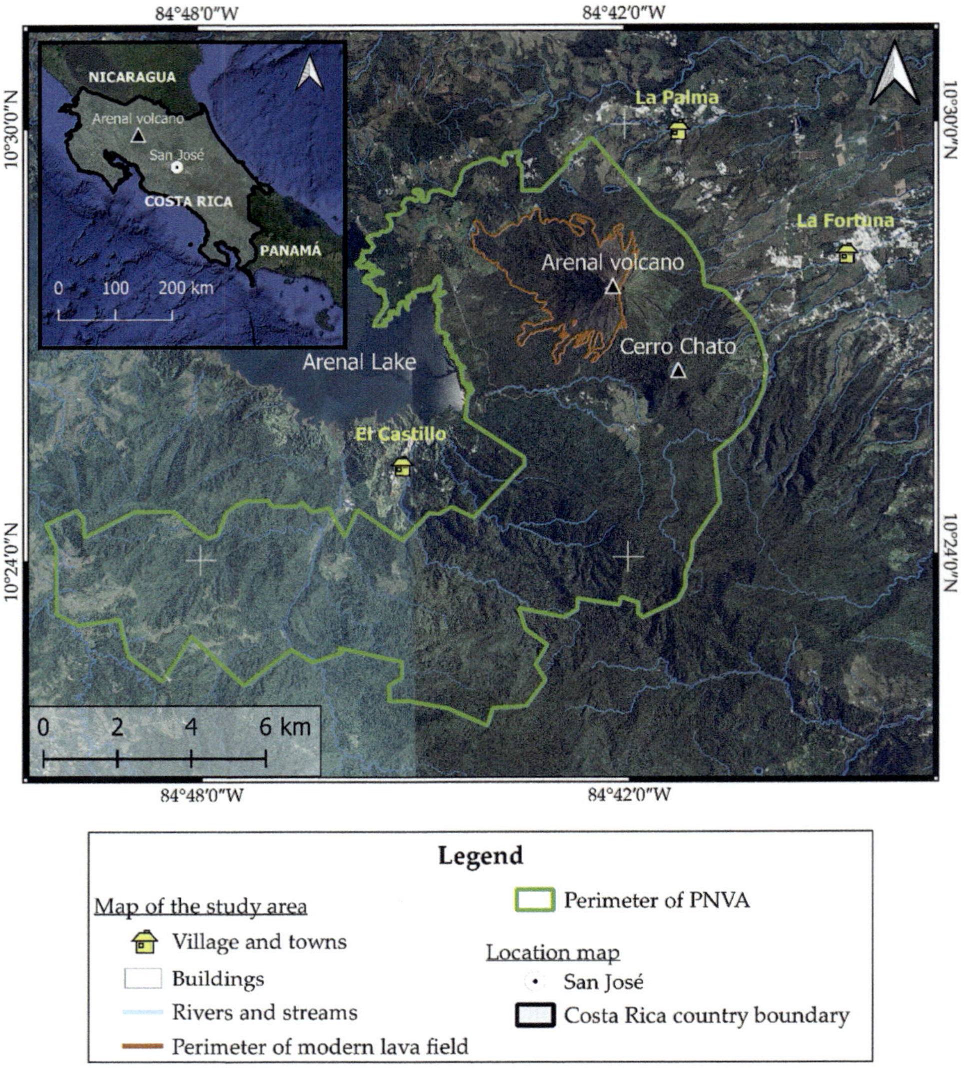

Fig. 14.1 Map of the study area showing Parque Nacional Volcán Arenal (PNVA), Arenal, and Chato volcano and the historic lava field of Arenal volcano (1968–2006, after Wadge et al. 2006). Elaborated from the cartography available at SNIT (2017a, b). Orthophoto of the location map obtained from Google 2015

can be also associated with erosion and fluvial processes. These events feed back into the problem, as there is a constant washing out of materials and nutrients that hinders the establishment of future seedlings that would recolonize the system. In tropical ecosystems, erosive effects are more pronounced due to heavy rainfall (Labrière et al. 2015). Charles Lyell (1834) realized that vegetation was one of the natural components capable of intervening in geomorphological processes. He considered that vegetation could retain loose soil in the face of erosive factors such as wind, rain, and rivers. Confirming Lyell's statements, modern studies support the ability of vegetation to reduce erosion, increase infiltration, decrease runoff, and reduce flow speed (e.g. Gyssels et al. 2005; Marston 2010; Zhang et al. 2014).

The first purpose of this chapter is to provide an overview and update on the volcanic geomorphology of Arenal, including field geology and providing relative ages of this small stratovolcano that was permanently active for more than four decades. There is currently a wide interest and concern about the environment of the Arenal National Park, and how the volcanic activity and the post-eruptive process (erosion, landslides, bioerosion, and other anthropic processes) work and affect or favour the development of biota, and how to conserve the vegetation community and manage geosites in the best way. Erosive processes are events that have a notable impact on ecosystem functioning. They give rise to feedback processes in which the vegetation community is one of the elements that receive the greatest impact. In disturbed systems, these erosive processes are intensified because of the loss of vegetation community. Therefore, another objective is to focus on quantifying the degree and trend of average erosion in lava fields of Arenal volcano considering plant recolonization, based on the thesis of Martínez-Atienza (2023).

For this purpose, average erosion rate maps were produced using the Revised Universal Soil Loss Equation (RUSLE) model using a geographical information system (GIS). The maps were produced after the cessation of the period of modern activity (1968–2010) of Arenal volcano, for the years 2010, 2012, 2015, 2017, 2019, and 2020. Modern methodologies support the ability to obtain the C factor or vegetation factor of the RUSLE formula through vegetation indices. For this reason, the C factor was elaborated through the Normalized Difference Vegetation Index (NDVI) obtained from satellite images.

Therefore, the study of vegetation succession and runoff erosion processes is crucial to understand how the ecosystem functions. Some authors even promote the term biogeomorphology, referring to the mixture of ecological and geomorphological science (Osterkamp et al. 2012). For this reason, the study of geomorphological processes must consider the biological components of the natural systems. In the present case, the study of recolonization of the Arenal volcano helps to understand the areas of greatest susceptibility to erosive processes.

14.2 Geological and Geomorphological Setting

14.2.1 Geological Background

The Chato and Arenal are small brother stratovolcanoes located between the two major volcanic ranges: the Cordillera de Guanacaste and the Cordillera Central. Chato volcano (~1140 m a.s.l., 12 km^2, 4 km^3) is about 50 ka old (Alvarado and Gans 2012) and has a truncated cone with an explosion crater (500 m × 415 m with a crater lake 346 m × 215 m), a volcanic dome, and several other secondary cones with rest of craters (Fig. 14.2). The profile of the cone has a break in the slope at about 900 m a.s.l. suggesting that the 240 m high cone hosting the crater lake grew over a previous cone (Alvarado et al. 2023). The lavas

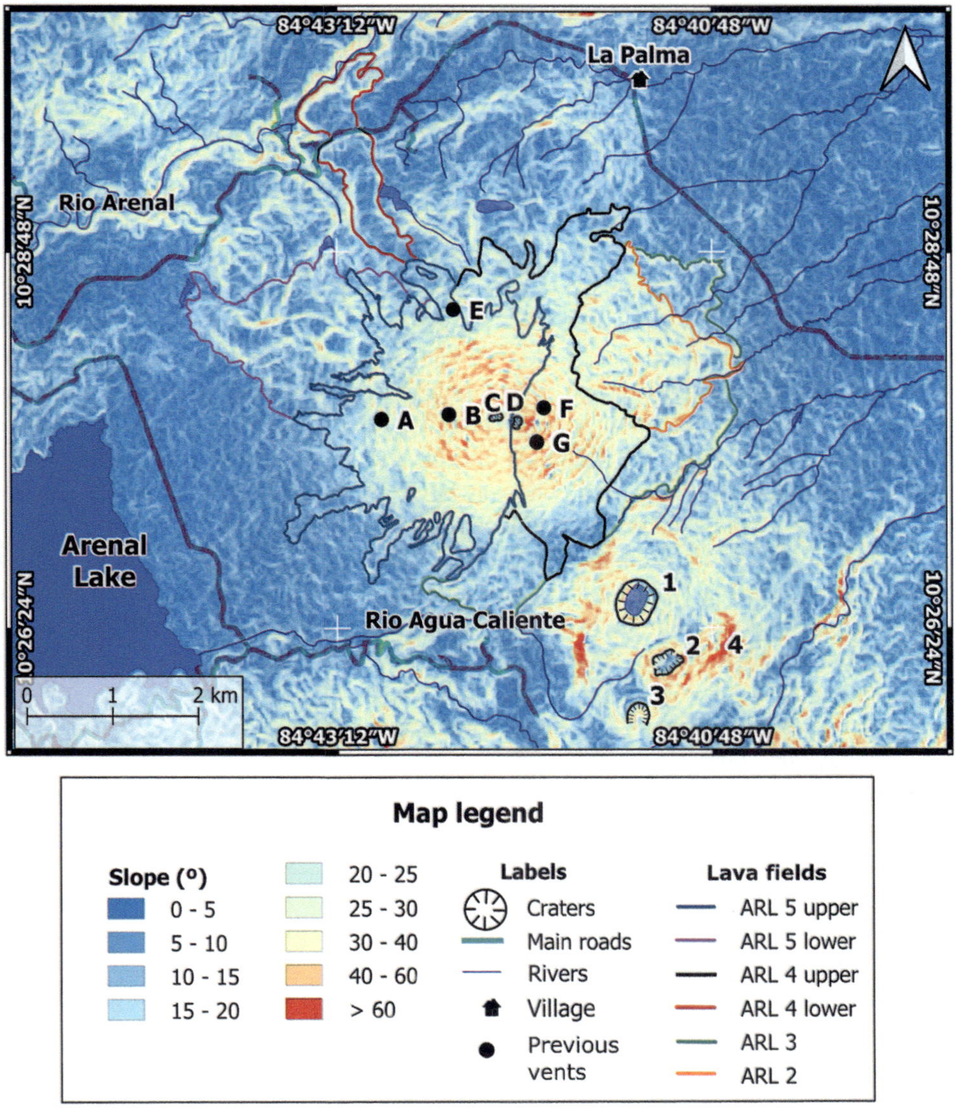

Fig. 14.2 Slope map of the study area with the perimeter of the lava fields (elaborated from DEM, USGS 2011). Rivers and roads obtained from SNIT (2017b). The craters of the last eruptive period (**A** and **B**, now covered by lava flows; **C** completely exposed), and the previous one (crater **D**, partially buried) are indicated. Also, the inferred position of craters **E**, **F**, and **G**) is shown (Borgia and Linneman 1990). The Chato crater (1), the Chatito cone (2), one other unnamed crater (3) and the La Espina dome (4) are also presented. Ages of the lava fields: ARL1 (>1250 B.C.), ARL2 (1250–350 B.C.), ARL3 (~650–700 A.D.), ARL4 lower and upper (<1400 A.D.), and ARL5 lower and upper (lower: 1968–1973 A.D.; upper: 1974–2010 A.D.) (Soto and Alvarado 2006a, b; Alvarado et al. 2023)

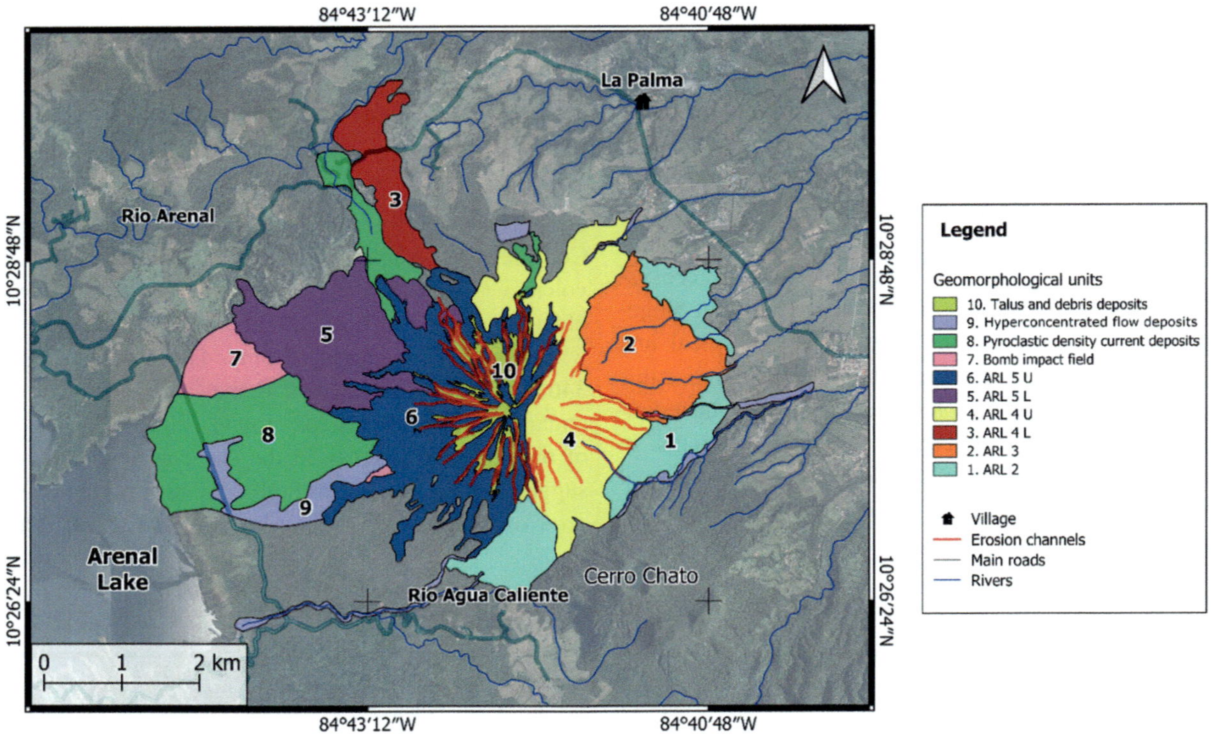

Fig. 14.3 Geomorphological map of the Arenal volcano. Orthophoto and main road network obtained from the SNIT (2017a, b). The map is modified after Borgia et al. (1988), Alvarado et al. (2006), Ebmeir et al. (2014), and Torres and Alvarado (2022), where erosional geomorphological updates have been included. ARL1 is not present because the outcrops are very small and have not generated any morphology

are basaltic to andesitic in composition (Borgia et al. 1988; Malavassi 1991).

Arenal volcano (~1684 m a.s.l., 33 km^2, 9 km^3) used to have an almost perfect conical shape, typical of a stratovolcano. As a result of eruptions of volcanic products during the recent period of activity, the volume of the volcanic edifice increased unevenly, and a new cone was built on crater C generating a twin cone together with cone D. Thus, if the volcano is viewed from Laguna Arenal or La Fortuna, it will appear to have a typical conical shape; however, if the volcano is observed from the N or S, the uneven profile (two cones with the same base) is visible.

The nature of Arenal volcano's eruptions has varied over time since its origin at least 7 ka. The Arenal volcano is known to have had phases of activity ranging from Plinian to Strombolian eruptions and is composed of tephra layers (ash, lapilli, blocks, and volcanic bombs), lava flows, and fluvial deposits (Soto and Alvarado 2006a, b). The 1968 eruption of Arenal volcano was one of the most violent eruptions in Costa Rica's recorded history, claiming at least 78 human lives (Melson and Sáenz 1968).

Arenal volcano is deforming its basement under the pressure of *ca.* 20×10^3 kPa, affecting it for several kilometres below the surface and about 5 km around the volcano base. The total settlement below the historical lava fields is 2 m or more, but the subsidence will be continuous for years. The results of numerical finite models show that the volcano is in an incipient stage of deformation by spreading of the basement (Alvarado 2003; Alvarado et al. 2010). The volcano's western flanks are moving downslope at an angle of ~55° below the horizontal plane and at a consistent rate of at least ~7 cm/yr (Ebmeir et al. 2010).

A geomorphological summary of the Arenal volcano is presented below. The main units can be summarized in the following units (Fig. 14.3), which are detailed in this local stratigraphic framework, including petrographic and petrologic aspects.

14.2.2 Geomorphological Units

Field geology, radiometric ages, and morphometric relationships between different geomorphic units, as well as the grade of erosional process related to the historical eruptive sequence allow one to provide a summary of the geological–geomorphological development of the volcano (Fig. 14.3). Tephra units, soils, and vegetation cover were efficient indicators of the relative age of older lava fields (Borgia et al. 1988; Soto and Alvarado, 2006a, b).

14.2.2.1 Lava Fields

Arenal volcano is composed of at least four well-developed lava fields (ARL 2, ARL 3, ARL 4, and ARL 5 in order of formation) and an old lava field that is currently covered by the four previously mentioned. Lava fields were recognized by geological field mapping, petrological and textural studies, geomorphologic features, grade of forest evolved, relative age, and historical evolution in the case of the last eruptive period; the whole of lava fields constitute the lava armour of the volcano. Because of the juxtaposition of different lava fields and their intersection, the topography of the volcano is not perfectly a symmetrical cone and, therefore, exhibits clear discontinuities in the slope (Borgia et al. 1988; Borgia and Linneman 1990). The slopes of the lava armour are not completely uniform and are dissected by a well-defined radial drainage pattern. Streams form ravines and gullies incised 10–30 m deep into the tephra and lava breccia, eroding down to the lava flows. The lava flows often initially formed surfaces but became blocky in the downslope direction as the flow was disrupted (Cigolini et al. 1984). Petrographically, they are basaltic andesites in composition (except for one old lava flow at the base, of andesite composition; Borgia et al. 1988). The lava fields were spatially and chronologically ordered with good approximation and renamed by Soto and Alvarado (2006a, b) as ARL1 (>1250 B.C.), ARL2 (1250–350 B.C.), ARL3 (~650–700 A.D.), ARL4 lower and upper (<1400 A.D.), and ARL5 lower and upper (lower: 1968–1973 A.D.; upper: 1974–2010 A.D.).

The lava field ARL3 and the older one is overgrown by a dense tropical rain forest, evolved into thick soils, and do not bear any evident structure of lava flows. Concave relief present at the NE flank near the main road and several hot springs suggest a continuity of the ARL2 lavas that are now covered by sediments. ARL4 prehistoric lava field can be recognized on aerial photographs and satellite images by means of typical features of lava flows such as lobules, pressure cracks (suture lines), central channels, and lava fronts. The forest is similar to a secondary forest, and soil development is scarce. In the prehistorical lava fields present near the top of the cone D, it is locally possible to recognize small and incipient *planèzes*. In the historical ARL5 lava field, the typical meso- and microfeatures are easy to recognize, and in addition to those already described in ARL4, there are micro-spines covered by scarce to almost no vegetation and soil. The lowest lava fields ($ARL4_L$ and $ARL5_L$) closed and dammed local streams and generated small lakes or small endorheic basins.

The first effusive phase (1968–1973) formed the lower ARL5 lava field from crater A, now all these lavas are highly recolonized. From August 1973 to March 1974, there was a 7-months long hiatus in lava effusion. The activity resumed and moved to crater C about 400 m higher up the volcano, and between 1974 and 2010 the upper lava field or $ARL5_U$ was built (Wadge 1983). Mapping of the historical lava field was carried out by Wadge et al. (2006) and included the perimeter of the lava field formed from 1968 to 2006 but did not cover the last four years of activity, during which the volcano activity was limited or negligible. During 1974–2010, crater C emitted many volcanic products that covered craters A and B, and therefore, only craters C and D are visible today.

14.2.2.2 Hummocky Morphology by Impact Craters

An undulating surface is still a recognized product of the 1968 Pelean explosion that generated a rain of ballistic bombs and blocks that impacted the ground producing several impact craters from less than one meter in diameter up to 60 m in diameter and 4 m deep (Melson and Sáenz 1968; Alvarado et al. 2006).

14.2.2.3 Depositional and Erosive Landforms: Hyperconcentrated Flows, Talus, and Debris Deposits

The degree of fluvial dissection produced by rainfall and slides and its associated deposits could be observed in a heterogeneous distribution on the flanks, where channels are deeper in the cone flanks. Tephra and lava breccia eroded from the steeper slopes of the lava armour are the source of mainly hyperconcentrated flow deposits present as fans and plains at the foot of the volcano, where branching and rill dissection are locally present, although in most of the pyroclastic fans the runoff rate is low and surface drainage is poor (Borgia et al. 1988; Soto 2008; Torres and Alvarado 2022; Paniagua, 2023). The modern talus and debris deposits are unconsolidated materials that are transported by gravity and hydrological events. These materials accumulate in the form of debris fans in the distal parts of the cone where the slope decreases abruptly (Fig. 14.4). This unit is the youngest one, still being formed by erosion of the upper lava field ($ARL5_U$).

Ebmeir et al. (2014) and Torres and Alvarado (2022) presented high-resolution satellite radar and Google imagery evidence, respectively, of frequent rock falls and slow-moving, shallow translational landslides. Landslides have faster-moving depletion zones at their heads and lower displacement rates in the accumulation zones at their toes (Ebmeir et al. 2014). In 2023, at least 16 slow-moving landslides were active, and about 7% of Arenal's surface area was covered by active landslides (estimated to be 5–11 m thick) where $\sim 3.8 \times 10^7$ m^2 of sliding material happened between 2011 and 2023 with a total volume of 35×10^6 m^3 (Table 14.1).

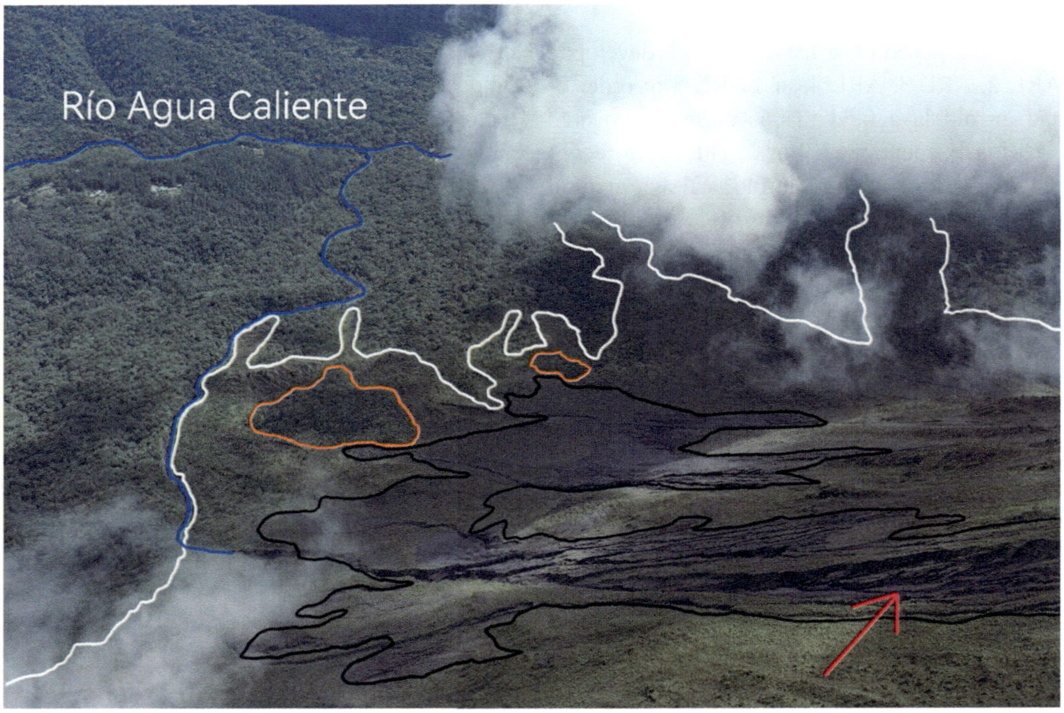

Fig. 14.4 Oblique image taken in 2022 from the summit of Arenal volcano during an aerial flight in 2022 (Prado 2023). The southern flank of the volcanic cone includes one of the largest gullies of the volcano (red arrow). The perimeter of $ARL5_U$ (white) can be seen with the presence of vegetation islands known as *kipukas* (small islands of life surrounded by lava flows drawn in orange). In the background the Agua Caliente River

14.2.2.4 Volcanotectonic Framework

Craters

The last eruptive period began on 29–31 July 1968, with the formation of three new explosion craters along a E-W line on the western flank (craters A, B, and C) and activity of the pre-existent crater D (Melson and Sáenz 1968). Two of these craters, C and D, are still preserved. The crater C is located on the summit and grows on a horseshoe scarp carved into the summit during the 1968 explosive phase (Fig. 14.5). Since the beginning of the effusive activity at crater C in 1974 (Wadge 1980), a cone has grown up with different crater sizes, small hornitos and contented partially molten lava (lava pool, and ephemeral domes) in the last decade of eruptive activity (Alvarado 2021).

From 1968, the volcano entered the phase of eruptive activity. Three new explosion craters were formed on its western flank, oriented in an E-W direction: (i) crater A, approximately at 1060 m a.s.l. and with a dimension greater than the others; (ii) crater B at 1170 m a.s.l., the smallest one, and (iii) crater C at 1460 m a.s.l., with intermediate dimensions with respect to the other two. The fourth and last is crater D, at 1633 m a.s.l., which predated the others (Melson and Sáenz 1968; Alvarado et al. 2006). At present, only craters C and D exist. Crater A was active from 1968 to 1973. From 1974 to 2010, cone C grew by 224 m (data updated to 2022) at a rate of between 4.5 and 9 m/year; averaging about 4 m/year from 1998 to 2005 (Hofton et al. 2006). By 1987 cone C was about 50 m lower than cone D, although by 1996 both had roughly the same elevation and by September 2000 crater C was about 20 m higher than crater D (Cole et al. 2005). Crater C at times had two or more outflow foci, or even involved the formation of domes and coulees (Alvarado 2021).

At present, crater C (altitude 1684 m) has an ellipsoidal shape, 154.7 m long by a variable width between 71.7 m and 80.8 m, with several internal depressions with a maximum depth of only 19 m. Crater D (altitude 1622 m, according to new topographic survey in 2022), on the other hand, is 129 m long, 87 m wide, and 45 m deep. Its circular shape was altered by volcanic processes between 1968 and 2010, as well as by the growth of cone C, which covered and partially filled part of crater D (Alvarado et al. 2023).

Based on the asymmetric distribution and different slopes of the lava fields, Borgia and Linneman (1990) estimated the position and elevation of the corresponding craters E, F, and G (Fig. 14.2). Crater E and the paleo-valley are aligned on the Chatito-Chato-Arenal axis, the northwestern trend of the Central American volcanic arc; thus, the paleo-valley probably formed by preferential erosion

Table 14.1 Geomorphological units and their characteristics (based on Borgia et al. 1988; Borgia and Linneman 1990; Alvarado et al. 2006, present work)

Unit	Lithology	Relief	Morpho-chronology	Morphogenesis
ARL2	Basaltic andesite lava flows	Smooth morphology exposed only at the southern and eastern flank	1250–350 B.C	Lava flows from crater G
ARL3	Basaltic andesite lava flows	Semi-cone with triangular shape presents only on the eastern flank	650–700 A.D	Lava flows from crater F
$ARL4_L$	Basaltic andesite lava flows	Thick and wide extended lavas with levees, lobules, pressure cracks, central channels, and lava fronts	< 1400 A.D	Lava flows from crater E
$ARL4_U$	Basaltic andesite lava flows	Thin and narrow lava flows with levees, lobules, pressure cracks, central channels, and lava fronts	< 1400 A.D	Lava flows from crater D
$ARL5_L$	Basaltic andesite lava flows	Thick and wide lava flows with levees, lobules, pressure cracks, central channels, and lava fronts	1968–1973	Lava flows from crater A
$ARL5_U$	Basaltic andesite lava flows	Thin and narrow, lava flows with levees, lobules, pressure cracks, central channels, and lava fronts. Surface inclination 30°-40° in the summit, and < 30° on the slope break	1974–2010	Lava flows from crater C
Impact craters	Basaltic andesitic blocks and bombs	Irregular morphology of small depressions (secondary craters)	July 29–32, 1968	Bombardment during the Pelean eruption by ballistic blocks and bombs
PDC	Basaltic andesitic pumice and scoria	Fan with relatively smooth topography	1968–2000	Pyroclastic flow deposits
HyCF	Poorly to moderate well-stratified and poor to well-sorted boulders and sand deposits	Smooth topography	1968–2010	Hyperconcentrated flow deposits by mechanical disaggregation of lava flows and erosion of pyroclastic and previous epiclastic deposits
T & D	Poorly to moderate well-stratified and poor to medium sorted boulders and sand deposits	Steep topography	1980–2023	Mechanical disaggregation of lava flows and erosion of pyroclastic deposits

Fig. 14.5 Oblique image of the summit of Arenal volcano during an aerial flight in 2022 (Prado 2023). One can see crater C (red), crater D (orange), the illegal trail (blue line) connecting crater C and D, and the boundary between the lava fields $ARL5_U$ and $ARL4_U$

along fractures. By analogy to crater A, crater E opened near the intersection of a northwestern regional fracture with the lava armour-tephra apron transition. Therefore, craters would tend to open on the northwestern or western flank of Arenal.

Faults and Other Structures

The lower and major crater (crater A) created by the July 1968 explosive activity opened within a deep valley near the transition between lava armour and the tephra apron. This valley was parallel to the regional north-trending faults and tangential to the cone, thus, possibly the crater A opened at the intersection between this north-trending fracture with the lava armour-tephra apron transition (Borgia et al. 1988).

Arenal is cut by several fractures and active faults (<4000 years old) in N-S, NE-SW, and NE-SW strike directions, which disturbed the Holocene volcanic sequence. Another peculiar erosive important structure, on the western and southern flanks and now covered by lava flows, has been the subject of debate over the years (Fig. 14.6a). It was interpreted as a caldera rim (Melson and Sáenz 1968; Bergoeing 1978), a normal fault scarp (Bravo 1982), or simply as an erosive border (lithological contact) at the front of the previous lava field (Borgia et al. 1988). However, this rim does not correspond to a caldera because there are no associated ignimbrites, nor it indicates the contact of lava armour-tephra apron since the erosive edge of the contact was dipping towards the most resistant lava rocks. After a review of the pre-1970 aerial photographs, it is possible to interpret this rim (Fig. 14.6b) as a reverse fault scarp produced by the continuous propagation of movement through the cone from the N-S dextral-normal Danta fault (Alvarado 2003; Alvarado et al. 2006), following the structural models of Lagmay et al. (2000). All the craters are parallel to an E-W maximum horizontal compressive stress direction (López 1999), favouring a radial fracture on the western flank of the volcano. Craters B and C probably originated after the crater A. Thus, the 1968 crypto-dome could intrude first through the low-angle thrust fault favouring the lateral blast eruption (Alvarado et al. 2006).

14.2.3 Erosional Processes After Volcanic disturbance—The Case of the Historic Lava Field (ARL5)

Studies on the vegetation of Arenal volcano are scarce and much localized. Vargas (1985) characterized the successional stages of vegetation of the Arenal volcano and surrounding areas. That study was carried out in 1985, and the historic eruptive period ended in 2010. It is useful and

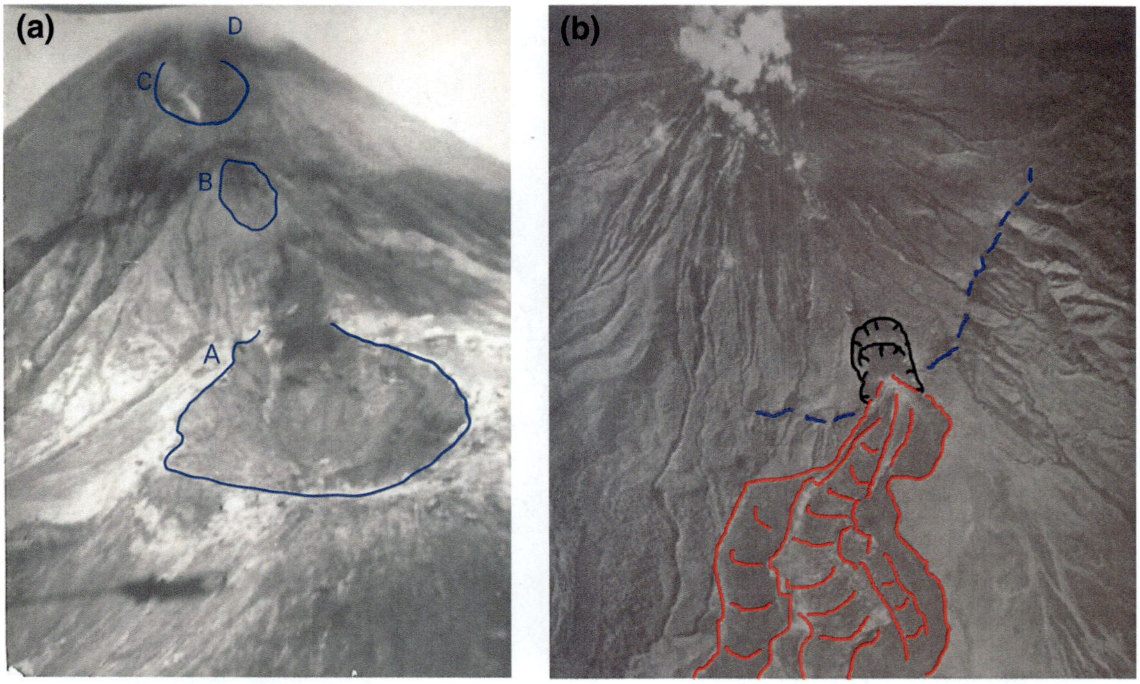

Fig. 14.6 Historical photographs of Arenal volcano: **a** Western flank with the craters **a**, **b**, **c** and **d**. Image taken by an ICE engineer on an aerial flight in 1968. **b** Western flank of Arenal with crater A (black lines) and the first lava flows (red lines). The pressure cracks are visible. The controversial morphological rim (blue dashed line in A), see text for discussion). Aerial photograph IGN, 4 September 1971

valuable, but also outdated. Another study was carried out by the same author (Vargas et al. 1993) and involved a zoned sampling in three regions. In the central and western flank, they quantified a decrease in species richness between 1985 and1993, in contrast to the NW sector, where the trend of recolonization proved to be progressively positive. The increase in the latter region could be explained by only limited disturbance during the period 1968–1973, as opposed to the other regions where acid rain and ash deposition events were still occurring. In any case, the study shows the importance of zoning the study area, as each region may have its dynamics with respect to the disturbance. For this reason, the zoning developed by Oramas-Dorta et al. (2012) was used in this study to delimit two geological areas with different ages after the last disturbance.

The $ARL5_L$ was created in 1968–1973. This lava field has been in plant recolonization since 1973, so the vegetation succession is at an advanced stage. The $ARL5_U$ (1974–2010) has been exposed to several recent disturbances (lava and pyroclastic flows, rock falls, erosion, ash fall, and acid rain). Therefore, vegetation is at a young recovery stage in most of the region. This delimitation allowed us to analyse the influence of time of vegetation recolonization on erosion events. Disturbed ecological systems are characterized by an immediate response to the event. In these systems, different transformation processes can be observed, but they occur continuously over time. The speed of transformation will depend on the capacity of the ecosystem to cope with the disturbance, which is known as ecological resilience (Holling 1973).

Due to steep slopes and heavy rainfall, large gullies have developed on the flanks of the Arenal (Fig. 14.7), especially in the steeper upper lava field ($ARL5_U$). Because of erosion processes, seedling and seed establishment is unfeasible in areas where disturbances such as gullies and sedimentary material deposition zones occur. However, seed establishment was observed in areas adjacent to gullies and deposition zones. The average NDVI value of the ARL5 in 2010 was 0.51, increasing to 0.66 in 2020. In 10 years, the vegetation community showed a high capacity for recolonization and reduced the efficacy of erosion processes. These changes have been monitored through the creation of erosion maps for different years. A distinction was made between the upper lava field ($ARL5_U$) and the lower lava field ($ARL5_L$) because the lower lava field started to become recolonized at different stages after 1974. Therefore, the $ARL5_L$ is expected to show lower average erosion rates.

14.2.3.1 General Erosion Trends Over the Historic Lava Field (ARL 5)

This work has studied the spatial–temporal changes in the average erosion rate of the historic lava field (1968–2010) after the cessation of volcanic activity in 2010. For this purpose, erosion susceptibility maps were obtained using the

Fig. 14.7 Gullies on the flanks of the upper lava field ($ARL5_U$) generated by water erosion processes (Prado 2023). In the gully and debris deposition areas, vegetation has not been able to establish. However, in the surrounding regions, vegetation is present. High-resolution photograph taken during an aerial flight

Revised Universal Soil Loss Equation or RUSLE (Renard et al. 1991). To calculate the C factor or vegetation cover factor, the Normalized Difference Vegetation Index (NDVI) (Rouse 1974) derived from 5 TM, Landsat 7 ETM+, and Landsat 8 OLI satellite images were used (Fig. 14.8). Satellite images were obtained from Google Earth Engine (Gorelick et al. 2017) following the methodology proposed by Quintero et al. (2019). This tool has proven to be very efficient for obtaining the relevant value. Specifically, Durigon et al. (2014) determined that in regions with high rainfall intensities the best way to calculate the C factor was through vegetation indices. Erosion maps were produced for the years 2010, 2012, 2015, 2017, 2019, and 2020, with which the trend of erosion processes could be represented as vegetation cover increased. The factors of the RUSLE formula were obtained from various data sources. Climate data (2022) was used to develop the R factor. The ratio between clays, sands, silts, and organic matter content present in the soil was considered to calculate the K factor. The data to calculate it were obtained through soil grids (Poggio et al. 2021), which was a project supported by the International Soil Reference and Information Centre (ISRIC). The LS factor considers slope and orientation of the slope and was calculated from the digital elevation model (DEM) obtained from the United States Geological Survey (USGS 2011). The P factor or support practice factor refers to soil conservation features implemented in the study region that affects the effects of erosion. No such element has been found in Arenal volcano, so the P factor has been estimated to be equal to 1. Finally, the erosion susceptibility maps had been drawn up, and the values of raster layers of erosion maps were extracted using QGIS 3.12 software (2022). The extraction of values of raster layers was carried out by systematic sampling for each $30 \text{ m} \times 30 \text{ m}$ per pixel within the historic lava field (1968–2006), as developed by Wadge (1983) and Wadge et al. (2006).

The RUSLE method is efficient in calculating erosion values. However, there is no generalized table of potential soil loss values, and therefore, in some studies, the value at which severe soil loss occurs is set as 10 t ha^{-1} y^{-1} (Mahala 2018). On the other hand, other studies indicate that the value of severe soil loss occurs when it exceeds 300 t ha^{-1} y^{-1} (Montalvo 1998). In the present work, it has been determined that the threshold value for a severe soil loss should be considered as 100 t ha^{-1} y^{-1} (Table 14.2).

At the end of 2010, the eruptive activity of the Arenal volcano ceased, and only fumarolic activity and acid rainfall have been recorded until the present day; thus, processes of ecological succession accelerated. In 2010, the

Fig. 14.8 Examples of NDVI values obtained at Arenal volcano. Photograph of the NW flank taken in April 2015 by Federico Chavarría, in Alvarado (2021)

Table 14.2 Classification of the values of potential soil loss used for this work

Soil loss classification	Potential soil loss (tonnes/ha/year)
Very low	0–10
Low	10–25
Moderate	25–50
High	50–100
Severe	>100

average erosion rate was estimated to be 50.7 t ha^{-1} y^{-1}, with a maximum erosion value of 576 t ha^{-1} y^{-1}. However, due to the Landsat 7 banding error, the 2010 composition had a large gap with no information on the upper lava field (Fig. 14.9a). The composition was valid for the lower lava field.

In 2012, the average erosion rate was estimated at 53.5 t ha^{-1} y^{-1}, reaching a maximum value of 566 t ha^{-1} y^{-1}. The median was lower, 27.5 t ha^{-1} y^{-1}. The median helps to understand that most of the data is clustered around low erosion values. This indicates that regions with high erosion values are rare in the study area. In 2015 there was a slight increase in the mean value of the average erosion rate (54.9 t ha^{-1} y^{-1}) compared to 2012. At the same time, the median value decreased (Fig. 14.10), meaning that the mean erosion rate decreased at most of the pixels. On the contrary, for some pixels located in the steeper regions of the volcanic cone, the mean erosion rate values increased. This becomes coherent if we observe that the maximum value in 2012 was 566 t ha^{-1} y^{-1}, compared to 2015 where the maximum value was 647 t ha^{-1} y^{-1}. At the same time, the effects of weathering and slope instability generated gullies and rock falls, which are the conditions in which the average erosion rate is at maximum. Therefore, the maximum values in 2015, higher than in 2012, were those located in these regions. This hypothesis can be supported by the fact that the trend (Fig. 14.11) between 2012 and 2015 in the lower lava field decreased, where there are no erosion channels. By contrast, in the upper lava field, where erosion channels and gullies are located, the average erosion rate value increased. This will be discussed in more detail in the following subsections.

For 2015 to 2020, a progressive decrease in the average and median erosion value was obtained (Fig. 14.10). This trend corroborates that the increase of vegetation cover in the lava field decreased the intensity of erosion processes. In 2020, the value of the average erosion rate was 40.9 t ha^{-1} y^{-1}, which is 9.7 t ha^{-1} y^{-1}less than in 2010, and 12.5 t ha^{-1} y^{-1} less than in 2012.

The erosion maps show the change in the erosion trend in the historic lava field of Arenal volcano. Most of the lava field suffers less from erosion as time progresses. However, there are regions in the upper lava field, especially in the erosion channels (gullies), where the value of the average erosion rate increases with time. Similarly, the areas surrounding the gullies show decreasing average erosion rates. Thus, the 2020 map (Fig. 14.9f) shows an asterisk-shaped pattern concerning the crater where the values are between 50 and 650 t ha^{-1} $y^{-1}$of the average erosion rate. These regions coincide with the areas of higher altitude and slope. In general, it has been observed that the points with the

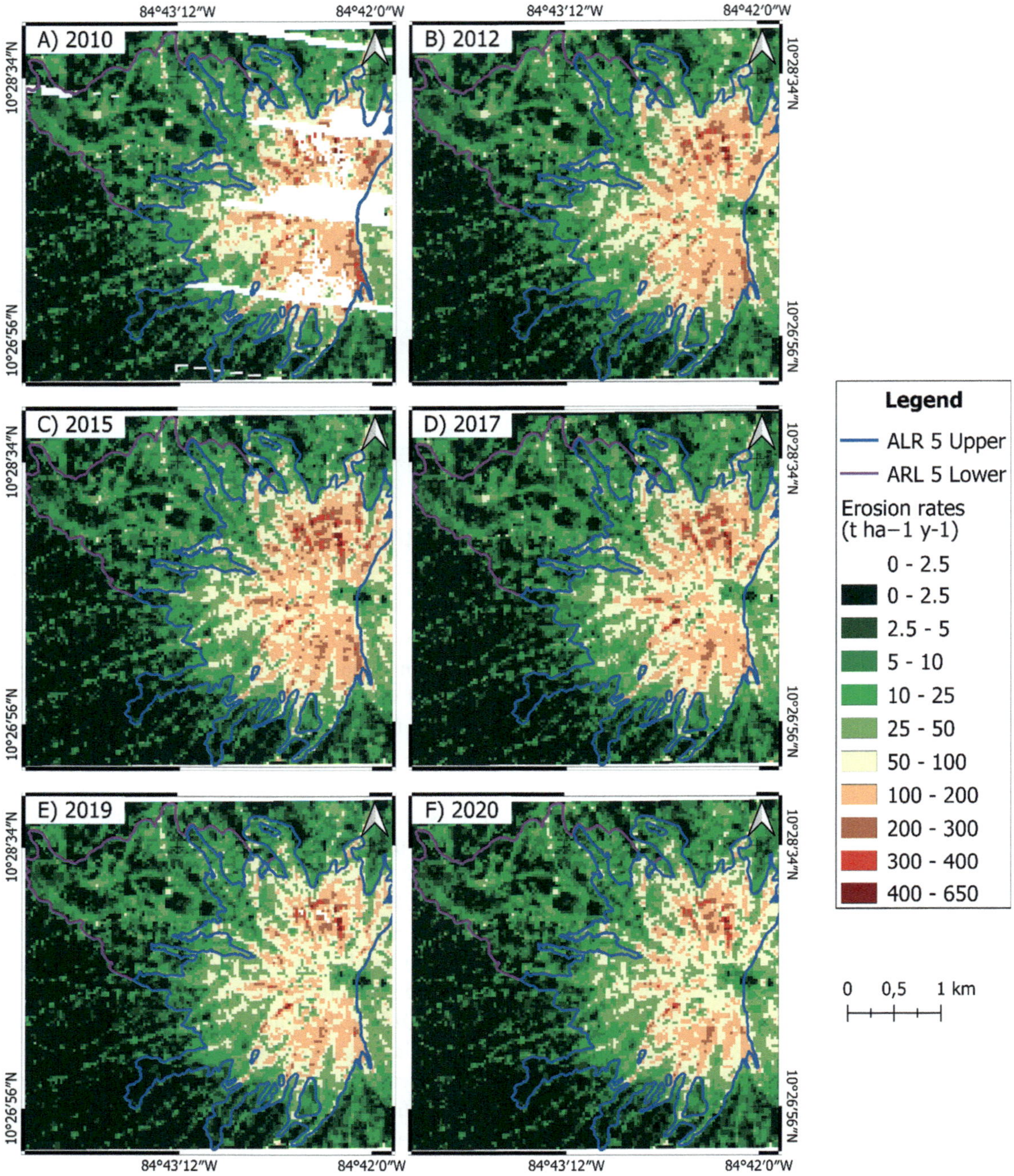

Fig. 14.9 Maps of average erosion rates obtained for the ARL 5 $_U$ and ARL 5 $_L$

highest average erosion rate values of both lava fields are linked to regions with high slopes.

14.2.3.2 Erosional Processes of ARL5$_L$ from 2010 to 2020

The lower lava field has shown a decreasing average erosion rate trend over the studied period (Fig. 14.11). In 2010, the average erosion rate presented a value of 14.64 t ha^{-1} y^{-1}, while in 2020 it diminished to 9.82 t ha^{-1} y^{-1}. Despite this decrease, this variability is of little relevance. The main reason why the trend has decreased a little over time is explained by the fact that the lower lava field has been undergoing a process of recolonization since 1973. For this reason, erosional processes have been progressively decreasing from 1973 to 2010. Today, the lower lava field shows a dense vegetation cover, with average NDVI values of 0.79. In general, erosional processes in the lower lava field are currently of low magnitude.

Within the lower lava field, the largest lava tongue of the entire lava field occurs. This tongue moved in NW direction

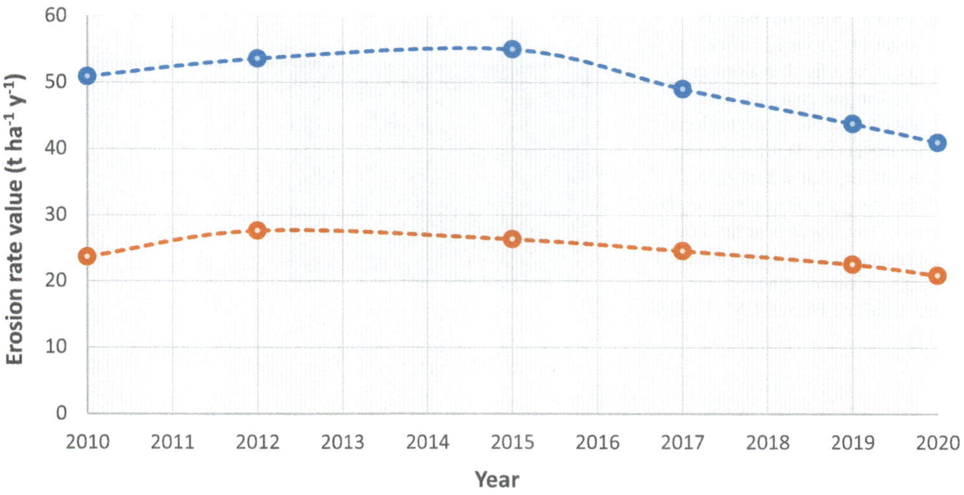

Fig. 14.10 Trend graph of average (blue line) and median (red line) erosion values (t ha^{-1} y^{-1}) of the ARL 5 (1968–2010) for the years studied

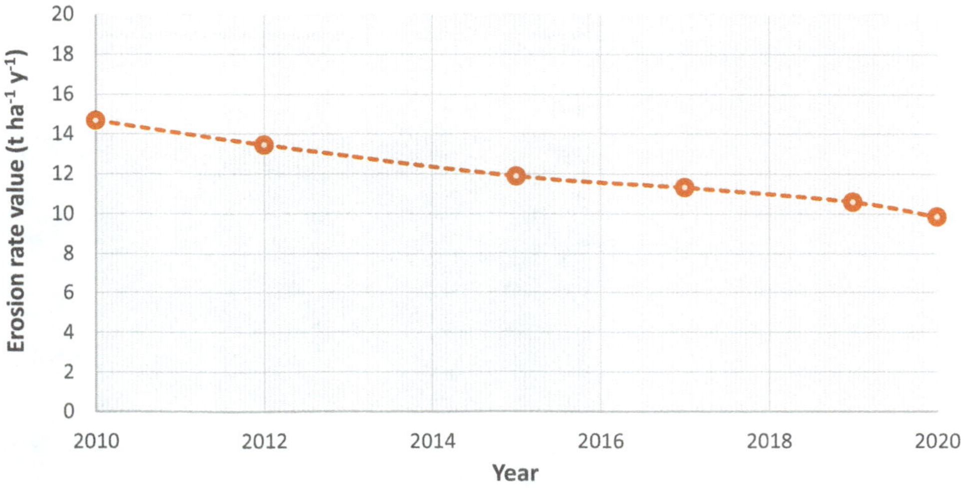

Fig. 14.11 Trend graph of the average erosion rate (t ha^{-1} y^{-1}) in the ARL 5_L during the period 2010 to 2020

during its formation. It has a meandering shape with a length of 2 km. This tongue can be divided into two zones (Fig. 14.12a). On the one hand, there is the top part, where an average erosion rate is mostly in the range of 0–5 t ha^{-1} y^{-1} in 2020. Here, parallel furrows can be observed that were formed during the cooling of the tongue. During the formation of this tongue, the melt front moved more slowly than the intermediate region of the tongue. For this reason, these easily visible furrows were formed, which have a high potential to retain moisture. At the same time, plant roots can penetrate more easily as the material is more fragmented due to pressures it has undergone during cooling. This is probably the reason why these furrows have been better recolonized by vegetation than the rest of the top part of the lava tongue. This top region of the lava tongue has a relatively flat morphology, with a slight slope in the direction of flow. When observing the lava tongue with contour lines, 2 further spatial units can be distinguished within the top region, mainly differentiated by the degree of slope. One can distinguish flat and steep units, which altogether give the lava tongue a terraced appearance. As expected, in the steeper units the average erosion rate increases markedly compared to the flat ones. In contrast, the flat units within the top region of the lava tongue have the lowest average erosion rate of the entire tongue.

On the other hand, the second region of the lava tongue (Fig. 14.12a) corresponds to the slopes, where the average erosion rate ranges between 5 and 50 t ha^{-1} y^{-1}in 2020. This region is steeper than the top one, which likely explains the difference in erosion rates. However, it is interesting to analyse this region in more detail as it should be considered that vegetation development has been notably higher than in the top section. Therefore, one would expect inverse correlation between vegetation cover and erosion rate. However, the opposite happens. This phenomenon is further evidence of the importance of slope in estimating the magnitude of erosion processes. At the same time, this result shows that slopes can notably reduce the capability

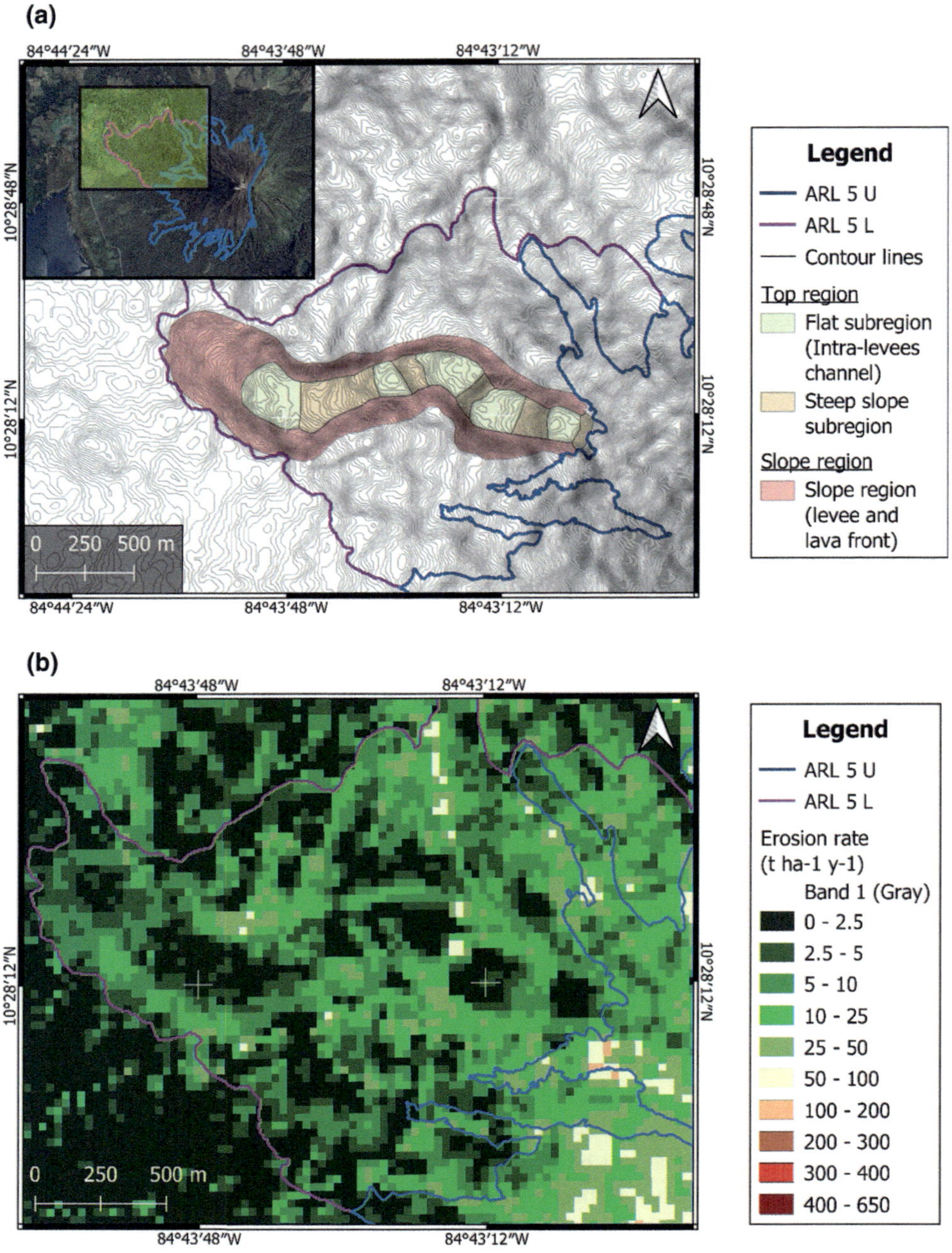

Fig. 14.12 Specific maps of the larger lava tongue formed in the lower lava field: **a** Zoning for lava tongue analysis. The top region and the slope regions can be observed. Within the top region are the flat sub-region and the steep slope sub-region. **b** Zoom of the lower lava field on the average erosion map obtained for 2020. Contour lines at 2 m were obtained from DEM (USGS 2011)

of vegetation to retain soil. Regarding the rest of the lower lava field, the region located to the N–NE presents a group of values that range between 10 and 50 t ha^{-1} y^{-1}in 2020. Again, this result coincides with an increase in the slope gradient of the N - NW facing slope. The relationship between the slope and the NDVI value can be seen by comparing Fig. 14.10a and b.

14.2.3.3 Erosional Processes Within the $ARL5_U$ from 2010 to 2020

The trend in the average erosion rate of the upper lava field has varied markedly from the beginning of the monitoring to the end. In 2012, the mean erosion rate value was 74.6 t ha^{-1} y^{-1}, while in 2020 the value was 57.4 t ha^{-1} y^{-1}. Monitoring of the upper lava field started in 2012 because of the Landsat 7 banding error discussed above. From 2012 to 2015, there was an increase in the average erosion rate (Fig. 14.13). This result may be associated with the creation of erosion channels after the volcanic activity ended. In contrast, from 2015 to 2020, there was a noticeable decrease in the average erosion rate, associated with an increase in vegetation cover. However, in regions where rock fall occurs, vegetation has not been able to establish successfully as the system is subject to constant disturbance. These regions have had the highest average erosion rate values in the entire Arenal volcano lava field.

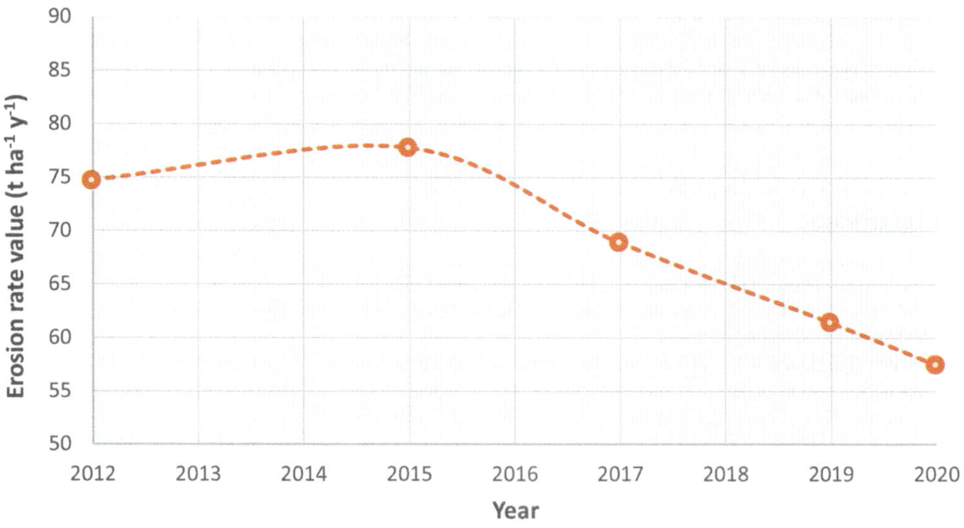

Fig. 14.13 Trend graph of the average erosion rate (t ha^{-1} y^{-1}) in the $ARL5_U$ during the period 2012 to 2020

The results obtained suggest that the upper lava field is composed of very heterogeneous patches regarding the average erosion rate. There are two main reasons for this heterogeneity. On the one hand, the upper lava field was formed over 36 years (1974–2010), although some areas were disturbed only during a single eruptive event. In general, the regions affected by only one event were located furthest away from the crater. In contrast, there are regions where the recurrence of volcanic disturbance has been the norm, especially close to the crater. Therefore, the result obtained is typical of a heterogeneous system where plant recolonization started at different times throughout the modern eruptive process. The second reason why the upper lava field shows high heterogeneity regarding the average erosion rate resides in the morphological variation of the terrain. There are regions within the upper lava field where the geomorphological characteristics of the terrain contribute to erosive events. The slope increases with the altitudinal range, and the most elevated parts will present higher erosion rates. In addition to the slope, the tropical climate of the region has a high rainfall. The combination of these two factors contributes to slope instability and gully formation.

14.3 Conclusions

In the last two decades, morphological studies of volcanic landforms have multiplied worldwide indicating the growing interest in the field very important for territorial planning (Carrión-Mero et al. 2020). The present chapter is additional evidence of this interest. Here the geomorphological aspects of different lava fields with different ages and degrees of forest development and erosion in at least the last 3 ka are presented, where the evidence of microrelief of the individual lava flows (central lava channel, levees, pressure cracks, and lava front) is gradually erased. Overall, the disappearance of volcanic morphological features is fast, but understandable given a rainy tropical climate.

During Arenal's last two eruptive periods, major craters were formed near intersections of regional fractures at the lava armour-tephra apron transition, and such intersections are potential sites for future major explosions, preferentially in the north flank followed in importance by the western flank (Borgia et al. 1988; Borgia and Linneman 1990; Alvarado et al. 2023). It was possible to quantify the magnitude of erosion processes in the historic lava field of Arenal and hence the degree of potential erosion in the region. Consequently, most vulnerable regions could have been identified, and these include gullies and rock fall-prone regions of the upper lava fields. In turn, these processes result in secondary disturbances on the vegetation community that have repercussions on the regions where there are depositional materials.

Based on the results obtained, the average erosion rate in the years following the writing of this work is expected to change over time. This is because the process of plant recolonization will continue to progress, and therefore, the value of the C factor will change. This variation depends on the region of the lava field. In the case of $ARL5_L$, it will be smaller than in the upper lava field because the recolonization started in 1974. In contrast, the variation in the $ARL5_U$ is expected to be larger because the recolonization started much later and is subject to strong geomorphological dynamics that show a noticeable landscape change over time. Considering that there is a large time difference between both lava fields concerning plant recolonization, it is consistent to assume that the trend plots cannot be compared. Therefore, it must be assumed that each component behaves differently, largely because of the successional stage of the vegetation community, and not only because of their geomorphological differences.

Acknowledgements We thank Natalia Quintero (University of Castilla-La Mancha) for her advice on remote sensing. We thank Kiersten Crye (Georgia State University) for proofreading the text. To the remarkable help of the rangers of the Parque Nacional Volcán Arenal.

References

Alvarado GE, Soto GJ, Schmincke H-U et al (2006) The 1968 andesitic lateral blast eruption at Arenal volcano, Costa Rica. J Volcanol Geoth Res 157:9–33

Alvarado GE, Gans PB (2012) Síntesis geocronológica del magmatismo, metamorfismo y metalogenia de Costa Rica, América Central. Rev Geológica América Cent 46:7–122. https://doi.org/10.15517/rgac.v0i46.1836

Alvarado GE, Carboni S, Cordero M et al (2010) Stability of the cone and foundation of Arenal volcano, Costa Rica. In: Olalla C, Hernandez LE, Rodriguez-Losada JA, et al. (eds) Volcanic rock mechanics. Rock mechanics and geo-engineering in volcanic environments. CRC Press, Taylor & Francis Group, London, pp 135–150

Alvarado GE, Brenes-André J, Núñez D, et al (2023) Actualización del análisis del peligro volcánico del Arenal, Costa Rica. ICE Internal report

Alvarado GE (2003) Diagnóstico de la estabilidad del cono y comportamiento de la fundación debido al crecimiento del edificio volcánico del Arenal (Costa Rica). M. Sc. thesis, University of Costa Rica

Alvarado GE (2021) Costa Rica y sus volcanes. San José, Costa Rica, EUCR, EUNA, ETCR

Bergoeing JP (1978) La fotografía aérea y su aplicación a la geomorfología de Costa Rica. Ministerio de Obras Públicas y Transportes, Instituto Geográfico Nacional, San José

Borgia A, Poore C, Carr MJ et al (1988) Structural, stratigraphic, and petrologic aspects of the Arenal-Chato volcanic system, Costa Rica: Evolution of a young stratovolcanic complex

Borgia A, Linneman SR (1990) On the mechanisms of lava flow emplacement and volcano growth: Arenal, Costa Rica. In: Fink JH (ed) Lava flows and domes: emplacement mechanism and hazard implications. Springer-Verlag, Berlin, pp 208–243

Bravo J (1982) Geomorfología de la Hoja Fortuna (Escala 1:50 000). Universidad Nacional, Sede Omar Dengo

Carrión-Mero P, Montalván-Burbano N, Paz-Salas N, Morante-Carballo F (2020) Volcanic geomorphology: a review of worldwide research. Geosciences 10(347). https://doi.org/10.3390/geosciences10090347

Cigolini C, Borgia A, Casertano C (1984) Intracrateric activity, aa-block lavas, viscosity and flow dynamics: Arenal volcano, Costa Rica. J Volcanol Geotherm Res 20:155–176

Cole PD, Fernández E, Duarte E, Duncan AM (2005) Explosive activity and generation mechanisms of pyroclastic flows at Arenal volcano, Costa Rica between 1987 and 2001. Bull Volcanol 67:695–716

Cotton CA (1969) Volcanoes as landscape forms, 2nd edn. Hafner Publishing Company, New York

Climate Data (2022) El clima de Tabacón (Costa Rica). https://es.climate-data.org/america-del-norte/costa-rica/alajuela/tabacon-344003/. Accessed 16 Jun 2022

Durigon VL, Carvalho DF, Antunes MAH, Oliveira P, Fernandes MM (2014) NDVI time series for monitoring RUSLE cover management factor in a tropical watershed. Int J Remote Sens 35:441–453. https://doi.org/10.1080/01431161.2013.871081

Ebmeir SK, Biggs J, Mather TA, Wadge G, Amelung F (2010) Steady downslope movement on the western flank of Arenal Volcano, Costa Rica. Geochem Geophys Geosyst 11(12). https://doi.org/10.1029/2010GC003263

Ebmeir SK, Biggs J, Muller C, Avard G (2014) Thin-skinned mass-wasting responsible for widespread deformation at Arenal volcano. Front Earth Sci 2:35. https://doi.org/10.3389/feart.2014.00035Google (2015) Map data 2015 Google. https://www.google.at/permissions/geoguidelines/attr-guide.html. Accessed 7 Aug 2022

Gorelick N, Hancher M, Dixon M, Ilyushchenko S, Thau D, Moore R (2017) Google earth engine: planetary scale geospatial analysis for everyone. Remote Sens Environ 202:18–27

Gyssels G, Poesen J, Bochet E, Li Y (2005) Impact of plant roots on the resistance of soils to erosion by water: a review. Prog Phys Geogr 29:189–217

Hofton MA, Malavassi E, Blair JB (2006) Quantifying recent pyroclastic and lava flows at Arenal Volcano, Costa Rica, using medium-footprint lidar. Geophys Res Lett 33:21

Holling CS (1973) Resilience and stability of ecological systems. Annu Rev Ecol Syst 4:1–23

Labrière N, Locatelli B, Laumonier Y, Freycon V, Bermouex M (2015) Soil erosion in the humid tropics: a systematic quantitative review. Agricul Ecosyst Environ 203:127–139. https://doi.org/10.1016/j.agee.2015.01.027

Lagmay AMF, van Wyk de Vries B, Kerle N, Pyle DM (2000) Volcano instability induced by strike-slip faulting. Bull Volcanol 62:331–346. https://doi.org/10.1007/s004450000103

Lopez A (1999) Neo and paleo stress partitioning in the SW corner of the Caribbean plate and its fault reactivation potential. PhD thesis, Tübingen Geowissenschaftliche Arbeiten 53:1–294

Lyell C (1834) Principles of geology: being an attempt to explain the former changes of the Earth's surface by references to causes now in operation. London

Mahala A (2018) Soil erosion estimation using RUSLE and GIS techniques—a study of a plateau fringe region of a tropical environment. Arab J Geosci 11:335

Malavassi E (1991) Magma sources and crustal processes at the southern terminus of the Central American Volcanic Front (Ph. D. thesis). Univ. de Santa Cruz, EE. UU

Marston RA (2010) Geomorphology and vegetation on hillslopes: Interactions, dependencies, and feedback loops. Geomorphology 116:206–217. https://doi.org/10.1016/j.geomorph.2009.09.028

Martínez-Atienza D (2023) Spatiotemporal changes on vegetation recovery in a post-disturbed ecosystem (Arenal volcano, Costa Rica). Universidad de Castilla-La Mancha, Spain, Master's Thesis

Melson WG, Saenz R (1968) The 1968 Eruption of Volcan Arenal, Costa Rica: preliminary summary of field and laboratory studies. Smithsonian Inst. Center for Short-lived Phenomena Trad al español (1977) La erupción del volcán Arenal, Costa Rica, en julio de 5–6

Montalvo N (1998) Capítulo 12: Determinación de áreas con riesgo potencial de erosión con sistema de información geográfica y percepción remota, sector: Quebrada La Solana, Río Quiroz, Margen izquierda Río Macara Piura-Región Grau. In: Navegando entre brumas: La aplicación de los sistemas de información geográfica al análisis de riesgo en América Latina. pp 303–326

Oramas-Dorta D, Cole PD, Wadge G, Alvarado G, Soto G (2012) Pyroclastic flow hazard at Arenal volcano, Costa Rica: scenarios and assessment. J Volcanol Geoth Res 247–248:74–92. https://doi.org/10.1016/j.jvolgeores.2012.07.015

Osterkamp WR, Hupp CR, Stoffel M (2012) The interactions between vegetation and erosion: new directions for research at the interface of ecology and geomorphology. Earth Surf Processes Landform 37:23–36

Paniagua N (2023) Reconstrucción geomorfológica del volcán Arenal (Costa Rica) pre y post erupción de la actividad que inició en 1968. Universidad de Costa Rica, Escuela Centroamericana de Geología, Thesis

Poggio L, De Sousa LM, Batjes NH, Heuvelink G, Kempen B, Ribeiro E, Rossiter D (2021) SoilGrids 2.0: producing soil information for the globe with quantified spatial uncertainty. Soil 7:217–240
Prado A (2023) Proyecto "Levantamiento de información Topográfica por medio de fotogrametría en el área de influencia de las erupciones del volcán Arenal, para actualizar los escenarios de riesgo". Internal report of the photogrametric survey GEOINN-CNE
QGIS (2022) QGIS Geographic information system. http://www.qgis.org. Accessed 30 Jul 2022
Quintero N, Viedma O, Urbieta IR, Moreno JM (2019) Assessing landscape fire hazard by multitemporal automatic classification of Landsat time series using the Google Earth engine in West-Central Spain. Forests 10(6):518. https://doi.org/10.3390/f10060518
Renard KG, Foster RG, Weesies GA, Porter GA (1991) C" RUSLE: Revised universal soil loss equation". J Soil Water Conserv 46:30–33
Rouse JW (1974) Monitoring the vernal advancement of retrogradation of natural vegetation. NASA/GSFC, Type III, Final Report, Greenbelt, MD 371
SNIT (2017a) Servicios OGC: Ortofoto 2014–2017 5k. https://www.snitcr.go.cr/ico_servicios_ogc_info?k=bm9kbzo6NDM=&nombre=Ortofoto%202014-2017%205k. Accessed 28 Jul 2022
SNIT (2017b) Servicios OGC: Cartografía 1:5 mil. https://www.snitcr.go.cr/
Soto GJ, Alvarado GE (2006a) Eruptive history of Arenal Volcano, Costa Rica, 7 ka to present. J Volcanol Geoth Res 157:254–269. https://doi.org/10.1016/j.jvolgeores.2006.03.041
Soto GJ, Alvarado GE (2006b) Eruptive history of Arenal Volcano, Costa Rica, 7 ka to present. J Volcanol Geotherm Res 157:254–269. https://doi.org/10.1016/j.jvolgeores.2006.03.041
Soto JG (2008) Coladas de lava formadoras de abanicos de bloques en el volcán Arenal (Costa Rica): mecanismos, facies e implicación para los estudios aplicados de terrenos volcánicos. IX Congreso Geol. Amér. Central y VI Congreso Geológico Nacional, 02–04 de julio 2008, San José, pp 179–180
Thouret JC (1999) Volcanic geomorphology-and overview. Earth Sci Rev 47:95–131
Torres O, Alvarado GE (2022) Cambios geomorfológicos por procesos erosivos y sedimentarios en el volcán Arenal, período post eruptivo 2010–2022. San José
USGS (2011) Earth explorer: digital elevation model from Costa Rica. https://earthexplorer.usgs.gov/. Accessed 14 May 2021
Vargas G (1985) La sucesión vegetal primaria en una región de vulcanismo reciente en el volcán Arenal y sus alrededores, Costa Rica. Rev Biol Trop 33:171–183
Vargas G, Cortés V, Zúñiga W (1993) Evolución de la dinámica vegetal entre 1985 y 1993 sobre las lavas recientes (1968–92) y flujos piroclásticos (1968) del volcán Arenal. Boletín Del Observatorio Vulcanológico Del Arenal 5:31–51
Wadge G (1983) The magma budget of Volcán Arenal, Costa Rica from 1968 to 1980. J Volcanol Geoth Res 19(3–4):281–302
Wadge G, Oramas Dorta D, Cole PD (2006) The magma budget of Volcán Arenal, Costa Rica from 1980 to 2004. J Volcanol Geoth Res 157:60–74. https://doi.org/10.1016/j.jvolgeores.2006.03.037
White PS (1979) Pattern, process, and natural disturbance in vegetation. Bot Rev 45:229–299
Zhang X, Yu GQ, Li ZB, Li P (2014) Experimental study on slope runoff, erosion and sediment under different vegetation types. Water Resour Managem 28:2415–2433. https://doi.org/10.1007/s11269-014-0603-5

Tectonic Geomorphology of the Fortuna Fault and the Venado Transtensive System, Northern Costa Rica

15

Alejandro Argüello-Sáenz, Walter Montero-Pohly, and Andrés Ulloa-Carmiol

Abstract

The northwestern dextral strike-slip Haciendas-Chiripa fault system crosses the backarc region of northern Costa Rica. This system defines the eastern border of the Guanacaste Volcanic Arc Sliver, which is moving northwest at ~ 11 mm year$-^1$ and nearly parallel to the Middle America Trench. The NW-oriented dextral strike-slip Fortuna Fault is a branch within the Haciendas-Chiripa fault system. A right bend along the Fortuna Fault in the Venado region originates the Venado Transtensive Zone. The Venado region is located within a backarc Miocene sedimentary basin, including sandstone, mudstone, and limestone (with karst), covered by Plio-Quaternary volcanic rocks. The Fortuna Fault and the Venado Transtensive Zone were defined using geomorphological and geological evidence. Tectonic landforms typical of strike-slip and normal faulting (e.g., scarps, linear valleys, and displaced drainages) were identified in the study region. The Venado dilatational stepover developed inside a tilted planation surface, causing Pleistocene to Holocene alluvial depocenters to locate in its eastern sector. Our work contributes to a better definition of the Haciendas-Chiripa fault system and its relationship with the tectonic sliver in northern Costa Rica.

Keywords

Active tectonics · Tectonic geomorphology · Fortuna fault · Venado transtensive system · Karst · Caves · Photogrammetry · Speleoseismology

A. Argüello-Sáenz (✉) · A. Ulloa-Carmiol
Escuela Centroamericana de Geología, Universidad de Costa Rica, San José, Costa Rica
e-mail: alejandro.arguello.cr@gmail.com

A. Argüello-Sáenz · W. Montero-Pohly · A. Ulloa-Carmiol
Centro de Investigación en Ciencias Geológicas, Universidad de Costa Rica, San José, Costa Rica

15.1 Introduction

Costa Rica is located in an active tectonic margin, where the subduction of the Cocos Plate under the Caribbean Plate produces important tectonic deformation. Tectonic blocks have formed as a result, which includes the Central American Forearc Sliver (CAFS), with northwestward displacement, parallel to the Middle America Trench (Fig. 15.1; DeMets 2001; Alvarado et al. 2011; Álvarez-Gómez et al. 2019). Three models have been proposed for the origin of CAFS motion: (a) oblique convergence of the Cocos Plate below the Caribbean Plate (Molnar and Sykes 1969; Harlow and White 1985; DeMets 2001), (b) Cocos Ridge collision (Montero 1994; Lewis et al. 2008; La Femina et al. 2009; Alvarado et al. 2011; Feng et al. 2012; Kobayashi et al. 2014), and (c) the dragging model, where the North American Plate pulls the sliver toward the northwest (Álvarez-Gómez et al. 2008).

Recent publications have aimed to define the geometry and kinematics of the bounding faults within the CAFS and the Caribbean Plate limit. For example, in Guatemala, Garnier et al. (2021) recognized the dextral-slip Jalpatagua Fault (JF in Fig. 15.1a) as the singular northern boundary.

In El Salvador, W to WNW-trending faults connected by dilatational stepovers have created a more complex northern limit (Martínez-Diaz et al. 2004; Corti et al. 2005; Alvarado et al. 2011; Alonso-Henar et al. 2014). The El Salvador Fault Zone is nearly parallel and close to the volcanic arc. In Nicaragua, the eastern boundary of the CAFS consists of sinistral northeast-striking faults which cause clockwise rotation as explained by a bookshelf faulting model (LaFemina et al. 2002; Álvarez et al. 2018) or by an orthogonal fault model including left-lateral northeast faults and right-lateral northwest faults (Suárez et al. 2015). In northwestern Costa Rica, Montero et al. (2017) proposed that the northwest dextral-slip Haciendas-Chiripa fault system defines the eastern side of the Guanacaste Volcanic Arc Sliver representing the final southeastern limit of the CAFS (Fig. 15.1b).

A. Quesada-Román (ed.), *Landscapes and Landforms of Costa Rica*, World Geomorphological Landscapes, https://doi.org/10.1007/978-3-031-64940-0_15

Geodetic-derived data also support northwest sliver motion of the Central America Forearc (e.g., Correa-Mora et al. 2009; Alvarado et al. 2011; Kobayashi et al. 2014) and of northwestern Costa Rica (Lundgren et al. 1999; Norabuena et al. 2004; LaFemina et al. 2009; Kobayashi et al. 2014; Hobbs et al. 2019). Montero et al. (2017) showed that the trench-parallel component of GPS velocities obtained from Feng et al. (2012) substantially decreases east of the Haciendas-Chiripa fault system.

This chapter describes the geometry of the dextral-slip Fortuna Fault (FF) as a newly recognized eastern branch of the Haciendas-Chiripa fault system. The FF crosses the northern part of Costa Rica, extends for ~ 45 km, and has an associated right-hand bend named the Venado Transtensive Zone (VTZ). Detailed geological mapping and identification of tectonic landforms were carried out to characterize the fault traces associated with the VTZ.

15.2 Tectonic Framework

Costa Rica and Central America sit on the northeast border of a subduction zone, where the Cocos Plate underthrusts the Caribbean Plate (Fig. 15.1). An anomalously thick and thermally buoyant Cocos ridge collides with southern Costa Rica, south of the Osa Peninsula (Kolarsky et al. 1995; Fig. 15.1b), which produces different tectonic regimes across Costa Rica.

The Central American active volcanic arc nearly parallels the Middle America Trench off the Pacific coast. In northern Costa Rica, the Quaternary volcanic arc is called the Guanacaste Volcanic Arc (CVG in Fig. 15.1b), while in central Costa Rica, the volcanic arc is called the Central Costa Rica Volcanic Front (CVC in Fig. 15.1b). The CVC end to the east with the Irazú-Turrialba volcanoes. Southeast of the Central Costa Rica Volcanic Front, the Talamanca Cordillera is located in a volcanic gap between Irazú-Turrialba and Baru volcanoes (Gazel et al. 2019).

Frequent, damaging shallow depth (generally < 20 km) earthquakes with intermediate magnitudes ($5.0 \leq M \leq 6.5$) occur along the entire length of the active volcanic arc of Central America, including Costa Rica (White 1991; White and Harlow 1993). Focal mechanism studies indicate that these events have a strike-slip motion with dextral northwest and sinistral east-northeast nodal planes (Montero and Dewey 1982; White and Harlow 1993; Montero 2001). The Haciendas-Chiripa fault system (HCFS) accommodates this seismogenic deformation in northwestern Costa Rica (Montero et al. 2017; Araya and Biggs 2020) and the northwest dextral displacement of the Guanacaste Volcanic Arc Sliver relative to the Caribbean Plate (Montero et al. 2017). The HCFS includes the north to northwest-striking, dextral Haciendas, Caño Negro, Cote-Arenal, and Chiripa faults (Alvarado 1989; Montero et al. 1998; Denyer et al. 2003; Camacho 2015; Montero et al. 2017).

The HCFS continues to the south and southeast with the Central Costa Rica Deformed Belt (CCRDB) (Marshall et al. 2000; Montero 2001). The CCRDB is a broad shear zone passing through central Costa Rica (Fig. 15.1b). Norwest dextral-slip faults predominate along the CCRDB. The principal examples are the Candelaria, Jaris, and Atirro-Río Sucio faults (CAF, JAF, and ARSF, respectively, in Fig. 15.1b; Montero et al. 2013; Montero and Rojas 2014). Additionally, E-NE-trending sinistral faults are present within the CCRDB. Finally, E to WNW-trending thrust faults borders the northern and southern sides of the Central Costa Rica Volcanic Arc (Montero et al. 2013). An example of a thrust fault is the Florencia-La Palmera Fault (FPF in Fig. 15.1b). The CCRDB defines the root of the Central America forearc sliver (Lewis et al. 2008; Montero et al. 2013). In the eastern part of central Costa Rica, the CCRDB joins the North Panama Deformed Belt (NPDB), a fold and thrust fault system bordering the Caribbean slope of the Cordillera de Talamanca and the northern Caribbean side of Panama (Fig. 15.1; Silver et al. 1995).

The geology of the northwestern Costa Rica outer arc includes an Upper Jurassic-Late Cretaceous basement covered by sedimentary and volcanic rocks with ages from Upper Cretaceous to the Quaternary (Dengo 1962; Denyer and Alvarado 2007; Fung et al. 2009). A Tertiary and Quaternary volcanic cordillera is located east of the outer arc. The Guanacaste Volcanic Arc is the presently active volcanic cordillera. The study area for this chapter is located between the southeastern slopes of the Tenorio Volcano and the eastern slopes of the Arenal Volcano, the two southeasternmost volcanic active structures of the Guanacaste Volcanic Arc (Fig. 15.1b).

15.3 Geologic Framework

The Venado region is characterized by the presence of siliciclastic to carbonated Miocene sedimentary rocks associated with the Venado Formation, which are partially covered by Quaternary volcanic rocks of the Monte Verde Formation (e.g., lavas, lahars, and tephras) and volcanic-landslide deposits (Fig. 15.2). Miocene-Pliocene volcanic rocks are also present; some are contemporaneous with the sedimentary sequence and are correlated with the Grifo Alto Formation (Aguacate Group). Multiple geological studies such as field geology, geophysics, drilling, and micropaleontology analyses were performed in the 1980s because of hydrocarbon resources (Echandi 1986; Obando 1986; Calvo and Bolz 1987; Pizarro 1993). The main geological formations are described below.

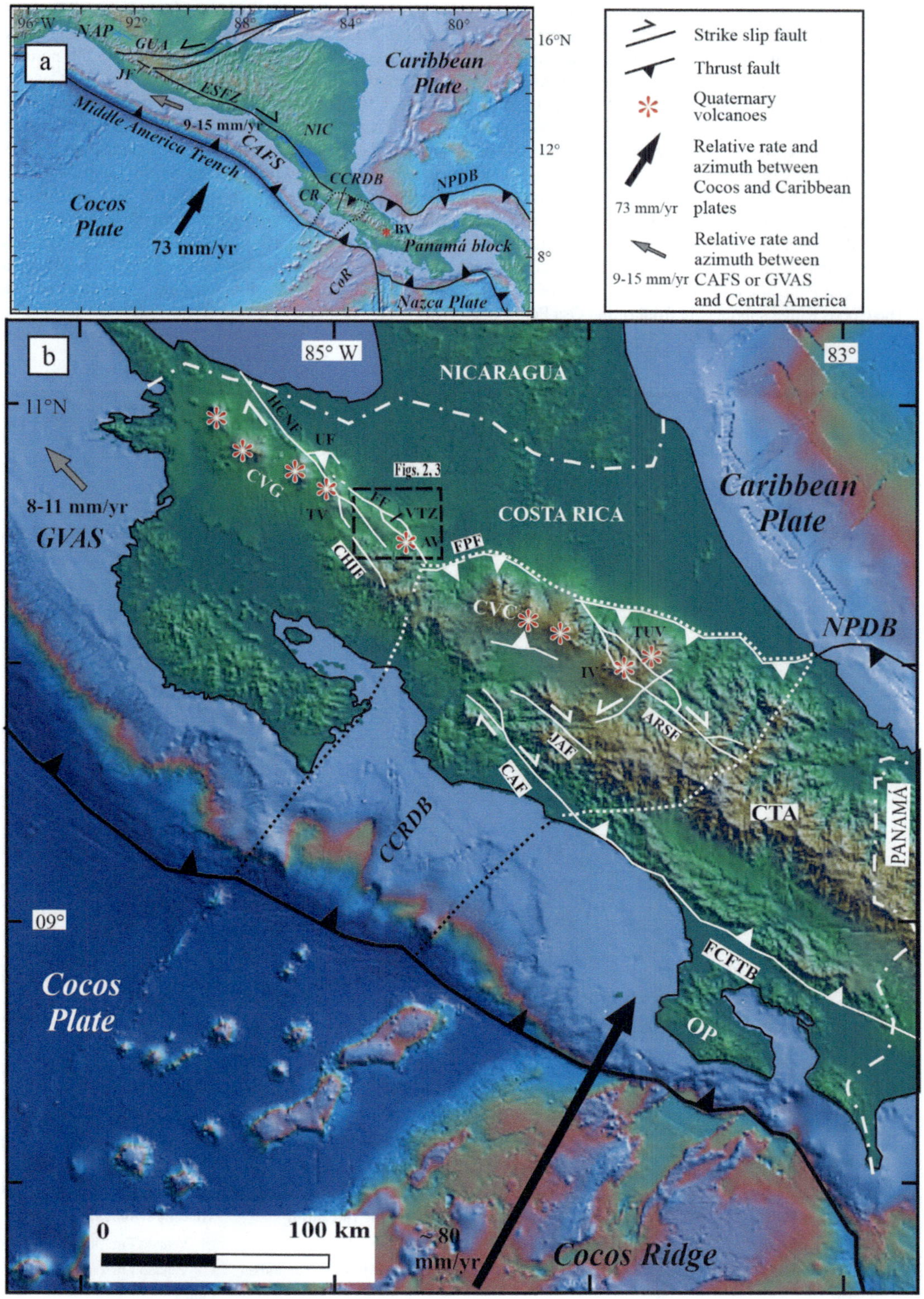
a
Caribbean Plate
Cocos Plate
73 mm/yr
9-15 mm/yr
Middle America Trench
CAFS
ESFZ
NIC
CCRDB
NPDB
CR
BV
Panamá block
Nazca Plate
CoR
NAP
GUA
JF
Strike slip fault
Thrust fault
Quaternary volcanoes
Relative rate and azimuth between Cocos and Caribbean plates
73 mm/yr
Relative rate and azimuth between CAFS or GVAS and Central America
9-15 mm/yr
b
85° W
83°
11°N
09°
NICARAGUA
COSTA RICA
Caribbean Plate
NPDB
8-11 mm/yr
GVAS
Figs. 2, 3
CVG
CVC
TUV
IV
UF
TV
VTZ
AV
FPF
CHIF
JAF
ARSF
CAF
CCRDB
CTA
PANAMÁ
FCFTB
OP
Cocos Plate
0 100 km
~80 mm/yr
Cocos Ridge

◀ **Fig. 15.1** Regional tectonic setting of Central America and Costa Rica: **a** The Cocos Plate underthrusts the Caribbean Plate along the Middle America Trench at a 73 mm/yr rate in front of northern Nicaragua (NIC; Argus et al. 2011). The Central American Forearc Sliver (CAFS) displaces horizontally to the northwest at about 9–15 mm/yr relative to the Caribbean Plate (Kobayashi et al. 2014). The Cocos Ridge (CoR) subducts below southern Costa Rica. The Central Costa Rica Deformed Belt (CCRDB) crosses central Costa Rica (boundaries defined by point lines). East of the CCRDB is the Panama Block, whose northern boundary is the North Panama Deformed Belt (NPDB). BV: Barú Volcano. JF: Jalpatagua Fault. GUA: Guatemala, ESFZ: El Salvador Fault Zone, CR: Costa Rica, **b** Tectonic setting of Costa Rica. The Cocos Plate underthrusts the Caribbean Plate at an 80 mm/yr rate in front of the Osa Peninsula (NP). The Cocos Ridge subducts in front of the Osa Peninsula (OP). The Fila Costeña Fold and Thrust Belt (FCFTB) rises along thrust faults in front of the Cocos Ridge subduction. The Guanacaste Volcanic Arc Sliver (GVAS) displaces horizontally to the northwest at a rate of 8–11 mm/yr (Feng et al. 2012). The Haciendas-Caño Negro (HCNF), Upala (UF), Chiripa (CHIF), and the Fortuna (FF) Faults and the associated Venado Transtensive Zone (VTZ) define the eastern limit of the GVAS. The dotted lines show the boundaries of the CCRDB. The principal dextral faults of the CCRDB are the Candelaria (CAF), Jaris (JAF), and Atirro-Rio Sucio (ARSF). The thrust Florencia-La Palmera Fault (FPF) is the northwestern limit of the CCRDB. CVG is the Cordillera Volcánica de Guanacaste, and CVC is the Cordillera Volcánica Central. CTA is the Cordillera de Talamanca. TV, AV, IT, and TU are the Tenorio, Arenal, Irazú, and Turrialba volcanoes, respectively. A dashed rectangle shows the location of Figs. 15.2 and 15.3

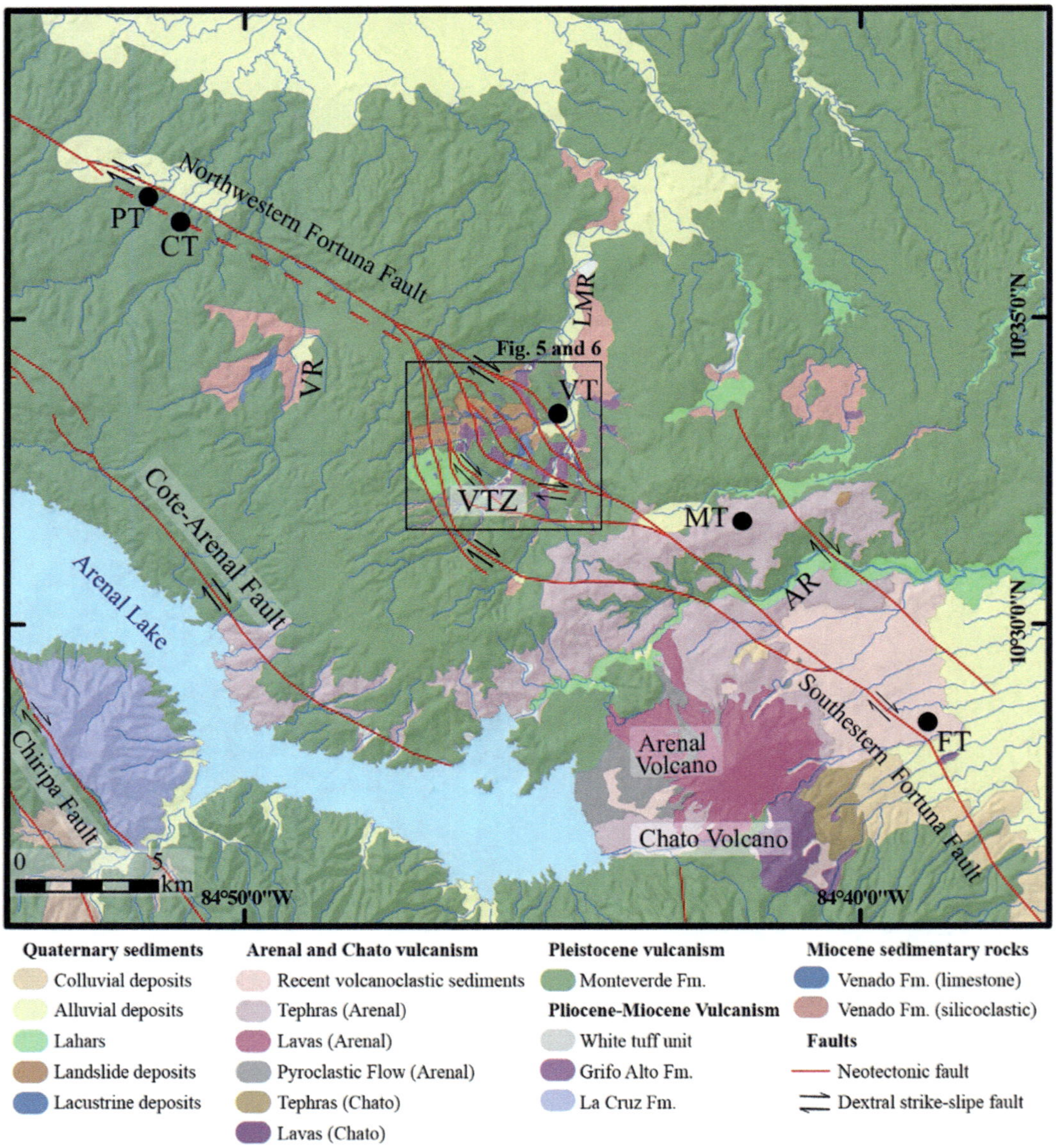

Fig. 15.2 Regional geologic map shows the different stratigraphic units present in the study area. VTZ is the Venado Transtensive Zone. Black circles mark the towns of Venado (VT), Fortuna (FT), Monterrey (MT), Cabanga (CT), and Pejiballe (PT). The black rectangle shows the locations of Figs. 15.5 and 15.6. Rivers: Arenal (AR), La Muerte (LMR), Nilo (NR). Location in Fig. 15.1. Geology is based on Estrada (1983), Echandi (1986), Alvarado (2009), Rojas (2019), Argüello (2021), and Solano et al. (2021)

15.3.1 Venado Formation

The Venado Formation is composed of lithofacies of limestones, shales, siltstones, sandstones, tephras, coal layers, and conglomerates (Malavassi and Madrigal 1970; Obando 1986; Calvo and Bolz 1987). It formed in littoral to infralittoral environments, associated with lagoonal estuaries with external sand bars (Sen Gupta et al. 1986; Calvo and Bolz 1987). According to different authors (Obando 1986; Calvo and Bolz 1987; Argüello 2021), the sedimentary rocks of Venado Formation have a significant volcaniclastic contribution (e.g., pumice levels, fragments of andesites, and vitroclasts), which is related to the acid volcanism of Nicaragua (Alvarado and Gans 2012). The Venado Formation is dated to the Middle to Upper Miocene (Malavassi and Madrigal 1970; Sen Gupta et al. 1986; Rojas 2019; Argüello-Sáenz et al. 2022). Limestone lithofacies of the Venado Formation are constituted by grainstone to packstone. These carbonated facies are well-exposed in the surroundings of El Túnel Creek, where karst and cave systems (~ 5 km length), dolines, sinks, and karst springs occur (Ulloa et al. 2021). Limestone sequences are also exposed in diverse localities near La Muerte River and Nilo River, but caves and karst expressions are scarce (Ulloa et al. 2021). Another locality where limestone is present is El Pato Creek, NW of the VTZ (Echandi 1986).

15.3.2 Grifo Alto Formation

Calco-alkaline magmatism correlated with the Grifo Alto Formation is exposed in the central part of the study area. This unit is composed of andesites to basalts and some scattered dacites. The most representative outcrops correspond to fault scarps associated with the VTZ and rounded hills related to intrusive bodies. This magmatism is contemporaneous with the upper part of the Venado Formation; two dates have been obtained in this unit, giving ages between 5.2 and 8.7 Ma (Sen Gupta et al. 1986; Rojas 2019). In specific areas, it is possible to recognize blocky peperites, which suggests the interaction of magma with unconsolidated sediments of the Venado Formation (Argüello 2021).

15.3.3 Pleistocene–Quaternary Volcanism

The Monteverde Formation includes andesitic to basaltic lavas, lahars, and tuffs. The lavas have been subject to significant weathering, generating a halo of reddish soil alteration with plagioclase crystals, a product of spheroidal weathering, and some relicts of the original rocks. This formation is dated from 2.1 to 1.1 Ma (Alvarado and Gans 2012) and locally covers an outcrop of Pliocene tuff (Rojas 2019; Argüello 2021), defined as a "*Toba Blanca*" unit. This formation also overlies the Miocene rocks of the Venado and Grifo Alto formations. The study area is also covered by decimeter to meter-thick tephra deposits, mainly covering the Monte Verde Formation. A regional tephra event that occurred 20,000 years ago is overlain by tephras associated with the Arenal volcano, ranging from 7000 years to the present (Soto and Alvarado 2006).

15.3.4 Alluvial and Mass Removal Deposits

The biggest landslide in the study area occurred in rocks correlated to the Monte Verde Formation. The deposit covers ~ 0.8 km^2 and is classified as a debris avalanche. The volcanoclastic deposits (clays and metric blocks of andesites) partially cover zones with limestone and karst around El Túnel Creek. Some rounded hills that resemble hummocks are associated with this event (Argüello 2021; Ulloa et al. 2021). Recent and active mass removal deposits are classified as earthflow-type landslides, as movement is similar to a plastic flow. Some landslide deposits caused local damming of channels, generating colluvial and alluvial deposits. Alluvial deposits are composed of gravel, sand, and decimeter to meter blocks (dominated by volcanic composition). The alluvial deposits occur in the margins of the La Muerte and Nilo rivers, along the channel, and on some multi-level alluvial terraces. Main depocenters occur near the confluence of the La Muerte and Nilo rivers, and some terraces are associated with traces of the VTZ (Argüello 2021).

15.4 Geomorphological and Geological Evidence of the Fortuna Fault

Argüello (2021) first referred to the dextral Fortuna Fault (FF) as a northwesterly-trending regional fault, slightly oblique to the Haciendas-Chiripa fault system (Montero et al. 2017). To the north, the FF continues with the Cote-Arenal Fault (Fig. 15.1; Camacho 2015; Montero et al. 2017). Southeast of the town of La Fortuna, the FF ends at the thrust Florencia-La Palmera Fault (FPF in Fig. 15.1; Montero et al. 2013). The Venado Transtensive Zone is located at the right bend along the FF. Southeast and northwest of the VTZ, the FF is characterized morphologically by scarps, dextral drainage offsets, linear valleys, fault saddles, and other tectonic landforms.

15.4.1 The Northwestern Branch of the Fortuna Fault

Previous regional and detailed geological maps of the northwestern part of the study region did not identify the

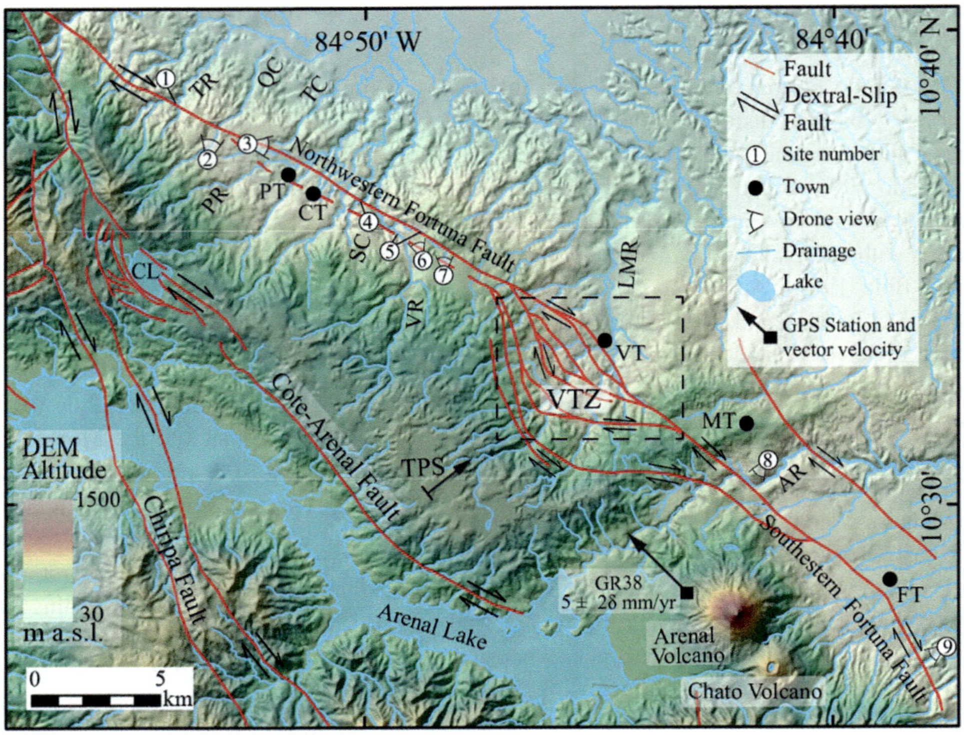

Fig. 15.3 Regional neotectonic map. The northwest-trending dextral Fortuna Fault and the related Venado Transtensive Zone (VTZ) are shown with the regional northwest-trending dextral Cote-Arenal and Chiripa faults (Camacho 2015; Montero et al. 2017). The northwestern part of the Fortuna Fault (NFF) and the southeast branch (SFF) are also shown. The location of numbered sites is discussed in the text. An eye shows the drone view. Towns: Pejibaye (PT), Cabanga (CT), Venado (VT), Monterrey (MT), and Fortuna (FT). Rivers: Triminio (TR), Pejibaye (PR), Venado (VR), La Muerte (LMR), and Arenal (AR). Creeks: Quebradón (QC), Ticamar (TC), and Salto (SC). Lake: Cote Lake (CL). TPS is the tilted planation surface crossed by the Fortuna Fault, with the arrow indicating the tilted NE direction

northwest branch of the Fortuna Fault (NFF; e.g., Sandoval et al. 1982; Denyer et al. 2003; Denyer and Alvarado 2007). Nevertheless, diverse tectonic landforms define the NFF, including dextral drainage offsets, water divide deviations, fault scarps, linear valleys, and fault saddles. The geomorphic evidence suggests a double fault trace for the FF between the northwest of Pejibaye and the VTZ. In the northwestern sector, the NFF deflects the Treminio River to the right by ~ 0.5 km along a linear valley (location 1 in Fig. 15.3). At location 2 (Fig. 15.3), the NFF shows various landforms typical of strike-slip faults. For example, we observed a right-stepping deviation of a tributary of the Quebradón Creek, west-facing scarps, an L-shaped spur, an interrupted divide, and a pond at the foot of the fault scarp (Fig. 15.4a). At site 3 (Fig. 15.3), the NFF follows a linear valley. Other evidence of dextral and dip-slip faulting includes a right offset of the Ticamar Creek, suspended drainage, an offset water divide, and a west-facing scarp (Fig. 15.4b). Near the town of Pejibaye, the Pejibaye River and a tributary trend subparallel to the NFF (Fig. 15.3).

Southeast of the town of Cabanga, the NFF dextrally offsets the Salto Creek by ~ 150 m (location 4 in Fig. 15.3). At location 5, two parallel and linear fault valleys follow the Venado River and a right-bank tributary (Fig. 15.3). These fault valleys define the northern and southern traces of the NFT. At location 6 (Fig. 15.3), a tilted sedimentary unit, including tuff deposits, crops out between the fault traces and terminates against the northern Fortuna Fault trace (Fig. 15.4c). On the eastern fault block, this sedimentary unit does not crop out. At location 7 (Fig. 15.3), the NFF follows a fault valley along the right-bank tributary of the Venado River (Fig. 15.4d).

15.4.2 The Venado Transtensive Zone (VTZ)

The VTZ corresponds to a right-hand bend along the FF. The VTZ structure includes eight fault traces. Fault traces were identified using geomorphic, detailed geological mapping (Fig. 15.4; Argüello 2021), and structural evidence. Drone photogrammetry allowed us to get detailed geomorphological features of the faults. Karstic cave topographic detail maps and structural cave data contributed to defining the fault system. The faults identified are described below from east to west.

15.4.2.1 La Muerte Fault

La Muerte Fault (F1) trends nearly N40°W and suggests a predominant dextral-slip movement. The dip-slip component of F1 is unclear, and diverse tectonic landforms have been identified along this trace. Near the town of Venado, the La Muerte River is rightly offset by ~ 300 m (Site 1 in Figs. 15.6 and 15.7a). Also, there are several levels of alluvial terraces downstream of the fault trace. South of Venado town, F1 follows the linear valley of the Nilo River (Site 2 in Fig. 15.6). Different terrace levels are also recognized upstream of the F1 fault trace (Fig. 15.6).

Fig. 15.4 Drone photographs of the northwestern Fortuna Fault branch (NFF): **a** View to the north of site 2 shows the NFF where it crosses the Treminio River. Different tectonic landforms support strike-slip kinematics. Also shown is a west-facing scarp (FS), **b** View to the east of site 3, showing different geoforms characteristic of strike-slip and dip-slip faults, **c** View to the NNW of location 6. A tilted sedimentary unit ends against the NFF, **d** View to the NNE of site 7, where the fault follows a linear fault valley and cuts across a water divide along the headwaters of an affluent to the Venado River

Geological evidence of the F1 was also found. For example, the hypabyssal units of the Grifo Alto Formation have rightly deviated at Sites 1 and 2 (Fig. 15.5). In addition, a secondary fault related to F1 displaces the Grifo Alto formation on the right bank of the Nilo River (Fig. 15.7b).

15.4.2.2 La Mina Fault

La Mina Fault (F2) is a northwest-trending fault with tectonic landforms suggestive of dextral-slip and dip-slip displacement components. Both components are present at the confluence of the El Túnel Creek with La Muerte

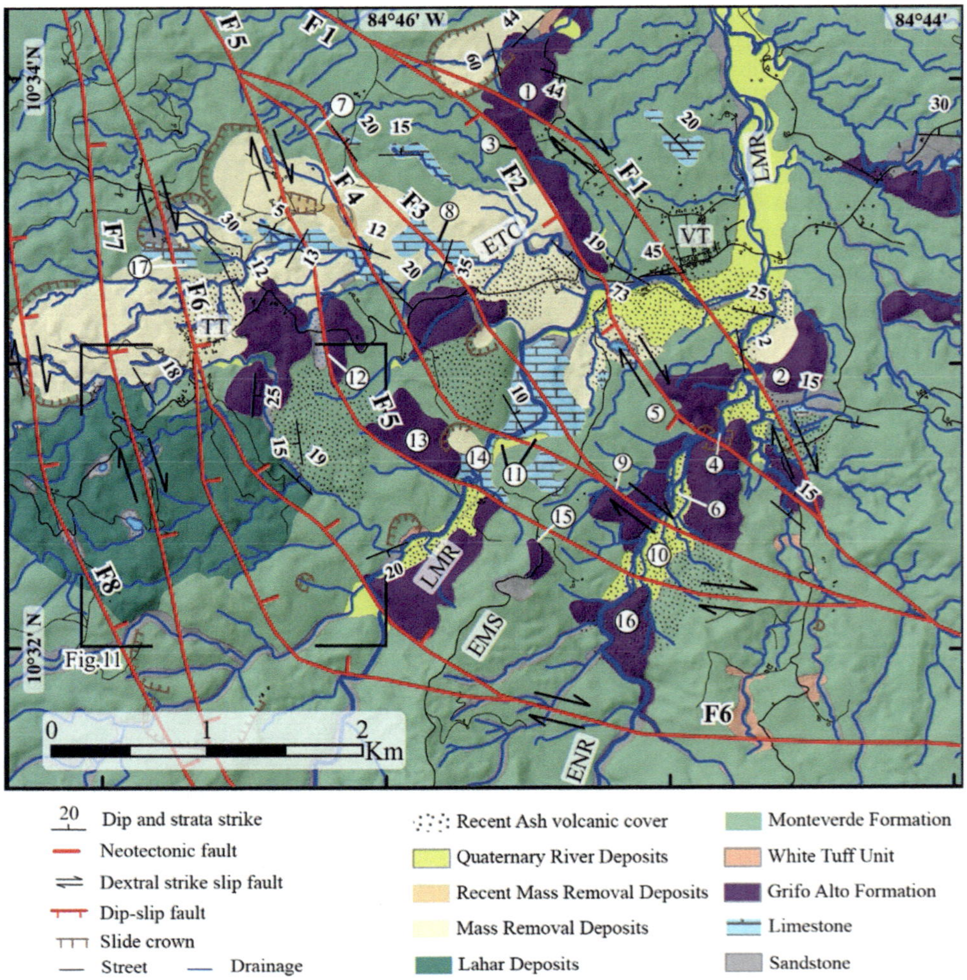

Fig. 15.5 Geological map over a hillshade model shows the Venado transtensive Zone (based in Argüello 2021). Fault names are La Muerte (F1), La Mina (F2), Santa Lucía (F3), La Cueva (F4), Túnel (F5), La Tigra (F6), Catanuria (F7), and Suampo (F8). White numbered circles are discussed in the text. Towns: Venado (VT) and La Tigra (TT). Rivers: La Muerte (LMR) and Nilo (ENR). ETC corresponds to El Túnel Creek, and EMS to Enmedio Road. Black corners show the location of Fig. 15.11. Location in Fig. 15.2

River (Site 3 in Figs. 15.6 and 15.8a). A prominent west-facing fault scarp reflects the dip-slip component of F2. F2 also displaces the El Túnel Creek ~ 400 m to the right before its confluence with the La Muerte River (Fig. 15.8a). Additionally, F2 raises the hypabyssal rocks of the Grifo Alto Formation, putting them in fault contact against the Monteverde and Venado formations (Site 3 in Fig. 15.5).

Southeast of the La Muerte River, on the west side of the fault trace, a small lake drains to the west (Site 4 in Fig. 15.6). South of the town of Venado, on the upthrown and eastern block of F2, the Nilo River follows an antecedent valley when crossing the Grifo Alto Formation (Site 4 in Fig. 15.5). At Site 5 (Fig. 15.5), F2 shows a fault contact between the Monteverde and Grifo Alto formations. Upstream of the fault, the Nilo River deposited several levels of terraces on the downthrown side of F2, originating a ponded alluvium (Site 6 in Fig. 15.5). A west-facing fault scarp is found on the left bank of the Nilo River (Site 5 in Fig. 15.6). Finally, F2 offsets dextrally a water divide on the right bank of the Nilo River (Site 6 in Fig. 15.6).

15.4.2.3 Santa Lucia Fault

The Santa Lucía Fault (F3) is a northwest-trending fault with a dextral offset and an unclear component of dip-slip offset. F3 has a weak morphotectonic expression in its northeastern sector but contains geological evidence of faulting. The Catano Cave has been dextrally offset by ~ 16 m, with dissolution occurring subparallel to the direction of the fault trace (Site 7 in Figs. 15.5 and 15.8b).

Southwest of Site 7, along the fault trace, the El Túnel Creek is dextrally displaced by ~ 100 m (Site 7 in Fig. 15.6). A right deviation is also observed in the Venado Formation (Site 8 in Fig. 15.5). South of the road between the towns of Venado and La Tigra, a tributary creek to La Muerte River is dextrally offset by ~ 250 m (Site 8 in Fig. 15.6). Based on geological evidence, the contact between the Monteverde and Grifo Alto Formations is dextrally offset where the F3 crosses both sides of the Nilo River valley (Site 9 in Fig. 15.5). Additionally, large alluvial deposits outcropping on the upstream side of the F3 fault suggest ponded alluvium and an uplift component (Site 10 in Fig. 15.5).

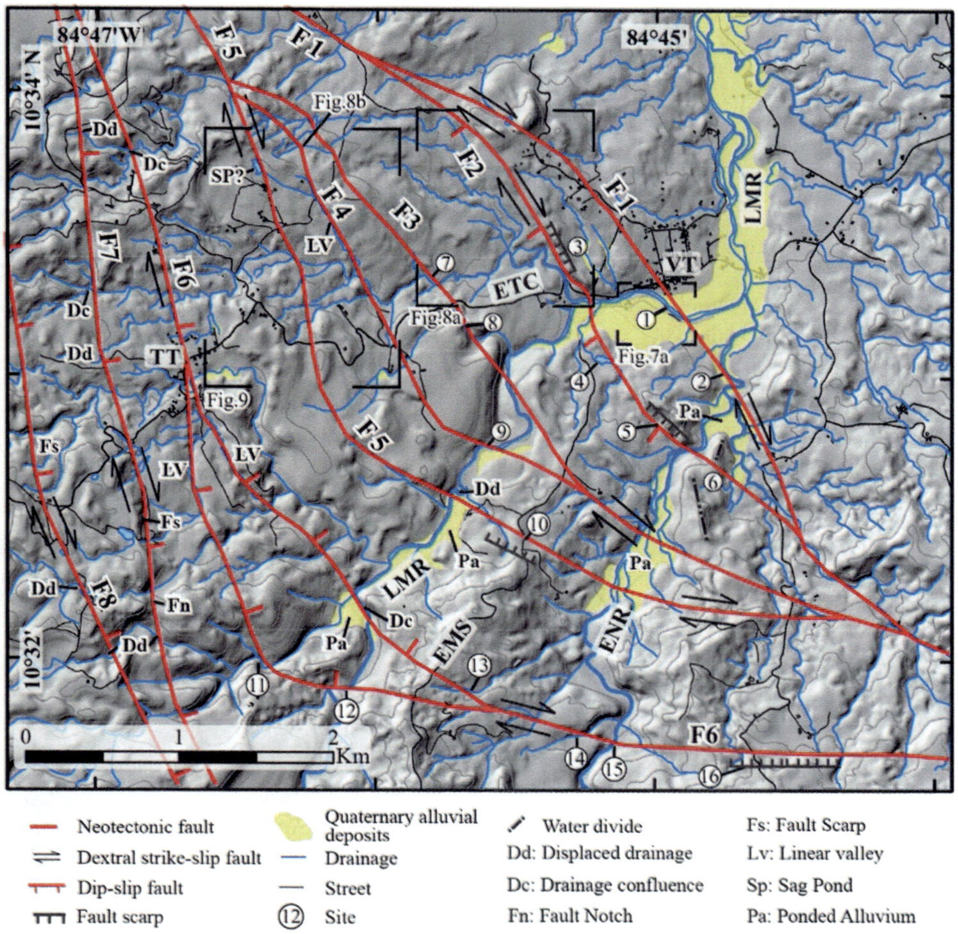

Fig. 15.6 Neotectonic map of the Venado Transtensive Zone overlapping a hillshade model. Fault names are La Muerte (F1), La Mina (F2), Santa Lucía (F3), La Cueva (F4), Túnel (F5), La Tigra (F6), Catanuria (F7), and Suampo (F8). Towns: Tigre (TT) and Venado (VT). Rivers: La Muerte (LMR) and Nilo (ENR). ETC is El Túnel Creek. EMS is the Enmedio road. The location of Figs. 15.7a and 15.8a, b, and nine are indicated by black corners. Numbered site locations 1 to 16 are discussed in the text

15.4.2.4 La Cueva Fault

La Cueva Fault (F4) is a northwest-trending fault with predominantly dextral movement. The dip-slip component is not clearly defined. In its northern sector, F4 follows the linear valley (LV in Fig. 15.9) of a tributary to El Túnel Creek. A west-facing scarp marks the east side of the fault. Along the western depressed area of the F4 fault, there are several drainage confluences and at least three dextral offset drainages (see Dd's in Fig. 15.9). Moreover, the Gabinarraca Cave has a subparallel trend to the fault (Fig. 15.9).

In the southeastern sector of F4, La Muerte River is dextrally offset by ~ 100 m (Site 9 in Fig. 15.6). Alluvial terraces suggest a dextral deviation on the south bank of the river (Site 11 in Fig. 15.5). Southeast of the La Muerte River, the F4 fault joins the F3 trace.

15.4.2.5 Túnel Fault

The Túnel Fault (F5) is an NW-WNW-trending fault with predominantly dextral-slip. On the northern side, the F5 fault appears to dextrally offset a water divide by ~ 200 m (Site 1 in Fig. 15.9). West of Site 1, a marshy area is elongated along the fault trace. It is interpreted as a sag pond or a swamp related to the landslide. A tributary of El Túnel Creek drains this sag pond to the southeast. Southeast from Site 1, F5 crosses Los Menonitas, and La Higuera caves (Fig. 15.9). Argüello (2021) mentioned that underground drainages in the Los Menonitas Cave are dextrally offset, and outcropping faults are sub-parallel to F5. La Higuera Cave has three subterranean channels with 5 to 15 m dextral offsets associated with three subparallel fault traces (Argüello 2021).

A west-facing fault escarpment defined the eastern F5 fault block, where the fault trace crosses the road between the towns of Venado and La Tigra (Fig. 15.9). The Grifo Alto Formation crops out on the eastern side of the fault (Site 12 in Fig. 15.5). A similar structural control occurs to the north of La Muerte River, where outcrops of the Grifo Alto Formation are present on the east side of the F5 fault (Site 13 in Fig. 15.5).

Limestones of the Venado Formation (east side) are in fault contact with alluvial deposits and the Grifo Alto Formation (west side), where F5 crosses La Muerte River (Site 14 in Fig. 15.5). Southeast of Site 14, F5 shows an

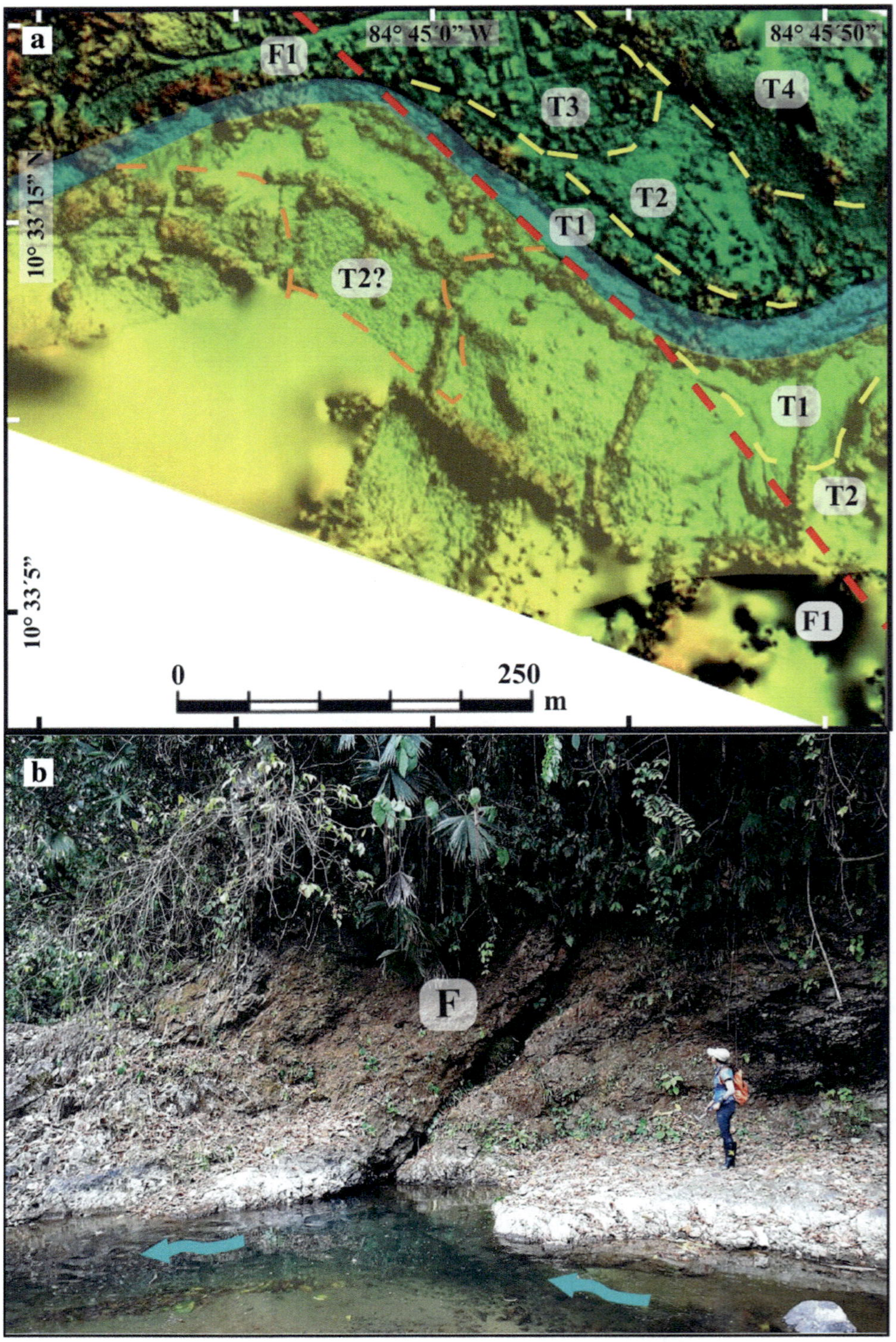

Fig. 15.7 Geomorphic and geologic evidence of F1: **a** Drone photogrammetry map of a ~ 300 m right offset of La Muerte River related to F1. Outer borders of terraces T1, T2, T3, and T4 are drawn with a yellow dashed line. The inner side of the suspected terrace T2 is shown with a dashed orange line. Location in Fig. 15.6, **b** Outcrop photography of intrusive rocks of the Grifo Alto Formation displaced by a secondary fault (F), related to the deformation zone of F1, right bank of the Nilo River. Solid light blue arrows show river flow. The person is shown for scale. Numbered site location 2 in Fig. 15.5

east-facing scarp where it crosses the Enmedio road (Site 10 in Figs. 15.6 and 15.10a). Slightly west and on the north side of the road, F5 displaces pyroclastic deposits and paleosoils overlaying lavas from the Grifo Alto Formation (Site 15 in Figs. 15.5 and 15.10b and c). At this location, two fault traces of F5 dipping 50° to 60° east have a dip-slip component (Fig. 15.10c). In the Nilo River, the Grifo Alto Formation crops out on the west side of the fault (Site 16 in Fig. 15.5).

15.4.2.6 La Tigra Fault

La Tigra Fault (F6) has a predominantly dextral motion and an NW-to-WNW trend. The F6 fault does not show prominent morphotectonic features in its northern part. Nevertheless, there is geological evidence for this fault. For example, to the north of La Tigra town, the fault dextrally displaces the Venado Formation (Site 17 in Fig. 15.5). Argüello (2021) also described a dextral displacement of the Venado Formation in

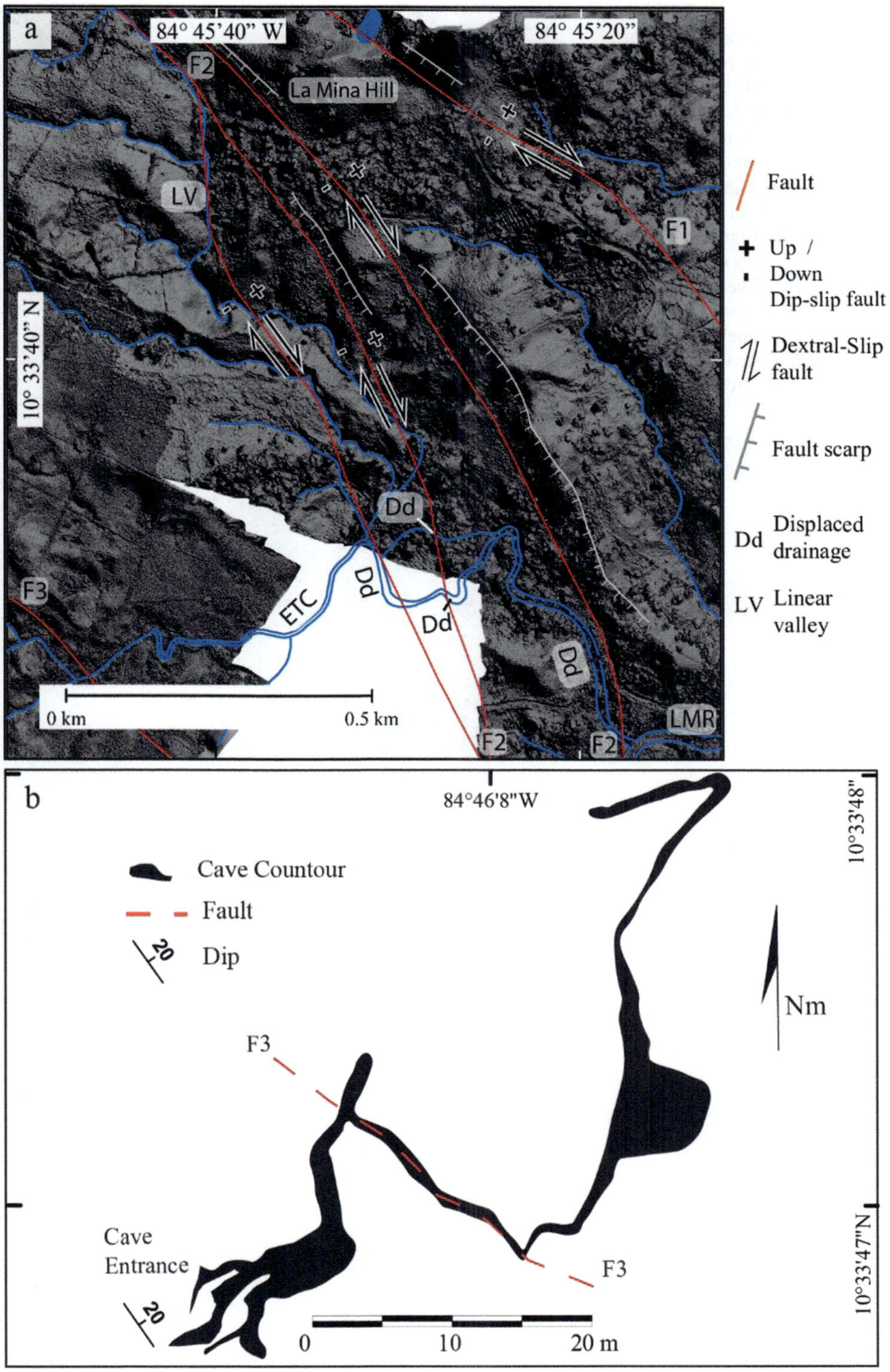

Fig. 15.8 Geomorphic and geologic evidence of F2 and F3: **a** Drone photogrammetry where La Mina Fault (F2) crosses El Túnel Creek (ETC) upstream from its junction with the La Muerte River (LMR). F2 shows west-facing scarps and ~400 m right offset of the El Túnel Creek. Location in Fig. 15.6b The map view of the Catano Cave outline shows a ~16 m right offset related to the Santa Lucía Fault (F3). Location in Figs. 15.6 and 15.9. Based on Vargas (2014)

different nearby creeks. South of La Tigra town, an alluvium patch is found where F6 splits into two fault traces (Site 1 in Fig. 15.11). South of this sector, both fault traces follow linear valleys of tributaries to La Tigra Creek (LV in Fig. 15.11) and also suggest dextral drainage offsets (Dd in Fig. 15.1). The region between both fault traces is depressed.

The east F6 fault trace shows a west-facing fault scarp and a rejuvenated valley on the eastern fault block where it crosses La Muerte River (FS and Rv, respectively, in Fig. 15.11). Upstream of the fault trace, a significant alluvial patch covers the F6 downthrown block (Pd in Fig. 15.11). Also, a fault contact occurs between the Grifo Alto Formation on the east

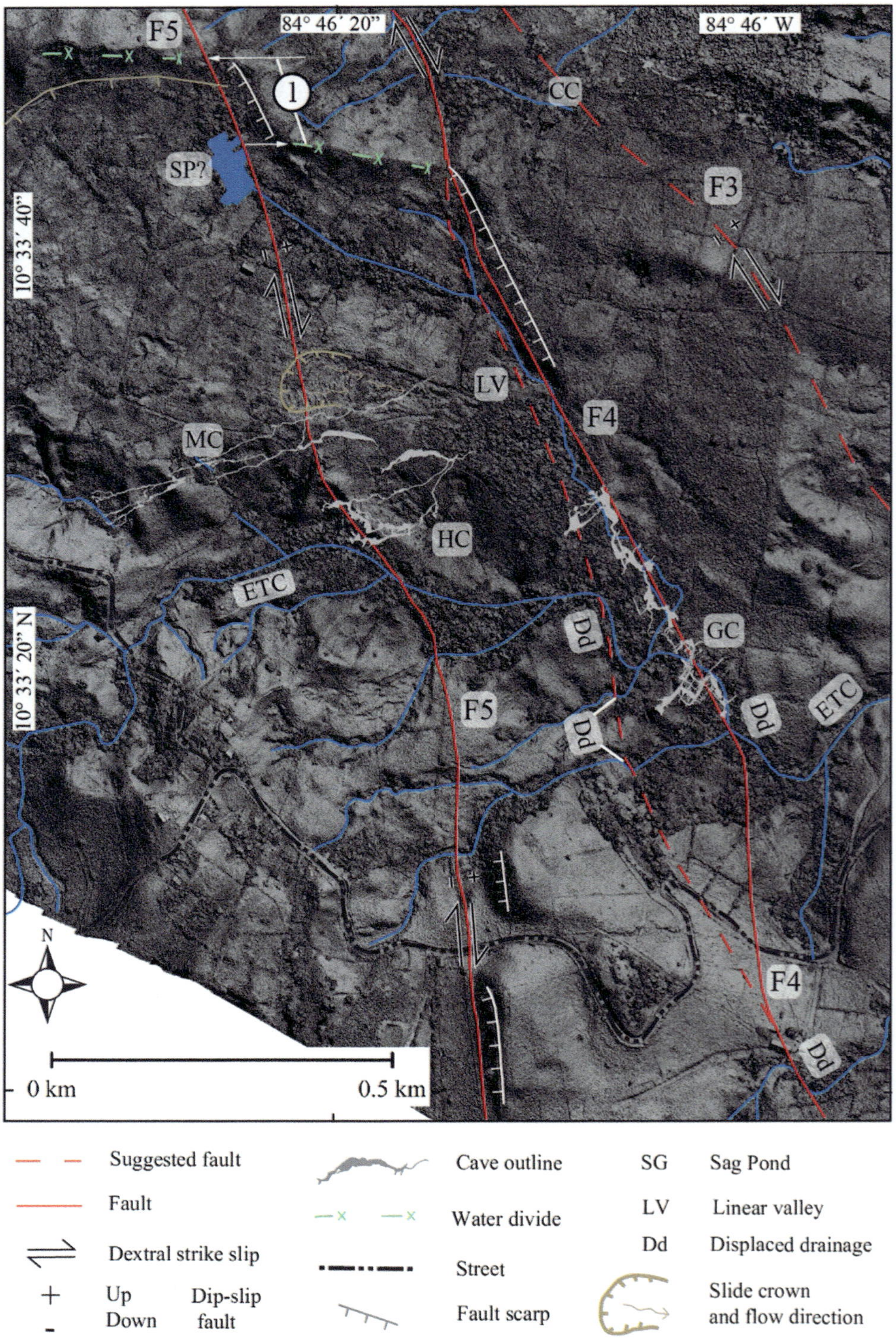

Fig. 15.9 Drone photogrammetry map where the Gabinarraca Cave (GC) follows the La Cueva (F4) fault trace. The cave contours of the Menonitas (MC) and La Higuera (HC) show the geometric relationship with the F5 fault trace. Also shown is a sector of the Santa Lucia fault trace (F3). Drainage and various tectonic landforms define F4 and F5 fault traces. The numbered site location 1 is discussed in the text. ETC is the El Túnel Creek. CC is the location of the Catano Cave displaced by F3 (see Fig. 15.8b)

side and the Monteverde Formation on the west side (site 2 in Fig. 15.11).

The F6 fault trends WNW where it crosses the La Muerte River. At this place, the river is dextrally offset by ~ 100 m (Site 11 in Fig. 15.6). F6 also shows various dextral deflections of tributary streams of the La Muerte and Nilo rivers (Sites 12, 13, 14, and 15 in Fig. 15.6). Finally, a north-facing scarp is found in its eastern sector termination (Site 16 in Fig. 15.6).

15.4.2.7 Catanuria Fault

The Catanuria Fault (F7) trend oscillates between NNW and WNW, showing a prominent geomorphic trace at its central part and less clear at its northern and southern sectors. Geological and geomorphic evidence of dextral-normal oblique movement characterizes the F7 fault trace. The northern sector of F7 has an NNW trend, shows an east-facing fault scarp, and causes displaced drainages (Dd in Fig. 15.6). In its central part, F7 trends NNW.

Fig. 15.10 Geomorphic and geologic evidence of F5: **a** Drone photograph looking north of F5. FS is a cross-section view of the east-facing fault scarp. The Enmedio road outcrop is shown in detail in Fig. 15.10b. The crown of a suggested landslide is shown cutting a farm road. **b** 3D photogrammetry model where fault traces of F5 cuts the Enmedio road, **c** Photography of the road cut outcrop showing two F5 fault traces cutting pyroclastic deposits and paleosoils. Both faults suggest dip-slip components. The person is shown for scale

Continuous east-facing scarps, with a height of ~ 60 m, characterize the Catanuria Fault (FS in Figs. 15.11 and 15.12a). West of La Tigra town, where F7 crosses the road, an outcrop on the uplifted block shows west-tilted lahar deposits, and pyroclastic deposits are covered by subhorizontal recent sediments (Fig. 15.12b). This angular discordance reflects the rotation of the uplifted block. Additionally, small ponds are found at the foot block (Pond P1 in Fig. 15.11) or behind the fault scarp (P2 and P3 in Figs. 15.11, 15.12c and d). They are tectonically related to perturbations of the drainage network due to the rotation of both F7 fault blocks. The F7 fault causes dextral drainage offsets along different sectors of the fault trace (Dd in Fig. 15.6).

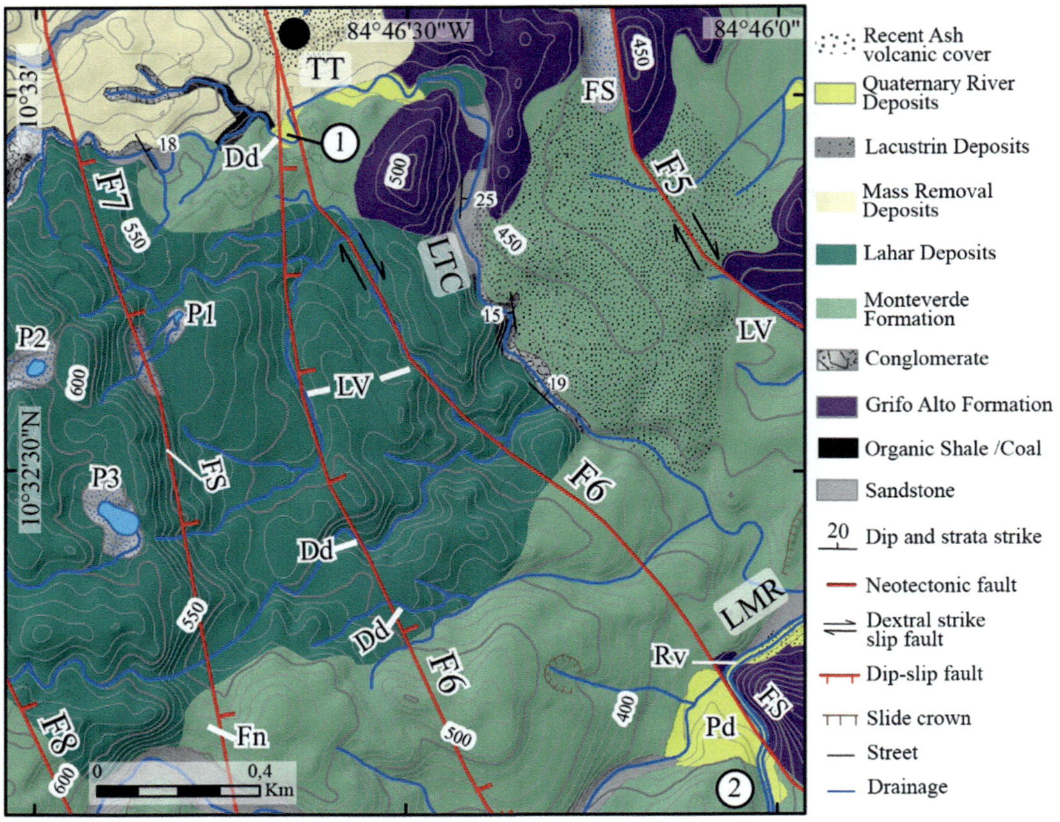

Fig. 15.11 Geologic and morphotectonic map of the F5, F6, and F7 fault traces south of the La Tigra town (TT). Tectonic landforms are P: pond, FS: fault scarp, Dd: displaced drainage, LV: linear valley, Pd: Ponded alluvium, and Rv is Rejuvenated valley. LMR is the La Muerte River. LTC is the La Tigra Creek. Numbered site locations 1 and 2 are discussed in the text

15.4.2.8 Suampo Fault

The Suampo Fault trace (F8) is an NNW-trending fault with a moderate to weak geomorphic expression. Geomorphic evidence suggests F8 is a dextral-normal oblique fault. The central sector of F8 shows an east-facing fault scarp west of the town of La Tigra, suggesting that the eastern block down-dropped relative to the western one (Fig. 15.12e). Small right drainage offsets suggest that F8 also has a strike-slip component (Dd in Fig. 15.6). The trace of this fault loses geomorphic prominence both north and south of its central sector. Geological evidence to support F8 is not presently available.

15.4.3 Southeastern Closure of the VTZ and the Southeastern Fortuna Fault Branch

The southeastern closure of the VTZ and the southeastern Fortuna Fault branch (SFF) continues between the northern bank of the Arenal River and the eastern slopes of the Arenal and Chato volcanoes (Fig. 15.3). Tectonic landforms such as dextrally offset or deflected drainages, displaced landforms, fault scarps, and linear valleys define the southeastern sector of VTZ and the Fortuna Fault. The southeastern VTZ closure includes two right offsets of the Arenal River. The eastern one suggests the river is dextrally displaced by about 1 km (location 8 in Figs. 15.3 and 15.13a). In addition to the morphotectonic evidence, thermal springs are reported in this section of the Fortuna Fault (Alvarado 2009). Northwest of the town of Fortuna, in the eastern foothills of Arenal Volcano, Alvarado (2009) identified several thermal springs. It is strongly suggested that the southeastern closure of the VTZ structurally controls the location of hot springs and thermal aquifers. A transtensional tectonic environment is favorable for the rise of hot water heated by the magma chamber of the Arenal Volcano (Fig. 15.3).

West of Fortuna, the FF crosses a Quaternary alluvial region composed of recent sediments eroded and transported from the Arenal and Chato volcanoes. This area is in an active depositional basin associated with the denudational processes of the two volcanoes; for this reason, the fault trace does not present unambiguous morphotectonic evidence. South of Fortuna town, the Fortuna Fault crosses the eastern slopes of the Cerros Los Perdidos (Fig. 15.13b). Volcanic products of the Monteverde Formation cover this region (Alvarado 2009). In this

Fig. 15.12 Geomorphic and geologic support of F7 and F8: **a** Photography of the F7 frontal east-facing fault scarp (FS). View looking WSW from near La Tigra town. The arrow indicated by **b** shows the location of the road outcrop shown in 12b, **b** Photography looking north of a road outcrop of the rocks units on the floor wall of F7 slightly west of the fault scarp. The photography shows a tilted lahar (Tla) and pyroclastic deposits (TPd) overlying by a subhorizontal volcanic sequence (Rd) covered by soil, **c** Photo looking SSE of the F7 fault scarp (FS) and its upthrown west side. Also shown is a small pond (Not visible in Fig. 15.11). A dashed red line shows the F7 fault trace at the foot of the east-facing scarp, **d** A creek flowing NE ends at the small pond (P2) on the west and upthrown F7 block, **e** Photography viewing NNW of the fault scarp (FS) related to F8. A dashed red line shows the fault trace. A house provides scale

area, FF presents several morphotectonic indicators of oblique displacement. Tectonic landforms include offset or deflected drainages where it crosses the Burrito River and a right bank tributary, a displaced planation surface, a fault scarp, and a wine-glass-shaped canyon (location 9 in Figs. 15.3 and 15.13b). South of the previous site, the SFF finishes at the thrust Florencia-La Palmera Fault (Montero 2001; Montero et al. 2013). Finally, we like to mention that Alvarado (2009) published a geological map for the Fortuna quadrangle showing various NW-trending faults on the eastern slopes of the Arenal and Chato volcanoes. Some of these faults have similar trends and dextral movement as our proposed faults related to the southeast termination of the VTZ and the SFF.

15.5 Discussion

15.5.1 Structure of the VTZ Deduced from Geological and Geomorphic Evidence

The F1 to F8 faults of the VTZ have different geomorphic and geological characteristics, which have significant structural and tectonic implications. The F1 and F2 eastern faults of the VTZ provide geomorphic and geologic evidence supporting strike-slip and dip-slip components. Although we did not mention any geomorphic features to deduce a dip-slip component for the La Muerte Fault (F1), geologic

Fig. 15.13 Drone photographs of the SW Fortuna Fault: **a** View WSW of the western branch of the Venado Transtensive Zone (WVTZ), where it crosses the Arenal River. The river has an estimated right displacement (Dd) of 1.0 km (defined between the white arrows). Image view is from site number 8 (Fig. 15.3), **b** View WSW showing the southeastern Fortuna Fault trace (SFF) where it crosses the eastern slopes of the Cerros Los Perdidos (CLP). CHV is the Chato volcano. The image view is from site number 9 (Fig. 15.3)

evidence supports it. The normal movement is inferred from an outcrop of limestones found on the eastern F1 block to the north of the town of Venado (Fig. 15.5). The limestones crop out at an elevation of around 270 m a.s.l and dip ~ 20° to the NE. We propose that the F1 fault uplifted the limestones outcropping in the eastern fault block. La Mina Fault (F2) shows geomorphic evidence of strike-slip and dip-slip components because F2 shows dextral drainage offsets and west-facing scarps. Geological evidence also supports both slip components are present on the F2 fault.

The fault traces F3, F4, and F5, located within the central sector of the VTZ, show various landforms suggestive of dextral faulting. However, consistent uplift on the eastern or western fault blocks along any of the previous three tectonic structures was not found. For example, the F5 fault shows a west-facing scarp at the road cross between the towns of Venado and La Tigra (Fig. 15.9). However, where F5 crosses the Enmedio road present an east-facing scarp (Fig. 15.10). The previous observations concerning the dip-slip component of the F3, F4, and F5 could be explained for two factors. First, their fault planes are nearly vertical and have minor west or east dip variations or release zones. A second explanation for the opposite direction of the facing scarps could be related to the scissor effect typical of strike-slip faults.

The F7 and F8 faults, located on the west side of the VTZ structure, show geomorphic and geologic evidence of strike-slip and dip-slip components. The principal landforms expressive of the dip-slip component are east-facing scarps along a significant extension of their central fault traces. Geologic evidence of a dip-slip was also found for the F7 fault trace. For example, where the fault crosses the road west of the town of La Tigra, older volcanic units are tilted to the west, and younger deposits above lie subhorizontally (Fig. 15.12b). Geomorphic and geological evidence of a dextral component is also found for F7 and F8, especially in the northern and southern fault trace sectors. The geologic and geomorphic evidence concerning F7 and F8 faults imply they have both strike-slip and dip-slip components and dip between 60° to 70° to the east. The last values are deduced from the relatively straight F7 and F8 fault traces. Therefore, F7 and F8 should be considered dextral-normal oblique faults. Concerning the F6 fault, it shows typical strike-slip landforms and a less clear dip-slip component. The dip-slip component is suggested from the depressed area between the east and west F6 fault traces to the south of La Tigra town. The western fault shows an east-facing scarp, while the eastern fault shows a less clear west-facing scarp. We infer the eastern fault trace is antithetical to the western synthetic fault trace.

Most of the F1 to F8 fault traces of the VTZ structure are prominent in the central sector. However, their northern and southern sectors show less clear geomorphic expression. In addition, the geological evidence in these sectors of the fault traces is also scarce. The reason partly relates to the fact that both regions have limited communication routes and dense vegetation. For the reasons above, the structural closure of the VTZ, both in the northern and southeast sectors, is inferred. However, essential evidence of the continuity of the fault traces in the southeast sector of the VTZ is the presence of two significant dextral offsets of the Arenal River (Figs. 15.6, 15.5 and 15.13a). The eastern trace has a displacement of about 1.0 km, whereas the western one shows ~ 0.5 km.

The previous geomorphic and geologic observations of the F1 to F8 fault traces are characteristics of a transtensional stepover showing inward west-dipping faults in the east sector of the VTZ tectonic structure (F1 and F2 faults dipping 60 to 70° to the west). In contrast, in the west sector, the F6, F7, and F8 fault traces incline 60° to 70° to the east. On the other hand, the central F3, F4, and F5 faults are relatively vertical. We interpret the west, central, and east VTZ faults to converge at depth with a central displacement fault zone related to the Fortuna Fault. Thus, F1 to F8 faults represent a negative flower structure (Fig. 15.14d). The maximum horizontal distance between the fault pairs also supports the convergence of the fault traces at depth. In the southern part of the VTZ, the maximum length is ~ 1.70 km between the fault pair F7-F6, while in the northern region, the fault pair F6-F5 has a maximum distance of ~ 1.15 km. In other words, the relative closeness of the different connected F1 to F8 fault traces in the plan view supports their connection at depth and the interpretation of a VTZ as a negative flower structure. Releasing bends are commonly related to negative flower structures (e.g., Dooley and McClay 1997; Cunningham and Mann 2007).

15.5.2 Caves of the Karstic System and Its Relationship with the VTZ Structure

The main caves of the karstic system were developed along the F4 and F5 fault traces (Fig. 15.9), which are considered predominantly strike-slip faults. Nevertheless, the caves show two different trends concerning the previous faults; some are nearly perpendicular to the F4 and F5 trends, and others are subparallel. The main passage of Gabinarraca Cave and the most extensive galleries have a northwest tendency, nearly parallel to a dilatational release zone related to the F4 fault trace. The last releasing structure is favorable for the development of the Gabinarraca Cave, where an underground river flows predominantly along the fault trace. On the other hand, the Menonitas and La Higuera caves trend nearly perpendicular to the F5 fault zone. These caves are dissolved parallel to the regional dip direction. Where the F5 fault zone crosses the Menonitas and Higuera caves, both caves show chamber displacement, rock striations, broken flowstone and stalagmites, and drainage offsets. La Higuera Cave is developed from the F5 fault trace to the northeast. Also, the western side of the Higuera Cave is subparallel to the F5 fault trace. The underground water that flows to the northeast in the Menonitas and Higuera caves follows the direction of the dip of the rock until the F4 fault zone and the Gabinarraca Cave. In the northwestern part of the Gabinarraca Cave, the water of both caves converges and changes its flow direction to the southeast, following the F4 fault zone.

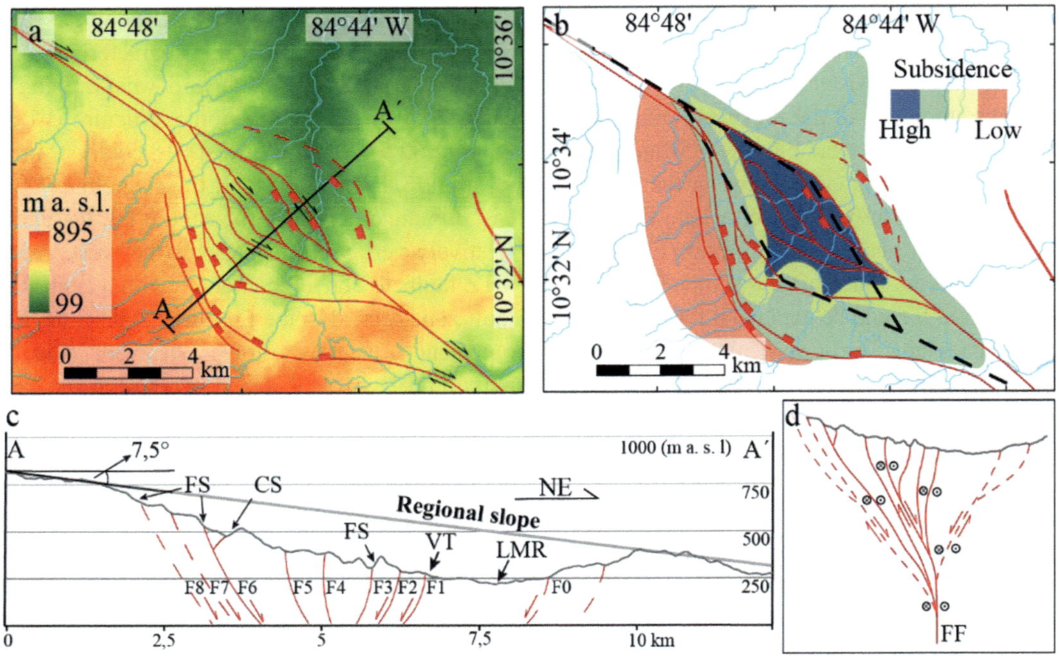

Fig. 15.14 Geomorphic and tectonic model of the VTZ: **a** Plan view of the Venado Transtensive Zone superimposed over a DEM. Profile AA' is shown in **c**. **b** Schematic subsidence of different sectors of the VTZ. See discussion in the text, **c** Profile AA' shows the depressed area between faults F0 to F8, defining the VTZ structure. Venado Town (VT), La Muerte River (LMR), fault scarp (FS), reverse scarp (CS), **d** Schematic profile model of the negative flower structure of the VTZ

15.5.3 Alluvial Deposits and Their Relation to the VTZ Tectonic Subsidence

The northwestern Fortuna Fault and the VTZ were developed inside a tilted planation surface (Figs. 15.6, 15.14a, and c). The planation surface shows an estimated tilt angle of 7.5° to NE along the A-A' profile (Fig. 15.14a and c). A drainage network whose principal collector is the La Muerte River has incised the VTZ structure. Headwater erosion along the different valleys of this drainage network was more significant than in other rivers that drain the tilted planation surface to the north (Figs. 15.3 and 15.14a). The reason is the tectonic subsidence related to the VTZ, which produced many small to medium-sized alluvial patches. Many represent different terrace levels located on the downthrown side of various faults of the VTZ and are considered, in many cases, ponded alluviums. The largest ponded alluviums are found along the La Muerte and Nilo headwaters, which have the greater power of the drainage network.

Nevertheless, the largest alluvium deposits occur downstream of the La Muerte Fault and outside the proposed VTZ structure. This could be explained in two ways. First, these terrace deposits are located slightly downstream of the La Muerte and Nilo junction. After receiving the waters of the Nilo River, the water increment of the La Muerte River means an increase in the capacity of alluvium transport downstream of the town of Venado. We also recognize a change in the course of the La Muerte River from the northeast to a more northern one. This course change means the La Muerte River did not follow the regional maximum slope angle the tilted planation surface represents. The shift to a more northern course means the La Muerte River diminishes the course gradient. The lower slope of the La Muerte River downstream from Venado town represents a change from erosion to aggradation. This last observation would explain the deposition of the most significant volume of alluvial deposits out and north of the VTZ structure. A second explanation that does not necessarily leave out the previous reasoning is that on the east side of F1, there is a geomorphic and geologic suggestion of another fault trace called Las Pendientes (F0) (Fig. 15.14; Argüello 2021). F0 would be supported geomorphologically by west-facing scarps, linear valleys, and dextral drainage offsets, among others. The F0 fault also explains the trend change course of the La Muerte River from the northeast to the north. F0 is geologically also supported by dextral offsets in the hypabyssal unit of the Grifo Alto Formation. F0 corresponds to the eastern limit of the alluvial deposits associated with the La Muerte River downstream of the junction with the Nilo River. F0 also explains the northern lobe of high subsidence present along the reach of the La Muerte River downstream of the town of Venado (Fig. 15.14b). Finally, if F0 is part of the VTZ, it will imply that this tectonic structure is of greater amplitude. In that case, the VTZ dilatational bend associated with the Fortuna Fault will cover a greater area.

15.5.4 The Haciendas-Chiripa Fault System and Its Relationship with the Fortuna Fault

The Fortuna Fault and the associated VTZ releasing bend is a newly defined eastern branch of the Haciendas-Chiripa fault system described by Montero et al. (2017). The FF branch explains the trench-parallel component of ~ 3 mm/yr GPS velocity found at the permanent station GR38 by Feng et al. (2012). This GPS station is located east of the Cote-Arenal Fault and west of the FF (Fig. 15.3). The east block of the Fortuna Fault is considered to displace with the Caribbean Plate. The Fortuna Fault connects to the north and slightly outside the study area with the Cote-Arenal Fault, which is part of the Haciendas-Chiripa fault system (Figs. 15.1, 15.2 and 15.3). To the south, it connects with the Florencia-La Palmera thrust fault (Montero et al. 2013) that borders the northwestern border of the Cordillera Volcánica Central (Fig. 15.1). This tectonic connection implies a tectonic link between the Haciendas-Chiripa fault system with the Central Costa Rica Deformed Belt as was suggested by Montero et al. (2013, 2017)(Fig. 15.1).

15.6 Conclusions

The regional Fortuna Fault and the associated VTZ dilatational bend is a newly defined eastern branch of the Haciendas-Chiripa fault system. This fault system is the eastern limit of the Guanacaste Volcanic Arc Sliver that is part of the Central American Forearc Sliver (CAFS). Along this fault system, the CAFS displaces around 11 mm/yr to the northwest, parallel to the Middle America Trench. The Fortuna Fault connects to the northwest with the Cote-Arenal Fault, which is part of the Haciendas-Chiripa fault system. To the south, the Fortuna Fault continues with the transpressive zone that limits the northern and southern borders of the Cordillera Volcánica Central, linking the Haciendas-Chiripa fault system with the Central Costa Rica Deformed Belt. The Fortuna Fault and the Venado Transtensive Zone are considered active based on geomorphic and geological evidence.

Various tectonic landforms typical of strike-slip faults have been recognized along the Fortuna Fault. The VTZ is defined using geomorphic evidence and detailed geological mapping. Right-lateral drainage offsets along the Fortuna Fault and the VTZ suggest meters to around one kilometer of horizontal displacement. The east or west-facing scarps show heights of tens up to sixty meters, but the accumulative dip-slip displacement along the different faults defining the VTZ would be greater than hundreds of meters. The VTZ is a closed structure defined by eight to nine faults. These faults join at depth, defining a negative flower structure.

Acknowledgements We thank Cristina Araya and Bridget Garnier for their helpful reviews and comments of the manuscript. This research is a result of projects "Reconocimiento del sistema kárstico de Venado de San Carlos y sus implicaciones espeleológicas, hidrogeológicas, geológicas y geo-turísticas" (B7297) as well as projects "Libro de cuevas y karst de Costa Rica" (B9707), "Análisis espeleogenético de cuevas clave en Costa Rica" (C0045), and "Avance en el Cartografiado Geológico y el léxico estratigráfico de Costa Rica" (No 830-C2-158) of the Vicerrectoría de Investigación de la Universidad de Costa Rica. Thank you also to the CeNAT-Conare Scholarship awarded to Alejandro Argüello for collaborating with the PRIAS Laboratory for the research. The Anthros Speleological Group is thanked for the support during the speleological explorations and fieldwork and for providing some information in the article. Thanks also to the Solis Cubero Family for their support during the field research and different expeditions.

References

Alonso-Henar J, Álvarez-Gómez JA, Martínez-Díaz JJ (2014) Constraints for the recent tectonics of the El Salvador fault-zone, Central America Volcanic Arc, from morphotectonic analysis. Tectonophysics 623:1–13

Alvarado GE (1989) Consideraciones neotectónicas recientes en los alrededores de la Laguna de Arenal, Costa Rica. Bol OSIVAM 3:6–14

Alvarado GE (2009) Geología de la hoja Fortuna, Alajuela, Costa Rica. Rev Geol Am Cent 41:117–122

Alvarado GE, Gans PB (2012) Síntesis geocronológica del magmatismo, metamorfismo y metalogenia de Costa Rica, América Central. Rev Geol Am Cent 46:7–122

Alvarado D, DeMets C, Tikoff B, Hernández D, Wawrzyniec TF, Pullinger C, Mattioli G, Turner HL, Rodriguez M, Correa-Mora F (2011) Forearc motion and deformation between El Salvador and Nicaragua: GPS, seismic, structural, and paleomagnetic observations. Lithosphere 3:3–21

Álvarez L, Rodríguez AM, González L, Moreno B, Cabrera A (2018) Seismotectonics of the Nicaraguan depression from recent seismicity. J Geol Geophys 7:1–13

Álvarez-Gómez JA, Meijer PT, Martínez-Díaz JJ, Capote R (2008) Constraints from finite element modeling on the active tectonics of northern Central America and the Middle America Trench. Tectonics 27:1–12

Álvarez-Gómez JA, Staller Vázquez A, Martínez-Díaz JJ, Canora C, Alonso-Henar J, Insua-Arévalo JM, Béjar-Pizarro M (2019) Push-pull driving of the Central America Forearc in the context of the Cocos-Caribbean-North America triple junction. Abstr Sci Rep 9(1). https://doi.org/10.1038/s41598-019-47617-3

Araya MC, Biggs J (2020) Deformation associated with sliver transport in Costa Rica: seismic and geodetic observations of the July 2016 Bijagua earthquake sequence. Geophys J Int 220:585–597

Argüello A (2021) Estudio tectónico y kárstico de la región Venado de San Carlos, Costa Rica. Universidad de Costa Rica, San José, Costa Rica (Lic. Thesis), pp 139

Argüello-Sáenz A, Laurito CA, Valerio AL, Quesada G, Ulloa A (2022) Primeros registros fósiles de Condrictios para la Formación Venado, Mioceno Superior, provincia de Alajuela, Costa Rica. Rev Geol Am Cent 67:1–9

Argus D, Gordon, DeMets C (2011) Geologically current motion of 56 plates relative to the no-net-rotation reference frame. Geochem Geophys Geosyst 12:1–13

Calvo C, Bolz A (1987) La secuencia de Venado, un estuario lagunar tropical del Mioceno Medio, Costa Rica. Rev Geol Am Cent 6:1–24

Camacho J (2015) Caracterización Geofísica y Neotectónica de la Falla Cote-Arenal, Costa Rica. Universidad de Costa Rica, San José, Costa Rica (Lic. Thesis), pp 158

Chaves R, Sáenz R (1974) Geología de la Cordillera de Tilarán. San José, Costa Rica (Technical Report), pp 48

Correa-Mora F, DeMets C, Alvarado D, Turner HL, Mattioli G, Hernández D, Pullinger C, Rodríguez M, Tenorio C (2009) GPS-derived coupling estimates for the Central America subduction zone and volcanic arc faults: El Salvador, Honduras, and Nicaragua. Geophys J Int 179:1279–1291

Corti G, Carminati E, Mazzarini F, Garcia MO (2005) Active strike-slip faulting in El Salvador, Central America. Geology 33:989–992

Cunningham WD, Mann P (2007) Tectonics of strike-slip restraining and releasing bends. Geol Soc London Special Public 290:1–12

DeMets C (2001) A new estimate for present-day Cocos-Caribbean plate motion: implications for slip along the Central American Volcanic Arc. Geophys Res Lett 28:4043–4046

Dengo G (1962) Estudio geológico de la región Guanacaste, Costa Rica. Impr. del Instituto Geográfico, San José, Costa Rica, pp 112

Denyer P, Montero W, Alvarado GE (2009) Atlas tectónico de Costa Rica, 2 edc. Editorial de la Universidad de Costa Rica, San José, Costa Rica, p 55

Denyer P, Alvarado GE (2007) Mapa Geológico de Costa Rica,Escala 1: 400 000, Librería Francesa, San José, Costa Rica

Denyer P, Montero W, Alvarado G (2003) Atlas tectónico de Costa Rica, 81 pp. Ed. UCR, San José, Costa Rica

Dooley T, McClay K (1997) Analog modeling of pull-apart basins. AAPG Bull 81(11):1804–1826

Echandi E (1986) Informe geológico de la región Venado-Monterrey. San José, Costa Rica. RECOPEs (Internal Report)

Feng L, Newman AV, Protti M, González V, Jiang Y, Dixon TH (2012) Active deformation near the Nicoya Peninsula, northwestern Costa Rica, between 1996 and 2010: Interseismic megathrust coupling. J Geophys Res Solid Earth 117:1–23

Funk J, Mann P, McIntosh K, Stephens J (2009) Cenozoic tectonics of the Nicaraguan depression, Nicaragua, and Median trough, El Salvador, based on seismic-reflection profiling and remote-sensing data. Bull Geol Soc Am 121:1491–1521

Garnier B, Tikoff B, Flores O, Jicha B, DeMets C, Cozensa-Muralles B, Hernández D, Marroquin G, Mixco L, Hernández W (2021) An integrated structural and GPS study of the Jalpatagua fault, Southeastern Guatemala. Geosphere 17:201–225

Gazel E, Hayes JL, Ulloa A, Alfaro A, Coleman DS, Carr MJ (2019) The record of the transition from an Oceanic Arc to a young continent in the Talamanca Cordillera. Geochem Geophys Geosyst 20:2733–2752

Harlow DH, White RA (1985) Shallow earthquake along the volcanic chain in Central America: evidence for oblique subduction (abstract). Earthquake Notes 55:28

Hobbs TE, Newman AV, Protti M (2019) Enigmatic upper-plate sliver transport paused by megathrust earthquake and afterslip. Earth Planet Sci Lett 520:87–93

Kobayashi D, LaFemina P, Geirsson H, Chachaco E, Abrego AA, Mora H, Camacho E (2014) Kinematics of the western Caribbean: collision of the Cocos ridge and upper plate deformation. Geochem Geophys Geosyst 15:1671–1683

Kolarsky R, Mann P, Montero W (1995) Island arc response to shallow subduction of the Cocos Ridge, Costa Rica. Geologic and tectonic development of the Caribbean plate boundary in Southern Central America. Geol Soc Amer Spec Pap 295:235–262

La Femina PC, Dixon TH, Strauch W (2002) Bookshelf faulting in Nicaragua. Geology 30:751–754

LaFemina P, Dixon TH, Govers R, Norabuena E, Turner H, Saballos A, Mattioli G, Protti M, Strauch W (2009) Fore-arc motion and Cocos ridge collision in Central America. Geochem Geophys Geosyst 10:1–21

Lewis JC, Boozer AC, López A, Montero W (2008) Collision versus sliver transport in the hanging wall at the Middle America subduction zone: constraints from background seismicity in central Costa Rica. Geochem Geophys Geosyst 9:1–12

Lundgren P, Protti M, Donnellan A, Heflin M, Hernández E, Jefferson D (1999) Seismic cycle and plate margin deformation in Costa Rica: GPS observations from 1994 to 1997. J Geophys Res Solid Earth 104:28915–28926

Malavassi VE, Madrigal R (1970) Reconocimiento geológico de la Zona Norte de Costa Rica. Inf Tec y Not Geol 38:1–18

Marshall JS, Fisher DM, Gardner TW (2000) Central Costa Rica deformed belt: Kinematics of diffuse faulting across the western Panama block. Tectonics 19:468–492

Martínez-Díaz JJ, Álvarez-Gómez JA, Benito B, Hernández D (2004) Triggering of destructive earthquakes in El Salvador. Geology 32:65–68

Molnar P, Sykes LR (1969) Tectonics of the Caribbean and Middle America regions from focal mechanisms and seismicity. Geol Soc Am Bull 80:1639–1684

Montero W (2001) El cinturón deformado del centro de Costa Rica: Frontera oeste de la microplaca de Panamá. Rev Geol Am Cent 24:29–56

Montero W, Dewey JW (1982) Shallow-focus seismicity, composite focal mechanism, and tectonics of the Valle Central of Costa Rica. Bull Seismol Soc Am 72:1611–1626

Montero W, Rojas W (2014) Las fallas Purires y Picagres, y su relación con la secuencia sísmica de Puriscal de 1990. Rev Geol Am Cent 50:39–69

Montero W, Lewis JC, Marshall JS, Kruse S, Wetmore P (2013) Neotectonic faulting and forearc sliver motion along the Atirro-Río Sucio fault system, Costa Rica, Central America. Geol Soc Am Bull 125:857–876

Montero W, Lewis JC, Araya MC (2017) The Guanacaste Volcanic arc sliver of Northwestern Costa Rica. Sci Rep 7:1–9

Montero W, Denyer P, Barquero R, Alvarado GE, Cowan H, Machette MN, Haller KM, Dart RL (1998) Map and database of quaternary faults and folds in Costa Rica and its Offshore Regions. USGS Open File Rep 98–481

Montero W (1994) Neotectonics and related stress distribution in a subduction-collisional zone: Costa Rica, In Seyfried H, Hellmann W (eds) Geology of an evolving Island Arc: The Isthmus of Southern Nicaragua, Costa Rica, and Western Panamá. Profil 7, pp 125–141

Norabuena E, Dixon TH, Schwartz S, DeShon H, Newman A, Protti M, González V, Dorman L, Flueh E, Lundgren P, Pollitz F, Sampson D (2004) Geodetic and seismic constraints on some seismogenic zone processes in Costa Rica. J Geophys Res Solid Earth 109:1–25

Obando LG (1986) Estratigafía de la formación Venado y rocas sobreyacientes (Mioceno-Reciente). Provincia de Alajuela. Costa Rica. Rev Geol Am Cent 5:73–104

Pizarro D (1993) Los pozos profundos perforados en Costa Rica: Aspectos litológicos y bioestratigráficos. Rev Geol Am Cent 15:81–85

Rojas MM (2019) Mapa Geológico de la Hoja Monterrey (3247-I), Escala 1:50.000. Dirección de Geología y Minas, San José, Costa Rica

Sandoval LF, Sáenz R, Acuña J, Castro JF, Gómez M, López A, Mederos B, Monge A, Vargas J, Fernández T, Ulate R, Ramírez C (1982) Mapa Geológico de Costa Rica, Escala 1:200 000, Instituto Geográfico Nacional, San José, Costa Rica

Sen Gupta B, Malavassi L, Malavassi E (1986) Late Miocene shore in northern Costa Rica: benthic foraminiferal record. Geology 14:218–219

Silver EA, Galewsky J, McIntosh KD, Mann P (1995) Variation in structure, style, and driving mechanism of adjoining segments of the North Panama deformed belt. Geologic and tectonic development of the Caribbean plate boundary in Southern Central America. Geol Soc Am Spec Pap 295:225–233

Solano JP, Soto GJ, Vargas Zúñiga JE, Denyer P (2021, November) Modelado geológico del cuadrante Tilarán (3247 - III), Costa Rica. Poster presented at the 5th UCR Geological Congress 2021. https://doi.org/10.13140/RG.2.2.20032.66566

Soto GJ, Alvarado GE (2006) Eruptive history of Arenal Volcano, Costa Rica, 7 ka to present. J Volcanol Geothermal Res 157:254–269

Suárez G, Muñoz A, Farraz IA, Talavera E, Tenorio V, Novelo-Casnova D, Sánchez A (2015) The 10 April 2014 Nicaraguan crustal earthquake: evidence of complex deformation of the Central American Volcanic Arc. In: Bandy WL, Dañobeitia J, Gutiérrez C, Taran Y, Bartolomé R (eds) Geodynamics of the Latin American Pacific Margin. Springer, pp 3305–3315

Ulloa A, Argüello A, Obando A, Vargas M (2021) Geomorfología y espeleogénesis del sistema kárstico de cuevas de Venado, Costa Rica. Rev Geol América Cent 64:1–31

Vargas M (2014) Pozo Catano [Escala 1:200]. San José, Grupo Espeleológico Anthros. (Cave topography)

White RA, Harlow DH (1993) Destructive upper-crustal earthquakes of Central America since 1900. Bull Seismol Soc Am 83:1115–1142

White RA (1991) Tectonic implications of upper-crustal seismicity in Central America. In: Slemmons DB, Engdahl ER, Zoback M, Blackwell D (eds). Neotectonics of North America. Geological Society of America, pp 323–338

Geomorphology and Morphometric Aspects of the Upper Region of the Barva Volcano, Costa Rica

16

Sebastián Carmona

Abstract

The Barva volcano consists of a volcanic stratigraphic sequence dating from the middle Pleistocene and extending to the present. The stratigraphy consists of the Colima, Tiribí, and Barva Formations. These formations are in a tropical environment, making them susceptible to landslide processes. Slope instability was evaluated by studying its geology, morphometry, and geomorphology. The use of LiDAR data was important for the acquisition of high-quality topographic databases to support the fieldwork. The application of LiDAR and field work allowed the establishment of detailed geomorphological cartography that classified the morphologies into two genetic groups, endogenous and exogenous. Map algebra made it possible to define that the Northwest and Central territory of the region presents the highest degree of susceptibility to mass movements, as well as the highest number of this type of events, verified using an inventory of mass removal processes. It was determined that the morphologies associated with the Early Lavas and Guararí Lavas and Fluvial Valleys are the geoforms that concentrate most of the slope instability, accounting for about 68% of the total number of events inventoried.

Keywords

Barva volcano · Geology · Morphometry · Geomorphology · LiDAR · Landslides

16.1 Introduction

The Barva volcano is an extensive andesitic shield complex located in the Central Volcanic Cordillera of Costa Rica (CVC), covering a total area of 1120–1500 km^2, which makes it the largest volcano in the country and indeed, in the entire Central American region (Alvarado 2011). Located 21 km northeast of the capital city of San José and with the existence of multiple socioeconomic developments on its slopes, it requires further research, such as detailed geomorphological assessment to determine possible areas susceptible to the impact of the exogenous activity. This volcano has lacked studies on the evaluation of landslides and their relationship with the morphologies with which they are associated. For this reason, the interest in implementing this type of analysis arose.

Due to the above, this type of analysis began with the development of a complementary detailed geomorphological model generated by Carmona (2021), which has an emphasis on the recognition of mass removal processes, based on a morphometric analysis. Figure 16.1 shows the area of study on a local scale, which includes the provinces of Alajuela and Heredia, specifically in the cantons of Alajuela, Santa Barbara, Barva, and Heredia, on the CVC. The main towns in the area are Sacramento, Cinco Esquinas, Porrosatí, Los Cartagos, and La Laguna. The study area is located on the Pacific Ocean slope, within the Virilla river basin; for this paper, the analysis was focused on a total area of 63 km^2.

The research is specifically limited to this area of interest, since there is a greater amount of population and vulnerable infrastructure works in comparison with the sites in the northern sector of the volcano, in addition to the fact that the available LiDAR was used as a tool for the determination of different variables, and likewise, field visits are necessary as a means for data collection and verification of the interpreted data. Therefore, those areas located in the

S. Carmona (✉)
Escuela Centroamericana de Geología, Universidad de Costa Rica, San José, Costa Rica
e-mail: sebastiancarmona135@gmail.com

Bioterra Consultores Ambientales, San José, Costa Rica

A. Quesada-Román (ed.), *Landscapes and Landforms of Costa Rica*,
World Geomorphological Landscapes, https://doi.org/10.1007/978-3-031-64940-0_16

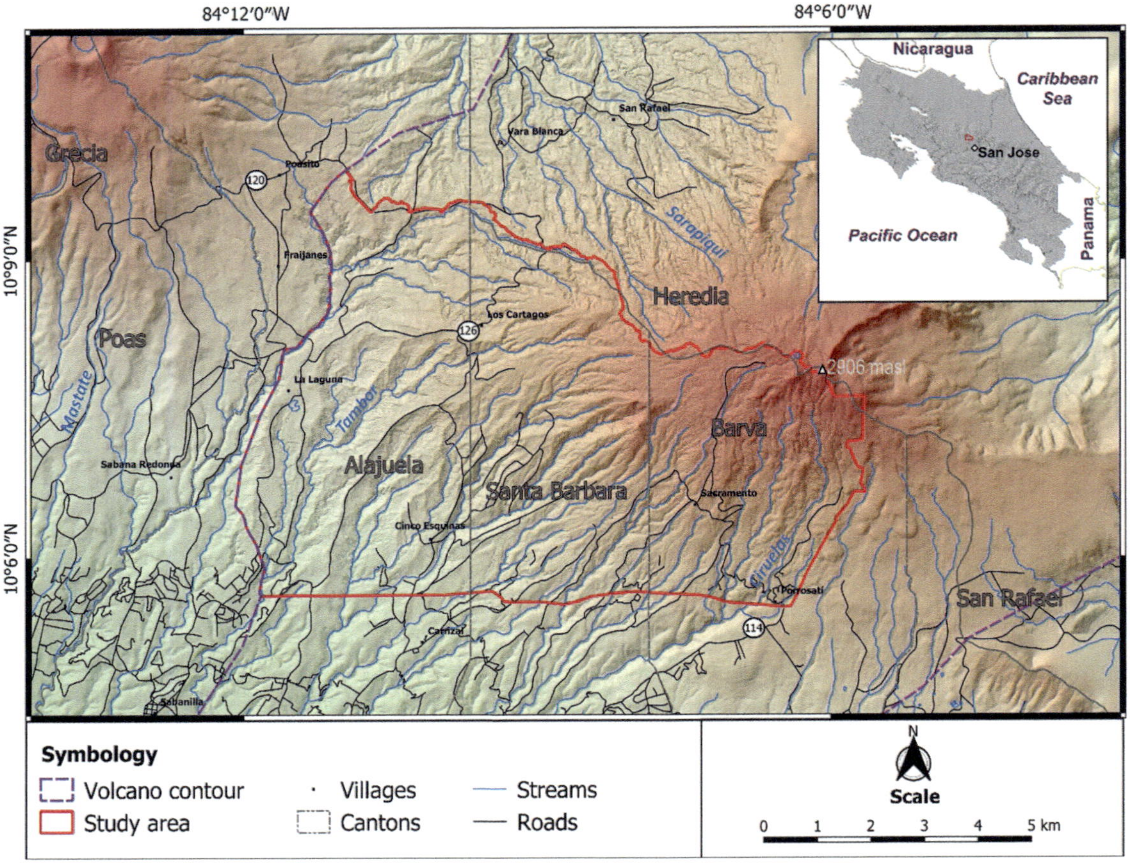

Fig. 16.1 General location map of the study area

Caribbean slope sector would be difficult to verify due to the high forest density and topographic conditions.

The surface materials of the Barva volcano are mostly tephra and soils with varying degrees of consolidation, present within steep slopes, which can affect the loss of stability. Furthermore, the intense precipitation, both seasonal and cyclonic, and historical seismicity, enhances the occurrence of slope movements, thus the interest in understanding the link between lithology and its competition with erosion.

16.2 Regional Context of the Volcano

The Barva volcano is located in the central region of Costa Rica forming part of the volcanic forearc, this territory is due to the interaction of the Cocos, Caribbean, and Panama microplate, where the Cocos plate underlies both. According to Montero and Rojas (2014), this process occurs at an annual rate of about 89 mm/yr. The volcano lies on the Panama microplate, which in the central territory of Costa Rica develops the so-called Central Costa Rica Deformed Belt (CCRDB), which corresponds to a wide belt of deformation that defines the western margin of the microplate and is characterized by active east to northeast sinistral and north to northwest dextral faulting, dating from 2 to 3 Ma ago (Montero 2001; Marshall et al. 2000; MacMillan et al. 2004). As for its stratigraphy, although the volcano covers a stratigraphic sequence that begins during the Lower Pleistocene (758 kyr), before its emplacement, there was a wide series of sedimentary sequences associated with the Central Valley basin dating from the Upper Eocene to the Upper Miocene (Denyer and Arias 1991). It is important to clarify that the stratigraphy of the Barva volcano can be classified into three temporal or evolutionary stages (Soto 1999), Proto-Barva, Paleo-Barva, and Neo-Barva, the last two being the ones described in this research because they have better surface exposure and are in the studied area.

16.3 Geology

The volcano stratigraphy was limited to the last two stages or evolutionary phases of the volcano proposed by Soto (1999), which, for this case, are Paleo-Barva (~800–270 kyr) and Neo-Barva (< 270 kyr), because they have evidence or surface outcrops, unlike Proto-Barva. Figure 16.2 shows the study area's geological map and stratigraphic column, which includes the geological units identified in the

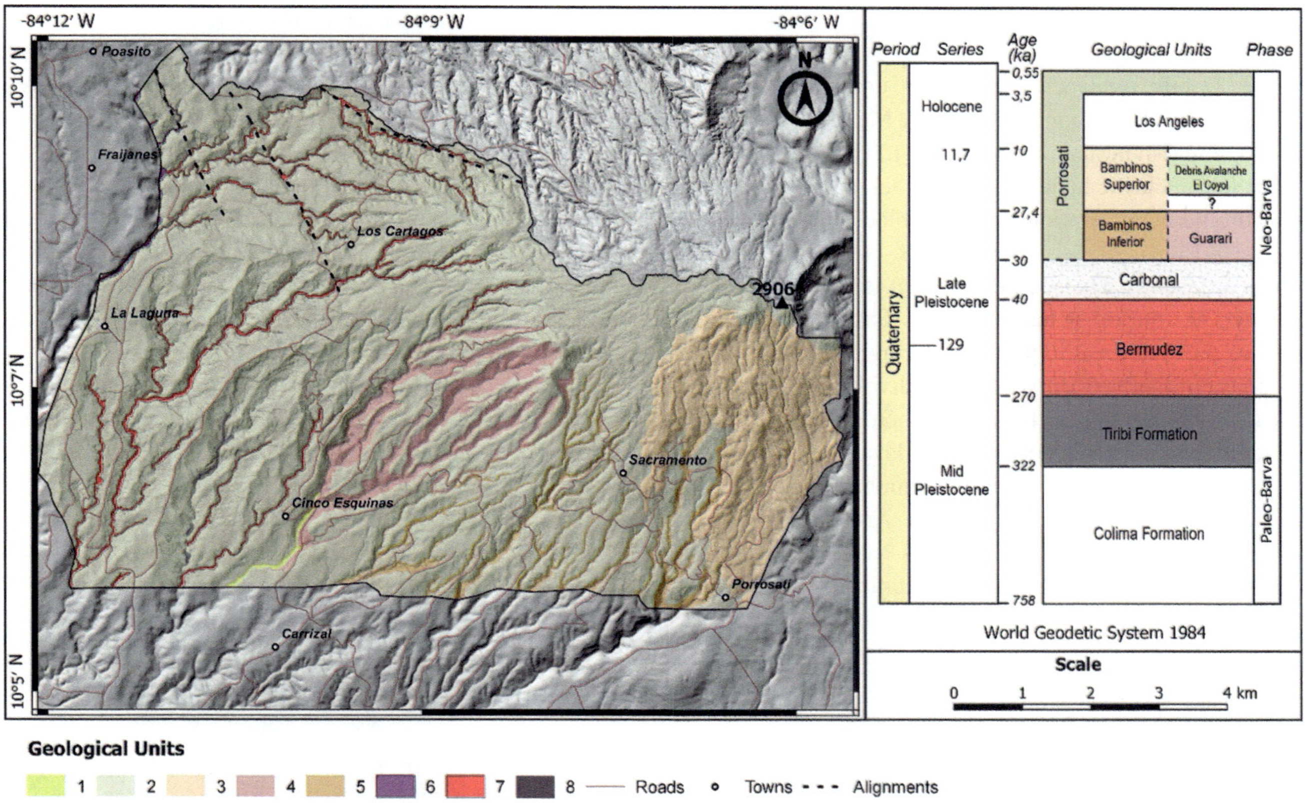

Fig. 16.2 Geological map and stratigraphic column of the study area. 1—Alluvial deposits; 2—Porrosatí unit, 3—Los Bambinos Superior unit; 4—Guararí unit; 5—Los Bambinos Inferior unit; 6—Poasito unit (Poás Volcano); 7—Bermudez unit; 8—Tiribí formation

field (Bermudez, Guararí, Los Bambinos, and Porrosati), as well as others identified in drilling logs or the literature.

As part of the Paleo-Barva evolutionary stage (< 270 kyr), we have the Colima and Tiribi formations. The first one corresponds to the volcano-stratigraphic floor of the volcanic sequence of the Central Valley of Costa Rica, dating from the Middle Pleistocene (758 –331 kyr) (Alvarado and Gans 2012), and is constituted by a series of lavas alternating with ignimbrites and a debris-avalanche deposit, where the source of origin is associated with two covered volcano-tectonic fractures (Thiele and Vargas 2017). The composition of the products varies from basaltic to andesitic, with moderate preservation.

The products of the Tiribí formation were studied by Perez (2000), who recognized at least four ignimbrite facies (Valle Central, Puente de Piedra, La Garita y Orotina). From the geochemical point of view, Perez et al. (2006) indicated that it presents a wide range of compositions, from basaltic–andesitic to rhyolitic, where trachytic is the predominant one. Its source is a pre-Neo-Barva caldera. Based on the $^{40}Ar/^{39}Ar$ dating of Perez et al. (2006), an age of 322 ± 2 kyr is defined for the ignimbrite event.

As part of the Neo-Barva phase, the Bermudez, Carbonal, Los Bambinos Inferior, Los Bambinos Superior, Guararí, Debris Avalanche El Coyol, Los Angeles, and Porrosatí units were defined, as well as the less studied Poza Azul and Cacho Negro units on the Caribbean slope.

The Bermudez unit consists of mostly andesite–basaltic lavas of porphyritic aphanitic texture with wide distribution toward the fluvial valleys of the volcanic piedmont, and these lavas correspond to the base of the recent volcanic sequence (Neo-Barva). Its maximum thickness is close to 100 m, and its age would be between 258 kyr for the most distal products, up to 40 kyr in the studied area (Carmona 2021).

The Carbonal unit was only recorded in stratigraphic sections of wells. It consists of a package of pyroclasts and epiclasts with a high degree of weathering in transition to soils, and the matrix is poorly welded pumiceous ash. Its maximum thickness reaches 15 m, and the age ranges from 40 to 27.4 kyr (Arredondo and Soto 2006; Vargas and Acosta 2007). The Guararí unit consists of blocky and massive lavas that crop out to the southwest of the Cerro Guararí cone. The lavas are classified as basaltic–andesites to andesites, with pyroclastic intercalations, the total sequence is estimated at 85 m thick (Carmona, 2021). In general, they present a porphyritic aphanitic texture, with plagioclase and pyroxene crystals, within a dark gray matrix. Based on the estimates of Arredondo and Soto

(2006), and Vargas and Acosta (2007), the age is close to 27.4 kyr.

The Debris Avalanche, El Coyol unit, was identified through some stratigraphic sections by Méndez and Hidalgo (2004), and Thiele and Vargas (2017). They describe polymict volcaniclastic materials of poor selection and affected by hydrothermal alteration, with lava blocks of large dimension (10–15 m), likewise, the matrix is gray to reddish-brown tuffaceous. The age of this event is between 27.4 and 13.4 kyr. On the other hand, the Los Bambinos unit is an effusive pulse of great importance during the last evolutionary phase of the Neo-Barva. It is subdivided into two subunits (Lower and Upper) by radiocarbon dating of organic matter in lahar deposits between both lava events, as well as by geomorphological expression (Carmona 2021).

The Los Bambinos Inferior (Lower) consists of a series of block flows, classified as andesites-basaltic–andesites to andesites, its texture is usually aphanitic porphyritic with dark gray matrix, in which only plagioclase and pyroxene crystals can be distinguished (Fig. 16.3). It has a wide extension in the southwestern riverbeds of the studied area; its age was estimated between 30 and 27.4 kyr (Arredondo and Soto 2006).

The Los Bambinos Superior unit (Upper) is composed of blocky lavas that, in contrast to the Lower unit (Lower), have a better surface expression due to the insignificant or no tephra coverage and a younger age range (< 27.4–10 kyr, Arredondo and Soto 2006). Furthermore, it has a low to moderate degree of weathering. These lava flows were classified as andesite–basaltic to andesite, usually, the matrix in macroscopy is dark gray, of porphyritic aphanitic texture, without vesicles, with crystals of plagioclase, pyroxene, and occasionally, some olivines.

The Los Angeles unit is a field of andesitic–basaltic to andesitic lavas located northeast of the town of Barva. Classified in two subunits, the first one is an

Fig. 16.3 **a** Lava field of the Los Bambinos Superior (upper) unit. **b** Contact between lower lavas of Los Bambinos Inferior (lower) and Porrosatí tephras. **c** Cross-section of a block lava flow, Guararí unit. **d** Tephra stratification of upper Porrosatí unit. **e** Massif lavas associated with Bermudez unit. **f** Block lava flow outcrop of Guararí unit. Photos were taken by the author

andesite–basaltic with scoriaceous blocks, and the second one comprises vesicular andesites (Rojas et al. 2017). A relative age between 10 and 3.5 kyr is associated with these flows (Arredondo and Soto 2006; Rojas et al. 2017). This unit, however, was not identified within the study area.

The Porrosati unit signifies the youngest eruptive stage of the volcano, whose products are mainly of pyroclastic genesis and are distributed on both, the Caribbean and Pacific slopes of the volcano, and cover most of the volcanic sequence (Fig. 16.3). Brenes (2003) dated the Porrosatí unit using ^{14}C in carbonized wood, which yielded an age of 11,380 ± 288 years for the pyroclastic sequence layer. However, this age would be higher when considering this unit as a contemporary event with the lavas of Los Bambinos Inferior (Lower) and Guararí, so age ranges from 27.4 to 0.55 kyr is assigned (Arredondo and Soto 2006; Vargas and Acosta 2007).

16.4 Morphometry of the Volcanic Slopes

Morphometry is the quantitative characterization of relief forms and provides information about the influence of exogenous processes, their magnitude, and associated morphology (Fig. 16.4). The morphometric variables used were density and depth of dissection, relief energy, as well as Altimetry and Terrain Inclination. These variables correlate in part with the dynamics of the slopes and, when analyzed as a whole, made it possible to identify terrains that are susceptible to mass movement processes.

This model was generated for the territory that has LiDAR coverage, from which contours were taken every 5 m. The altimetric map simplified the information contained, visualizing valley floors and slopes through altitudinal intervals. With the terrain slope index, slopes were classified into slope grades established under a geomorphological criterion, reviewing which of the results visualized or clearly represented the relief forms. Thus, it was possible to recognize the relationship between the dynamics of the slopes, their morphology, and their spatiality.

The dissection density quantifies the intensity of fluvial erosion, indirectly associating the susceptibility of the rock to erosion, permeability, and type of cover. For this analysis, the concentration of riverbeds in each area (1 km/km^2) was calculated; the highest concentration of streams reaches values close to 5 km/km^2. Dissection depth evaluated the vertical erosion of rivers by measuring the distance between the talweg and the nearest slope break on the lower slope of the valley. This analysis made it possible to quantify the erosive intensity in the vertical component of the relief, in which the maximum values were close to 120 m.

Relief energy identifies the erosive potential to which the relief may be subject by measuring the topographic slope. This was done by measuring the maximum and minimum height, where 340 m was the greatest difference. Once the values of the three morphometric indices were determined, and prior to generating the synthesis map (Fig. 16.5), it was necessary to transform the metric units of each parameter to a numerical range between 0 and 1 using the Omega standardization (Velazquez and Celemin 2011; Buzai 2013):

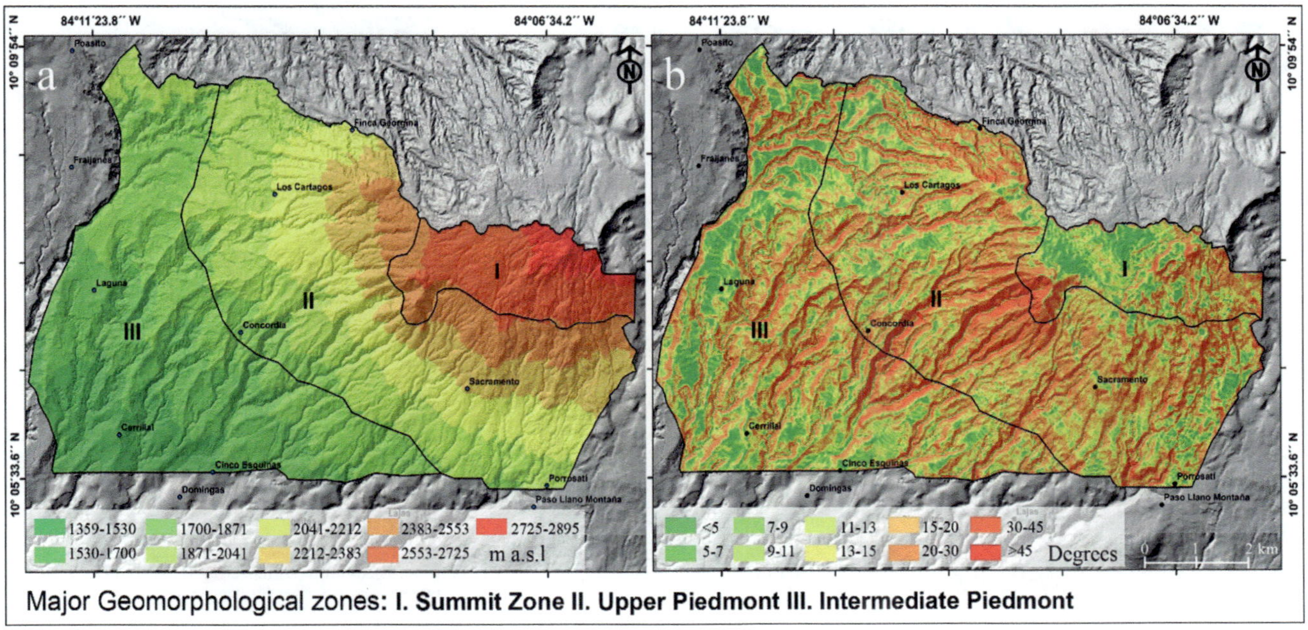

Fig. 16.4 Delimitation of the major geomorphologic regions based on altimetry and slope indices. I Summit zone, II Upper piedmont, III Intermediate piedmont

$$\text{Susceptibility} = \frac{\text{Density of dissection} + \text{Depth of dissection} + \text{Energy of Relief}}{3}$$

The most intense gravitational processes would be expected in areas where the morphometric indices are high. Figure 16.5d shows that the Upper Volcanic Piedmont and Intermediate Volcanic Piedmont are the regions with the highest incidence of these values and therefore concentrate the highest expected instability. The influencing factors are sites with intense fluvial dynamics, organized in a dense dendritic pattern, fluvial valleys with depths > 80 m, and thicker weathered rocks. The summary map shows the terrain susceptible to gravity-driven processes, for which the susceptibility is classified as low, medium, and high. The correlation between the susceptible areas of landslides and the inventory of slope movements (Fig. 16.6) showed that the areas of high susceptibility comprise 76% of the landslides, 23.9% for medium susceptibility, and < 0.1% for low susceptibility.

16.5 Geomorphology

The geomorphological units were classified by genetic hierarchy into two categories: endogenous and exogenous, according to the criteria of Simonov (1985) and Zamorano (1990). The first group is associated with forms resulting from volcanism and its derivatives; the second group is associated with morphologies resulting from erosive and depositional processes (Fig. 16.7).

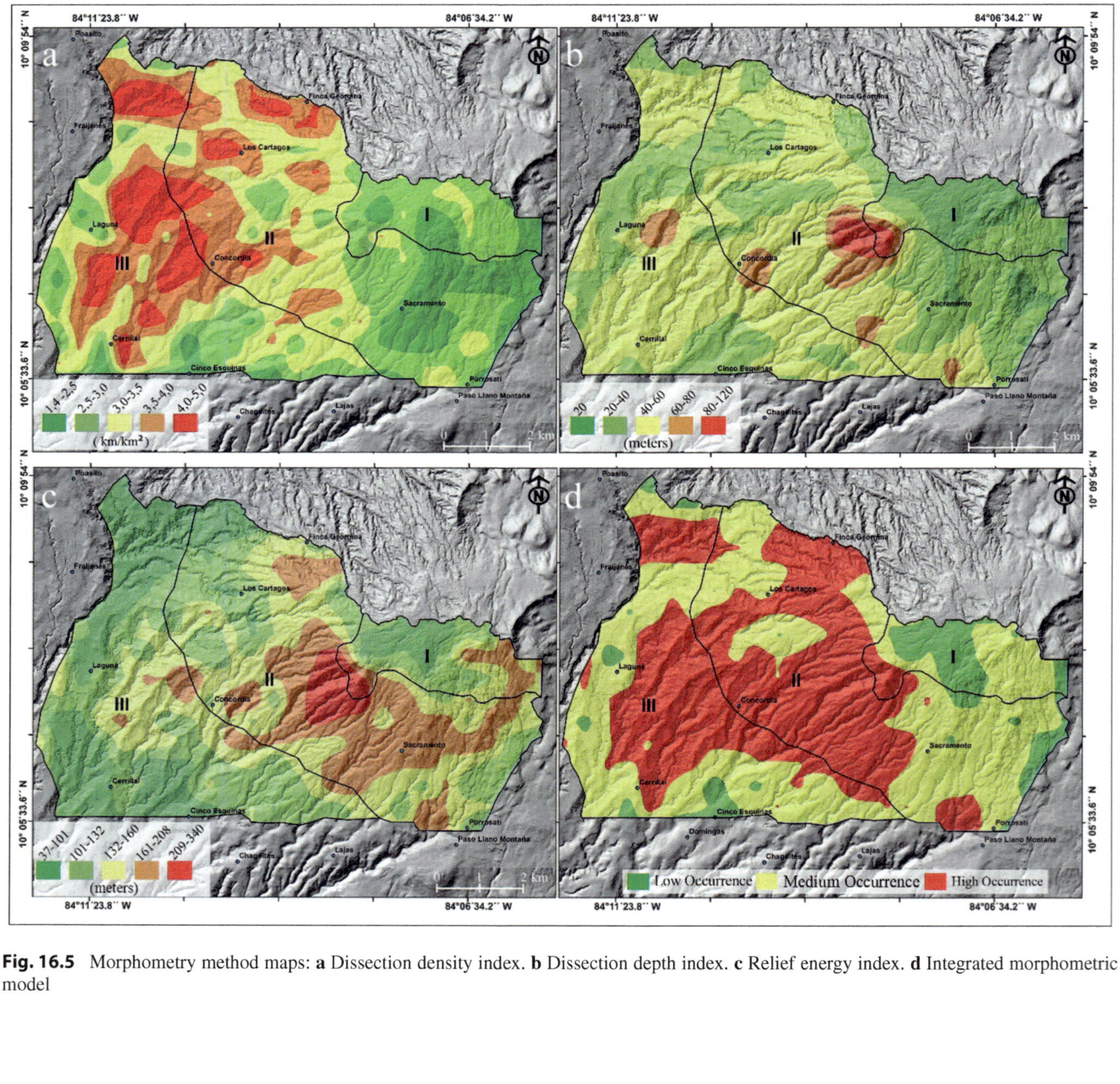

Fig. 16.5 Morphometry method maps: **a** Dissection density index. **b** Dissection depth index. **c** Relief energy index. **d** Integrated morphometric model

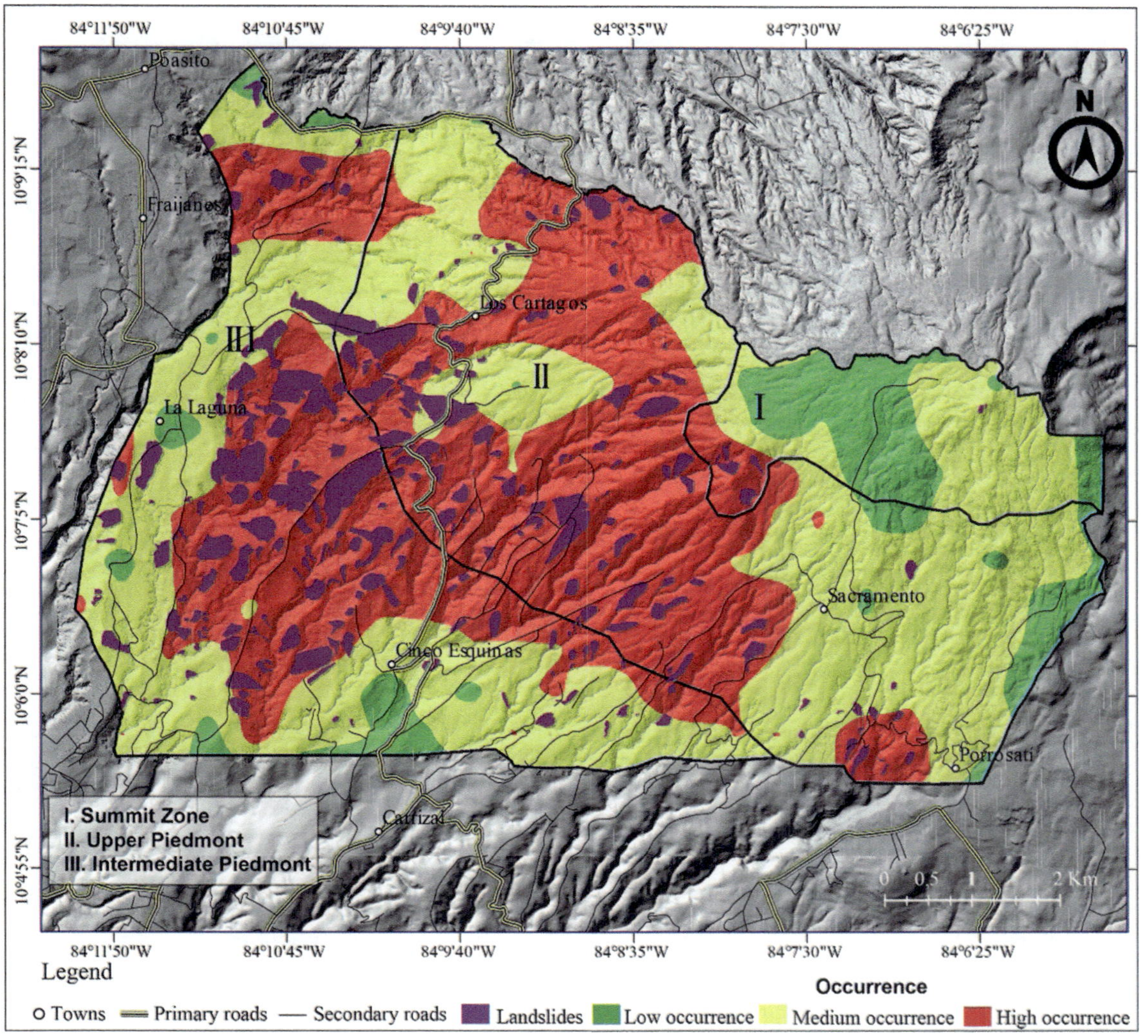

Fig. 16.6 Morphometric model and mass movement distribution

16.5.1 Endogenic Relief

16.5.1.1 Barva Volcano

The Barva volcano is an edifice culminating at 2906 m asl, formed by the accumulation of rocks of basaltic–andesitic composition, which have reached the surface through multiple effusive structures that have formed preferentially on the northern and southern flanks. The geometry of the volcano is convex with a wide base, totaling an approximate area of 1400 km^2. On its slopes, monogenetic and adventitious cones responsible for extensive lava fields on the eastern slopes of the structure and powerful layers of pyroclasts ejected in the last eruptive stages (> 100 m thick) developed in the Middle Pleistocene—early Holocene.

16.5.1.2 Barva Volcano Lava Flows

For the study of the lava flows, a relative temporal criterion was used. Thus, two groups were established, Early Lavas and Intermediate Lavas. All of them are associated with the main structure, and in the study, the area occupies the total area of 41.31 km^2 (Fig. 16.8). The Early Lavas are the oldest flows in the area (Neo-Barva Phase). This type of flows outcrops on the W, SW, SE, and N slopes, and their surface area is 33.82 km^2. Their primary formation features are unclear, due in part to being covered by potent pyroclastic mantles (upto 100 m, BA-333 SENARA well). This type of flows integrates the middle and low sectors of the volcanic piedmont, where the general inclination is > 15° and they are associated with the Bermudez geological unit (~270–40 kyr) at the base of the Neo-Barva volcanic sequence (Soto 1999). These lavas usually outcrop in deep valley bottoms.

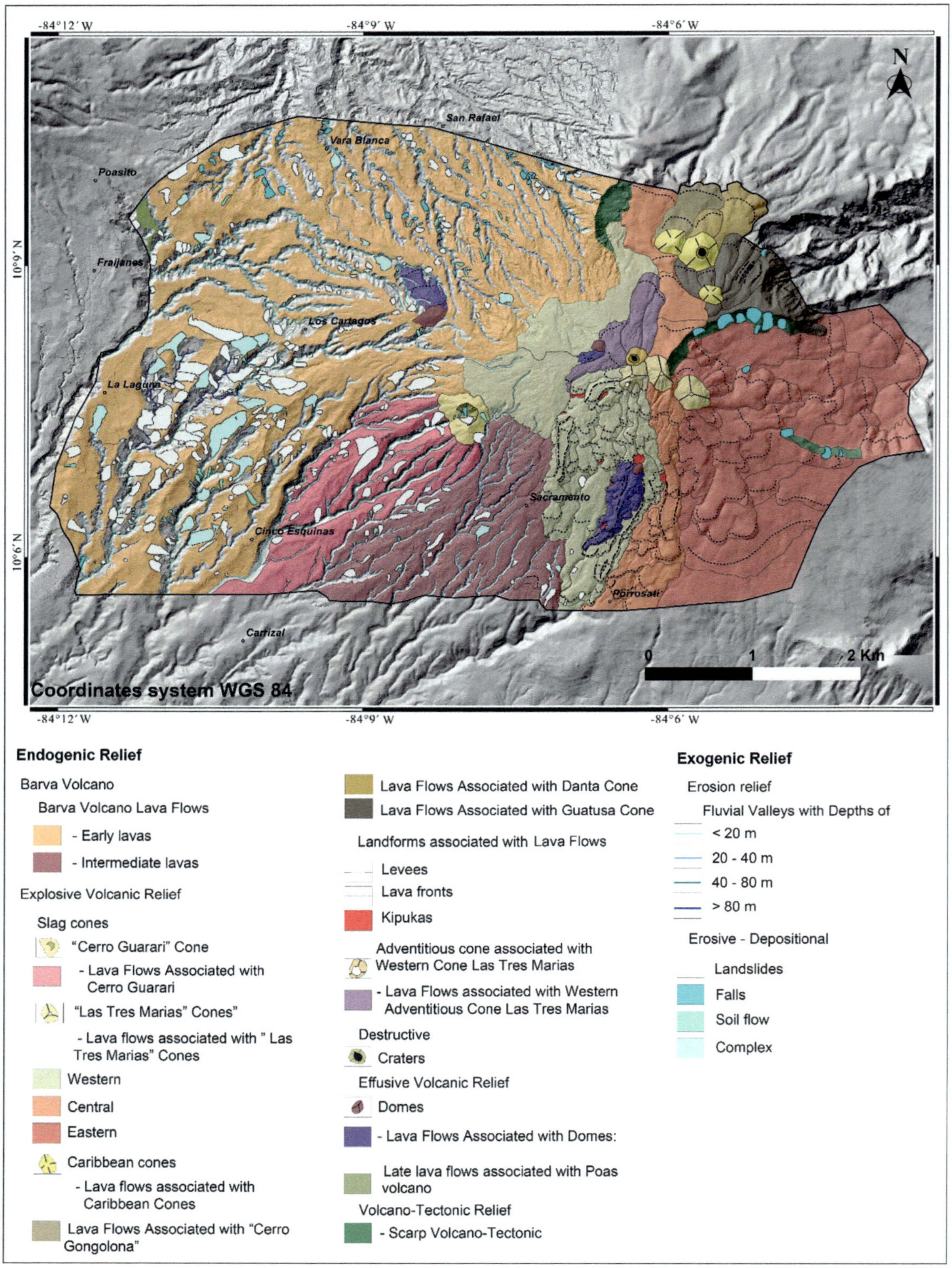

Fig. 16.7 Geomorphological units present in the study area

The Intermediate Lavas overlie the previous ones, corresponding to flows of the geological unit of Los Bambinos Inferior (~30–27.4 kyr), which are covered by pyroclastic packages between 10 and 40 m thick, as well as by more recent flows of the unit Los Bambinos Superior. The slope in general is > 20° and their primary flow morphologies are not clear. They are covered by pyroclastic layers of variable thickness (10–40 m: SENARA wells BA-727, BA-760 and

Fig. 16.8 **a** Barva volcano seen from the south. **b** Smooth relief of Older Lavas unit covered by tephras. **c** Thickness of tephras covering Older Lavas unit. **d** Intermediate lavas lying in river valleys. Photos taken by the author

BA-916). Their composition varies from andesite–basaltic to andesite, and outcrops occur in the bottom and on slopes of the valleys. This unit comprises at least 7 main lava flows oriented to the southwest, covering a total area of 7.5 km^2.

16.5.1.3 Explosive Volcano Relief

This classification includes morphologies related to violent explosive events, generated by acidic and `intermediate magmas. This eruptive style is responsible for creating positive (constructive) and negative (destructive) reliefs.

Slag Cones

A total of seven slag cones were identified in the study area. They are < 350 m high and have cratered or convex summits. These are located both on the summit and on the upper piedmont, in addition to these structures multiple lava flows originate. The "Cerro Guararí" Cone is located in the central–eastern region of the study area, between the altitudes of 2285 and 2605 m asl. The edifice has a semicircular shape, is inclined to the southwest, and has a major axis of 1115 m and a minor axis of 710 m. The maximum height is 320 m, and its area is 0.55 km^2. The Cerro Guararí is formed by the intercalation of lavas and tephras of the Guararí geological unit (Arredondo and Soto 2006), and recent pyroclasts of the Porrosatí unit were also observed. The slopes are steeper than 30°, and at the top, there is an open crater. From this cone, five primary flows of andesitic composition derive, which are covered by thick pyroclastic covers, making it impossible to recognize their original features. In general, the morphology of these structures is lobular, with broad summits and smoothed surface. They are bounded by deeply incised (> 45°), well-integrated fluvial valleys. The length of the Cerro Guararí lava flows reaches 3 km, and they are up to 110 m thick, all of them covering in total 6.83 km^2, maintaining a preferential 135° direction.

Las Tres Marias cones are three slag cones located in the summit zone (Fig. 16.9). They present a truncated cone morphology and are responsible for the emission of important lava flows and pyroclasts, which are known as the Los Bambinos Superior (Upper) and Porrosatí geological units (Arredondo and Soto 2006). These structures, according to their position, were named as Occidental, Central, and Oriental. The first one covers an area of 0.2 km^2; it has a height of 115 m and is semicircular; its major axis is oriented 45°; and its length reaches 530 m. The minor axis is oriented 180° and is 379 m long. The central cone covers 0.2 km^2; its height is 120 m, and it has a convex summit. It has a slightly elliptical shape where the major axis reaches 659 m and is oriented 180°. The minor axis measures 389 m and is oriented from east to west. For its part, the eastern cone is the largest, covering an area of 0.4 km^2; its height is 80 m, its plan is semicircular, its axis is oriented

Fig. 16.9 **a** Guararí cone seen from the Pacific side. **b** Las Tres Marias Cones. **c** Caribbean cones. **d** Open crater associated with the adventitious cone "Laguna Barva." Photos taken by the author

from 315°, and its length is 671 m. The minor axis is 655 m long and runs from east to west, parallel to the central one.

These cones supplied a lava field originated from the superposition of andesitic–basaltic to andesitic flows. This set of lavas correlates with the geological unit of the Los Bambinos Superior. The primary formation features are more noticeable, due to the reduced pyroclastic coverage and less time of exposure to exogenous processes. Associated with the Occidental Cone, elongated lava flows of upto 1.5 km were identified, arranged semi-radially, and covering an area of 10.06 km^2. In general, the southern section has a smoother morphology than the northern one due to the existence of a greater pyroclastic cover (< 20 m). In the unit, the presence of levees is notorious, with lava flow fronts and lobular shapes. From the Central Cone, 315° and 135° trending lava flows originated, with a morphology of overlapping lobes in a staggered manner, covering an area of 4.82 km^2. The pyroclastic cover is usually reduced (< 10 m), so it is possible to observe some of its primary formation features (levees and escarpments); the general inclination varies between 5°and 20°, and the average thickness is 41 m. As for the Oriental Cone, the lava field has a wide extension. Within the study area, the flows cover an area of 17.01 km^2. These lavas have courses that vary between 165°, 90°, and 45°; they have sub-horizontal and inclined surfaces between 2° and 30°. The flows overlap in a staggered manner and are lobular shaped. The thickness of the flows averages 36 m.

On the volcano's Caribbean side, three cones were delimited: Gongolona, Danta, and Guatusa. The first one is the westernmost and has a height of 140 m, its shape is semicircular and has a major axis of 710 m and a minor of 525 m, its area is 0.21 km^2, and its top is convex, and to the northwest, some lava flows occur. The second one is central and the highest among the three (160 m), covering an area of 0.38 km^2; its shape is also semicircular, its major axis measures 742 m and the smaller 654 m, its top is concave and includes the crater of Laguna Danta, and its lava flows have a 225° direction of emplacement. The last one is named Guatusa, and it is the easternmost and of smaller extension (0.12 km^2). It reaches a maximum height of 100 m, its shape is oval, the major axis is 460 m, and the minor one is 345 m. It has a concave top and multiple flows radially derive from it. Generally, they have a short length and width, the largest one comes from the Guatusa cone (1070 m and 630 m, respectively).

Associated with the Caribbean Cones, there is a lava field with a preferential emplacement direction of 157°. Its compositional characteristics are not clear due to the inaccessibility of the terrain and it was not possible to define primary formation features; however, they maintain a stepped lobular shape. Those coming from the Gongolona hill cover an area of 0.45 km^2, from the Danta cone—0.87 km^2 and those from the Guatusa cone extend over 2.35 km^2. Three forms associated with primary formation features were identified. Some correspond to levees, and the presence of these structures is associated with recent eruptive events. In the study area, levees were recognized on the lava flows from the Las Tres Marias and the Caribbean Cones. The other landforms identified are the Kipukas, which consist of a land area that is surrounded by younger lava flows, creating an "island" shape surrounded by lavas, were only recognized in the western and central lava fields.

Associated with the western cone of "Las Tres Marias" is an adventitious cone. It is located to the north of the western cone. It covers an area of 0.1 km^2; from the base, the height is 30 m. The largest diameter is 405 m, arranged 45°, while the smallest is 307 m, oriented 315°. Its slopes have an inclination that varies between 7° and 20°, and at the top, there is a closed crater.

An andesitic flow located southeast of the adventitious cone was identified, covered by the last tephras emitted by the volcano. The age range of this lava varies between 27.4 and 10 kyr. The flow covers an area of 1.2 km^2; its shape is an elongated lobe with a length of 1.4 km and an approximate thickness of 24 m.

Craters

Three craters were defined, which were separated into open and closed craters. The first one is located on the summit of Cerro Guararí; it is horseshoe-shaped, with an opening to the 225°, and it covers an area of 0.1 km^2. The largest diameter is 384 m, while the smallest diameter is 205 m, and its depth is 145 m. The closed crater is located on the cusp of the adventitious cone (Laguna Barva) and cone Danta; it has a semicircular shape, with the length and width of 210 m and 163 m, respectively. It covers an area of 0.02 km^2, and the depth of the crater is < 50 m. The second one presents an almost circular shape, where the radius is 240 m, the depth of the crater is 26 m, and it covers an area of 0.04 km^2.

Domes and Related Lavas

Three domes were defined as the north dome, south dome, and northwest dome, which together cover 0.3 km^2. The north dome is in the summit zone of Barva volcano; its morphology is asymmetric, slightly inclined 315° and convex. The major axis reaches 250 m and is oriented 270°–180°; the minor axis measures 235 m and is oriented 315°. The height from the base is 55 m. This southern dome is located near the town of Montaña La Isla, its height being 45 m. Its morphology is also asymmetric, and it has a well-defined convex top. The dome form is inclined toward the southwest, the major axis measures 235 m, oriented 45°, while the minor one is 172 m long and arranged 315°. The northwest dome is located to the west of Los Cartagos, its major axis is 45° and measures 651 m, while the minor axis measures 360 m in 315° direction. It has the shape of an asymmetric dome, with a convex top and its height is 50 m.

The presence of lava flows was identified in each of these domes. Only one flow originates from the north dome, which is located to the southwest of this structure. The flow is of short length and width (0.5 km and 215 m), has a lobe shape with 129° orientation, and has a thickness of 15 m. In the south dome, two main andesitic lava flows were identified, which are distributed subradially to the southeast of the emitting center and overlie the lavas from the eastern cone. They present a superimposed lobular morphology, where the lower flow has a greater length of ~ 1780 m and 209° orientation. On the other hand, the upper lava is shorter 1230 m and its emplacement direction changes to 190°. In the northwest dome, a flow covers the north slope, having a length of 1 km and a thickness of approximately 67 m. There is no evidence of primary features, due to the pyroclastic cover of the Porrosatí geological unit, and the preferential direction of flow is 320°.

Within the study area, only one lava flow associated with the Poás volcano was recognized, which covers a small area of 0.2 km^2. It is located at the distal end of the volcanic piedmont, to the northwest of the study area. In this part, it does not present primary formation features, since it is covered by pyroclastic packages of up to 20 m.

16.5.1.4 Volcano-Tectonic Relief

This section includes the escarpment caused by a volcano-tectonic event that the literature associates with a caldera (Perez 2000; Hannah 2000; Alvarado and Gans 2012; Bergoeing 2015), where the approximate age is 322 kyr. In general, it has a maximum height difference of 180 m present to the east of the Las Tres Marias Cones and represents part of some of the vestiges that managed to be preserved, its slopes are steep (> 45°) and face 45° and 225°. Within this structure, the intense superposition of recent lava flows of the Neo-Barva stage, mainly the Los Bambinos unit, is evident.

16.5.2 Exogenic Relief

16.5.2.1 Erosive Relief

This category groups together landforms that owe their origin to the erosive action of rivers. The fluvial valleys located in the northwestern territory are deeper and wider than in the eastern sector. These were classified into: < 20 m, 20–40 m, 40–80 m, and 80 m deep. The valleys < 20 m are minor forms of fluvial erosive relief and represent the stage prior to the development of valleys and ravines. Those in the 20–40 m range correspond to narrow, sinuous valleys with steep slopes (> 45°), located in the volcanic piedmont regions. Their origin and configuration are due to geomorphological contacts between lavas and eruptive sources.

Valleys between 40 and 80 m deep are concentrated in the western and northern regions of the volcanic piedmont, developed over older deposits, and along geological contacts between Early Lavas or in the presence of disjunctive structures. The average width is 405 m, V-shaped and with steep slopes on the flanks. In the > 80 m depth category, there are erosive structures of greater amplitude and depth; they owe their origin to the erosion within little consolidated pyroclastic materials of great thickness, and their setting is between the geomorphologic contacts of Early Lavas.

16.5.2.2 Erosive Depositional

This category corresponds to landforms originated by the downslope movement of materials that occur by the action of gravity, due to the loss of cohesion in the rocky body or deposit, triggered by lithological, disjunctive, climatic, tectonic, anthropic, etc. factors. According to Alcantara (2000), these elements were classified into landslides, falls, soil flows, and complex gravitational movements, which together cover an area of 7.45 km^2 (Fig. 16.10).

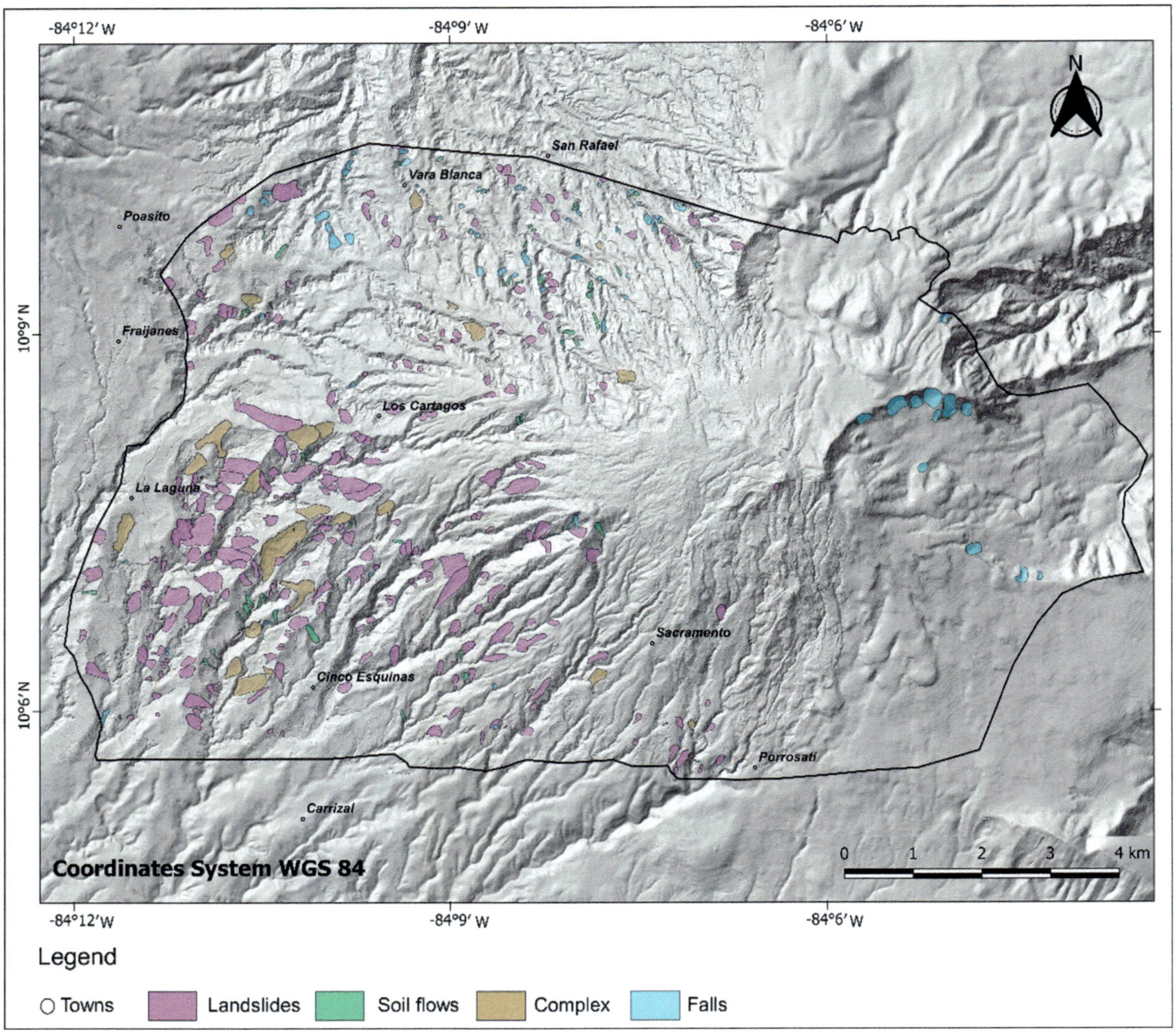

Fig. 16.10 Distribution of mass movement processes

Most of the landslides are in the central–western region of the studied area (Fig. 16.11), within the volcanic piedmont regions, although they are also found in the northern region, which is the largest unit (4.92 km^2). Landslides develop in river basins and along narrow interfluves, where the slope is steep. It should be noted that their occurrence is preferentially associated with granular lithologies that depend more on their low degree of compaction than the age of the deposit.

The falls (debris or soils) are strictly located in the volcanic piedmont regions, mostly in river basins with deep valleys and steep slopes. The displaced material corresponds to soils and tephra detritus and covers an area of 0.8 km^2. According to Alcantara (2000), soil flows correspond to rapid displacements of short duration and limited thickness of the displaced material, triggered by intense periods of precipitation. Within the study area, they are located mainly in the central–western sector of the volcanic piedmont, have an elongated shape and steep slope, occur on sites with potent pyroclastic mantles, and cover an area of 0.29 km^2. As for the category of complex gravity-driven movements, their genesis is associated with events where there is at least a combination of two types of displacements. They occur mostly in the western territory, on the banks of rivers with steep slopes and in the interfluves with high pyroclastic coverage, covering an area of 1.43 km^2.

16.6 Conclusions

The analysis of the local stratigraphy focused on the sources that in some way defined the local geological units of the Barva volcano (British Geological Survey and SENARA 1985; Protti 1986; Brenes 2003; Obando and Peraldo, 2004; Arredondo and Soto, 2006); however, such data were complemented with field verification information generated in Carmona (2021), to give validity to the descriptions that were not clear. The units that make up the local stratigraphy (Neo-Barva) are defined by Bermudez, Carbonal, Guararí, Debris Avalanche El Coyol, Los Bambinos, and Porrosatí units. Some of these units were only described at the correlation level of well correlation reports since they outcrop further south of the volcanic edifice. It was not possible to verify the geology of the

Fig. 16.11 Some mass movements identified in the western region of the study area

Caribbean slope due to the difficulty of access. It is important to note that there are few absolute age data for the Neo-Barva phase, in total there are four. Two are by $^{40}Ar/^{39}Ar$ for the Bermudez lavas for an age between 270 and 258 kyr (Perez et al. 2006); the remaining by ^{14}C, one for Porrosatí resulting in 11.3 kyr (Brenes 2003) and 27.4 kyr for the lahar between the lower and upper Los Bambinos (Arredondo and Soto). Therefore, this condition makes that the other ages are by relative correlations.

Five variables were analyzed from the morphometric analysis: altimetry, terrain inclination, density and depth of dissection, and relief energy. This requires an adequate topographic information, which was only available through LiDAR on the Pacific slope of the volcano, which is why this region was considered in the morphometric analysis. Regarding the indexes that integrate the final morphometric map, it was obtained that the Density of Dissection allowed the identification of areas of greater concentration of drainage, which have a spatial distribution correlatable with the regions with greater dissection of the relief and gravitational processes. From the elaboration of the use of the Dissection Depth, the areas associated with a greater erosive intensity in the vertical component were defined, which are mostly concentrated in the tephras (Porrosatí unit) and geomorphological contacts (Bermudez and Guararí lava flows). The relief energy quantified the potential difference that exists in a given territory, since this is a determining factor for terrain instability, the highest values are located in the central–western sector (Guararí cone).

From the morphometric method, the areas were classified into three categories (low, mid, and high occurrence), the moderate zones were the most extensive with 49%, followed by high at 46% and low at 5%. When comparing the results with the inventory of slope processes, it was obtained that, in both, the high degree zones comprise the greatest amount of gravity movements, 76%, while those of medium occurrences are 23.9% and low occurrence $<0.1\%$. Therefore, the coincidence of the method and the reality can be evidenced. However, in addition to the substratum component, the morphometric section showed that the areas with the highest presence of slope processes coincide with the highest susceptibility categories obtained through the morphometric method. Morphometry showed that the territories of the lower piedmont with a high frequency of landslides occur toward the center–northeast, due to a developed fluvial network with wide and deep valleys, partly due to the spatial arrangement of the forms (contacts), high rainfall regimes, structural features, and lithological characteristics and slope inclination $> 20°$.

With the elaboration of the detailed geomorphological model by Carmona (2021) and with the extension of the present work, unlike the geomorphological models proposed in previous works, it was possible to define detailed cartography that shows the different geomorphological units that make up the area, in turn, this map gives information about the evolution of the relief, the relationships with the geology, climate, topography, and exogenous dynamics. In total, 324 gravitational processes were identified in the territory comprising the Pacific slope, which together cover an area of 5.93 km^2. The geomorphological units that presented the largest slid surface were the Early Lavas and Fluvial Valleys for a total of 4.1 km^2 (68% of the total slid surface). As can be seen, this condition of greater apparent instability is found in the western sector of the volcano, while in the eastern sector, the occurrence of these events was lower than in the remaining area of the study area, which could be intrinsically associated with a shallower rocky substrate.

16.7 Concluding Remarks

The stratigraphy of the study area, from the base to the top, is conformed by the Colima and Tiribí Formations, followed by several units of the Barva Formation, including Bermudez, Carbonal, Guararí, Los Bambinos Inferior and Superior, Debris Avalanche El Coyol, Los Angeles, and Porrosatí. The use of LiDAR, as well as high-resolution orthophotos, were essential for the determination of geological and geomorphological features, especially at sites of difficult access (Caribbean sector), due to the dense vegetation cover, high precipitation rates, and inaccessible topography. In addition, the quality of the LiDAR data allowed for the identification of those mass movement processes that were not evident during fieldwork or by photointerpretation of larger-scale aerial images.

Morphometry made it possible to establish the intensity of erosive processes on the surface of the volcano, and when correlating these susceptibility zones with the catalog of mass removal processes, it was clearly observed that the spatial pattern of mass movements coincides with the results of the numerical model. In this sense, the geomorphological regions of the Summit Zone and the Upper Piedmont of the eastern sector are the least susceptible to mass removal processes.

In general, the total area considered unstable on the Pacific slope was 5.93 km^2. The forms that concentrate the largest number of landslides are the Early Lavas, the Guararí Lavas, and the fluvial valleys, which together cover a total of 68% (4.1 km^2) of the area affected by rock-mass or soil displacements.

Geomorphological mapping allowed for the delimitation of seven slag cones, as well as three domes, which had not been reported in previous works. The entire eastern sector (Bambinos Superior) is composed of three sources associated with the three cones located in the summit zone (Las

Tres Marias cones) and a dome to the south of them. The early flows were classified according to their location as Eastern, Central, and Western ones, while those overlying the previous ones were classified as Lavas Associated to the South Dome. The lavas of the Caribbean sector were associated with three slag cones called Gongolona, Danta, and Guatusa.

References

Alcántara I (2000) Landslides: ¿deslizamientos o movimiento del terreno? Definición, clasificaciones y terminología. Boletín De Investigaciones Geográficas, Instituto De Geografía-UNAM 41:7-25

Alvarado GE (2011) Volcanes de Costa Rica, Geología, historia y riqueza natural. Editorial UNED

Alvarado GE, Gans P (2012) Síntesis geocronológica del magmatismo, metamorfismo y metalogenia de Costa Rica, América Central. Revista Geológica De América Central 46:7–122

Arredondo S, Soto G (2006) Edad de las lavas del Miembro Los Bambinos y sumario cronoestratigráfico de la Formación Barva, Costa Rica. Revista Geológica De América Central 34:59–71

Bergoeing J (2015) Geomorphology of Central America: a syngenetic perspective. Elsevier, Amsterdam

British Geological Survey, SENARA (1985) Mapa hidrogeológico del Valle Central de Costa Rica escala 1:50000. E.S.R. Limited

Brenes J (2003) Geología y peligro volcánicos del Flanco Oeste y Suroeste del volcán Barva (Undergraduate thesis). University of Costa Rica, San Jose

Buzai G (2013) El mapa social de la ciudad de Luján, 2010. Modelo socioespacial basado en cluster analysis. Tiempo y Espacio, 31. http://revistas.ubiobio.cl/index.php/TYE/article/view/1989

Carmona S (2021) Análisis geomorfológico de la vertiente pacífica del Volcán Barva, aplicado a la remoción de procesos de remoción en masa (Undergraduate thesis). University of Costa Rica, San Jose

Denyer P, Arias O (1991) Estratigrafia de la región central de Costa Rica. Revista Geológica De América Central 12:1–59

Hannah R (2000) Origin of the chemical variation in the Valle Central Tuff. Costa Rica. (Master´s thesis). University of Michigan

Macmillan I, Gans P, Alvarado G (2004) Middle Miocene to present plate tectonic history of the southern Central American Volcanic Arc. Tectonophysics 392:325–348

Marshall J (2000) Active tectonics and Quaternary landscape evolution across the western Panama block, Costa Rica, Central America (PhD thesis). University of Pennsylvania, Pennsylvania

Méndez J, Hidalgo P (2004) Descripción del depósito de Debris Avalanche El Coyol, Formación Barva, Costa Rica. Revista Geológica De América Central 30:199–202

Montero W (2001) El Cinturón Deformado del Centro de Costa Rica: Frontera oeste de la microplaca de Panamá. Revista Geológica De América Central 24:29–56

Montero W, Rojas W (2014) Las fallas Purires y Picagres, y su relación con la secuencia sísmica del Puriscal de 1990. Revista Geológica De América Central 50:39–69

Obando L, Peraldo G (2004) Geología, geomorfología, geotecnia, hidrogeología y consideraciones ambientales del flanco Sur del Volcán Barva. University of Costa Rica

Perez W (2000) Vulcanología y petroquímica del evento ignimbrítico del Pleistoceno Medio (0,33 Ma) del Valle Central de Costa Rica (Undergraduate thesis). University of Costa Rica, San Jose

Perez W, Alvarado GE, Gans PG (2006) The 322 kyr Tiribi tuff: stratigraphy, geochronology, and mechanisms of deposition of the largest and most recent ignimbrite in the Valle Central, Costa Rica. Bull Volcanol 69:25–40

Protti R (1986) Geología del flanco sur del volcán Barva. Boletín De Vulcanología. National University 17:23–31

Rojas V, Barahona D, Alvarado G (2017) Geomorfología y petrografía de la colada Ángeles y del cono Monte de la Cruz, volcán Barva, Costa Rica. Revista Geológica De América Central 56:17–35

Simonov Y (1985) Morphometric analysis. Moscow State University, Editorial (in Russian)

Soto G (1999) Geología Regional de la Hoja Poás (1:50 000). In: Alvarado GE, Madrigal L (eds) Estudio Geológico-Geotécnico de Avance a la Factibilidad. San Jose, Costa Rica

Thiele K, Vargas A (2017) Actualización volcanoestratigráfica del piso del Valle Central, de Costa Rica, específicamente la denominada Formación Colima y sus miembros (Undergraduate thesis). University of Costa Rica, San Jose

Vargas C, Acosta P (2007) Tephrostratigraphy and volcanic hazards for the upper part of Barva volcano, Costa Rica. Workshop to integrate subduction factory and seismogenic zone studies in Central America. Heredia, Costa Rica

Velazquez G, Celemin J (2011) Elaboración y aplicación de un índice de calidad ambiental para la Región de Cuyo, Argentina, 2010. Cuadernos Geográficos 49:179–197

Zamorano, J (1990) Análisis ingeniero-geomorfológico de la cuenca de México (doctoral thesis). Moscow State University

17 Geomorphology and Slope Instability in a Densely Populated County, Escazú, San José, Costa Rica

Diana Quirós Maroto and Rolando Mora Chinchilla

Abstract

In the Escazú county, geoforms of denudational, structural, and fluvial origin occur. Denudational morphologies comprise 94% of the study area and their characteristics are largely controlled by the lithological units. These units generate moderately stable soils and mass wasting morphologies such as old and active landslides, block falls, incipient and intense creep-affected areas, and areas with concentrated erosion. The areas of sedimentary rock with steep slopes represent 51% of mapped landslides, where the probability of landslide occurrence (PLO) is moderate to high. There are 31 inactive landslides, which are sites with slopes greater than 15 ° and covering large areas. About 44 active landslides were mapped that are located mainly on the slopes near riverbeds. Seven reactivation sectors were identified within old landslides. By integrating this geomorphological characterization with the PLO, the suggested land use for the study area is defined.

Keywords

Slope stability · Structural landforms · Landslides · Probability of landslide occurrence · Land use

17.1 Introduction

The Escazú country presents factors such as changes in land use, climatic variations, active seismic sources, and population growth toward areas with steep slopes that favor unstable slope conditions. In addition, this condition of instability has generated a negative impact on the population or on the infrastructure (Rivier 1979; Hernández and Vargas 1992; Chaves et al. 2012; Ruiz and Soto 2014; González 2014). Therefore, there is a need to carry out various studies to obtain useful information for decision-making on issues related to territorial planning and risk management. That is why a detailed study of the geomorphology was carried out using LiDAR information to characterize the various morphologies associated with landslides present in the study area.

The study area is located in the districts of San Antonio and Escazú of Escazú county, in the province of San José; between the coordinates CRTM05 1097,000/1099,000 N, 482,000/487,000 E, and covers an area of 25 km^2. The Pacacua, Peña Negra, Escazú Intrusive, and Escazú Hornfels formations outcrop in the region, with the superficial mantle of recent alluvial and colluvial deposits (Castillo 1969; Obando 1983; Denyer and Arias 1991; Hernández and Vargas 1992; Ruiz & Soto, 2014). The Pacacua Formation has three lithological units, among them medium-grained gray sandstones, gray breccias, and medium-grained sandstones with clasts. The Peña Negra Formation is composed of fine-grained sandstones, and the Escazú Intrusive is made of igneous rocks from granodiorites to gabbro. The Escazú Hornfels is made of silicified sandstones (Quirós, 2019) (Fig. 17.1).

Regarding structural geology, in the northern part of the study area, there are NW-striking reverse faults, while in the southern sector, the faults strike to the NE. The main structures are Escazú, Bello Horizonte, and Aserrí faults, which are classified as neotectonic faults (Montero, 2001; Fernández and Montero, 2002; Montero et al. 2005). Additionally, the dips of the sedimentary formations show a trend to SW and an average dip of 39 °.

D. Q. Maroto (✉) · R. M. Chinchilla
Escuela Centroamericana de Geología,
Universidad de Costa Rica, Apartado: 2142060 UCR,
San Pedro, Montes de Oca, San José, Costa Rica
e-mail: diana.quirosmaroto@ucr.ac.cr

A. Quesada-Román (ed.), *Landscapes and Landforms of Costa Rica*,
World Geomorphological Landscapes, https://doi.org/10.1007/978-3-031-64940-0_17

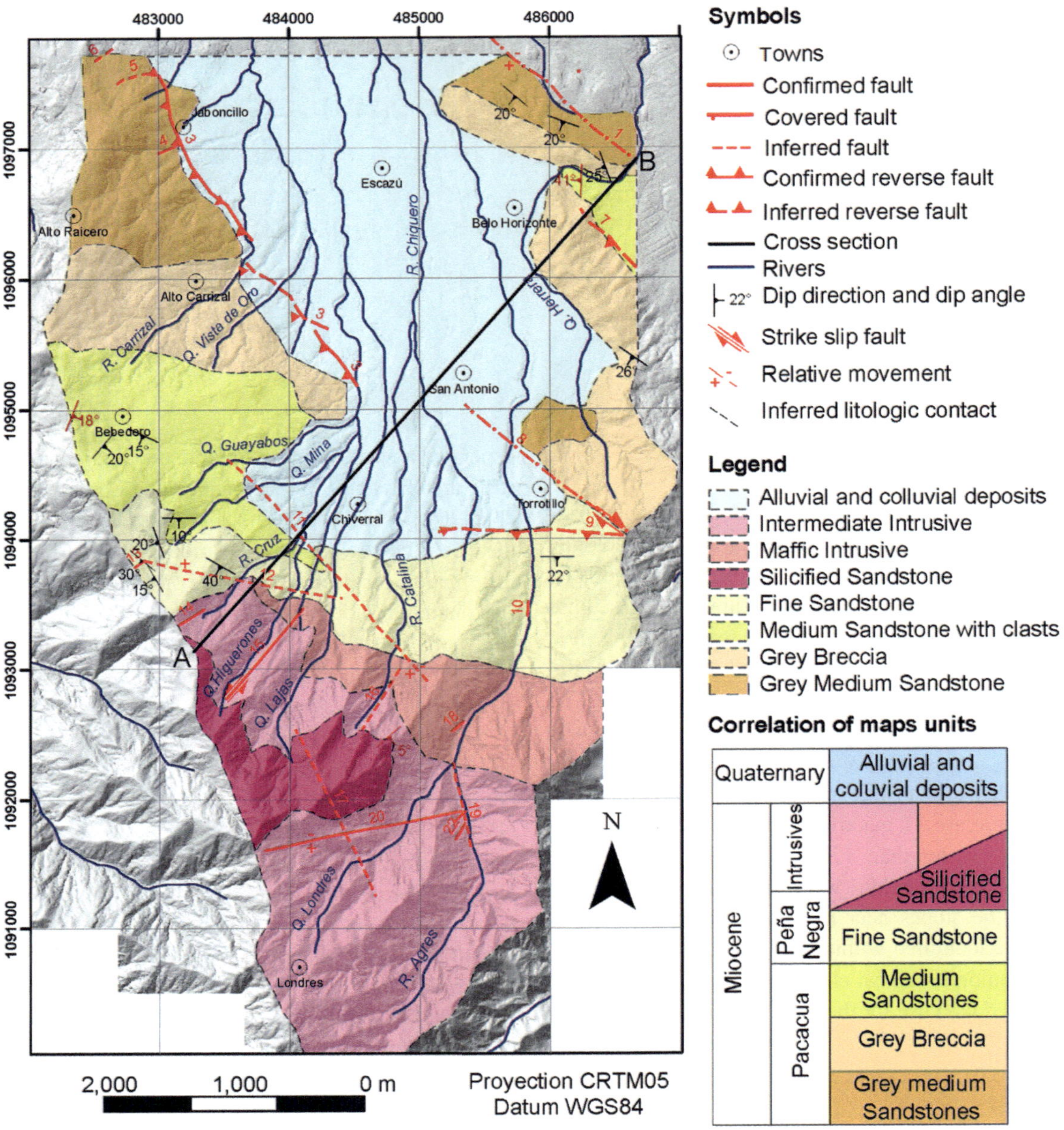

Fig. 17.1 Geological framework of the area under study

17.2 Geomorphological Units

Geoforms of denudational, structural, and fluvial origin are foundin the study area (Fig. 17.2). Slopes of structural origin are associated with reverse faults, while riverbeds are units of fluvial origin. The denudational geoforms are of igneous, metamorphic, and sedimentary. Sedimentary lithologies generate denudational morphologies related to massive wear processes, such as old and active landslides, rockfalls, incipient and intense landslides, and areas with concentrated erosion. These geoforms are described below.

17.2.1 Structural Geoforms

17.2.1.1 Slopes of Structural Origin

This geoform is located near the town of Bello Horizonte (Fig. 17.2) and corresponds to a structural chain linked to the Pacacua Formation, specifically to the Gray Breccia lithological unit. It presents a strike to the northwest–southeast, with slopes that vary from 20 to 43 ° and heights between 25 and 40 m. This geoform represents 2% of the total area and is associated with the Bello Horizonte reverse fault (Fig. 17.3a). It includes geomorphological evidence

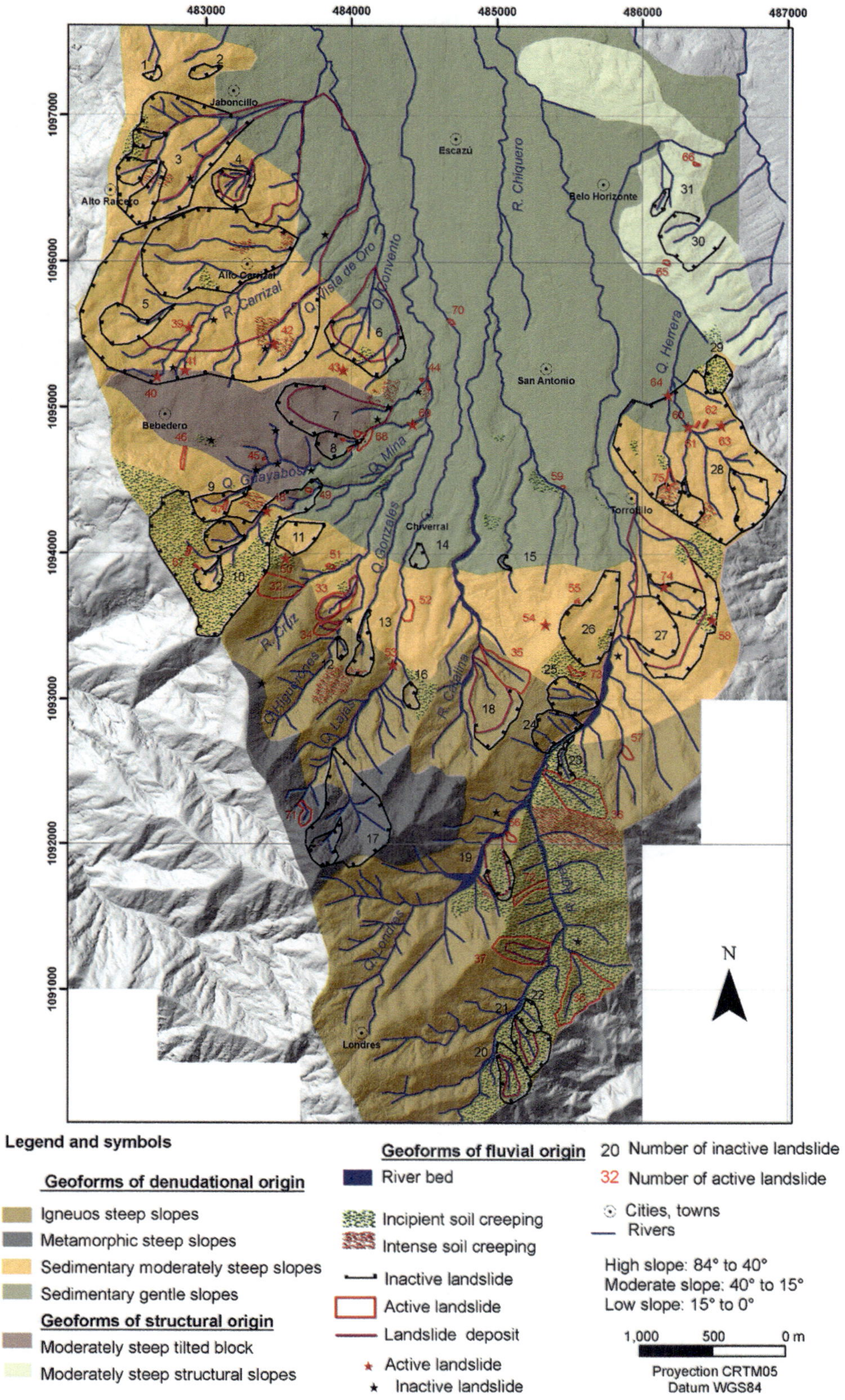

Fig. 17.2 Geomorphological map of the study area

of the dextral displacement of Herrera creek (Fig. 17.3b) and amphitheater-shaped morphology (Fig. 17.3c). The offset and the amphitheater are interpreted to be linked by the alignment (Fig. 17.3a) oriented in the same direction as the Bello Horizonte fault, which could be understood as a prolongation of this fault. The Herrera creek offset could

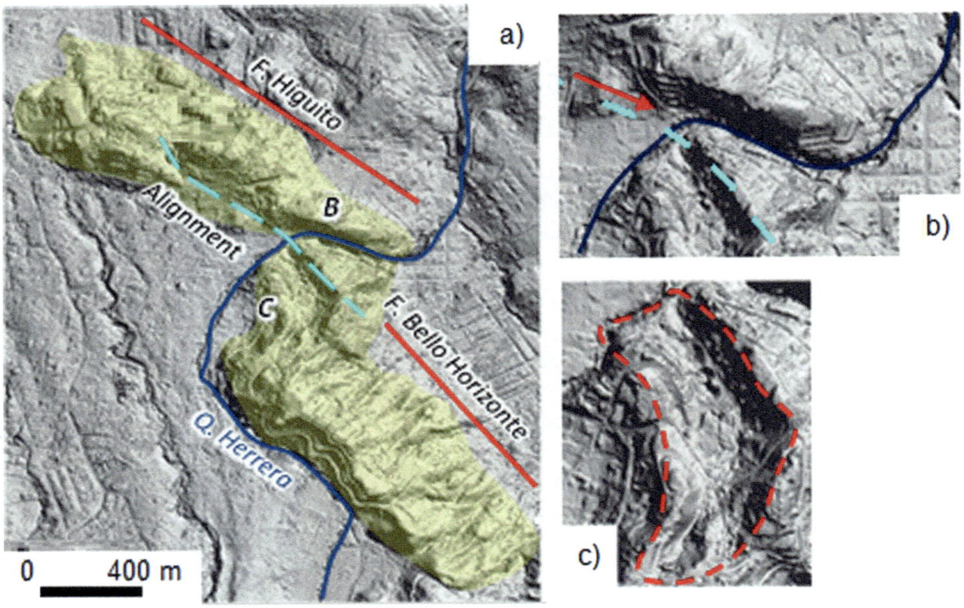

Fig. 17.3 Structural slope at Bello Horizonte, **a** General view with the location of geomorphological features. In yellow the structural morphology and in red the traces of the faults, **b** Displaced Herrera creek, **c** Amphitheater morphology

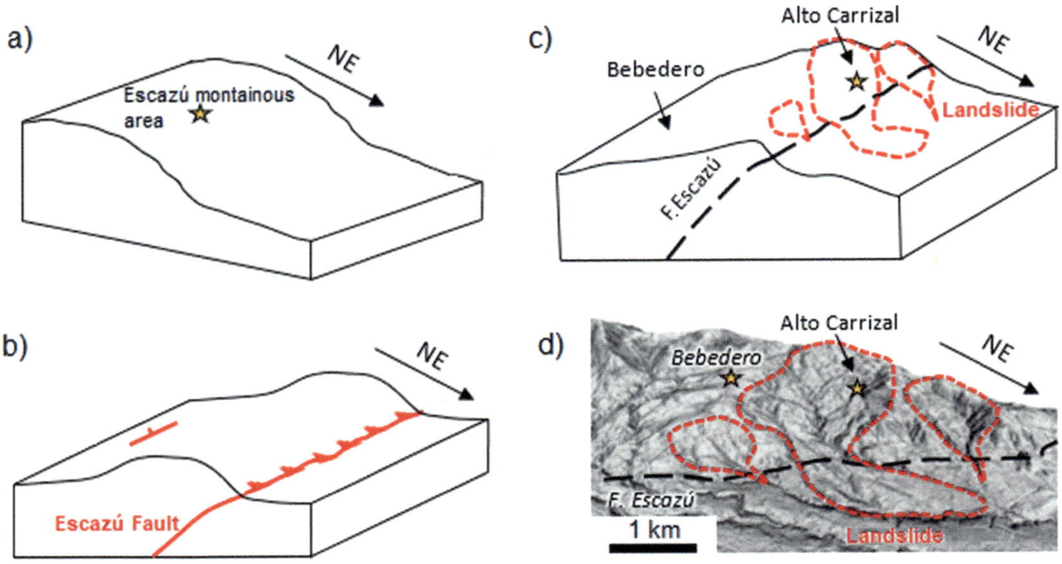

Fig. 17.4 Interpretation of the evolution of Bebedero tilted block

be related to the Bello Horizonte fault or to Higuito fault as well, the latest being a dextral displacement fault.

17.2.1.2 Tilted Block

This geoform is located around the town of Bebedero. It has a flat-undulating topography with slopes lower than 15 ° and is tilted to the southeast. The geoform is associated with the Pacacua Formation, specifically with the medium-grained sandstone with larger clasts. It is proposed that its origin is linked to the Escazú reverse fault (Fig. 17.4) and its evolution is explained in four stages. A first stage, where the fault has not generated any type of affectation. In the second stage, as a result of the activity of the Escazú reverse fault, the SW block was uplifted, generating a tilt in the same direction (Fig. 17.4b). In the third stage, the fault scarp provided instability conditions that generated landslides (Fig. 17.4c). The fourth and last stage represents the current situation where the Bebedero settlements are found in the tilted block and the Alto Carrizal village is located in the old landslide morphology (Fig. 17.4d).

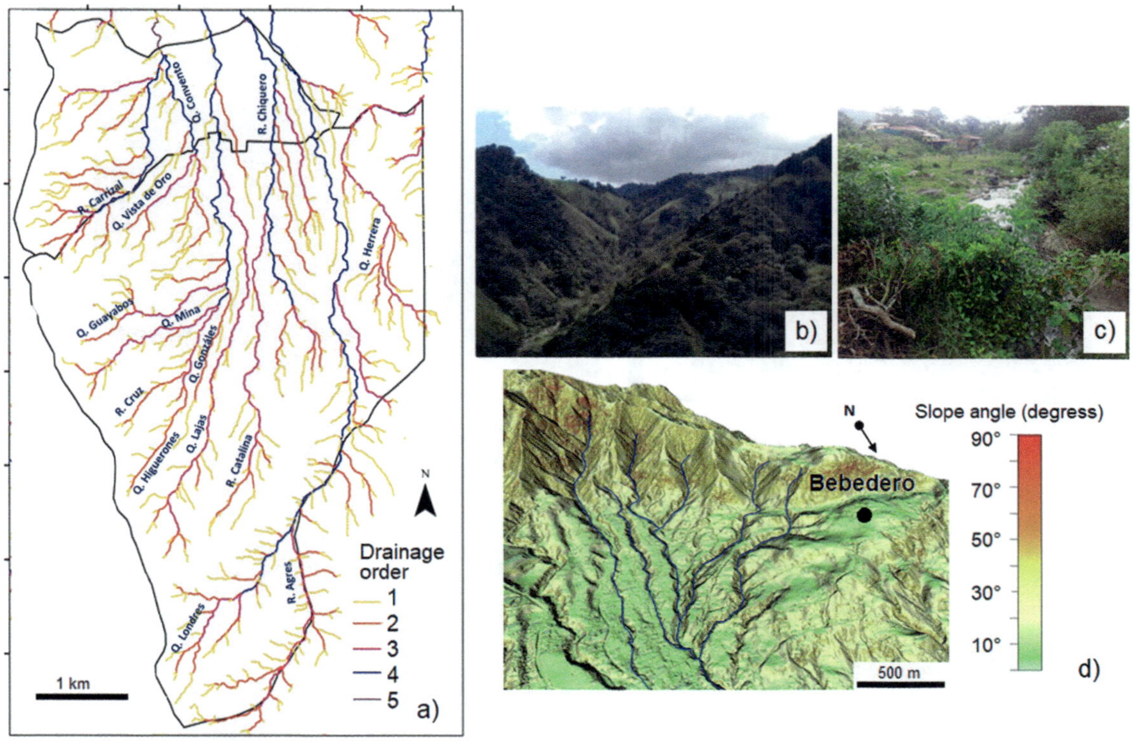

Fig. 17.5 **a** Drainage network and drainage order of the main streams, **b** Morphology of the streams in the upper part of the basin, **c** Morphology of the streams in the lower part of the basin, **d** 3D view of the streams in the Bebedero

17.2.2 Fluvial Geoforms

Fluvial geoforms correspond to the riverbeds of Carrizal, Catalina, Agres, and Cruz rivers and Vista de Oro, Convento, Mina, Higuerones, Gonzáles, Lajas, Londres, and Herrera creeks (Fig. 17.5). The drainage direction of these streams is toward the northeast in the upper parts of the basins, while in the lower parts, they drain toward the north. All streams show a moderate sinuosity. In the northern sector, where the topography becomes flat-undulating, they show a narrower channel, and most of them are artificially confined due to urban development.

17.2.3 Denudational Geoforms

The study area comprises denudational slopes developed in igneous, metamorphic, and sedimentary rocks. They vary in terms of slope and degree of concentrated erosion, as well as host medium-scale landforms associated with mass wasting processes such as ancient landslides, active landslides, and creep-affected surface.

17.2.3.1 Denudational Igneous and Metamorphic Steep Slopes

These slopes are found in the southern sector of the study area, mainly in the Escazú range. They cover 28% of the slopes, where the Escazú Intrusive and the silicified sandstones of the Escazú Hornfels crop out (Figs. 17.1, 17.2, and 17.6). These geological formations are associated with strongly eroded steep slopes, with slope angles between 40 and 84 °, and with elevations varying from 1500 to 2420 m asl. The drainage basins in which they are located are oriented to the northeast and show an abrupt and steep topography, with ridge tops between 5 and 21 m wide. The drainage pattern is dendritic to sub-dendritic, and the current land use is mainly forests and grasslands.

Denudational sedimentary moderate slopes are found in the lower section of the Escazú range, around the villages of Alto Carrizal, Alto Raicero, and Torrontillo. They cover 26% of the slopes, where the Peña Negra and Pacacua formations crop out. They are moderately eroded, with slope angles between 15 and 40 ° and elevations varying from 1100 to 1500 m asl. The water divides of this unit coincide with the border of the county. They are oriented to the north–northeast, show smooth and rounded shapes, with ridge tops less than 8 m wide. The drainage pattern is dendritic, trending northeast and northwest (Fig. 17.7). The land use is devoted for housing, agriculture, and pastures.

Denudational sedimentary gentle slopes are found in the northern sector of the study area, around the settlements of San Antonio, Bebedero, Escazú, Bello Horizonte, and Jaboncillo. They cover 36% of the sloping ground and are underlain by recent alluvial and

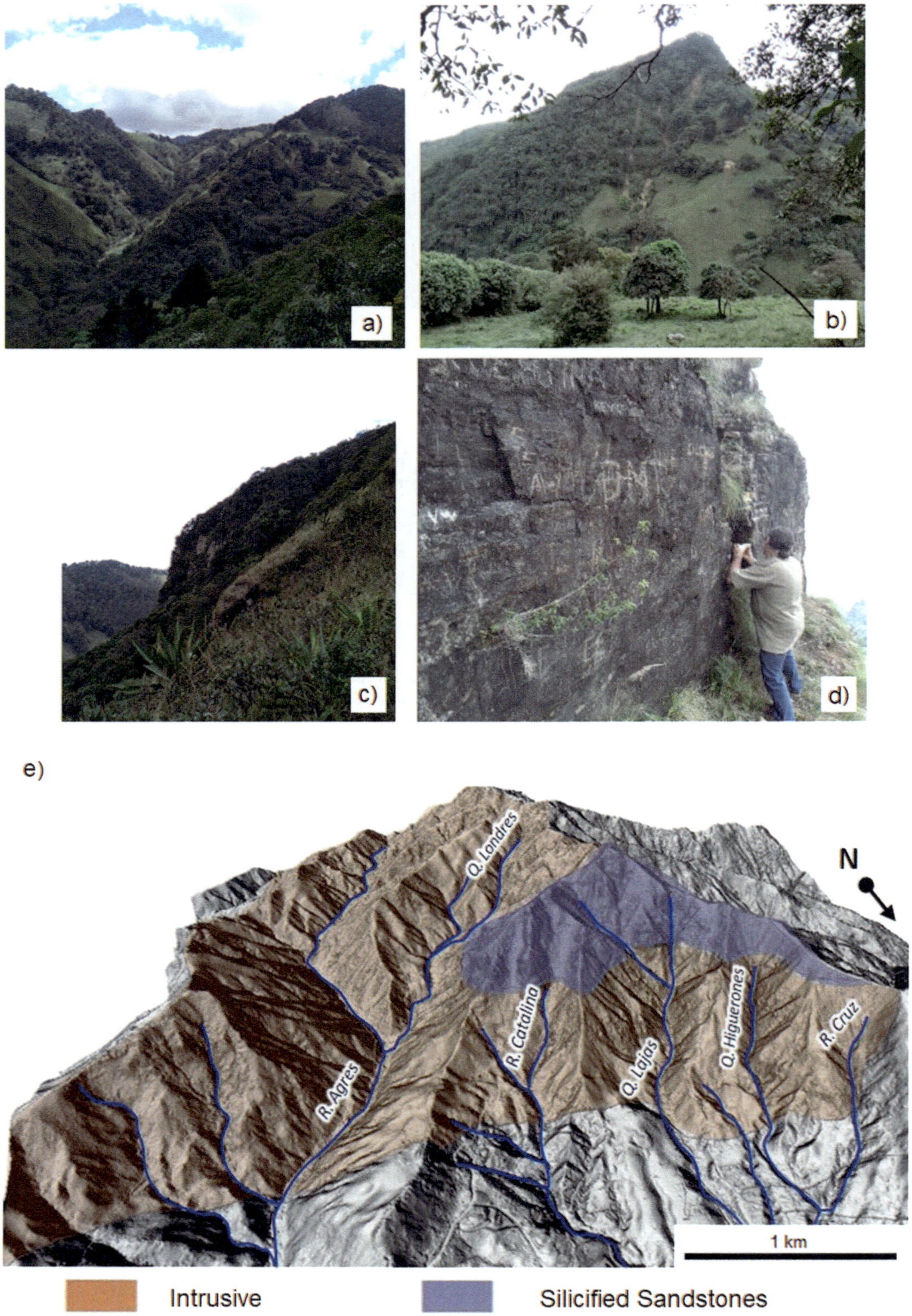

Fig. 17.6 Denudational igneous and metamorphic steep slopes. **a** and **b** Igneous slopes, **c** and **d** Metamorphic slopes, **e** 3D view

colluvial deposits, with slope angles between 0 and 15 ° and elevations that vary from 1020 to 1100 m asl. The drainage pattern is parallel, and the main rivers and creeks drain to the north (Fig. 17.7). The current land use is for housing mainly. Geoforms associated with mass wasting are old or inactive landslides, active landslides, and creep-affected surfaces (Fig. 17.2). The observed landslides are mostly of rotational type. The main

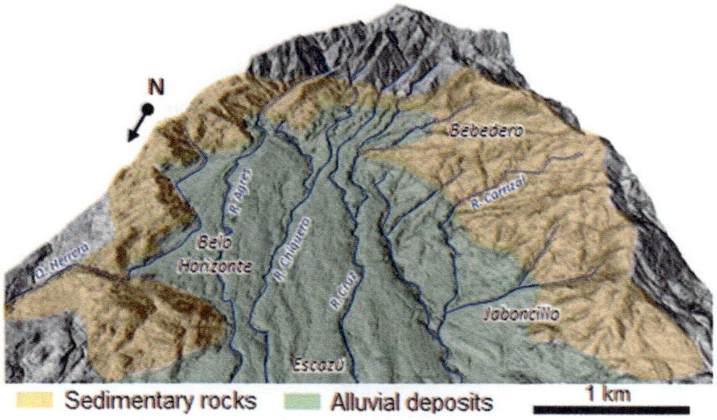

Fig. 17.7 Denudational sedimentary moderate and gentle slopes

geoforms for each of the categories mentioned above are described below:

17.2.3.2 Ancient Inactive Landslides

Ancient inactive landslides are mass wasting geoforms having the largest dimensions. They are located in the upper part of the slopes, where slope angles typically exceed 15 °. These landslides probably were triggered by seismic activity and/or ground saturation. They are linked mainly to denudational sedimentary moderate slopes, and their main features are shown in Table 17.1 and Figs. 17.8, 17.9, and 17.10.

17.2.3.3 Active Landslides

Active landslides have smaller dimensions compared to the ancient inactive landslides and they are mainly located on the slopes in the upper and middle parts of the basins. The characteristics of some of these landslides are shown in Table 17.2. In addition, seven reactivation zones were identified within old landslides, which correspond to sectors of intense creeping, gullying, exposed soil, and intense erosion. They are located in the highest parts of Escazú range and linked to the denudational unit of igneous steep slopes. Some of these landslides are described below, and the landslide number corresponds to the location on the geomorphological map (Fig. 17.2).

The Higuerones creek landslide (33) occurs within a tributary stream on the left bank of the Higuerones stream (Fig. 17.11a) and is linked to the Escazú Intrusive lithological unit. Its dimensions are approximately 147 m wide and 405 m long. The main crown has slopes between 20° and 35° and a length of 690 m. The deposit covers an area of 2 ha, with a general slope of 20 ° and the main movement direction toward N52°E. In the depositional are, due to Nate tropical storm in 2017, a debris flow was triggered (Fig. 17.11a), which overflowed the banks of the creek and affected the road that runs parallel to the right bank of it. The landslide in a tributary valley of the Catalina River (35) is linked to the fine-grained sandstones lithological unit of the Peña Negra Formation. Its dimensions are 161 m wide,

Table 17.1 Features of ancient inactive landslides on denudational sedimentary moderate slopes

Landslide		Alto Raicero (3)	Palo Campana (27)	Torrontillo (29)	Alto Granadilla (5)	Alto El Tajo (10)	Bebedero (7)
Dimensions	Width [m]	669	943	752	1300	650	280
	Length [m]	1400	582	1000	2600	1190	760
Main crown	Slope angle	20°-35°,>35° northward	20°	10°–25°	25°	25°	15”
	Length	2700 m	1300 m	3600 m (perimeter)	3600 m	3900 m (perimeter)	1000 m
Deposit	Area	24 ha	39 ha	Not calculated	1,4 km^2	Not calculated	15 ha
	Slope angle	15°	15°	–	15°	–	10”
	Direction of movement	N60°E	N72°W	N37°W	N60°E, N04°W at the toe	N55°E	S75°E
Current land use		Pasture and forest	Croplands and housing	Croplands and housing	Pasture, forest, and housing	Pasture, forest, and housing	Croplands and housing
Internal events[a]		3	2	3	1	2	1
Infrastructure at risk		Houses at its toe	Road 105 (Pacayas road). Uncommunication of houses possible	Houses at its toe	Alto Carrizal village	Houses at its toe	Houses at its toe
Figure 17		8a	8b	8c	8d	8e	9

Note [a] The landslide number corresponds to the location on the geomorphological map (Fig. 17.2), and internal events corresponds to the number of smaller landslide morphologies within the main landslide

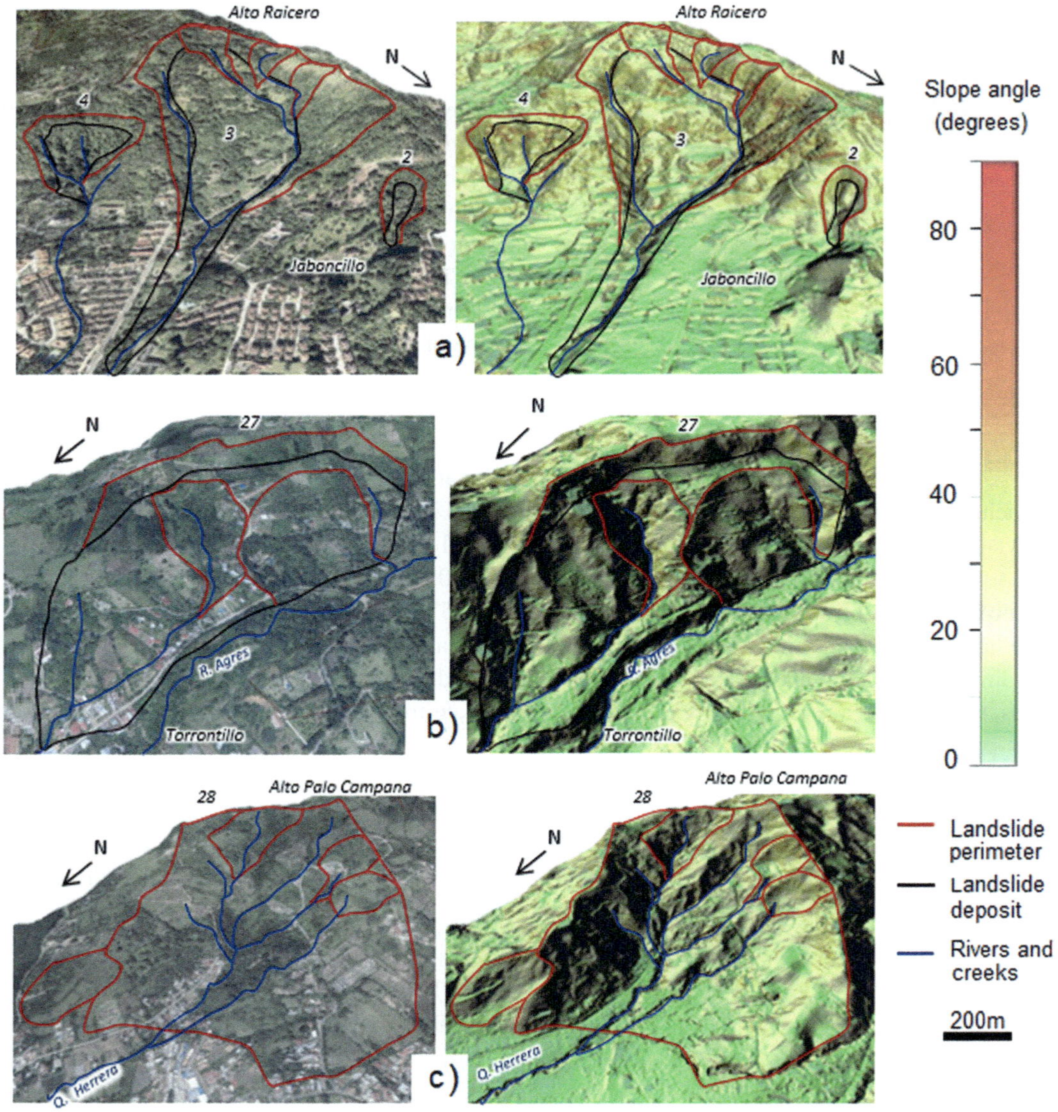

Fig. 17.8 Ancient inactive landslides: **a** Alto Raicero, **b** Palo Campana, **c** Torrontillo

422 m long, with a perimeter of 1000 m. The direction of movement is toward N51 W. The landslide shows evidence of reactivation of two active landslides (Fig. 17.11b). The first (Fig. 17.12b) is 60 m long and 15 m wide, shows tilted vegetation (trunks and branches) in its depositional part, and exposes orange soil with 30% of sub-angular clasts of 10 cm size. The second slide (Fig. 17.13b) exposes orange soil and tilted vegetation and shows 20 m of vertical displacement at its crown. The material released from these landslides is conveyed through a tributary creek toward the Catalina River. The current land use is forest.

The San Miguel peak landslide (36) is located on the flanks of San Miguel peak, at the right bank of Agres River and is linked with the Escazú Intrusive lithological unit. It's approximate dimensions are 230 m wide, 426 m long, with a perimeter of 1250 m. The direction of movement is toward N55°W. It shows signs of reactivation evident in exposed soil sectors, which are the result of mud and/or debris flows (Fig. 17.11c). The material released from these landslides falls through a tributary to the Agres River. The current land use in the landslide corresponds to forested land.

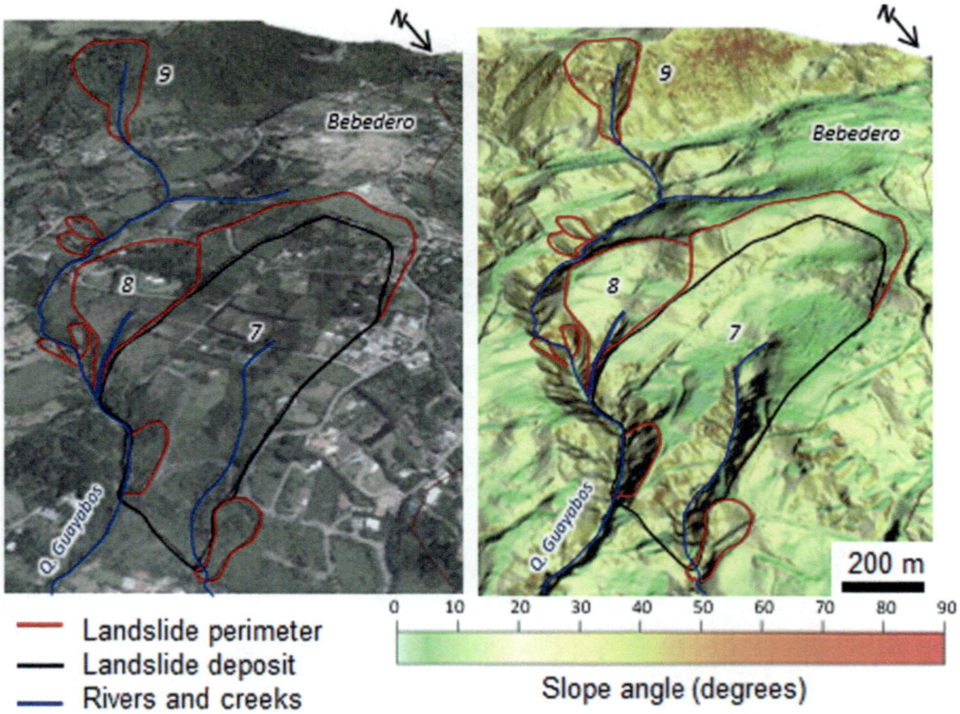

Fig. 17.9 Ancient inactive landslides: Bebedero (7) and Quebrada Guayabos (9)

17.3 Relationship Between Geomorphology and Slope Instability

In order to know the relationship between the geomorphological units and the slope instability problems, the morphogenesis, morphography, and morphometry of each unit were established. In addition, geotechnical studies were carried out to know the behavior of the soils generated by the different lithologies and finally integrating all the data. The probability of landslide occurrence (PLO) was calculated using PISA (probabilistic infinite slope analysis) code (Haneberg 2007).

The geoforms associated with mass wasting correspond to old or inactive landslides, active landslides, and creep-affected surfaces. Thirty-one old landslides are mapped, which are located on slopes with angles greater than 15 ° and at the same time are the evidence of the biggest events that occurred in the area. There are 44 mapped active landslides, which are of small dimensions and are located mainly on the banks of fluvial erosional incisions. Seven of these landslides correspond to ancient landslide geoforms that show some evidence of reactivation. The creeping areas are found on steep slopes (greater than 40 °) or in sectors associated with ancient landslides.

Regarding the geotechnical characteristics, the soils of the lithological units involved in landsliding are moderately stable and behave geotechnically similar. Present characteristics include intermediate to high plasticity and moisture content very close to the liquid limit and very low or almost zero cohesion in saturated soil conditions, which means problems related to mudflows due to their relatively low liquid limit. Considering all the geotechnical features and evaluating different scenarios with PISA code, the result is that unstable slopes are located mainly in the southern sector, in the upper parts of the watersheds, on the banks of the riverbeds and within old landslides. The most stable slopes are located in the northern sector where the urban center of the county is located. About 51% of the landslides are mapped on the slopes underlain by sedimentary rocks with moderate angles, and the PLO is moderate to high, except for the lithological unit medium sandstone with clasts of the Pacacua Formation, where PLO is low. On the other hand, steep denudational slopes in igneous or metamorphic rocks are associated with rock masses of medium quality and show evidence of rockfalls and incipient and intense creeping, and the PLO is high to very high.

Bandera peak, San Miguel peak, and surroundings of Pico Blanco peak are the most unstable zones, with a PLO greater than 50%. These zones have slopes steeper than 40 °, which is the main factor governing instability processes. Other areas with very high to high PLOs are associated with old landslide geoforms, as in the case of Alto Raicero and Alto Granadilla landslides. Finally, integrating geological, geotechnical, and geomorphological features together with the current land use (housing, forest, and pasture for agriculture and livestock), it is possible to establish a recommended land use zoning, considering categories such as conservation, controlled development, and urban

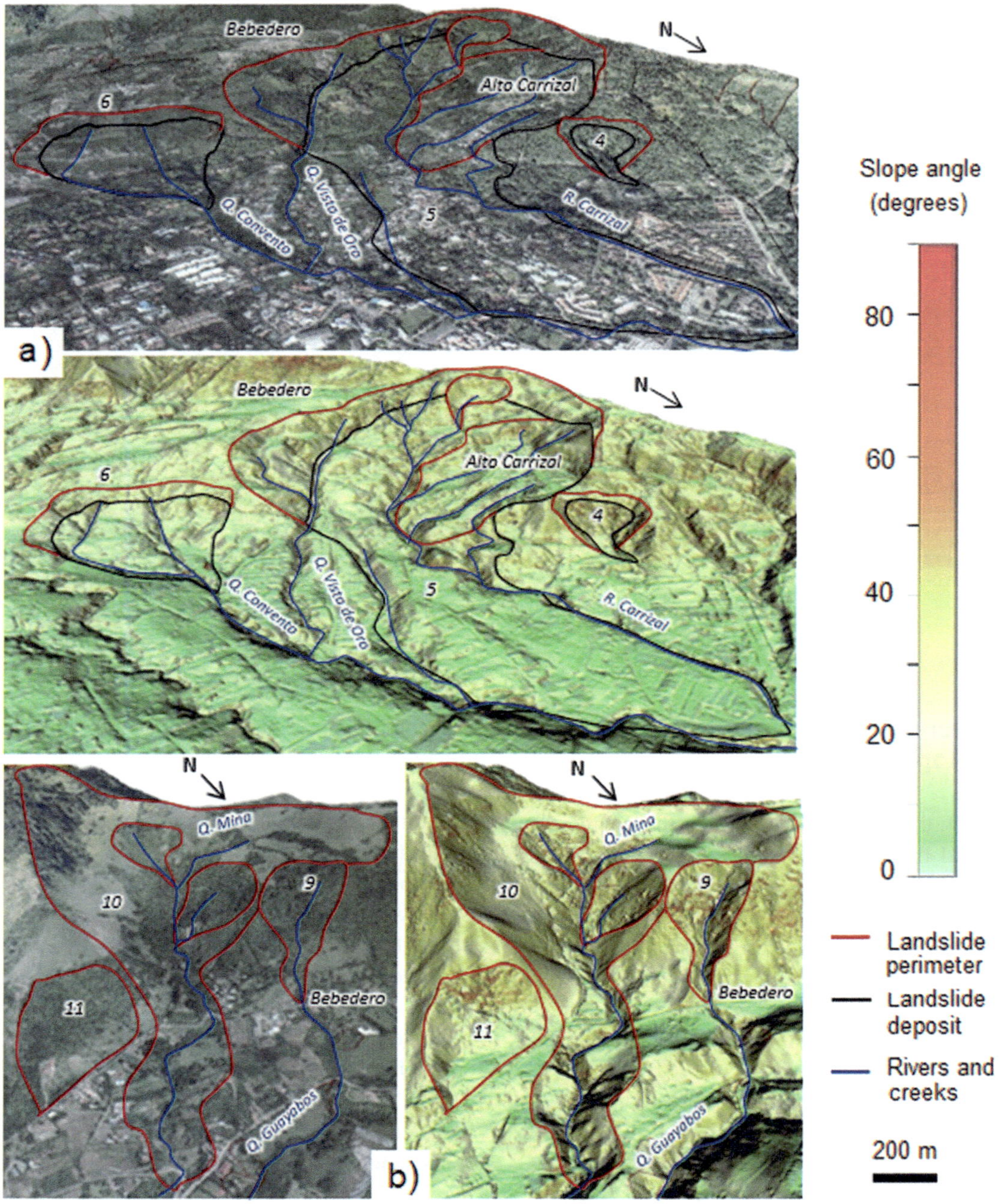

Fig. 17.10 Ancient inactive landslides: **a** Alto Granadilla (5) y Quebrada Convento (6), **b** Alto El Tajo (10)

development (Fig. 17.14), where the current land use differs by 23% from the recommended land use.

17.4 Conclusions

The detailed study of mass wasting processes and their relationship with the geological, geotechnical, and geomorphological conditions allows for the modeling of scenarios of probability of landslide occurrence and recommendations regarding land use, which are important inputs for decision-making on issues of territorial planning and risk management since it allows for the characterization of present threats and the subsequent hazards. The types and mechanisms that trigger landslide processes and the areas with the highest PLO were established; based on this, the sites that require more specific studies were identified and a land use zoning was established. Using this information, it is possible to identify zones that require a stability control and monitoring systems, such as active landslides and

Table 17.2 Main features of the active landslides

Active landslide	Landslide number				
	40	41	44	66	68
River in which it is located	Carrizal	Tributary of Carrizal river	Confluence of Guayabos and Mina creeks	It is not located on or near a riverbed	Middle basin of Guayabos creek
Lithological unit	Medium sandstone with clasts (Pacacua Fm.)	Medium sandstone with clasts (Pacacua Fm.)	Alluvial and colluvial deposits	Medium sandstone with clasts (Pacacua Fm.)	Medium sandstone with clasts (Pacacua Fm.)
Dimensions	120 m wide 171 m long	29 m wide 35 m long	Right bank: 5 m wide and 9 m long Left bank: 20 m wide and 10 m long	Steps of 50 cm high	26 m wide 37 m long
Evidence of activity	Tilted vegetation, exposed soil 15 m thick and open cracks 2 – 3 cm wide and 14 – 35 cm deep	10 m drop at the crown with exposed soil in the first 3 m	Exposed soil. Left bank shows intense creeping up to 20 cm deep	Slope showing creeping and multiple landslides in upper layer of soil	Unstable zone with cracking and exposed floor
Direction of movement	N56 E	NE	Toward the river	NE	
Figure 17	10A	10C	10B	10D	11

Note The landslide number corresponds to the location on the geomorphological map (Fig. 17.2)

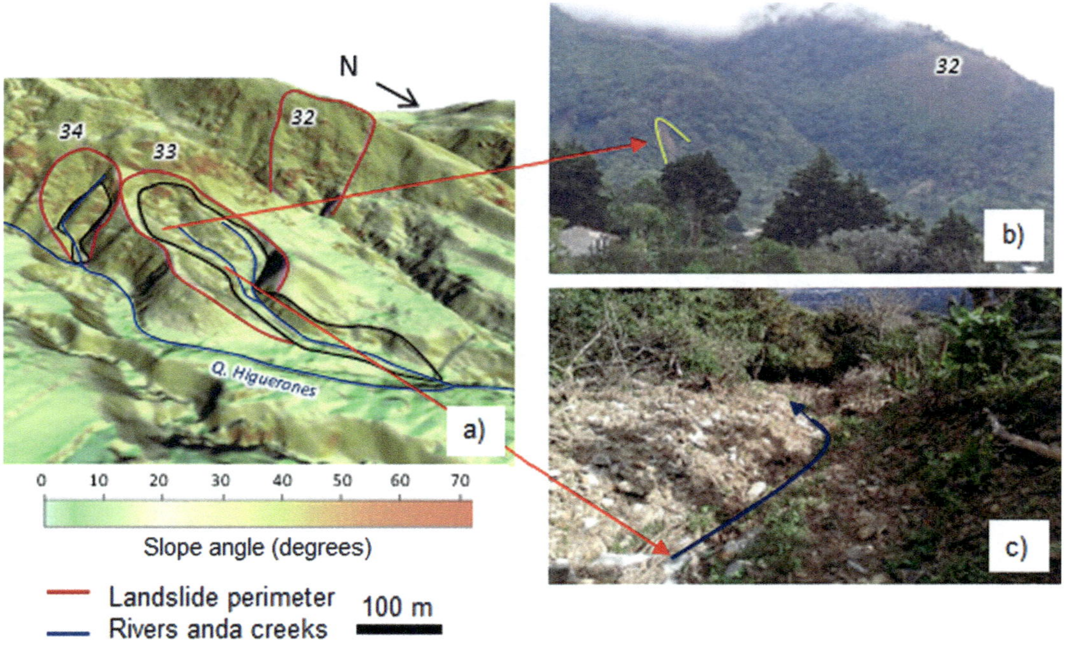

Fig. 17.11 **a** Higuerones creek landslide, **b** Earthflow, **c** Debris flow

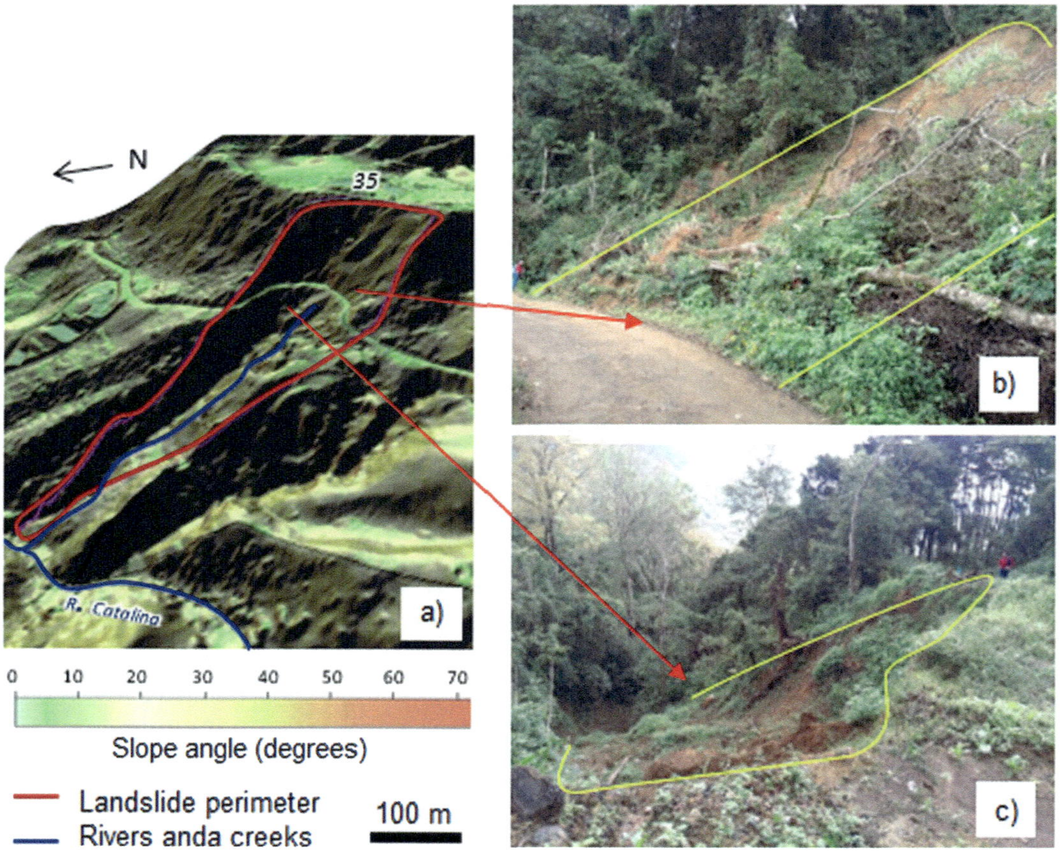

Fig. 17.12 **a** Landslide on a tributary of Catalina River, **b** and **c** Earthflow

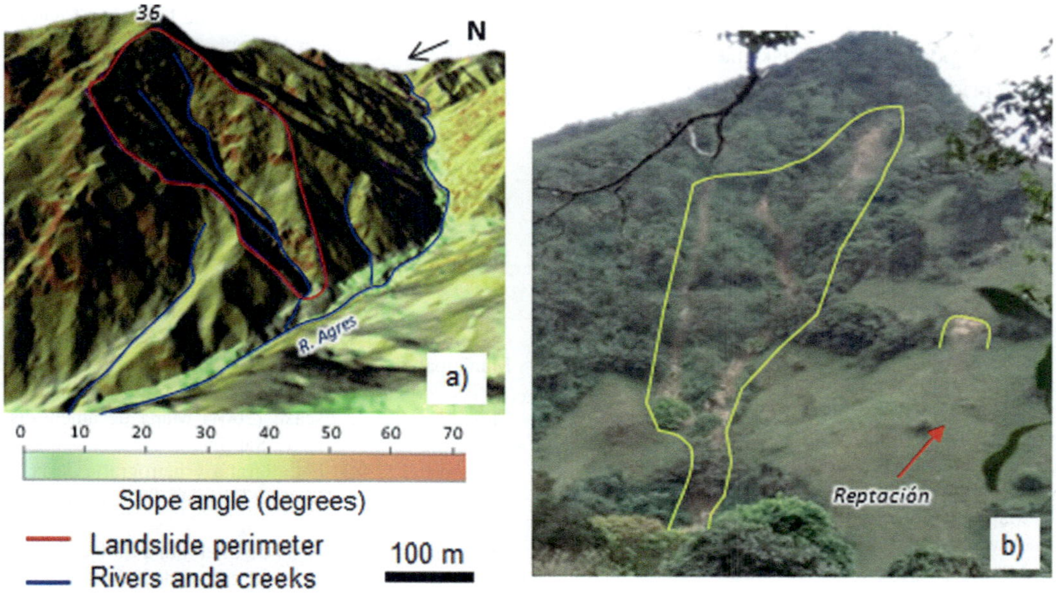

Fig. 17.13 **a** San Miguel peak landslide, **b** Earthflow

Fig. 17.14 Recommended land use

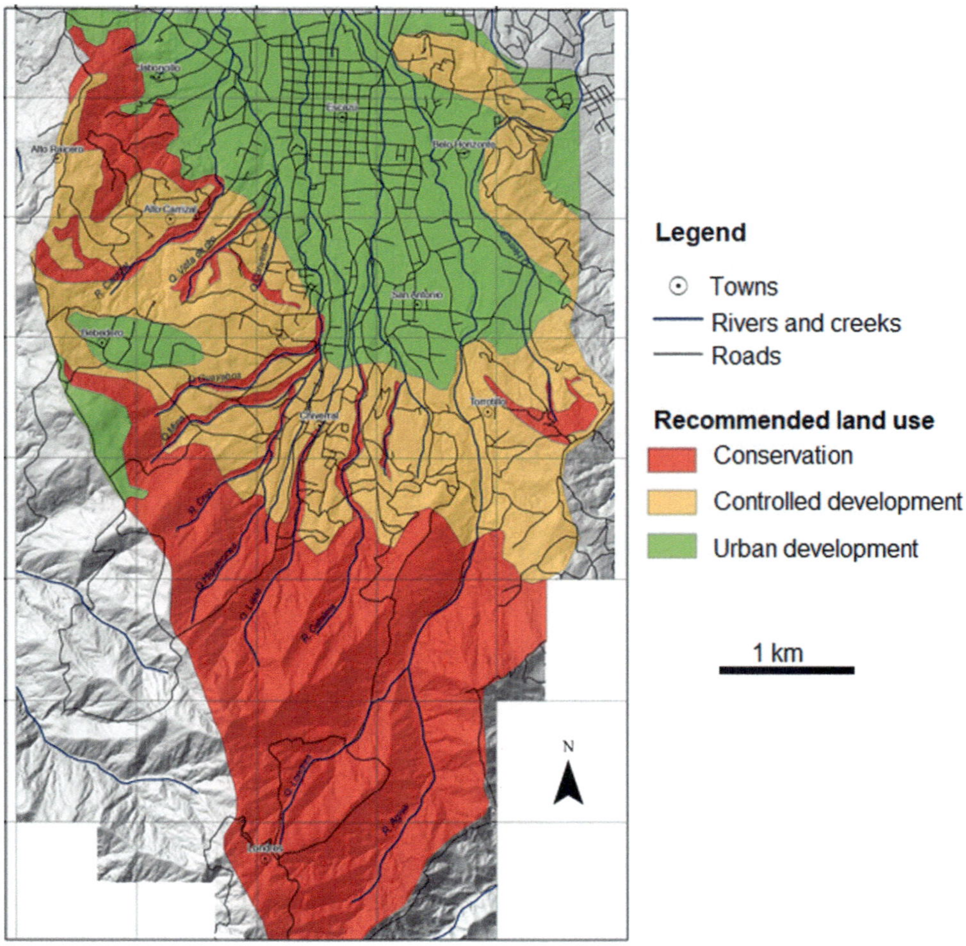

reactivation zones. It is also possible to establish priority sites for training the population on risk management issues and the sites with priority in case of emergency.

References

Alvarado G, Gans P (2012) Síntesis geocronológica del magmatismo, metamorfismo y metalogenia de Costa Rica, América Central. Revista Geológica De América Central 46:7–122

Castillo R (1969) Geología de los mapas básicos Abra y partes de Rio Grande, Costa Rica. Informe MEIC, mapa 1:50000, p 40

Chaves SI, Duran MM, Herrera HD, Soto AD (2012) Susceptibilidad de deslizamiento del cantón de Escazú y análisis a detalle de cinco casos específicos: Palo Campana. Universidad de Costa Rica, San José. Seminario de graduación, Salitrillo, Curio y la cuenca de los ríos Agres y Guayabos, p 130

Cole RA (2013) Map-based probabilistic infinite slope analysis of the Stephens Creek watershed. Portland State University, Tesis M.Sc, Portland, Oregon, p 76

Denyer P, Arias O (1991) Estratigrafía de la región central de Costa Rica. Revista Geológica De América Central 12:1–59

Fernández M, Montero W (2002) Fallamiento y sismicidad del área entre Cartago y San José, Valle Central de Costa Rica. Revista Geológica De América Central 26:25–37

Gonzales LA (2014) Análisis de caída de bloques de roca con microzonificación de la amenaza en el sector de Bebedero. Universidad de Costa Rica, San José, Tesis Lic, San Antonio de Escazú, p 162

Haneberg WC (2007) PISA-m: Map-based probabilistic infinite slope analysis, version 1.0.1 user manual. Haneberg Geoscience, p 20

Haneberg WC (2005) PISA: probabilistic infinite slope analysis, version 1.0 user manual. Haneberg Geoscience, p 15

Hernández J, Vargas A (1992) Geología urbana del cantón de Escazú, provincia de San José. Universidad de Costa Rica, San José, Informe de Campaña Geológica, Costa Rica, p 172

Montero W (2001) Neotectónica de la Región Central de Costa Rica: frontera Oeste de la Microplaca de Panamá. Revista Geológica De América Central 24:29–56

Montero W, Barahona M, Rojas W, Taylor M (2005) Los sistemas de falla Agua Caliente y río Azul y relevos compresivos asociados, Valle Central de Costa Rica. Revista Geológica De América Central 33:7–27

Mora R, Granados R (2012) Estabilidad y probabilidad de ocurrencia de deslizamiento en la cuenca del río Cruz. Informe FUNDEVI, p 25

Mora R, González LA, Campos F, Castro J (2015) Determinación de la probabilidad de ocurrencia de deslizamientos en condiciones estáticas y dinámicas en el cantón de San José, Costa Rica. Informe FUNDEVI, p 45

Municipalidad, de, Escazú (2006) Plan regulador del cantón de Escazú. Universidad Nacional, Heredia, San José, p 61

Municipalidad de Escazú (2004) Mapa de amenazas potenciales. – Escala 1:10 000, CNE, San José

Obando LG (1983) Estratigrafía y petrografía de las rocas aflorantes al Sur del Valle Central (Tabarca). Universidad de Costa Rica, San José, Tesis Lic, p 136

Quirós D (2019) Análisis de estabilidad de laderas en los distritos de Escazú y San Antonio del cantón de Escazú. Una contribución para una futura gestión del riesgo. Universidad de Costa Rica, San José. Tesis Lic. p 206

Rivier F (1979) (1979) Geología del Área norte de los Cerros de Escazú, Cordillera de Talamanca, Costa Rica. Informe Semestral 1:99–137

Ruiz P, Soto G (2014) Preparación del mapa de susceptibilidad a deslizamientos utilizando imágenes LIDAR en los Cerros de Escazú, cantones Aserrí, Desamparados, Alajuelita, Santa Ana y Escazú, Costa Rica. Informe CNE: p 222

Suárez J (2009) Deslizamientos: Análisis Geotécnico. Instituto sobre erosión y deslizamientos, Bucaramanga, p 582

18 Caves in Volcanic Terrains in Costa Rica with the Description of New Volcano-Speleogenetic Mechanisms

Andrés Ulloa and Guillermo E. Alvarado

Abstract

The continental territory in Costa Rica is composed in 50% of Cretaceous to Quaternary volcanic rocks. However, volcanic caves *sensu strictu* (formed only by volcanic processes) are very rare due to the geotectonic conditions of the Pre-Oligocene submarine volcanism, and rheological conditions of Neogene-Quaternary sub-alkaline lava flow in a subduction zone. A total of 90 volcanic caves are reported in the country, with a total length of passages of ~ 2.2 km. A few reported lava tubes (pyroducts) are related to Middle-Upper Pleistocene basaltic to basaltic-andesitic aa lava flows. However, there are an enormous number and variability of caves of different origin and hosted in volcanic terrains like mold caves in lahars, piping caves, erosion caves in fluvial settings (lateral erosion of a river or at the base of a waterfall) and littoral systems, including those in which the cave formed along fault zones or a stratigraphic contact in ignimbrites. Less common are dissolution caves in active hydrothermal systems, or caves in ignimbrites probably formed in a similar way as tumuli or pressure ridges in lava flows. Caves can form in fluvial systems when erosion happens at a stratigraphic contact within an ignimbrite package or when large blocks fall, and landslides occur in a narrow valley followed by further erosion at the base. Sometimes, the rock above the cave can be used as a natural bridge when the length is short. Several caves are the result of complex processes or are of mixed origin. A few other caves were apparently enlarged by anthropogenic interventions (or possibly by Plio-Pleistocene megamammals). More studies are required in order to better understand the diversity of the given volcano-speleogenetic processes.

Keywords

Costa Rica · Vulcanospeleology · Volcanic caves · Pyroducts · Lava tubes · Mold caves · Erosion caves · Weathering caves · Piping

18.1 Introduction

The origin and formation of pyroducts (often called lava tubes) in pahoehoe lava flows, widely distributed throughout the world, has been studied since the eighteenth century (Kempe 2012; 2019). On the contrary, caves developed in volcanic rocks having a different origin and genesis (e.g., mold caves, pit craters, and different types of erosion caves) are much less studied. The most common definition of cave in the scientific community is "a natural void below the surface that is accessible to humans" (Gillieson 1996; Ford and Williams 2007; De Waele and Gutiérrez 2022). According to Halliday (2004b), a volcanic cave is strictly formed by volcanic processes, including lava tubes, hollow tumuli, lava rise caves, hollow hornitos, flow lobe caves, mold caves, volcanic fissure caves, and open vertical volcanic conduits (Kempe et al. 2021). Since the term "volcanic cave" is used for caves formed by volcanic processes *sensu strictu*, in this study we prefer to use the general term "caves in volcanic rocks" or "caves in volcanic terrains", as some of the caves described in this chapter are formed by complex processes and are probably of mixed origin. These processes can be erosion, embedding, acidic weathering, and tectonics with no certainty that only volcanic processes were involved in their formation.

Almost 50% of the continental territory of Costa Rica is volcanic (Alvarado 2021), with

A. Ulloa (✉) · G. E. Alvarado
Centro de Investigación en Ciencias Geológicas, Universidad de Costa Rica, San José, Costa Rica
e-mail: andres.ulloa@ucr.ac.cr

A. Ulloa
Escuela Centroamericana de Geología, Universidad de Costa Rica, San José, Costa Rica

A. Quesada-Román (ed.), *Landscapes and Landforms of Costa Rica*, World Geomorphological Landscapes, https://doi.org/10.1007/978-3-031-64940-0_18

volcanic-geomorphological features dominating in an important part of the country. Thus, it is expected that volcanic caves should be abundant in Costa Rica. However, andesitic magmas dominate over basaltic ones. All the lavas in Costa Rica´s continental territory (excluding Cocos Island) is aa or blocky lava flows, and Quaternary pahoehoe lava fields are completely missing. Consequently, lava tubes (or better, pyroducts), which are developed predominantly in pahoehoe lavas, are rarely reported. In the vulcanospeleological literature, only a few pyroducts have been reported in aa andesitic lavas.

In addition to the rheological behavior, both the tropical climate and weathering are common factors that affect the conservation of volcanic caves in tropical countries such as Costa Rica. Few studies reported volcanic caves in Costa Rica (Mora-Amador et al. 2004; Ulloa 2009; Alvarado and Vega, 2013; Ulloa et al. 2013, 2015, 2018; Rojas et al. 2017; Alvarado 2021). The current study presents, for the first time, a comprehensive review of the known caves in volcanic rocks found in Costa Rica, as well as their speleogenetic features, and different origins. This review most likely underestimates the number of caves in volcanic terrains of Costa Rica, but we hope it can be a starting point for future discoveries and studies.

In this paper, the caves in volcanic rocks are subdivided in: (A) pyroduct (lava tube) caves, (B) mold caves in volcanoclastic rocks, (C) caves in tuff, (D) sulfuric-acid weathering caves, (E) caves of erosive origin, (F) other volcanic caves, and G) volcanic caves intercepted by drilling (Fig. 18.1). Table 18.1 lists the location of the caves, their origin, their host rock, presence of cave map and length of the cave (measured or estimated) and contains speleogenetic information. Table 18.2 presents the information of the Cocos Island's wave-cut caves with the location and dimension reported by Benumof and Lockwood (2000).

18.2 Caves in Volcanic Rocks

18.2.1 Pyroduct (Lava Tube) Caves

Pyroducts, traditionally known as lava tubes, have for centuries been the most studied and explored volcanic caves around the world, especially but not exclusively in hot spot

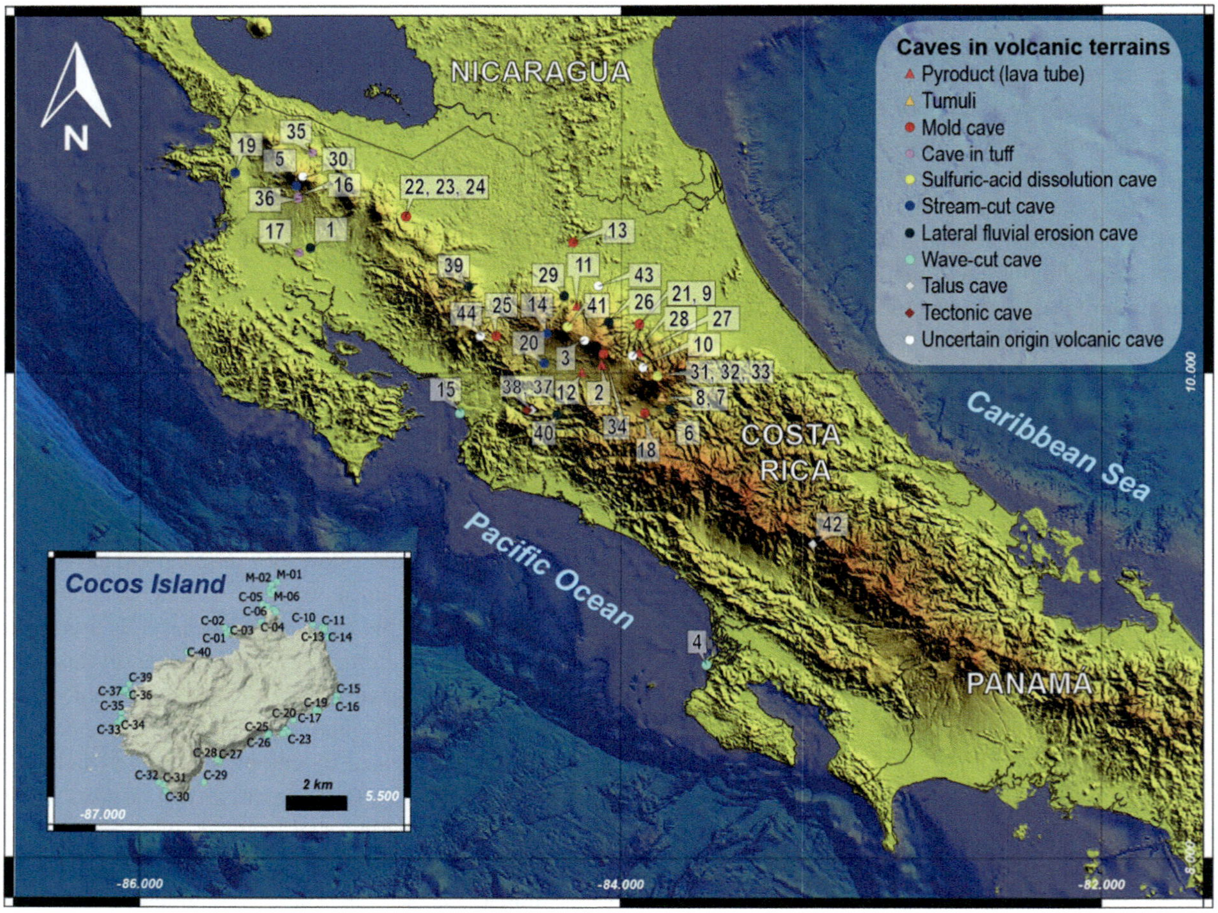

Fig. 18.1 Location of caves formed in volcanic rocks in Costa Rica territory, including caves in Cocos Island

Table 18.1 Costa Rica's volcanic caves showing the location, origin, host rock, cave map, and length (measured or estimated)

Cave ID	Cave name	Lat	Long	Cave origin	Host rock	Cave map	Length (m)
1	Llanos de Cortés	10.52406	– 85.29728	Lateral fluvial erosion	Tuff (Pan de Azúcar Member)	No	~5 to 10
2	Fogón	10.03639	– 84.08555	Pyroduct (lava tube)	Basaltic andesitic lava flow	No	<2
3	Cueva del Indio	10.13841	– 84.15796	Uncertain	Volcanoclastic deposits	Yes	27.4
4	Playa Violín	8.79321	– 83.64985	Wave-cut	Tholeiitic basalt flow	Yes	14
5	Quebrada Escondida	10.78117	– 85.35670	Stream-cut	Landslide of weathered lava flow	No	~300
6	Cueva Cachí	9.85039	– 83.80280	Lateral fluvial erosion	Andesite lava flow	No	~5 to 10
7	Cipreses	9.90304	– 83.83263	Pyroduct (lava tube)	Andesitic basaltic lava flow	Yes	7.7
8	Oratorio	9.90318	– 83.83707	Pyroduct (tumuli)	Andesitic basaltic lava flow	Yes	11.4
9	General	10.20520	– 83.93300	Pyroduct (lava tube)	Andesitic lava flow	No	15
10	Locos por el Bosque	10.02410	– 83.91590	Uncertain	Andesitic lava flow and tuff	Yes	37.5
11	Cariblanco	10.28220	– 84.19140	Pyroduct (lava tube)	Basaltic andesite lava flow	No	<3
12	Ribera	10.00610	– 84.16931	Pyroduct (lava tube)	Basaltic lava flow	No	>10
13	Cerro Blanco	10.54660	– 84.20700	Mold cave?	Block and ash flow deposit	Yes	7.5
14	El Encanto	10.16570	– 84.31540	Stream-cut	Debris avalanche deposit	Yes	28.2
15	Cuevas en Guacalillo	9.83419	– 84.67540	Wave-cut	Debris avalanche deposit	No	6
16	Torre 14	10.73799	– 85.34942	Cave in tuff	Ignimbrite	Yes	103
17	Ciruelas	10.50490	– 85.34520	Cave in tuff	Ignimbrite	No	~5 to 10
18	Agua_Caliente	9.83501	– 83.90729	Mold cave	Ignimbrite	No	5
19	El Duende	10.83650	– 85.61137	Stream-cut	Ignimbrite	Yes	41
20	Puente de Piedra	10.04160	– 84.32700	Stream-cut (natural bridge arc) cave	Ignimbrite	No	5.5
21	Ceibo	10.20640	– 83.93100	Mold cave	Lahar deposit	No	15
22	Cueva 1	10.65365	– 84.90093	Mold cave	Lahar deposit	Yes	27
23	Figuras	10.65337	– 84.90111	Mold cave	Lahar deposit	Yes	39.5
24	Ramales	10.65382	– 84.90088	Mold cave	Lahar deposit	Yes	116
25	El Piquete	10.15480	– 84.52590	Mold cave	Lahar deposit	Yes	56.9
26	Cueva del río Puerto Viejo	10.21379	– 84.05763	Lateral fluvial erosion	Lava flow?	No	25
27	Cueva del río Sucio	10.08757	– 83.93156	Pyroduct (lava tube)?	Lava flow?	No	30
28	Cueva Río Blanco	10.07318	– 83.95959	Uncertain	Lava flow?	No	~5 to 10
29	De la Muerte	10.32103	– 84.24339	Lateral fluvial erosion	Not reported	No	<2
30	Rincón del Jilguero	10.82244	– 85.33033	Uncertain	Not reported	No	~10
31	Los Minerales	9.98640	– 83.85390	Sulfuric-acid weathering	Pyroclastic rocks with hydrothermal alteration	Yes	50
32	Los Mucolitos	9.98720	– 83.85420	Sulfuric-acid weathering	Pyroclastic rocks with hydrothermal alteration	Yes	38
33	El Pizote Espantado	9.98800	– 83.85430	Sulfuric-acid weathering	Pyroclastic rocks with hydrothermal alteration	No	15
34	La Condesa	10.08144	– 84.07800	Mold cave?	Tephras	Yes	8

(continued)

Table 18.1 (continued)

Cave ID	Cave name	Lat	Long	Cave origin	Host rock	Cave map	Length (m)
35	Los Araya	10.92502	− 85.28936	Cave in tuff (piping)	Tuff (contact of Bagaces and Liberia Formations)	Yes	171.8
36	Viscoyola	10.72833	− 85.35050	Cave in tuff	Tuff (Bagaces Fm.)	Yes	3
37	Puente de Piedra (Puriscal)	9.84807	− 84.37664	Talus cave	Volcanic breccias (Grifo Alto Fm.)	No	~20
38	Cueva del Tigre	9.84861	− 84.38004	Tectonic	Volcanic breccias of Fm. Grifo Alto	No	~5 to 10
39	Peñas Blancas	10.36015	− 84.64132	Lateral fluvial erosion	Volcanic rocks with hydrothermal alteration	No	17.25
40	De los Murciélagos	9.82760	− 84.27300	Lateral fluvial erosion	Volcanic rocks with hydrothermal alteration	Yes	35
41	Cueva del volcán Poás	10.19502	− 84.22990	Sulfuric-acid weathering?	Volcanic rocks with hydrothermal alteration	No	8
42	El Zoncho	9.29300	− 83.20720	Talus cave	Volcaniclastic colluvium	Yes	40
43	Bijagual	10.36422	− 84.10152	Uncertain	Volcanoclastic deposits or in highly altered lava	No	35
44	San Ramón	10.15498	− 84.59107	Uncertain	Not reported	No	~250 to 300

regions (e.g., Hawai'i, Azores, Easter Island, Galápagos, New Mexico, Australia, Saudi Arabia, Canary Islands, Italy, and Iceland). They develop in very fluid, often basaltic lavas (pahoehoe type), and usually have a short preservation time, because they usually develop at only a few meters of depth below the topographic surface. A limited number of lava caves have been described in aa lava flows, e.g., in Nishiyuura Ana in Japan and on Etna volcano in Italy (Calvari and Pinkerton 1998; Halliday 2004b). In Costa Rica, although there are several volcanoes, only a few pyroducts has been encountered so far. The scarce presence of pyroducts is caused by higher viscosity of the Quaternary lavas in subalkaline lava flows (andesitic to rhyolitic compositions). In most cases, aa and block lava flows inhibit the development of pyroducts, which usually are formed in pahoehoe lava flows (basaltic in composition).

The Cervantes lava flow field, located at Irazú volcano in central Costa Rica, has a basaltic to andesitic-basaltic composition and consists of two different lava flow fields. The older western Cervantes lava field (age of 57 ka), and the younger eastern Cervantes lava field age (~ 17 ka) (Alvarado and Vega 2013). Alvarado and Vega (2013) reported a 100 m^2 depression in the lava flow field, associated with a collapsed lava tube roof. This depression has two small entrances, the largest corresponds to the Cipreses cave which presently reaches 7.7 m in length (Fig. 18.2a). Local reports indicate that this cave was penetrable for at least 200 m until some decades ago. However, the landowner filled the entrance with soil, rocks, and agricultural waste, in order to prevent vampire bats to roost in the cave thus causing damage to livestock at that time. Figure 18.3a shows the accessible part of the cave. A tumuli structure is located to the west-northwest of the Cipreses cave, where an entrance to a second cave (Oratorio cave) is located at the top of the structure. Oratorio cave presents an entrance of ~ 3 m in diameter, with a vertical-dominated development of 11.4 m depth (Fig. 18.2b) and a volume of the cave of ~ 406 m^3. The Oratorio cave is located in a smooth dome with less than 10 m elevation above the surrounding terrain (Fig. 18.2c), which is interpreted as a tumuli (lava pressure ridge). For the formation of the Oratorio cave, the lava of the tumuli must have drained before crystallizing. Unfortunately, the Oratorio cave has been used to deposit garbage and the continuation of possible cave passages is blocked.

The Cariblanco cave was reported in 2010 in the distal lava flows of the Von Frantzius cone, which represents a secondary cone of the Poás volcano forming a basaltic-andesitic lava field. The entrance to the cave was opened during mining activity in the area (Fig. 18.3b). However, the origin of this cave appears to be associated with folds in high viscosity lava flows, rather than to flowing lava.

Other two small pyroducts were found during the construction of the General hydroelectric project in Horquetas of Sarapiquí. These pyroducts developed in Pleistocene andesitic lava flows (Fig. 18.3c, d) originating the General caves. Unfortunately, these caves were covered with concrete for the development of the project. The existence of these caves is known from pictures, and estimations that suggest that the biggest cave was at least 15 m long. Several small lava caves have been reported in the lava fields associated with the Barva volcano, but most of the reports lack

Table 18.2 Location and dimensions of the wave-cut caves of Coco Island according to (Benumof and Lockwood 2000)

Cave id	Latitude	Longitude	Height	Width	Depth	Length
C-01	5.54846	– 87.0611	5	3	7.5	91
C-02	5.55007	– 87.0616	2.5	10	8.5	4.1
C-03	5.54928	– 87.0599	3.5	7	4	6.5
C-04	5.55181	– 87.0508	1	5	4	31.5
C-05	5.55713	– 87.0503	4	12	4	3.5
C-06	5.55653	– 87.049	2.5	6	11	42
C-07	5.55553	– 87.0471	1	2	2.5	10
C-08	5.55535	– 87.0467	0.5	2	3	5
C-09	5.55474	– 87.0464	3.5	11	2.5	25
C-10	5.55103	– 87.0354	2	3	2	7
C-11	5.55017	– 87.0326	2	1.5	0.5	2.5
C-12	5.55085	– 87.0319	3	5	1	2
C-13	5.54946	– 87.0318	1	5	2	14
C-14	5.54725	– 87.0307	2	4	1	4
C-15	5.52997	– 87.0282	1	2	1	6
C-16	5.52829	– 87.0285	1.5	2	2.5	7
C-17	5.52547	– 87.0336	0.5	1.5	2	5
C-18	5.52547	– 87.0345	5	1	3	5
C-19	5.52587	– 87.0344	1.5	5	ND	8
C-20	5.52276	– 87.0412	0.5	3.5	0.5	4
C-21	5.51987	– 87.0435	3	7	1	2.5
C-22	5.51987	– 87.0435	3	1	2.5	10
C-23	5.51919	– 87.0427	5	3	2	25
C-24	5.51876	– 87.0443	5	4	4	4
C-25	5.51873	– 87.0484	6.5	6	0	5
C-26	5.51816	– 87.0487	5.5	3	3	12
C-27	5.51059	– 87.0632	4	1.5	0	8
C-28	5.51102	– 87.0638	5	1.5	5	20
C-29	5.50445	– 87.0677	3.5	3	0	6.5
C-30	5.5017	– 87.079	4	3	2	9.5
C-31	5.50377	– 87.0798	4	5	8	2.5
C-32	5.50438	– 87.0812	3	1.5	8	15
C-33	5.52194	– 87.0934	3	3.5	3	6
C-34	5.52283	– 87.0924	2	3.5	0.5	2
C-35	5.52483	– 87.0924	2.5	2	3	10
C-36	5.53008	– 87.0901	2	2	0	2
C-37	5.53118	– 87.0922	1.5	3.5	2	2.5
C-38	5.53151	– 87.0913	0.5	1	4	10
C-39	5.53318	– 87.0901	2	2	3.5	6
C-40	5.54271	– 87.0727	5	8	6	35
M-01	5.56377	– 87.0461	10	13.5	10	30
M-02	5.56281	– 87.0477	2.5	0.5	0.5	8
M-03	5.56134	– 87.0483	4	3.5	3.5	15
M-04	5.56049	– 87.0483	0.5	1	2	10
M-05	5.55999	– 87.0479	11	9.5	0	35
M-06	5.56092	– 87.0469	N.R	N.R	N.R	N.R

N.R. Not Reported.

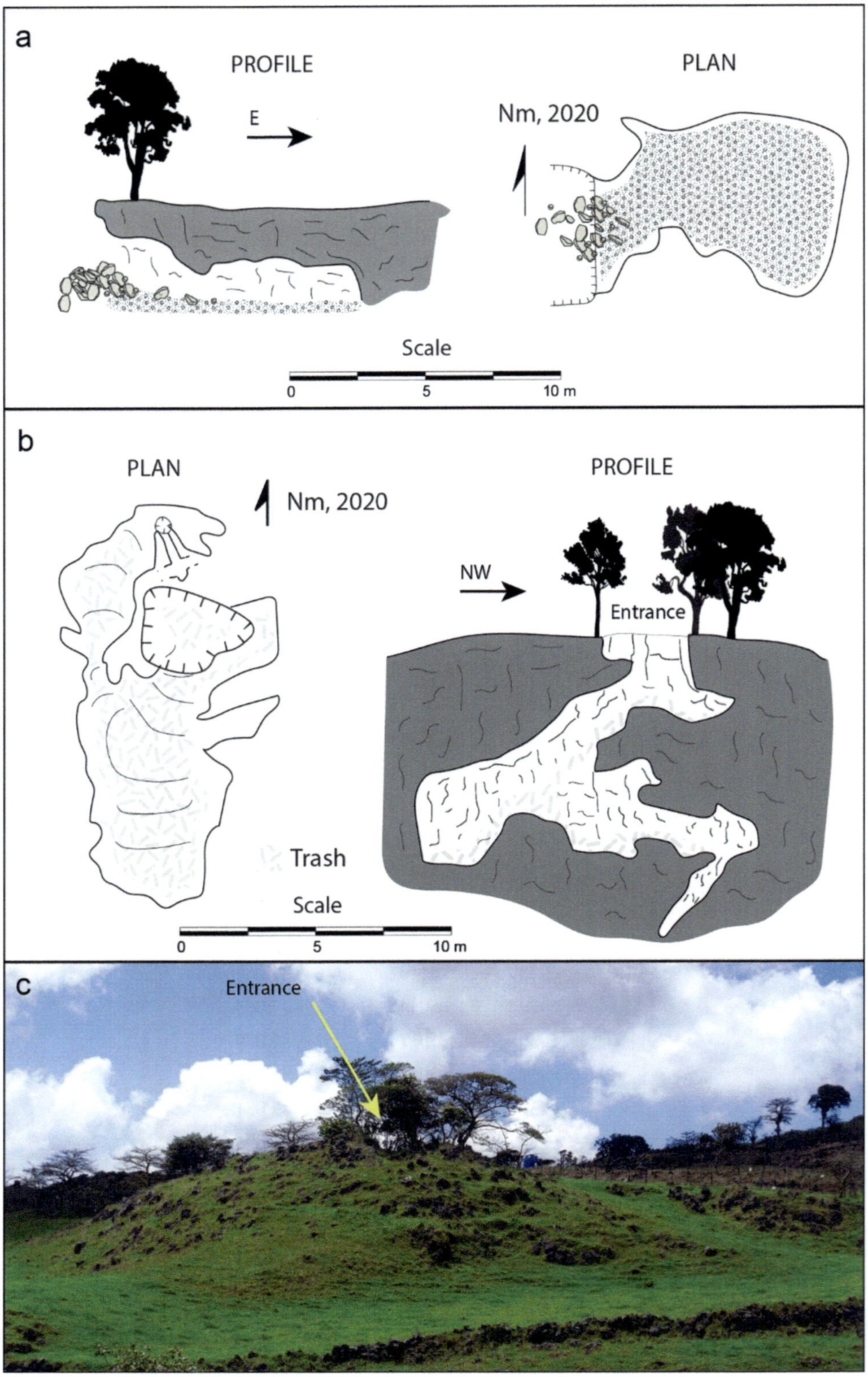

Fig. 18.2 Maps in plan view and profile of the two volcanic caves found in Cervantes Lava Field. **a** Cipreses cave map. **b** Oratorio cave map. **c** Photography showing the main view of Cervantes lava field in the area. The central part corresponds to a dome interpreted as a tumuli, and the arrow indicates the entrances of Oratorio cave

description of detailed locations and characteristics of the caves (Tschinkel 1984; Protti 1986). The El Fogón cave is the only lava tube recently found in Los Angeles (Rojas et al. 2017), but it is only 75 cm wide, 45 cm tall, and 1 m deep, really too small to be considered a "true cave". Finally, in the Central Valley, some basaltic lava flows of fissure origin are reported in the Barva Formation, in the Bermúdez Member. In la Ribera de Belén (Heredia), a pyroduct of some meters in size has been reported (water concession number 1394 in Dirección de Aguas, MINAE).

Fig. 18.3 Photographs of pyroduct lava caves. **a** Photography of Cipreses cave. **b** Cave located in Cariblanco, photography by Edwin Garita. **c, d** Two lava tubes found in General Hydroelectric project. Photos from Leonel Rojas

This cave is referred to as the Ribera Cave, which is partially flooded and considered one of the most important springs of the area with a flow rate of 40 L/s. There are no maps of this cave and access is not allowed as it is located within a private property.

18.2.2 Mold Caves in Volcanoclastic Rocks

Pyroclastic deposits are not the common type of rock for the development of caves. However, small caves formed in pyroclastic rocks have been reported from Greece, Argentina, United States, Czech Republic, Hungary, Slovakia, Japan, Kenya, and Costa Rica (De Waele et al. 2008; Ulloa 2009; Bauman et al. 2022; De Waele and Gutiérrez 2022). Bella and Gaal (2007) divided tree-mold caves, according to the disintegration of wood, into syngenetic pyrogenetic caves (with wooden material burnt out in the volcanic active period), and epigenetic mold caves (associated with mechanical or biogenic destruction of the wood some time after the volcanic activity).

During the building of the General Hydroelectrical project, a mold cave was encountered in a lahar deposit (Fig. 18.4a) and was named the Ceibo cave. This cave was 2 m in diameter, having a depth of around 15 m (Ulloa 2009). The cave diameter could be associated to the presence of a large tropical tree (i.e., a Ceibo tree) trapped in the lahar, which was degraded afterward, forming an epigenetic tree-mold cave.

La Condesa cave is a small cave with a length of 8.6 m, which is located in the foothills of the Chompipe hill, north of Heredia. It corresponds to a cave developed in pyroclastic rocks, and according to its morphology, it could correspond to a small mold cave, however, it requires further characterization. El Piquete cave is another possible mold cave, located in the lahar deposits of the Monteverde Formation, in the La Paz de San Ramón area in Alajuela, in the western part of Valle Central. This cave has semi-circular cross sections (Fig. 18.4b). The semi-rectilinear morphology, semi-circular cross sections, and the presence of small molds of branches observed in the cave suggest that a tree-mold could be a feasible origin. Presence of runoff water could be responsible of secondary erosion and cave enlargement.

In the surroundings of Treminio de Guatuso (Alajuela), three potential tree-mold caves (Las Figuras, Los Ramales and Cueva 1) are reported in lahar deposits. These caves seem to be tree-mold caves due to their morphology, nature of the host rock, and the semi-circular cross section of some passages (Fig. 18.4c). The lahar material in the area is poorly sorted, matrix-supported, with a brown–red matrix, abundant hydrated plagioclase crystals, and sub-angular

Fig. 18.4 Examples of mold caves in lahars with semi-circular sections. **a** Ceibo cave in El General Hydroelectrical Project photography from Leonel Rojas. **b** Photo of the entrance of El Piquete Cave in San Ramón. Photography by Gustavo Quesada. **c** Photo of the Cross-section of Ramales cave in Treminio of Guatuso. **d, e** Mold cave in ignimbrites related to Irazú volcano, in Aguacaliente of Cartago

lava blocks ranging in size from a few to 50 cm. However, some complex morphologies detected in the caves are difficult to explain only by a mold cave origin. Moreover, the area presents anthropogenic activity. A possibility is that these caves were enlarged by natives of the area (Guatusos Amerindians) to use them as shelter or hideouts, since this culture was oppressed in the nineteenth century and before, but further anthropological research needs to be done in the area. Another possibility is that the caves were enlarged by Plio-Pleistocene megamammals. Cave Las Figuras has a length of 39.5 m and a difference in elevation from the entrance of+5.5 m (Fig. 18.5a). The name of the cave derives from the presence of "aboriginal drawings" carved in the rock. However, archeologists of the National Museum of Costa Rica confirmed that these are not pre-Columbian petroglyphs. Cueva 1 (27.2 m long and -7.8 m deep) is the smallest of the three caves (Fig. 18.5b). The longest is Cueva Los Ramales (116.2 m long and − 11.2 m deep); the unusual length of this cave could be explained by the union of multiple tree-molds. The longitudinal profile and plan view of Cueva Los Ramales in addition to circular-shaped cross sections (Fig. 18.5c) indicates a pattern similar to other tree-mold caves such as examples from Mt. St. Helens (Bauman et al. 2022).

Another small mold cave has been found in an ignimbrite deposit in Aguacaliente river in Cartago. This cave is approximately 5 m long, and its diameter ranges from 0.5 to 1 m (Fig. 18.4d, e). This cave is aligned in the N–S direction. Cerro Blanco is located in Chaparrón hills (San Carlos), a relic of the Sarapiquí volcanic arc of Miocene age. In terms of lithology, this outcrop is constituted by a 60 m thick sequence of tuff and volcanic breccias (block

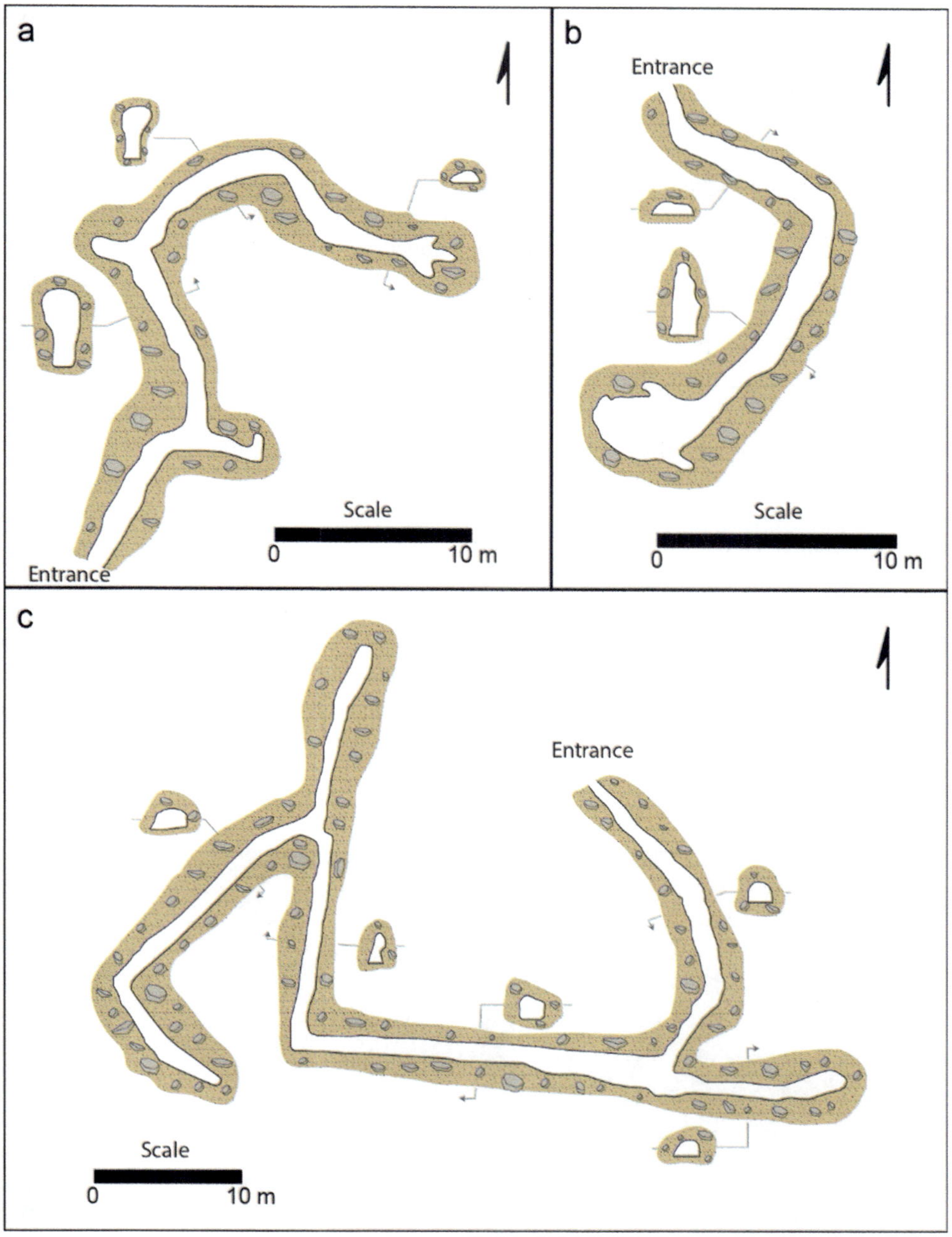

Fig. 18.5 Cave maps (plan view) of three presumed mold caves in Treminio of Guatuso. **a** Cueva las Figuras, **b** Cueva 1, and **c** Cueva los Ramales

and ash flow deposits), with an age of 16.10 ± 0.25 Ma ($^{40}Ar/^{39}Ar$ dating method; Alvarado and Gans 2012). According to local reports, at least three caves have developed in the 4 m thick lapilli and tuff deposit, but only the Cerro Blanco cave has been measured (also known as *Cueva de la India*). This cave is small, 7.5 m in length and 2.5 m to 0.8 m in diameter. The origin of this cave is uncertain, but the morphology of the cave suggests it to be a mold cave. The cave has an excellent visibility over the plains and can be accessed following a ledge running at middle height along the cliff, without needing vertical caving equipment. Consequently, anthropogenic enlargement of the cave is feasible due to its accessibility and panoramic location. In addition, some local reports suggest that it was used as a surveillance point during colonial times, but this has not been confirmed yet.

18.2.3 Caves in Tuffs

According to De Waele et al. (2008), the main processes involved in the genesis of caves in pyroclastic rocks are groundwater sapping, chemical and physical weathering, breakdown and geophagy (i.e., elephants). In Guatemala (Central America), big pseudokarstic sinkhole collapses up to ~ 30 m diameter and 60–120 m deep have been reported in pumice in Ciudad Guatemala (Hermosilla 2012; Bergoeing 2015) associated with piping (or suffosion) processes within thick deposits of tuffs. In this section, we describe the caves developed in pyroclastic ash flow deposits, which are not related to mold origin.

The Guanacaste Volcanic Range is characterized by Quaternary volcanoes and an extensive ignimbrite platform of Neogene to Quaternary age (Alvarado and Denyer 2019). The Torre 14 cave is located 13 km northeast of the town of Liberia (Guanacaste). This cave is developed along a fault plane and a lithological contact between two tuff packages. The cave is 102.9 m long and 13.6 m deep (Ulloa and Quesada 2010). It has four accesses, three of them aligned with a fault (one blocked by collapsed material). The access aligned with the fault plane presents elliptical to semi-circular morphology in cross section (Fig. 18.6a), and the cave has a main gallery which is 4 m high on average (Fig. 18.6b). This cave presents branch molds on the walls (~ 15 cm diameter), but the general morphology of the cave does not point to a tree-mold origin. Mineral and thermal springs have been reported in the surroundings of the caves. Speleogenesis of the cave is not clear, but the presence of thermo-mineral waters in the surroundings, semi-circular morphologies in some galleries and the presence of a fault suggest that acid weathering and subsurface erosion may have been involved in the formation of the cave. After the formation of the main passages, the cave could also have been enlarged by erosion. Detailed studies on the genesis of this cave are required. Other cave entrances have been identified in some nearby outcrops, but they remain unexplored. In addition to geotechnical research, two geoelectrical profiles were investigated in the area, and caves were correlated with high resistivity values, > 200 Ω-meter (Ávila 2010). Near the Torre 14 cave, the small erosional (3 m long) Viscoyola cave is located.

The Ciruelas cave is developed in ignimbrites found in Hacienda Ciruelas (Guanacaste). This cave is ~ 9 m wide, ~ 4 m deep, and ~ 1.5 m high (Fig. 18.6c). The origin of the cave can be associated with ductile deformation during the ignimbrite formation (Alvarado and Denyer 2019), similar to a pressure ridge or tumulus in lava flows.

The Los Araya cave developed in a tuff sequence, along the geological contact of Bagaces and Río Liberia tuffs. This contact is characterized by the regional geological "Green layer" located under the Río Liberia tuff and over the Bagaces formation. The "Green layer" is composed of vitroclastic tuffs with stratified dacitic pumice in centimetric-sized layers (Alvarado and Denyer 2019). This cave has underground streams in each gallery, the map of this cave reports a length of 172 m (Trescott 2018), and shows

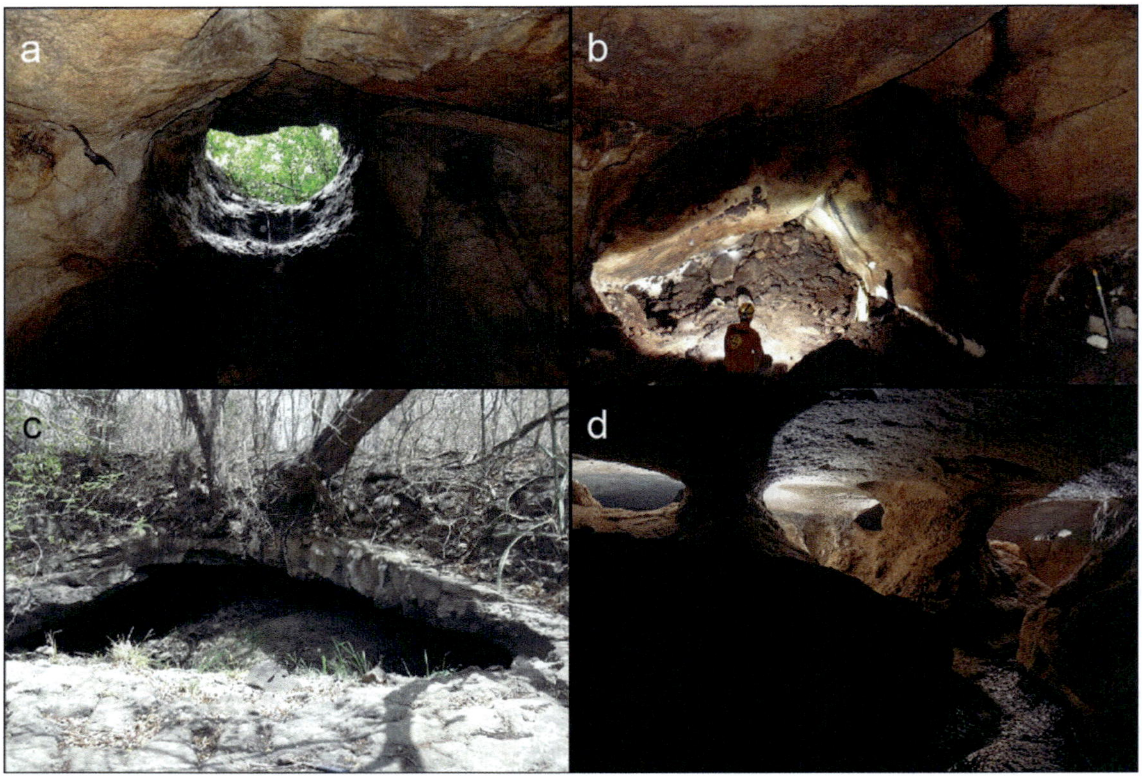

Fig. 18.6 Caves formed in pyroclastic rocks. **a** Semi-circular shape entrances of cave Torre 14. Photography by Scott Trescott. **b** Main gallery of the cave Torre 14. Photography by Scott Trescott **c** Small cave in Hacienda Ciruelas associated with ductile deformation **d** Cave los Araya in the north foothills of Rincón de la Vieja volcano, this cave is formed by piping of highly weathered tuff and presents some characteristic columns. Photography by Ferdinand Salazar

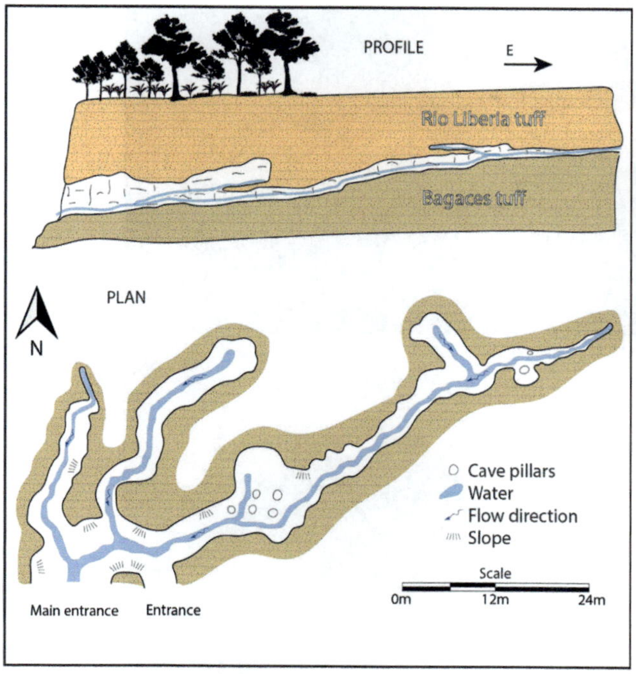

Fig. 18.7 Cave map of Los Araya cave, modified from Trescott (2018)

a branchwork pattern with at least five different tributaries that join into a single stream at the cave entrance. The Los Araya cave is characterized by planar passages developed along the geological contact, and by the presence of cave pillars (Fig. 18.6d). The cave pillars are similar to those in quartzite caves reported by Sauro (2014) and Wray and Sauro (2017), which are formed by stress and erosion. The size of the passages increases toward the entrance and decreases progressing upstream in the tributaries (Fig. 18.7). The origin of the cave is associated with erosion by piping along the contact with the "Green layer", which has different resistance to erosion. Cave development follows the lithological contact, which is slightly dipping to the SW.

Weathering is more intense in the Caribbean region of Rincón de la Vieja volcano, affecting the physical resistance of the tuff and making it subject of efficient piping. This could explain why these types of caves have not been reported in the Pacific sector, where tuffs are widely distributed and are more accessible, but less affected by weathering.

18.2.4 Sulfuric-Acid Weathering Caves: The Case of Irazú Volcano

Caves in the NW sector of Irazú volcano are located in an area of difficult access due to steep and unstable terrain. Three caves have been identified in this area, namely

Fig. 18.8 Photography of the NW section of the Irazú volcano showing the location of the Irazú volcanic caves in a highly unstable and hydrothermal-altered area. Photography by Iván Salazar

the Cueva los Minerales (CMI), the Cueva los Mucolitos (CMI), and the Cueva Pizote Espantado (CPE). These caves have developed in pyroclastic rocks with intense hydrothermal alteration (Fig. 18.8). A volcanic debris avalanche and related debris flows occurred on the night of December 8, 1994, uncovering the site where the caves are located (Alvarado et al. 2013).

These caves have been the object of multiple scientific investigations due to the presence of high mineral diversity. Close to fifty different minerals have been identified in the CMI cave, mainly hydrated sulfates (with Fe, Al, Ca, etc.), native sulfur, oxides, and halides (Fig. 18.9b). Thirteen new cave minerals have been reported in this cave environment (Ulloa et al. 2013, 2018; Ulloa, 2019). It is very likely that the CMI presents the richest cave mineral diversity in the world (Ulloa 2019).

The caves in the Irazú volcano are characterized by positive thermal anomalies, acidic environment (dripping water has a pH 1–2), high mineral diversity, and the presence of extremophiles. The CMI has a wide entrance (Fig. 18.9a, d), following a stratum of highly hydrothermally altered pyroclastic rocks. This cave presents some thermal anomalies (12 °C in the entrances, and 30 °C in some chambers), where volcanic gas at high temperature flows into the cave from the floor and walls near to faults (Ulloa et al. 2013, 2018; Ulloa 2019).

These caves are formed in an acidic environment and are characterized by the presence of extremophiles (Ulloa 2019). Particular white mucous-like extremophiles are present in the CMU and the CMI, hanging from the walls. Those biofilms are known as "snottites", and they are reported for the first time in volcanic-acidic caves. These biofilms were analyzed using molecular techniques, and taxonomic findings identified by 16S rRNA

Fig. 18.9 Caves of Irazú volcano. **a** Entrance of the CMI, where it is possible to observe a significant number of collapsed rocks. Photography by Víctor Carvajal. **b** Example of crust minerals along the walls of the cave. Photography by Víctor Carvajal. **c** Cylindrical hole in the roof of the cave, which could be evidence of sulfuric-acid weathering. **d** Entrance of the CMI where it is possible to recognize crust minerals in the walls and floor of the cave. Photography by Scott Trescott

gene sequencing yielded that Nitrospirae, Proteobacteria-Actinobacteria, Chlamydiae, Euryarchaeota, and OD1 were the dominant taxa. Also, negative values of $\delta^{34}S$ isotopes measured in sulfate minerals are suggestive of microbial sulfur transformation. Finally, due to these particular conditions, these volcanic caves can be considered as analogs to aqueous acidic paleo-Mars environments (Ulloa 2019).

The caves in the Irazú volcano have a complex origin. In the proximity, at least three traces of the active Río Sucio Fault System have been reported (Montero et al. 2013) and detailed observations suggest that the Río Sucio Central active fault is present in the CMI and the CMU. The scarp located above the entrance of CMU may correspond to one trace of the Río Sucio Central fault which is also present in the CMI. Weathering-dissolution of andesitic glass, pyroxene, and plagioclase is a feasible process in hyper-acidic environments such as the Poás Volcano (Rowe and Brantley 1992). Recent hydrogeochemical analyses of the Irazú volcano estimated a rock removal rate of ~ 3000 m^3/year in the Irazú-Turrialba system by chemical leaching drained by the springs (Rouwet et al. 2021). This new evidence of strong chemical weathering (including dissolution of selected rock-components), in addition to the extremely acidic environment of the cave, suggests that sulfuric-acid weathering of volcanic rocks by strong acidic waters along active faults in the Irazú caves area is an important process in the formation of the caves. Moreover, the presence of a cylindrical hole in the roof of the cave, with ~ 1 m diameter, similar to feeders found in sulfuric-acid caves (Fig. 18.9c), is suggestive of sulfuric-acid weathering in some areas of the cave and rising air. Another important process involved in the speleogenesis is the formation of gypsum crusts or alteration rims where mineral precipitation in the cracks can produce extensive forces and promote rock-falls within the caves (Ulloa 2019). Various factors, such as hydrothermal fluids, fumaroles, and volcanic gases, influence the formation of these caves. These elements come together to trigger a combination of sulfuric-acid weathering (including dissolution) and erosion, contributing to the speleogenesis of those caves. Probably the materials are transported out of the cave by the hydrothermal system (fines and some dissolved minerals), seeping water and even air currents. Therefore, the Irazú volcanic caves could be used as a unique model for sulfuric-acid weathering and speleogenesis in volcanic rocks.

18.2.5 Caves of Erosive Origin

18.2.5.1 Stream-Cut Caves in Volcanoclastic Rocks

Flowing water can preferentially remove softer rocks and produce caves. For instance, caves formed by river erosion are quite common. Rarer examples occur when flowing water is able to preferentially remove sections of the basal breccia in an aa lava flow, leaving the solid lava core, as observed in some caves in Jordan (Kempe 2012, 2019). A particular example is the kilometer-long and 100 m-deep Kuka'iau cave located in Hawai'i, which was formed by fluvial erosion without following a pre-existing lava cave (Kempe and Werner 2003). The best-known example in Costa Rica is the Puente de Piedra, a natural bridge arc cave, located in Grecia, at the southern foothill of Poás volcano. This cave has cross sections of 12.3–8.1 m wide, 4.6–6.3 high, and roof thickness of 7 m (Fig. 18.10a), and has been developed in Middle Pleistocene ignimbrites (Mora 2004). The cave was formed by erosion of the Poró river, and differential removal of the ignimbrite lithofacies (Mora 2004; Ruiz and Suárez 2017).

Approximately 14 km north of Puente de Piedra, another stream-cut cave is located in the locality of Alto Palomo,

Fig. 18.10 Photographs of selected caves with erosive origin. **a** Puente de Piedra cave, which consists of a natural bridge developed in Puente de Mulas ignimbrite. **b** El Encanto cave, which consists of a stream-cut cave developed in debris avalanche deposits **c** Entrance of Duende cave, in Santa Rosa National Park. Photography by Stanimira Deleva. **d** Cueva de los Murciélagos

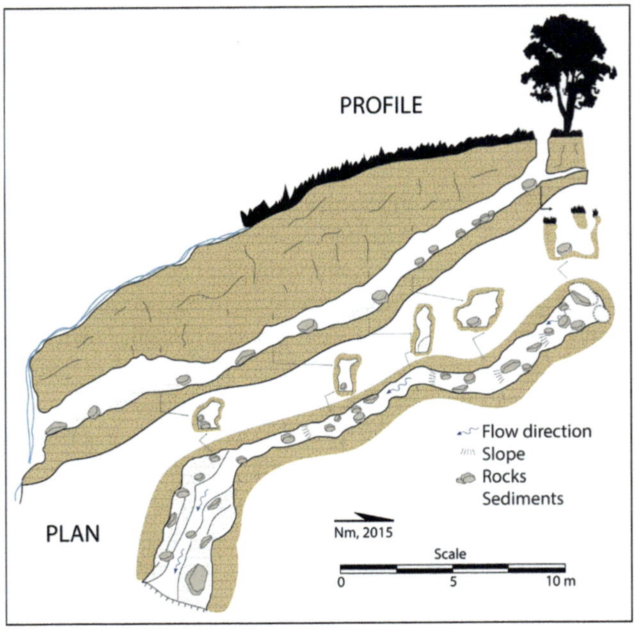

Fig. 18.11 Cave map of El Encanto cave showing the plan view and profile of the cave, with a development of 28 m. Modified from Ulloa and Madrigal (2015)

Sarchí Norte (Alajuela). This cave is known as El Encanto and was developed in a volcanic debris avalanche deposit. Water from a nearby spring has been eroding and enlarging the cave (Fig. 18.10b). The development of the cave occurs along the contact of two facies with different resistance to erosion. The cave has a length of 28.2 m, and there is 11 m difference in elevation between the two entrances (Fig. 18.11).

In the Santa Rosa National Park (Guanacaste), a small cave developed in ignimbrites is located in El Duende creek (Madrigal 1982), hence named cave El Duende. The cave is less than 10 m wide and 41 m long (Deleva et al. 2018), and developed in Pliocene ignimbrites of the Río Colorado flow deposit (Denyer et al. 2019). The origin of this cave is associated with water erosion between layers of differential consistency within the volcanic deposit.

Another cave found in a volcanic landslide is located in the southern flank of Rincón de la Vieja volcano. This cave is known as Quebrada Escondida and has not been formally explored. A creek was naturally channeled through the cave for about 300 m, suggesting an erosive origin. During periods of heavy rain, this cave is temporarily blocked, and the water of the creek returns to the surface riverbed. In addition, volcanic gas smell is often perceived in the area. The origin of this cave could be associated to stream-cutting, but further speleological exploration is needed.

18.2.5.2 Lateral Fluvial Erosion Caves in Volcanic Rocks

Caves created by lateral fluvial erosion are quite common in volcanic rocks. In order to be formed, it is necessary to have rocks with different physical resistance (due to physical or chemical conditions), and flowing water able to remove rock particles as the caves developed on stacked series of lava flows (Kempe 2012, 2019). The erosion can be lateral along the margin of the river or vertical and headward in the case of waterfalls. In this section, we describe some of the most interesting lateral erosion fluvial caves in volcanic rocks discovered in Costa Rica.

Various caves have been reported along the banks of Peñas Blancas river in San Ramón of Alajuela. These caves are located in areas with intense hydrothermal alteration where most of the original minerals have been replaced by secondary clay minerals (Vargas 2001). In the area, it is common to find epithermal alteration, with caves located in hydrothermally altered volcanic rocks of the Aguacate Group, according to the geological map of Fortuna Quadrangle (Alvarado 2009). Quesada and Goicoechea (2021) mapped the Cueva del río Peñas Blancas, with a length of 17.25 m and a width of 12 m at the entrance.

The Cueva de la Muerte is located in the left margin of the Toro River (Venecia, San Carlos). Volcanic gas emissions are present in this cave (e.g., CO and CO_2), generating a toxic environment for most living organisms (Zimmer et al. 2004), hence receiving its name (Cave of Death). The origin of this small cave is probably associated with preferential river erosion along Quaternary volcanic and fractured rocks (i.e., faults) that also explains the accumulation of gasses detected inside the cave. McFarlane et al. (2019) indicated that CO_2 stratifies inside the Cueva de la Muerte, emerging from the cave in a density flow at ground level. The term "mofette cave" was coined for caves that emit geogenic CO_2 (McFarlane et al. 2019).

The Cueva de los Murciélagos is another volcanic cave originated by fluvial erosion in a hydrothermally altered area. It is located in Floralia of Puriscal (San José) where the host rock consists of tuffs and lavaflows. The rock sequence is tilted and differential erosion in weak areas (such as interbeds, fracture zones, and intense hydrothermal alteration zones) are responsible for the formation of the cave. The Cueva de los Murciélagos is 35 m long and 19 m deep (Quesada 2009), while it is composed just by one chamber (Fig. 18.10d). It was historically used by locals as a hideout to avoid recruitment during the civil war of 1948. Another lateral erosion cave has been identified in the northwest of the Cacho Negro volcano. It is a small cave at the base of a waterfall developed in lavas. In addition, other caves have been reported in Llanos de Cortés, Liberia and Reventazón river valleys, but are not described here due to the lack of detailed information or small size.

18.2.5.3 Wave-Cut Caves

Costa Rican marine erosion caves in volcanic rocks are restricted to the Pacific coast and nearby islands. In contrast, the Caribbean coast lacks outcrops of volcanic rocks. Hitherto, there is no systematic work to identify marine caves in the Pacific coast of Costa Rica, so the number of wave-cut caves is certainly highly underestimated. It is common to find caves close to a cliff or developed in oceanic ophiolites, which are generally Cretaceous-Eocene in age in the littoral Pacific (e.g., Punta Islita Beach), but neither formal geolocation nor surveying has been performed for these caves. The best well-known wave-cut caves have been reported in the Cocos Island and are described below.

In addition, littoral caves are found in Playa Guacalillo cliffs (Pacific coastline), where small caves are carved in volcanoclastic rocks (Middle Pleistocene volcanic debris flows/avalanche deposits; Alvarado and Gans 2012). Along the south Pacific coast of Costa Rica, in Violines Beach, close to the Sierpe River mouth there is a wave-cut cave of 14 m length, reported by Stanimira and Ivanov (2018). The Playa Violín cave is developed in Paleocene basalts related to the Rincón Block, which corresponds to accreted seamount fragments (Denyer and Gazel 2009).

The Cocos Island is the highest point of the Aseismic Cocos Ridge (a submarine volcanic range) in the Pacific, originated on the Galapagos Hot Spot. Lavas are basaltic to basaltic-andesitic. According to Hey (1977), the basement rocks of Cocos Island are between 10 and 15 Ma old (Middle Miocene), but there are some anomalous ages of 1.5–2.2 Ma (Harpp et al. 2005; O'Connor et al. 2007) that suggest local volcanism at much later date (Lower Pleistocene). An updated synopsis of the geology and tectonics of the Cocos Island was presented by Rojas and Alvarado (2012).

The Cocos Island is the only subaerial exposure of the volcanic Cocos Ridge and 46 caves (Fig. 18.1) have been reported there (Benumof and Lockwood 2000). These caves are associated with fracture enlargement by wave action. Table 18.2 lists the location and dimensions of the caves, with an average cave length of 12.7 m, and a total reported length for the 46 caves of 574.6 m. These caves are of interest to conduct microbiological and mineralogical studies due to the isolated and pristine character of these volcanic caves, similar to the Galapagos (Espinasa et al. 2015; Miller et al. 2020, 2022). Even though marine caves have been recognized, from a geological and speleological point of view these caves require a vulcanospeleological exploration. For instance, informal communications suggest the existence of volcanic caves (lava tubes) around Wafer Bay (north section of the island).

18.2.6 Other Volcanic Caves

In this section, we describe volcanic caves that have been previously reported, but for which additional studies are necessary to elucidate their origin and speleogenesis. One of these caves is located in Monserrat de Coronado, 11 km NW of the Irazú volcano. The cave is developed in rocks affected by hydrothermal alteration within an active faulting area. Genesis of this cave could be associated with tectonic-erosive processes. The cave is named Locos por el Bosque and is located next to the Cajón River. It has a 37.5 m longitudinal development and a maximum height of 5 m. Also, some secondary mineral deposits have been recognized in the cave (Fig. 18.12a, b). A second cave has been reported in the surroundings but has not been formally surveyed.

Another cave located in the Caribbean margin of the Irazú volcano, alongside Río Sucio, has been reported in the walls of the canyon. According to photographic evidence, this cave could be associated with a pyroduct (Fig. 18.12c, d). A particularity of this cave is the presence of yellow to orange microbial mats in a small flow of water inside the cave. Another cave was observed in the margins of Río Blanco, in the Braulio Carrillo National Park. This cave was observed during a helicopter flyover Fig. 18.13a, and is presumed to be formed in lavas, where differential fluvial erosion may have been an important factor in its formation. Unfortunately, access to the site requires a multi-day trek through the dense jungle of the Braulio Carrillo National Park.

A small cave has been reported in the locality of Los Cartagos in a volcanic area located between Barva and Poás volcanoes. This cave is called Cueva del Indio and is developed in volcanoclastic deposits associated to the Paleo-Barva unit. According to Quesada and Quesada (2020), this cave is 27.4 m long and has a depth of 11.5 m, with two entrances and a 11.5 m pit. Most likely, the origin of the cave is associated with piping of the tuff of the paleo-Barva unit, but additional speleogenetic interpretation is required for this cave.

The Rincón del Jilguero, located ~ 1.1 km SE from the main crater of Rincón de la Vieja volcano, is a small and partially flooded cave (Fig. 18.12e). The cave has biofilms hanging from the walls.

Talus caves consist of voids between boulders, normally located at the base of cliffs or steep slopes. Generally, this type of caves are roofed by overlapping blocks and demonstrate complex three-dimensional development (Palmer

Fig. 18.12 **a, b** Locos por el Bosque cave with some details of secondary mineral deposits. **c, d** Possible pyroduct in the margins of Río Sucio. Photography by Ivan Salazar. **e** Main view of the entrance of Rincón del Jilguero cave in Rincón de la Vieja volcano. Photographs by Scott Trescott

Fig. 18.13 **a** Aerial photography of a cave in the margins of Río Blanco cave, Braulio Carrillo National Park. **b** A talus cave in La Máquina creek, in Barbacoas of Puriscal

2007; De Waele and Gutiérrez 2022). El Zoncho cave, located in Olán of Buenos Aires (Puntarenas), classifies as a talus cave; specifically, it is found in a talus at the foothills of Arbolado Mountain (Talamanca Range). The cave is composed mainly of Pliocene pyroclastic flow deposits (Ulloa and Delgado 2010), is 40 m long, and shows 5 m of difference in elevation. Another talus cave has been recognized in Barbacoas of Puriscal, where the Máquina creek is covered by a big rock related to rock-fall deposits forming a natural bridge used by 4WD cars (Fig. 18.13b). The cave is known as Puente de Piedra (Puriscal) and extends for approximately 20 m along the river. In this location, lithology consists of Pliocene volcanic breccias of the Grifo Alto Formation. In the same area, another cave located in a fault-trace is known as Cueva del Tigre and has a tectonic origin.

The San Ramón cave, reported in the surroundings of Cerro Azahar, is developed in volcanoclastic deposits, with a length between 250 and 300 m. According to a regional geological map of the area (Žáček et al. 2012), the cave is located in the Monteverde Formation within an area of intense laterization and close to some fault traces. Probably laterization of the volcanoclastic deposits and structural influence are factors that controlled the origin of the cave. Another cave, approximately 30 m long, was developed apparently in volcanoclastic deposits or in a highly altered lava flow close to la Virgen of Sarapiquí within an ecological reserve called Bijagual.

18.2.7 Volcanic Caves Intercepted by Drilling

Costa Rica is a country where groundwater is an important source for freshwater supply, with the presence of more than 17,000 wells reported in the country according to the Servicio Nacional de Aguas Subterráneas, Riego y Avenamiento (SENARA) wells database. During the drilling of wells, some have reported loss of drilling fluids and/or the interception of metric-sized cavities in volcanic materials. We correlated some of these cavities with volcanic caves, some of which are totally flooded.

Particularly, in the Central Valley, there are multiple places where caves have been intersected by drilling. In Heredia, drilling operations at several wells have found cavities, however, in many cases, the "anomaly" is not presented in the drilling reports. During the construction of a piezometer by Empresa de Servicios Públicos de Heredia (ESPH), a 1–2 m cavity was reported in Barreal de Heredia. In la Valencia, located in Santo Domingo of Heredia, cavities were reported during the drilling process of big-diameter water-production wells in the 1970s by Servicio Nacional de Aguas Subterráneas (SENAS). In Coyol de Alajuela, the well RG-312 reports a cave at ~ 68 m depth, developed in lavas. According to the drilling report, this well has an exceptionally high transmissibility and was pumped at 14.8 L/s with well drawdown less than 2 cm. According to the driller manager (Mr. Sergio Fonseca), at least three more wells in the surroundings of Coyol intersect volcanic caves and are high-capacity wells. These caves are probably developed in the Colima Superior, a geological formation, which is composed of andesitic lava flows. Another example in the andesitic lava flows related to the Paraíso Formation (Sojo 2018), at the Paraíso de Cartago location, where the well TP-47 for water supply reported a cave at 78 m of depth, under phreatic level, most likely corresponding to a flooded lava tube. A piezometer placed in tuff sequences, in the location of Curubandé at Liberia, also reported a cave-intersection of a few meters.

The Llano Grande de Cartago mine is composed of pyroclastic highly-silicified rocks (>80% SiO_2), used as natural *pozzolana*. Here a cave *sensu strictu* has not been found, but we have evidence that strongly supports the presence of cavities in these materials. The cavities are associated to paleo-fumaroles, and some are filled with native sulfur. These paleo-fumaroles possibly follow old paths of weathering channels formed by paleo-acid waters of the Irazú hydrothermal system. Those channels have been found filled with sulfur in the mine and "empty" in the drillholes related to the mine. A similar cavity of 3 m was found at 55.5 m of depth during the drilling of an exploratory well. A ~ 2.5 m void was intercepted during the drilling at 75 to 80 m deep in the locality of Alto las Palomas in Santa Ana. According to the driller manager, this well had no water losses during the drilling process, and showed the presence of well-formed crystals, suggesting this cavity to be a geode.

18.3 Speleogenetic Remarks

Costa Rica has a high diversity of magma compositions, from felsic to mafic. Also, in a relatively small territory, a wide variety of volcanic deposits are exposed (e.g., lava fields, ignimbrite deposits, accreted oceanic basalts, lahars, debris avalanches). Although there are more than 400 volcanic centers (from relict to active volcanoes, Alvarado 2021), volcanic caves are little known and studied compared to caves of karstic origin. Diverse volcano-speleogenetic mechanisms are described in this chapter, with some particular, or unique cases, described in this section.

18.3.1 Caves in Blocky Aa Lava Flows

Figure 18.14a illustrates a volcano-speleogenetic process of cave formation in a blocky lava flow. During the lava flow, the lava can present a depression or *jameo* (a local Canary Island term in Spain). After lava crystallization, the lava flow can be partially drained (central part) and a pyroduct elongated along the lava flow direction can be formed. This scenario can explain the formation of the Cipreses cave.

Another mechanism of cave formation is when a tumuli is formed by internal pressure of the lava flow (Shick 2009). During the tumuli formation, the internal pressure of the lava flow can be modified by the morphology of the flow and lava rises by piston effect. Fluid-lava can be drained afterward, subsequently crystallization of lava occurs, and a cave can be formed with open fissures at the top of the tumuli (Fig. 18.14b). The Oratorio cave is the only example of tumuli cave in a blocky aa lava flow described in Costa Rica so far.

18.3.2 Mold Caves

Mold caves can be formed in different volcanoclastic deposits (e.g., pyroclastic flow, lahar, debris avalanche), or even lava flows. The most common object for the formation of mold caves corresponds to tree trunks, but molds can be diverse: for example, a rhinoceros created the Blue Lake Rhino lava mold cave (Palmer 2007). The mold caves described in this chapter probably originated with trunks as mold. Figure 18.14c demonstrates a possible origin of a mold cave in a lahar deposit. Hereby first a tree trunk is trapped in the deposit, afterward, the trunk is disintegrated and the cavity is formed. Due to the tropical forest and the presence of relatively big trees in Costa Rica's jungles (i.e., ceibo tree), the Ceibo cave (2 m in diameter, ~ 15 m long) could be one of the biggest single-tree-mold caves reported in the world. Other caves such as Piquete (56 m long) and Ramales (116 m long) are associated with mold cave origin, but more than one tree trunk was likely involved in the genesis, due to the complex morphology. Also, other factors such as erosion or anthropogenic modification could be involved in the final shaping of some of these caves.

18.3.3 Piping Caves in Tuff

Piping caves are formed in unconsolidated and poorly soluble rocks by different processes such as groundwater sapping, seepage face erosion, "tunnel scouring", and other erosive mechanisms (Halliday 2004a). Figure 18.14d shows an example of piping cave formation in two poorly consolidated tuffs, along a geological contact. For this model, two tuff deposits are required, with different welding and

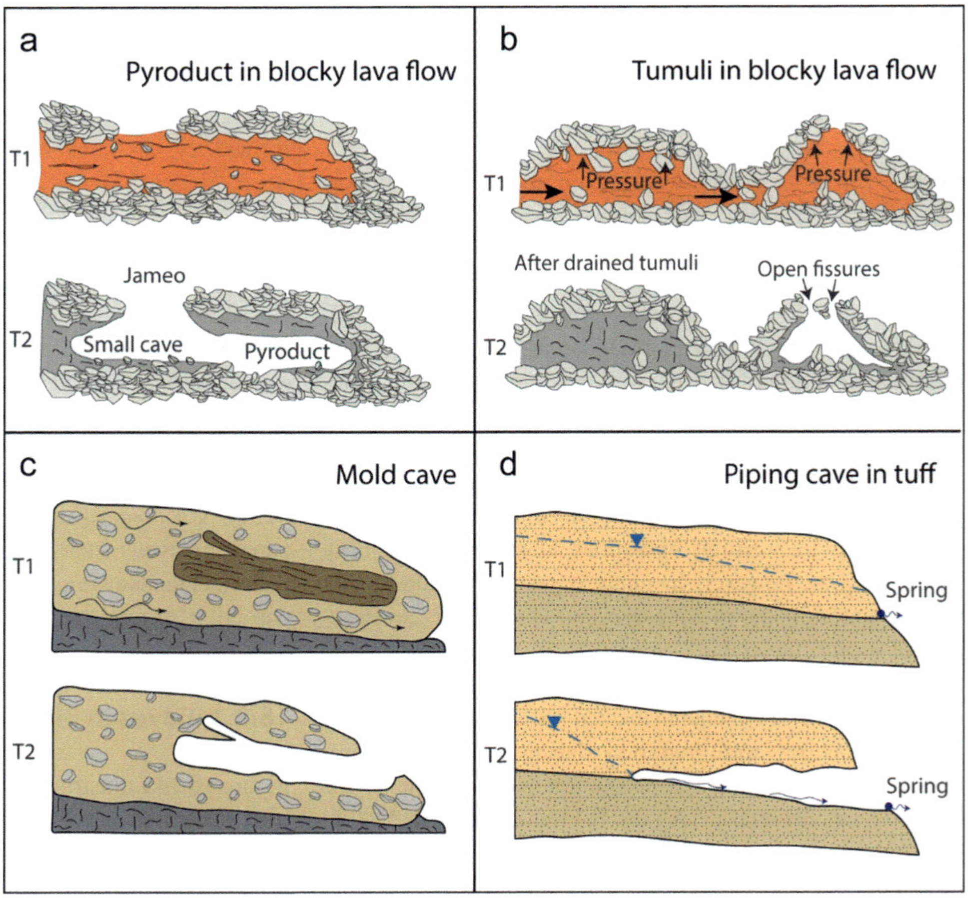

Fig. 18.14 Diagrams showing diverse volcano-speleogenetic mechanisms for cave formation. Detailed explanations are in the text. T1: Cave formation period and T2: cave after cooling, erosion and/or piping formation time. **a** Pyroduct formation in the blocky lava flow. **b** Formation of tumuli cave in the blocky lava flow. **c** Origin of a mold cave in lahar. **d** Formation of caves by horizontal piping along the contact between two different tuff layers

consequent intergranular permeability. The clastic particles of the tuff are removed by groundwater sapping and erosion, preferentially along the contact between the two layers. When the conduit is large enough, water erosion increases the size of the conduit. With more tributaries, the flow increases the hydraulic capacity, so the cave becomes larger in size further downstream. The upstream erosion process is associated with the enlargement of the cave and catchment area of the river. Piping caves can quickly collapse if the diameter of a gallery becomes too big, so they are not common in the geological record. An exceptional case is Los Araya Cave, which is 172 m long.

18.3.4 Caves Formed by Volcanic Landslide

On rare occasions, natural bridges and consequent caves are formed in landslide deposits in confined valleys in volcanic terrains. In such cases, when creeks reestablish again after the landslide, they cut the slide through its base, mainly consisting of an accumulation of blocks and boulders, favoring the erosion at the base of the landslide. Rare, but excellent examples are Puente de Piedra (Puriscal) and Quebrada Escondida (Fig. 18.15).

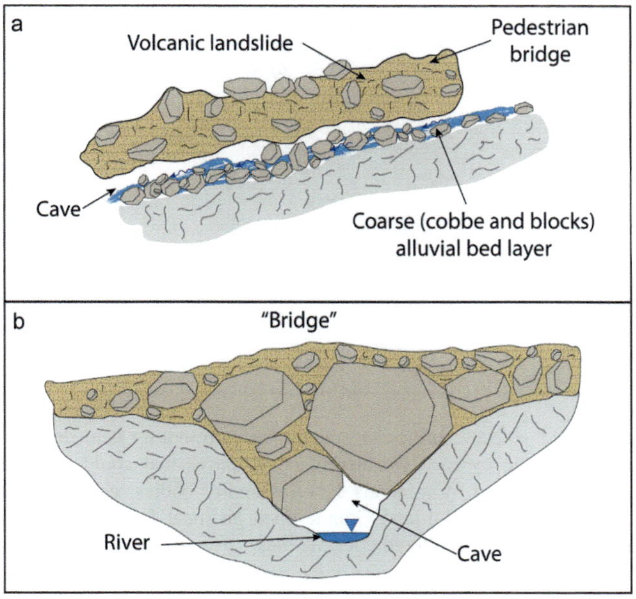

Fig. 18.15 Two diagram examples showing how caves can be formed by volcanic landslides in a narrow valley associated with river flow, such as Puente de Piedra (Puriscal) cave and Quebrada Escondida (Cave). **a** Schematic profile along river flow, covered by the volcanic landslide. **b** Schematic profile perpendicular to river flow, showing how a river can erode and flow under the landslide

18.3.5 Sulfuric-Acid Weathering in Volcanic Environment

Caves formed by sulfuric-acid speleogenesis are well-known in carbonate rocks (e.g., Lechuguilla, Villa de la Luz, Frasassi caves; Hose and Pisarowicz 1999; Engel 2007; Jones et al. 2014; Galdenzi and Jones 2017), but this process is not described for "poorly soluble" volcanic rocks. Based on weathering-dissolution of pyroxene, plagioclase and volcanic glass in acidic environments (Rowe and Brantley 1992), hydrogeochemical analyses that suggest chemical leaching from the thermal springs (Rouwet et al. 2021), direct morphological observations in the Irazú volcanic caves, and evidence of a cavity during drilling in Llano Grande (correlated hydrothermal-altered volcanic rocks of Irazú volcano), it is proposed that very acidic (pH ~ 0–2) and hot water from the hydrothermal volcanic system is capable of weathering volcanic rocks, thus constituting a volcano-speleogenetic mechanism. Certainly, the acidic weathering (including dissolution) of volcanic rocks is not an alone-standing process. It formed caves combined with other common processes such as collapse, high-temperature degasification, and erosion, which occur along local faults. Ulloa (2019) describes the formation of gypsum crusts or alteration rims the Cueva los Minerales (CMI) the Cueva los Mucolitos (CMI) at Irazù volcano. The alteration rims occur when the sulfuric acid reacts with the calcium of the host rock producing gypsum minerals. The gypsum precipitation in the cracks is able to produce extensive forces and promotes the collapse and enlargement of the cave.

Figure 18.16a shows a volcanic environment where sulfuric-acid weathering could occur, based on a simplified model of Irazú volcano. The Irazú volcanic caves (CMI, CMU, and the Cueva Pizote Espantado (CPE)) are partially formed by sulfuric-acid interaction with the host rock. This complex origin for volcanic caves is a new approach in the volcano-speleological literature. A schematic cartoon (Fig. 18.16b) illustrates a simplified model of the pervasive weathering of the volcanic rocks in aggressive hydrothermal environments (temperature > 30 °C, pH < 5, strong ionic force) through aqueous fluids (fumaroles, acid/hyper-acid volcanic brines in crater lakes). This process promotes acid drainage in highly permeable layers, generating open caves, often filled with secondary minerals. Additionally, the Llano Grande open pit presents intense hydrothermal alteration with sulfur-filled dissolution channels (Fig. 18.16c), correlated to the sulfuric-acid weathering process in volcanic environments.

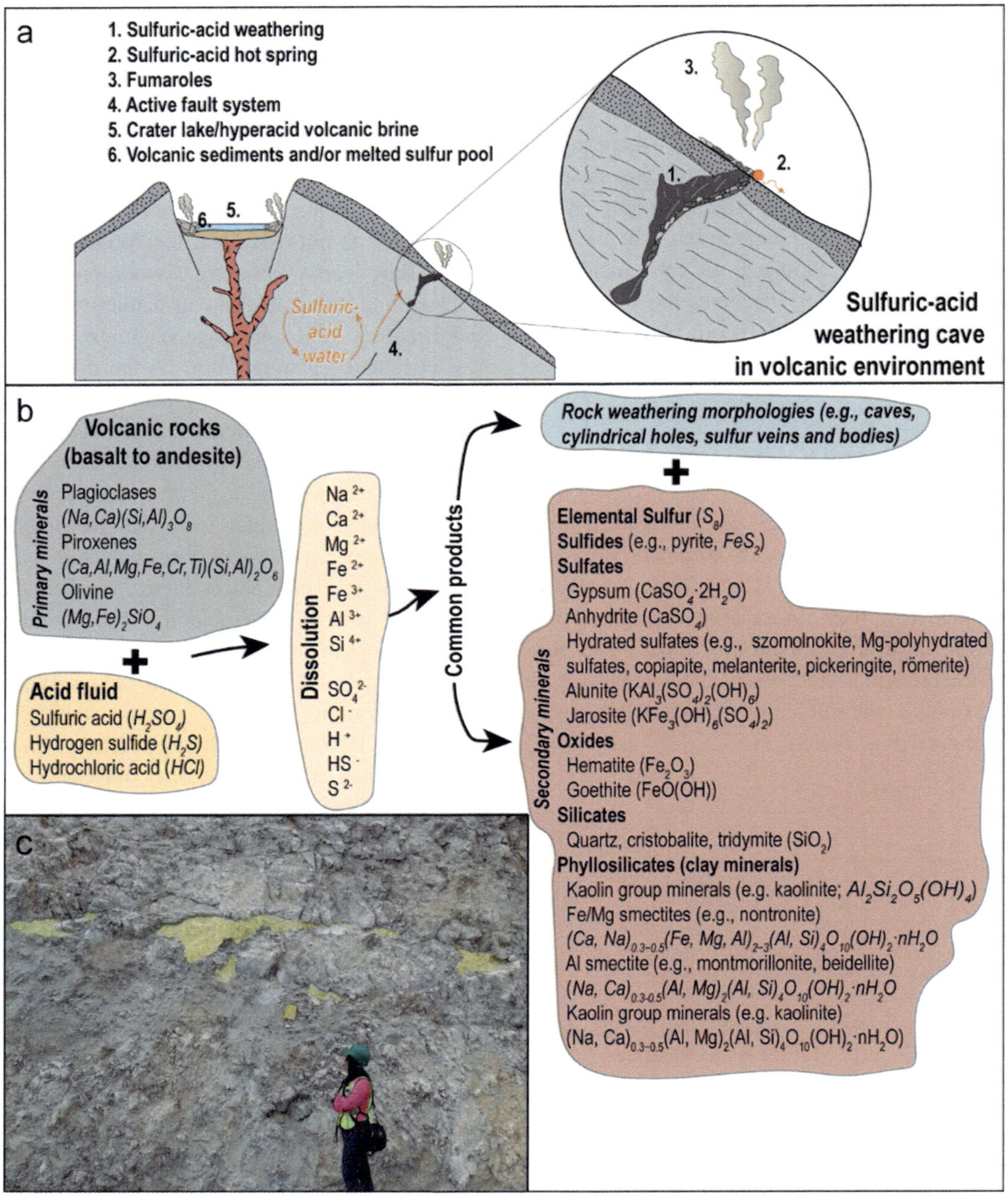

Fig. 18.16 **a** Diagram showing a volcanic environment with the presence of sulfuric acid, where the weathering of volcanic rocks is a feasible origin for volcano-speleogenesis. **b** Cartoon with a simplified model of the pervasive dissolution of volcanic rocks in aggressive volcanic environments through aqueous fluids (fumaroles, acid/hyper-acid volcanic brines in crater lakes) promoting acid drainage and producing by rock weathering open spaces that could generate caves or be filled by minerals as veins and/or bodies. **c** Photograph of sulfur channels in Llano Grande open pit (near Irazú volcano) in highly hydrothermal-altered volcanic rocks

18.3.6 Caves Formed by Plio-Pleistocene Megamammals?

Recently, in South America (e.g., Brazil, Argentina), thousands of caves of tens of meters long, and 1.5–4 m wide have been attributed to biogenic activity (considered paleoburrows). Sub-elliptical cross sections characterize these caves, they are formed in "non-soluble" Plio-Pleistocene rocks, sometimes with evidence of water flow, and some present scratches (interpreted as digging marks). The origin of these caves (considered ichnofossils) is attributed to giant armadillos and ground sloths (Frank et al. 2015; Lopes et al. 2017). Such animals have also been reported in the fossil record of Costa Rica (Lucas et al. 1997). Therefore, the probability exists that some reported caves with semi-elliptical cross sections, smooth walls, and non-water or non-tectonic origin (e.g., three caves in Terminio of Guatuso) were formed by ancient digging species. However, this is only a suggestion, and more detailed studies focused on a more accurate characterization of these caves of uncertain origin need to be conducted to elucidate this hypothesis.

18.4 Conclusions

A total of 90 Costa Rican caves in volcanic rocks are described in the present study. These caves are genuinely heterogeneous in origin. The total length of cave passages reaches ~ 2.2 km. Nevertheless, this number needs to be considered a first approximation because just 23% of the caves in volcanic rocks have a cave map. Water-supply drills have intercepted several volcanic caves in diverse localities in Costa Rica (e.g., Barreal, Santo Domingo, Coyol, Paraiso). Most of the reported cave lithologies in these boreholes are associated with lavas and intercepted voids usually have dimensions less than 5 m. Flooded lava tubes present exceptionally high transmissibility, so they are likely the best sites for water supply of fractured volcanic aquifers. The Rivera cave is the only partially flooded lava tube that might be accessible. This cave is considered one of the most important springs in the area.

We described two caves in aa lava flow of the Cervantes lava field. The Cipreses cave has a pyroduct-like origin. This cave was penetrable for several tens of meters but was blocked with soil, rocks, and agricultural waste by the landowner. The Oratorio cave, on the other hand, is associated with a tumuli cave, developed in a blocky lava flow. Tumuli caves in blocky lavas are not common and, probably, this cave is the first reported in the volcano-speleological literature. We described six mold caves for lahar deposits and one in ignimbrites. All the described mold caves are probably associated with disintegrated tree trunks.

The Los Araya cave is a unique example of a piping cave developed in volcanic tuff. This cave is 172 m long and is formed along the geological contact between two tuff layers. The origin of the cave is associated with groundwater sapping and erosion. Sulfuric-acid weathering in a volcanic environment is presented as a new volcano-speleogenetic model for the Irazú volcanic caves. This speleogenetic mechanism occurs in conjunction with other cave formation mechanisms including tectonism, erosion, high-temperature degasification, and alteration rims with gypsum precipitation associated with extensive forces and collapse of the rock.

In general, most of the caves still need more detailed studies to establish their dimensions and geometry as well as their speleogenesis. For caves developed in soft volcanic rocks, where there is no evidence of formation by water flows, it is possible that they were excavated by prehistoric megamammals and/or anthropic activities (Pre-Columbian and post-Columbian).

Acknowledgements Centro de Investigación en Ciencias Geológicas de la Universidad de Costa Rica supported this work through the projects C0045 and B9707. Our hydrogeologist colleagues: Sandra Arredondo, Ingrid Vargas, Carlos Vargas, Franz Ulloa, Ernesto Echandi, and the driller Sergio Fonseca "Mingo" are thanked for valuable information about volcanic caves intercepted by wells. Thanks to Grupo Espeleológico Anthros for facilitating reports and cave maps of some of the described volcanic caves. The reports and photographs from Ivan Salazar of Cueva del Río Puerto Viejo and Cueva del Río Sucio are highly appreciated. Leonel Rojas is thanked for the report and pictures of volcanic caves in General Hydroelectric project. Amanda Vicent is acknowledged for bringing detailed information of San Ramón and Bijagual caves. Special thanks to Scott Trescott for bringing the location, pictures, and description of Rincón del Jilguero cave, in Rincón de la Vieja volcano. We thank Andrea Hidalgo, Yosselyn Alvarez, and Karina Cerdas for their collaboration in different stages of this work. The reviewers Paulina Vega, Jo De Waele, and Theofilos Toulkeridis are also acknowledged for reviewing the manuscript.

References

Alvarado GE (2021) Costa Rica y sus volcanes, Editorial Tecnológica de Costa Rica, San José, Costa Rica, pp 933

Alvarado GE (2009) Geología de la hoja Fortuna, Alajuela, Costa Rica. Rev Geol Am Cent 41:117–122

Alvarado GE, Denyer P (2019) Estratigrafía volcánica del Neógeno y Cuaternario. In: Denyer P (eds) Perspectiva geológica del noroeste de Costa Rica, Editorial UCR, San José, Costa Rica, pp 101–238

Alvarado GE, Gans PB (2012) Síntesis geocronológica del magmatismo, metamorfismo y metalogenia de Costa Rica, América Central. Rev Geol Am Cent 46:7–122

Alvarado GE, Mora MM, Ulloa A (2013) La caída de "ceniza" proveniente del volcán Irazú (Costa Rica) el 8 de diciembre de 1994: ¿Una explosión freática? Rev Geol Am Cent 48:159–168

Alvarado GE, Vega AE (2013) La geomorfología de la colada de Cervantes, Volcán Irazú (Costa Rica): Descripción de uno de los campos de lava más grandes de América Central. Rev Geol Am Cent 48:99–118

Ávila M (2010) Estudio geofísico de resistividad eléctrica para la Torre 14 P.G. Pailas, ICE, San José, Costa Rica, pp 12 (Internal report)

Bauman A, Bauman O, Petrie G (2022) Tree Mold Caves of Mt. St. Helens. NSS News 80:12–14

Bella P, Gaal L (2007) Tree mould caves within the framework of cave genetic classification. Nat Conserv 63:7–11

Benumof BT, Lockwood JP (2000) Investigaciones de las cuevas marinas de la isla del Coco. Datos de campo, pp 47 (Report)

Bergoeing JP (2015) Geomorphology of Central America: a syngenetic perspective. Elsevier, pp 162

Calvari S, Pinkerton H (1998) Formation of lava tubes and extensive flow field during the 1991–1993 eruption of Mount Etna. Journal of Geophysical Research: Solid Earth 103(B11):27291–27301

De Waele J, Gutiérrez F (2022) Karst hydrogeology, geomorphology, and caves. Wiley, Chichester, p 912

De Waele J, Sanna L, Rossi A (2008) Pseudokarstic cavities in pyroclastic rocks: some examples from north Sardinia. In: Proceedings of the 10th international symposium on pseudokarst. pp 53–62

Deleva S, Mitchel M, Vicente A (2018) Cueva El Duende (Escala 1:200)

Denyer P, Aguilar T, Gazel E, et al (2019) Mapa geológico de la hoja Murciélago [Escala 1:50 000]. In: Denyer P (eds) Perspectiva geológica del noroeste de Costa Rica, Editorial UCR, San José, Costa Rica, pp 360

Denyer P, Gazel E (2009) The Costa Rican Jurassic to Miocene oceanic complexes: origin, tectonics and relations. J South Am Earth Sci 28:429–442. https://doi.org/10.1016/j.jsames.2009.04.010

Engel AS (2007) Observations on the biodiversity of sulfidic karst habitats. J Cave Karst Stud 69:187–206

Espinasa L, Garvey R, Espinasa J, Fratto CA, Taylor S, Toulkeridis T, Addison A (2015) Cave dwelling Onychophora from a Lava Tube in the Galapagos: 16s rRNA sequence and relationships. Subterranean Biol 15:1–10

Ford DC, Williams PW (2007) Karst Hydrogeology and Geomorphology. Wiley, Chichester, UK, p 563

Frank HT, Althaus CE, Dario EM, Tramontina FR, Adriano RM, Almeida ML (2015) Underground chamber systems excavated by cenozoic ground sloths in the state of Rio Grande do Sul, Brazil. Rev Bras Paleontol 18:273–284. https://doi.org/10.4072/rbp.2015.2.08

Galdenzi S, Jones DS (2017) The Frasassi caves: a "classical" active hypogenic cave. In: Klimchouk AB, Palmer AN, De Waele J, Aler AS, Audra P (eds) Hypogene Karst Regions and Caves of the World. Springer, Cham, pp 143–159

Gillieson D (1996) Caves: processes, development and management, Blackwell Publisher Inc., Oxford, UK, pp 324

Halliday WR (2004a) Piping caves and badlands pseudokarst. In: Gunn J (ed) Encyclopedia of caves and karst science. Taylor and Francis Group, New York, pp 1260–1268

Halliday WR (2004b) Volcanic caves. In: Gunn J (ed) Encyclopedia of caves and karst science. Taylor and Francis Group, New York, pp 1624–1633

Harpp KS, Wanless VD, Otto RH, Hoernle K, Werner R (2005) The Cocos and Carnegie aseismic ridges: a trace element record of long-term plume-spreading center interaction. J Petrol 46:109–133. https://doi.org/10.1093/petrology/egh064

Hermosilla RG (2012) The Guatemala city sinkhole collapses. Carbonates Evaporites 27(2):103–107

Hey R (1977) Tectonic evolution of the Cocos Nazca spreading center. Geol Soc Am Bull 88:1404–1420. https://doi.org/10.1130/0016-7606(1977)88<1404:TEOTCS>2.0.CO;2

Hose LD, Pisarowicz JA (1999) Cueva de Villa Luz, Tabasco, Mexico: reconnaissance study of an active sulfur spring cave and ecosystem. J Cave Karst Stud 61:13–21

Jones DS, Schaperdoth I, Macalady JL (2014) Metagenomic evidence for sulfide oxidation in extremely acidic cave biofilms. Geomicrobiol J 31:194–204. https://doi.org/10.1080/01490451.2013.834008

Kempe S (2012) Volcanic rock caves. In: White WB, Culver D (eds) Encyclopedia of Caves, Elsevier Academic Press, San Diego, California, pp 865–873

Kempe S (2019) Volcanic rock caves. In: White WB, Culver DC, Pipan T (eds) Encyclopedia of Caves, Elsevier Academic Press, San Diego, California, pp 1119–1127

Kempe S, Werner MS (2003) The Kuka'iau Cave, Mauna Kea, Hawaii, created by water erosion: a new Hawaiian cave type. J Cave Karst Stud 65:53–67

Kempe S, Middleton G, Addison A, Toulkeridis T, Hoese G (2021) New Insights into the Genesis of Pyroducts of the Galápagos Islands, Ecuador. Acta Carsologica 50(1):53–73

Lopes RP, Frank HT, Buchmann FS, Caron F (2017) Megaichnus igen. nov.: Giant Paleoburrows Attributed to Extinct Cenozoic Mammals from South America. Ichnosan Int J Plant Anim 24:133–145. https://doi.org/10.1080/10420940.2016.1223654

Lucas SG, Alvarado GE, Vega E (1997) The Pleistocene mammals of Costa Rica. J Vertebr Paleontol 17:413–427. https://doi.org/10.1080/02724634.2011.546742

Madrigal R (1982) Geología de Santa Rosa, UNED, San José, Costa Rica, pp 56

McFarlane DA, van Rentergem G, Lundberg J, Umaña Ramirez C, Christenson K (2019) Carbon dioxide emissions from a 'mofette cave', Costa Rica. Cave Karst Sci 46(3):118–120

Miller AZ, Garcia-Sanchez AM, Coutinho ML, Pereira MFC, Gazquez F, Calaforra JM, Forti P, Martínez-Frías J, Toulkeridis T, Caldeira AT, Saiz-Jimenez C (2020) Geomicrobiology of secondary mineral deposits coating lava cave walls from Galapagos Islands (Ecuador). Coatings 10:1134

Miller AZ, Jiménez-Morillo NT, Coutinho ML, Gazquez F, Palma V, Sauro F, Pereira MFC, Rull F, Toulkeridis T, Caldeira AT, Forti P and Calaforra, J. M. (2022). Organic geochemistry and mineralogy suggest anthropogenic impact in speleothem chemistry from volcanic show caves of the Galapagos. iScience, 104556

Montero W, Lewis JC, Marshall JS, Kruse S, Wetmore P (2013) Neotectonic faulting and forearc sliver motion along the Atirro-Río Sucio fault system, Costa Rica, Central America. Bull Geol Soc Am 125:857–876. https://doi.org/10.1130/B30471.1

Mora-Amador R, Alvarado GE, Ramírez C (2004) Caverns in volcanic terrains in Costa Rica, Central América. In: Espinasa-Pereña R, Pint J (eds) Proceedings of the XI International Symposium on Vulcanoespeleology, Pico Island, Azores, pp 78–79

Mora R (2004) El puente de piedra de grecia : ¿ un arco o un puente natural ? Rev Geol Am Cent 31:61–66

O'Connor JM, Stoffers P, Wijbrans JR, Worthington TJ (2007) Migration of widespread long-lived volcanism across the Galápagos Volcanic Province: evidence for a broad hotspot melting anomaly? Earth Planet Sci Lett 263:339–354. https://doi.org/10.1016/j.epsl.2007.09.007

Palmer AN (2007) Cave Geology. Cave Books, Dayton, Ohio, p 502

Protti R (1986) Geología del flanco sur del Volcán Barva: Heredia, Costa Rica. Boletín De Vulcanología 17:23–31

Quesada G (2009) Cueva De Los Murciélagos (Escala 1:250)

Quesada G, Goicoechea C (2021) Informe de Exploración, cueva del río Peñas Blancas. San José, Costa Rica (Report)

Quesada G, Quesada F (2020) La Cueva del Indio, en "Los Cartagos" (CR391). San José, Costa Rica (Report)

Rojas V, Barahona D, Alvarado GE (2017) Geomorfología y petrografía de la colada Ángeles y del cono Monte de la Cruz, volcán Barva, Costa Rica. Rev Geol Am Cent 56:17–35. https://doi.org/10.15517/rgac.v0i56.29240

Rojas W, Alvarado GE (2012) Marco geológico y tectónico de la Isla del Coco y la región marítima circunvecina, Costa Rica. Rev Biol Trop 60:15–32

Rouwet D, Mora-Amador R, Ramírez C, González G, Baldoni E, Pecoraino G, Inguaggiato S, Capaccioni B, Lucchi F, Tranne CA (2021) Response of a hydrothermal system to escalating phreatic unrest: the case of Turrialba and Irazú in Costa Rica (2007–2012). Earth Planets Space 73:1–26. https://doi.org/10.1186/s40623-021-01471-8

Rowe GL, Brantley SL (1992) Estimation of the dissolution rates of andesitic glass, plagioclase and pyroxene in flank aquifer of Poás Volcano, Costa Rica. Chem Geol 105:71–87. https://doi.org/10.1016/0009-2541(93)90119-4

Ruiz P, Suárez C (2017) Evaluación integral de "Puente de Piedra", Grecia, Costa Rica y su potencial turístico. Rev Geol Am Cent 121–147. https://doi.org/10.15517/rgac.v0i57.30151

Sauro F (2014) Structural and lithological guidance on speleogenesis in quartz–sandstone: evidence of the arenisation process. Geomorphology 226:106–123

Shick H (2009) Understanding lava tubes and lava caves, 2nd ed. Trafford Publishing, pp 130

Sojo D (2018) Geología de la hoja paraíso (1:10 000), Costa Rica. Rev Geol Am Cent 97–120. https://doi.org/10.15517/rgac.v59i0.34161
Stanimira D, Ivanov A (2018) Túnel Marino Playa Violin (Escala 1:200)
Trescott S (2018) Cueva De Los Araya (Escala 1:800). Unpublished report
Tschinkel W (1984) Zophobas Atratus (Fab.) and Z. Rugipes Kirsch (Coleoptera: Tenebrionidae) Are Same Species. Coleopt Bull 38:325–333
Ulloa A, Madrigal H (2015) Cueva El Encanto (Escala 1:100). Unpublished report
Ulloa A (2009) Caves of Costa Rica (Central America) and their geologic origin. In: White WB (eds) 15th International congress of speleology. International Union of Speleology, Kerrville, Texas, pp 1930–1935
Ulloa A (2019) Mineralogy and geomicrobiology in active volcanic cave environments in Central America. Ph.D. thesis, University of Nova Gorica, Nova Gorica, Slovenia
Ulloa A, Delgado C (2010) Características geoquímicas y mineralógía del flanco Pacífico Central de la cordillera de Talamanca: Implicaciones para la evolución de una corteza continental. Thesis, Escuela Centroamericana de Geología, Universidad de Costa Rica, San José, Costa Rica
Ulloa A, Quesada G (2010) Informe de exploración de la cueva "Torre 14." San José, Costa Rica. Unpublished report
Ulloa A, Campos-Fernández CS, Rojas L (2013) Cueva los Minerales, Volcán Irazú, Costa Rica: descripción, mineralogía y origen. Rev Geol Am Cent 48:169–187
Ulloa A, Haydock A, Trescott S (2015) Explorations in the Caves of Irazú Volcano and Their Amazing Minerals. NSS News 73:4–11
Ulloa A, Gázquez F, Sanz-Arranz A, Medina J, Rull F, Calaforra JM, Alvarado GE, Martínez M, Avard J, De Moor JM, De Waele J (2018) Extremely high diversity of sulfate minerals in caves of the Irazú Volcano (Costa Rica) related to crater lake and fumarolic activity. Int J Speleol 47:229–246
Vargas R (2001) Geología, Hidrogeología y Modelo Conceptual de Reservorio para la prefactibilidad del Campo Geotermico Pocosol, San Ramon - San Carlos, Costa Rica. Ph.D. thesis, Universidad de Costa Rica, San José, Costa Rica
Wray RAL, Sauro F (2017) An updated global review of solutional weathering processes and forms in quartz sandstones and quartzites. Earth-Science Rev 171:520–557. https://doi.org/10.1016/j.earscirev.2017.06.008
Žáček1 V, Vorel T, Kycl P, Huapaya S, Mixa P, Grygar R, Havlíček P, Čech S, Hrazdíra P, Metelka V, Ševčík J, Pécskay Z (2012) Geología y estratigrafía de la Hoja 3246-II Miramar, Costa Rica. Rev Geol Am Cent 47:7–54
Zimmer MM, Fischer TP, Hilton DR, Alvarado GE, Sharp ZD, Walker JA (2004) Nitrogen systematics and gas fluxes of subduction zones: Insights from Costa Rica arc volatiles. Geochem Geophys Geosyst 5:2–19. https://doi.org/10.1029/2003GC000651

19 Landslides of the Middle Reventazón Catchment—Lithological and Topographic Controls

Giovanni Peraldo Huertas and Elena Badilla Coto

Abstract

Many methodologies developed to establish susceptibility or probability to slope instability prioritize the slope as the main passive factor that triggers movement. However, landslides along the Reventazón River, between Angostura of Turrialba and Siquirres, have been observed on low-angle slopes. These slopes show movements even in areas of less than 20°. They were influenced by the passive factor of lithology. Based on a bibliographic review and field work, the Uscari Formation, composed of sandstones and fissile shales, has been identified as a relevant factor in the generation of many of the landslides that occur on the slopes of the Reventazón River. This lithology, presenting clays and high friability, moves downslope in the presence of water. On the other hand, there are volcanic lithologies and denser colluviums on top of the Uscari Formation, exerting weight on the unstable mass and, therefore, allowing for mass displacements. The areas of the San Antonio, Jesús María, Torito, or Tres Equis landslides represent intensive mass movement processes on gentle slopes, the result of which is very important in the modeling of the relief. The associated destruction has been extreme for the case of the Jesús María landslide, which destroyed in a timespan of two decades, the homonymous town.

Keywords

Geomorphology · Landslide · Lithology · Reliefs · Turrialba · Reventazón

19.1 Introduction

The traditional concept to define landslides refers to a mass of soil, rock, or both moving along one or more rupture surfaces, and this refers to the geological structure of the landslide area, as well as its lithological and mineralogical characteristics. In addition, the moving mass responds to weathering, hydrothermal processes, as well as seismic shocks, hydrometeorological events, hydrogeological conditions, and land use. This article describes the relationship between geology and geomorphology that develops very complex relief patterns associated with processes of instability of the slopes located along the Reventazón River, between the cantons of Turrialba and Siquirres, Costa Rica (Fig. 19.1). In this area, landslides appear on relatively gentle slopes (5–20° on average), which makes landslide recognition difficult through slope instability zonation methods that include slope as a weighting factor.

Historical, geological, geomorphological, and geophysical information was collected to characterize the study area and describe, as best as possible, the areas of complex mass movement. The analysis of the landslides was carried out through remote sensing methods such as aerial photo interpretation, which covers considerable areas to interpret large features of slope instability. It was complemented by field observations focused on small-scale features related to very local conditions and that can hardly be observed in aerial photos or in LIDAR images. In addition, they are ephemeral forms that must be analyzed almost at the moment of their formation. For some cases, the deformations in the landslide area were recorded by taking photos at different moments of the landslide progress.

G. Peraldo Huertas · E. Badilla Coto (✉)
Central American School of Geology and Geological Research Centre (CICG), University of Costa Rica, San José, Costa Rica
e-mail: elena.badillacoto@ucr.ac.cr

A. Quesada-Román (ed.), *Landscapes and Landforms of Costa Rica*,
World Geomorphological Landscapes, https://doi.org/10.1007/978-3-031-64940-0_19

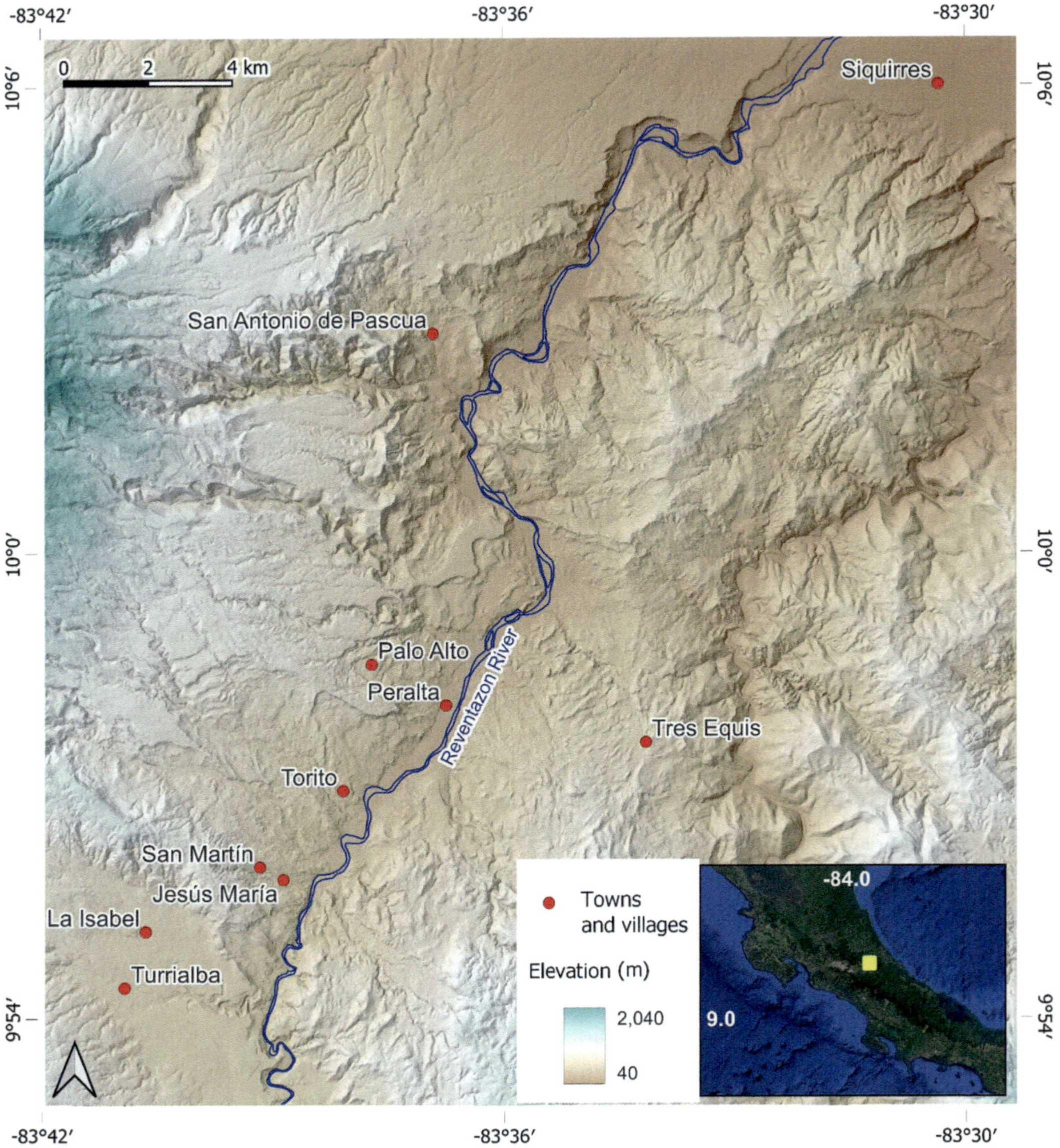

Fig. 19.1 Location of the study area

19.2 Background—Previous Landslide Studies

Segura et al. (2011) carried out a landslide susceptibility zonation between Turrialba and Siquirres along the Reventazón River, based on aerial photointerpretation and bibliography. They considered relative relief and a landslide inventory map in the evaluation of susceptibility to slope instability. The study showed low susceptibility (33%) areas associated with Tuis and Suretka Formations and part of the Andesitas Poás Formation; medium susceptibility (25.3%) developed in the Andesitas Poás, Uscari, and Guayacán Formations; and high susceptibility (13.4%) associated with the Uscari, Río Banano, Andesitas Poás, and Brechas La Unión Formations.

Peraldo et al. (2012) studied the complex mass movement of the village called San Antonio de Pascua, which had affected the socioeconomic development of the territory, including plantations, grazing areas, the railroad to the Caribbean coast, homes, pipelines, and rural and national roads. The associated deposits exhibited gently sloping morphology with small undulations, some of them fan-shaped. Barahona et al. (2013) carried out an analysis of complex landslides of Tres Equis and Chitaría, with a surface area of 32 km^2, and the Pacuare landslide, with a surface area of 28 km^2. They classified them, according to Varnes (1978), as lateral displacements. Geologically, they are characterized as the movement of volcanic rocks of the Doán Formation over the Miocene sequence of sandstones and shales of the Uscari Formation. Through diffractometry tests on sandstones and shales of the Uscari Formation, they identified expansive clays such as montmorillonite and illite. These complex landslides are not appropriately recognized by landslide susceptibility zonation methods because of their low slope.

Calderón (2015) studied the San Antonio de Pascua landslide, located near La Alegría, Siquirres. The local stratigraphy corresponds to the Río Banano Formation (fine friable gray sandstones, and intercalations of grayish sandstones and shales with little fossiliferous content) and the Doán Formation (lithologies of volcanic origin), which overlies the Río Banano Formation. This author compiles historical events on slope instability that have affected the area and recognized the landslide mechanism as the movement of colluvial deposits over the sedimentary strata (translational sliding). Peraldo and Badilla (2015) studied an area of complex mass movement located near Río Torito, Turrialba, where a sector of approximately twenty hectares reactivated in January 2011. They highlighted historical evidence of a reference from 1838, which described ditches on the slopes, suggestive of slope instability. Years later, these areas caused disruption to the railroad service. The geology involved in slope instability are sandstones and shales of the Uscari Formation and volcanic rocks of the Doán Formation.

Peraldo et al (2015) studied the Palo Alto landslide. They found that the north crown is only slightly eroded, contrary to the strongly eroded east-south-east crown, which suggested various phases of landslide formation. Evidence of stream capture was also reported. This landslide currently shows no activity. Caravaca et al. (2017) analyzed slope instability along La Suiza district, especially the Pacuare landslide, using probabilistic and landslide susceptibility zonation methods. They conclude that gently sloping unstable areas are not detected by either of the two methods.

Abarca et al. (2018) studied La Isabel district. They detailed its geology and geomorphology with emphasis on the Jesús María landslide, which destroyed the town of the same name. They pointed to the sandstones of the Uscari Formation as an important lithological factor triggering the movement, even when they do not outcrop in the entire study area. These authors estimated that Uscari Formation underlies the units of volcanic origin and due to its lithological and mechanical characteristics, they cause the volcanic formations to move. By interpretation of geoelectrical profiles, they determined that the center of the population is located on top of a limestone block, possibly in situ, which prevented impact on the houses located on it and which produces differential movements, generating a process of complex instability.

Arroyo-González (2021) analyzed the geomorphology of the Siquirres canton through photo interpretation, map review and field work, to apply it to the cantonal regulatory plan. This author proposed a zonation indicating that 70% of the forms of the canton are of fluvial origin and the remaining 30% correspond to forms of denudational origin. An area of 47 km^2 called "semi-active landslide", has the greatest number of restrictions on urban development, and is suggested for forest conservation. ITC-RAPCA (2000) analyzed natural hazards of the Turrialba River basin, including mass movement hazard. The methodology they used involved the survey of a series of parameters and possible factors, to determine the relationship between relief conditions and landslides; they designated this relationship as indirect. However, the information used is difficult to locate or calculate, such as groundwater flow, or groundwater loss in rock. They classified landslides located as rotational *slumps* and rotational slides.

19.3 Historical Review of Slope Instability

The first references to instability mentioning "derrumbos" (landslides), date from the first and second half of the nineteenth century, due to the opening of a road to the Caribbean coast and the railroad track, which caused reactivation of landslides. They appeared in newspapers from different periods (Peraldo and Rojas 1998, 2000). In 1838, the surveyor Henry Cooper, described the area where it was thought possible to build a road to the Caribbean. He refers to an opening in the mountain, near the source of the Torito River, occupied by a 1200 m long wetland covered by "gamalotillo" (type of vegetation), and cracked land. Peraldo and Rojas (1998) interpreted it as areas of instability near Alto Churuca. As of 1899, there were reports in the press about landslides in Las Lomas and other points close to the railway line, being the main cause of delays in the railway service and the constant destruction of sections of the railway line (Peraldo and Rojas 1998). In the documents of the Northern Railway Company, by October 11, 1899, the line is reported in good condition, except in "*Lomas slide section …*" (Peraldo and Rojas 2000).

The complex mass movement areas along the Reventazón River have caused economic losses in housing and agricultural activities, such as the destruction of the Jesús María village by the homonymous landslide; its inhabitants, mostly farmers, migrated (Abarca et al. 2018). This means family and community uprooting. Table 19.1 shows historical information on slope instability for the study area. Mora (1989) characterized the Reventazón-Parismina basin and refers to mass movement processes for various sectors but does not consider the areas of strong mass movements of Turrialba-Siquirres, despite the historical damage to the railway line, roads, and homes. The Cantonal Human Development Plan (Municipality of Turrialba 2016) did not characterize mass movements that

Table 19.1 Chronology of landslides in the study area

Month-Day-Year	Description	Approximate location
1838	Cracks, swamps associated with an opening in the mountain	Alto Churuca, Torito River
09-08-1866	Landslides cover the road	Bóvida, Angostura, Calvario
12-18-1903	Landslides obstruct the road, Peralta, and Las Lomas	Las Lomas—Peralta
07-03-1908	Landslide crushed a house, one injured	Las Lomas
12-27-1908	"…slide fell on the machine and the driver was injured."	Las Lomas
09-01-1914	"A lot of the mountain opened. In 5 days, the train service will be restored"	Torito River
07-28-1920	Derailment, the driver, and the stoker died because of the rains. … landfills and part of the terrain will fail	Mile 45, near Las Lomas
11-27-1956	The railway service has been interrupted due to landslides; the problem is the groundwater that causes internal washes that instabilize the hillside	Piedras de Fuego, Las Ánimas
10-1957	Subsidence of one meter in the railway due to landslides. Water coming down the hills washes away the railway	Piedras de Fuego, Las Ánimas
06-09-1960	Landslides between Jesús María and Santa Teresita	Jesús María, Santa Teresita

Source Peraldo and Rojas (2000)

exist in various sectors of the canton either, even though they constitute permanent hazard. On the respective map, there are just few landslides and, for example, Jesús María landslide or those existing in the Torito River sector, are not included.

The Costa Rican Electricity Institute (ICE) reported that in July 2012, a significant flood occurred, and by the end of that year, a landslide, previously stable in the Reventazón dam spillway, was reactivated due to the presence of weak material lenses. This necessitated an excavation of 1 million cubic meters to rehabilitate the ground. Landslides pose a geological risk to the project (Instituto Costarricense de Electricidad 2017).

19.4 Relationship Between Geology and Geomorphology

19.4.1 Brief Geological Context

The study area belongs to the South Limón back-arc basin of south-east Costa Rica. Starting in the Campanian, the basin received various oceanic sediments until the Pliocene, when it began to fill with continental sediments (Fig. 19.2) (Bottazzi-Basti et al. 2021).

During the Paleocene to the middle Eocene, thick hemipelagic and volcaniclastic sediments, associated with submarine volcanism, were deposited. In the Late Eocene, coral structures were built and the exoskeletons of foraminifera and macroforaminifera accumulated, giving rise to Las Ánimas limestone. The construction of carbonate structures continued during the upper Oligocene to lower Miocene with the Jesús María limestones of Punta Pelada Formation (Fig. 19.3a, b). Mora (1989) brought together Fila de Cal limestones with Jesús María limestones. During Early Miocene to Pliocene, oceanic siliciclastic sands and muds were deposited and formed Uscari and Río Banano formations (Fig. 19.3c–h). Uscari Formation includes semi-rounded decimetric concretions of brown tones. Río Banano Formation is similar to the Uscari Formation in terms of lithology but presents a large number of fossils and rounded igneous extraclasts, suggesting a littoral depositional environment influenced by fluvial activity.

In the Uscari Formation, Barahona et al. (2013) observed facies of fine sandstones and brittle calcareous shales with trace fossils and parallel lamination (Fig. 19.3c, d). Mora (1989) describes this formation as the intercalation of clastic sedimentary rocks with fossiliferous calcarenites. He indicates that they are quite weathered and generate montmorillonite clays.

The Río Banano Formation can be associated with a shallow depositional environment. It includes fine, medium, and coarse sandstones, interbedded with shales and carbonate rocks. It presents fossiliferous levels with well-preserved exoskeletons of gastropods (Fig. 19.3e–h, 10° 1′ 11″ N 83° 36′ 24″ W), where there are also levels of rounded decimetric blocks of andesitic lavas. Well-preserved oysters also occur.

During the Pliocene, shallow littoral sedimentation formed sandbars, reef banks, and paralic zones. At the end of the Pliocene and during the Pleistocene, the basin captured continental sediments derived from andesitic and basaltic lavas, which were poorly sorted conglomerates with sandy matrix (Sojo 2018) that may contain lenses of medium to coarse sandstones with siliceous cement (Colima River, 9° 59′ 4″ N 83° 36′ 57″ W). These deposits are the

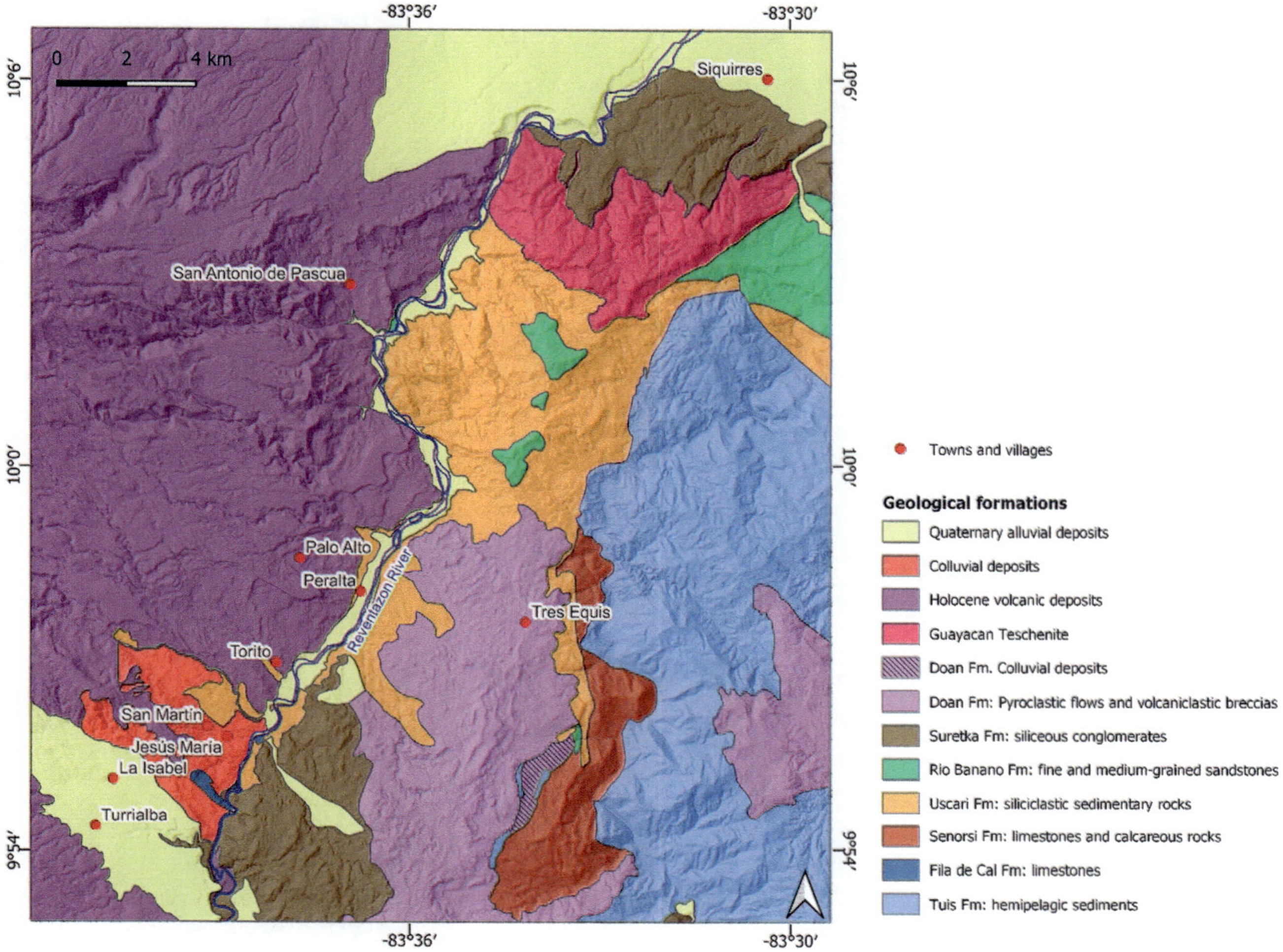

Fig. 19.2 Geological map of the area between La Angostura and Siquirres. Integrated map based on geological information provided by Fernández (1987), Barahona et al. (2013), Calderón (2015), Caravaca et al. (2017), Abarca et al. (2018), Segura (2009), Linkimer (2003) and own observations

product of the Talamanca Mountain range uplift (Suretka Formation) and intense volcanic activity. These materials are typical of the activity and destruction of the Pliocene volcanic arc; they are flows of blocks and ash possibly from volcanic avalanches (Doán Formation) (Fig. 19.3i, j).

Barahona et al. (2013) indicated that in the Tres Equis district, the Doán Formation is widely distributed and described it as chaotic breccia deposits and associated pyroclastic flows. It has a possible lateral stratigraphic relationship with the Suretka Formation (Sojo 2018). However, in the study area it is difficult to recognize both formations, because the epiclastic deposits associated with the Doán Formation are similar to the Suretka Formation deposits.

From the Pleistocene to recent, the unit containing andesites, scarce basalts, and lahar deposits (Fernández 1987) derived from the Central Volcanic Cordillera has been emplaced. The Miocene sedimentary lithologies crop out in small patches within the study area, whereas the volcanic and Plio-Pleistocene lithologies have much greater coverage.

19.4.2 Landforms Related to Slope Instability Processes

The geomorphology of the area is broadly characterized by two zones: (1) in the western sector there are moderately steep volcanic slopes, slightly to moderately dissected, associated with recent volcanic deposits (andesitic lava flows, avalanche deposits, lahars, and pyroclastic fall deposits) (2) toward the Reventazón River, relief is rougher due to intense erosive processes. Well-defined escarpments surround morphologically complex areas of moderate to gentle slopes with large landslides that constantly transform the landscape (Fig. 19.4).

Within the relief, ancient crowns of different sizes are recognized as the limit between slopes that have not yet been moved and those that have slipped (Fig. 19.5a). These areas show recent movements affecting deposits of the old landslides. In the Río Torito complex mass wasting area, recent crowns and deposits are observed (Fig. 19.5b).

Fig. 19.3 Lithologies present in the study area: **a** Jesús María limestone quarry. **b** Corals of Jesus María. **c** Uscari Formation, appearance of the rock mass and rock fracture plane. **d** Uscari Formation, concretions. **e** Vertical sequence, Río Banano Formation, Bonilla River, road 412 bridge. **f** Rounded lava clasts, Bonilla River, road 412 bridge **g** Gastropods, Río Banano Formation. **h** Turritella, Río Banano Formation. **i** Detail of the Doán Formation, Río Torito landslide. **j** Medium to fine sand lenses in the Doán Formation. *Photos* G. Peraldo

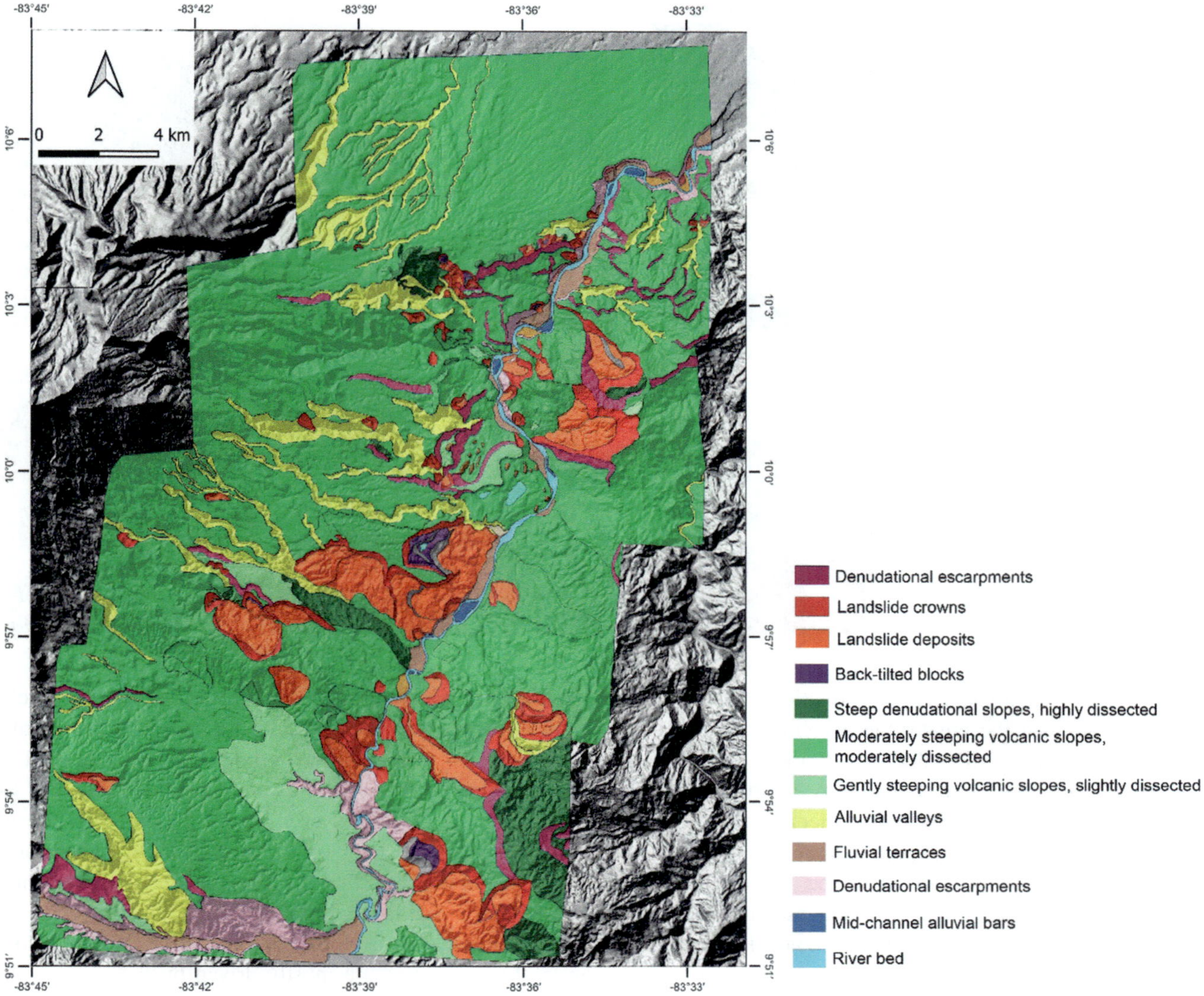

Fig. 19.4 Geomorphological photointerpretation of the sector of the study area most strongly affected by landslides

Back-tilted sectors and gently sloping flat-undulating terrains occur, associated with terrains affected by movements along concave failure surfaces, suggesting a rotational type of landslide; they are present in different sizes (Fig. 19.5c), from decametric to metric (Fig. 19.6a, b). The size of these tilted blocks can provide an indication of the depth of the failure surface. Strongly eroded and rounded micro-reliefs also stand out (Fig. 19.5d).

Depressions created by differential movements in a complex mass wasting area can be filled by water from shallow groundwater levels intersected by block movements in the landslide or by accumulated precipitation. Figure 19.6c shows a small pond formed in the Torito landslide, and Fig. 19.6d, the Palo Alto lagoon, formed in the interior of a moved and slightly tilted block. The microrelief also suggest stress patterns exerted on the unstable mass, partly due to the relief of the sliding surface. In Fig. 19.6e, a region subjected to tensional stress is observed, forming graben-type structures, and in Fig. 19.6f, a wall of soil lifted by compression in the San Antonio landslide is observed. The deposits of landslides can appear as wavy reliefs with gentle hills, such as the shape seen in Fig. 19.6g. Strong creeping may also occur as shown in Fig. 19.6h.

19.5 Possible Landslide Mechanisms

There are three geological formations that allow for movement on gentle and moderately steep slopes in the study area. Caravaca et al. (2017) recognized an important factor of movement in the Uscari Formation. Barahona et al. (2013) applied the RMR (Rock Mass Rating) criteria (Bieniawski 1989) and assigned a lithological susceptibility index of 5 to the Uscari Formation, and a value of

Fig. 19.5 Morphological features on landslides of the study area: **a** Palo Alto landslide, north crown. **b** Torito landslide crown. **c** Back-tilted block in the Palo Alto landslide. **d** Gently sloping flat-undulating terrain within the unstable area of Río Torito, formed by old deposits. *Photos* G. Peraldo

4 to the Río Banano and Doán formations, meaning the higher the number, the greater the instability. Abarca et al. (2018) assigned an RMR value of 2 to the Andesitas Poás Formation, indicating a regular quality of the rock mass. Lithologies of the Uscari Formation incorporate highly friable clays, which, in the presence of water, can expand and deform plastically. The mounds of the Uscari Formation observed in the Torito landslide suggest such behavior. These have an almost symmetrical conical shape (Fig. 19.7a, b), with moderate to steep flanks and small fringes, which may be an evidence of such expansion. These forms can be confused with the anthills built by a specific type of ants called "zompopas"; however, they show significant differences (Fig. 19.7e). Mora (1989) suggested that the sandstones described for the Uscari Formation generate montmorillonite. Barahona et al. (2013) found montmorillonite and illite by X-ray diffractometry in the Uscari Formation. These clays are expansive, as verified by Peraldo (2000) by a diffractometric analysis of a clay from the Llano de la Piedra slide in Los Santos area, Costa Rica; it turned out to be montmorillonite with a expansion rate of 19%.

Recent volcanic lithologies cover large parts of the study area compared to sedimentary formations. However, the latter extend over an unknown area below the Plio-Pleistocene to recent deposits, forming a lithological cushion susceptible to instability. Regarding density, the volcanic colluvium has a unit weight of 18 KN/m^3, against 23 and 26 kN/m^3 for sandstones and 22 and 26 kN/m^3 for shales of the Río Banano Formation (Calderón, 2015). The average value of unit weight for the andesites is 27 kN/m^3 (https://geologiaweb.com/rocas/andesita/). This stratigraphic relationship between denser rocks on top of less dense lithologies is similar to that when a group of bricks is placed on top of moldable clays. Stresses are generated that allow for the movement of slopes, even when low-angle slopes are predominant. Landslides on gently inclined slopes demonstrate, in these cases, that the slope angle is not the main passive factor that favors instability.

Therefore, with a lower influence of gravity, slope instability processes are mostly driven by the passive factor of lithology, rock geotechnical properties, stratigraphic arrangement, and the amount of water in the subsoil. Barahona et al. (2013) recognized the stratigraphic characteristics as a factor in the movement of slopes. If we consider that the lithologies of Suretka, Doán and recent volcanic deposits lie on top of the Uscari and Río Banano Formations, they exert stress on the unstable mass of these sedimentary lithologies, allowing for lateral movement for two reasons: either 1. Higher density of the rock that overlies the Uscari and Río Banano Formations, or 2. Well-defined planes in the sedimentary rock that allow its most weathered part to flow, moving the overlying lithological cover by default.

When observing the boulders at the confluence of the Torito River with the Reventazón River, clear differences can be made regarding the resistance of lithologies associated with the Uscari Formation and those associated with recent lavas. Figure 19.7c, d compare the weathering of fine- to medium-sandstone cobbles from the Uscari Formation that present fibrosities that grew on their surface or fissures, and recent lava cobbles that only show oxidation

Fig. 19.6 Microrelief of landslide terrains: **a** Back-tilted block, Torito landslide. **b** Jesús María landslide back-tilted blocks. **c** Small pond due to damming of exposed groundwater and accumulated precipitations, Torito landslide. **d** Palo Alto Pond, possibly due to a rupture of water tables. **e** and **f** Areas subjected to stresses: E shows tensional stress that cause the formation of grabens (Torito landslide) and F shows compression sites that lift blocks (San Antonio landslide). **g** Torito landslide deposit. **h** Creeping processes in San Antonio landslide crown. *Photos* G. Peraldo

patinas of their ferromagnesian minerals. The presence of expansive clays in the Uscari Formation leads us to think that the effect of clay expansion is another factor to consider in the movement mechanism.

When analyzing the triggering mechanism, fluvial dynamics and fluvial erosion of the fronts of unstable areas should not be neglected. Swaying of a river bed can leave areas inactive for decades (Fig. 19.8a). Due to climatic activity or simply due to changes in its sedimentation pattern, the river can resume its erosive activity in front of a landslide, causing its reactivation. These fluctuations are usually recurrent. In the case of the river bank shown in Fig. 19.8a, the concrete walls and iron structure are evidence of strong erosion caused by the Reventazón River in past times, since they were built to protect the railway line (Fig. 19.8b). A flood in 2012 caused this river to resume undercutting of the front of the Torito landslide area, destroying these old concrete walls (Fig. 19.8c–e). The

Fig. 19.7 Features related to weathering of lithologies with expansive clays: **a** and **b** Mounds of the Uscari Formation, associated with the expansion of montmorillonite clays. **c** and **d** Comparison between clasts of lavas (not weathered) and rocks of the Uscari Formation. Note the degree of destruction of the Uscari clasts in relation to the lavas. **e** Mounds build by "zompopas" in the Torito landslide. *Photos* G. Peraldo

consequences of this flood were discussed in a session of the Municipality of Siquirres, according to agreement 1760-22-10-2012 (Municipality of Siquirres 2012). The Guayabo river has also changed its course, leaving the railway bridge abandoned (Fig. 19.8f).

Another factor that contributes to the shifting of a riverbed is mudflow deposition associated with the activation of a landslide. Such was the case of the Torito River, which moved its active channel due to the deposition of material originated by the activation of the Torito landslide (Fig. 19.8g). This forced the river to move toward the front of the active landslide. It undermines the frontal part of the landslide deposit, forming cracks that delimit blocks that fall into the river, causing dams that later break, generating mudflows toward the Reventazón River (Fig. 19.8h). Another factor is the behavior of groundwater, which can be a very important factor in causing ground movement, as it acts as a lubricant on rupture surfaces and weakens the cohesion of soil particles (Mora 1985).

19.6 The Slope Factor

The slope factor has been weighted in landslide zoning methodologies, however, in the case of landslides located along the Angostura-Siquirres corridor, it is relatively less important among the triggers. In studies that have been carried out in the area, landslide areas remain invisible on maps showing results of the application of methods such as landslide susceptibility or probability; even morphometric methods do not recognize such areas, although the use of photogrametric models generated from drone images might be extremely helpful. In the slope map of the study area (Fig. 19.9), a limited representation of steep or very steep slopes (in red tones) can be observed, whereas low- and medium-angle slopes are represented more extensively (in green and yellow tones).

Figure 19.10 shows the location of the landslides, whose hillshade 3D views can be seen in Fig. 19.11 and 19.12. The

Fig. 19.8 Morphological and erosive evidence of changes in the Reventazón Riverbed **a** Abandoned sector of the Reventazón Riverbed next to the confluence with the Torito River. Vegetation growth is observed. **b** Concrete structure to support the railway line. **c** Riverbank erosion reactivated due to July 2012 flooding. **d** View of the affected riverbank and partial destruction of protective concrete walls. **e** Fragments of the protective wall in the river. **f** Guayabo River bridge and channel in 2014. **g** Mound of material deposited during the activation of the Torito landslide in 2010. **h** Front of the Torito landslide affected by erosion of the river of the same name. *Photos* G. Peraldo

predominance of low-angle slopes within the landslides (between 10 and 15°) is clearly shown in the topographic profiles that accompany each 3D view. These landslides are also very large areas within which complex movement is recognized. Tres Equis (Fig. 19.11) is the one with the lower slopes, so when the landslide is visited, almost flat slopes of 10° on average are observed, as reported by Barahona et al. (2013). The same is true about the landslides of San Antonio (Fig. 19.10), Torito and Jesús María (Fig. 19.11).

The profile of the Palo Alto landslide (Fig. 19.11) shows breaks in the relief and back-tilted blocks that, because of the movement, become practically horizontal. In the first break, the Palo Alto Pond is located. Thus, in the particular case of the study area, slopes are no longer a fundamental factor in the movement. Rather, the lithological factor, which ranges from mineralogy to density and stratigraphic relationships, becomes of primer importance for landslides located in low-sloping areas.

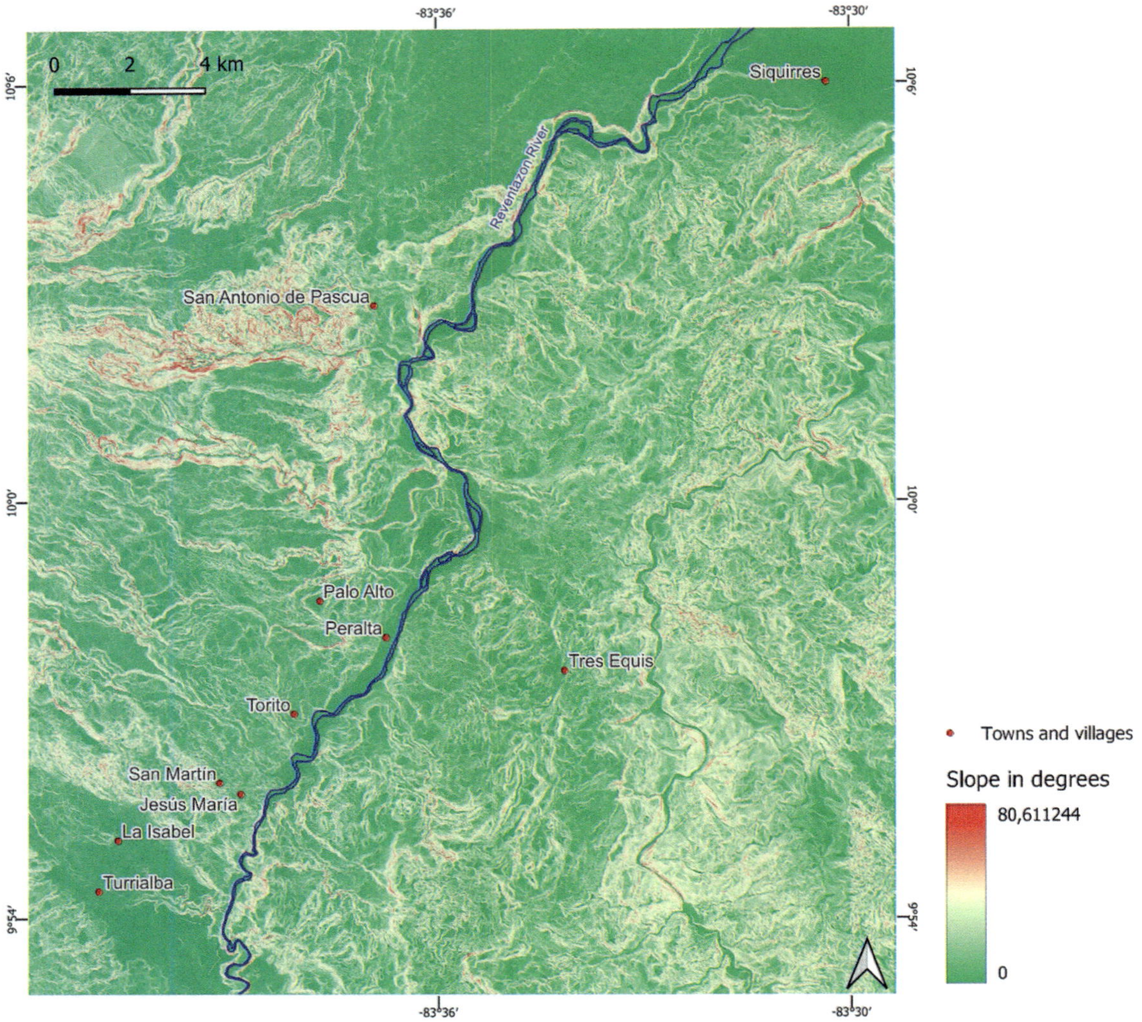

Fig. 19.9 Slope map of the study area. Low and medium slopes dominate over steep slopes

19.7 Classification

Following Varnes (1978), these forms of movement should be classified as lateral spreads, where extensive efforts predominate. Alcántara (2000) points out that these processes occur when there is liquefaction or fluidization of the underlying material, where the shear surface is not clear, and indicates that these processes are common in lacustrine and shallow marine areas or in areas of former ice caps. In the case of the landslides studied, the cause is the fluidization of the sedimentary substrate of the Uscari Formation.

Peraldo et al. (2012) defined the San Antonio de Pascua area as a case of complex landslides that involve debris deposits. Alcántara (2000) defined complex movements when the type of initial movement is transformed into another one as the failed mass moves down the slope. In this chapter, it is preferred to call them complex mass wasting processes, since it is not only complexity when the type of movement is transformed during displacement, but also in terms of the movement itself, because it produces various micro-reliefs that testify to different directions of movement, multiple surfaces of rupture and internal blocks of independent movement.

19.8 Case Study: Jesús María Complex Mass Wasting Area

19.8.1 History and Background

Since 2007, the area of Jesús María presents significant mass wasting processes due to the destruction and morphological changes that have occurred. This area has a history of activity dating back to the opening of the railroad to the Caribbean coast. The Jesús María landslide has had important movements in the past, especially in the sector of

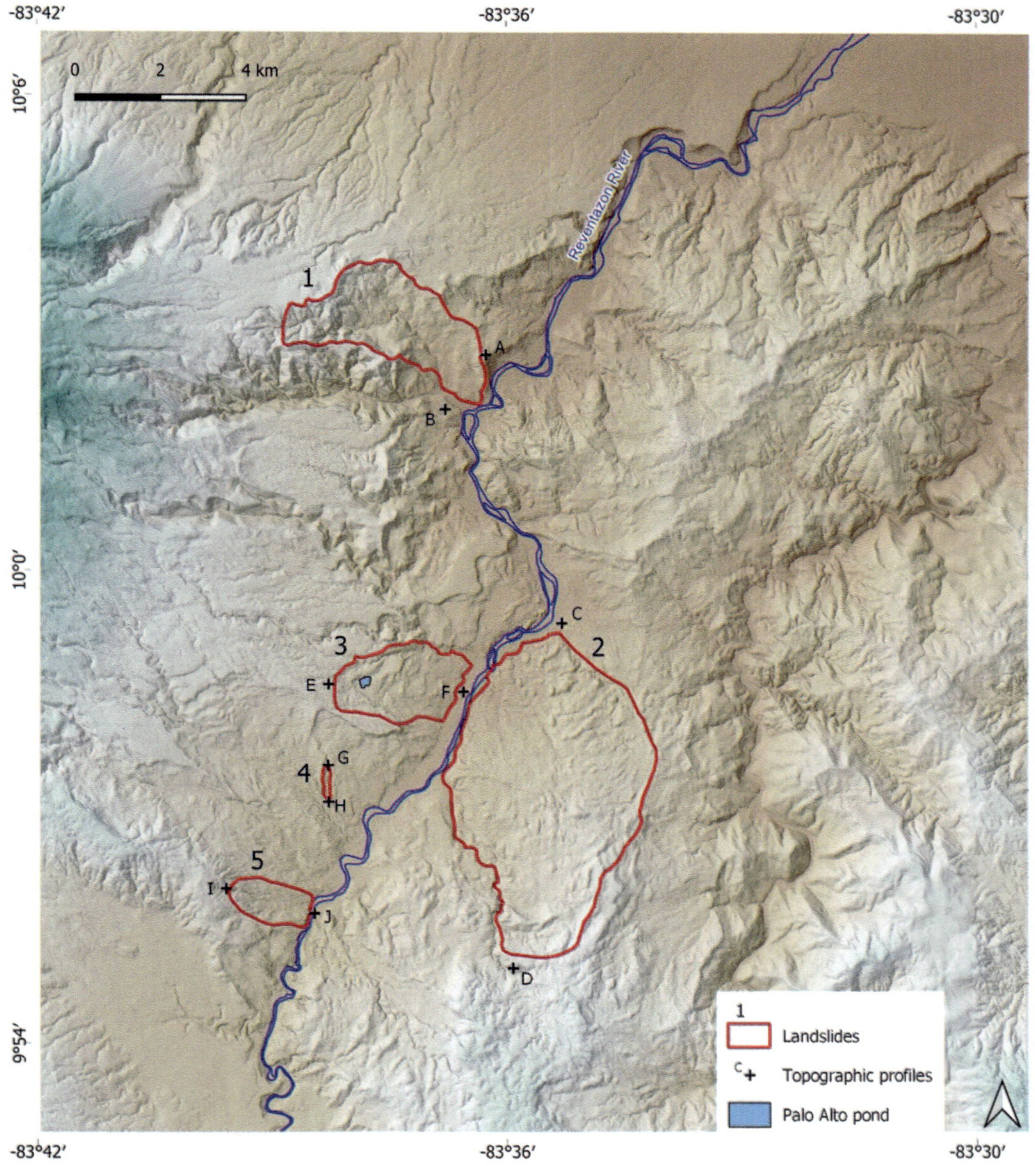

Fig. 19.10 Locations of landslides and topographical profiles detailed in Figs. 19.11 and 19.12. (1) San Antonio de Pascua landslide, (2) Tres Equis landslide, (3) Palo Alto landslide, (4) Torito landslide, and (5) Jesús María landslide

the old railway line along the Jesús María river, as demonstrated by some decametric structures of concrete and rails, possibly built by the Northern Railway Company in order to hold the ground near the railway line, currently removed by the landslide. A concrete structure was also built to support route 415, in the town of Las Brisas, as well as a cement wall, located outside the currently active area, which was built to stop landslide processes obstructing the rail road (Fig. 19.13a–c). In 2007, the landslide was reactivated, destroying houses, which led to the intervention of the National Commission for Risk Prevention and Emergency Response (CNE, by its historical acronym), through an inspection to find out and analyze the situation in the town. A thickness of 20 m is currently known for this landslide at the site of Jesús María. The slide has an irregular shape and covers an area of approximately 1.3 km^2 (Abarca et al. 2018), although not all of it remains active. Segura et al. (2011) calculated a volume of 9,848,454 m^3. Abarca et al. (2018) apply deterministic and probabilistic methods to the area they studied, particularly to the Jesús María landslide, and conclude that they did not observe a direct relationship between the results of these methods and the instability present in the area. The probabilistic method shows a high probability of landslides in the crowns; however, the instability in areas such as Jesús María involves movement in the deposit. It is important to indicate that Barahona et al. (2013) reached similar conclusions, using the same methodologies, for the Tres Equis landslide and Caravaca et al.

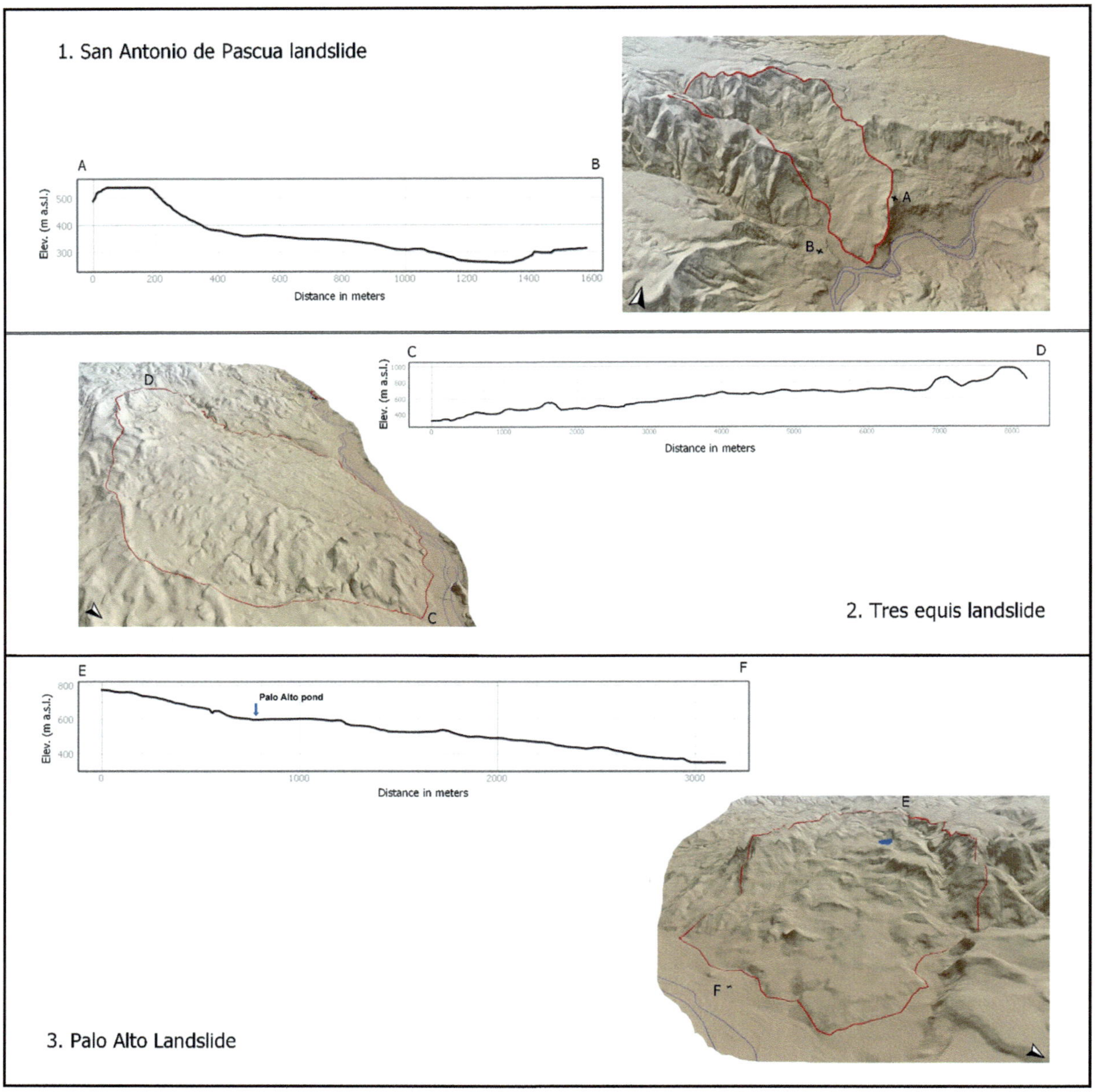

Fig. 19.11 Hillshade 3D views and topographical profiles of the San Antonio de Pascua, Tres Equis, and Palo Alto landslides

(2017) for the Pacuarito landslide, both in the canton of Turrialba.

19.8.2 Movement Direction

In a small slide, the direction of movement is usually the same for the entire slide area. However, in areas of complex mass wasting, it is preferable to speak of directions of movement. The irregularities in the slip surfaces result in the existence of blocks that move in different directions, and this is determined according to the differences in the direction of the cracks formed. The direction of the slipped mass can be observed in striations preserved in the crack planes and in tension in natural elements such as taut roots (Fig. 19.14).

Striations on the surface of the crack planes show the direction and sense of movement of the sector where they are located. In Fig. 19.14c, the striations have a small vertical component, with a rake angle of 10°, this means that the initial movement shown by the striations is preferably horizontal along the fracture. The tense root is an element that has not been given the importance it requires; it is useful as a temporal criterion for crack formation (Peraldo 2000); it is stretched and has a 45° inclination (Fig. 19.14d), which indicates a movement with a larger vertical component that

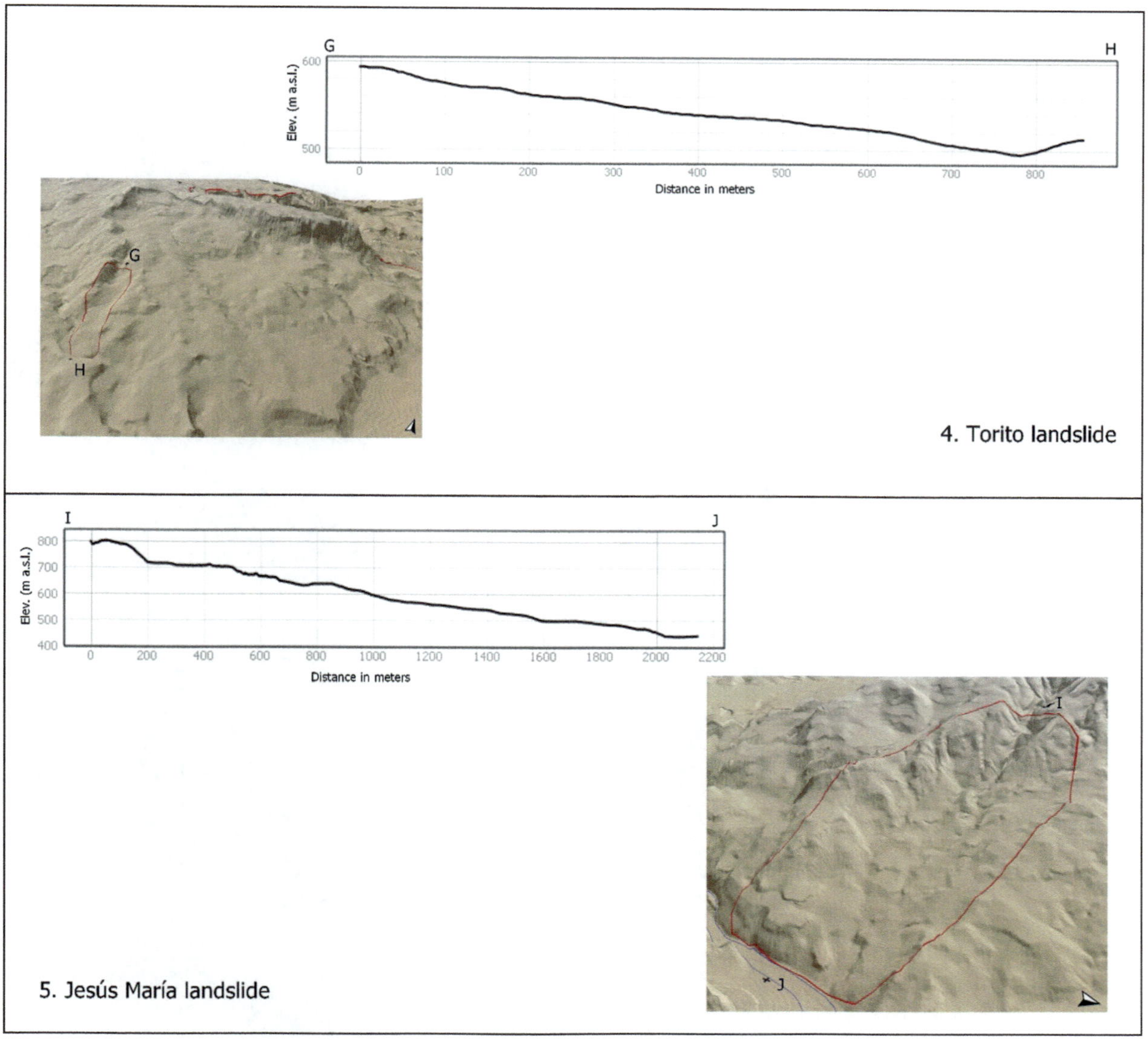

Fig. 19.12 Hillshade 3D views and topographical profiles of the Torito and Jesús María landslides

explains the collapse of the houses affected by an oblique movement and gave rise to a 4 m high escarpment heading E-W (Fig. 19.14c, d). These observations are important when the movement vectors could not be measured by geodetic methods or striations, so the direction of the tense roots show the complex movements in a complex mass wasting area such as Jesús María.

19.8.3 Effects on Infrastructure and Microrelief Associated with the Geology of the Landslide

The surface dynamics present in the Jesús María landslide allow the relief to also become dynamic, meaning it changes constantly. As a result, depressions, among other forms, are formed and transformed over time due to rotational movement (tilted blocks) or translational movement (grabens): these depressions are occupied by small bodies of water. Since some of this geoforms have been formed during the years of observation of the landslide, they serve to calibrate similar microrelief and thus help in the recognition of currently inactive mass removal areas.

Abarca et al. (2018) carried out two geoelectric profiles in a downtown area not affected by the landslide and recorded resistivities greater than 25 Ωm, whose interpretation assumes a block from Jesús María limestone that is located downtown. Angular blocks of this limestone can be seen in the center of the town of Jesús María. Also, in the Jesús María River, there is an outcrop of a colluvium formed by subrounded blocks of limestone with metric diameters, in a section of the river affected by the landslide. The landslide moves around this limestone block, allowing

Fig. 19.13 Structures built to halt the movement of the Jesús María landslide. **a** Part of the structure built to hold active areas of the Jesús María landslide so that they would not affect the railway. **b** Structure that supports the Las Brisas section of route 415. **c** Retaining wall built in a currently stable landslide area, to save the railway line from ground movements. *Photos* G. Peraldo

for the observation of areas of distension and compression, which results in the type of existing microrelief.

A first microrelief that was observed since its formation is a block generated by compression when the western block of the landslide moved toward the immobile central block. This causes the lifting of the west block, generating a shape similar to a back-tilted block (Fig. 19.15g–i). Another microrelief is an escarpment formed from an E-W strike fracture, generated by the movement of the north block toward the east in relation to the immobile block.

The damage to infrastructure is by sectors, since the temple and several surrounding houses have not been affected because they are located on top of the immobile block of limestone (Fig. 19.15d). A row of houses located to the north of the immobile block was destroyed by a combination of eastward horizontal movements (about 20 m) and vertical movements (about 5 m) (Fig. 19.15a–c). The infrastructure to the west of the immobile block was destroyed in situ, and the side street was affected by a movement of the ground that deformed and tilted the street (Fig. 19.15f,

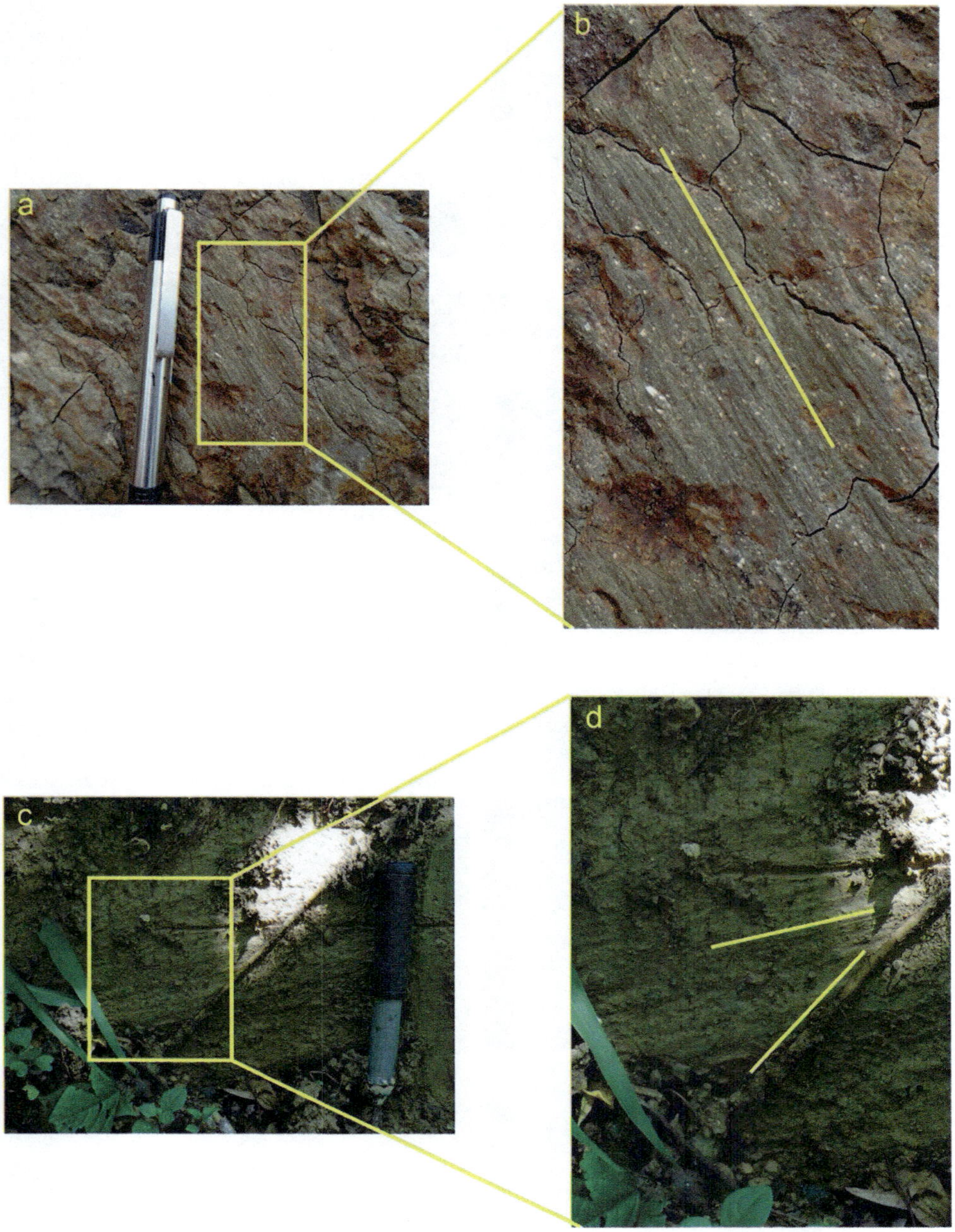

Fig. 19.14 Cinematic evidence of the movement of the Jesús María landslide. **a** Landslide plane in Uscari, road cut in Las Brisas. **b** Detail of the surface showing movement striations. **c** Escarpment plane located in the center of the former population of Jesús María. A tense root is observed due to the movement of blocks. **d** Detail of the surface of the plane, faint striations are observed

g) and the soccer field in the center of the town. This occurred because the movement of the ground met the immobile block and this generated tilting (Fig. 19.15g–i). A decompression sector adjacent to the tilted block formed open fissures that affected the Jesús María River, leaving the channel dry, and also contributed to the destruction and movement of the houses (Fig. 19.15a–c).

In 2016, the landslide was reactivated in the Community Hall sector (Fig. 19.15j–l) and this generated stresses in the central sector of the village. As a result of this, starting in 2017, started to be evident the tilting of the soccer field and the adjoining street (Fig. 19.15g–i). Recently, in 2020, the landslide was again reactivated in the Community Hall sector and generated a displacement of the internal crown of about 20 m from where it was located in 2017. This implies that the previous relief disappeared (Fig. 19.15m–o). Figure 19.15k, n show, through large radius ovals, the current place where the internal crown is positioned and its location with respect to the situation of 2014.

A vital line affected is the polyduct of the Costa Rican Oil Refinery (RECOPE), locally completely deformed (Fig. 19.16a, b). It was first affected by an important fissure, then subsidence of the ground occurred and since then it has remained hanging. It is interesting that previous studies of the pipeline did not analyze the morphology of the Jesús María landslide, since Aguilar (1995) lists the

Fig. 19.15 Effects of landslide displacement in the populated area of the Jesús María landslide: **a** 2010, destroyed house, the arrow points to the fissure that was beginning to form, and which originated the escarpment observed in Fig. 19.14c, d. **b** 2014, the houses had been moved horizontally. **c** The arrow indicates the displacement of the house with respect to the entrance sidewalk. **d** Temple without appreciable damage, located on top of the in situ limestone block interpreted by geophysics. **e** Street heading E-W in front of the immobile sector and the sector of destroyed and displaced houses. **f** Street heading N-S, facing north, in the background you can see the police station that served as a jail. The white block is the remains of one of the houses destroyed without any horizontal displacement. **g** Same street, but facing south, with remains of a house in the background. **h, i** significant tilting of the central football field and the street (view to the South) When compared with views f and g, the tilting that raised the street and shortened its width is observed. **j** On the way to the town of Jesús María, fine fissures were observed that cut the road perpendicularly. **k** A broader view of the landscape, the path with an oval indicates the position of the fissures of j, and in the background the Community Hall. **l** Community Hall completely destroyed, the back wall and the floor can be observed. **m** Same driveway to the town of Jesús María. People are in front of the local school. n) Recent escarpment formed in the Community Hall sector. The oval shows where it is based on the historical photo of K. **o** Continuation of the escarpment formed in 2020 and that affects the football field near the Community Hall. *Photos* G. Peraldo

Fig. 19.16 Effects of Jesus Maria landslide on infrastructure. **a** RECOPE polyduct pipeline completely deformed by the movement of the landslide. A crack that can potentially damage the pipeline can be observed. **b** Same sector of a), where the pipe has no support whatsoever, the double arrow shows the support structure of the pipe, in both cases it became obsolete. **c** Football field to the W of the town, near the Community Hall, with an important fissure that deforms it. **d** Formation of a pond. **e** Internal crown that has already reached the field. **f** Deformed football field. The double arrow indicates the place where the internal crown is currently positioned, and the small arrow indicates the elevation of the central sector of the football field. *Photos* G. Peraldo

threats that affect the pipeline such as the Piedras de Fuego, Santa Marta and Guayabo landslides, but it does not refer to the morphology of the Jesús María landslide. Another sector affected close to the pipeline is a soccer field that was deformed, rising in the center and presenting cracks (Fig. 19.16c–f).

19.9 Discussion

In general terms, the complex mass wasting areas represent intense processes on gently to moderately steep slopes, which were historically confirmed through newspapers, documents or infrastructure works. Its study should be essential to prevent social development in similar areas. One aspect that must be considered is the integration of different scales of observation and analysis, from remote sensing, digital elevation models and other digital techniques, to field analysis with the observation of microrelief. The latter provides crucial information for the interpretation of the type of lithology present, the mechanism of movement and, in the case of the tilted blocks, according to their size, the depth of the associated rupture or the results of the collision of a mobile block with an immobile one are suggested. Also, detailed microrelief mapping can inform the shape and complexity of the landslide rupture surface relief. These observations favor detailed knowledge of the geological and geometric elements that led to better understanding of dynamic behavior of areas subjected to slope instability. Another aspect to consider in the understanding and interpretation of unstable areas is the dynamic relationship between lithologies. Zonation methodologies consider each rock unit independently. However, the relationships between lithological units (stratigraphy) must be quantified in an integrated manner based on the nature and mechanical characteristics of the interacting lithologies. The study of the clay content cannot be neglected, as it is important in the movement mechanism, especially the identification of expansive clay mineral species, for which clay diffractometry is relevant.

The geomorphological study is made up of two integral parts: 1. The analysis of aerial photos (vital for a temporal analysis of instability processes), LIDAR data in order to refine details, and other useful images, as well as the use of GIS for the creation of maps. 2. Field work, vital to describe and analyze microrelief, which provides details to refine the analysis of the process. It is important to consider especially in landslides where the slope does not seem to be the main movement parameter.

In particular, the Jesús María landslide was chosen as a case study since, due to its low slope, it is not revealed if deterministic and probabilistic methods of landslide recognition are used. This landslide has been observed continuously since 2007. This has made it possible to document changes observed in Fig. 19.17 This landslide destroyed, within a span of a decade, the homonymous population and generated a breakdown of its community character, given the loss of their homes and farmland; the constant deterioration of the infrastructure was a severe blow to the community. A series of morphological observations showed the complexity of movement in the affected areas. Thanks to the internal conformation of the landslide, known from geoelectric profiles (Abarca et al 2018), it was possible to identify sectors of the landslide areas subjected to compressional and tensional stresses. Figure 19.17a shows a detailed view of the activity of the landslide. Around the stationary block marked with 2, movement occurs. To the north, the fissure that destroyed and mobilized several houses was formed, the arrow indicates the direction of movement to the east–northeast. This movement generated a distensive effect that formed the series of N–S trending fissures that dried up a section of the Jesús María River; the flow manages to come out about 80 m downstream. The mobile block where the square shown with number 1 in the figure was located, generated a compressive stress and formed a block tilted toward the east due to collision with the immobile block where the temple was (2 in the sketch). Recently, part of the tilted block has been deposited on top of the immobile block because the landslide continues to move east in that sector. This is known through an interview with a resident of the area, who indicates that a structure still standing is being pushed by the ground that comes from what was the street (interview with Mr. Ronaldo Obando, 8-10-22). Figure 19.17b–e show the process at different observation periods. It is inferred from the figure that the process began in a distal part of the landslide, close to the Reventazón River and ran toward the highest parts. The Reventazón River could have influenced the movement by eroding the front of the old landslide.

19.10 Conclusions

It is relevant to write about the examples of slope instability occurring in Turrialba, along the Reventazón River, Costa Rica, in a book on geomorphology. The significance lies in the fact that this is a territory with a high percentage of large-scale complex landslides, generating intricate, active, and therefore changing morphologies. The resulting geomorphology exhibits distinct features that stand out compared to other geomorphic units in the area.

The climate and interaction between sedimentary and volcanic lithologies in the study area are crucial factors for denudational processes, especially in a region with historical reports of significant damage to civil structures such as the railway and the pipeline. Towns like Jesús María have been completely destroyed by the eponymous landslide, leading to a socio-economic crisis for their inhabitants.

Understanding the geomorphology of the study area contributes significantly to land use planning and risk management in a region of strategic importance for regional and local development, driven by socio-economic activities such as agriculture or tourism. The latter is based on the variety of environments, ranging from urban and agricultural to secondary and primary forested areas. The construction of the reservoir for the Reventazón hydroelectric plant by the Costa Rican Institute of Electricity (ICE) for large-scale energy generation is boosting the local economy. It attracts investors for the creation of a capital base for tourism projects, necessitating hotels, residences, restaurants,

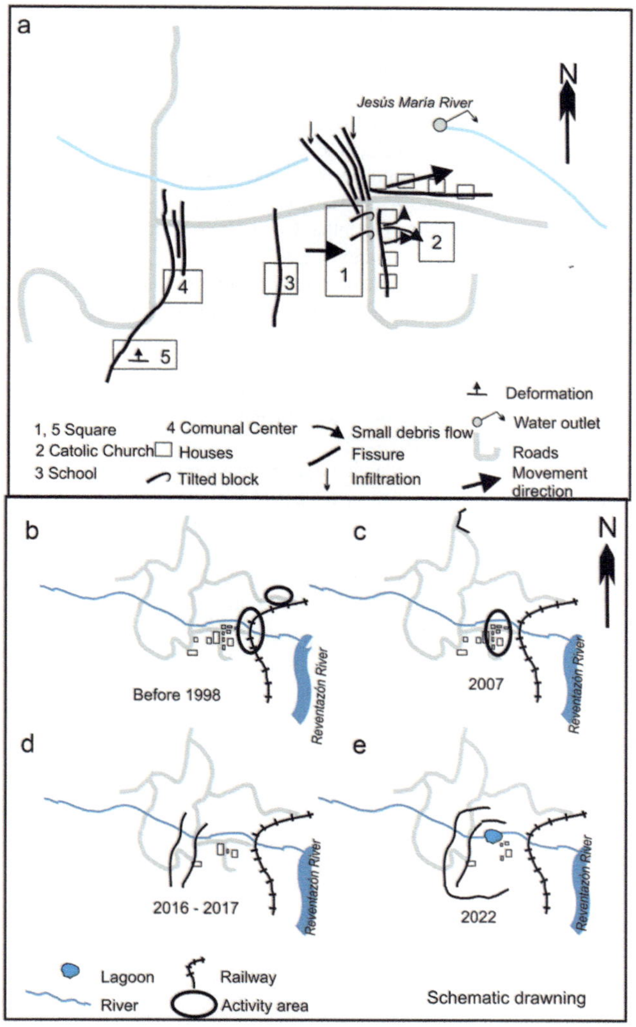

Fig. 19.17 Summary of the evolution of the areas affected by the landslide in the vicinity and in the town of Jesús María. **a** Elements of movement observed in the town of Jesús María. **b** Affected areas before 1998. **c** The movement of the landslide affects the town center. **d** The area to the east of the town is incorporated into the movement. **e** A crown is formed and affects the entire town area

recreational sites, among others. However, this development must be based on territorial planning that considers detailed studies of geomorphological instability processes; otherwise, the risk in the area could increase.

It is essential to emphasize that probabilistic and susceptibility methods tested in the area do not recognize low to medium slope instability sectors, leading to territorial management issues. This happens because these methodologies prioritize slope as a factor over others and consider lithology independently. Observations in the study area suggest that lithological and geotechnical characteristics should be considered integratively, creating a geotechnical stratigraphy. This approach incorporates stratigraphic relationships of lithologies in the area and their geotechnical characteristics to improve the accuracy of methods for recognizing unstable low-slope areas.

Additionally, it is advisable to analyze the concept of compound landslides and include it in their definition. Intrinsic differences in movement directions expressed on the surface by generated morphologies, indicating compression-decompression sectors in the sliding mass, highlight the complexity of rupture surfaces. Finally, the geomorphology of slope instability should be considered in the teaching of science at the secondary and university levels. It should be approached in a transdisciplinary manner, as the instability process needs to be studied from multiple university disciplines for better understanding and integration into the socio-economic development processes of areas like the one under study.

Acknowledgements We want to thank the geologist Mauro Fallas for his collaboration in the geomorphological analysis of the study area. In addition, we want to thank Ph.D. Maximiliano Garnier Villareal, for the technical and language revision of the text. To the University of Costa Rica for financing this research through the projects 113-96-221 of the Central American School of Geology and 830-B0-055 and 830-B5-121 of the Center for Research in Geological Sciences.

References

Abarca S, Marín C, Gutiérrez V (2018) Análisis de procesos de remoción en masa en el distrito La Isabel, Turrialba, con énfasis en el área de Jesús María. Graduation Seminar Report, University of de Costa Rica

Aguilar Á (1995) Diagnóstico general de amenazas naturales y su impacto en la infraestructura de RECOPE. In: I Taller latinoamericano de los efectos de los desastres naturales en la infraestructura energética. San José

Alcántara I (2000) Landslides: ¿deslizamientos o movimientos del terreno? Definición, clasificaciones y terminología. Investigaciones Geográficas 41:7–25

Arroyo-González LN (2021) Geomorfología del cantón de Siquirres, Costa Rica. Revista Geográfica de América Central 1:217–239 https://doi.org/10.15359/rgac.68-1.8

Barahona D, Méndez J, Sjöbohm L (2013) Análisis de la susceptibilidad a deslizamientos en el distrito de Tres Equis: una base para la gestión del riesgo y ordenamiento territorial, Turrialba, Costa Rica. Graduation Seminar Report, University of Costa Rica

Bieniawski ZT (1989) Engineering rock mass classifications: A complete manual for engineers and geologists in mining, civil, and petroleum engineering. Wiley

Bottazzi-Basti G, Obando-Acuña LG, Fernández-Solorzano JA (2021) Episodios tectonosedimentarios de la cuenca Limón Sur. Revista Geológica de América Central 65:1–17. https://doi.org/10.15517/rgac.v0i66.46648

Calderón M (2015) Definición preliminar del modelo geológico – geotécnico del deslizamiento San Antonio de Pascua, Siquirres, Limón. Unpublished tesis, University of de Costa Rica

Caravaca D, Chamorro C, Miller R (2017) Análisis de la estabilidad de laderas en el sector Noreste del distrito La Suiza, Turrialba, Cartago. Graduation Seminar Report, University of de Costa Rica

Fernández-Solorzano JA (1987) Geología de la hoja topográfica Tucurrique, 1:50000 IGN. Unpublished thesis, University of de Costa Rica

ICE (2017) Hydropower sustainability assessment report. Reventazón, Costa Rica. Internal report, Instituto Costarricense de Electricidad

ITC-RAPCA (2000) Estudio preliminar de amenazas naturales en la cuenca del río Turrialba, cantón de Turrialba, Costa Rica. Capacity building for natural disaster reduction (CBNDR) and Regional Action program for Central America (RAPCA)

Linkimer L (2003) Neotectónica del extremo oriental del Cinturón Deformado del Centro de Costa Rica. Unpublished tesis, University of de Costa Rica

Mora S (1985) Las laderas inestables de Costa Rica. Revista Geológica De América Central 3:129–161

Mora S (1989) Análisis de la susceptibilidad geológico – geomorfológica a la erosión de la cuenca del Río Reventazón, Costa Rica. FAO-ONU-MAG. Unpublished report

Municipalidad de Turrialba (2016) Plan cantonal de desarrollo humano local (PCDHL) 2016–2026. Ministerio de Planificación Nacional y Política Económica (MIDEPLAN). Internal report

Municipalidad de Siquirres (2012) Ordinary session number 129, 22 October 2012. Municipality of Siquirres

Peraldo G, Rojas E (1998) La deslizable historia del ferrocarril al Caribe. Costa Rica. Anuario De Estudios Centroamericanos 24(1–2):97–128

Peraldo G, Rojas E (2000) Catálogo de deslizamientos históricos para Costa Rica, período 1772–1960. Informe Semestral Instituto Geográfico Nacional 36:123–171

Peraldo G, Badilla E, Camacho J, Morera M, Chávez I, Valverde W, Sánchez D (2012) Deslizamientos complejos que afectan a la población de San Antonio de Pascua, Siquirres, Costa Rica. Letras Verdes 11:4–26

Peraldo G, Badilla E (2015) El área de deslizamiento de Torito: un proceso histórico de amenaza. En Torno a La Prevención 14:7–21

Peraldo G, Badilla E, Soto D, González M, Herra D, Ramos J, Méndez H, Ceciliano A, Vega A, van der Laan E (2015) El deslizamiento de Palo Alto, Turrialba, Costa Rica: apuntes para su estudio. En Torno a La Prevención 15:7–23

Segura G (2009) Análisis mediante aplicaciones SIG de la susceptibilidad al deslizamiento en el corredor Siquirres-Turrialba como factor de riesgo del poliducto de RECOPE. Unpublished tesis, University of de Costa Rica

Segura G, Badilla E, Obando L (2011) Susceptibilidad al deslizamiento en el corredor Siquirres—Turrialba. Revista Geológica De América Central 45:101–121

Sojo D (2018) Geología de la hoja Paraíso (1:10000), Costa Rica. Revista Geológica De América Central 59:101–124

Varnes DJ (1978) Slope movement types and processes. In: Schuster R, Krizek R (Eds) Landslides analysis and control, vol 17, pp 62–33. Transportation Research Board, National Academy of Sciences Washington D.C. Special Report

20 Geomorphology of the Caribbean Coastal Plains of Northern Costa Rica: The Realm of Avulsive Fluvial Systems

Jorge P. Galve, Guillermo E. Alvarado, Iván Sanabria, Sebastián Granados-Bolaños, Patricia Ruano, José Vicente Pérez-Peña, and José Miguel Azañón

Abstract

The Caribbean coastal plains of northern Costa Rica have drawn little attention from geomorphologists even though this region possesses singular landforms resulting from exogenous and endogenous processes that shape the Costa Rican landscape. The Quaternary plain is the surface of a back-arc sedimentary basin called North Limón Basin that is fed by tropical rivers flowing down from the Central and Talamanca cordilleras. These rivers create spectacular fluvial sceneries, their channels changing from pure braided to anabranching and meandering types; they form meander necks and leave oxbow lakes and paleo-channels scattered across the floodplain. The western sector of the plain is covered by a complex volcanogenic megafan (Late Pleistocene to recent), which is the first modern landform of this type described in the international scientific literature. The eastern sector is a coastal plain where active thrust faults divert and strangle rivers as the movements of these faults slowly uplift the ground. The southern limit of the plain is also remarkable in that it shows unique tectonic ridges in alluvial deposits, where the deformation produced by active folding can be observed clearly in tilted and faulted sediments. All the landforms that can be admired in this region contribute to an understanding of how a landscape changes due to climatic, volcanic, tectonic, and marine processes.

Keywords

Alluvial plain · Avulsion · Volcaniclastic sediments · Megafan · Active tectonics

20.1 Introduction

The Caribbean coastal plains of northern Costa Rica are the region of the country that has received the least attention from a geomorphological point of view. Costa Rica's geomorphology is noteworthy for the exuberance of volcanic landscapes; the paradigmatic landforms associated with active tectonics that are clearly visible throughout the country; the frequent and varied forms of landslides, and associated social and economic problems, and the world-renowned uplifted coasts of the Pacific margin. Nevertheless, amidst the multitude of breathtaking natural wonders mentioned above, the significance of the Caribbean coastal plains as home to some of the most remarkable landforms in the Costa Rican landscape has been inadvertently overlooked.

The combined action of tectonics and volcanism has configured the Caribbean coastal plains, an area of active sedimentation covered by ever-changing fluvial landscapes largely modeled by endogenous processes. On the one

J. P. Galve (✉) · P. Ruano · J. V. Pérez-Peña · J. M. Azañón
Departamento de Geodinámica, Universidad de Granada, Granada, Spain
e-mail: jpgalve@ugr.es

G. E. Alvarado
Comisión Nacional de Prevención y Atención de Emergencias (CNE), Unidad de Investigación y Análisis del Riesgo, San José, Costa Rica

Centro de Investigación en Ciencias Geológicas, Universidad de Costa Rica, San José, Costa Rica

S. Granados-Bolaños
Escuela de Geografía, Universidad de Costa Rica, San José, Costa Rica

Department of Geosciences, University of Padova, Padova, Italy

I. Sanabria
Instituto Geográfico Nacional, San José, Costa Rica

P. Ruano · J. M. Azañón
Instituto Andaluz de Ciencias de La Tierra, Armilla, Spain

J. V. Pérez-Peña · J. M. Azañón
Instituto Andaluz de Geofísica, Granada, Spain

A. Quesada-Román (ed.), *Landscapes and Landforms of Costa Rica*,
World Geomorphological Landscapes, https://doi.org/10.1007/978-3-031-64940-0_20

hand, the recurrent explosive eruptions of the Cordillera Central volcanoes (Alvarado et al. 2020, 2021) contribute with great amounts of volcanic sediments. These materials end up in the coastal plain, transported by coarse lahars in the piedmont and diluted lahars in the distal parts that often transform the landscape of the plain (Galve et al. 2016). On the other hand, the recent uplift of the Cordillera de Talamanca during Quaternary (Morell et al. 2012) has produced piedmont coarse sediments that feed the plain. These sediments, in turn, are deformed by the activity of thrust faults affecting the piedmont of both cordilleras (Denyer et al. 2003). Furthermore, the deformation occurs in a transitional geodynamic region, between the fold-and-thrust belt located on the Costa Rican shelf and an incipient subduction zone in northern Panama, making this area a unique example of this type of transition (Alvarado 2016; Arroyo and Linkimer 2021).

Below, our brief description of the geomorphology of the Caribbean plains focuses on the interaction of endogenous processes with the landscape, pointing out the most distinctive landforms of the region and the mechanisms behind their formation. In view of the great potential for geoscientific research in this area, we also indicate research efforts expected to be developed in the mid- to long-term.

20.2 Background

The Caribbean coastal plains of northern Costa Rica (Fig. 20.1) were considered a remote region just 150 years ago. For example, the western sector of the plain was

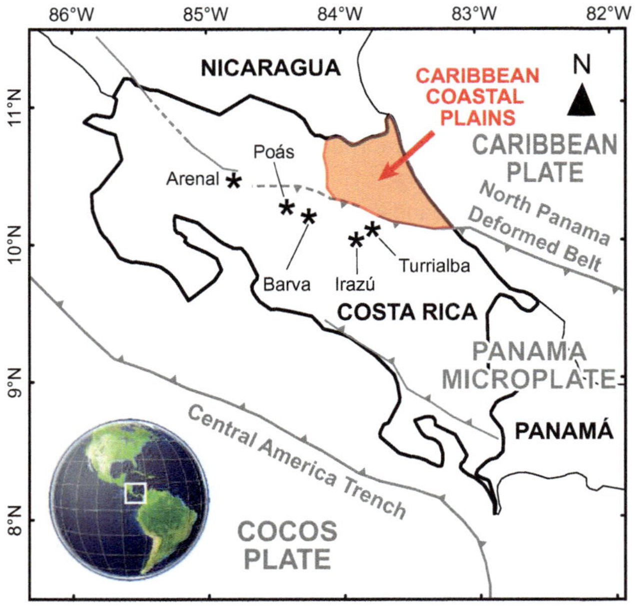

Fig. 20.1 Location and extension of the Caribbean coastal plains of northern Costa Rica

defined in the mid-nineteenth century as "almost entirely unknown" (see historic map of Frantzius of 1861). At that time, only the piedmont of the Central and Talamanca Cordilleras, and the routes to Limón, were mapped in detail, because the population was concentrated in these areas. One of the first geological and edaphic recognizant maps —by Sapper (1905)—depicts volcanic sands, diluvial debris, and fluvial soils in this region. It was not until well into the twentieth century that the entire region of the coastal plain was explored and mapped in sufficient detail to reflect shapes and dynamics in the landscape and impel geomorphological studies. The pioneer publications on geomorphology of the plains were in Spanish. Bergoeing (1978, 1998) commented on specific geomorphological aspects of the Parismina river near the Caribbean coast such as floodplain, meanders, oxbow lakes, lagoons, mangroves, and beaches. Madrigal and Rojas (1980) published the first comprehensive geomorphological work to focus on this region, recognizing a series of alluvial fans and providing a general description of the floodplains. Alvarado (1990) provided details regarding the four fluvial terraces (the oldest, T4; the most recent, T1) in the confluence of Sarapiquí and Puerto Viejo rivers, at Puerto Viejo town (Fig. 20.2), highlighting the first radiocarbon age in the floodplain region in T2, with an age over 43 ka.

In the realm of international literature, a major step was the paper by Kesel and Lowe (1987); then again, the early 1990s witnessed several studies and projects developed by the University of Wageningen (Netherlands) in collaboration with the Tropical Agronomic Center of Research and Teaching (*Centro Agronómico Tropical de Investigación y Enseñanza*, CATIE) and the Ministry of Agriculture and Livestock of Costa Rica (*Ministerio de Agricultura y Ganadería de Costa Rica*, MAG). Yet Kesel and Lowe (1987) were the first to describe in detail the geomorphology and sedimentology of this region, revealing a tropical fan through their description of a sector within the foothills of the Cordillera Central. In the wake of Kesel and Lowe's paper, Dutch researchers carried out geomorphological and edaphological investigations in the region, providing useful radiocarbon dating. They underlined the decisive influence of explosive eruptions on changes in the landscape of the Caribbean plains (van Seeters 1993; Wielemaker and Vogel 1993; Dechesne 1994; Nieuwenhuyse and Kroonenberg 1994; van Ruitenbeek 1994; Niewenhuyse 1996) and inspired the recent investigations of Galve et al. (2016, 2022). New publications in this century include the descriptions by Marshall (2007) and the maps of Bergoeing et al. (2010). The latter authors elaborated maps at a 1:100,000-scale to register piedmont, alluvial, colluvial, glacis and delta fans, floodplains with flat-shaped valleys, as well as Holocene transgression areas and littoral strings. The maps were re-published internationally through the book

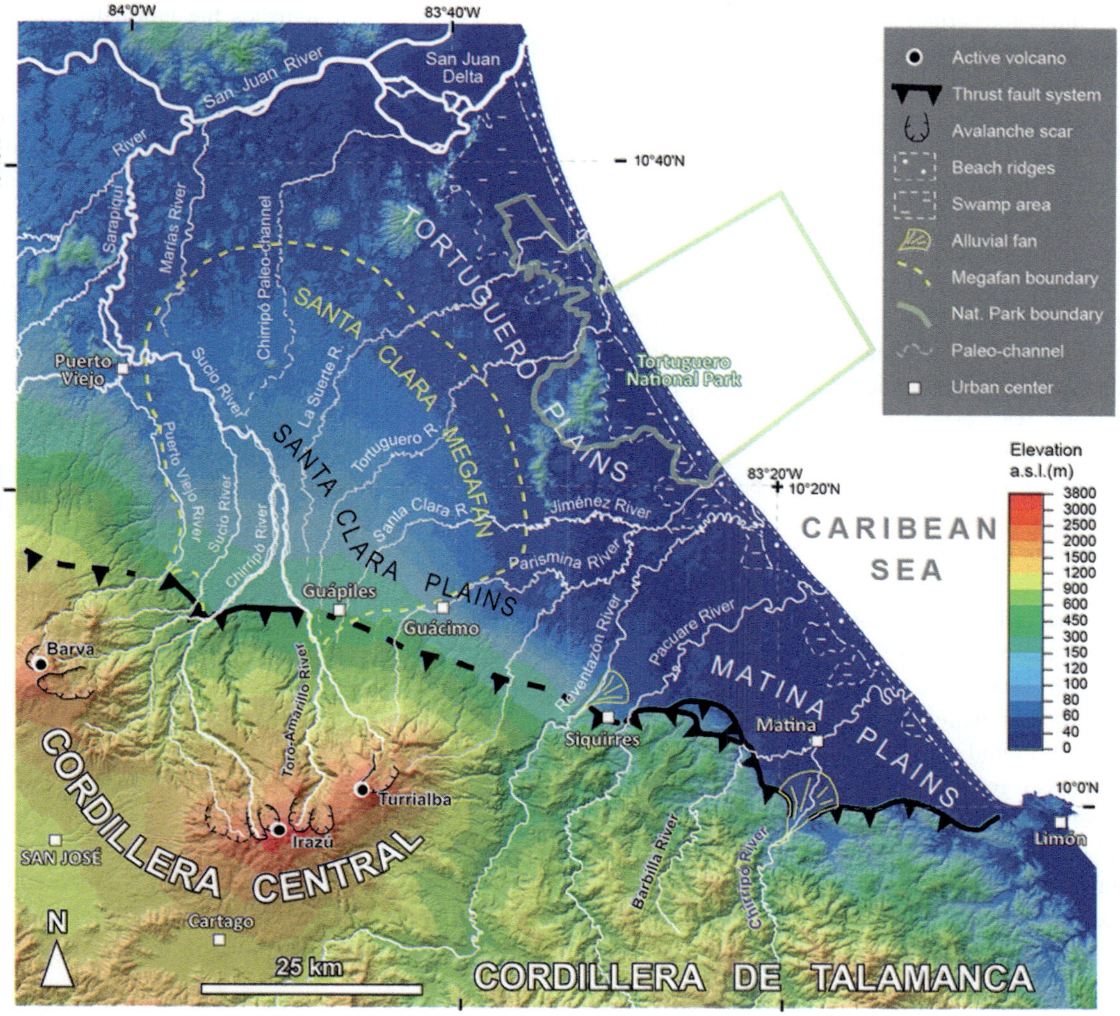

Fig. 20.2 Topography and main geographical features of the analyzed coastal plains

"Geomorphology and Volcanology of Costa Rica" by Bergoeing (2017). Some features represented in these geomorphological maps were recently revised by Galve et al. (2022), and the authors of this chapter are currently updating the maps of several sectors of the Caribbean coastal plains. Hence, the region has been revisited to delve deeper into the driving forces behind its fluvial landscape. The data gathered through diverse projects developed in the past two decades have been fully reviewed in this process. Chronological and geomorphological information sheds new light on the origin and evolution of the most singular landforms in the plains, including the isolated volcanic hills (Alvarado and Gans 2012; Galve et al. 2016, 2022; Oconitrillo and Sanabria 2020). Recent research highlights unique characteristics of the Caribbean plains and raises key questions about the origin and evolution of its fluvial systems.

20.3 Geographical Setting

The Caribbean coastal plains of northern Costa Rica cover about 4500 km^2 of the lowlands between the mouth of San Juan River to the northwest and the Limón headland to the southeast (Figs. 20.1, 20.2). The plain extends for 40–70 km, from the Puerto Viejo and Sarapiquí rivers and the foothills of the cordilleras Central and Talamanca, to the Caribbean Sea. From a geographic standpoint, these coastal plains can be divided into three areas: Santa Clara, Tortuguero, and Matina plains (Bergoeing 1998; Santillana 2008). Furthermore, the region is crossed by several important rivers, most notably the Reventazón-Parismina, Chirripó-Sucio, Pacuare and Matina. Before the construction of roads and railways, these rivers were the usual routes of communication between coastal communities and population centers located at the foot of cordilleras. The use of double names for certain rivers is unique to this region, where river names may change over their course or history. We inferred that such changes are related to frequent diversions and avulsions affecting their courses.

The towns of Guápiles, Siquirres and Guácimo make the region near the cordilleras the most populated and best communicated area of the plains (Fig. 20.2). In contrast, the land between this area and the sea is one of the least densely populated regions of Costa Rica. From an ecological point of view, it is worth noting that the coastal strip of the region and part of the offshore area are home to one of Costa Rica's best-known national parks. The Tortuguero National Park was established in 1975 and has been extended on several occasions. It initially aspired to protect the green sea turtle (*Chelonia mydas*), but nowadays the objectives are broader. The park preserves very high biodiversity, harboring tropical rainforests, and coastal wetlands.

The Caribbean coastal plains have a tropical rainforest climate according to the Koppën classification. The mean annual temperature is 25 °C, and annual precipitation is around 3500–4000 mm (ICE 2016). Two seasons can be discerned in the region: a dry season from December to April, and a humid season from May to November, yet variation between the two is not very significant. Intensive rainfall events are exceptional. Hurricanes impact only indirectly on this region, for instance Hurricane Otto, November 2016—the first hurricane that crossed Costa Rica since records have been kept (Quesada-Román et al. 2019).

20.4 Geological Setting

The geological origin of the Caribbean coastal plains of northern Costa Rica is related to the formation of a sedimentary basin (i.e., Nicaragua tectonic depression) in the back arc of the Central Costa Rica volcanic arc (Malavassi and Madrigal 1970; Fernández et al. 1994). Specifically, the basin to which we refer has two sectors: the South Limón Basin, affected by active deformation of the North Panama Deformed Belt (Fig. 20.1); and the largely undeformed sector called the North Limón Basin, embracing the coastal plains of reference.

North Limón Basin started in the Late Cretaceous with deep-marine deposits, then evolved from pelagic to continental conditions until the present (Brandes et al. 2008). The younger fill of the basin—that is, the materials on which the plains developed—is composed of volcanic coarse and fine sediments from the Central Cordillera in the Tortuguero and Santa Clara plains, along with sediments comprising intrusive, sedimentary, hornfels rock components from the Talamanca Cordillera in the Matina Plain. As indicated above, most of the basin fill is undeformed, an aspect that distinguishes the North Limón Basin from the South Limón Basin. However, the western and southern boundaries of the basin present incipient deformation in the form of a narrow and segmented fold-and-thrust belt centered in the already defined active faults of Guápiles and Siquirres-Matina (Fig. 20.3; Denyer et al. 2003). These faults appear to be interconnected, forming a singular fault system; they would pertain to the Coastal Fault System of the Caribbean (i.e., *Falla Frente de Montaña del Caribe*), in liaison through a strike-slip fault called the Río Blanco Fault, which limits the Limón promontory to the west and serves as a southeastern margin for the North Limón Basin (Fig. 20.3). The Coastal Fault System of the Caribbean is responsible for the 1991 Limón Earthquake of 7.7 Mw, the most serious earthquake in the world that year (Fig. 20.3; Denyer et al. 2003). This fault system represents the western end of the North Panama Deformed Belt. Eastward from there a transitional zone begins, from thin-skinned tectonic deformation to a thick-skinned sector that grades to an incipient subduction zone (Fig. 20.1; Alvarado 2016; Arroyo and Linkimer 2021).

At several locations along the Tortuguero and Santa Clara plains, and the Tortuguero coast, the low-relief landscape is interrupted by a series of isolated and abrupt volcanic hills composed of old Pliocene to Pleistocene alkaline volcanic rocks, and rare Neogene subalkaline lava flows (Fig. 20.3; Alvarado and Gans 2012). These volcanic hills, however, represent only a minor departure from the overall monotonous topography of the Tortuguero lowlands (Marshall 2007).

20.5 Geomorphology of the Caribbean Coastal Plains of Northern Costa Rica

The Geomorphology of the coastal plain is dominated by fluvial and wetland landscapes, Pliocene volcanic reliefs (Cocori and Lomas de Sierpe-Azules) and an isolated Late Pleistocene or Holocene volcanic center, the Tortuguero cinder cone (Fig. 20.4). Three sedimentary regions can be differentiated from a geomorphological viewpoint: the cordilleras piedmont, alluvial plains, and littoral (Fig. 20.4). These regions would correspond to the three classes of floodplains already recognized by Narson and Croke (1982) in the Costa Rican Caribbean plains: (1) high-energy non-cohesive in the proximal part; (2) medium-energy non-cohesive in the middle part; and (3) low-energy cohesive floodplains including the subtype organic-rich floodplains.

The Cordilleras piedmont comprises a juxtaposition of volcanic aprons and alluvial and fluvial fans. The proximal parts of these fans are crossed by braided, straight and anabranching alluvial rivers carrying coarse sediments composed of boulders to sands (Fig. 20.5a, b, c, and d).

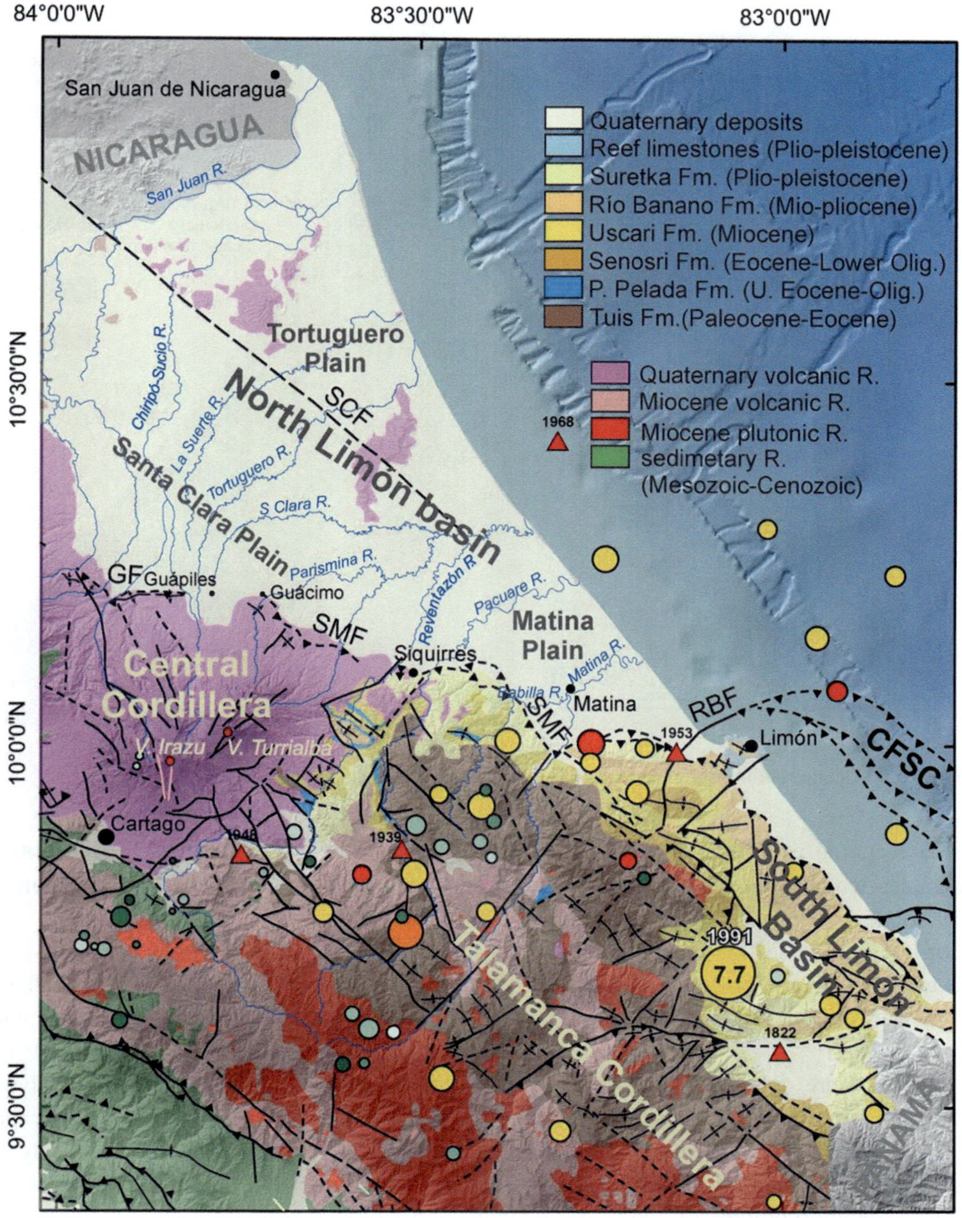

Fig. 20.3 Geological map of the Caribbean coastal plains. M 7.7 1991 Limón earthquake seismic series is shown: Green dots, earthquakes till 6 days after main shock (darker shades signaling greater depth, Barquero and Rojas 1994); Orange dots, M>4 seismic series four months after main shock (USGS). The historic seismicity is indicated with red triangles (Borchini and Montero 1994). CFSC: Coastal Fault System of the Caribbean; RBF: Río Blanco Fault; SMF: Siquirres-Matina Fault; GF: Guápiles Fault; SCF: San Carlos Fault

The fans grade to a band of alluvial plains, and the rivers transform their planform to a meandering style as their gradient decreases (Fig. 20.5e and f). Sinuosity of the channels varies as they pass through the plains, changing from highly sinuous segments to low sinuosity segments, then back to high sinuosity ones. Such alterations are meaningful for an understanding of blind (covered) active tectonic structures that affect the plains, as will be explained below. Furthermore, the fluvial channels show high dynamism, changing their courses or abandoning segments within a few years. These dynamics are related to laterally unconfined meandering with an active floodplain, expressed in the abundance of oxbow lakes and paleochannels along the plains (Fig. 20.6).

In the distal part of the alluvial plains there is a band of swamps and marshes, fed by the rivers and groundwater of the plains, that connects with a littoral formed by a succession of beach ridges and lagoons, generating an organic-rich floodplain sensu Narson and Croke (1982). The beach ridges are formed primarily by volcanic sands, dated within the last 5 ka to 4 ka years based on Nieuwenhuyse and Kroonenberg (1994). These authors pointed out that the formation of these beach ridges could be episodic and controlled by sand supply, probably related to eruptions and erosion associated with the volcanic source area of the sediments. The beach ridges appear to have grown in favor of the main north–south ocean currents identified in this sector of the Caribbean Sea (Andrade et al. 2003).

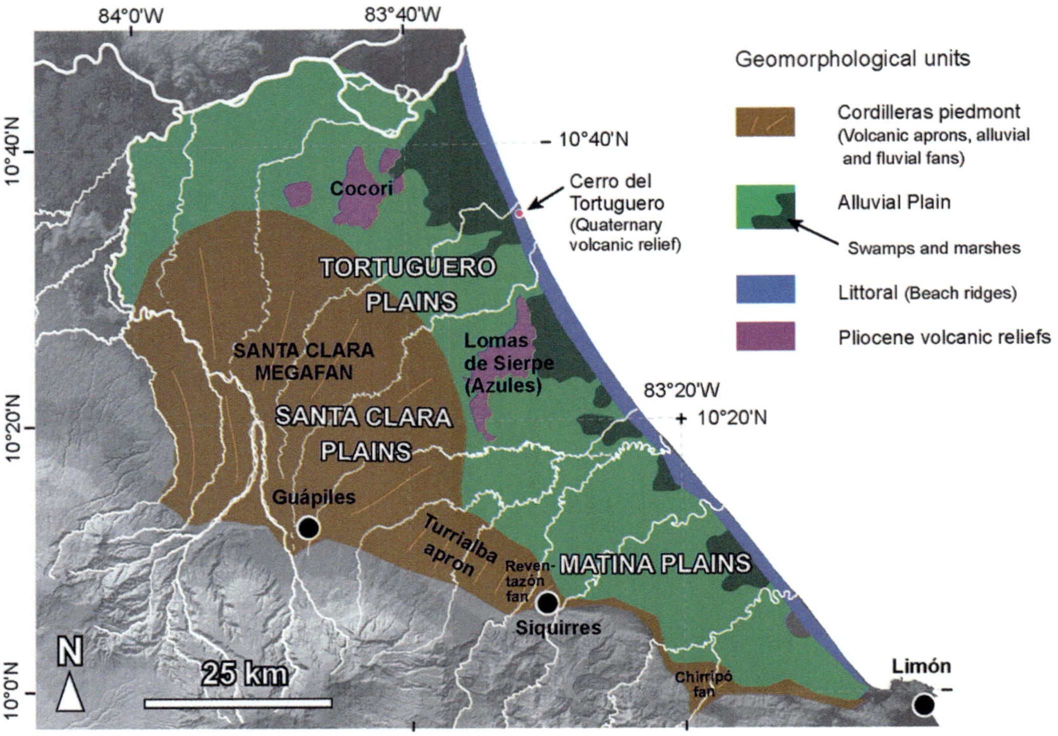

Fig. 20.4 Geomorphological units of the Caribbean coastal plains

However, the evolution of the coast observed in the last decades would indicate that the coast is being configured by south-north littoral drifts that make the sand bars grow northward (Fig. 20.7). This general picture contains some exceptions, where bi-directional currents do not allow for a clear evolution in one direction, e.g., in the Matina river mouth area.

For a more detailed description of the geomorphology of this Costa Rican coast stretch, we recommend the chapter by Denyer and Cárdenas (2000) and the paper and map of Quesada-Román and Pérez-Briceño (2019). As for the inland geomorphology of the coastal plains, the following sections describe certain particularities and the most remarkable landforms related to the avulsive nature of their fluvial systems.

20.6 The Avulsive Fluvial Systems of the Caribbean Plains

In the following sections we focus on the most geomorphologically outstanding features of the coastal plains—the fluvial systems. They belong to two main geomorphological settings: (1) the Santa Clara Megafan; and (2) the Matina Plain (Fig. 20.2).

20.6.1 The Santa Clara Megafan: An Avulsive Fluvial System Driven by Active Vulcanism

The Santa Clara Megafan is a fan-shaped mega-landform, 40 km long and 50 km wide, that covers 1440 km^2 of the Santa Clara Plains, at the piedmont of the Cordillera Central (Fig. 20.2; Galve et al. 2016). Previous authors classified this kind of large fan according to different aspects, hence they may be referred to as fluvial fans or Distributed Fluvial Systems (DFSs) (e.g., DeCelles and Cavazza 1999; Weissmann et al. 2010; Ventra and Clarke 2018). Generally speaking, megafans are large fans mostly composed of sandy and muddy sediments with very low gradients and commonly generated by perennial rivers at the foot of mountain ranges that limit inland sedimentary basins. The physiography and components of the Santa Clara Megafan would fit these characteristics. It covers more than 10^3 km^2, most of the sediments being sandy, silty, and clayey, and shows an average gradient of about 6 m/km (<0.35°). Because megafans also have sub-types, the Santa Clara Megafan would be a "bending swinging-type" according to the classification of Zhang et al. (2020), since (1) it is formed by two sub-fans or sedimentary lobes, the Sucio-Marías and Tortuguero lobes; and (2) the fluvial

Fig. 20.5 Fluvial landscapes of the Caribbean plains: **a** Braided channel of the Sucio River just upstream of the Cordillera Central mountain front. **b** Straight plane-bed channel of the Toro-Amarillo River at the apex of the Santa Clara Megafan. **c** Braided channel of Talamancas' Chirripó River at the apex of its alluvial fan when it flows into the Matina Plains. **d** Minor watercourse crossing agricultural and livestock areas of the Santa Clara Megafan's proximal zone. Note the boulders deposited in this area of the megafan. **e** Meandering creek and gallery forest formed along the paleochannel of the Marías River in the distal zone of the megafan. **f** High sinuosity segment of the Sucio River just before flowing into the Sarapiquí River

system that covers the megafan is formed mostly by meandering rivers (Fig. 20.8). These rivers constitute a divergent fluvial network in which Sucio-Chirripó, Toro-Amarillo, La Suerte, Tortuguero and Santa Clara rivers stand out. The Sucio-Chirripó and Toro-Amarillo rivers still form the main active channel of the fan in our days and show braided and anabranching channels in the megafan proximal zone (Fig. 20.8).

The course of the megafan's main fluvial channels has undergone many changes over time, alterations that confer the megafan one of its principal characteristics. Rivers in megafans divert their courses recurrently through processes called avulsions. Avulsions allow rivers to radiate from the apex of the fan, cyclically filling the different sectors of the megafan with sediments, as is observed in conventional alluvial fans. The Santa Clara Megafan is a space where avulsions can be easily explained to the public because of the paradigmatic examples observed there. For example, in 1970 a major avulsion of the Chirripó River suddenly diverted its course, forcing the abandonment of 55 km of

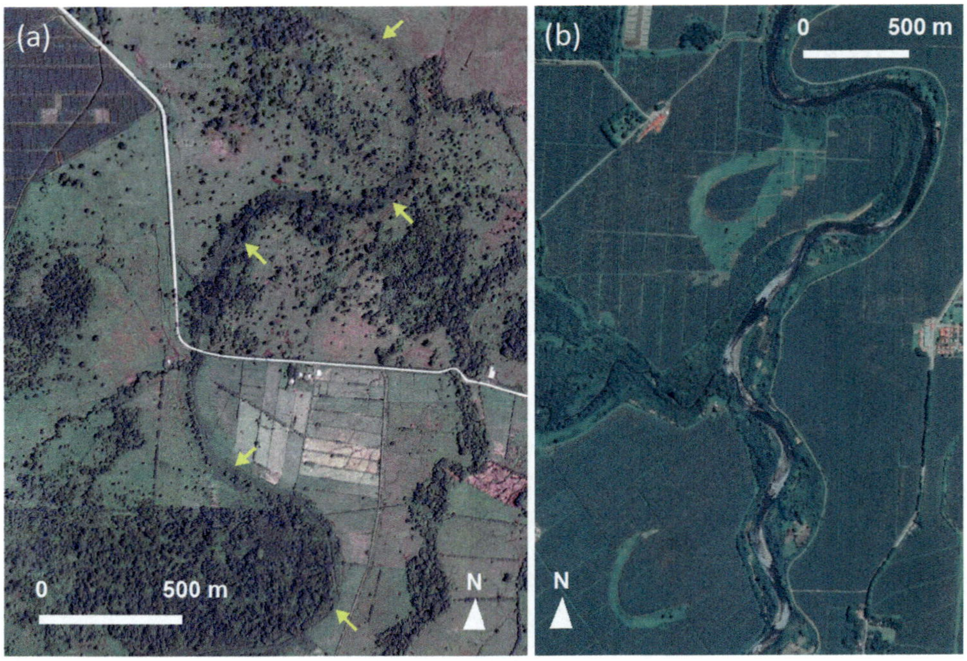

Fig. 20.6 High-resolution satellite image from Google Earth™ where paleochannels of the Caribbean plains clearly illustrate some of the main terms used to describe the landscape elements of rivers. **a** Paleochannel of the Marías River. **b** Remnants of oxbow lakes in the Pacuare River alluvial plain

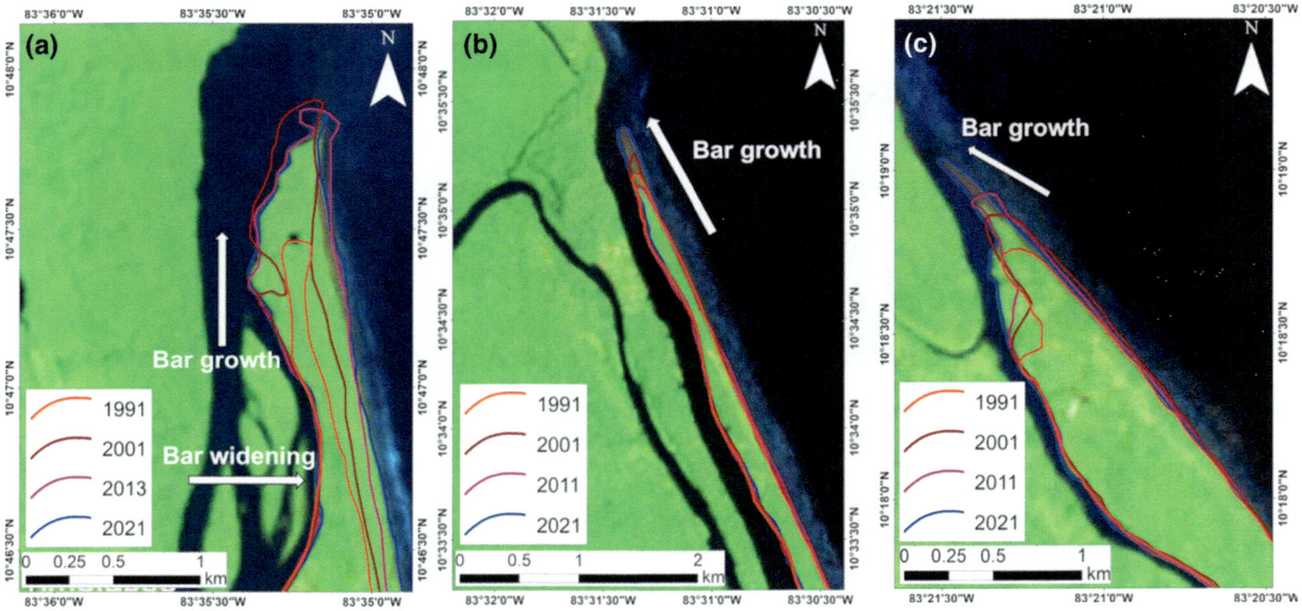

Fig. 20.7 Evolution of the Tortuguero, Parismina and Colorado sand bars, 1991–2021. Analysis performed over a collection of images generated through the app "Timelapse" from streamlit.geemap.org selecting the Landsat TM-ETM-OLI Surface Reflectance collection and the SWIR1/NIR/Red band combination. **a** The Tortuguero sand bar has a south-north-growing tendency on the point bar, while the width of the bar remained relatively stable during the timelapse. **b** Despite the fact that the Parismina bar width remained stable overall, the north-west side of the bar evidences an increase in width of around 300 m during the timelapse. **c** The Colorado sand bar tends to grow in a south-north direction in terms of length. When compared to the other sand bars, this bar is seen to show substantial width variations, augmenting its width by approximately 500 m during the timelapse

its former channel and displacing its mouth by 90 km — from the San Juan River in the north to the Sarapiquí River in the west (Fig. 20.9; Galve et al, 2016). Nowadays, the only remnants of Chirripó River old channel are meandering depressions several kilometers long and around 50 m wide. Furthermore, the fact that the border between the

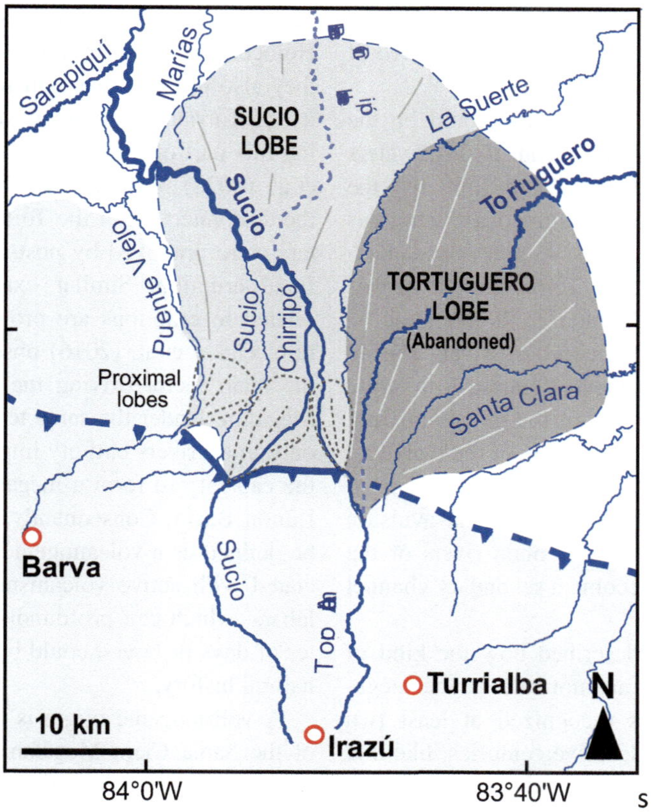

Fig. 20.8 Parts and main channels of the Santa Clara Megafan

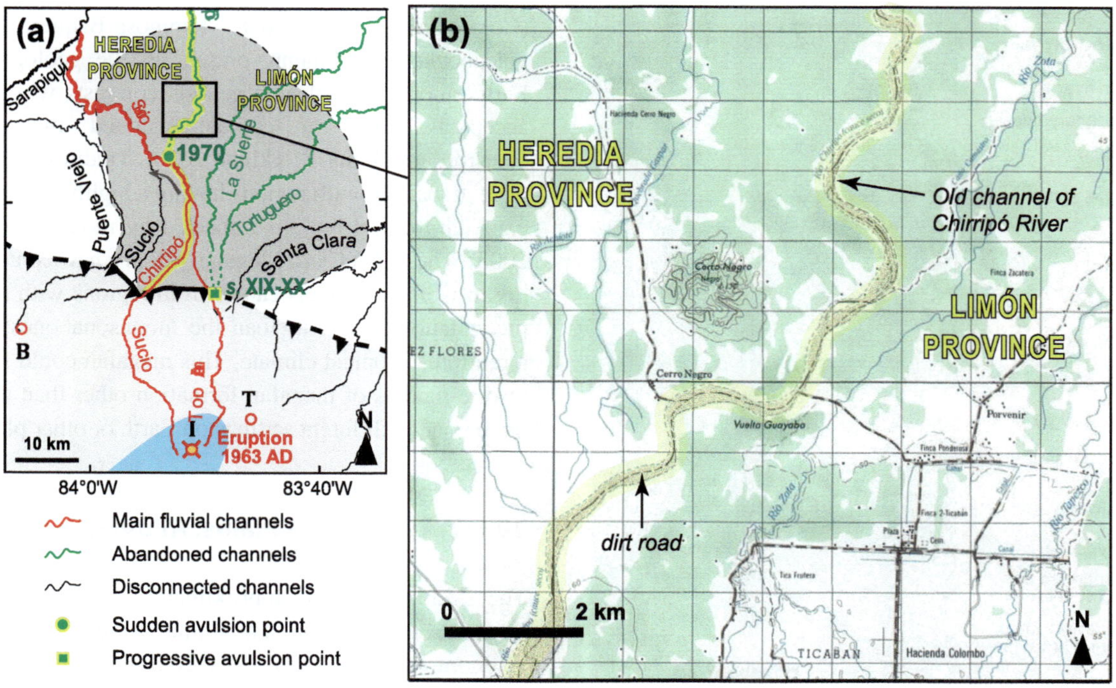

Fig. 20.9 **a** Scheme of the disconnection of Tortuguero and La Suerte rivers from their source area during the last 150–200 years and the avulsion of the Chirripó River in 1970. These avulsions —the first progressive and the second sudden—resulted in the abandonment of the eastern depositional lobe of the Santa Clara Megafan. **b** Extract of the topographic map in the area indicated in panel a. It illustrates where the border between the provinces of Heredia and Limón was traced in the nineteenth century along a river that no longer exists. Now it is a meandering depression where a dirt road was built to cross this area of the plains. The border between the provinces of Heredia and Limón is highlighted in yellow in the two panels

provinces of Heredia and Limón was delimited in the nineteenth century along the course of this river is indicative of its previous situation (Fig. 20.9b).

Another distinctive avulsion in the area would be the disconnection of the Tortuguero River from its headwaters formed by the Toro-Amarillo River Basin. Initially, the Toro-Amarillo and Tortuguero rivers formed a continuous fluvial channel. Toro-Amarillo was the stretch of the channel located in the mountains and the Tortuguero River was the course that crossed the alluvial plain to its mouth at the sea. Their disconnection took place progressively during a period of 150–200 years. The channel started to migrate from the east to the west in the area of the mountain front until the final cut-off occurred in the middle of the twentieth century, when the Toro-Amarillo River finally connected with the Chirripó-Sucio River (Fig. 20.10). This avulsion made the Tortuguero River, one of the main rivers of the megafan in previous centuries, become a secondary channel due to the loss of its discharge.

Galve et al. (2022) recently described how the kind of major changes described above are not rare in the megafan environment. These authors recognized at least two more sudden avulsions over the last five centuries, like that of the Chirripó, and another progressive displacement of a river course that caused the abandonment of a river channel, as in the case of the Tortuguero (~one major avulsion/

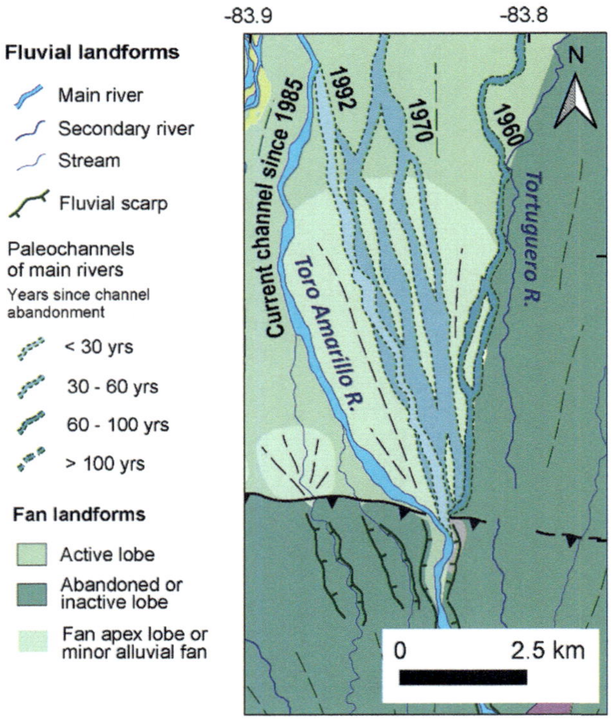

Fig. 20.10 Scheme of the progressive migration of Toro-Amarillo River to the west during the second half of the twentieth century, causing the disconnection of this channel from the Tortuguero River (modified from Galve et al. 2022)

century). In view of new chronological information on Holocene volcanic eruptions (Alvarado et al. 2020, 2021) they also inferred that avulsions were favored by the activity of Turrialba and Irazú volcanoes, most likely impacting this part of the Caribbean margin of Costa Rica. Galve et al. (2022) surmised that the 1970 Chirripó avulsion and the disconnection of the Tortuguero River from its headwaters were provoked by post-eruptive lahars of the 1963–65 Irazú eruption. Similar examples of previous avulsions related to eruptions are provided by the same authors. In turn, Galve et al. (2016) observed that active volcanism is the main factor driving the formation of the Santa Clara Megafan. Under the same tectonic and climatic conditions, only those rivers entirely impacted by active volcanism had the capacity to form a megafan in the setting of the North Limón Basin. Consequently, the Santa Clara Megafan can be defined as a volcanogenic megafan; and processes associated with active volcanism, such as flank collapses and lahars, which can profoundly change landscapes in a matter of days or hours, could be at the heart of this megafan's natural history.

A volcanogenic origin is the most relevant characteristic of the Santa Clara Megafan because it represents the only modern and active megafan of this type identified on Earth so far. This makes it the prime current model to understand the formation and evolution of other volcanogenic megafans, whether active in the recent past, such as the megafans of Jakarta (Indonesia) (Hosono et al. 2011) or Quindio-Risaralda (Colombia) (Guarín 2008; Espinosa-Baquero 2020); or volcaniclastic sediments present in the geological record. Furthermore, the volcanic genesis contrasts with the widely accepted model of climate-generated megafans (e.g., Hansford and Plink-Björklund 2020). Traditionally, megafans were associated with regions having a highly variable climate, e.g., sub-tropical regions impacted by monsoon (Leier et al. 2005). The Santa Clara Megafan offers a new context, according to which megafans can form in regions with much lower precipitation variability than the moonsonal ones, such as in a rainforest tropical climate. This megafan could furthermore inspire models of megafan formation other than the climate-driven one in different settings on Earth or other planets.

20.6.2 The Matina Plains: Alluvial Rivers

The Matina Plains approximately cover the lowlands located between the Parismina River and the Limón Promontory (Fig. 20.2). This coastal plain of nearly 900 km^2 is crossed by the Reventazón, Pacuare, Barbilla, Chirripó (not to be confused with the Chirripó River of the Santa Clara Plains) and Matina rivers. To the south it is bordered by the mountain front of the Talamanca Cordillera. This mountain front is an active tectonic boundary of the

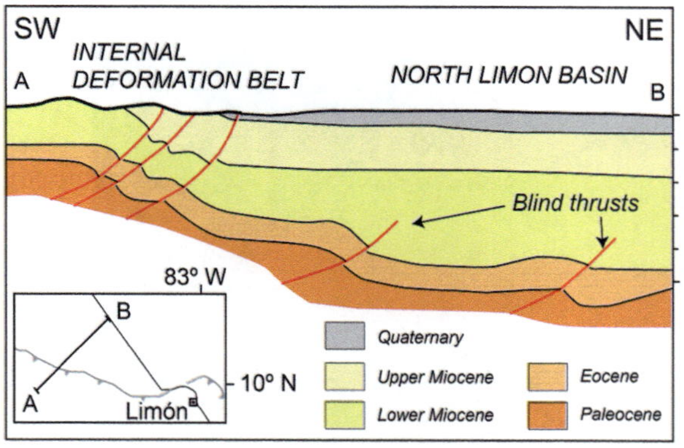

Fig. 20.11 Geo-seismic profile of the SW sector of the Tortuguero plain (modified from Fernández et al., 1994). Further research should review this interpretation by taking into account the deformation observed at the surface

North Limón Basin, formed by a fold-and-thrust belt that is currently deforming and uplifting the proximal sediments of the basin. Seismic profiles acquired in the 1980s show that several buried thrust fronts affect the Miocene bedrock of the area (Fig. 20.11, Fernández et al. 1994) and, as surface deformation shows, also affect the Quaternary cover. The deformation of sediments caused slight changes in the topography of the southern sector of the coastal plain, modifying the directions of the fluvial network. Thus, the Matina Plains offer cases that nicely illustrate river diversions induced by active tectonism; indeed, they may be even considered as paradigmatic examples in tectonic geomorphology textbooks. Nevertheless, the area has not attracted the attention of specialists in active tectonics (unlike the Pacific Coast of Costa Rica) and it has not been studied in detail despite suggesting promising outputs in this field. Here we summarize some preliminary observations on the geomorphology of this area, highlighting the most prominent landforms and some historical notes on the dynamics of the region.

The rivers that cross the Matina Plains have their headwaters in the Talamanca Cordillera, a mountain range that undergoes active uplift at present (<1 m/ka; Morell et al. 2012). The uplift results in high erosion rates in the cordillera and an important supply of detrital sediments to the North Limón Basin. Galve et al. (2016) estimated long-term denudation rates of ~ 0.2 m/ka in the western sector of the Cordillera de Talamanca, and long-term incision rates of ~ 0.6 m/ka in the Pacuare, Barbilla and Chirripó rivers. Such rates explain the creation of alluvial fans in the river outlets. Moreover, the dynamics of these fans determine the path taken by the rivers on their way to the sea (Fig. 20.2). The natural migration of rivers from margin-to-margin of the fans leads the rivers to change their courses periodically through avulsions, which means that the rivers can cross different sectors of the plain in different periods. This process also gives rise to the confluence of rivers in the plain for some time until the next avulsion and channel diversion. Changes in the position of river mouths are usual in such an ever-changing fluvial system, and the rivers themselves show stretches with anomalous orientations or/and sudden changes in planform.

As mentioned earlier in this chapter, the modification of river names (or segments of them) is common in this region of Costa Rica. It evidences the migration and abandonment of fluvial channels throughout history. Recent analyses of historical chronicles from the fourteenth and fifteenth centuries suggest that the rivers used by Spaniards when exploring and colonizing this area have since changed their courses. Apparently the Suerre River, well known since 1529 and upon which Diego Gutiérrez explored the region in 1544, was a channel formed by the confluence of the actual Reventazón (formerly known as Grande de Orosi River) and Pacuare rivers (De Peralta 1890) (Fig. 20.12a). The connection of these two rivers was through a branch of the Reventazón River that still exists as a secondary intermittent stream (Fig. 20.12d). In other words, the actual Reventazón and Pacuare rivers converged in a fluvial channel named Suerre River in the fourteenth century. The historical confluence of these two rivers afforded the fluvial channel sufficient discharge to move vessels such as the brigantines used by Diego Gutiérrez in his endeavors (De Peralta 1890).

In 1576, some decades after the arrival of Diego Gutiérrez to the Suerre River, a permanent port was established at its mouth, named Suerre Port (González 1906; Payne-Iglesias 2022; Fig. 20.12a). The location of this port, and consequently the Suerre River mouth, is a topic of debate due to the absence of any discovered traces or remains of its facilities. The port was active roughly until 1659 (Fig. 20.12b). Historical documents suggest that the Suerre Port was abandoned much beyond that date, probably related to a drop in the river flow (Calvo 1887; Fonseca et al. 2001; Montes de Oca 2005; Payne-Iglesias 2022). It appears that problems with the draft of the port began when several changes occurred in the fluvial channels that fed the Suerre mouth. The first, dated in 1633, was the bifurcation of the Grande de Orosi River at a place called "La Junta", seven leagues (i.e., ~ 33 km) from the sea according to the chronicles (Fig. 20.12b; González 1906; Obregón 1932; Montes de Oca 2005; Hilje 2017). This bifurcation was sudden and created the actual Reventazón River, while the previous fluvial channel of the Grande de Orosi River remained as a secondary branch connected to the former Suerre River, nowadays the Pacuare River. It is reasonable to imagine that this event was tied to a major flash flood, as the word "reventazón" used to name the new river comes from the Spanish word "reventón" or "blow-out". Even though the bifurcation would have developed in a sudden

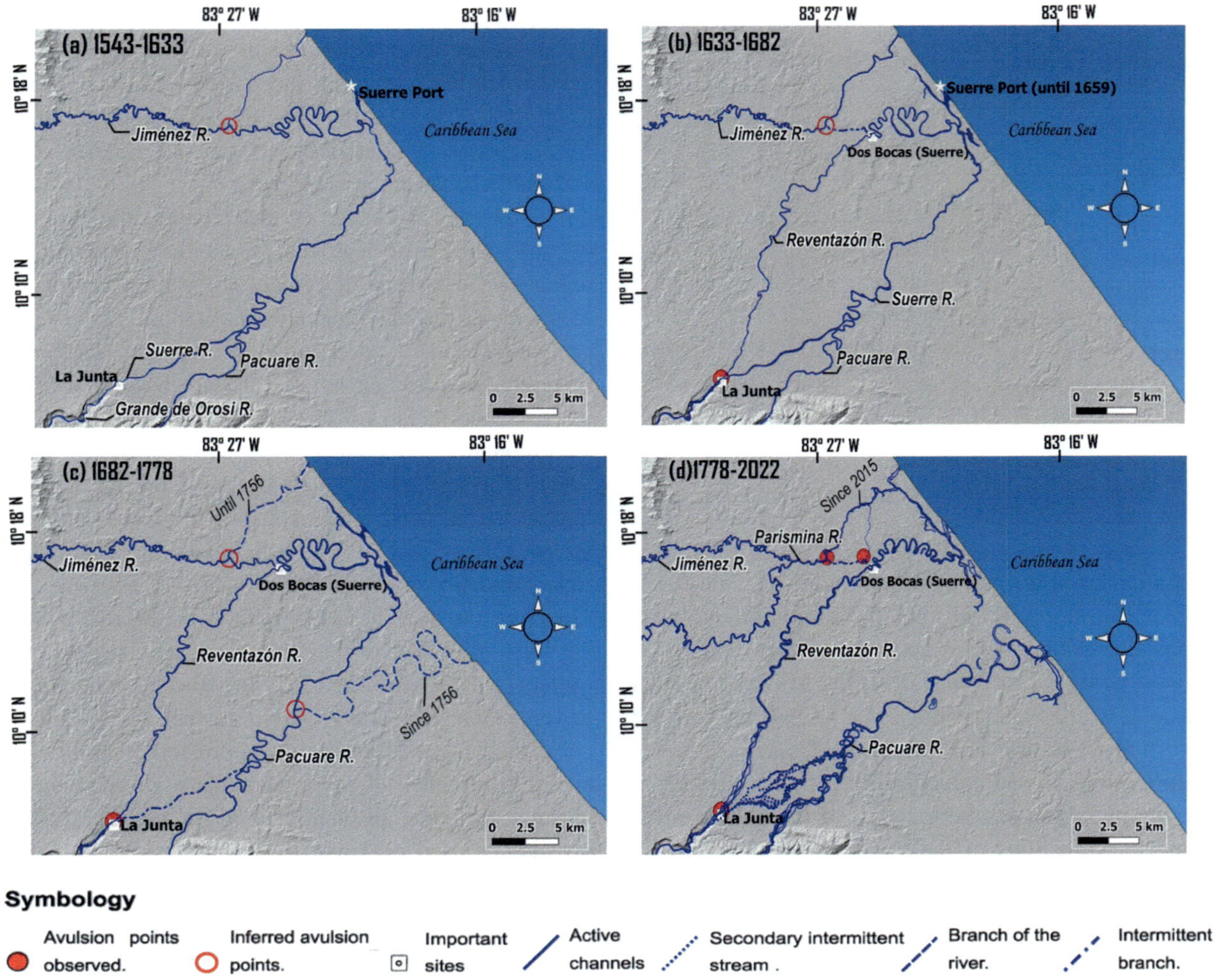

Fig. 20.12 Evolution of the Reventazón and Pacuare rivers from the sixteenth century to the present

manner, the abandonment of the secondary branch connected with the Pacuare river channel appears to have been progressive.

Further fluvial changes probably related to a decreased draft of the Suerre Port can be envisaged. Two possible avulsions in the sixteenth century may have begun to develop in the previous century: (1) the confluence of the Jimenez-Parismina River with the Reventazón River at "Dos Bocas"; and (2) the diversion of the Pacuare River, causing disconnection between this river and the Reventazón (Fig. 20.12c). The historic documentation of Gonzalez (1906) underlines two particular facts that point to avulsions possibly developed between 1756 and 1778. Accordingly, (1) the Pacuare diverted from the Reventazón, generating its own mouth to the sea and conforming its actual course; and (2) the Jimenez-Parismina River became a tributary of the Reventazón River and abandoned its outlet channel into the sea (recovered in recent times; see below).

Although this narrative is supported by an exhaustive review of historical documents and geomorphological assessments, it is worth stressing that considerably more research is needed to determine the consistency of this argument. The present study raises new questions about the origin of avulsion processes in the setting of the Matina Plains. For instance, the area of the Reventazón bifurcation coincides with the apex of its alluvial fan when it arrives at Matina Plains. Therefore, the bifurcation is most likely related to the sedimentary dynamics of the fan. Conversely, the narrative has both historical and archaeological significance, which motivates further research. The search to locate and corroborate the existence and position of the port of Suerre may be facilitated by further geomorphological investigations based on the interpretations outlined above. In addition, other archaeological sites related to the people of Suerre and early colonial settlements in this area can be linked to historical chronicles taking into account the evolution of the river channels in the region.

Other inferred avulsions in the Matina Plains call for alternative explanations to those of the Reventazón-Parismina-Pacuare fluvial systems. These other avulsions

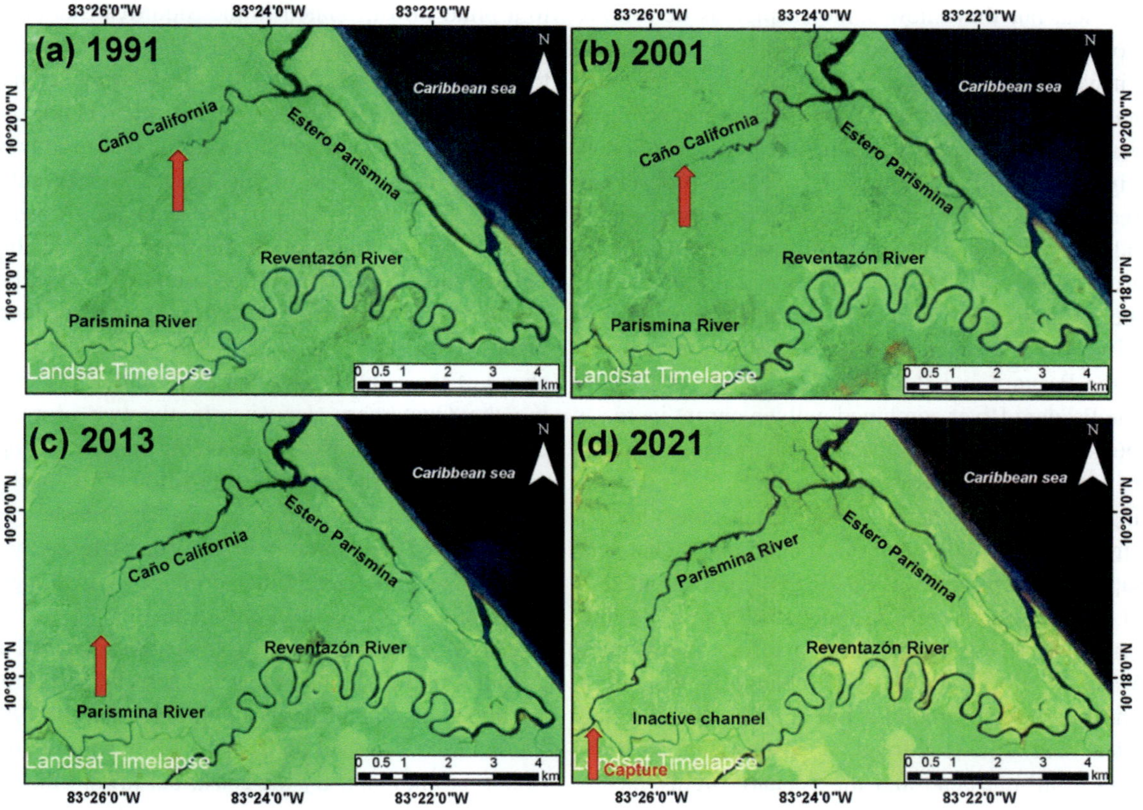

Fig. 20.13 Evolution of the Parismina-Reventazón rivers 1991–2021. Images generated using the app "Timelapse" from streamlit.geemap.org selecting the Landsat TM-ETM-OLI Surface Reflectance collection and the SWIR1/NIR/Red band combination. **a** The 1991 image does not show a clear connection between the Caño California creek and the Parismina river, yet this could be to the spatial resolution of the imagery used and the dense vegetation that characterizes the area. **b** The detectable flooded area of the channel increased by approximately 2 km, as the red arrow shows the channel width has varied. **c** Despite the low spatial resolution of the imagery it is possible to detect a connection between the Caño California creek and the Parismina river. **d** The 2021 image shows a clear connection between streams, leaving an abandoned channel in the southern part of the figure. By 2021, the Parismina had changed its channel approximately 5 km to the north. There might have been a connection between the two rivers before 1991 but with low streamflow. After 2013, the channel width grew sufficiently to be detected by Landsat imagery, with increased streamflow and a loss of vegetation along the riverbanks

probably are linked to active tectonics that bore an impact on the coastal plain fluvial dynamics. In conjunction with the above framework, the last progressive avulsion identified at human timescale is the recent diversion of the Parismina River just at its confluence with the Reventazón River near "Dos Bocas" (Figs. 20.12d and 20.13), disconnecting the former river from the latter. There is no register of the moment that marks the beginning of this avulsion or the event that triggered the process. A recent field survey and analysis of historical aerial photographs and maps by Oconitrillo and Sanabria (2020) suggest that the avulsion could have started before 1953 and was completed around 2015, leaving the linking branch from the Parismina to the Reventazón rivers as an inactive channel (Fig. 20.13). This avulsion produced the migration of the Parismina River mouth 5 km to the NW. The authors of the geographical/photo-interpretative analysis hypothesize that the avulsion is linked to the San Carlos Fault (Fig. 20.3). This is a regional strike-slip fault, to which the dextral displacement with normal component of the Lomas de Sierpe-Azules (a Pliocene volcanic relief located to the northwest of the avulsion area, Fig. 20.4) is attributed. If the hypothesis is correct, the fault may continue to the Parismina and Reventazón area. Again, further research efforts are needed to contrast this hypothesis and demonstrate modern activity of this fault.

20.6.3 The Matina Plains: Active Tectonics

Apart from the avulsion processes observed historically and at present, the perturbation of active tectonism is patent in certain features of the rivers that cross the Matina Plains. The main ones are a change in the general direction of the meander belt of the fluvial channels and an associated change in their sinuosity, the latter owing to a very slight variation in the slope of the channel. Initially, the rivers of the coastal plain are expected to follow a direction roughly toward the sea, but they show some suspicious deviations likewise observed in alluvial rivers altered by active tectonics. For instance, Galve et al. (2015) pointed out how the Parismina and Reventazón rivers show convergence and then divergence just before the avulsion point mentioned above (Fig. 20.14b). This type of behavior is also observed in the Secchia and Panaro rivers in the Po alluvial plain, in coincidence with subsiding and uplifting terrains due to active thrust folding (Fig. 20.14a; e.g., Burrato et al. 2003).

Another example of river diversions and sudden changes in sinuosity, also associated with channel abandonments, is found in the Pacuare and Barbilla rivers. There is a slight shift in the Pacuare River combined with an abrupt increase of its sinuosity along 8 km in the Galicia area (Fig. 20.15), to later return to its previous direction downstream, decreasing in sinuosity. The high sinuosity segment shows a wide strip of oxbow lakes and abandoned paleo-channels that disappear downstream (Fig. 20.15). Ten kilometers southeast of this stretch we find the Barbilla river's bend, a landform that reflects the disturbing effect of tectonics in the Matina Plain. There is a 70º change in the direction of the river to get around a hill (Fig. 20.16a) formed by uplifted and tilted alluvial deposits (Fig. 20.16b). Between the Barbilla bend and the Pacuare highly meandering stretch, another key landform is the now extinct lake of Cuatro (*Laguna Cuatro*) (Fig. 20.15). Originally, the lake had an elongated and meandering morphology with three branches. Its shape reflects that it constituted the last remains of an abandoned fluvial channel, now totally obliterated by agricultural activities. The three features described above define an area where the plain topography is being modified with enough magnitude to disturb fluvial channels that cross it. First, the Barbilla's bend is a clear example of a river diverted by active thrust folding. The uplift of the terrain caused by the deformation linked to a blind thrust, related to the Siquirres-Matina fault system, prevented the river from crossing the growing tectonic ridge. Second, the highly meandering segment of the Pacuare on one hand and, on the other hand, the Cuatro Lake would indicate a zone of uplift that produces a slight decrease in the slope of the fluvial profile, sufficient to modify the sinuosity and type of channel. Considering the tectonic setting, the zone of uplift is most probably caused by growth of the fault-propagation fold associated with another blind thrust front. This fault, not yet identified or mapped, may be a new thin-skin thrust sheet developed ahead of the Siquirres-Matina Fault and at present deforming the Quaternary deposits. In the two aforementioned cases, some small marsh areas and undulated topography are observed in the theoretical position of the back-limbs of the cited thrust sheets (Fig. 20.17). The marshes developed as a result of the damming of rainwater caused by the now-growing tectonic ridges that produce W-E elongated hills and depressions (Fig. 20.17).

Lastly, the most remarkable features legated to active tectonics in the Matina Plains and surroundings are the synseismic topographic changes in the wake of the 1991 Limón Earthquake. A coastal uplift up to 1.8 m was recorded in the Limón promontory. By the coast, but toward the northwest, at the Matina River mouth, 0.5 m of subsidence was calculated (Denyer et al. 2003). Yet, the most impressive topographic changes were measured by the Costa Rican Electric Institute (*Instituto Costarricense de Electricidad*, ICE) along a power line. Data acquired in surveys before and after the earthquake allowed personnel from ICE to determine that the base of the pylons had been uplifted as much as 4.46 m in the central sector of the powerline, whereas they had settled up to 0.7 m in some parts of the westernmost segment. These data were divulged in a map of the "Tectonic Atlas of Costa Rica" (*Atlas Tectónico de Costa*

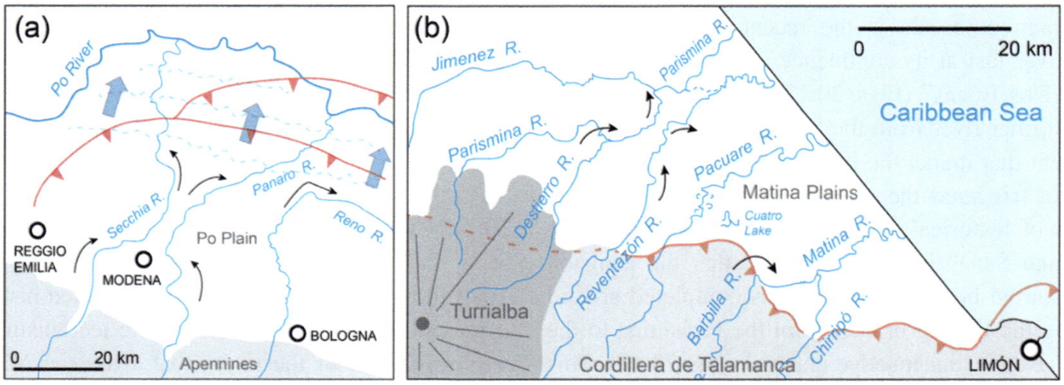

Fig. 20.14 Scheme to compare the fluvial tectonic disturbance in the Po Plain (Italy) and the river diversions observed in the Matina Plains. **a** Convergence and consecutive divergence of the Secchia and Panaro rivers and bends of the Po and Reno rivers due to the uplift produced by blind thrust faults affecting the central sector of the Po Plain (Burrato et al., 2003). **b** Convergence and consecutive divergence of the Parismina and Reventazón rivers and bend of the Barbilla River. In the case of the Matina Plains, further geophysical research must be developed to contrast these surficial observations with the subsurface active structures, as has been done in the Po Plain

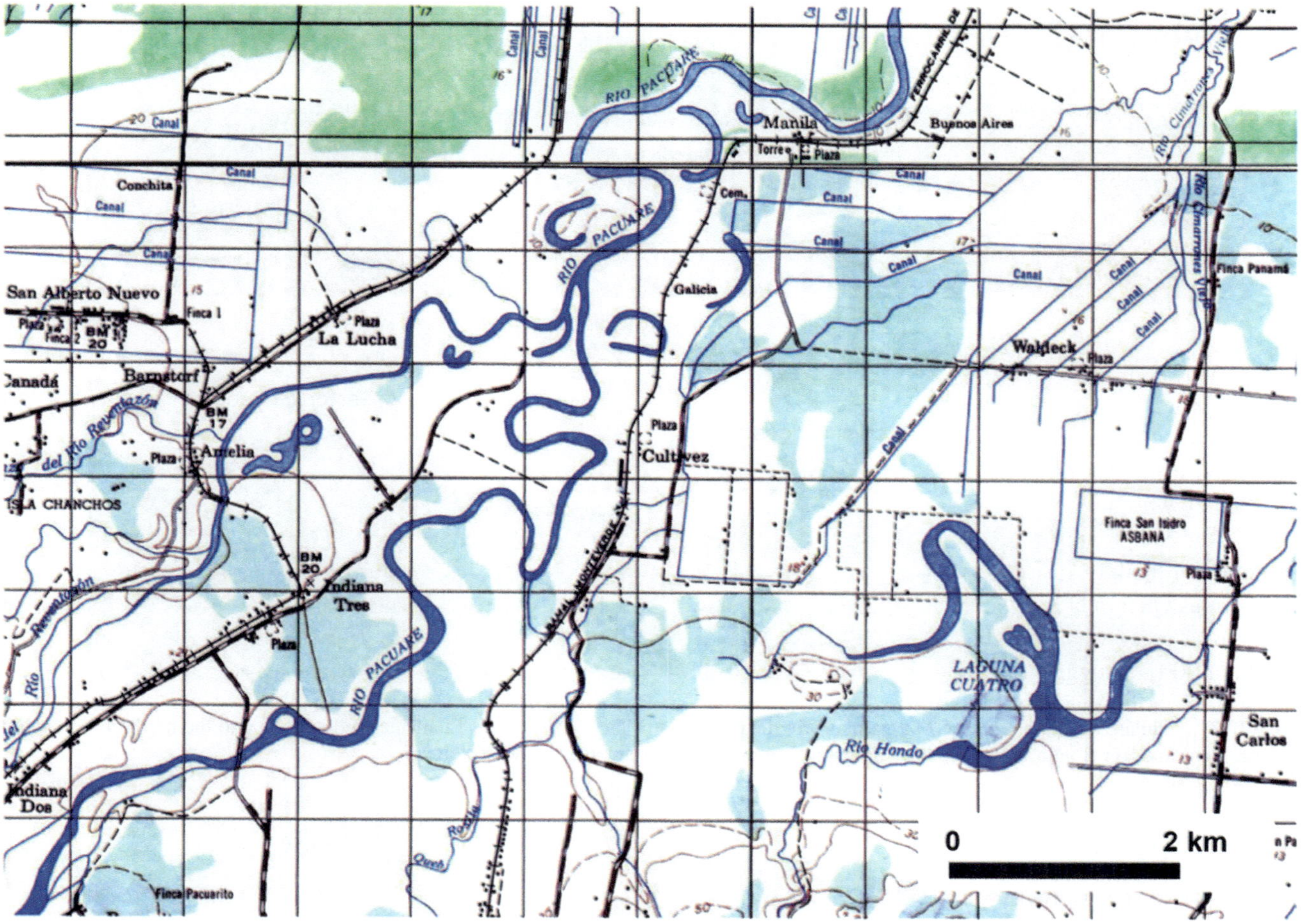

Fig. 20.15 Topographic map of the National Geographical Institute of Costa Rica (*Instituto Geográfico Nacional de Costa Rica, IGN*) where the anomalous segment of the Pacuare River and the Cuatro Lake is highlighted

Fig. 20.16 **a** Aerial oblique photograph of the Barbilla's bend. The black arrow indicates flow direction. The red arrow indicates the location of photograph in b. **b** Alluvial deposits uplifted and tilted observed in the hill that causes diversion of the Barbilla River

Rica, Denyer et al. 2003), a local Spanish-language publication. Figure 20.18 depicts this information with more geographical references than the original map, showing in greater detail how the area was affected by deformation associated with the earthquake. Note that the zone disturbed by topographic changes lies within the coastal plain, precisely by the piedmont of the Talamanca Cordillera, hence in the area of influence of the Siquirres-Matina Fault. The magnitude of the changes is significant, taking into account that the elevation of most of the coastal plain is between 15 and 20 m asl. Such vast modifications in surface morphology can explain how river choking and diversions developed in the Matina Plains after major earthquakes, e.g., that of 1991.

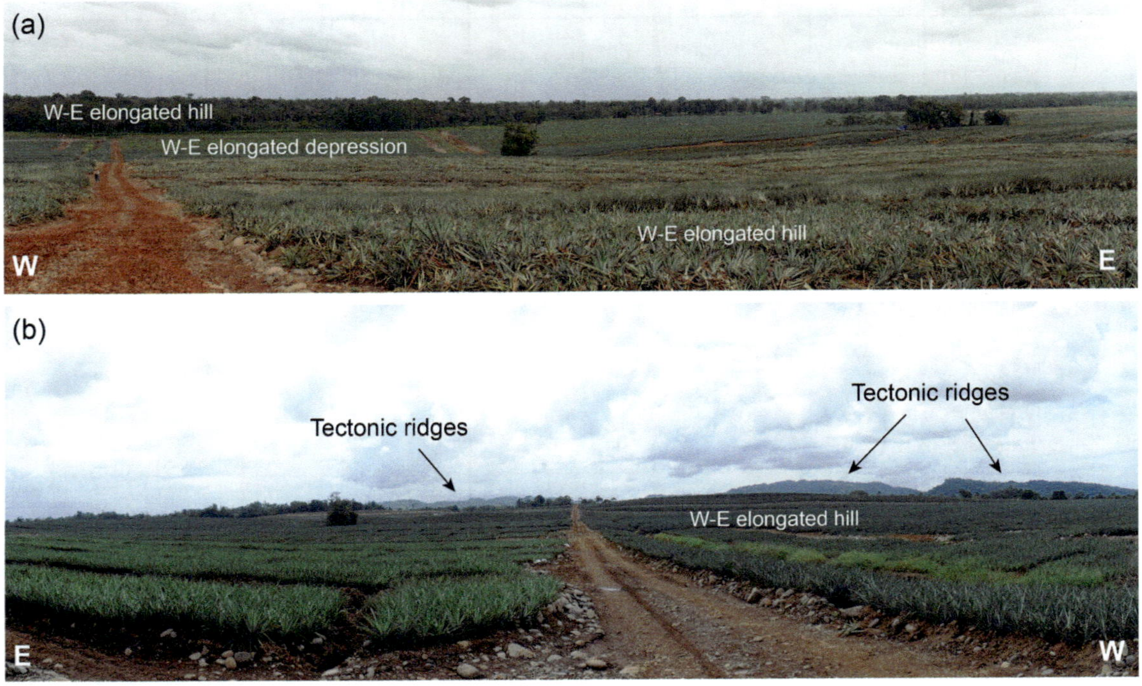

Fig. 20.17 Undulated landscape of the Matina plains in the piedmont of the Cordillera de Talamanca. **a** Landscape to the north in the vicinity of Santa Rosa and Río Hondo. **b** View to the south at the same point where image "a" was taken

Fig. 20.18 Topographic changes measured in the eastern end of the Matina Plains after the 1991 Limón Earthquake. The squares mark measurements along the coast (Denyer et al. 1994), while the circles are measurements by ICE personnel published in Denyer et al. (2003)

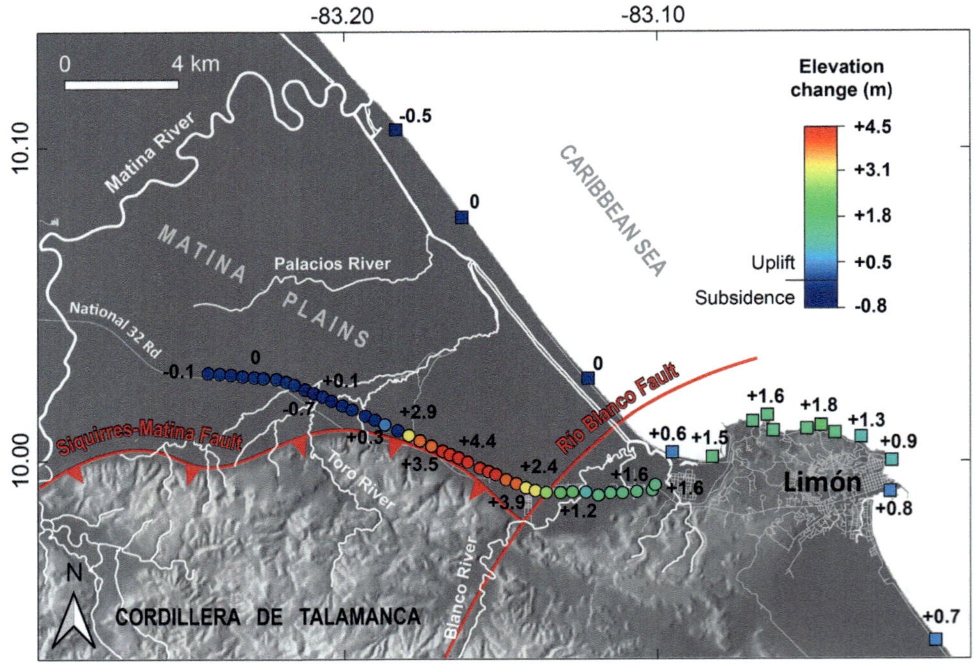

20.7 Relevance of Costa Rica's Caribbean Coastal Plains and Further Research

The Caribbean coastal plains of northern Costa Rica were left disregarded for years in spite of their outstanding geomorphological richness. Recent research has brought to light old and new data on the characteristics, diversity and complexity of a landscape long overlooked (e.g., Galve et al. 2016, 2022; Oconitrillo and Sanabria 2020). Still, geomorphological research focusing on this region is in early stages. Future investigations might determine the pace of changes in the fluvial system, the rates at which the processes act on the relief, the relationship between landforms and subsurface structures, or the detailed evolution of the landscape. The influence of the inferred sea-level change on coastal plain geomorphology during the Late Quaternary is a key question to resolve. Sea-level changes during glacial and interglacial periods most certainly impacted the profiles of the rivers described in our study, but also conditioned the dynamism of the geomorphic systems associated with them. Thus, with a background of tectonic and volcanic processes acting at decadal or centennial temporal scales, the piedmont and floodplain landscape resulted from millennial sea-level changes that influenced relief dynamics, then and even now. The region would be a fruitful area for more geomorphological work.

Another important aspect to consider is that geomorphology and other earth and social sciences converge in these coastal plains. On the one hand, seismological and tectonic investigations continue to be developed taking the 1991 Limón Earthquake as a reference; and on the other hand, archaeological and historical investigations are being carried out with the support of geomorphological evaluations. A critical point in this respect would be the search for the port of Suerre, work which could benefit from further geomorphological analysis in the area. It is also worth mentioning that future volcanic hazard assessments for some of the most dangerous volcanoes in Costa Rica (i.e., Barva, Irazú and Turrialba) could be constructed using the geological data obtained from the Santa Clara plains. The volcanoclastic deposits that form the Santa Clara Megafan record the strongest eruptions of these volcanoes and may constitute the most reliable source of data for estimating the hazard of future damaging eruptions.

So as to conclude, the alluvial and tectonic landforms described in this chapter create a special landscape worth disseminating to the geomorphological community as a fine example of a plain environment formed by interactions between endogenous and exogenous processes. The region harbors a unique modern volcanogenic megafan and offers good examples of alluvial rivers disturbed by active faults, plus several prehistoric and historic avulsions. No matter how far one looks over the world, it is difficult to find all these features stemming from different geological origins in a space as small as the Caribbean coastal plains of northern Costa Rica.

References

Alvarado GE (1990) Características geológicas de la Estación Biológica La Selva. Tecnología en Marcha 10(39):11–22

Alvarado GE, Gans PB (2012) Síntesis geocronológica del magmatismo, metamorfismo y metalogenia de Costa Rica, América Central. Rev Geol Amér Central 46:7–122

Alvarado GE, Esquivel L, Sánchez BE, Matamoros G (2020) Peligro Volcánico del Turrialba, Costa Rica. Comisión Nacional de Prevención de Riesgos y Atención de Emergencias (CNE) de Costa Rica

Alvarado GE, Campos D, Brenes-André J, Alpizar Y, Nuñez S, Sojo D, Esquivel L (2021) Peligro Volcánico del Irazú, Costa Rica. Comisión Nacional de Prevención de Riesgos y Atención de Emergencias (CNE) de Costa Rica

Alvarado GE (2016) El cinturón deformado del norte de Costa Rica-Colombia: ¿Desde una faja de empuje-plegamiento hasta una zona de subducción incipiente dentro de un bloque o microplaca? IASPEI Regional Assembly Latin-American and Caribbean Seismological Commission—LACSC, June 20–22, San José, pp 39

Andrade CA, Barton ED, Mooers CNK (2003) Evidence for an eastward flow along the Central and South American Caribbean Coast. J Geophys Res 108. https://doi.org/10.1029/2002jc001549

Arroyo IG, Linkimer L (2021) Modelo unidimensional de velocidades sísmicas y características tomográficas tridimensionales del Caribe Sur de Costa Rica. Rev Geol Amér Central 65:1–15

Barquero R, Rojas W (1994) Sismicidad inducida por el terremoto de Limón. Rev Geol Amér Central Vol. Esp Terremoto de Limón, pp 111–120

Benzoni C (1565) La historia del Mundo Nuevo. Apresso Francesco Rampazetto, Venecia

Bergoeing JP (2017) Geomorphology and volcanology of Costa Rica. Elsevier

Bergoeing JP, Brenes LG, Protti R, Arce R, Artavia LG, Salas D, Carrillo M (2010) Atlas geomorfológico del Caribe de Costa Rica. San José, UCR-IGN

Bergoeing JP (1978) La fotografía aérea y su aplicación a la geomorfología de Costa Rica. Ministerio de Obras Públicas y Transportes, IGN, San José

Bergoeing JP (1998) Geomorfología de Costa Rica. Inst. Geográf. Nacional, San José

Boschini I, Montero W (1994) Sismicidad histórica e instrumental del Caribe de Costa Rica. Rev Geol Amér Central 18:65–72

Brandes C, Astorga A, Littke R, Winsemann J (2008) Basin modelling of the Limón Back-arc Basin (Costa Rica): Burial history and temperature evolution of an island arc-related basin-system. Basin Res 20:119–142

Burrato P, Ciucci F, Valensise G (2003) An inventory of river anomalies in the Po Plain, Northern Italy: Evidence for active blind thrust faulting. Ann Geophys 46:865–882

Calvo JB (1887) Apuntamientos geográficos, estadísticos é históricos, San José de Costa Rica, Imprenta Nacional

Curray JR (1961) Late Quaternary sea level: a discussion. Gel Soc Amer Bull 72:1707–1712

DeCelles PG, Cavazza W (1999) A comparison of fluvial megafans in the Cordilleran (Upper Cretaceous) and modern Himalayan foreland basin systems. Bull Geol Soc Amer 111:1315–1334

Dechesne M (1994) Island-arc volcanism and episodic fluvial sedimentation in the Atlantic Zone of Costa Rica. CATIE/AUW/MAG. Atlantic Zone Programme. Report N°. 91. Field Report No. 137. Costa Rica. sp

Denyer P, Montero W, Alvarado GE (2003) Atlas tectónico de Costa, Rica. Univ, Costa Rica, San José

Denyer P, Cárdenes G (2000). Costas marinas. In: Denyer P, Kussmaul S (eds) Geología de Costa Rica, Ed. Tecnológica de Costa Rica, Cartago pp 185–218

Denyer P, Arias O, Personius S (1994) Efecto tectónico del Terremoto de Limón, Rev Geol Amér Central Vol. Esp Terremoto de Limón, pp 39–52

De Peralta MM (1890) Límites de Costa Rica y Colombia. Nuevos documentos para la historia de su jurisdicción territorial con notas, comentarios y un examen de la cartografía de Costa Rica y Veragua. Imprenta de Manuel Ginés Hernández, Madrid

Espinosa-Baquero A (2020) A model of the Quindío and Risaralda Quaternary deposits. The Geol Colombia 4:333–352
Fernández JA, Bottazzi G, Barboza G, Astorga A (1994). Tectónica y estratigrafía de la Cuenca Limón Sur. Rev Geol Amér Central Vol Esp Terremoto de Limón, pp 15–28
Fonseca E, Alvarenga P, Solórzano JC (2001) Costa Rica en el siglo XVIII/(Colección historia de Costa Rica; t.8): EUCR, San José
Galve JP, Alvarado G, Azañón JM, Giaconia F, Pérez Peña JV, Notti D, Mora Fernández M, Booth-Rea G (2015) Analysis of the river anomalies in the Tortuguero plain (Costa Rica). In: Proceedings of the World Multidisciplinary Earth Science Symposium (WMESS), Prague, Czech Republic, 7–11 September 2015
Galve JP, Alvarado GE, Pérez-Peña JV, Mora MM, Booth-Rea G, Azañon JM (2016) Megafan formation driven by explosive volcanism and active tectonic process in a humid tropical environment. Terra Nova 28:427–433
Galve JP, Alvarado GE, Pérez-Peña JV, Pérez-Consuegra N, Ruano P, Becerril L, Devoto S, Reyes-Carmona C, Azañón JM (2022) River avulsions and megafan development triggered by high recurrent explosive volcanism. Geomorphology 419:108466
González C (1906) Apuntes sobre geografía histórica de Costa Rica. San José de Costa Rica, Imprenta de Avelino Alsina, pp 102
Guarín F (2008) Etude sédimentologique du fan volcanoclastique du Quindío-Risaralda (Colombia) et sa relation avec la morphonéotectonique. University of Geneva
Hansford MR, Plink-Björklund P (2020) River discharge variability as the link between climate and fluvial fan formation. Geology 48:952–956
Hilje L (2017) Los puentes en Angostura, Turrialba. Comunicación 26(2):93–127. https://doi.org/10.18845/rc.v26i2-17.3447
Hilje L (2018, comp.) Turrialba en la mirada de los viajeros. Ed Tecnológica de Costa Rica
Hosono T, Delinom R, Nakano T, Kagabu M, Shimada J (2011) Evolution model of δ34S and δ18O in dissolved sulfate in volcanic fan aquifers from recharge to coastal zone and through the Jakarta urban area, Indonesia. Sci Total Environ 409:2541–2554
Instituto Costarricense de Electricidad (ICE) (2016) Meteorological database from the "Área de Hidrología". [Database]. C.S. Estudios Básicos de Ingeniería: Pronóstico, Procesamiento y Análisis Meteorológico, Área de Hidrología. San Jose, Costa Rica (Retrieved 23 Febr 2016)
Kesel RH, Lowe DR (1987) Geomorphology and sedimentology of the Toro Amarillo Alluvial fan in a humid tropical environment, Costa Rica. Geogr Ann 69:85–99
Leier AL, DeCelles PG, Pelletier JD (2005) Mountains, monsoons, and megafans. Geology 33:289–292
Madrigal R, Rojas E (1980) Manual Descriptivo del Mapa Geomorfológico de Costa Rica (escala 1:200 000). San José, Costa Rica: SEPSA, Imprenta Nacional
Malavassi E, Chaves R (1970). Estudio geológico regional de la Zona Atlántica Norte de Costa Rica. Inf Técnico y Notas Geol 9(35):1–16; Direc Geol Min y Petról, San José
Malavassi E, Madrigal R (1970). Reconocimiento Geológico de la Zona Norte de Costa Rica. Inf Técnico y Notas Geol 9(38):1–12; Direc. Geol Min y Petról, San José
Marshall JS (2007) Geomorphology and physiographic provinces. In: Bundschuh J, Alvarado GE (eds) Central America: geology, resources and hazards. Taylor & Francis, London, vol 1. pp 75–122
Montes de Oca C (2005) Y las mulas no durmieron…Los arrieros en Costa Rica, siglos XVI al XIX. EUNED, San José
Morell KD, Kirby E, Fisher DM, Van Soest M (2012) Geomorphic and exhumational response of the Central American Volcanic Arc to Cocos Ridge subduction. J Geophys Res [solid Earth] 117:B04409
Narson GC, Croke JC (1982) A genetic classification of floodplains. Geomorphology 4(6):459–486
Nieuwenhuyse A, Kroonenberg SB (1994) Volcanic origin of Holocene beach ridges along the Caribbean coast of Costa Rica. Mar Geol 120:13–26
Nieuwenhuyse A (1996). Landscape formation and soil genesis in volcanic parent materials in humid tropical lowlands of Costa Rica. Universiteit Wageningen
Obregón MR (1932) Geografía general de Costa Rica. Lecturas Geográfica 3° serie. Tomo I geografía Física. Unión Panamericana Washington, San José
Oconitrillo G, Sanabria IJ (2020). Nuevo cauce del Río Parismina Estudio cartográfico/fotointerpretativo, inspección de campo y contexto geológico/tectónico. Materia Registral. Rev Registro Nacional Costa Rica. V2 (16):23–31
Payne-Iglesias E (2022) "Robos, ultrajes y cautiverios": los puertos del Caribe de Costa Rica (siglos XVI al XIX.). Universidad de Costa Rica, CIHAC, San José, Costa Rica
Quesada-Román A, Pérez-Briceño PM (2019) Geomorphology of the Caribbean coast of Costa Rica. J Maps 15:363–371
van Ruitenbeek F (1994) Holocene, arc-volcanism controlled, episodic sedimentation in the atlantic lowland of Costa Rica. CATIE/AUW/MAG. Atlantic Zone Programme. Report N°. 74. Field Report No. 120. Costa Rica. sp
Santillana (2008) La Enciclopedia del Estudiante. Geografía de Costa Rica (Tomo 20). Peru, Santillana S.A.
Sapper K (1905) Gebirgsbau und Boden des südlichen Mittleamerika. Justus Perthes, Gotha, 151. 1–82 + 3 mpas + 2 profiles
van Seeters RJM (1993) A study on geomorphology, mineralogy and geochemistry of the Toro Amarillo-Tortuguero and the Chirripo-Matina river system in the Atlantic Zone of Costa Rica. CATIE/AUW/MAG. Atlantic Zone Programme. Report N°. 47. Field Report No. 95. Costa Rica. sp
Ventra D, Clarke LE (2018) Geology and geomorphology of alluvial and fluvial fans: Current progress and research perspectives. Geol Soc Spec Publ 440. https://doi.org/10.1144/SP440.16
von Frantzius (1861) Originalkarte des nördlichen Theiles von Costarica, nach einer Original-Zeichnung. Gotha, Justus Pethes
Weissmann GS, Hartley AJ, Nichols GJ, Scuderi LA, Olson M, Buehler H, Banteah R (2010) Fluvial form in modern continental sedimentary basins: distributive fluvial systems. Geology 38:39–42
Wielemaker WG, Vogel AW (1993). Un Sistema de Información de Suelos y Tierras para la Zona Atlántica de Costa Rica. Atlantic Zone Program Report 22: Wageningen Agricultural Univ, Wageningen
Zhang Y, Dai X, Wang M, Li X (2020) The concept, characteristics and significance of fluvial fans. Pet Explor Dev 47:1014–1026

Highly Weathered Soil Landscapes of Costa Rica

21

Manuel E. Camacho-Umaña, Adolfo Quesada-Román, Mario Villatoro-Sánchez, Bryan Alemán-Montes, Rafael Mata, Carlos Henríquez-Henríquez, Jesús Céspedes-Rivera, and Alfredo Alvarado

Abstract

Highly weathered soils, formerly known as lateritic soils, are currently classified according to the Soil Taxonomy as Oxisols and Ultisols. These soils have been identified in various landscapes of Costa Rica, developing under contrasting climatic conditions indicated by annual precipitation values between 1749 and 5778 mm. The chapter provides a concise description of five specific landscapes of Costa Rica, where Oxisols have been identified as dominant. The parent materials were grouped within four general categories: (1) Basaltic rocks from Jurassic-Miocene Oceanic Complexes, (2) Igneous materials of Paleocene-Pliocene ages, (3) Sedimentary materials from the Pliocene–Pleistocene ages, and (4) Volcanic materials from the Pleistocene. Despite this diversity in parent materials and climatic conditions, the stability of the geomorphic units is the common factor that allowed for the development of Oxisols and some great groups of Ultisols, corroborating the fundamental role of geomorphology in the origin of highly weathered soils.

Keywords

Costa Rica · Oxisols · Ultisols · Weathered soils · Geomorphology

M. E. Camacho-Umaña (✉) · M. Villatoro-Sánchez · B. Alemán-Montes · R. Mata · C. Henríquez-Henríquez · J. Céspedes-Rivera · A. Alvarado
Centro de Investigaciones Agronómicas, Universidad de Costa Rica, San José, Costa Rica
e-mail: manuel.camachoumana@ucr.ac.cr

A. Quesada-Román
Laboratorio de Geografía Física, Escuela de Geografía, Universidad de Costa Rica, San José, Costa Rica

21.1 Introduction

Highly weathered soils are usually understood as deep, red soils with high pedogenetic development and low activity clays. In Central America, soil characterization studies started in the twentieth century, with notable publications by Anderson and Byers (1930), Dóndoli (1943), Jaramillo (1969), Martini (1970), Macías (1969), Eswaran (1972). Previously, highly weathered soils were classified as "laterites" (Hardy 1970) until Soil Taxonomy came out (Soil Survey Staff 1975). Further, international committees were established to deal with taxonomic studies and aspects of these weathered soils, designating them as Oxisols (Buol and Eswaran 1988, 2000).

Soil-forming process includes factors that act together and independently over a parent material to modify it into a soil body. Jenny (1994) developed a mathematic function named "Universal equation for soil-forming factors" based on the combined effects of five forming factors: parent materials, relief (landforms), organisms, climate, and time, which depending on the magnitude and specific combinations between each other yield in different soils. The concept of these five forming factors remains valid up today, and they are the cornerstone for the main soil-forming processes: transformations, translocations, additions, and losses (Alvarado 1985; Buol et al. 2011; Targulian and Krasilnikov 2007), providing fundamental information to explain and understand the genesis of a certain soil. Within these soil-forming factors, landscape morphology becomes fundamental to understand the pedogenetic pathways forming highly weathered soils in the tropics. For instance, stability in geomorphologic units is required for Oxisols-forming processes, especially when dealing with materials and landforms formed during the Quaternary (Thomas 1994; Camacho et al. 2020, 2021).

Hence, the proper study and characterization of geomorphological landscapes and their features becomes essential in soil survey, genetic considerations, and classification of

A. Quesada-Román (ed.), *Landscapes and Landforms of Costa Rica*, World Geomorphological Landscapes, https://doi.org/10.1007/978-3-031-64940-0_21

highly weathered soils. The present chapter focuses on five landscapes from Costa Rica, where climatic conditions and landform features allowed for the formation of Oxisols and Ultisols. Forming factors involved in pedogenesis will be also described. Our results are key to other tropical landscapes where highly weathered soils are present and which face similar environmental challenges as the ones presented in this chapter.

21.2 Landscapes with Highly Weathered Soils Development in Costa Rica

21.2.1 Mountains of Peninsula de Nicoya

Located in the Pacific Northwestern region of Costa Rica, these mountains were formed from the Nicoya Oceanic Complex (Jurassic–Lower Cretaceous), a geological formation mainly composed by igneous and sedimentary rocks, with some materials dated from 72 ± 4 to 30 ± 15 Ma (Alvarado et al. 1992; Denyer and Gazel 2009). This geological formation is considered among the oldest within the Costa Rican Territory. These mountains and their associated landforms are included within the large geomorphologic unit "Península de Nicoya". This unit (Fig. 21.1) is mainly composed by small hills, with evidence of intense weathering cycles of igneous and sedimentary materials (Bergoeing 2007). These weathering processes produce clay-rich weathering mantles, modeled by severe erosion processes which on steep slopes include solifluction, soil reptation, and landslides (Mora 1985).

The climate in this region (Fig. 21.2) is characterized by annual precipitation values of 1853 mm, with a bimodal distribution and peaks in May–June and August–October. This rainfall pattern includes a marked dry period between December and April (almost 0 mm), when the monthly average air temperature reaches the highest values (30 °C).

These climatic conditions restraint the number of plant species that cover this land, only allowing for the growth of species with drought resistant mechanisms, for instance coffee (*Coffea arabiga* L.) and orange (*Citrus sinensis* L.) have been successful crops. In addition, grasslands and cattle farms are possible within this area, and more recently orange production has become a cash crop (Fig. 21.3). However,

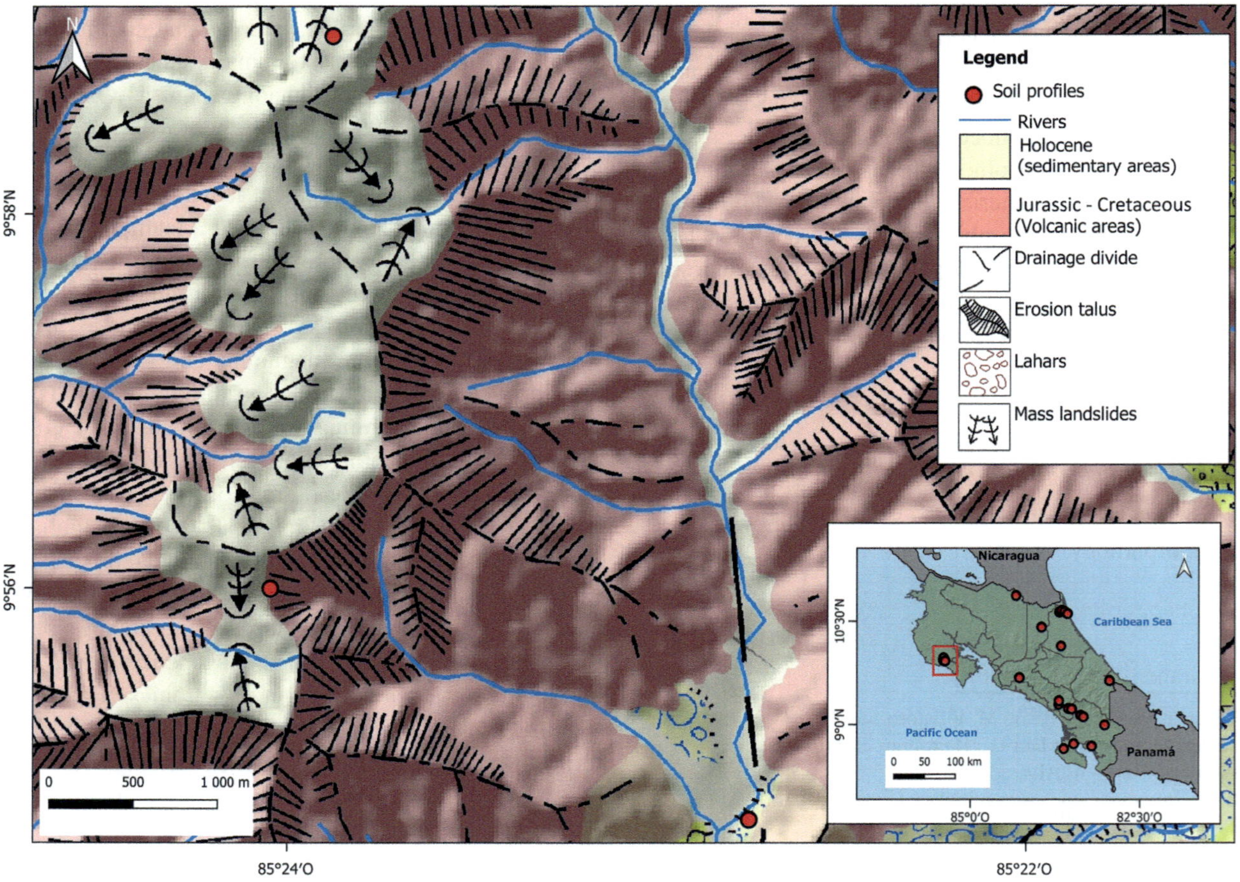

Fig. 21.1 Location, main geomorphologic features, and soil profiles observed in the Mountains of Peninsula de Nicoya. Adapted from Bergoeing and Brenes (2017)

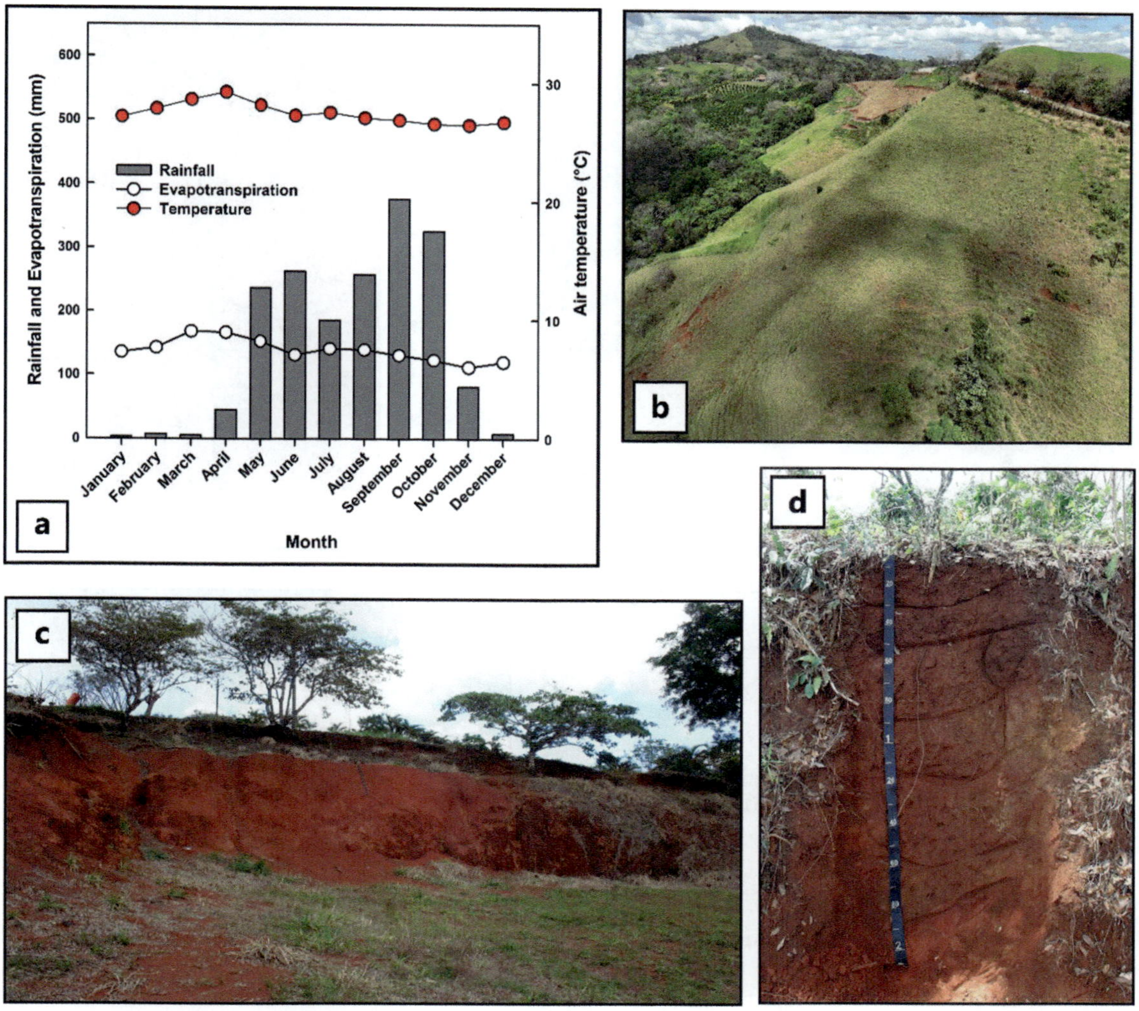

Fig. 21.2 Climate, landforms, and soils in the Mountains of Peninsula de Nicoya: **a** Atmospheric water balance estimated for the town of Hojancha, showing six months of water deficit; **b** Steep slopes and land denudation showing clayey reddish materials; **c** Red soil exposed in a soccer field, providing a complete view of the soil profile; **d** Typical soil observed in the highlands cultivated with coffee (*Coffea arabiga* L.)

the plant cover observed in this landscape can be classified as Lowland Semi deciduous Wet Forest, with tall plant species with clear defoliation during the dry season such as *Enterolobium ciclocarpum* and *Scheelea rostrata* (Vargas 1992). Likewise, the vegetation found in this region can be classified according to Holdridge's Life Zones as Premontane Wet Forest (wf-P) (Bolaños and Watson 1993), with secondary forest species such as *Cordia alliodora*, *Dalbergia retusa*, *Tabebuia rosea*, and *Lonchocarpus parviflorus*. (Soudre 2004; Granda 2015). In addition, quality tropical hardwood species like teak (*Tectona grandis* L.f.) has become an economic alternative due to its adaptability to dry conditions and timber high productivity (Fernández-Moya et al. 2014).

The soils in this landscape (Fig. 21.2) can be classified within the orders of Alfisols and Oxisols. Camacho et al. (2021), while studying soils in the highlands of Hojancha, classified the soils as *Kandiustalfic Eutrustox,* highly weathered soils with enough contents of exchangeable cations (> 35% base saturation) along all horizons and under an ustic soil moisture regimen. This soil order was previously described in this region by Wilcke et al. (1998), but these authors classified the soils as *Kadiudalfic Eutruduox* (Fig. 21.3), inferring a short drought period that is unlikely to occur under the aforementioned climatic conditions for this landscape. Interestingly, these Oxisols seem to be related with Alfisols, representing a transitional state between the orders Alfisols and Oxisols (as stated in the taxonomic Subgroup *Kandiustalfic*). This is consistent with the *Typic Rhodustalfs* found by Camacho (2016) just 4 km away from the Oxisols, and with other reports on Alfisols-Oxisols associations for other tropical and subtropical regions (Buol and Eswaran 2000). Despite those soils were classified as Oxisols (the most weathered soils within Soil Taxonomy), the exchangeable bases saturation was higher than 85%, in agreement with that found by Moura-Filho and Buol (1972), who described and

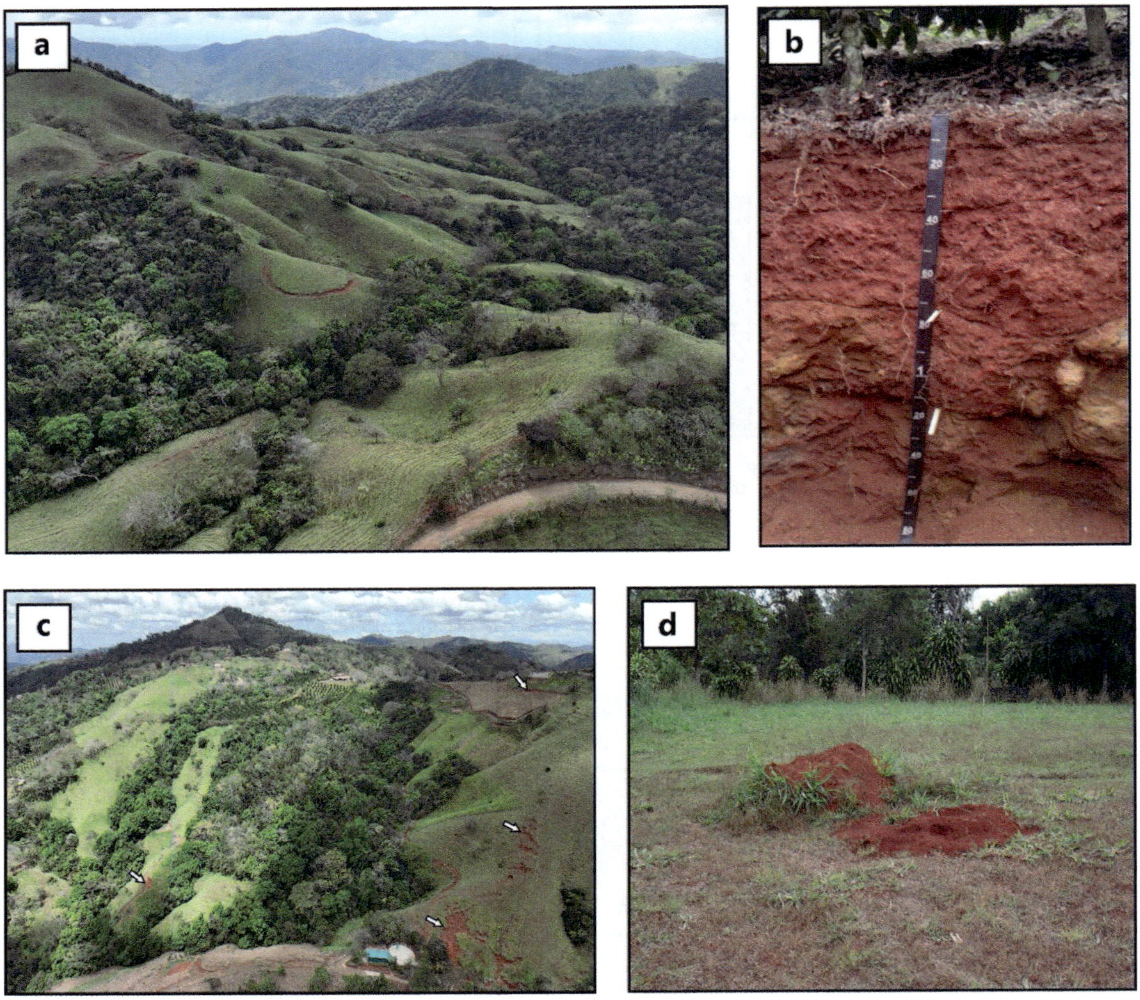

Fig. 21.3 Scenic views, landforms, and soils observed in the Mountains of Peninsula de Nicoya: **a** The mountains of Hojancha, with typical dissected relief and evidence of human activities like coffee (*Coffea arabiga* L.) and orange (*Citrus sinensis* L.) plantations and cattle farms; **b** The profile of a *Kandiustalfic Eutrustox* cultivated with coffee; **c** Steep slopes, road cuts and erosional scars showing clayey reddish materials (highlighted with white arrows); **d** Ant mounds of leaf-cuttler ants (*Atta cephalotes* L.) which show a reddish subsoil taken to the surface

classified "Latosolos Roxos" (*Eutrustox*) in Southern Sao Paulo and Brasilia in Brazil.

21.2.2 Alluvial Fans from Valle del General

Situated in the Central South region of Costa Rica, this landscape extends across the whole Río General Watershed, including the cities of San Isidro del General (Pérez Zeledón) and Buenos Aires de Puntarenas (Fig. 21.4). The current landforms observed developed from the progressive deposition of highly weathered old materials, deposited as alluvial fans during the Pliocene and the Quaternary (Kesel and Spicer 1985; Bergoeing 2011). These old alluvial fans present a clayey-red matrix, with some sandstones, andesitic rocks, and iron stones and duricrusts.

These old alluvial fans (Fig. 21.4) were formed due to extensive processes of deposition, associated with rapid deglaciation of the extensive glaciers that existed in the Cordillera de Talamanca during the Last Glacial Maximum (LGM). In addition, resulting alluvial fans have been modeled by high rainfall rates, high temperatures, and intense weathering processes (Camacho et al. 2020), which progressively have been molding the current fluvial landforms (Quesada-Román and Zamorano-Orozco 2019). In this regard, the rainfall distribution pattern observed in this landscape (Fig. 21.5) is characterized by annual precipitation values of 3050 mm, mainly from May to November. This rainfall pattern allows for a water deficit period (around −380 mm) from December to March, which represents an ustic soil moisture regime for most of the soils. The highest rainfall values occur in August, September, and

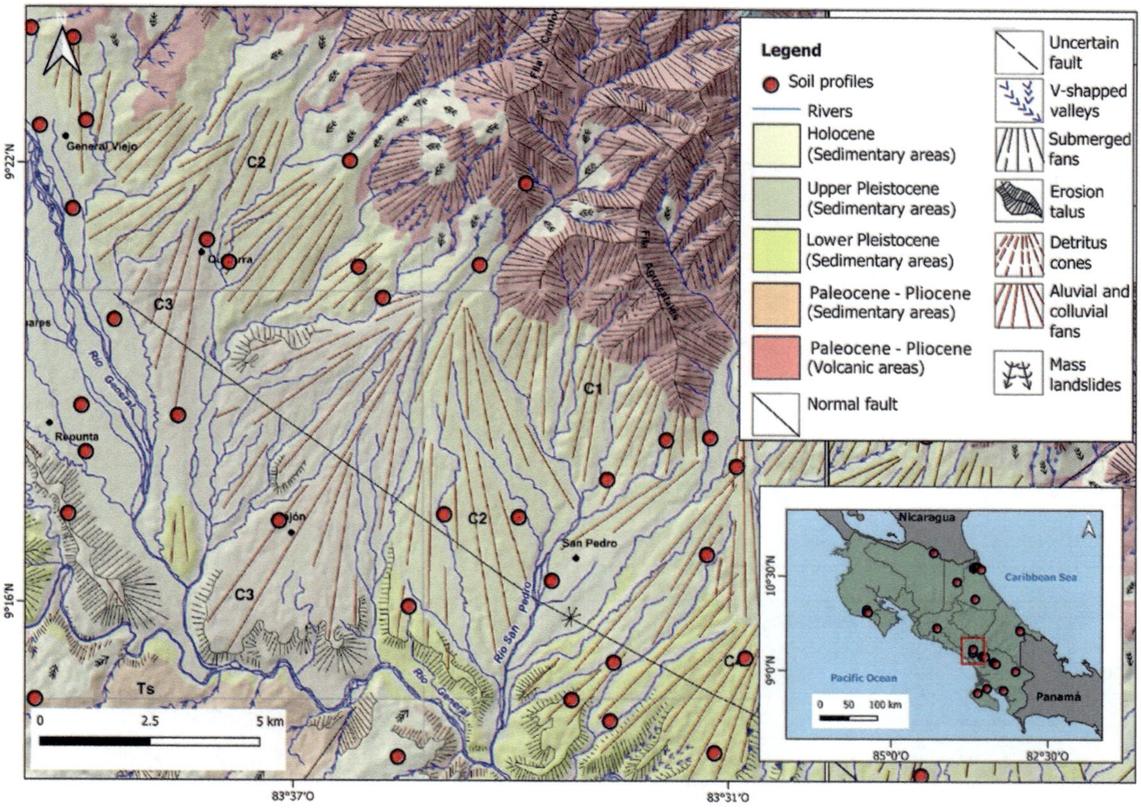

Fig. 21.4 Location, main geomorphologic features, and soil profiles observed in El Valle del General. Adapted from Bergoeing and Brenes (2017)

October (with average monthly values > 350 mm). In addition, the average air temperature reaches the highest values (25 °C) in March, and the lowest in November (21 °C), in agreement with the highest evapotranspiration value (Fig. 21.5).

Those climatic conditions allowed for the development of cash crops like sugarcane (*Saccharum officinarum* L.) (Fig. 21.6), or pineapple (*Ananas comosus* (L.) Merr.) (Fig. 21.7), which in addition to the population growth have been responsible for landscape transformation with time, extending the crop land across the whole area, enhancing land fires and even further formation of savannas (Budowsky 1963; Miranda 1983). Both cash crops represent a fundamental economic activity for this region, based on the high yields observed and the amount of people involved.

The soils of these valleys have been classified as Ultisols and Oxisols. Several literature reports indicate that highly weathered soils are dominant within these alluvial geomorphologic units (Kesel and Spicer 1985; Wilcke et al. 1998). For this region, Camacho et al. (2021) classified the soils as *Plinthic Kandiustox* and *Anionic Acrustox,* the latter being the most weathered soils of Costa Rica. These *Anionic Acrustox* are characterized by quite low values of effective cation exchange capacity (< 1.5 $cmol_c$ kg $clay^{-1}$) within 150 cm from the soil surface, an ustic soil moisture regimen, pH dependent charge, and presence of lateritic bodies (Fig. 21.7).

Interestingly, those *Acrustox* developed over older alluvial fans (formed 85,000–45,000 y.a.), whereas the *Kandiustox* can be found over intermediate alluvial fans (formed 45,000–7000 y.a.). This difference in time of formation within the parent materials produces changes among the soil-forming processes, which results in different degrees of weathering within these soils classified as Oxisols (Camacho et al. 2020).

21.2.3 Lowlands of San Carlos

This landscape is located in the northern sector of Costa Rica, more precisely in the Central American Physiographic Province of "Nicaraguan depression" (Fig. 21.8; Marshall 2007). This region is characterized by *debris flows* and old alluvial fans deposition from the Guanacaste and Central volcanic ranges in the western sector, with the presence of red and highly weathered matrices (Fig. 21.9), mostly composed by kaolinitic clays (Bergoeing and Protti 2006).

The eastern sector of this landscape presents a diversity of volcanic materials, including some dacitic porphyries mounts, pyroclastic-felsic block deposits, and vitric welded

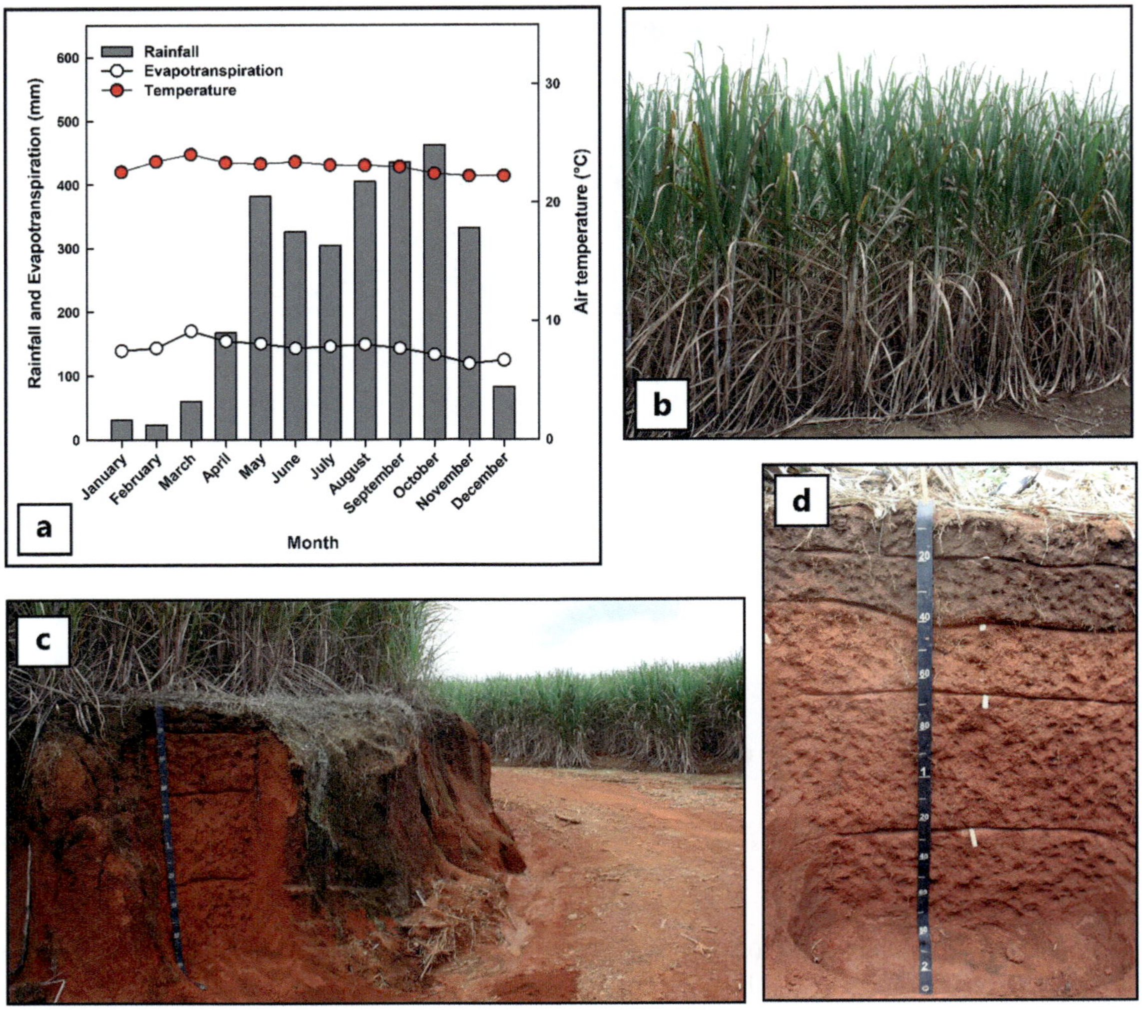

Fig. 21.5 Climate, crops, and soils observed in El Valle del General: **a** Atmospheric water balance estimated for the region of Valle del General, showing 4 months of water deficit; **b** Sugarcane (*Saccharum officinarum* L.); **c** Sugarcane growing over clayey reddish and highly weathered soils within old alluvial fans; **d** The profile of a *Anionic Acrustox* cultivated with sugarcane

tuffs limited by two quartz-stripes (Vargas and Alfaro 1992; Gazel et al. 2005). All these materials have been suffering from intense weathering processes, which are enhanced due the climatic conditions. Solano and Villalobos (2001) and Korpela (2014) described the region "Llanuras de San Carlos" as very rainy, with annual precipitation values around 3020 mm, with two months without rainfall (considered as a short drought period) (Fig. 21.9). This region presents an average air temperature of 23 °C, with minimum and maximum values of 20° and 31°, respectively. These climatic conditions promote the growth of a number of plant species typical of Tropical wet forest (wf-T), with heights that could reach 40–60 m. Among them are *Anacardium excelsun*, *Ceiba pentandra,* and other important forestry species like *Vochysia guatemalensis, V. ferruginea, Gmelina arborea,* and *Terminalia amazonia* (Fagan et al. 2015). More recently, the crop pineapple (*A. comosus*) has been extending within the northern side of this landscape, covering marginal areas, where cattle and forestry plantations were formerly established (Figs. 21.10 and 21.11). Other important crops (mostly in low input agriculture) found in this landscape are common red and black beans (*Phaseolus vulgaris* L.,), cassava (*Manihot esculenta* Crantz), and some extensive crops like orange (*C. sinensis* L.).

Soil development in this landscape reflects intense weathering processes acting over igneous materials, enhanced by hot and rainy environments which promote leaching and neoformation of sesquioxic and kaolinitic clay minerals. As a result, soils have been classified as *Plinthic Kandiudox, Typic Rhodudults,* and *Typic Haplohumults* (Fig. 21.9). Highlighted are these Oxisols, which present the development of plinthite within their bottom subsoil

Fig. 21.6 Crops, soils, and special features observed on Quaternary alluvial fans: **a** Extensive areas cultivated with sugarcane (*Saccharum officinarum* L.); **b** A typical soil profile observed in the plantations of sugarcane; **c** Deposits of magnetite grains (red arrows), and iron-rich duricrust (white arrows); **d** Zoom picture of the iron-rich duricrust observed

horizons. This feature indicates (i) potential fluctuation of the groundwater table, allowing the differentiation of iron forms and further hardening, and (ii) changes within the landscapes which prevent erosion and degradation of plinthic materials (Eze et al. 2014).

In general, these weathered soils have low fertility (high soil acidity, low content of nutrients) which present a constraint for most of agricultural crop's development, except the pineapple crop which is tolerant to soil acidity (Figs. 21.9 and 21.10). Nonetheless, there are some areas where the plinthite formation close to the soil surface represents a serious restriction for root growth and water movement, providing conditions for redoximorphic features development like glei colors (Fig. 21.11).

Consequently, it is not surprising that economic activities like agriculture and cattle production are limited in this landscape, constraining human settlement and therefore population growth and development. Nevertheless, in 1992, gold ores were discovered within the subsoil in the northern part of this area (Crucitas de Cutris), which could represent a potential economic activity. Indeed, Crucitas Mine Project (Fig. 21.11) was intended to start gold extraction in 1999, but the project was shut down due to a resolution from the Costa Rican Constitutional Court (Korpela 2014).

21.2.4 Mountains of Peninsula de Osa

This landscape is located in the southeastern part of Costa Rica, extending along the inner shore of Golfo Dulce and inward of the Peninsula (Fig. 21.12). This landscape is described as a litho-stratigraphic unit, composed by a sedimentary layer overlying basaltic materials, formerly associated with the Nicoya Oceanic Complex. Subsequently, these materials were reclassified as: (a) Golfito Terrain (90–64 Ma) mostly composed of basalts, breccias, and

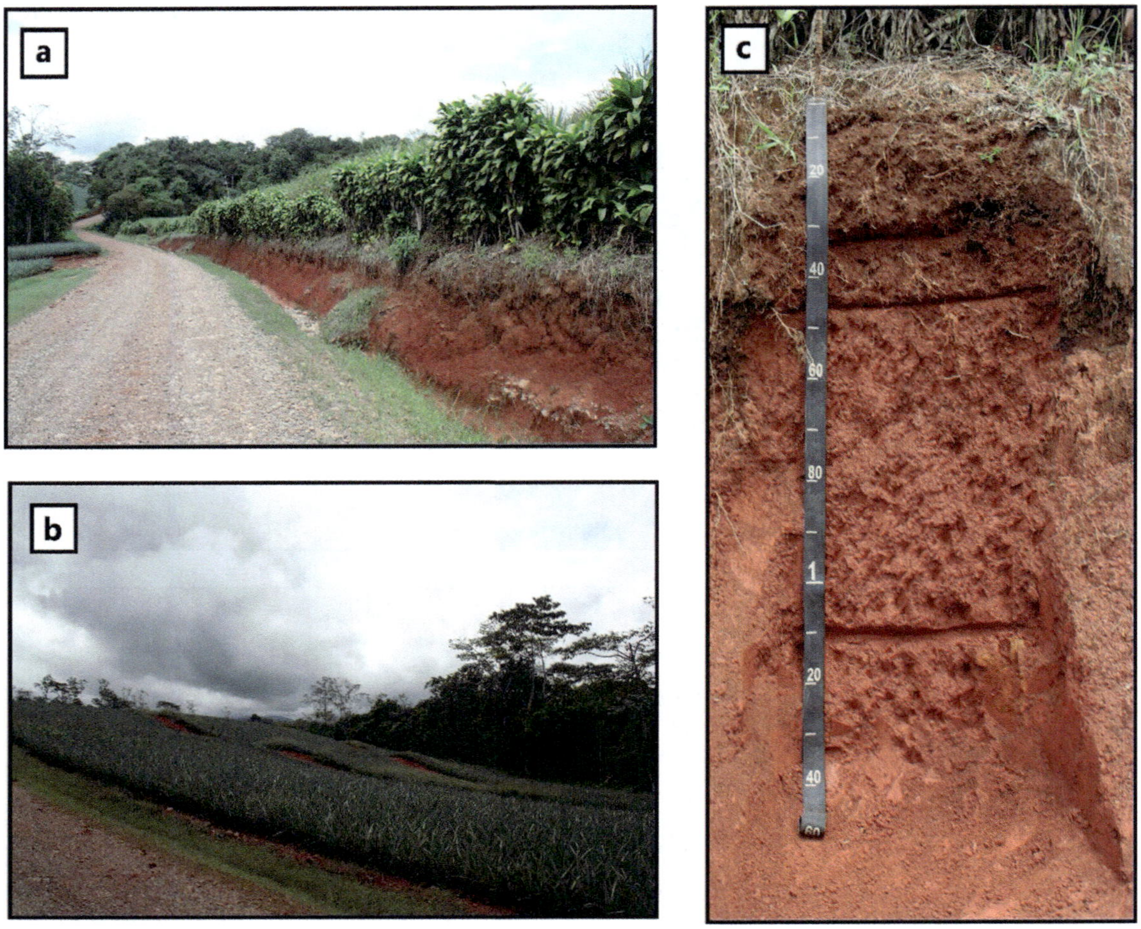

Fig. 21.7 Typical landscape with pineapple fields on Quaternary alluvial fans: **a** Road slope that extends along the whole field, exhibiting a distinct red soil; **b** Field planted with pineapple (*Ananas comosus* (L.) Merr.), in specific sowing beds arrangement; **c** The soil profile observed in this pineapple field, described as deep, reddish, and highly weathered soil

dolerites, with interbedded pelagic sediments from the upper Maastrichtian; and (b) Rincón Block (62–54 Ma) with accreted seamounts composed by basaltic materials (Denyer and Gazel 2009). According with Marshall (2007), the geomorphologic unit "Mountain range of Peninsula de Osa" belongs to the Central American Physiographic Province "Chorotega fore arc". This mountain range crosses the Peninsula from NW–SE, with rivers draining the mountains toward the flatlands and separating among tectonic and erosive geoforms, with highly weathered clay materials and saprolite exposed by landslide activity (Fig. 21.13).

Climatic conditions observed in this region are described as warm and wet (Solano and Villalobos 2001). For instance, total annual precipitation is 4282 mm (Fig. 21.14), with the highest monthly average values during August–October (more than 500 mm month^{-1}). This rainfall pattern also suggests a very short dry period (January and February), with a clear decrease of precipitation (< 200 mm month$-^{1}$). The annual average, minimum, and maximum temperature values are 27 °C, 23 °C, and 30 °C, respectively.

These particular climatic conditions allow for the growth of Tropical wet forest (wf-T) species according with Holdridge's Life Zones System (Bolaños and Watson 1993). Among these species, *Peltogyne purpure* Pittier (aka "Nazareno" or Purpleheart wood) is highlighted, a precious wood that currently is forbidden to cut down, being an endangered species. Despite this, the forest has been submitted to fractioning and deforestation, mostly associated to emerging development of extensive crops like oil palm (*Elaeis guineensis* JACQ) and cattle (Rosero-Bixby et al. 2002).

Within this landscape, the mountains and other elevated landforms present clayey deep soils with a characteristic reddish pattern (Fig. 21.15). These soils were previously described as Oxisols (Cleveland et al. 2003), but that study did not provide enough data to support this taxonomic classification. More recently, Camacho et al. (2021) studied and characterized the soils of Rincón de Osa, classifying

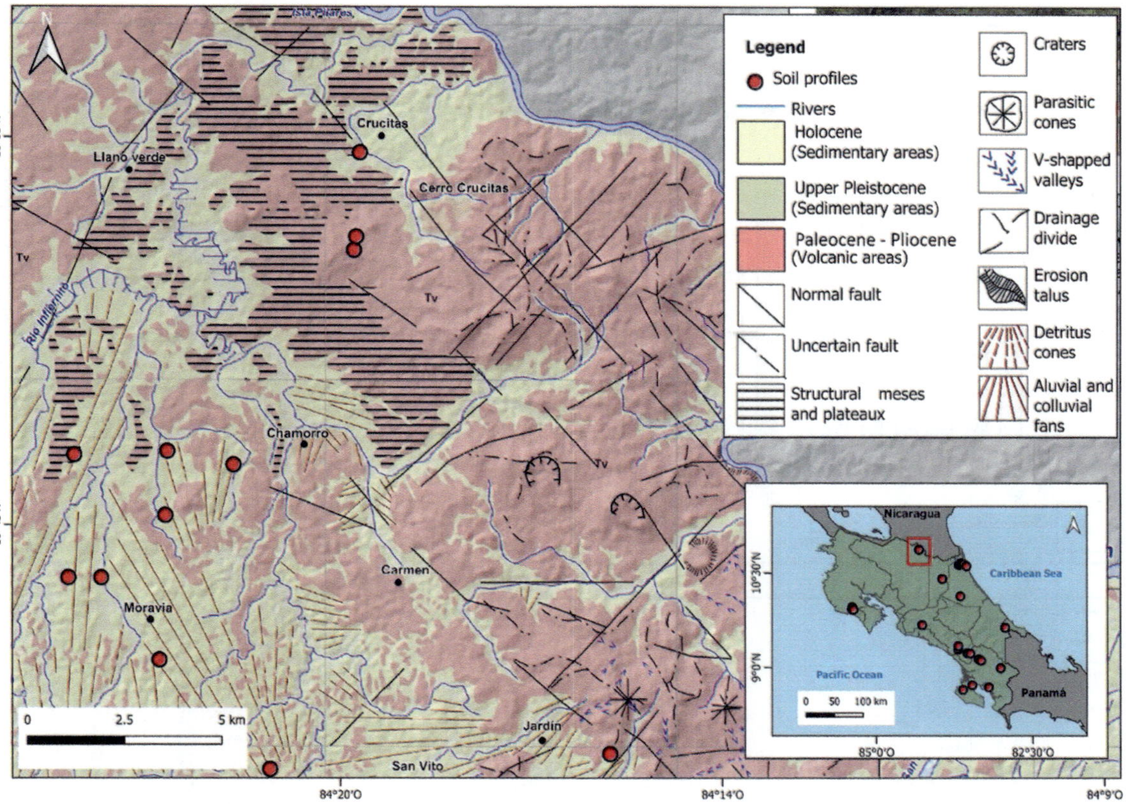

Fig. 21.8 Location, main geomorphologic features, and soil profiles observed in the Lowlands of San Carlos. Adapted from Bergoeing and Brenes (2017)

these soils as *Rhodic Kandiudox*, and confirming the presence of these highly weathered soils at Península de Osa (Fig. 21.15).

Rhodic Kandiudox is typically dark reddish soils (Hue 10R 3/6 or redder), with a kandic horizon where the content of weatherable minerals is less than 10%. These soils have normally a low exchangeable bases content (lower than 5 $cmol_c\ kg^{-1}$) within the subsoil horizons, but present excellent soil physical properties (Camacho et al. 2021). These soil morphological patterns (clayey soils with reddish colors) were observed along the whole landscape (Figs. 21.14 and 21.15).

21.2.5 Mountains of Tortuguero-Barra del Colorado

The mountains of Tortuguero-Barra del Colorado are located in the northeastern part of Costa Rica, bordering the San Juan River delta with the Caribbean Sea (Fig. 21.16). These mountains are surrounded by lowlands composed by marine sediments and fluvial-volcanic deposits from the Quaternary (Nieuwenhuyse 1996). In general, these mountains are described as Quaternary conic-basalts with a dated age of 1.2 ± 0.4 Ma (Alvarado et al. 1992; Gazel et al. 2011). These basaltic cones are lower than 300 m asl and represent fissure eruptions (Fig. 21.17).

This landscape is included within the Central American Physiographic Province of "Chorotega back arc", described as an alluvial plain dissected by a network of rivers (Fig. 21.16) that drain waters from the elevations of the Central Volcanic Range to the Caribbean (Marshall 2007). The flatness of the coastal plain is abruptly interrupted by low-elevation mounts. For instance, highlights two mounts: (1) Cerro Cocorí (an ancient volcanic cone, dated from the Pliocene) and Cerro Coronel (an eroded volcanic cone, dated from the Pleistocene) (Nieuwenhuyse 1996). These mounts reaching elevations lower than 400 m asl, modifying the distribution of vegetation patterns and the course of the rivers that drain into the Caribbean coast providing differentiating conditions for soil formation (Figs. 21.17 and 21.18).

Climatic conditions observed in this region are particularly complex, described as hot and very wet (Solano and Villalobos 2001). The total annual precipitation is around 4860 mm (Fig. 21.17), where the highest monthly average values are observed from July to November (more than 450 mm month$-^{1}$). Because the rainfall surpasses the monthly average evapotranspiration during all twelve months, a clear dry period does not occur. Typically, there is a clear decrease in rainfall (< 200 mm month^{-1}) in March.

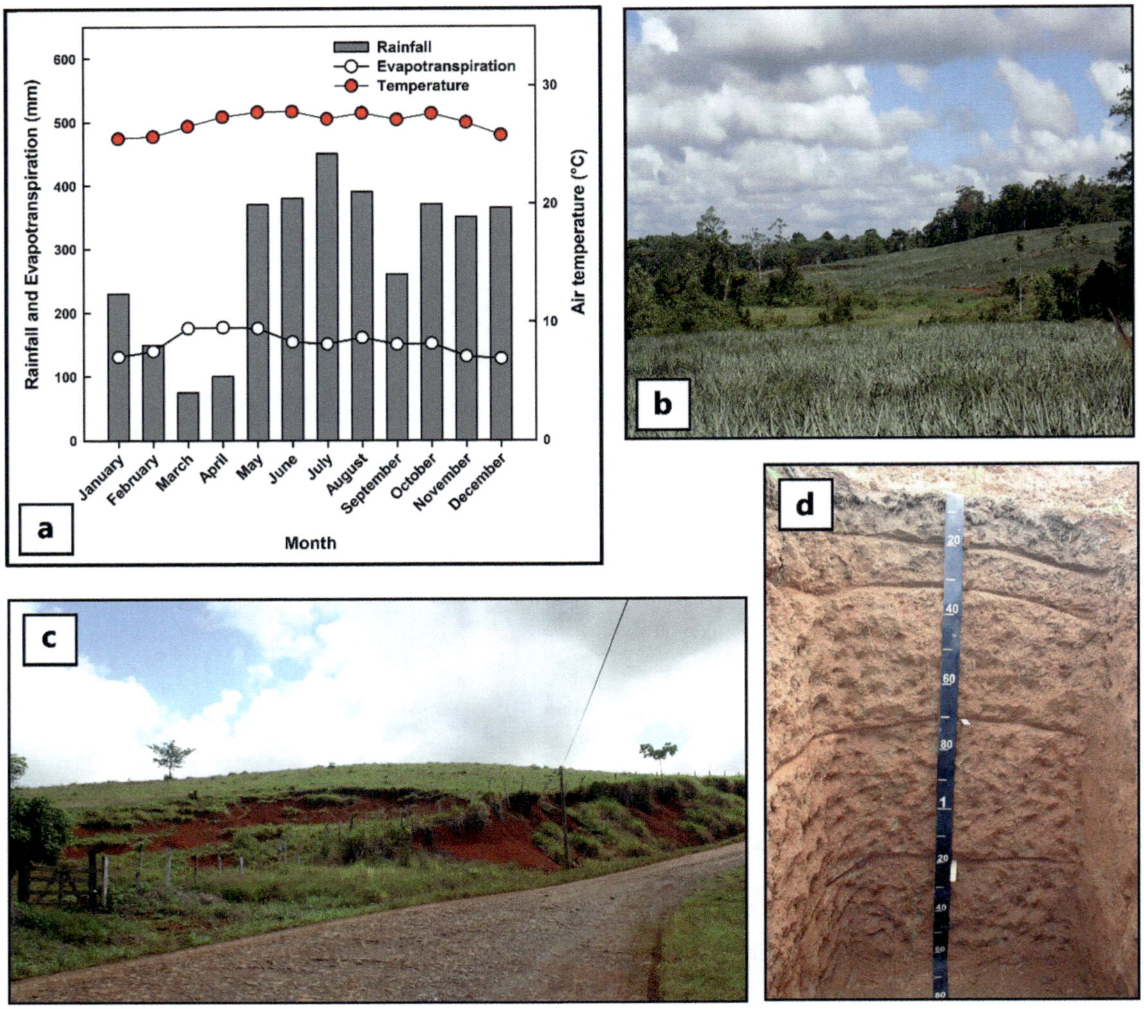

Fig. 21.9 Climate, crops, and soils observed in the Lowlands of San Carlos: **a** Atmospheric water balance estimated for the town of Crucitas de Cutris, with just two months of water deficit; **b** Field planted with pineapple (*Ananas comosus* (L.) Merr.), **c** Grassland with clayey reddish soils exposed in erosional scars; **d** The profile of a *Plinthic Kandiudox* studied in Crucitas de Cutris

In addition, the annual average, minimum, and maximum temperature values are 26°, 24°, and 28 °C respectively.

These climatic conditions allow to classify the vegetation within this landscape as Tropical wet forest (wf-T) according to Holdridge´s Life Zones System (Bolaños and Watson 1993). On the other hand, Vargas (1992) described this vegetation as low-height evergreen wet forest species, typical of lowlands forests. The key species include *Terminalia chiriquensis* Pittier, *T. amazonia* (Gmel.) Excell, and *Calophyllum brasiliense* Cambess (Figs. 21.18 and 21.19).

The soils within this landscape are contrasting. For instance, young alluvial soils (Inceptisols) have developed along the coastal plain, while highly weathered soils (Oxisols) have been observed in the elevations of lomas del Cerro Cocorí and Cerro Coronel (Fig. 21.17). The soils developed from the alluvial deposits have been classified as the following soil suborders: Aquepts, Udepts, and Aquents (Mata et al. 2022), all with very low degree of pedologic development, and saturation issues. More recently, Camacho et al. (2021) revisited the weathered soils previously described in the highlands of Coronel and reclassified as *Andic Haploperox* (Fig. 21.19). These soils developed within a perudic soil moisture regime, this meaning no soil water deficit (Fig. 21.19), but the soil does not saturate enough (due to high hydraulic conductivity) to get reduced iron forms. In addition, these soils present certain andic properties, like low bulk density values and high $PO_4{-}^{1}$ retention.

This landscape offers limited opportunities for agricultural activity, focused mainly on subsistence crops and small cattle farms, mostly due the climatic conditions and low fertility of predominant soils. However, scenic qualities of the area are worth emphasizing, including the delta of the San Juan River and the contrasts made by differentiated vegetation within the highlands and the lowlands (Fig. 21.17), in addition to high biodiversity in both flora

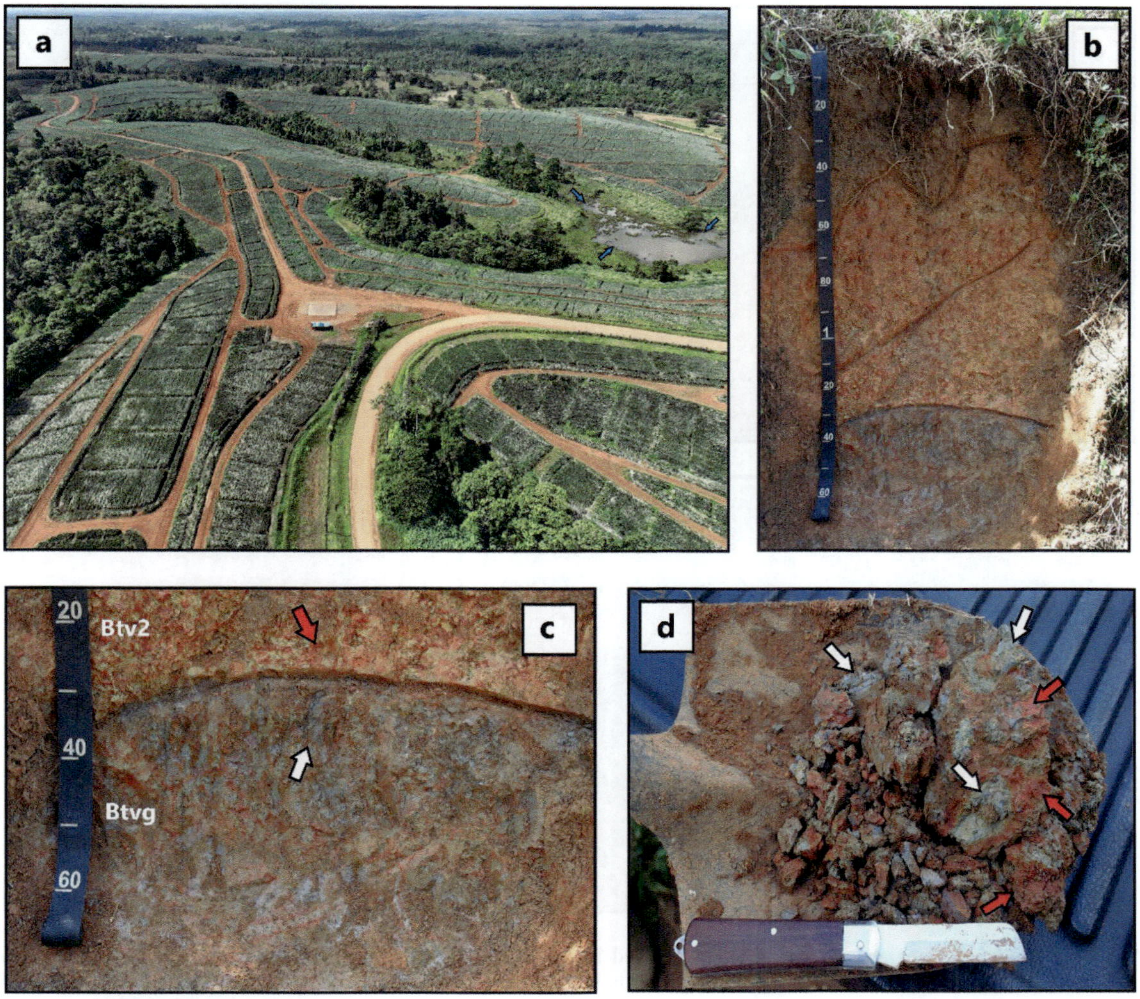

Fig. 21.10 Agricultural activities and typical soil features within the Lowlands of San Carlos: **a** Field planted with pineapple (*Ananas comosus* (L.) Merr) in El Concho de Cutris, with wetlands located in low-lying areas (blue arrows); **b** A soil profile observed in Crucitas de Cutris; **c** Boundary between contrasting soil horizons with plinthite Btv2 (red arrow) and bluish reduced iron conditions Btvg (white arrow) indicating possible fluctuation of water table; **d** Zoom picture of redoximorphic features observed in the horizon Btvg, showing reduced iron (white arrows) and oxidated iron forms (red arrows)

and fauna. These characteristics could be exploited through eco-touristic projects with a minimum environmental impact and considerable positive social impact, especially for this region with high poverty rate and low human development index.

21.3 Genesis of Highly Weathered Soils in Costa Rica

21.3.1 Parent Materials

Parent materials observed in landscapes presented above can be grouped according to their geologic features and ages within four main general categories: (1) Basaltic rocks from Jurassic-Miocene Oceanic complexes; (2) Igneous materials of Paleocene-Pliocene age; (3) Sedimentary materials of Pliocene–Pleistocene age, (4) Volcanic materials of Pleistocene age.

21.3.1.1 Basaltic Rocks from Jurassic-Miocene Oceanic Complexes

Bedrock presented in the mountains of Península de Nicoya and Península de Osa was originally classified within a great geological unit of "Nicoya Ophiolite Complex" (Berrangé and Thorpe 1988), but more recently, basaltic materials within these two landscapes were separated and reclassified, based on rock dating, into two main geologic units (Denyer and Gazel 2009): Nicoya Complex (Jurassic to Cretaceous, 139–83 Ma) and Rincón Block (Paleogene,

Fig. 21.11 Other human activities and associated soils observed in the Lowlands of San Carlos: **a** Road cut that extends along the whole road to Crucitas de Cutris, exhibiting a weathered reddish deep soil; **b** Headquarters of former Crucitas Mine Project, where red soils can be observed (red arrows); **c** The soil profile observed in the road cut (A)

55–62 Ma), respectively. This differential exposure of easily weatherable materials (e.g., basaltic, and ultramafic rocks) in time and space to non-uniform conditions has been described as a suitable background for the development of Oxisols and associated Alfisols and Ultisols (Buol and Eswaran 2000), in agreement with the taxonomic Subgroups observed in both landscapes (Camacho et al. 2021).

21.3.1.2 Igneous Materials of Paleocene-Pliocene Ages

The Paleocene and Pliocene in Costa Rica (65–2 Ma) were characterized by intense volcanic activity throughout the emerging regions, creating extensive areas of depositions of pyroclastic rocks, lahars, and further sedimentary materials (Bergoeing and Brenes 2017). In this regard, parent materials described within the landscape Lowlands of San Carlos are essentially pyroclastic rocks, dacitic and rhyolitic lavas, and welded tuffs (Alvarado et al. 1992; Gazel et al. 2005).

These old volcanic materials are highly susceptible to weathering processes, especially under the climatic conditions of the tropics. For instance, rhyolitic rocks include amphiboles, green hornblende, and plagioclase phenocrystals (Gazel et al. 2005), all of them easily weatherable, so that under the conditions of protracted exposure to humid tropical the development of Oxisols (Buol and Eswaran 2000), or other soil taxonomic orders like Ultisols (West et al. 1997), will take place.

21.3.1.3 Sedimentary Materials of Pliocene–Pleistocene Ages

These materials have been exclusively identified in Valle del General and linked with the glaciation during the LGM, which took place in this specific region of Costa

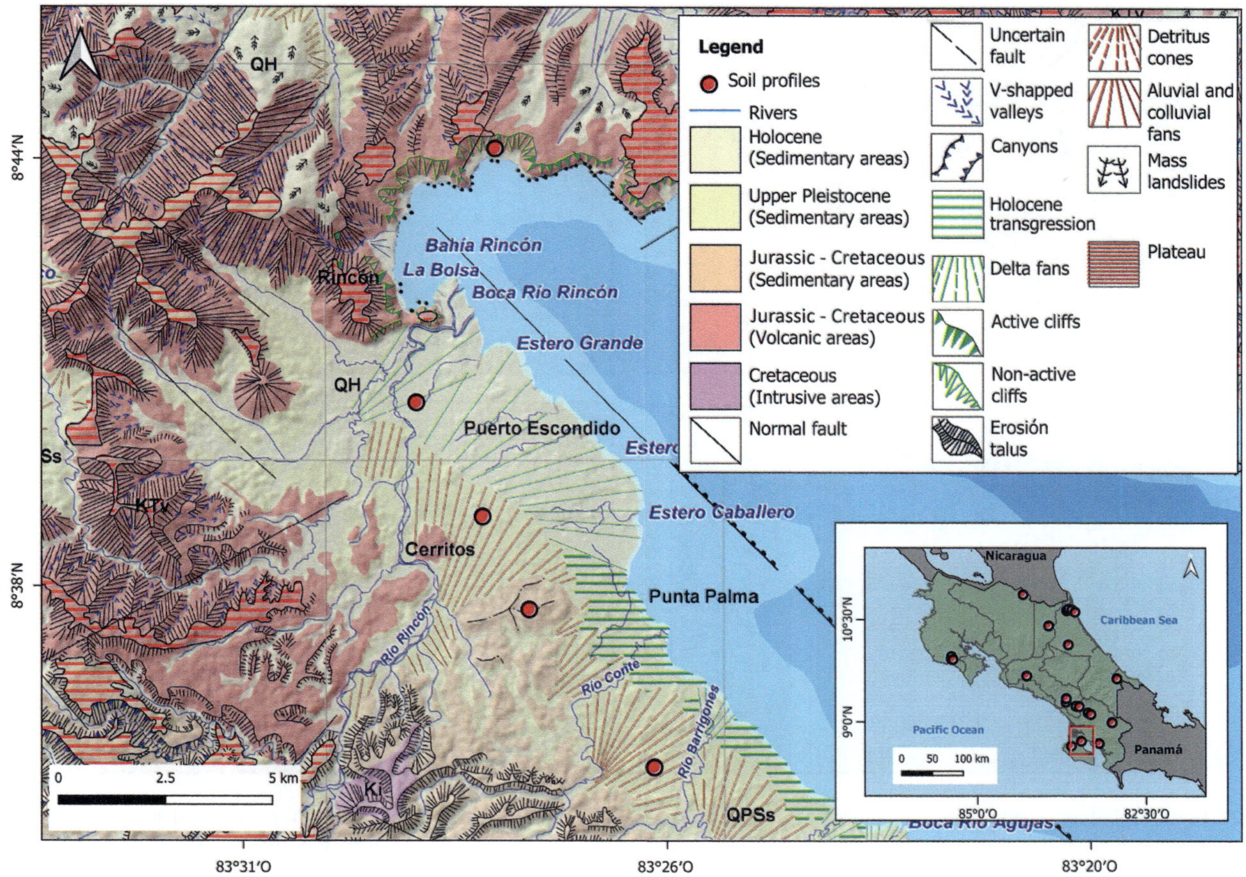

Fig. 21.12 Location, main geomorphologic features, and soil profiles observed for the landscape Mountains of Peninsula de Osa. Adapted from Bergoeing and Brenes (2017)

Rica (Quesada-Román et al. 2020, 2021). After accelerated deglaciation in the Talamanca Mountain Range (Quesada-Román et al. 2019), high volumes of sediments were deposited as massive alluvial fans along the whole piedmont (Camacho et al. 2020). These extensive geomorphic units are unique in Costa Rica (Quesada-Román and Zamorano-Orozco 2019).

Because these sediments were previously weathered and reworked during the transport, and they form very stable and extensive landforms, they allowed for the formation of highly weathered soils like Oxisols and Ultisols (Camacho et al. 2020), even within timespans shorter than 100,000 years. This relative geomorphic stability could have led to a feasible genetic process that gave rise to highly weathered soils such as Oxisols or Ultisols. Kesel and Spicer (1985) highlighted large clastic elements observed within these alluvial fans that have been turned into clay, which could indicate that weathering processes in both rocks and soils occurred in situ. Indeed, these clayey and red matrices commonly observed in old alluvial fans have become a pattern along this great landscape, where highly weathered soils like Oxisols and Ultisols have been identified (Kesel and Spicer 1985; Camacho et al. 2021). This is consistent with observations by Chen et al. (2015), who described Oxisols and Ultisols in Taiwan, genetically associated with Quaternary terraces formed from old alluvium.

21.3.1.4 Volcanic Materials of Pleistocene Age

The Quaternary history in Costa Rica (2.58–0 Ma) also included volcanic activity and associated deposition events (Bergoeing and Brenes 2017). For instance, Cerro Coronel and the corresponding mountain range are described as remnants of volcanic cones built of alkaline basaltic materials formed approximated 1.2 Ma and potentially extended up to Holocene age (Alvarado et al. 1992). These geologic materials support pedogenetic processes that result in highly weathered soils such as Oxisols and Ultisols. In this regard, Buol and Eswaran (2000) asserted that the development of Oxisols is likely when old volcanic areas within the Tropics are subjected to intensive climatic conditions, in agreement with the assessment provided by Camacho et al. (2021) for this region of Costa Rica.

Fig. 21.13 Morphology and some landform features in the Mountains of Peninsula de Osa: **a** Slope next to the highway, covered by abundant vegetation which hides the parent material (visible after the landslide, and highlighted with red arrows); **b** A soil profile studied within these mountains, **c** River meanders behind a small elevation, where two landslides (30 m height) display the presence of a clayey matrix and reddish soils (red arrows); **d** A highway across the mountains, providing perspective about the elevations within this landscape in contrast with the Talamanca Mountain Range, with red soils observed in the top (red arrows)

21.3.2 Rainfall Distribution Within Landscapes and Neoformation Processes

Costa Rica has contrasting climatic conditions across its territory, which are indirectly modified by volcanic and mountain ranges, and the influence of both Caribbean and Pacific coast (Solano and Villalobos 2001). For instance, the province of Guanacaste is considered the driest region, with less than 1900 mm of annual rainfall distributed in 7 months, meanwhile the rainfall pattern in the mountains of Tortuguero-Barra del Colorado is such that more than 4500 mm of precipitation is distributed within 12 months.

Therefore, it is expected to find variations in the pedogenic processes, resulting in different soil taxonomic features. Indeed, different taxonomic subgroups (within Oxisols) were assigned to soils studied within these five landscapes. For example, when comparing *Kandiustalfic Eutrustox* from Península de Nicoya with *Andic Haploperox* observed in the mountains of Tortuguero-Barra del Colorado (Cerro Coronel), one can infer marked differences in fertility, in addition to soil water availability (Camacho et al. 2021).

These variations are associated with differences in the bases and Si leaching rates, and differential neoformation and ferratilization processes associated with changes in rainfall amount and distribution (Nieuwenhuyse and van Breemen 1997; Camacho et al. 2021). This last statement agrees with those results obtained by Pincus et al. (2017), who found in the Pacific Southeastern of Costa Rica that weathering rates could reach up to four times fold under annual rainfall values (higher than 4000 mm) when compared with rates in drier sectors (lower than 2800 mm).

21.3.3 The Fundamental Role of Geomorphology

When comparing these five landscapes, and despite variations in geologic materials and climatic conditions, there is a common factor that allowed for the development of highly weathered soils with similar morphologic features: *very stable geomorphologic units*. Buol and Eswaran (2000) explained the role of stable geomorphic surfaces in the genesis of Oxisols, where regardless the parent material (rock weathering or polycyclic sediments previously weathered)

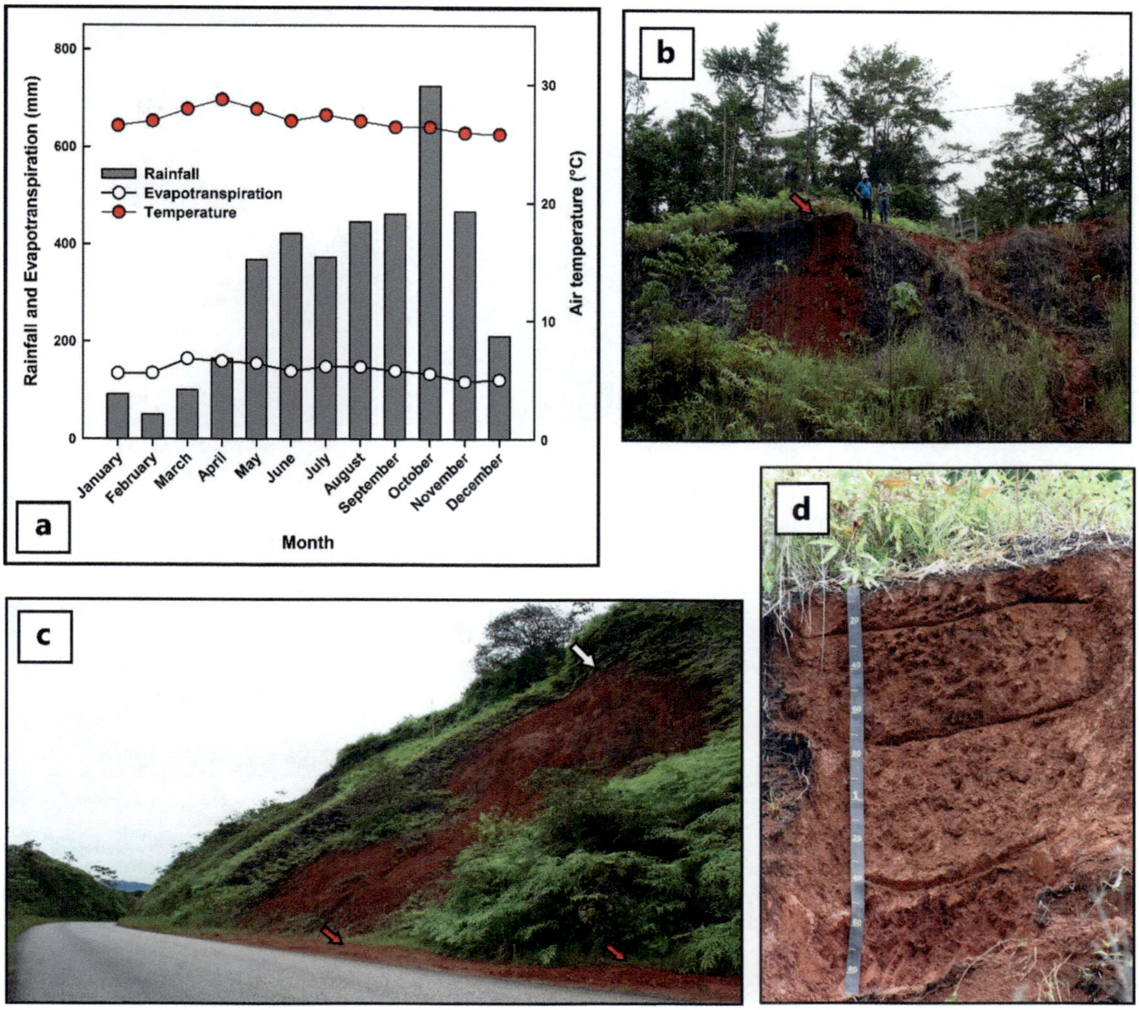

Fig. 21.14 Climate and soils observed in the Mountains of Peninsula de Osa: **a** Atmospheric water balance estimated for the town of Rincón de Osa, with three months of water deficit and high precipitation values from May to November; **b** Top of the mountain with an exposed soil profile (red arrow), and evidence of pluvial erosion along the slope **c** Deposits of red materials (red arrows) along the highway, which are products of landslides and further pluvial erosion from the top of the slope (white arrow); **d** Zoom picture of the soil profile assessed in the top of this mountain presented in panel A

and climatic conditions, this stability can control the occurrence of Oxisols within the landscape. Landscapes assessed in the present work represent diverse examples of stable geomorphic surfaces such as old alluvial fans from the Quaternary within El Valle del General (with higher occurrence of highly weathered soils), the mountains in both Península de Nicoya and Península de Osa, or lowlands in the northern part of Costa Rica (Bergoeing 2007).

Some of the Oxisols developed within the landscape "Alluvial fans from El Valle del General" are considered the most weathered soils of Costa Rica, classified as *Anionic Acrustox* (Camacho et al. 2021). Despite ages between 200,000 and 7000 years ago, these materials were weathered before their deposition as massive alluvial fans (Quesada-Román and Zamorano-Orozco 2019). Then, landform stability allowed for the genesis of highly weathered soils even within younger materials.

Conversely, in landscapes with older geologic materials under higher precipitation regimens, soils are consider highly weathered, but taxonomically they seem to be less weathered than those mentioned above. For instance, within the mountains of Peninsula de Osa, Oxisols developed from basaltic rocks (around 60 Ma) were classified as *Kandiudox*, which compared with the aforementioned *Acrustox, and* are considered less weathered soils (Camacho et al. 2021).

Fig. 21.15 Panoramic views and distribution of red soils within the Mountains of Peninsula de Osa: **a** Scenic view of the elevations in Rincón de Osa (foreground), diverging with Talamanca Mountain Range (background) and red soils (red arrows); **b** A *Rhodic Kandiudox*, studied in the elevation of Rincón de Osa; **c** The upper coast of Dulce Gulf contrasting the elevations with the presence of red soils (red arrow); **d** Cleared area within the mountains, where red soils are visible (red arrows)

21.4 Conclusions

Highly weathered soils (Oxisols and Ultisols) have been identified in diverse landscapes of Costa Rica. Despite this country has a "young geological age", there are other fundamental features that allowed for the development of highly weathered soils, similar to regions which are geologically much older. Diversity in parent materials (various Cenozoic igneous and sedimentary rocks), contrasting climatic conditions and specific geomorphologic features provided the stage for different genetic pathways that resulted in different taxonomic Subgroups found within the five landscapes studied in present work, ranging from *Eutric* Oxisols to the most developed taxonomic groups among the Oxisols: *Acrustox*.

The comparative analysis highlights the fundamental role of geomorphologic features in the development of weathered soils, where highly stable landforms are required

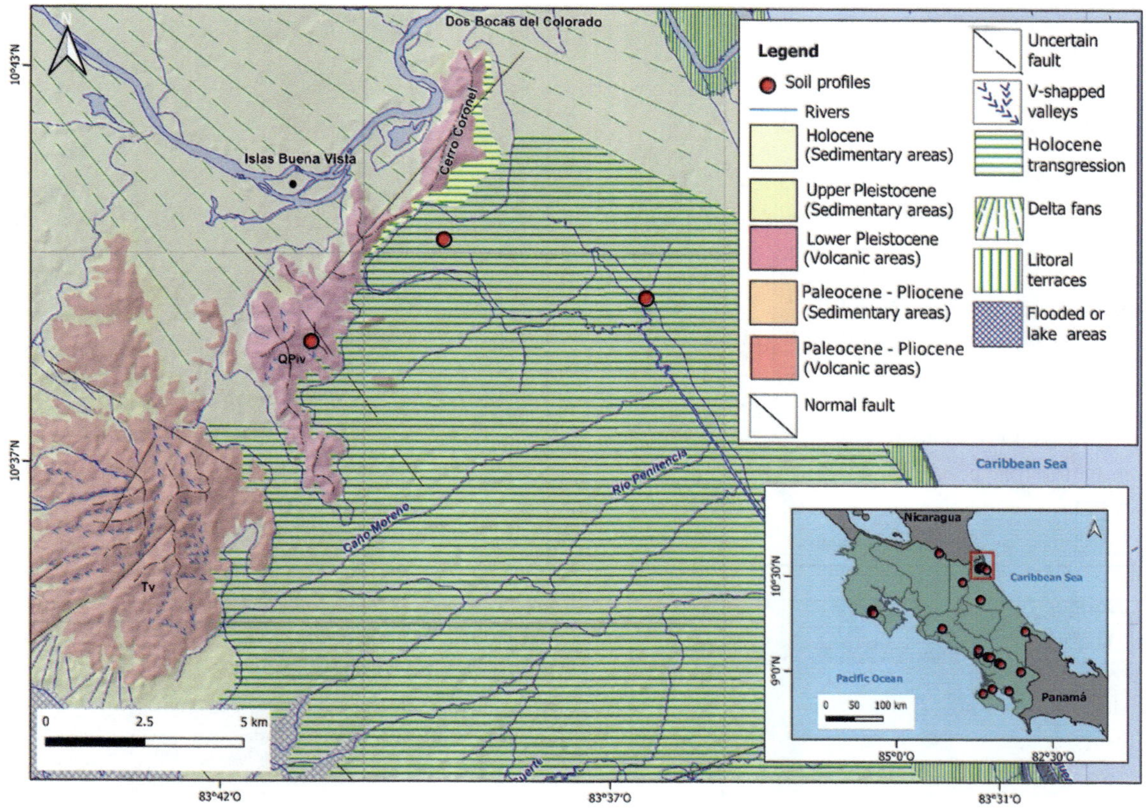

Fig. 21.16 Location, main geomorphologic features, and soil profiles observed in the mountains of Tortuguero-Barra del Colorado. Adapted from Bergoeing and Brenes (2017)

to achieve diverse forming processes (additions, losses, transformations, and translocations) even though the other forming factors are favorable. From economic perspective, these stable landforms allow important cash crop establishment, including coffee and pineapple, which becomes profitable under a proper soil management.

A concise review of soil-forming factors becomes fundamental to understand the origin of soils and their further fate, allowing not only for their taxonomic classification but also better soil management and conservation. All this information should be considered in land-use planning, considering the limitations, trade-offs, and advantages of soil taxonomic subgroups within the landscapes assessed, aiming to optimize the use of available resources with minimum ecological impact.

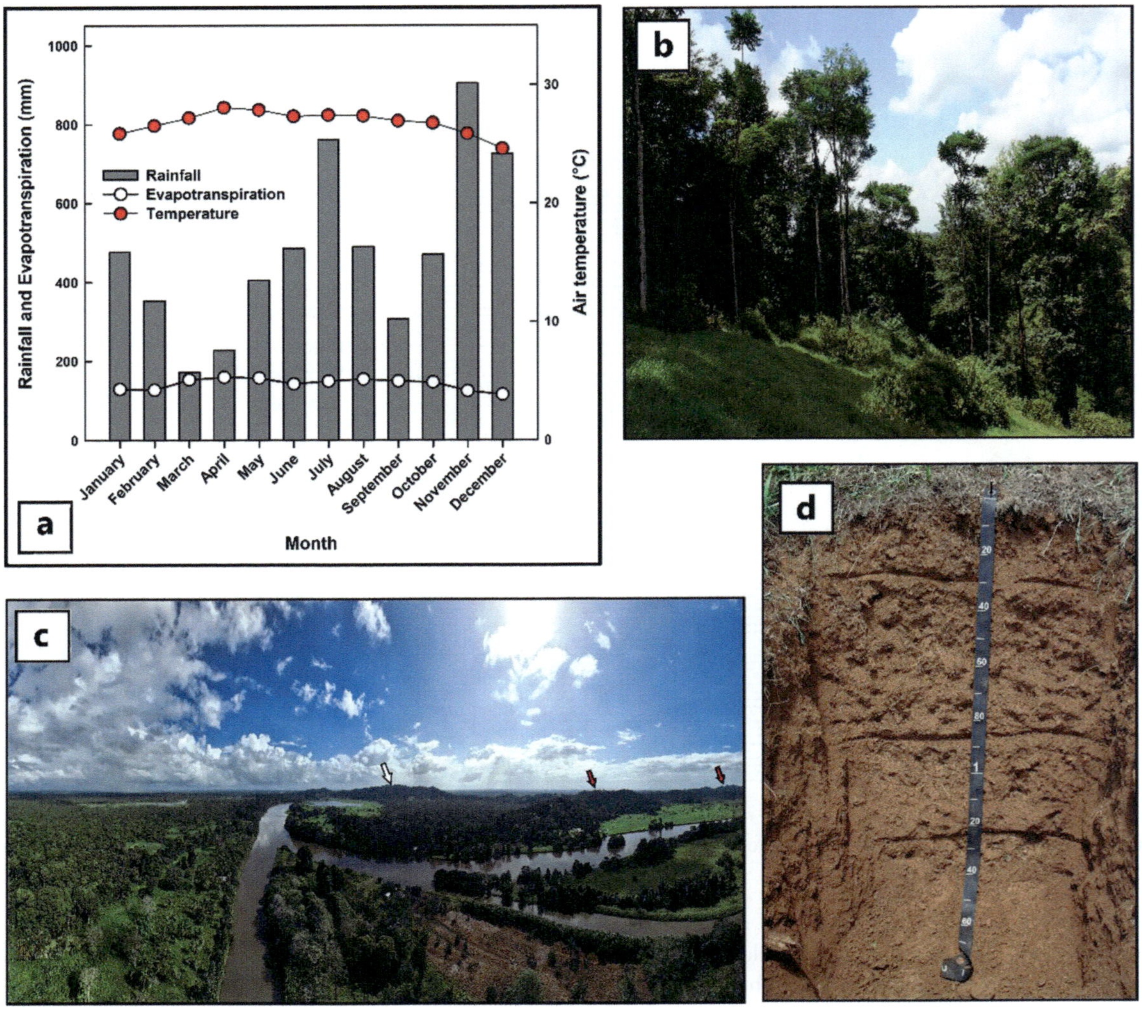

Fig. 21.17 Climate, vegetation, and soils observed in the Mountains of Tortuguero-Barra del Colorado: **a** Atmospheric water balance estimated for the Barra del Colorado, showing no water deficit, and high values of monthly precipitation in July, November, and December; **b** Evergreen vegetation corresponding to Tropical wet forest in Cerro Coronel; **c** Panoramic view of Barra del Colorado, with distant elevations of Cerro Coronel (white arrow) and Lomas del Cerro Cocorí (red arrow); **d** Typical soil observed on the top of Cerro Coronel

Fig. 21.18 Landscapes and landforms observed in the region of Tortuguero-Barra del Colorado: **a** Mountains covered by abundant vegetation (white arrows), and associated elevations that have been cleared to grasslands, with the presence of deep clayey reddish soils (red arrows); **b** A reddish soil profile studied at the same elevation as shown in panel A, **c** Lowlands punctuated by low hills (white arrows) with contrasting vegetation due difference in land-use (forest versus grassland), and distant view of Cerro Tortuguero (red arrow); **d** Terracettes and lanes (black lines and red arrows) observed in grasslands due to cattle movement (cattle in white circles)

Fig. 21.19 Highly weathered soils in the Mountains of Tortuguero-Barra del Colorado: **a** Road cut that extends along the road to Barra del Colorado, exhibiting weathered reddish deep soils (white arrows); **b** A soil profile observed in these same road cut (2 m depth between the red arrows); **c** Zoom picture of the soil profile observed in the road cut (panel B), and classified as *Andic Haploperox*

References

Alvarado GE, Kussmaul S, Chiesa S, Gillot PY, Appel H, Wörner G, Rundle C (1992) Resumen cronoestratigráfico de las rocas ígneas de Costa Rica basado en dataciones radiométricas. J S Am Earth Sci 6(3):151–168

Alvarado A (1985) El origen de los suelos. Serie materiales de enseñanza No. 24. Centro Agronómico Tropical de Investigación y Enseñanza (CATIE). Turrialba, Costa Rica, p 54

Anderson MS; Byers HG (1930) Character of the colloidal materials in the profiles of certain major soil groups. United States Department of Agriculture. Technical bulletin. 228, p 24

Bergoeing JP, Brenes LG (2017) Atlas Geomorfologico de Costa Rica: Escala 1: 100000. Primera Edición. Escala 1:100 000. Heredia, Costa Rica. EUNA

Bergoeing JP, Protti R (2006) Geomorfología Paleo-Lacustre del Sur del Lago de Nicaragua. Revista Geográfica 139:27–38

Bergoeing JP (2007) Geomorfología de Costa Rica. 2 ed. Librería Francesa. p 328

Bergoeing JP (2011) Los conos de deyección del Valle de El General, Costa Rica. Revista Geográfica de América Central. Número Especial 47E(2):1–16

Berrangé JP, Thorpe RS (1988) The geology, geochemistry and emplacement of the Cretaceous—Tertiary ophiolitic Nicoya Complex of the Osa Peninsula, southern Costa Rica. Tectonophysics 147:193–220

Bolaños RA, Watson V (1993) Mapa ecológico de Costa Rica: según el sistema de clasificación de zonas de vida del mundo de L.R. Holdridge. San José, Costa Rica. Centro Científico Tropical. 1:200 000 Color

Budowsky G (1963) Forest succession in tropical lowlands. Turrialba 13(1):42–44

Buol SW, Eswaran H (2000) Oxisols. Adv Agron 68:151–195

Buol SW, Southard RJ, Graham RC, McDaniel PA (2011) Soil genesis and classification, 6th edn. John Wiley and Sons, Chichester, UK, p 543

Buol SW, Eswaran H (1988) International committee on Oxisols. Final report. Technical monograph no 17. Soil management support services and North Carolina State University. Washington DC and Raleigh NC, p 157

Camacho ME, Quesada-Roman A, Mata R, Alvarado A (2020) Soil-geomorphology relationships of alluvial fans in Costa Rica. Geoderma Reg 21:e00258

Camacho ME, Mata R, Barrantes-Víquez M, Alvarado A (2021) Morphology and characteristics of eight Oxisols in contrasting landscapes of Costa Rica. CATENA 197:104992

Chen ZS, Hseu ZY, Tsai CC (2015) Oxisols and Ultisols. In: Chen ZS, Hseu ZY, Tsai CC (eds) The soils of Taiwan. Springer, Netherlands, Dordrecht, Netherlands, pp 95–108

Cleveland CC, Townsend AR, Schmidt SK, Constance BC (2003) Soil microbial dynamics and biogeochemistry in tropical forests and pastures, southwestern Costa Rica. Ecol Appl 13(2):314–326
Denyer P, Gazel E (2009) The Costa Rican Jurassic to Miocene oceanic complexes: origin, tectonics and relations. J S Am Earth Sci 28(4):429–442
Dóndoli C (1943) La región del General: condiciones geológicas y geoagronómicas de la zona. Revista Del Instituto De La Defensa Del Café De Costa Rica 13(106):513–528
Eswaran H (1972) Micromorphological indicators of pedogenesis in some tropical soils derived from basalt from Nicaragua. Geoderma 7:15–31
Eze PN, Udeigwe TK, Meadows ME (2014) Plinthite and its associated evolutionary forms in soils and landscapes: a review. Pedosphere 24(2):153–166
Fagan ME, DeFries RS, Sesnie SE, Arroyo-Mora JP, Soto C, Singh A, Chazdon RL (2015) Mapping species composition of forests and tree plantations in Northeastern Costa Rica with an integration of hyperspectral and multitemporal landsat imagery. Remote Sens 7:5660–5696
Fernández-Moya J, Alvarado A, Miguel-Ayanz S, Marchamalo-Sacristán M (2014) Forest nutrition and fertilization in teak (*Tectona grandis* Lf) plantations in Central America. NZ J Forest Sci 44(1):1–8
Gazel E, Alvarado GE, Obando J, Alfaro A (2005) Geología y evolución magmática del arco de Sarapiquí, Costa Rica. Revista Geológica De América Central 32:13–31
Gazel E, Hoernle K, Carr MJ, Herzberg C, Saginor I, Van den Bogaard P, Hauff F, Feigensson M, Swisher C III (2011) Plume–subduction interaction in southern Central America: mantle upwelling and slab melting. Lithos 121(1–4):117–134
Granda VA (2015) Caracterización ecológica y del potencial forestal de bosques secundarios en la Península de Nicoya, Costa Rica, y sus relaciones con factores ambientales. Tesis Maestría. Centro Agronómico Tropical de Investigaciones Tropicales CATIE. Turrialba, p 65
Hardy F(1970) Edafología Tropical, Edit. Herrero Hermanos, Sucesores S.A. México, p 420
Jaramillo LR (1969) Caracterización de algunos "Latosoles" de Mesoamérica. Tesis M.Sc. IICA. Turrialba, Costa Rica, p 233
Jenny H (1994) Factors of soil formation. A system of quantitative pedology. Dover Press, New York (Reprint, with Foreword by R. Amundson, of the 1941 McGraw-Hill publication). pdf file format
Kesel RH, Spicer BE (1985) Geomorphologic relationships and ages of soils on alluvial fans in the Rio General Valley, Costa Rica. CATENA 12(1):149–166
Korpela D. (2014). A social and environmental impact assessment of the Crucitas gold mining project in Costa Rica. Master Degree Thesis. Findland. University of Jyväskylä, p 81
Macías M. (1969). Propiedades morfológicas, físicas, químicas y clasificación de ocho "Latosoles" de Costa Rica. Tesis Mg. Sc. Turrialba, Costa Rica. IICA, p 193
Marshall JS (2007) Geomorphology and physiographic provinces of Central America. In: Bundschuh J, Alvarado G (eds) Geology, resources, and natural hazards. Balkema, Leiden, The Netherlands, pp 75–122
Martini JA (1970) Allocation of cation exchange capacity to soil fractions in seven surface soils from Panama and the application of a cation exchange factor as a weathering index. Soil Sci 109(5):324–331
Mata R, Rosales A, Sandoval D, Vindas E, Alemán B (2022) Mapa de Órdenes de Suelos de Costa Rica, 2022. Esc. 1:200.000. Universidad de Costa Rica, Centro de Investigaciones Agronómicas. San José, Costa Rica
Miranda M (1983) Cambio en el uso del suelo en General Viejo de Pérez Zeledón. Revista Geográfica De América Central 17–18:91–121
Mora S (1985) Las laderas inestables de Costa Rica. Revista Geológica De América Central 3:129–161
Moura-Filho W, Buol SW (1972) Studies of a Latosol Roxo (*Eutrustox*) in Brazil. Experientiæ 13:201–247
Nieuwenhuyse A, Van Breemen N (1997) Quantitative aspects of weathering and neoformation in selected Costa Rican volcanic soils. Soil Sci Soc Am J 61(5):1450–1458
Nieuwenhuyse A (1996) Landscape formation and soil genesis in volcanic parent materials in humid tropical lowlands of Costa Rica. Ph.D. thesis, Wageningen University, Wageningen, The Netherlands, p 131
Pincus LN, Ryan PC, Huertas FJ, Alvarado GE (2017) The influence of soil age and regional climate on clay mineralogy and cation exchange capacity of moist tropical soils: a case study from Late Quaternary chronosequences in Costa Rica. Geoderma 308:130–148
Quesada-Román A, Zamorano-Orozco JJ (2019) Geomorphology of the upper general river basin, Costa Rica. J Maps 15(2):94–100
Quesada-Román A, Ballesteros-Cánovas JA, Stoffel M, Zamorano-Orozco JJ (2019) Glacial geomorphology of the Chirripó National Park, Costa Rica. J Maps 15(2):538–545
Quesada-Román A, Campos N, Alcalá-Reygosa J, Granados-Bolaños S (2020) Equilibrium-line altitude and temperature reconstructions during the last glacial maximum in Chirripó National Park, Costa Rica. J S Am Earth Sci 100:102576
Quesada-Román A, Campos N, Granados-Bolanos S (2021) Tropical glacier reconstructions during the last Glacial Maximum in Costa Rica. Rev Mex Cienc Geol 38(1):55–64
Rosero-Bixby L, Maldonado-Ulloa T, Bonilla-Carrión R (2002) Bosque y población en la Península de Osa, Costa Rica. Rev Biol Trop 50(2):585–598
Soil Survey Staff (1975) Soil Taxonomy. A basic system for making and interpreting soil surveys. soil conservation service. Washington DC, USDA Handbook No. 436, pp 459
Solano J, Villalobos R (2001) Aspectos fisiográficos aplicados a un bosquejo de regionalización geográfico climático de Costa Rica. Tópicos meteorológicos y Oceanográficos 8(1):26–39
Soudre M (2004) Factores que influyen sobre las características del suelo y la vegetación secundaria regenerada en pasturas abandonadas de Hojancha, Guanacaste, Costa Rica. Tesis Maestría. Centro Agronómico Tropical de Investigaciones Tropicales CATIE. Turrialba, p 110
Targulian VO, Krasilnikov PV (2007) Soil system and pedogenic processes: Self-organization, time scales, and environmental significance. CATENA 71(3):373–381
Thomas MF (1994) Geomorphology in the tropics: a study of weathering and denudation in low latitudes. West Sussex, UK. John Wiley & Sons Ltd., pp 460
Vargas G (1992) Las Formaciones Vegetales de Costa Rica: un Estudio Ecológico Regional. Revista Geográfica 116:113–136
Vargas F, Alfaro A (1992) Presencia de serpentinitas, basaltos alcalinos y rocas volcánicas ácidas en la zona norte-Atlántica de Costa Rica. Revista Geológica De América Central 14:105–107
West LT, Sumner FBM, Kang BT (1997) Ultisols: Characteristics and impacts on society. Adv Agron 63:179–239
Wilcke W, Kretzschmar S, Bundt M, Saborío G, Zech W (1998) Aluminum and heavy metal partitioning in a horizons of soils in Costa Rican coffee plantations. Soil Sci 163(6):463–471

Index

A. Quesada-Román (ed.), *Landscapes and Landforms of Costa Rica*,
World Geomorphological Landscapes, https://doi.org/10.1007/978-3-031-64940-0

GPSR Compliance

The European Union's (EU) General Product Safety Regulation (GPSR) is a set of rules that requires consumer products to be safe and our obligations to ensure this.

If you have any concerns about our products, you can contact us on ProductSafety@springernature.com

In case Publisher is established outside the EU, the EU authorized representative is:

Springer Nature Customer Service Center GmbH
Europaplatz 3
69115 Heidelberg, Germany

Batch number: 10755972

Printed by Printforce, the Netherlands